EFFECTS OF PERSISTENT AND BIOACTIVE ORGANIC POLLUTANTS ON HUMAN HEALTH

EFFECTS OF PERSISTENT AND BIOACTIVE ORGANIC POLLUTANTS ON HUMAN HEALTH

Edited by

David O. Carpenter
University at Albany
Institute for Health and the Environment

Published by John Wiley & Sons, Inc., Hoboken, New Jersey
Published simultaneously in Canada

Library of Congress Cataloging-in-Publication Data:
Carpenter, David O.
Effects of persistent and bioactive organic pollutants on human health / David O. Carpenter, University at Albany, Institute for Health and the Environment.
pages cm
Includes index.
ISNB 978-1-118-15926-2 (cloth)
1. Bioactive compounds–Toxicology. 2. Organic compounds–Toxicology. 3. Persistent pollutants–Health aspects. I. Title.
RA1235.C37 2013
615.9'5–dc23

2013004573

Printed in the United States of America

10 9 8 7 6 5 4 3 2 1

CONTENTS

CONTRIBUTORS

James S. Brown, Jr., MD, MPH, MS, Department of Psychiatry, Virginia Commonwealth University School of Medicine, Midlothian, VA

Kristopher K. Burnitz, MA, Department of Anthropology, University at Albany, Albany, NY

David O. Carpenter, MD, Institute for Health and the Environment, University at Albany, Rensselaer, NY

Mariano E. Cebrián, MSc, MD, PhD, Department of Toxicology, Center for Research and Advanced Studies (Cinvestav-IPN), Instituto Politécnico Nacional, Mexico D.F., Mexico

Chi-Hsien Chen, Department of Environmental and Occupational Medicine, National Taiwan University (NTU) College of Medicine and NTU Hospital, Graduate Institute of Occupational Medicine and Industrial Hygiene, NTU College of Public Health, Taipei, Taiwan

Richard W. Clapp, DSc, MPH, Lowell Center for Sustainable Production, University of Massachusetts Lowell, Lowell, MA

Adrian Covaci, PhD, Toxicology Centre, University of Antwerp, Belgium

Eveline Dirinck, MD, Departments of Endocrinology, Diabetology and Metabolic Disease, Antwerp University Hospital and University of Antwerp, Belgium

José L. Domingo, PhD, Laboratory of Toxicology and Environmental Health, School of Medicine, IISPV, Universitat Rovira i Virgili, Reus, Catalonia, Spain

Nina Dutton, MPH, Oak Ridge Institute for Science and Education (ORISE) Research Participation Program Fellow, ATSDR/CDC, Atlanta, GA

Mia V. Gallo, PhD, Department of Anthropology, University at Albany, Center for the Elimination of Minority Health Disparities, Albany, NY

Stefanie Giera, PhD, Molecular and Cellular Biology Program, University of Massachusetts Amherst, Amherst, MA

Howard P. Glauert, PhD, Graduate Center for Nutritional Sciences, University of Kentucky, Lexington, KY

Samuel M. Goldman, MD, MPH, The Parkinson's Institute, Sunnyvale, CA

Janelle Graham, MPH, School of Public Health, Curtin Health Innovation Research Institute, Curtin University, Perth, Western Australia

Yueliang Leon Guo, Department of Environmental and Occupational Medicine, National Taiwan University (NTU) College of Medicine and NTU Hospital, Graduate Institute of Occupational Medicine and Industrial Hygiene, NTU College of Public Health, Taipei, Taiwan

David R. Jacobs, Jr., PhD, Division of Epidemiology and Community Health, University of Minnesota School of Public Health, Minneapolis, MN

Molly M. Jacobs, MPH, Lowell Center for Sustainable Production, University of Massachusetts Lowell, Lowell, MA

Duk-Hee Lee, MD, PhD, Department of Preventive Medicine, School of Medicine, Kyungpook National University, Jung-gu, Daegu, Korea

Lizbeth López-Carrillo, MSc, Dr. Ph, Center of Population Health Research, National Institute of Public Health, Col. Santa María Ahuacatitlán, Cuernavaca, Morelos, Mexico

Gabriele Ludewig, PhD, Department of Occupational and Environmental Health, Interdisciplinary Graduate Program in Human Toxicology, The University of Iowa, Iowa City, IA

Rachel I. Massey, Toxics Use Reduction Institute at UMass Lowell, Wannalancit Mills, Lowell, MA

Juliana W. Meadows, PhD, National Institute for Occupational Safety and Health, Division of Applied Science and Technology, Biomonitoring and Health Assessment Branch, Cincinnati, OH

Martí Nadal, PhD, Laboratory of Toxicology and Environmental Health, School of Medicine, IISPV, Universitat Rovira i Virgili, Reus, Catalonia, Spain

Marian Pavuk, CDC Atlanta, ATSDR/CDC, Atlanta, GA

Susan R. Reutman, PhD, National Institute for Occupational Safety and Health, Cincinnati, OH

Anna Rignell-Hydbom, Division of Occupational and Environmental Medicine, Lund University, Sweden

Larry W. Robertson, PhD, MPH, ATS, Department of Occupational and Environmental Health, Interdisciplinary Graduate Program in Human Toxicology, The University of Iowa, Iowa City, IA

Krassi Rumchev, PhD, MSc, School of Public Health, Curtin Health Innovation Research Institute, Curtin University, Perth, Western Australia

Jérôme Ruzzin, Department of Biology, University of Bergen, Bergen, Norway

Lars Rylander, Division of Occupational and Environmental Medicine, Lund University, Sweden

Lawrence M. Schell, PhD, Department of Anthropology, University at Albany; Department of Epidemiology and Biostatistics, School of Public Health, Center for the Elimination of Minority Health Disparities, Albany, NY

Jeff Spickett, PhD, School of Public Health, Curtin Health Innovation Research Institute, Curtin University, Perth, Western Australia

Luc Van Gaal, MD, PhD, Department of Endocrinology, Diabetology and Metabolic Disease, Antwerp University Hospital and University of Antwerp, Belgium

Stijn Verhulst, MD, PhD, Department of Pediatrics, Antwerp University Hospital and University of Antwerp, Belgium

R. Thomas Zoeller, PhD, Department of Biology, Molecular and Cellular Biology Program, University of Massachusetts Amherst, Amherst, MA

CHAPTER 1

Introduction: Why Should We Care about Organic Chemicals and Human Health?

DAVID O. CARPENTER

ABSTRACT

Background: The last several decades have seen an enormous increase in the development and manufacture of different organic chemicals that have proven useful for many aspects of contemporary life. The question is the degree to which some of these chemicals cause harm to human beings.

Objective: This book is directed at the goal of identifying organic chemicals that, while useful in many regards, pose risks to human health because of their biological activity and often their persistence.

Discussion: The various chapters in this book are directed at the effects of organic chemicals on the various organ systems.

Conclusions: While recognizing the wonderful benefits that have come from the development and use of many organic chemicals, serious adverse human health effects have occurred because of inadequate testing prior to use and ineffective steps to prevent release of the chemicals into air, food, water, and the environment, resulting in exposure and disease in humans. It is urgent that more effective ways be found to ensure the safety of organic chemicals, no matter how useful they may be, before they are produced and released into the environment.

Effects of Persistent and Bioactive Organic Pollutants on Human Health, First Edition.
Edited by David O. Carpenter.

Organic chemicals are a major part of everyday life in the modern world. Without question, chemicals have made our lives much easier. But at the same time, it is important to recognize that there have been some downsides to the chemical revolution. This book is focused on the downsides, but that is not to indicate that the benefits of chemicals are ignored. The use of chemicals has resulted in increased food production and safety of food, safer drinking water, improvements in life expectancy from development of pharmaceuticals and antibiotics, and greater convenience to everyone.

It is quite remarkable how much has changed in our daily lives after the development of synthetic chemicals. In the past, our carpets, draperies, and clothes were all made from natural fibers such as wool, linen, or cotton. Today, many are made from synthetic products all derived from petroleum. Most carpets, draperies, and many clothes are treated with organic flame retardants. In the past, our cookware was made of glass, pottery, and various metals. Today, we store foods in plastic, and our cookware is lined with perfluorinated compounds to prevent food from sticking. We drive in cars that may have a metal motor and frame and have glass windows, but everything else is made from plastic and petroleum products. We spray our homes with pesticides and air fresheners. We bathe our bodies with personal care products (creams, cosmetics, deodorants, perfumes, polish for nails, etc.) containing many different chemicals, and often we have no idea what they are or what they might do to alter our health, no matter how beautiful they make us look and how good they make us smell. We dye our hair with chemicals and treat our hair with shampoos and conditioners that contain a variety of chemicals, often not even identified on the bottle because the mixture is proprietary.

We eat food that is often raised at distant places and depend on fossil fuels to get them to our local supermarket. Because we all like our fruits and vegetables to look perfect, they must be grown heavily treated with pesticides and fungicides, with herbicides added to keep the weeds under control. Since foods spoil over time, many fresh foods are treated with preservatives to make them look fresh even if they are not. Food additives are in almost every prepared product to reduce rate of spoilage and to improve color and flavor. There are some 3000 food additives in common usage. While our canned foods used to be in bare aluminum cans, we now line these cans with bisphenol A to avoid any metallic taste, assuming that the bisphenol A stays on the can. When we freeze our foods, we almost always place them in plastic, and we drink from plastic bottles and cups and assume that the plasticizers there, usually various phthalates or bisphenol A, do not leach into the food or drink.

It is not just fruits and vegetables that now contain chemicals that were not in them in earlier times. Now our meats come from animals treated with antibiotics and growth hormones. Our fish come from waters contaminated with persistent organic pollutants, such as bis[*p*-chlorophenyl]-1,1,1-trichloroethane (DDT) and its breakdown product, 2,2-bis(*p*-chlorophenyl)-1,1-dichloroethylene (DDE), other pesticides, polychlorinated biphenyls (PCBs), methyl mercury, and even pharmaceuticals that are discharged into the waste water through human excretion and deposition of unused pharmaceuticals down the toilet. Many of the fish we eat come from fish farms, where fish are caged and fed food that often is contaminated with chemicals (Hites et al. 2004). In addition, in order to prevent infectious and fungal diseases in the enclosed, concentrated environment, antibiotics and fungicides must be used. Even the wild fish from lakes, streams, and the ocean contain organic chemicals, especially those that are lipophilic and persistent. The same contaminants, albeit usually at a lower concentration, are in our meats, eggs, and dairy products as a result of the contemporary practice of adding waste animal fats and products into the food fed to domestic farm animals. The feeding of waste animal fats to domestic animals that are not naturally carnivorous has resulted in the recycling of dangerous persistent chemicals like DDT and PCBs, which have not been produced in developed countries for more than 30 years, back into our food supply (IOM 2003).

Most people assume that the chemicals in carpets, in plastic food containers, and in drink bottles, and those sprayed under the kitchen sink to deal with insects stay put. However, it is clear that this is often not the case. Furthermore, most people assume that governments would not allow chemicals that might pose a hazard to health to be used. However, this also is often not the case. Unfortunately, chemicals volatilize from carpets and under-the-sink pesticide applications. They leach out of food and drink containers. Even before reaching the kitchen, there are chemicals in the food reflecting what the food animal ate or was treated with, and there are chemicals on the fruits and vegetables that are only partially removed by washing. So, a variety of organic chemicals are in the food and water we eat and drink and in the air we breathe, and are also absorbed through our skin.

Because infants and children are particularly vulnerable to harm from exposure to contaminants, there is special concern about the impact of pesticides in the diets of infants and children (NRC 1993). However, the mother's body is the first environment for the child, and the contaminants in the mother's body are passed to the fetus. Thus, efforts to reduce exposure to dangerous organics should focus on all women of reproductive age, not just infants and children.

Governments struggle to balance the promotion of new chemicals that will be useful to humankind with the protection of the public from hazards. The development and marketing of organic chemicals has increased enormously in a relatively brief period of time after World War II. In the United States, the Toxic Substance Control Act of 1976 (TSCA) is the law that presently regulates new chemicals. At present, there are more than 84,000 chemicals in this inventory, most of them organics. When the law was passed, most existing chemicals (62,000) were grandfathered into the inventory and were allowed to remain on the market without further study. Some chemicals were specifically identified to no longer be manufactured and used, as was the case with PCBs. New chemicals continue to be added to the inventory, but most of the testing of safety is dependent on the manufacturer. Figure 1.1 shows the

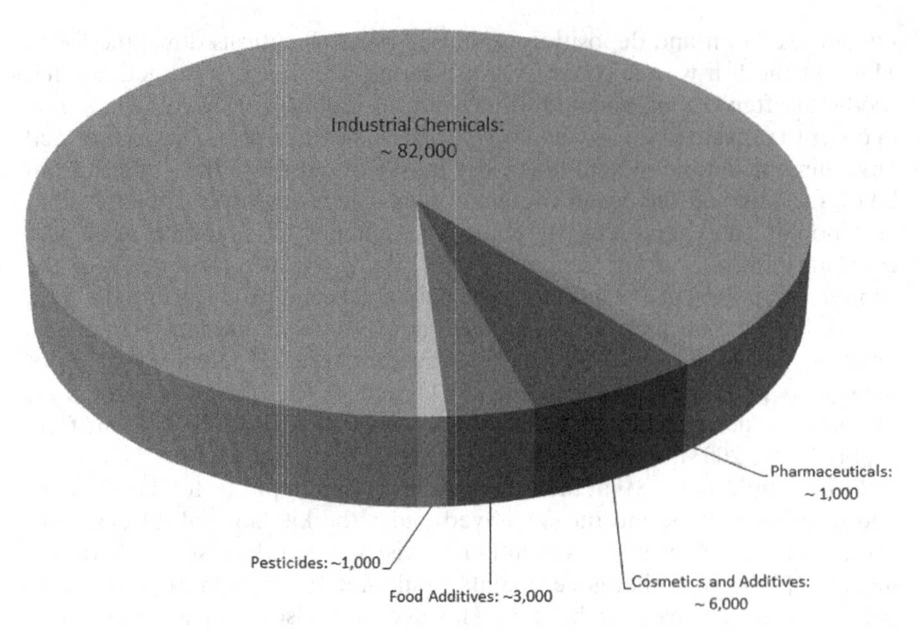

Figure 1.1. Approximately 100,000 individual chemicals have been registered for commercial use in the United States over the past 30 years. Chemical classes that receive the majority of public attention (e.g., pharmaceuticals, cosmetics and food additives, and pesticides) constitute only a small percentage of this inventory. Analytical methodologies are currently limited to several hundred of these nonregulated chemicals. Adapted from Muir and Howard (2006) with permission.

distribution of chemicals currently on the market. Most are organics, although there are also some metals. To date, only about 250 chemicals have been rigorously tested independent of the industry by the Environmental Protection Agency (EPA), and only 5 have been regulated. In addition, TSCA (and thus EPA) does not have regulatory authority over pesticides, tobacco and tobacco products, radioactive materials, foods, food additives, drugs, and cosmetics, all of which are regulated by different government agencies. While new legislation is needed, a number of steps have been taken to prioritize chemicals of high use and those that are the most worrisome in terms of impacts on public health.

In 1999, the Canadian government implemented a tiered approach to address chemicals of concern in their inventory under the Canadian Environmental Protection Act. They evaluated 23,000 chemicals with a screen including physicochemical properties that might relate to persistence and bioaccumulation, measures of toxicity to various organ systems with consideration of acute, subchronic, and chronic endpoints. They identified 500 chemicals of high priority and 193 that required regulatory action. The government is continually reviewing the high-priority chemicals.

In late 2008, the European Chemical Agency, in preparation for the implementation of Registration, Evaluation, and Authorization of Chemicals (REACH), preregistered about 150,000 substances (http://www.echa.europa.eu/). The stated goal of REACH is "to improve the protection of human health and the environment through the better and earlier identification of the intrinsic properties of chemical substances." It gives greater responsibility to industry to manage risks from chemicals and to provide safety information. It also has a goal of obtaining progressive substitution of the most dangerous chemicals when less dangerous alternatives are available. The provisions of REACH are to be phased in over a period of 11 years.

These actions by various governments are all intended to prevent chemicals, especially organic chemicals, from being produced and used before it is certain that they will not escape into the environment, lead to exposure to animals and people, and pose significant hazards to human health. However, the reality is that to do so is very difficult. Premarket tests usually look at acute lethality in animal models or study animal or human cells in culture. Investigation of the subtle effects on the nervous or immune systems and the delayed elevated risk of developing cancer is much more difficult and much more expensive. Even if this long-term testing is done in animal models, there is no certainty that humans will respond exactly the same. Thus, we all become guinea pigs for the effects of exposure to chemicals.

Another major problem is that most testing and understanding of the hazardous effects of chemicals in animal and cellular models are done one chemical at a time. But in the real world, each of us is constantly exposed to a very great mixture of chemicals. There is a mixture of chemicals in the air we breathe, a different mixture in the water or other fluids we drink, yet a different mixture in the food we eat and then we put yet other chemicals on or in our body through medications, lotions, shampoos, and other personal care products. However, interactions between the effects of two or more chemicals have been very poorly studied. There are three major possibilities—the effects of two chemicals may be additive, less than additive, or synergistic (Carpenter et al. 1998). Of particular concern is when there are synergistic effects.

To make things even more complex, the above-mentioned discussion assumes that one chemical has only one site of action. DDT, for example, kills insects by blocking the action potential in insect nerves and causing paralysis. This is the mechanism of action that kills pests. However, in humans, DDT does not block action potentials but increases the risk of a great variety of human diseases, including cancer, cardiovascular disease, diabetes, nervous systems effects, and changes in immune system function (detailed in the various chapters in this book). These different effects are certainly not mediated by actions at the neuronal sodium channel! And it is very unlikely that the effects on the different organ systems are mediated by the same mechanisms. This may involve different receptor binding sites or induction of different genes. Kiyosawa et al. (2008b) found that technical-grade DDT in rats induced genes associated with drug metabolism, cell proliferation and oxidative stress, and the nuclear receptors constitutive androstane receptor and pregnane X receptor. In another study, Kiyosawa et al. (2008a) reported that the pattern of gene induction in the mouse was significantly different from that in the rat to the same exposure. So one must conclude that any chemical that can induce genes regulating many different physiological functions has the potential to cause a great variety of different effects, but that there may be significant species differences which make extrapolation from animals to humans subject to errors.

These actions at different receptors and induction of a great variety of different genes likely explain the increasing frequency of demonstration of low-dose effects, nonlinear dose–response curves and what is commonly called "hormesis" (Calabrese 2008; Lee et al. 2010; Welshons et al. 2003). It has always been a tenant of toxicology that "the poison is in the dose." This may well be true if the poison has a single binding site that leads to a single action, but it is clearly not true for the actions of many organics that have both multiple binding sites in different organ systems and also induce genes that alter many different physiological functions.

One book cannot hope to cover all organic chemicals or all possible biological effects. However, in this book, we have tried to consider effects on the major organ systems and the actions of representative chemicals for which there is at least some information. In many cases, the focus is on the persistent organic pollutants for the very practical reason that, because of their persistence, we have better exposure assessment and more information than is available for less persistent organics. As will be clear, our knowledge on the range of human health effects of organic chemicals is incomplete and much more research is needed.

REFERENCES

Calabrese EF. 2008. Hormesis: 2008. Why it is important to toxicology and toxicologists. Environ Toxicol Chem 27:1451–1474.

Carpenter DO, Arcaro KF, Bush B, Niemi WD, Pang S, Vakharia DD. 1998. Human health and chemical mixtures: An overview. Environ Health Perspect 106 (Suppl 6):1263–1270.

Hites RA, Foran JA, Carpenter DO, Hamilton MC, Knuth BA, Schwager SJ. 2004. Global assessment of organic contaminants in farmed salmon. Science 303: 226–229.

IOM (Institute of Medicine). 2003. Dioxins and Dioxin-Like Compounds in the Food Supply: Strategies to Decrease Exposure. Washington, DC: National Academies Press.

Kiyosawa N, Kwekel JC, Burgoon LD, Dere E, Williams KJ, Tashiro C, et al. 2008a. Species-specific regulation of PXR/CAR/ER-target genes in the mouse and rat liver elicited by o,p′-DDT. BMC Genomics 9:487.

Kiyosawa N, Kwekel JC, Burgoon LD, Williams KJ, Tashiro C, Chittim B, et al. 2008b. o,p′-DDT elicits PXR/CAR-, not ER-, mediated responses in the immature ovariectomized rat liver. Toxicol Sci 101:350–363.

Lee DH, Steffes MW, Sjödin A, Jones RS, Needham LL, Jacobs DR, Jr. 2010. Low dose of some persistent organic pollutants predicts type 2 diabetes: A nested case-control study. Environ Health Perspect 118:1235–1242.

Muir DC, Howard PH. 2006. Are there other persistent organic pollutants? A challenge for environmental chemists. Environ Sci Technol 40:7157–7166.

NRC (National Research Council). 1993. Pesticides in the Diets of Infants and Children. Washington, DC: National Academies Press.

Welshons WV, Thayer KA, Judy BM, Taylor JA, Curran EM, vom Saal FS. 2003. Large effects from small exposures. I. Mechanisms for endocrine-disrupting chemicals with estrogenic activity. Environ Health Perspect 111:994–1006.

CHAPTER 2

Sources of Human Exposure

MARTÍ NADAL and JOSÉ L. DOMINGO

ABSTRACT

Background: Persistent and bioactive organic pollutants may reach the human body through different pathways, which usually determine subsequent health effects. Although occupational exposure has a prominent role, the environmental/dietary contact with these substances may be also very important. Therefore, it is critical not only to identify but also to estimate the contribution of each one of the exposure pathways.

Objectives: This chapter presents current calculation methods to estimate the main pathways of exposure to organic pollutants. Information regarding a few chemicals (persistent organic pollutants, pesticides, benzene, and perfluoroalkyl substances) is also summarized.

Discussion: Direct (or nondietary) exposure can be estimated as the sum of pollutant intake through air inhalation (air concentration related), as well as soil ingestion and dermal absorption (both dependent of soil concentration). In turn, dietary exposure can be calculated by considering food intake and water consumption. Dietary intake seems to be the main human exposure route to organic contaminants such as POPs or pesticides, with only a few exceptions. To a lesser extent, other pathways may have some notable contribution, especially for particular subgroups of population characterized by being more vulnerable to environmental pollutants, such as children or aged people.

Conclusions: Some basic tools to perform a first-tier screening for human health risk assessment, focusing on human exposure, are provided here. Food consumption seems to be the most important contributive route to the total intake of persistent and bioactive organic pollutants.

Effects of Persistent and Bioactive Organic Pollutants on Human Health, First Edition.
Edited by David O. Carpenter.

INTRODUCTION

During normal life, people may be exposed to a broad range of chemicals through different pathways. Many contacts with those substances occur in an unconscious and/or involuntary manner during usual and daily activities. Indoor spaces are environments where the potential exposure to chemicals is especially significant. Moreover, occupational exposure to chemicals is also important for some adults during the working day. However, foodstuffs play a key role in the uptake of contaminants by humans. As it has been largely confirmed in recent years, dietary intake is the most critical pathway of exposure for many pollutant substances.

The effects of persistent and bioactive organic pollutants on human health are often dependent on the exposure routes through which those contaminants enter the human body. Therefore, it is critical to identify the main entrance pathways, as well as to estimate the contribution of each one. This information is essential to undertake actions to minimize the human exposure to organics, especially in those subpopulation groups for which the potential adverse health effects are more notable, such as children or the elderly.

This chapter is divided into two basic sections. The first one highlights current methods to estimate the main pathways of exposure to organic pollutants, while the second one compiles information for some specific chemicals, which are contemplated in subsequent chapters.

HUMAN EXPOSURE PATHWAYS

The U.S. National Research Council (NRC 1983), in its so-called Red Book, established a series of principles to be considered for human health risk assessment, defining it as a process in which information is analyzed to determine if an environmental hazard might cause harm to exposed persons and ecosystems. Human exposure was identified as a critical step in the original four-step risk assessment process. In recent years, scientists and governmental organizations have been encouraged to derive quick, easy, but robust mathematical tools to assess human exposure to environmental pollutants, considering that there exist diverse potential routes (dietary and nondietary) through which chemicals can enter the human body.

Direct or Nondietary

Air Inhalation Inhalation occurs when chemical, radioactive, or physical pollutants enter the respiratory system, reaching the lungs. This may be a very important route of exposure, especially for some volatile chemicals and semivolatile organic compounds (SVOCs). This pathway has been found to be the most significant for volatile organic compounds (VOCs), such as benzene and formaldehyde, among others.

The U.S. Environmental Protection Agency (EPA) developed a specific methodology to assess exposure through inhalation (U.S. EPA 2009b). This approach, consistent with the inhalation dosimetry methodology, involves the estimation of exposure concentrations (ECs), instead of doses, for each receptor exposed to contaminants via inhalation in the risk assessment. ECs are time-weighted average concentrations derived from measured or modeled contaminant concentrations in air. The estimation of ECs is a prior step to the evaluation of noncancer risks (hazard quotient) or cancer risks. The recommended process for obtaining a specific EC value is the following: (1) to assess the duration of the exposure scenario, (2) to assess the exposure pattern of the exposure scenario, and (3) to estimate the scenario-specific EC. In the first step, the duration of the exposure scenario is chosen among three possibilities: acute, subchronic, or chronic. The second step entails comparing the exposure time and frequency at a site to that of a typical subchronic or chronic toxicity test. The third and final step involves estimating the EC for the specific exposure scenario based on the decisions made in steps 1 and 2. For subchronic and chronic exposures, EC is calculated according to the following equation:

$$\mathrm{EC} = (\mathrm{CA} \times \mathrm{ET} \times \mathrm{EF} \times \mathrm{ED})/(\mathrm{AT} \times 365),$$

where EC is the exposure concentration (mg/m^3), CA is the concentration in air (mg/m^3), ET is the exposure time (h/day), EF is the exposure frequency (day/year), ED is the exposure duration (years), and AT is the averaging time (years). Specific values of the parameters can be obtained from the scientific literature, including U.S. EPA reports. In case of acute exposure, EC would be equivalent to CA.

Soil Ingestion Contact with contaminated soils may become an important pathway of exposure to organic chemicals, posing large and long-lasting health risks, through different activities (e.g., through hand to mouth by young children, gardening by adults, and tracking of soil and dust into the home) (Kimbrough et al. 2010). In addition, for some classes of organic pollutants, such as persistent organic pollutants (POPs), incidental ingestion of contaminated soil has been pointed out as the major nondietary exposure pathway (Rostami and Juhasz 2011).

The U.S. EPA (1989) developed specific formulations for the estimation of the contribution of each nondietary pathway. The expression used to evaluate the exposure through ingestion ($\mathrm{Exp}_{\mathrm{ing}}$, in mg/kg/day) is the following:

$$\mathrm{Exp}_{\mathrm{ing}} = (\mathrm{CS} \times 10^{-6} \times \mathrm{EF} \times \mathrm{IFP})/(\mathrm{BW} \times 365),$$

where CS is the concentration in soil (mg/kg), EF is the exposure frequency (day/year), IFP is the soil ingestion rate (mg/day), and BW is the body weight (kg).

Oral bioavailability is the fraction of an ingested contaminant that reaches the systemic circulation from the gastrointestinal tract. In turn, bioaccessibility, in relation to human exposure via ingestion, is defined as the fraction of a toxicant in soil that becomes soluble in the gastrointestinal tract, being then available for absorption (Guney et al. 2010). When data of bioavailability and/or bioaccessibility are unknown, worst-case scenarios are generally considered by assuming a value of 100%. In fact, a fraction of the contaminant may only be bioavailable, and therefore, this assumption may grossly overestimate the chemical daily intake, thereby influencing risk assessment (Rostami and Juhasz 2011).

Dermal Absorption Exposure to some indoor organic compounds through the dermal pathway is sometimes underestimated. Transdermal permeation can be substantially greater than is commonly assumed (Weschler and Nazaroff 2012).

When assessing exposure to organic pollutants through the dermal pathway, two different subroutes must be considered, as dermal contact may be relevant for chemicals contained in both water and soil (Ferré-Huguet et al. 2009; U.S. EPA 2009a). A generic formula is given for estimating the exposure through dermal contact (Exp_{derm}, in mg/kg/day):

$$Exp_{derm} = (CS \times 10^{-6} \times AF \times ABS \times EF \times SA)/(BW \times 365),$$

where CS is the concentration in soil (mg/kg), AF is the adherence factor soil (mg/cm), ABS is the dermal absorption fraction (unitless), EF is the exposure frequency (day/year), SA is the surface area (cm^2/day), and BW is the body weight (kg).

A summary of calculation equations to assess the human exposure through nondietary pathways is shown in Figure 2.1.

Dietary

Food A number of studies have shown that dietary intake is the main entrance route of POPs and other organic chemicals to the human body (Cornelis et al. 2012; Domingo 2012b; Martí-Cid et al. 2008a; Perelló et al. 2012b), accounting for more than 90% of the total exposure (Linares et al. 2010; Noorlander et al. 2011). Therefore, the calculation of the total ingestion of pollutants through food consumption is essential to estimate the total amount of chemicals to which humans are exposed.

The ingestion of pollutants (Exp_{diet}, in mg/kg/day) through food consumption is generally calculated as follows:

$$Exp_{diet} = \sum FIR \times CF/BW,$$

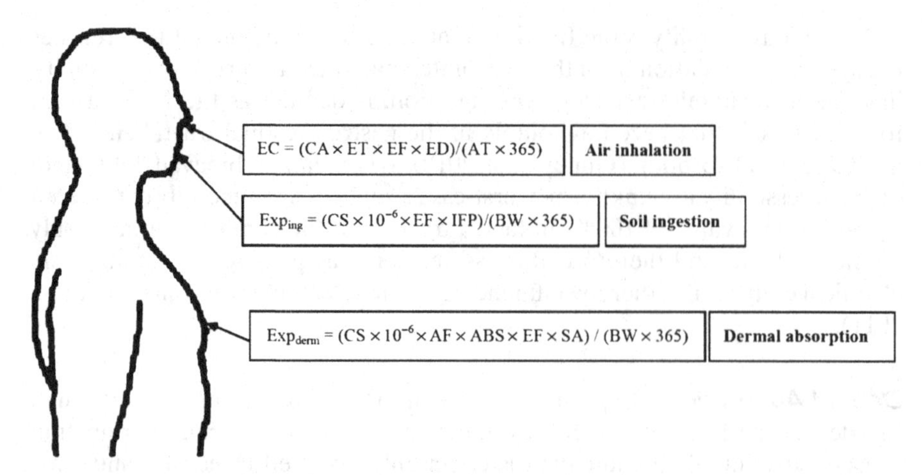

Figure 2.1. Main nondietary exposure pathway routes to persistent and bioactive organic pollutants. Calculation equations.

where FIR is the food ingestion rate (in kg/day), CF is the concentration in food (mg/kg), and BW is the body weight (kg). Thus, the daily intake of a chemical by a food group is estimated by multiplying the average concentration by the daily consumption of the food group. Finally, the estimated total dietary intake of each chemical is obtained by summing the respective intakes from each food group and dividing by the body weight.

Water Indoor exposure through the use of contaminated tap water is an issue of great concern (López et al. 2008). For certain chemicals, the water pathway may be especially significant, considering that adults may consume more than 2L daily. Furthermore, water is a part of the nutritional basis of food ingestion by babies, as many baby foods are prepared by using drinking water, either tap or bottled. In any case, exposure to organic substances through water consumption must not be underestimated.

Similar to food, the intake of chemicals through water ingestion (Exp_{water}, in mg/kg/day) is calculated by applying the following equation:

$$Exp_{water} = \sum WIR \times CW/BW,$$

where WIR is the water ingestion rate (L/day), CF is the concentration in water (mg/L), and BW is the body weight (kg).

CHEMICALS OF CONCERN

POPs

POPs are organic substances that may persist a long time in the environment, may present a high bioaccumulation potential through the food web, and may pose a high degree of toxicity for human health and the environment. Furthermore, POPs are characterized by their long-range transport capacity (LRTC); that is to say, they are able to travel long distances and to be deposited in territories where they have never been used or produced, posing then an important risk for the global community. Under the framework of the Stockholm Convention on POPs signed in 2001, a list of chemicals whose production, use, and storage must be eliminated, or seriously restricted, was developed. Among these, polychlorinated dibenzo-*p*-dioxins and dibenzofurans (PCDD/Fs) and polychlorinated biphenyls (PCBs) were included in the initial list of chemicals, commonly known as the "dirty dozen." However, in recent years, a number of other chemicals have also been catalogued as POPs, enlarging that list, while pollutants such as polychlorinated naphthalenes (PCNs) or polycyclic aromatic hydrocarbons (PAHs), already listed in the United Nations Economic Commission for Europe (UNECE) Protocol, have been also proposed (Nadal et al. 2011). Given that the treaty has been entering into force in many countries in the course of the 2000 decade, the number of studies to monitor the environmental levels of POPs has progressively increased. Furthermore, these investigations have been used to evaluate human exposure to those organic pollutants, as well as to compare the percentage of total exposure contributed by food intake. A number of studies has identified food consumption as the most important pathway of exposure to POPs (especially PCDD/Fs and PCBs), with contributions of >95% (Linares et al. 2010). Moreover, a number of those studies were focused on rather reduced groups of foodstuffs, mainly fish and seafood (Storelli et al. 2011; Yu et al. 2010), as this was the most contributive food group (Figure 2.2). In Catalonia (northeast of Spain), a wide surveillance program focused on measuring the levels of a number of chemical contaminants (including PCDD/Fs and PCBs) in various groups of foodstuffs is being performed, as requested by the Catalan Agency of Food Safety. Three campaigns have been carried out between 2000 and 2012 (Llobet et al. 2003, 2008; Perelló et al. 2012a). In the framework of these investigations, the dietary intake of these pollutants was subsequently estimated for various age and sex groups of the population of the country using deterministic and probabilistic methodologies (Perelló et al. 2012a). An important decreasing trend in the dietary exposure to PCDD/Fs and PCBs for the population living in Catalonia was noted. The authors associated this finding with the general decreasing trend in the atmospheric PCDD/F and PCB levels, which has also been observed in a number of countries in recent years. The intake of these pollutants was generally lower in Catalonia than those recently found in various other regions and countries over the world. With respect to the health risks derived from dietary exposure to PCDD/Fs and dioxin-like polychlorinated biphenyls (dl-PCBs), it must be remarked that the current total daily intake is lower (even considering the individuals in the extreme of the exposure distribution) than the tolerable daily intake (TDI) established by international organizations. In relation to this, for comparative purposes, the tolerable intake established by the World Health Organization (WHO) for dioxin-like compounds, including PCDD/Fs and dl-PCBs is within the range 1–4 pg WHO-TEQ/kg

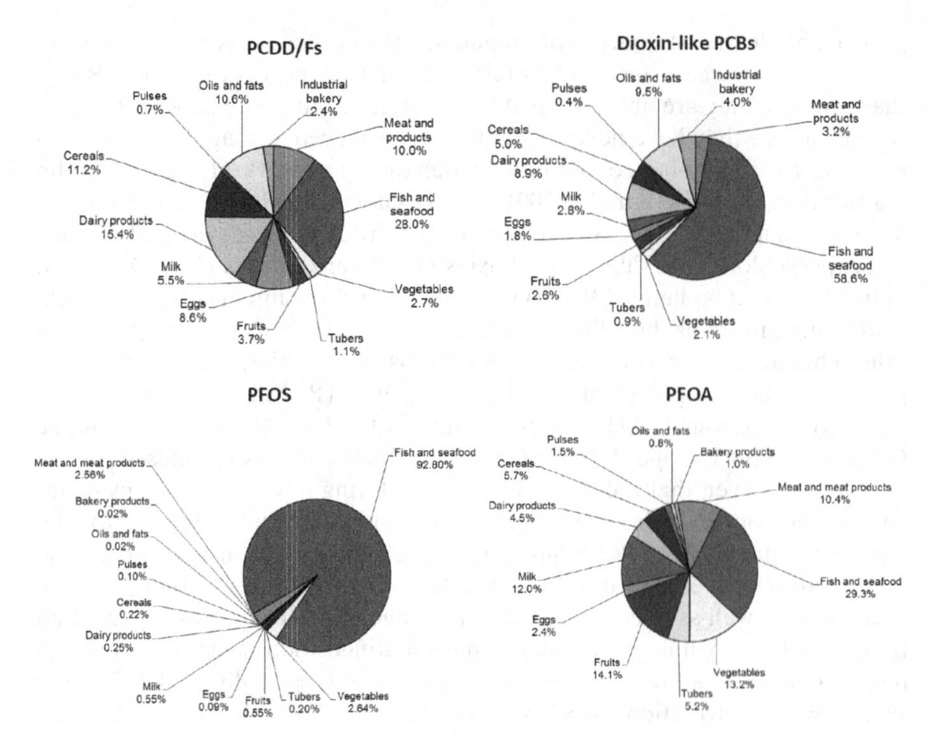

Figure 2.2. Percentages of contribution from each food group to the total dietary intake of several organic pollutants by the adult population of Catalonia (Spain).

of body weight per day (van Leeuwen et al. 2000), while the Scientific Committee on Food (SCF) of the European Commission set a value of 14 pg WHO-TEQ/kg of body weight on a weekly basis.

In parallel, the direct exposure to PCDD/Fs and PCBs by sources other than diet has been estimated for the population of Catalonia. In the most recent study (Linares et al. 2010), the direct exposure to PCDD/Fs ranged between $5.00{\cdot}10^{-6}$ and $9.69{\cdot}10^{-6}$ ng WHO-TEQ/kg·day. Dermal absorption was the main entrance route of PCDD/Fs to the human body (54%), while the lowest contribution corresponded to soil ingestion (15%). These values are in contrast with the results obtained in previous studies (59%, 28%, and 13% for inhalation, dermal contact, and soil ingestion, respectively) (Nadal et al. 2004). The decrease of importance of the inhalation route could be linked to the notable reduction of air PCDD/F levels observed in the area under evaluation, with up to 10-fold diminutions. Other investigations performed in industrial areas of Catalonia, where cement factories are operating, indicate that inhalation is still the most important route of direct contact (40–60%), compared to dermal absorption and soil ingestion (Rovira et al. 2010, 2011). Concerning other POPs, such as PAHs and PCNs, inhalation has also been pointed out as the most contributive among the direct exposure routes (Nadal et al. 2011), but still minor in comparison to total dietary intake, whose estimative value was at least 95% of the total intake (Martí-Cid et al. 2008a,b).

Pesticides

Although several pesticides are considered as POPs (i.e., aldrin, chlordane, DDT, among others), the group of pesticides includes a longer list of chemical agents of different physicochemical characteristics. Pesticides are actually defined as "chemical substances used to prevent, destroy, repel or mitigate any pest ranging from insects (i.e., insecticides), rodents (i.e., rodenticides) and weeds (herbicides) to microorganisms (i.e., algicides, fungicides or bactericides)" (Alavanja 2009). People may be exposed to pesticides through the same pathways, being mainly dietary intake, followed by inhalation, the two most influent exposure routes. The populations at risk of developing adverse health effects associated with pesticide exposure include subjects such as farm workers, workers in the pesticide production industry, pest control workers, and individuals environmentally exposed, such as farm residents and those using household pesticides (Ndlovu et al. 2011). In fact, the potential exposure of pesticides for the agricultural sector is very important since about 1800 million people in the world are involved in agriculture, most of this population being exposed to these products (Alavanja 2009). The potential health effects of agricultural pesticide exposures are of particular interest as these chemicals are designed to have adverse biological effects on target organisms (Weichenthal et al. 2010). An increasing incidence of cancer, chronic kidney diseases, suppression of the immune system, sterility among males and females, endocrine disorders, and neurological and behavioral disorders, especially among children, has been attributed to chronic pesticide poisoning (Abhilash and Singh 2009). The application of pesticides for pest control means an important entrance route of these chemicals to the human body through the dietary intake, with fish and seafood being the group with the most significant contribution (Törnkvist et al. 2011). However, when considering long assessment periods, it has been observed that intake estimations for organochlorine pesticides have apparently decreased (Fromberg et al. 2011).

Recent findings have demonstrated that inhalation and dermal absorption of pesticides in general, and DDT in particular, may be important routes of exposure (Sereda et al. 2009). In fact, specific studies on highly exposed populations living in tropical houses where there was indoor residual spraying (IRS) estimated that inhalation could account for 70% of total exposure to DDT, 2.5 times the estimated median exposure through diet (Ritter et al. 2011). However, the general population is primarily exposed to pesticides through food intake (Beamer et al. 2012; Dirtu and Covaci 2010). In turn, the percentage contribution of nondietary exposure routes seems to be lower, especially when professional activities do not include any contact with pesticides. Ferré-Huguet et al. (2009) found that the uptake of DDTs, their metabolites—dichlorodiphenyldichloroethylenes (DDEs) and dichlorodiphenyldichloroethanes (DDDs)—as well as other organochlorine compounds through soil ingestion and consumption of drinking water did not increase either noncarcinogenic nor cancer risks for the population living in the Catalan stretch of the Ebro River basin. This is an agricultural zone where industries and sewage treatment plants are of notable concern, taking into account the potential adverse impacts on water quality and local soils. The same research group also assessed the dietary intake of the same chemicals and estimated the health risks associated with food consumption (Martí-Cid et al. 2010). Although the consumption of local foods for different population age/gender groups should not mean an increase in noncancer and cancer risks, as all indices were less than the safety values, the dietary intake of DDTs, DDEs, and DDDs was found to be much higher than the environmental exposure to the same chemicals. Among the food items analyzed, fish and seafood were important contributors to the dietary exposure of DDT derivatives (DDE and DDD) in the adult population, while consumption of vegetables was especially notable for the parental compound (DDT). A review of the scientific literature indicates that ingestion of house dust may be also a major route of exposure to pesticides for infants and toddlers. The role of house dust as an exposure source is gaining more attention over the years.

However, several open questions related to health remain to be resolved. Pesticides applied outside or within the household, which are absorbed and preserved by house dust, can lead to an increased exposure through the everyday activities of children and infants. Residential exposure including house dust residues contribute to combined exposure from dietary and nondietary sources (Butte and Heinzow 2002).

Benzene

The chemical structure of benzene (C_6H_6) is a ring with six carbon atoms and a hydrogen atom attached to each carbon atom. Benzene is a natural constituent of crude oil, being one of the most basic aromatic petrochemicals, together with toluene, ethylbenzene, and *m,p,o*-xylenes. It is generally accepted that benzene is a risk factor for childhood acute leukemias and breast cancer, among other adverse health effects (McNally and Parker 2006; Rennix et al. 2005).

More than 99% of the intake of benzene is through the air, where it may originate from natural sources (e.g., forest fires) or from human activities such as smoking or exhaust fumes (Van Poucke et al. 2008). Other sources of benzene are drinking water and food, both through environmental contamination. Regarding inhalation, indoor air plays an important role in the total inhalation of VOCs. In addition to smoking, incense burning and emissions from consumer products are also sources of benzene and other VOCs in indoor spaces. Recently, Sarigiannis et al. (2011) reviewed bibliographic data on the occurrence of major organic compounds and evaluated cancer and noncancer risks posed by indoor exposure in dwellings and public buildings in European Union (EU) countries. The results indicate that significant differences in indoor air quality exist within and among the countries where data were available. Another important emission source of benzene and other VOCs is municipal solid waste (MSW). Handling and treatment of MSW are known to generate benzene. Particularly, composting facilities are known to release odorous VOCs due to biodegradation of waste (Domingo and Nadal 2009). Therefore, not only individuals working at composting plants but also residents living nearby may be potentially exposed to benzene (Nadal et al. 2009; Vilavert et al. 2012). Human exposure studies are usually focused on the inhalation pathway, as this has been identified as the most contributive route (Vilavert et al. 2011).

Unlike other organic chemicals, such as POPs, food has not been identified as the leading route of exposure to VOCs in general, and benzene in particular. Since the intake of benzene from the diet is usually about 1000 times lower than that derived from heavy cigarette smoking (estimated average = 1.8 mg/day), it is less likely that dietary sources are the major contributors to elevated levels of benzene and metabolites in the general population (Johnson et al. 2007). On the other hand, drinks can contain significant amounts of benzene, as benzoic acid is used as a preservative in some beverages and can react with ascorbic acid (vitamin C), either added or naturally occurring, to form benzene. The benzoate salts are preservatives that are added to beverages to inhibit growth of bacteria, yeasts, and mold but may also occur naturally in some fruit juices. Vitamin C may be added as a preservative, as a vitamin supplement, or may also be naturally present in some fruit juices (Haws et al. 2008). Recently, the content of pollutants, such as arsenic or benzene, is being carefully studied in baby foods, taking into consideration that infant exposure is critical for the further development of children. With respect to this, benzene has been detected not only in certain beverages and soft drinks but also in baby food, specifically in carrot juices intended for infants (Lachenmeier et al. 2010). These juices contain higher concentrations of benzene than any other beverage group, with an average content above the EU drinking water limit, which is 1 μg/L (Lachenmeier et al. 2008). In this sense, Lachenmeier et al. (2010) detected trace (μg/kg) levels of benzene in canned foods, jarred baby food, and juices containing carrots, showing that the level of exposure to benzene through food products could be currently underestimated.

Perfluoroalkyl Substances (PFASs)

PFASs are molecules made up of carbon chains to which fluorine atoms are bound. Due to the strength of the carbon/fluorine bond, these molecules are chemically very stable and are highly resistant to biological degradation, therefore being persistent in the environment (Stahl et al. 2011). Furthermore, these compounds are extremely bioaccumulative and of toxicological concern (Andersen et al. 2008; D'Hollander et al. 2010; Domingo 2012a). Because of their chemical properties, PFASs have been widely used in a broad range of applications, such as inks, varnishes, waxes, firefighting foams, metal plating and cleaning, coating formulations, lubricants, water and oil repellents for leather, paper, and textiles (Paul et al. 2009). In recent years, a number of studies have reported a ubiquitous distribution of PFASs in human tissues (Sturm and Ahrens 2010), as well as in invertebrates, fish, reptiles, and marine mammals worldwide (Houde et al. 2011). Perfluorooctane sulfonate (PFOS) is the predominant compound detected in humans and animals, as well as in environmental samples (Jogsten et al. 2009). The industrial production of PFOS and some of its derivatives was phased out by the major producer, 3M, in 2002, while the EU banned most uses of this compound since 2008. PFOS has been very recently included in the list of priority substances in the field of water policy, which includes the chemicals identified among those presenting a significant risk to or via the aquatic environment at the EU (according to the Water Framework Directive) (EC 2012). In 2009, PFOS was included in Annex B of the Stockholm Convention list of POPs (Buck et al. 2011). Unlike other organic chemicals, PFASs do not typically accumulate in lipids. In humans, exposure levels and pathways leading to the presence of PFASs have been better characterized by monitoring these chemicals in blood. In recent years, the concentrations of various PFASs in human blood have been determined in individuals from a number of regions and countries (Fromme et al. 2009). Although the relative importance of the routes of human exposure to these compounds is not yet well established, it has been suggested that food intake and packaging, water, house dust, and air are all potentially significant exposure sources (Domingo 2012a).

Since 2006, the Laboratory of Toxicology and Environmental Health, Universitat Rovira i Virgili (Catalonia, Spain), has been performing periodical studies to find out the contribution of each pathway to total exposure to PFASs. Biomonitoring investigations included analysis of the concentrations of 13 PFASs in the blood of Catalan residents (Ericson et al. 2007). In general terms, reported values were lower than those found in human blood and serum of subjects from different countries, with PFOS showing the highest concentration of all PFASs. Breast milk samples from primiparae mothers were also collected, and the content of PFASs was determined. Milk concentrations were similar to reported levels from other countries (Kärrman et al. 2010). Finally, human liver samples were collected from subjects who had lived in different areas of Tarragona County (Catalonia, Spain). Liver samples were found to contain more PFASs above quantification limits and higher PFOS concentrations compared to reports from the scientific literature. Interestingly, perfluorooctanoic acid (PFOA) levels from Catalan males were significantly higher ($p < 0.05$) than those from females in both liver and blood (Kärrman et al. 2010). Because it is highly consumed by humans, water could be an important contributor to the exposure of PFOA, PFOS, and other PFASs (Kim et al. 2011). The concentration of the same PFASs was analyzed in water samples from Tarragona Province (Catalonia, Spain), coming from diverse origins: tap and bottled drinking water, as well as river water. The samples of bottled water were found to be the least contaminated by PFASs (Ericson et al. 2008). Although values were similar or lower than those reported in the literature for surface water samples from a number of regions and countries, it was concluded that, in certain cases, drinking water could be a source of exposure to PFASs as important as the dietary intake of these pollutants (Ericson et al. 2008).

Considering these outstanding results, new environmental monitoring studies were carried out in 2008 and 2009 to evaluate the temporal evolution of PFASs in tap water, as well as to assess the trends in human exposure to PFASs by the Catalan population. The same PFASs were analyzed in municipal drinking water samples collected at 40 different locations from five different zones of Catalonia. The most contaminated water samples were found in Barcelona Province, which is the most industrialized area of Catalonia (Ericson et al. 2009). Notwithstanding, a decreasing tendency was observed for some PFASs in the last period of assessment (2008–2009), with reductions of 48% and 47% for PFOS and PFOA, respectively (Domingo et al. 2011).

Inhalation of air and dust ingestion is also believed to be important, especially since people spend more than 90% of their time indoors (Shoeib et al. 2011). In fact, indoor air levels of PFASs may be substantially higher than outdoor values (Harrad et al. 2010). Furthermore, dust ingestion may be of notable concern for toddlers as this population subgroup may ingest large quantities of dust through hand-to-mouth contact (Shoeib et al. 2011). The contribution of air sources to total human exposure to PFASs was estimated by determining their concentrations in both house dust and indoor air from selected homes in Catalonia (Ericson Jogsten et al. 2012). It was confirmed that these two routes contribute significantly less to PFAS exposure in this population. Food intake, and particularly fish and seafood, has been suggested as the main exposure route of PFASs in general populations. Some studies indicate that there is an association between fish consumption and PFAS concentrations in serum (Château-Degat et al. 2010; Halldorsson et al. 2008; Haug et al. 2010).

In order to estimate the human exposure to PFAS through food consumption, the levels of 14 PFASs were determined in 36 composite samples of foodstuffs randomly purchased in 2006 in various locations of Tarragona County (southern Catalonia). The dietary exposure to PFOS and other compounds was estimated for various age/gender groups. For the adult population, PFOS intake through food was estimated to be 62.5 (assuming undetected values as zero, ND = 0) or 74.2 ng/day (assuming that nondetected concentrations were one-half of the detection limit, ND = ½ LOD). Fish, followed by dairy products and meats, were the main contributors to PFOS intake (Figure 2.2). Similar results were also found very recently, as fish and seafood was identified as the largest contributor to total exposure to PFASs, in general, and PFOS, in particular (unpublished data). Moreover, those data confirmed that PFOS was the PFAS with the highest concentrations in marine species, mainly in blue fish, the species presenting the major lipid content (Domingo et al. 2011).

CONCLUSIONS AND RECOMMENDATIONS

As it has been shown, humans are regularly exposed to organic chemicals during their daily activities. The contact with these substances may be especially significant in occupational environments, although for nonoccupationally exposed populations, the contact to those pollutants must not be disregarded. Dietary intake seems to be the main exposure route of many organic contaminants, such as POPs or pesticides, to the human body, with only a few exceptions (e.g., benzene). To a lesser extent, other pathways may have some notable contribution, especially for particular subgroups of population characterized by being more vulnerable to environmental pollutants, such as children or elderly people. Therefore, it is critical to understand the diverse potential exposure pathways and to identify the contribution of each one in order to minimize the subsequent adverse effects for human health.

This second chapter reviews some methodologies to calculate exposure to organic pollutants through different pathways. The most notable direct routes were air inhalation, soil ingestion, and dermal absorption, while intake through water and food consumption was also considered. We provided some basic tools to perform a first-tier screening for human health risk assessment, focusing on the third step of the process: human exposure. In turn, information regarding some particular pollutants has been provided by using POPs, pesticides, benzene, and PFASs as examples. Subsequent chapters will go deeper on the potential health effects of these and other persistent and bioactive organic pollutants once they have reached the human body.

REFERENCES

Abhilash PC, Singh N. 2009. Pesticide use and application: An Indian scenario. J Hazard Mater 165:1–12.

Alavanja MCR. 2009. Introduction: Pesticides use and exposure extensive worldwide. Rev Environ Health 24:303–309.

Andersen ME, Butenhoff JL, Chang SC, Farrar DG, Kennedy GL, Jr., Lau C, et al. 2008. Perfluoroalkyl acids and related chemistries—Toxicokinetics and modes of action. Toxicol Sci 102:3–14.

Beamer PI, Canales RA, Ferguson AC, Leckie JO, Bradman A. 2012. Relative pesticide and exposure route contribution to aggregate and cumulative dose in young farmworker children. Int J Environ Res Public Health 9:73–96.

Buck RC, Franklin J, Berger U, Conder JM, Cousins IT, de Voogt P, et al. 2011. Perfluoroalkyl and polyfluoroalkyl substances in the environment: Terminology, classification, and origins. Integr Environ Assess Manag 7:513–541.

Butte W, Heinzow B. 2002. Pollutants in house dust as indicators of indoor contamination. Rev Environ Contam Toxicol 175:1–46.

Château-Degat ML, Pereg D, Dallaire R, Ayotte P, Dery S, Dewailly T. 2010. Effects of perfluorooctanesulfonate exposure on plasma lipid levels in the Inuit population of Nunavik (Northern Quebec). Environ Res 110:710–717.

Cornelis C, D'Hollander W, Roosens L, Covaci A, Smolders R, Van Den Heuvel R, et al. 2012. First assessment of population exposure to perfluorinated compounds in Flanders, Belgium. Chemosphere 86:308–314.

D'Hollander W, De Voogt P, De Coen W, Bervoets L. 2010. Perfluorinated substances in human food and other sources of human exposure. Rev Environ Contam Toxicol 208:179–215.

Dirtu AC, Covaci A. 2010. Estimation of daily intake of organohalogenated contaminants from food consumption and indoor dust ingestion in Romania. Environ Sci Technol 44:6297–6304.

Domingo JL. 2012a. Health risks of dietary exposure to perfluorinated compounds. Environ Int 40:187–195.

Domingo JL. 2012b. Polybrominated diphenyl ethers in food and human dietary exposure: A review of the recent scientific literature. Food Chem Toxicol 50: 238–249.

Domingo JL, Nadal M. 2009. Domestic waste composting facilities: A review of human health risks. Environ Int 35:382–389.

Domingo JL, Ericson-Jogsten I, Nadal M, Perelló G, Bigas E, Llebaria X, et al. 2011. Exposure to perfluorinated compounds through drinking water, and fish and seafood by the population of Catalonia (Spain). Organohalogen Compd 73:969–972.

EC. 2012. Proposal for a Directive of the European Parliament and of the Council amending Directives 2000/60/EC and 2008/105/EC as regards priority substances in the field of water policy.

Ericson I, Gómez M, Nadal M, van Bavel B, Lindström G, Domingo JL. 2007. Perfluorinated chemicals in blood of residents in Catalonia (Spain) in relation to age and gender: A pilot study. Environ Int 33:616–623.

Ericson I, Nadal M, Van Bavel B, Lindström G, Domingo JL. 2008. Levels of perfluorochemicals in water samples from Catalonia, Spain: Is drinking water a significant contribution to human exposure? Environ Sci Pollut Res 15:614–619.

Ericson I, Domingo JL, Nadal M, Bigas E, Llebaria X, Van Bavel B, et al. 2009. Levels of perfluorinated chemicals in municipal drinking water from Catalonia, Spain: Public health implications. Arch Environ Contam Toxicol 57:631–638.

Ericson Jogsten I, Nadal M, van Bavel B, Lindström G, Domingo JL. 2012. Per- and polyfluorinated compounds (PFCs) in house dust and indoor air in Catalonia, Spain: Implications for human exposure. Environ Int 39:172–180.

Ferré-Huguet N, Bosch C, Lourencetti C, Nadal M, Schuhmacher M, Grimalt JO, et al. 2009. Human health risk assessment of environmental exposure to organochlorine compounds in the Catalan stretch of the Ebro River, Spain. Bull Environ Contam Toxicol 83:662–667.

Fromberg A, Granby K, Højgård A, Fagt S, Larsen JC. 2011. Estimation of dietary intake of PCB and organochlorine pesticides for children and adults. Food Chem 125:1179–1187.

Fromme H, Tittlemier SA, Völkel W, Wilhelm M, Twardella D. 2009. Perfluorinated compounds—Exposure assessment for the general population in Western countries. Int J Hyg Environ Health 212:239–270.

Guney M, Zagury GJ, Dogan N, Onay TT. 2010. Exposure assessment and risk characterization from trace elements following soil ingestion by children exposed to playgrounds, parks and picnic areas. J Hazard Mater 182:656–664.

Halldorsson TI, Fei C, Olsen J, Lipworth L, McLaughlin JK, Olsen SF. 2008. Dietary predictors of perfluorinated chemicals: A study from the Danish National Birth Cohort. Environ Sci Technol 42:8971–8977.

Harrad S, De Wit CA, Abdallah MAE, Bergh C, Björklund JA, Covaci A, et al. 2010. Indoor contamination with hexabromocyclododecanes, polybrominated diphenyl ethers, and perfluoroalkyl compounds: An important exposure pathway for people? Environ Sci Technol 44:3221–3231.

Haug LS, Thomsen C, Brantsæter AL, Kvalem HE, Haugen M, Becher G, et al. 2010. Diet and particularly seafood are major sources of perfluorinated compounds in humans. Environ Int 36:772–778.

Haws LC, Tachovsky JA, Williams ES, Scott LLF, Paustenbach DJ, Harris MA. 2008. Assessment of potential human health risks posed by benzene in beverages. J Food Sci 73:T33–T41.

Houde M, De Silva AO, Muir DCG, Letcher RJ. 2011. Monitoring of perfluorinated compounds in aquatic biota: An updated review. Environ Sci Technol 45: 7962–7973.

Jogsten IE, Perelló G, Llebaria X, Bigas E, Martí-Cid R, Kärrman A, et al. 2009. Exposure to perfluorinated compounds in Catalonia, Spain, through consumption of various raw and cooked foodstuffs, including packaged food. Food Chem Toxicol 47:1577–1583.

Johnson ES, Langård S, Lin YS. 2007. A critique of benzene exposure in the general population. Sci Total Environ 374:183–198.

Kärrman A, Domingo JL, Llebaria X, Nadal M, Bigas E, van Bavel B, et al. 2010. Biomonitoring perfluorinated compounds in Catalonia, Spain: Concentrations and trends in human liver and milk samples. Environ Sci Pollut Res 17: 750–758.

Kim SK, Kho YL, Shoeib M, Kim KS, Kim KR, Park JE, et al. 2011. Occurrence of perfluorooctanoate and perfluorooctanesulfonate in the Korean water system: Implication to water intake exposure. Environ Pollut 159:1167–1173.

Kimbrough RD, Krouskas CA, Leigh Carson M, Long TF, Bevan C, Tardiff RG. 2010. Human uptake of persistent chemicals from contaminated soil: PCDD/Fs and PCBs. Regul Toxicol Pharmacol 57:43–54.

Lachenmeier DW, Reusch H, Sproll C, Schoeberl K, Kuballa T. 2008. Occurrence of benzene as a heat-induced contaminant of carrot juice for babies in a general survey of beverages. Food Addit Contam 25:1216–1224.

Lachenmeier DW, Steinbrenner N, Löbell-Behrends S, Reusch H, Kuballa T. 2010. Benzene contamination in heat-treated carrot products including baby foods. Open Toxicol J 4:39–42.

van Leeuwen FX, Feeley M, Schrenk D, Larsen JC, Farland W, Younes M. 2000. Dioxins: WHO's tolerable daily intake (TDI) revisited. Chemosphere 40:1095–1101.

Linares V, Perelló G, Nadal M, Gómez-Catalán J, Llobet JM, Domingo JL. 2010. Environmental versus dietary exposure to POPs and metals: A probabilistic assessment of human health risks. J Environ Monit 12:681–688.

Llobet JM, Domingo JL, Bocio A, Casas C, Teixidó A, Müller L. 2003. Human exposure to dioxins through the diet in Catalonia, Spain: Carcinogenic and non-carcinogenic risk. Chemosphere 50:1193–1200.

Llobet JM, Martí-Cid R, Castell V, Domingo JL. 2008. Significant decreasing trend in human dietary exposure to PCDD/PCDFs and PCBs in Catalonia, Spain. Toxicol Lett 178:117–126.

López E, Schuhmacher M, Domingo JL. 2008. Human health risks of petroleum-contaminated groundwater. Environ Sci Pollut Res 15:278–288.

Martí-Cid R, Llobet JM, Castell V, Domingo JL. 2008a. Evolution of the dietary exposure to polycyclic aromatic hydrocarbons in Catalonia, Spain. Food Chem Toxicol 46:3163–3171.

Martí-Cid R, Llobet JM, Castell V, Domingo JL. 2008b. Human exposure to polychlorinated naphthalenes and polychlorinated diphenyl ethers from foods in Catalonia, Spain: Temporal trend. Environ Sci Technol 42:4195–4201.

Martí-Cid R, Huertas D, Nadal M, Linares V, Schuhmacher M, Grimalt JO, et al. 2010. Dietary exposure to organochlorine compounds in Tarragona Province (Catalonia, Spain): Health risks. Hum Ecol Risk Assess 16:588–602.

McNally RJQ, Parker L. 2006. Environmental factors and childhood acute leukemias and lymphomas. Leuk Lymphoma 47:583–598.

Nadal M, Schuhmacher M, Domingo JL. 2004. Probabilistic human health risk of PCDD/F exposure: A socioeconomic assessment. J Environ Monit 6:926–931.

Nadal M, Inza I, Schuhmacher M, Figueras MJ, Domingo JL. 2009. Health risks of the occupational exposure to microbiological and chemical pollutants in a municipal waste organic fraction treatment plant. Int J Hyg Environ Health 212:661–669.

Nadal M, Schuhmacher M, Domingo JL. 2011. Long-term environmental monitoring of persistent organic pollutants and metals in a chemical/petrochemical area: Human health risks. Environ Pollut 159:1769–1777.

Ndlovu V, Dalvie MA, Jeebhay MF. 2011. Pesticides and the airways—A review of the literature. Curr Opin Allergy Clin Immunol 24:212–217.

Noorlander CW, Van Leeuwen SPJ, Te Biesebeek JD, Mengelers MJB, Zeilmaker MJ. 2011. Levels of perfluorinated compounds in food and dietary intake of PFOS and PFOA in the Netherlands. J Agric Food Chem 59:7496–7505.

NRC. 1983. Risk Assessment in the Federal Government: Managing the Process. Washington, DC: National Research Council, National Academies Press.

Paul AG, Jones KC, Sweetman AJ. 2009. A first global production, emission, and environmental inventory for perfluorooctane sulfonate. Environ Sci Technol 43: 386–392.

Perelló G, Gómez-Catalán J, Castell V, Llobet JM, Domingo JL. 2012a. Assessment of the temporal trend of the dietary exposure to PCDD/Fs and PCBs in Catalonia, over Spain: Health risks. Food Chem Toxicol 50:399–408.

Perelló G, Gómez-Catalán J, Castell V, Llobet JM, Domingo JL. 2012b. Estimation of the daily intake of hexachlorobenzene from food consumption by the population of Catalonia, Spain: Health risks. Food Control 23:198–202.

Rennix CP, Quinn MM, Amoroso PJ, Eisen EA, Wegman DH. 2005. Risk of breast cancer among enlisted Army women occupationally exposed to volatile organic compounds. Am J Ind Med 48:157–167.

Ritter R, Scheringer M, MacLeod M, Hungerbühler K. 2011. Assessment of non-occupational exposure to DDT in the tropics and the north: Relevance of uptake via inhalation from indoor residual spraying. Environ Health Perspect 119: 707–712.

Rostami I, Juhasz AL. 2011. Assessment of persistent organic pollutant (POP) bioavailability and bioaccessibility for human health exposure assessment: A critical review. Crit Rev Environ Sci Technol 41:623–656.

Rovira J, Mari M, Nadal M, Schuhmacher M, Domingo JL. 2010. Partial replacement of fossil fuel in a cement plant: Risk assessment for the population living in the neighborhood. Sci Total Environ 408:5372–5380.

Rovira J, Mari M, Nadal M, Schuhmacher M, Domingo JL. 2011. Levels of metals and PCDD/Fs in the vicinity of a cement plant: Assessment of human health risks. J Environ Sci Health A 46:1075–1084.

Sarigiannis DA, Karakitsios SP, Gotti A, Liakos IL, Katsoyiannis A. 2011. Exposure to major volatile organic compounds and carbonyls in European indoor environments and associated health risk. Environ Int 37:743–765.

Sereda B, Bouwman H, Kylin H. 2009. Comparing water, bovine milk, and indoor residual spraying as possible sources of DDT and pyrethroid residues in breast milk. J Toxicol Environ Health A 72:842–851.

Shoeib M, Harner T, M Webster G, Lee SC. 2011. Indoor sources of poly- and perfluorinated compounds (PFCS) in Vancouver, Canada: Implications for human exposure. Environ Sci Technol 45:7999–8005.

Stahl T, Mattern D, Brunn H. 2011. Toxicology of perfluorinated compounds. Environ Sci Eur 23:38.

Storelli MM, Barone G, Perrone VG, Giacominelli-Stuffler R. 2011. Polychlorinated biphenyls (PCBs), dioxins and furans (PCDD/Fs): Occurrence in fishery products and dietary intake. Food Chem 127:1648–1652.

Sturm R, Ahrens L. 2010. Trends of polyfluoroalkyl compounds in marine biota and in humans. Environ Chem 7:457–484.

Törnkvist A, Glynn A, Aune M, Darnerud PO, Ankarberg EH. 2011. PCDD/F, PCB, PBDE, HBCD and chlorinated pesticides in a Swedish market basket from 2005—Levels and dietary intake estimations. Chemosphere 83:193–199.

U.S. EPA. 1989. Risk Assessment Guidance for Superfund Volume I: Human Health Evaluation Manual: EPA/540/1-89/002. Washington, DC: United States Environmental Protection Agency.

U.S. EPA. 2009a. Risk Assessment Guidance for Superfund Volume I: Human Health Evaluation Manual (Part E, Supplemental Guidance for Dermal Risk Assessment): EPA/540/R/99/005. Washington, DC: United States Environmental Protection Agency.

U.S. EPA. 2009b. Risk Assessment Guidance for Superfund Volume I: Human Health Evaluation Manual (Part F, Supplemental Guidance for Inhalation Risk Assessment): EPA-540-R-070-002. Washington, DC: United States Environmental Protection Agency.

Van Poucke C, Detavernier C, Van Bocxlaer JF, Vermeylen R, Van Peteghem C. 2008. Monitoring the benzene contents in soft drinks using headspace gas chromatography-mass spectrometry: A survey of the situation on the Belgian market. J Agric Food Chem 56:4504–4510.

Vilavert L, Nadal M, Figueras MJ, Kumar V, Domingo JL. 2011. Levels of chemical and microbiological pollutants in the vicinity of a waste incineration plant and human health risks: Temporal trends. Chemosphere 84:1476–1483.

Vilavert L, Nadal M, Figueras MJ, Domingo JL. 2012. Volatile organic compounds and bioaerosols in the vicinity of a municipal waste organic fraction treatment plant. Human health risks. Environ Sci Pollut Res 19:96–104.

Weichenthal S, Moase C, Chan P. 2010. A review of pesticide exposure and cancer incidence in the agricultural health study cohort. Environ Health Perspect 118: 1117–1125.

Weschler CJ, Nazaroff W. 2012. SVOC exposure indoors: Fresh look at dermal pathways. Indoor Air 22:356–377.

Yu HY, Guo Y, Zeng EY. 2010. Dietary intake of persistent organic pollutants and potential health risks via consumption of global aquatic products. Environ Toxicol Chem 29:2135–2142.

CHAPTER 3

The Burden of Cancer from Organic Chemicals

MOLLY M. JACOBS, RACHEL I. MASSEY, and RICHARD W. CLAPP

ABSTRACT

Background: The majority of the industrial chemicals and drugs that have been identified as carcinogens by the International Agency for Research on Cancer are organic chemicals. Exposures to these organic chemicals occur in the workplace; in the outdoor and indoor environments; through air, water, and food; and through products.

Objectives: This chapter reviews the paths of exposure through which organic chemicals contribute to the global burden of cancer; summarizes the links between individual organic chemicals and specific cancer sites; examines selected individual chemicals, including both well-known carcinogens and emerging chemicals of concern; and provides a brief discussion of the methodological and conceptual difficulties associated with the effort to define the percentage of cancers attributable to occupational and environmental exposures.

Discussion: This review highlights a number of areas of concern, including rising rates of certain cancers, including children's cancers; and ongoing exposure to organic chemical carcinogens in the workplace, in the ambient environment, and through products and food. Individual chemicals highlighted as case studies include 2, 4-D, benzene, styrene, trichloroethylene (TCE), 2,3,7,8-tetrachlorodibenzo-*p*-dioxin (TCDD), bisphenol A (BPA), and n-propyl bromide (nPB).

Conclusions: Taking into account all the factors reviewed here, it is clear that reducing human exposure to organic chemical carcinogens in the workplace, the home, and the ambient environment is a key component of a comprehensive cancer prevention strategy.

Effects of Persistent and Bioactive Organic Pollutants on Human Health, First Edition.
Edited by David O. Carpenter.

INTRODUCTION

Cancer is a group of diseases that are "characterized by uncontrolled growth and spread of abnormal cells" (ACS 2011). Cancers evolve through a complex web of multiple factors involving multiple steps, often over the course of many years. Causal risk factors that can initiate or promote cancer include genetic predisposition, infectious agents such as viruses, and "lifestyle" factors such as smoking and diet, as well as occupational and environmental exposures to carcinogenic agents. Estimates suggest that over half of all cancers worldwide are preventable (ACS 2011). The World Health Organization (WHO) has noted that primary prevention efforts to eliminate or reduce exposure to recognized risk factors are "by far the most cost-effective and sustainable intervention for reducing the burden of cancer globally" (WHO 2011).

According to estimates from the International Agency for Research on Cancer (IARC), in 2008, there were 12.7 million new cases and 7.6 million deaths due to cancer around the world (ACS 2011). The total number of cancer cases worldwide continues to grow with the growth and aging of the world population, combined with reductions in childhood mortality and deaths from infectious diseases in developing countries. Taking these factors into account, the global burden of cancer is expected to nearly double to 21.4 million cases by 2030 (ACS 2011). If industrialization leads to increasing human exposure to chemical carcinogens in developing countries, this future burden could be even higher.

This chapter examines the role of organic chemicals in contributing to the global burden of cancer. To date, IARC has evaluated 941 agents and exposure circumstances for carcinogenicity. Of these, IARC has identified over 400 as carcinogens (Group 1, Group 2A, and Group 2B). Of these carcinogens, approximately 65% are industrial chemicals or drugs. (The remainder are other agents, such as viruses; "lifestyle" factors such as tobacco and diet; radiation; and conditions of exposure, such as aluminum production or shift work.) Of these industrial chemicals and drugs, approximately 88% are organic chemicals.

The chapter begins, in the section "The Global Burden of Cancer," with an overview of global trends in the burden of cancer. The section "Exposure to Organic Chemicals: Contributions to the Global Burden of Cancer" reviews the paths of exposure through which organic chemicals contribute to the global burden of cancer, with examples of exposures in the workplace, in the outdoor and indoor environments, through air, water, and food, and through products. This chapter focuses on involuntary exposures. While drugs are an important category of known or suspected organic chemical carcinogens, these primarily voluntary exposures are not reviewed in this chapter. The section "Linking Chemicals to Specific Cancer Sites" briefly summarizes the links between individual organic chemicals and specific cancer sites. The section "Examples of Individual Organic Chemicals" provides a more detailed information on selected individual chemicals, including both well-known carcinogens and emerging chemicals of concern. Finally, the section "Estimating the Percentage of Cancers Attributable to Occupational and Environmental Exposures: Methodological and Conceptual Difficulties" provides a brief discussion of the methodological and conceptual difficulties associated with the effort to define the percentage of cancers attributable to occupational and environmental exposures. Taking all of these factors into account, there is significant scope for preventing cancer by reducing human exposure to organic chemical carcinogens.

THE GLOBAL BURDEN OF CANCER

According to global cancer statistics from IARC, the highest overall cancer rates are in the most economically developed countries. For example, in Western Europe in 2008, the estimated overall yearly age-adjusted incidence rate for all cancers combined (excluding nonmelanoma skin cancers) was 335.3 per 100,000 in males and 250.5 per 100,000 in females. Similarly, in North America, the rate was 334.0 per 100,000 in males and 274.4 per 100,000 in females. In contrast, age-adjusted cancer rates in the less economically developed, highly populous countries of India, Brazil, and China were significantly lower in the same time period. Age-adjusted cancer incidence in males was 92.9, 190.4, and 211.0 per 100,000 in India, Brazil, and China, respectively. In females, the rates were 105.5, 158.1, and 152.4 per 100,000 in these countries, respectively. It should be noted that these rates are calculated based on far fewer cases than the Organization for Economic Cooperation and Development (OECD) country rates because there are limited cancer registry data covering only a small percentage of the population in these countries (IARC 2012b).

One recent analysis of cancer trends views the changing global pattern of cancer in the context of the globalization trends in the 21st century (Sasco 2007, 2008). This analysis suggests that the patterns of cancer in many low- and middle-income countries are beginning to reflect the patterns found in the wealthy countries. For example, rates of breast and lung cancer have risen in recent years in many developing countries (ACS 2011).

Overall, cancer mortality rates are beginning to level off or decline in the OECD countries but continue to rise in the low- and middle-income countries (IARC 2008). IARC notes that deaths from occupationally induced cancers have declined somewhat in high-income countries but not in low- or middle-income countries. This pattern can be expected to continue as hazardous materials and industrial processes formerly concentrated in the OECD countries are increasingly located in developing countries and in countries with economies in transition. It is projected that by 2020, 31% of global chemical production and 33% of global chemical use and consumption will be in developing countries (OECD 2001). In addition, the growing burden of chemicals in developing countries is compounded by the continued illegal trafficking of hazardous waste. For example, concern remains high regarding the millions of metric tons of electronic waste (e-waste) that are exported to countries such as Nigeria, Ghana, Pakistan, India, and China, among others. This e-waste produces an array of organic chemical carcinogens when processed and disposed of through burn piles, including dioxins and polyclic aromatic hydrocarbons (PAHs) (Robinson 2009). An estimated 75% of e-waste generated in the European Union and 80% of similar waste generated in the United States goes unaccounted for (Basel Action Network 2005).

Childhood Cancers

Childhood cancers are a particular source of concern. In industrialized countries such as the United States, Australia, the United Kingdom, and many countries in Europe where long-term trend data for childhood cancers are available, surveillance data document a steady rise in cancer among those under age 20 since the 1960s and 1970s (Baade et al. 2010; Cancer Research UK 2011; Kaatsch et al. 2006; Linabery and Ross 2008; Ward et al. 2006). For example, in the United States, the incidence of childhood cancers (under age 20 years) has increased 27% from 1975 to 2008, from 12.9 per 100,000 in 1975 to 17.7 per 100,000 in 2008 (Figure 3.1). While improved diagnostics explain

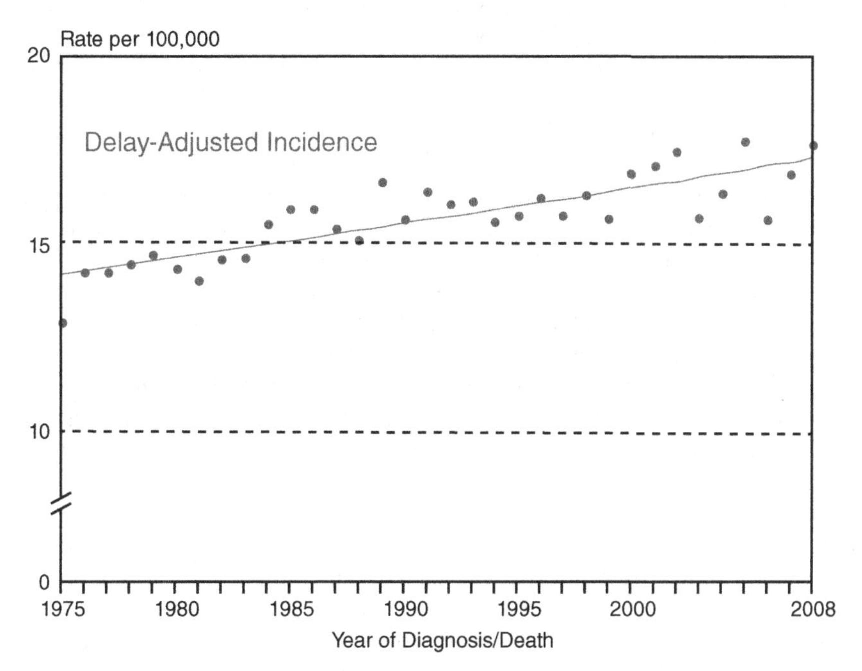

Figure 3.1. Childhood cancer incidence (0–19), all cancers, United States, 1975–2008. *Source*: National Center for Health Statistics, U.S. Centers for Disease Control. SEER Cancer Statistics Review, 1975–2008. Delay Adjusted Data from figure 28.1. Available at http://seer.cancer.gov/csr/1975_2008/index.html [accessed November 10, 2011].

some of the rise for some cancers, such as childhood brain cancer (Ward et al. 2006), environmental factors are thought to play an important role. Genetic factors influencing cancer risk are unlikely to change over the course of a few decades and thus cannot explain these trends (Clapp et al. 2006).

EXPOSURE TO ORGANIC CHEMICALS: CONTRIBUTIONS TO THE GLOBAL BURDEN OF CANCER

Exposure to organic chemicals in a variety of settings contributes to the global burden of cancer. Exposures occur in the workplace, in the outdoor and indoor environments, through air, water, and food, and through products. This section reviews each of these routes of exposure, including examples of chemicals used in industry, agriculture, and consumer products.

Occupational

Occupational exposure to organic chemical carcinogens occurs in a wide variety of work settings and is a significant source of preventable cancers. A recent study reviewed the IARC monographs and found that nearly 40% of Group 1 and Group 2A carcinogens (168 agents) are considered occupational carcinogens, meaning they have been linked to cancers in workers (Siemiatycki et al. 2004). A subset of those agents classified as occupational carcinogens is organic chemicals; these include solvents, paints, dyes, gasoline, and other petroleum products (Belpomme et al. 2007). Common solvents used in a wide variety of industries that are either known or strongly suspected of causing a range of cancers include benzene, trichloroethylene (TCE), and perchloroethylene (PCE), as well as a variety of other organic solvents (Clapp et al. 2008). A number of aromatic amines or their by-products, which are used in dye and pigment manufacturing as well as in the rubber industry, are strongly associated with bladder cancer (Belpomme et al. 2007). Mineral oils and lubricants used among metal workers, in pharmaceutical and cosmetic preparations, and in the printing industry, among other applications, are associated with skin cancers. Emerging evidence also links these chemicals to a range of additional cancers, including cancers of the larynx, lung, nasopharynx, rectum, stomach, and bladder (Clapp et al. 2008). Phthalates, used as plasticizers in polyvinyl chloride (PVC) plastic products, are a source of concern both in the workplace and in consumer products; for example, animal studies link diethylhexyl phthalate (DEHP) with cancers of the liver and testes (Grosse et al. 2011; Kluwe et al. 1982; Voss et al. 2005). There are several seminal examples of cancer clusters in occupational settings, including a cluster of angiosarcoma of the liver caused by exposure to vinyl chloride monomer, a finding subsequently confirmed in multiple studies (Boffetta et al. 2003; Creech and Johnson 1974). Vinyl chloride is also associated with hepatocellular carcinoma (Kielhorn et al. 2000). Parental occupational exposure to organic chemical carcinogens is a significant concern for cancers in children. For example, evidence indicates an increased risk of childhood leukemia associated with parents' exposure to benzene and other solvents, paints, and pigments (Colt and Blair 1998).

Outdoor Air Pollution

A wide variety of outdoor air pollutants are associated with increased cancer risk (Clapp et al. 2006, 2008). Carcinogenic organic air pollutants include by-products of fossil fuel combustion such as diesel exhaust, benzo[a]pyrene and other PAHs, 1,3-butadiene, benzene, and formaldehyde (Cohen 2000). A number of studies show that outdoor air pollution can increase cancer risk. For example, occupational exposure studies show increased risk of lung cancer among a number of occupations exposed to diesel exhaust, railroad, bus garage, trucking, and dock workers (Cohen and Pope 1995) as well as compelling new evidence for miners (Attfield et al. 2012; Silverman et al. 2012). PAHs and other organic chemicals can adhere to fine particulate matter (particles less than 2.5 microns in size) that can readily evade the defenses of the respiratory system and, when inhaled, are carried deep into the lung. Multiple, large cohort studies on air pollution have shown a significant increase in mortality from lung cancer associated with exposure to fine particulate matter (Pope and Dockery 2006). These studies clearly demonstrate that the higher the concentration of fine particulates in the air, the higher the number of deaths due to lung cancer (Pope and Dockery 2006). Studies have also linked childhood leukemia with living near areas of high traffic density (Belpomme et al. 2007).

Indoor Air Pollution

Indoor air can also be a significant source of exposure to organic chemical carcinogens. Indoor air pollutants can include pesticides, and volatile organic compounds (VOCs), including formaldehyde, benzene, and 1,3-butadiene, among others. Studies have shown that exposure to indoor VOCs can increase the risk of leukemia and lymphoma as can the indoor use of insecticides. Again, children's exposures are of particular concern as several epidemiological studies have shown that indoor air pollution can be an important contributing risk factor for childhood cancers. (Belpomme et al. 2007)

Water Pollution

Public and private water supplies can be contaminated from a variety of sources, including commercial, agricultural, and residential sources as well as hazardous waste sites. Exposure can occur by drinking contaminated water or by bathing, showering, or swimming in it. Studies of water contamination have linked a number of organic chemical pollutants to elevated risks of cancer, including PCE, TCE, and chlorophenols, among others (Cantor et al. 2006).

While disinfecting public water supplies with chlorine reduces illness and death associated with waterborne microbes, resulting disinfection by-products are linked to cancer. Strong evidence from epidemiological studies suggests that long-term exposure to disinfection by-products in drinking water—including trihalomethanes such as chloroform and bromoform, among others—elevates the risk of bladder cancer and possibly colon, rectal, and esophageal cancers (Cantor et al. 2006). Given the vast numbers of people globally who receive their drinking water from systems that rely primarily on chlorination for disinfection, even modest elevations in cancer risk from disinfection by-products can create a significant cancer burden.

Consumer Products

A variety of known or suspected organic chemical carcinogens are found in consumer products. Examples include formaldehyde in pressed wood products, methylene chloride in paint strippers, TCE in craft and home repair products, and DEHP in plastic household items. The evidence on human carcinogenicity of these chemicals in consumer products comes primarily from studies of workers exposed during product manufacturing. Few studies have examined links between cancer risk and use of consumer products containing carcinogens, in part because of the difficulty of tracking product use and understanding the range of exposures experienced by consumers through the variety of products they use each day. For example, human studies examining the carcinogenic effects of bisphenol A (BPA), used in producing polycarbonate plastics and for preparing epoxy resins, are virtually absent. However, animal studies demonstrate that exposure to low doses of BPA *in utero* can increase susceptibility to breast and prostate cancers later in life (Durando et al. 2007; Ho et al. 2006; Markey et al. 2001). Studies that have examined the risk of human cancers associated with the use of specific consumer products include studies of home and garden pesticide products as well as hair dyes. As reviewed further, studies examining the use of home and garden pesticide products have shown an increased risk of cancer, especially for childhood cancers. Hair dyes contain an array of organic chemicals, including aromatic amines, that are known to increase cancer risk. Some, though not all, studies have shown an increased risk of multiple myeloma and non-Hodgkin's lymphoma among hair dye users, particularly permanent hair dyes among long-term users (Altekruse et al. 1999; Brown et al. 1992; Zahm et al. 1992; Zhang et al. 2004). A recent meta-analysis found an association between bladder cancer and working as a hair dresser, especially among people who worked as hair dressers for 10 or more years (Harling et al. 2010).

Food Contaminants

An array of known or suspected organic chemical carcinogens have been found as food contaminants. A few key examples include dioxin-like compounds, pesticides, and emerging chemicals of concern such as BPA, 4-methylimidazole, and DEHP.

Certain persistent bioaccumulative compounds, such as polychlorinated dibenzodioxins (PCDDs), polychlorinated dibenzofurans (PCDFs), polychlorinated biphenyls (PCBs), and chlorinated insecticides such as dichlorodiphenyldichloroethene (DDT), have declined in the environment since the late 1970s (except where DDT is still used for malaria control). However, concern for exposure lingers because of their accumulation in the food chain, particularly in animal fat. Results of monitoring dioxins and dioxin-like PCBs in food in the European Union from 1999 to 2008 revealed that the highest mean levels of dioxins and dioxin-like PCBs in food were found in liver and liver products of terrestrial animals and fish, with 8% of samples exceeding various maximum allowable levels (European Food Safety Authority 2010). Evidence linking these organic chemical food contaminants with cancer is not based primarily on studies of chemical residues in food, but rather on occupational studies as well as human cohorts exposed to high levels from industrial accidents or insecticide sprays. IARC has classified 2,3,7,8-tetrachlorodibenzo-*p*-dioxin (TCDD), the most toxic dioxin, as a known carcinogen with links to all cancers combined as well as specific sites, such as soft tissue sarcoma and lung cancer. PCBs are considered known human carcinogens according to IARC with links primarily to melanoma, while DDT and its breakdown products such as dichlorophenyldichloroethylene (DDE) are considered possible carcinogens with studies suggesting links with a range of cancers (Clapp et al. 2005; Cogliano et al. 2011; Lauby-Secretan et al. 2013).

Evidence from animal carcinogenicity studies now suggests a few ubiquitous chemical food contaminants may be associated with cancer. For example, 4-methylimidazole, which is used as caramel colorant, particularly in popular soft drink beverages, was recently classified by IARC as a possible carcinogen based on increased incidences of lung cancers in mice and leukemia in rats (Grosse et al. 2011). DEHP, a plasticizer that can contaminate packaged food, is also considered a possible carcinogen by IARC based on laboratory studies consistently finding increased liver cancers in rats and mice, as well as supporting mechanistic studies (Grosse et al. 2011). Animal evidence suggests that BPA, which can leach into drinks and food from polycarbonate plastic containers, can linings, and other sources, can increase susceptibility to breast and prostate cancers later in life (Durando et al. 2007; Ho et al. 2006; Markey et al. 2001).

Pesticides

Since the 1950s, organic chemicals have been used as pesticides in both agriculture and domestic applications. The total value of the global agricultural pesticide market (including herbicides, insecticides, and fungicides) was nearly $38 billion in 2009 and is projected to reach $52 billion by 2014 (Croplife International 2010; Freedonia Group 2010).

Approximately 40 pesticides classified by IARC as known, probable, or possible human carcinogens are currently registered for use in the United States (IARC 2012a). While some of these chemicals are inorganic, such as chromium trioxide, the majority are organic chemicals such as organochlorines, organophosphates, carbamates and carbinols, among others. However, not all pesticides with either toxicological and/or epidemiological evidence of carcinogenicity have been reviewed by IARC. Examples of organic chemical pesticides that have not been reviewed by IARC include several pesticides such as fonofos, dicamba, and glyphosate (Alavanja et al. 2007).

Studies of exposed farmers or their families, pesticide applicators, crop duster pilots and pesticide manufacturers have reported elevated risk for several specific types of cancer. These cancer types include breast, colorectal, Hodgkin's disease, leukemia, lung, multiple myeloma, non-Hodgkin's lymphomas, prostate, soft tissue sarcoma, skin, stomach, and testicular cancers (Alavanja et al. 2007; Bassil et al. 2007; Brody et al. 2007; Clapp et al. 2006; Dich et al. 1997; Mills and Yang 2005). Evidence regarding the carcinogenicity of specific pesticides is often complex, with studies documenting both positive and negative findings. For example, studies throughout several decades have been mixed regarding the risk of breast cancer associated with exposure to DDT (Brody et al. 2007). However, a recent case-control epidemiological study observed that girls exposed to DDT during the time of puberty—when mammary cells are more susceptible to the deleterious effects of endocrine-disrupting chemicals—were five times more likely than controls to develop breast cancer when they reached middle age (Cohn et al. 2007). Such a study reveals how exposure to an endocrine-disrupting agent during critical windows of development significantly influences disease risk, and suggests avenues for future study design to understand these risks.

In children, several epidemiological studies suggest an increased risk of cancer among those with direct exposure to pesticides or parental exposure prior to and during pregnancy. Specific childhood cancers associated with pesticides include leukemias, central nervous system tumors, Wilms's tumor, Ewing's sarcoma, germ cell tumors, and non-Hodgkin's lymphoma (Cordier et al. 2001; Feychting et al. 2001; Ma et al. 2002; Zahm and Ward 1998). The risk of leukemia is consistently elevated among children whose parents used pesticides in the home or garden, including those who grew up on farms, as well as children of pesticide applicators (Meinert et al. 2000; Menegaux et al. 2006; Monge et al. 2007; Zahm and Ward 1998). Yet because these chemicals are often applied as mixtures, studies cannot clearly distinguish the risk associated with specific pesticides.

2,4-Dichlorophenoxyacetic Acid (2,4-D): Health Impacts of a Widely Used Organic Chemical Herbicide

2,4-D is a synthetic auxin (plant growth hormone) that is the most widely used herbicide in the world. It is contained in over a thousand products that have commercial, agricultural, and residential applications. It has been used to control weeds on wheat and corn farms, on lawns, turfs, and roadsides. In some areas, it is used to control weeds in lakes. The U.S. Environmental Protection Agency (EPA) estimated that 46 million pounds were used in all applications in 2005, with two-thirds of that being used in agriculture (U.S. EPA 2005).

2,4-D was one of the two main components of the herbicide Agent Orange, used as a defoliant in the Vietnam war during the years 1965–1970. It is estimated that 45 million liters of Agent Orange were used during this period (NRC 2009). Agent Orange use was discontinued when it was determined that it was contaminated with dioxin. Substantial dioxin contamination in several parts of Vietnam continues to the present (Schecter et al. 2006).

Evidence of the carcinogenicity of 2,4-D began to accumulate in the 1980s in the United States with the publication of studies by Shelia Zahm and colleagues in Kansas and Nebraska (Hoar et al. 1986; Zahm et al. 1990). The most convincing early study showed an increased risk of non-Hodgkin's lymphoma in those who mixed or applied pesticides in farm work, with increased risk associated with increased opportunity for exposure. Prior studies in Scandinavia and subsequent studies of workers exposed to 2,4-D in manufacturing or application of herbicides in the United States, Canada, several European countries, Australia, and New Zealand provided evidence that this chemical causes non-Hodgkin's lymphoma.

Studies of Vietnam veterans exposed to Agent Orange also showed increased risk of non-Hodgkin's lymphoma in those who were exposed during their service. The largest case-control study of selected cancers in Vietnam veterans was published in 1990 and found increased non-Hodgkin's lymphoma, especially in those in specific branches of service and specific heavily sprayed regions. As a result, the Agent Orange Act of 1991 specifically listed non-Hodgkin's lymphoma as one of the conditions for which Vietnam veterans should be compensated as service related.

Corroborating evidence of the carcinogenicity of 2,4-D included increased risk of canine lymphoma in dogs whose owner used 2,4-D (Hayes et al. 1991) and increased brain tumors in laboratory mice exposed to 2,4-D. Another study of deaths from non-Hodgkin's lymphoma in employees of a lawn care company (Zahm 1997) reported a large increase in those who had been employed three or more years. As with many studies of workers, the exposures are often mixtures of multiple substances, so it is difficult to disentangle the effect of a particular chemical from that of other herbicides or substances to which the study subject may have been exposed.

Industry spokespeople challenged the designation of 2,4-D as a possible human carcinogen (Industry Task Force II 2012), and IARC still considers the evidence in humans to be insufficient to determine whether 2,4-D causes cancer. This is partly because of the difficulty in examining the risk of the specific chemical, so IARC categorizes exposure to phenoxy herbicides, including methyl chlorophenoxyacetic acid (MCPA), 2,4-D, and 2,4,5-T as a group to be possibly carcinogenic to humans (Group 2B). The Industry Task Force has also argued strenuously against increasing restriction of 2,4-D by the EPA (Ibrahim et al. 1991).

An intriguing piece of evidence that phenoxy herbicides have caused population-level increases in non-Hodgkin's lymphoma in Sweden was reported by Hardell and Eriksson (2003). These authors noted that phenoxyacetic acid herbicides and chlorophenols were used heavily in the 1960s and 1970s, but were banned in Sweden in 1977 and 1978. Subsequently, population exposure to these and other persistent organic pollutants declined and there was a leveling off of the rate of non-Hodgkin's lymphoma in the years 1991–2000. They suggest that this reduction in exposure to the general population may have been responsible for the observed pattern in non-Hodgkin's lymphoma. Other authors have reported similar trends in other countries (Sandin et al. 2006; Viel et al. 2010), and recent trends in the U.S. SEER data follow the same pattern.

This example illustrates the difficulty in interpreting the evidence that organic chemicals cause cancer in humans. Often, the epidemiological data are inconclusive because of the inevitable mixture of exposures involved. Furthermore, vested interests may assemble panels to criticize the existing literature and delay regulatory decision making. Occasionally, people exposed in specific circumstances such as the war in Vietnam are compensated even in the absence of definitive cause–effect evidence. Or, when examined thoughtfully, population-level trends in cancer may suggest that precautionary reduction in exposure has reversed a trend in a related cancer.

LINKING CHEMICALS TO SPECIFIC CANCER SITES

A recent study presented an overview of chemicals that have been reviewed by IARC, and summarized the links between individual chemicals and specific cancer sites within the human body (Cogliano et al. 2011). Table 3.1, adapted from this larger study, shows the industrial organic chemical carcinogens linked to specific cancer sites in humans.

It should be noted that this information provides only a limited window into the larger set of organic chemical carcinogens due to the narrow criteria for inclusion used in the study. The researchers limited their review to data with evidence of cancer in humans (e.g., primarily evidence from epidemiological studies) and did not include data where only animal studies or *in vitro* laboratory studies were available. It is worth noting, for example, that while the table identifies just one industrial organic chemical, ethylene oxide, as clearly linked to breast cancer in humans, a study that defined the data universe based on positive findings in animal carcinogenicity studies identified more than 200 mammary carcinogens (Rudel et al. 2007). Despite these limitations, this overview provides an illustration of some of the many opportunities that exist for the primary prevention of specific cancers by preventing exposures to organic chemicals.

TABLE 3.1. List of Organic Chemical Carcinogens by Cancer Site (*Sufficient* or *Limited* Evidence in Humans, Based on IARC Reviews)

Cancer Site	Carcinogenic Agents with Sufficient Evidence in Humans	Agents with Limited Evidence in Humans
Lip, oral cavity, and pharynx		
Pharynx		Printing processes
Nasopharynx	Formaldehyde	
Digestive organs		
Esophagus		Dry cleaning Rubber production industry Tetrachloroethylene
Stomach	Rubber production industry	
Liver and bile duct	Vinyl chloride	Polychlorinated biphenyls Trichloroethylene
Respiratory organs		
Nasal cavity and paranasal sinus	Isopropyl alcohol production	Formaldehyde
Larynx		Rubber production industry Sulfur mustard
Lung	Bis(chloromethyl)ether, chloromethyl methyl ether (technical grade) Painting Rubber production industry Sulfur mustard	*alpha*-Chlorinated toluenes and benzoyl chloride (combined exposures) Creosotes Insecticides, nonarsenical (occupational exposures in spraying and application) Printing processes 2,3,7,8-Tetrachlorodibenzo-*para*-dioxin
Bone, skin, and mesothelium, endothelium, and soft tissue		
Skin (other malignant neoplasms)	Coal tar distillation coal tar pitch Mineral oils, untreated or mildly treated Shale oils Soot	Creosotes Nitrogen mustard petroleum refining (occupational exposures)
Mesothelium (pleura and peritoneum)	Painting	
Soft tissue		Polychlorophenols or their sodium salts (combined exposures) 2,3,7,8-Tetrachlorodibenzo-*para*-dioxin
Breast and female genital organs		
Breast		Ethylene oxide
Uterine cervix		Tetrachloroethylene

(*Continued*)

TABLE 3.1. (*Continued*)

Cancer Site	Carcinogenic Agents with Sufficient Evidence in Humans	Agents with Limited Evidence in Humans
Male genital organs		
Prostate		Rubber production industry
Urinary tract		
Kidney		Printing processes
Urinary bladder	4-Aminobiphenyl Auramine production Benzidine Magenta production 2-Naphthylamine Painting Rubber production industry *ortho*-Toluidine	4-Chloro-*ortho*-toluidine Coal tar pitch Dry cleaning Engine exhaust, diesel Printing processes Soot
Lymphoid, hematopoietic, and related tissue		
Leukemia and/or lymphoma	Benzene 1,3-Butadiene Formaldehyde Rubber production industry Thiotepa Treosulfan	Ethylene oxide Nitrogen mustard Painting (childhood leukemia from maternal exposure) Petroleum refining (occupational exposures) Polychlorophenols or their sodium salts (combined exposures) Styrene Tetrachloroethylene Trichloroethylene 2,3,7,8-Tetrachlorodibenzo-*para*-dioxin
Multiple or unspecified sites		
Multiple sites (unspecified)		Chlorophenoxy herbicides
All cancer sites (combined)	2,3,7,8-Tetrachlorodibenzo-*para*-dioxin	

Source: This table is adapted from a table provided on the IARC website (http://www.iarc.fr), accessed February 2012. The table on the IARC website, in turn, is adapted from Cogliano et al. 2011. The present version of the table includes only organic chemical carcinogens that are used in industry, omitting other chemicals as well as other carcinogenic agents.
Note: The information shown here represents the most recent information reviewed by IARC at the time of the original study (Cogliano et al. 2011). However, scientific information continues to accumulate on the chemicals shown here. For example, the level of concern for TCE has been upgraded in the most recent review by the U.S. EPA in 2011 and at the time of this writing is under review by IARC. More generally, the information shown here provides only a narrow window into the larger set of organic chemical carcinogens due to the narrow specifications used in the study. The researchers limited their review to data on human cancers and did not include data collected through animal studies or *in vitro* laboratory studies. Many more organic chemicals would appear on this list if these other data were included.

EXAMPLES OF INDIVIDUAL ORGANIC CHEMICALS

In the discussion that follows, we consider a few selected organic chemicals and their relationship to cancer: benzene, styrene, TCE, TCDD, BPA, and n-propyl bromide (nPB). These examples serve to illustrate some of the specific ways in which people are exposed to carcinogens at work and in the ambient environment, and point to specific opportunities for prevention. Moreover, the examples demonstrate the breadth of our economy's dependence on using these known and suspected carcinogens. Benzene and styrene are two basic, high-volume industrial chemicals used to make an enormous variety of downstream chemicals. TCE is a chlorinated solvent and is considered a lower-volume industrial chemical, yet still used in a variety of industrial applications. TCDD is a known carcinogen that is also a persistent organic pollutant. BPA and nPB are emerging sources of concern with regard to cancer.

The data available on these chemicals are somewhat variable. To the extent possible, for each chemical, we briefly describe what is known about the chemical's links to cancer, the amount used globally, and the applications for which the chemical is used, and available information on human exposures.

Sources of Exposure Information

Sources of information on occupational and ambient environmental exposure to organic chemical carcinogens are limited. Two sources provide the basis for the present discussion. The U.S. National-Scale Air Toxics Assessment (NATA) provides estimates of ambient environmental exposures to a variety of toxic air pollutants within the United States. The Canadian CAREX system provides estimates of both occupational and ambient environmental exposures to a variety of toxic chemicals in Canada. Equivalent data systems in Europe are not sufficiently up to date to be useful, and national-level estimates of this kind are completely lacking in developing countries.

Benzene

Links to Cancer IARC classifies benzene in Group 1 (carcinogenic to humans) (IARC 1987). Benzene is strongly linked to leukemia, multiple myeloma, and non-Hodgkin's lymphoma. It also has suspected links to cancer of the brain and the central nervous system, lung cancer, and nasal and nasopharyngeal cancer (Clapp et al. 2008).

Production and Use Benzene is one of the world's major commodity chemicals. It is a basic petrochemical that is used to make a wide variety of downstream chemical products with a global production of 40 million metric tons in 2010 (Davis 2011). Benzene is commonly used as an intermediate in the production of plastics, resins, and some synthetic and nylon fibers. It is used to make some types of rubbers, lubricants, dyes, detergents, drugs and pesticides, and it is found in crude oil, gasoline, and cigarette smoke. Benzene is among the most abundantly produced chemicals in the United States (ATSDR 2007).

Human exposure to benzene occurs primarily by inhaling benzene in ambient air (CDC 2009). Automobile sources and gasoline filling stations are significant industrial sources of benzene in outdoor air, while indoor sources include the off-gassing of building materials as well as exposure to environmental tobacco smoke (ATSDR 2007). The consumption of food, drinking water, and beverages is considered a negligible source of exposure unless benzene contamination has occurred (ATSDR 2007). In the United States, benzene is among the 10 chemicals most frequently found at hazardous waste sites (Reese 2008). Workplace exposure to benzene may come from a variety of sources but often is the result of the production, use, or transportation of petroleum products (CDC 2009).

Occupational and Environmental Exposure Estimates: Canada Occupational and environmental benzene exposure information is available for Canada through the carcinogen exposure surveillance program, CAREX Canada. CAREX estimates that nearly 300,000 Canadian workers (approximately 1.8% of the working population) are exposed to benzene. Workers most highly exposed include automotive service technicians (mechanics), delivery and courier drivers, taxi drivers, and firefighters. Additional exposure occurs among petroleum and chemical process workers, foundry workers, construction workers, and service station attendants (CAREX Canada 2011e). Studies have found that workers exposed to gasoline fumes, including garage mechanics, drivers, and street vendors as well as workers exposed to solvent fumes, have blood levels of benzene 10-fold higher than levels in the general population (CDC 2009).

As for ambient environmental exposure, CAREX estimates that just over half of Canadians (52.5% or 16,582,994 people) are exposed to benzene in outdoor air at levels corresponding to an estimated lifetime excess cancer risk of 1/100,000 and 1/50,000. CAREX estimates that a small percentage of Canadians are exposed to measurable levels above or below these levels. For most of the rest of the population (44.4%), exposures are simply unknown (CAREX Canada 2011a).

CAREX has not estimated ambient environmental exposures through indoor air, drinking water, or food. CAREX classifies benzene exposures in indoor air as a "moderate" priority, and exposures through drinking water and food as "low" priority (CAREX Canada 2011b).

Environmental Exposure Estimates: United States In the United States, the NATA provides estimates of benzene exposure in air, and resulting cancer risk, for the U.S. population as a whole. The lifetime excess cancer risk associated with exposure to benzene in 2005 is estimated at 7.4 cases per million people, or approximately 2100 excess cancer cases in Americans resulting from exposure to benzene in air in 2005. These represent lifetime risks for people exposed to benzene released to air from both point and nonpoint sources as well as "background" exposure, which is defined as exposure from natural sources, emissions from past years that have persisted in the air, or "long-range transport from distant sources" (U.S. EPA 2011a).

Styrene

Links to Cancer The IARC classifies styrene in Group 2B (possibly carcinogenic to humans) (IARC 2002). As summarized by the U.S. National Toxicology Program's (NTP) *Report on Carcinogens*, 12th Edition, studies indicate links between occupational styrene exposure and lymphohematopoietic cancers. There is also some evidence supporting an increased risk of esophageal and pancreatic cancer among workers exposed to styrene (NTP 2011b).

Production and Use Worldwide, 24 million metric tons of styrene were produced in 2008 (Davis 2009). Styrene is manufactured using benzene and ethylene as raw materials. It is used, in turn, to make a variety of downstream chemical products. These include polystyrene (used in consumer products such as disposable food containers and insulation); styrene acrylonitrile resins (used to make products including instrument lenses and housewares); styrene butadiene rubber (used in products such as tires, footwear, and sealants); and styrene butadiene latex (used in products such as carpet backings) (ACC 2011). It is also used in fiberglass production (CAREX Canada 2011c).

Styrene is commonly detected in urban air, especially near industrial facilities where it is produced and used as well as in areas with significant motor vehicle traffic. However, air concentrations of styrene may be greater indoors than outdoors due to emissions from photocopiers and laser printers, cigarette smoke, and consumer products in the home (CDC 2009).

Occupational and Environmental Exposure Estimates: Canada According to CAREX estimates, approximately 41,000 Canadians are potentially occupationally exposed to styrene. The sectors with the largest number of exposed workers are "plastic products manufacture, wood products manufacturing (from styrene-based glues, adhesives, and varnishes), and ship and boat building." CAREX notes that most styrene use in Canada is for polystyrene production and that occupational exposures "can occur during the manufacture and use of styrene and styrene based varnishes, coatings, adhesives, and other plastic parts and products." Fiberglass production, which is included in the broader category of plastic products manufacture, is estimated to have the highest exposures (CAREX Canada 2011c). Studies have found that workers exposed to styrene can have blood levels that are 25 times higher than those in the general population (CDC 2009).

CAREX has also developed estimates of styrene exposure in outdoor air for a portion of the Canadian population, but lifetime excess cancer risk levels associated with these exposures have not been calculated. CAREX has not developed estimates for exposure through food or indoor air, though it flags these as being of high and medium priority, respectively (CAREX Canada 2011d).

TCE

Links to Cancer IARC classifies TCE as a Group 2A carcinogen (probably carcinogenic to humans) (IARC 2012a). However, the U.S. NTP classifies TCE as reasonably anticipated to be a carcinogen and the U.S. EPA classifies TCE as a known human carcinogen based on more recent reviews of the evidence. In 2011, the U.S. EPA's review found that exposure to TCE can lead to kidney and liver cancer and non-Hodgkin's lymphoma, and it may also be linked to bladder, esophageal, prostate, cervical, and breast cancers, as well as leukemia (U.S. EPA 2011b).

Production and Use In 2007, 256,000 metric tons of TCE were produced globally (Glauser and Ishikawa 2008). TCE is primarily used as an intermediate in the production of hydrofluorocarbons and as a degreaser for metal parts. As a degreaser, TCE is used in industrial sectors including furniture and fixture production, fabricated metal products, electrical and electronic equipment, and transport equipment. TCE is also used as a solvent in the rubber industry, adhesive formulations, dyeing and finishing operations, printing inks, paints, lacquers, varnishes, adhesives, and paint strippers. In the past, TCE was used in food processing as a dry cleaning agent and as a general anesthetic. TCE is also used in some consumer products used for hobbies, crafts, and home maintenance. For example, it is listed as a major ingredient of 12 household products in the United States, constituting 80–100% in five products intended for use primarily in hobbies, crafts, and home maintenance (NTP 2011b).

TCE has been detected in urban and ambient air and in soils and drinking water, primarily as a result of contamination by industrial discharge (CDC 2009). It is among the 10 chemicals most frequently found at U.S. hazardous waste sites (Reese 2008).

Occupational Exposure Estimates: Canada To date, CAREX Canada has estimated occupational exposures to TCE but has no estimates for ambient environmental exposures. Initial data compiled by CAREX Canada show that approximately 13,000 Canadians are exposed to TCE in their workplaces. The largest number of exposures occurs among workers who perform metal degreasing. Other important occupational groups at risk from exposure to TCE include workers in textile processing as well as printing press operators. For the general public, indoor air is the principal source of TCE exposure; other sources include "food, drinking water and outdoor air" (CAREX Canada 2012).

Environmental Exposure Estimates: United States According to NATA estimates, the lifetime excess cancer risk associated with exposure to TCE in 2005 is estimated at 0.17 cases per million people, or about 50 excess cancer cases in Americans resulting from exposure to TCE in air in 2005. These figures represent lifetime risks for people exposed to TCE released to air in 2005 as well as TCE present in air from earlier emissions or long-range transport from distant sources (U.S. EPA 2011).

TCDD TCDD is formed as an unintended by-product of a variety of industrial processes, including smelting, chlorine bleaching of paper pulp, and the manufacturing of some herbicides and pesticides. TCDD is also formed from the incineration of products containing chlorinated compounds, such as the open burning of household municipal trash, landfill fires, solid waste incinerators, as well as agricultural fires (CDC 2009; WHO 2010).

IARC classifies TCDD as a Group 1 carcinogen (carcinogenic to humans). While IARC's classification in 1997 was primarily based on mechanistic evidence, strong evidence from epidemiological studies links TCDD to cancer at multiple sites based on observed increased rates for all cancer sites combined associated with dioxin exposure (Cogliano et al. 2011). Soft tissue sarcomas, leukemias and lymphomas (particularly non-Hodgkin's lymphoma), and lung cancer are also particular cancers of concern (Cogliano et al. 2011). While few studies have examined the risk of breast cancer and dioxin exposure, a compelling follow-up study of the Seveso, Italy, industrial disaster in 1976 provides evidence for concern (Pesatori et al. 2009). In addition, recent studies of specific military troops in Vietnam (the Ranch Hand veterans who were responsible for handling and spraying Agent Orange, the herbicide contaminated with dioxins) also link prostate cancer to dioxin exposure (Pavuk et al. 2006).

TCDD as well as other dioxins and furans are persistent organic pollutants that degrade slowly in the environment and concentrate as they go up the food chain. Approximately 90% of the general population exposure to TCDD and other dioxins and furans is through the ingestion of high-fat foods, including dairy products, and animal fats (WHO 2011). In many parts of the world, releases of dioxins from regulated industrial sources have decreased dramatically over the last few decades (see, e.g., U.S. EPA 2006). However, most soil and water samples reveal trace amounts of dioxins and furans (CDC 2009). Breast milk is a substantial source of exposure for infants, though breast milk levels of dioxins and furans have been decreasing in countries across the globe (CDC 2009; Ulaszewska et al. 2011; Van Leeuwen and Malisch 2002). Despite these promising trends, dioxin levels remain a significant concern, especially in some regions. For example, contamination with dioxins associated with e-waste recycling in Guiyu, China, has resulted in significantly elevated levels of dioxins measured in human milk, placentas, and hair (Chan et al. 2007). Moreover, widespread dioxin-contaminated animal-based food events, including a 1999 incident in Belgium and a 2008 incident in Ireland, demonstrate continued potential for exposure among the general population, even in developed countries. Ongoing exposures in Vietnam also continue to be a source of concern (Schecter et al. 2002, 2006).

Chemicals of Emerging Concern The chemicals described earlier are a few of the many organic chemical carcinogens that have been studied in depth over a significant period of time. However, there are many other organic chemicals that have not been evaluated fully for potential carcinogenicity. Two examples of organic chemicals that are a source of emerging concern from the perspective of carcinogenicity include BPA and nPB.

BPA BPA is a high-production volume chemical, with about 4 million metric tons produced and used globally in 2009 (Greiner and Funada July 2010). Globally, the use of BPA is driven by the polycarbonate plastic market with products that include baby bottles, food containers, toys, and automobile parts (CDC 2009; Greiner and Funada July 2010). BPA is also used heavily in the manufacture of epoxy resins for products including the protective linings of some food can containers, some dental composites, and wine vat linings, and is also used in the production of thermal paper products (Biedermann et al. 2010; CDC 2009). Human exposure can occur as BPA leaches into canned foods, infant formula, and other food products as well as through the handling of thermal paper (Biedermann et al. 2010; Noonan et al. 2011; Schecter et al. 2010). Compared with the U.S. general population, urinary levels of BPA have been observed to be higher in children, females, and those of lower-income levels. Retail workers also have high levels of BPA in urine, probably due to their exposure to thermal paper cash receipts (Calafat et al. 2008; Lunder et al. 2010).

A large number of studies have shown a wide variety of health effects associated with the ability of BPA to disrupt estrogen signaling. Possible carcinogenic effects of BPA have not been reported to date in humans, but a number of animal studies indicate that early-life exposure to BPA may increase susceptibility to prostate and breast cancers later in life. For example, transient developmental exposure to environmentally relevant doses of BPA in rats increases the susceptibility of the prostate gland to precancerous lesions following a second dose of hormonal exposure in adulthood (Ho et al. 2006). Several studies of BPA show effects on the mammary gland. Several studies using both rat and mouse models demonstrate that exposure to BPA *in utero* leads to changes in mammary tissue structure that are predictive of tumor development later in life (Maffini et al. 2006; Markey et al. 2001; Muñoz-de-Toro et al. 2005). Studies also show that *in utero* exposure to BPA increases both the number of precancerous lesions and *in situ* tumors (Murray et al. 2007). In addition, studies have shown that exposure to BPA *in utero* also increases the number of mammary tumors following adulthood exposures to lower doses of BPA than those generally needed to induce tumors (Durando et al. 2007; Jenkins et al. 2009). Studies using cultures of human breast cancer cells show that along with its many other effects on cell growth and proliferation, BPA has been shown to mimic estradiol in causing direct damage to DNA (Iso et al. 2006). In sum, these studies indicate that there is a basis for concern about BPA and cancer, although the chemical has not yet been reviewed and evaluated by IARC.

n-Propyl Bromide (nPB) nPB use began primarily in the 1990s, first as an intermediate in the production of pesticides, quaternary ammonium compounds, flavors and fragrances, and pharmaceuticals (NTP 2011a). In the mid to late 1990s, it was introduced as a replacement for methylene chloride (a suspected carcinogen according to both the U.S. NTP and IARC) and TCE (a known carcinogen according to the U.S. NTP and U.S. EPA) that were more strictly regulated in applications where emissions can be significant, such as vapor and immersion degreasing and cleaning of electronics and metals (NTP 2011a). It has also been used as a replacement for ozone-depleting refrigerants and as a solvent for adhesives (NTP 2011a). The United Nations Environment Programme, Ozone Secretariat (2010) estimates the global production of nPB to be around 40,000 metric tons.

Early claims that nPB was less toxic than the chlorinated organic chemicals it was replacing were based on the fact that the range of chronic toxicity tests had not been conducted (Jacobs 2011). In 2011, NTP published the first carcinogenicity studies of nPB in experimental animals. The studies found clear evidence of carcinogenicity in female rats and mice, as well as some evidence of carcinogenicity in male rats (NTP 2011a). A recent study of nPB levels in dry cleaning facilities found high levels of exposure (up to 54 ppm) (Blando et al. 2010), indicating that there is a serious basis for concern about ongoing occupational exposures. nPB serves as a drop-in substitute for chlorinated solvents in a number of industrial applications, and in the United States at least, it is not currently subject to the federal regulations controlling the use of these substances. Given recent, albeit preliminary, findings about its possible carcinogenicity, trends in use of nPB are a source of concern.

In summary, BPA and nPB are two examples of a much larger set of organic chemicals that are widely used in commerce and have not yet been fully evaluated for carcinogenicity, but which may be carcinogenic in humans. For chemicals such as benzene and TCE, authoritative bodies such as CAREX Canada and U.S. EPA are able to develop estimates of the burden of cancer that results from their use and emissions into the environment. For other, widely used chemicals of concern, even rough estimates of this kind are unavailable. Rather than wait for the evidence to accumulate to a point where such estimates might be possible, it would be prudent to act now to reduce exposures.

ESTIMATING THE PERCENTAGE OF CANCERS ATTRIBUTABLE TO OCCUPATIONAL AND ENVIRONMENTAL EXPOSURES: METHODOLOGICAL AND CONCEPTUAL DIFFICULTIES

A number of researchers have attempted to estimate the percentage of cancers that can be attributed to occupational and environmental exposures. However, it is increasingly clear that cancers arise through a complicated web of multiple causes. In this context, it is counterproductive to attempt to calculate the percentage of cancers attributable to specific exposures. At the same time, it is clear that preventable exposures are fueling many excess cancer cases and deaths each year.

Efforts to Estimate Attributable Fractions

As described elsewhere (Clapp et al. 2006), the first and most frequently cited estimate of attributable fractions was published in a 1981 monograph by Sir Richard Doll and Sir Richard Peto (Doll and Peto 1981). These authors estimated that in the United States, pollution was responsible for 2% of all cancer deaths, and occupation was responsible for 4%. They estimated the attributable fraction for tobacco at 30% and for diet at 35%, and assigned smaller percentages to a wide variety of other factors (Doll and Peto 1981).

Doll and Peto acknowledged that many of these factors can interact with one another so that it is misleading to estimate a set of percentages that add up to 100% (Doll 1998; Doll and Peto 1981). However, this key limitation of their estimates has largely been overlooked. Doll and Peto's estimates are cited in thousands of scientific articles. They have also been cited by commentators who argue that improving occupational and environmental standards will not have a significant effect on cancer rates. Later reviews of Doll and Peto have pointed out specific flaws in their estimates, including their failure to consider exposures in small workplaces (Landrigan and Baker 1995), indirect contact with carcinogenic substances such as asbestos in maintenance operations (Landrigan and Baker 1995), and cancers in people over the age of 65 (Davis 1990; Landrigan and Baker 1995; Landrigan et al. 1995).

A recent study estimates the number of people exposed to carcinogens in the U.K. workplaces and estimates the number of cancers that may result from these exposures. This effort, while likely to underestimate the burden, does estimate figures higher than those developed by Doll and Peto (Hutchings and Rushton 2011; Rushton et al. 2010). These estimates suggest that the overall percentages of U.K. cancer deaths in 2005 attributable to occupational exposures were 8.2% for men and 2.3% for women (with the higher percentage for men reflecting men's greater exposure to occupational carcinogens) (Rushton et al. 2010).[1] The study determined that for certain cancer sites, a particularly large percentage of total deaths can be attributed to occupational exposures. In men, for example, occupation is identified as the source of nearly all cases of mesothelioma (97%), nearly half of sinonasal cancers (46%), lung cancers (21%), and bladder and nonmelanoma skin cancers (7%) (Rushton et al. 2010). For women, the cancer sites at the top of the list for occupationally attributable cancer deaths are mesothelioma (83%), sinonasal cancers (20.1%), lung cancers (5.3%), breast cancers (4.6%), and nasopharyngeal cancers (2.5%). For a number of additional cancers, occupation is responsible for at least 2% of cases (Rushton et al. 2010).

Also, in 2011, researchers from the WHO estimated the deaths and burden of disease (calculated as disability-adjusted life years [DALYs]) attributable to selected chemical exposures in 2004, including both inorganic and organic chemicals. They calculate 4.9 million deaths and 86 million DALYs, with the largest contribution from exposure to air pollutants. The authors acknowledge the incompleteness of the data on which they base these estimates but nevertheless state that their review "shows that the currently known disease burden from chemicals is large, and that the yet unknown burden may be considerable" (Prüss-Üstün et al. 2011). Critical chemicals not able to be included in the analysis include dioxins, organic chlorinated solvents, PCBs, and chronic pesticide exposures as well as health impacts from exposure to local toxic waste sites (Prüss-Üstün et al. 2011). Thus, the figures developed by WHO represent just a fraction of the full health impact of chemical exposures.

Causes: Genes or Environment?

Current knowledge of the mechanisms of cancer suggests that all cancers are both environmental and genetic—meaning that there are multiple causes that involve exposures originating outside the body as well as hereditary or genetic changes that converge to produce the disease. Cancer researchers have identified at least six essential alterations that may overwhelm the natural defenses built into human cells and tissues over time to produce a tumor (Hanahan and Weinberg 2011). Although researchers are beginning to examine interactions among causal factors, the vast majority of epidemiological and toxicological studies continue to investigate cancer risk associated with individual factors. While the previous sections provide examples of the wide variety of ways investigators examined how environmental and occupational exposure to organic chemicals can cause cancer, even these recent studies are limited by our current level of knowledge and by current epidemiological methods. There are undoubtedly other interacting factors, such as prenatal and early childhood

[1]This analysis uses IARC Group 1 and Group 2a carcinogens. The figure is lower if only Group 1 carcinogens are considered and would be higher if Group 2b carcinogens were included as well.

exposures, nutrition, physical activity, genetics, and psychosocial factors such as stress, which together may be responsible for the development of cancer in ways not yet fully understood.

CONCLUSION: OPPORTUNITIES FOR PREVENTION

Large numbers of people are exposed to organic chemical carcinogens in the workplace; through air, water, soil, and food; and in the indoor environment, including through use of consumer products. Each of these exposure scenarios represents an opportunity for cancer prevention.

Some researchers have used the metaphor of a pie to represent the conditions required for cancer to develop (Rothman 2002). Each slice of the pie is a component cause contributing to the development of cancer. One slice could represent an inherited genetic trait; another slice could represent a "lifestyle" risk factor, such as smoking; another slice could represent an occupational exposure to one or more carcinogens; and so on. In most cases, no individual slice is sufficient to cause disease; disease develops only when all the components are present. Removing one slice of the pie (e.g., smoking or an occupational exposure) may protect the individual from developing cancer.

The U.S. President's Cancer Panel, in its 2008–2009 Annual Report, made a strong statement about the role of environmental factors in cancer etiology and recommended addressing environmental causes of cancer as a key component in the broader cancer prevention strategy. In the words of the Panel,

> A precautionary, prevention-oriented approach should replace current reactionary approaches to environmental contaminants in which human harm must be proven before action is taken to reduce or eliminate exposure. Though not applicable in every instance, this approach should be the cornerstone of a new national cancer prevention strategy that emphasizes primary prevention, redirects accordingly both research and policy agendas, and sets goals for reducing or eliminating toxic environmental exposures implicated in cancer prevention. (U.S. President's Cancer Panel 2010)

Taking into account all the factors touched upon in this chapter, it is clear that reducing human exposure to organic chemical carcinogens in the workplace, in the home, and in the ambient environment is a key component of a comprehensive cancer prevention strategy.

REFERENCES

ACC (American Chemistry Council). 2011. 2011 Guide to the Business of Chemistry. Washington, DC: ACC.

ACS (American Cancer Society). 2011. Global Cancer Facts and Figures, 2nd edition. Atlanta, GA: ACS.

Alavanja MC, Ward MH, Reynolds P. 2007. Carcinogenicity of agricultural pesticides in adults and children. J Agromedicine 12:39–56.

Altekruse SF, Henley SJ, Thun MJ. 1999. Deaths from hematopoietic and other cancers in relation to permanent hair dye use in a large prospective study (United States). Cancer Causes Control 10:617–625.

ATSDR (Agency for Toxic Substances and Disease Registry). 2007. Toxicological Profile for Benzene. Atlanta, GA: ATSDR. Available at http://www.atsdr.cdc.gov/toxprofiles/tp.asp?id=40&tid=14 [accessed May 28, 2012].

Attfield MD, Schleiff PL, Lubin JH, Blair A, Stewart PA, Vermeulen R, et al. 2012. The diesel exhaust in miners study: A cohort mortality study with emphasis on lung cancer. J Natl Cancer Inst 104:869–883.

Baade PD, Youlden DR, Valery PC, Hassall T, Ward L, Green AC, et al. 2010. Trends in incidence of childhood cancer in Australia, 1983–2006. Br J Cancer 102: 620–626.

Basel Action Network. 2005. The digital dump: Exporting reuse and abuse to Africa. Available at http://ban.org/BANreports/10-24-05/documents/TheDigitalDump_Print.pdf [accessed May 28, 2012].

Bassil KL, Vakil C, Sanborn M, Cole DC, Kaur JS, Kerr KJ. 2007. Cancer health effects of pesticides: Systematic review. Can Fam Physician 53:1704–1711.

Belpomme D, Irigaray P, Hardell L, Clapp R, Montagnier L, Epstein S, et al. 2007. The multitude and diversity of environmental carcinogens. Environ Res 105:414–429.

Biedermann S, Tschudin P, Grob K. 2010. Transfer of bisphenol A from thermal printer paper to the skin. Anal Bioanal Chem 398:571–576.

Blando JD, Schill DP, De La Cruz MP, Zhang L, Zhang J. 2010. Preliminary study of propyl bromide exposure among New Jersey dry cleaners as a result of a pending ban on perchloroethylene. J Air Waste Manag Assoc 60:1049–1056.

Boffetta P, Matisane L, Mundt KA, Dell LD. 2003. Meta-analysis of studies of occupational exposure to vinyl chloride in relation to cancer mortality. Scand J Work Environ Health 29:220–229.

Brody JG, Moysich KB, Humblet O, Attfield KR, Beehler GP, Rudel RA. 2007. Environmental pollutants and breast cancer: Epidemiologic studies. Cancer 109 (12 Suppl):2667–2711.

Brown LM, Everett GD, Burmeister LF, Blair A. 1992. Hair dye use and multiple myeloma in white men. Am J Public Health 82:1673–1674.

Calafat AM, Ye X, Wong LY, Reidy JA, Needham LL. 2008. Exposure of the U.S. population to bisphenol A and 4-tertiary-octylphenol: 2003–2004. Environ Health Perspect 116:39–44.

Cancer Research UK. 2011. Childhood Cancer Statistics. Available at http://info.cancerresearchuk.org/cancerstats/childhoodcancer/incidence/#source11 [accessed February 20, 2012].

Cantor KP, Ward MH, Moore LE. 2006. Water contaminants. In: Cancer Epidemiology and Prevention (Schottenfeld D, Fraumeni JFJ, eds.). New York: Oxford University Press, p. 382.

CAREX Canada. 2011a. Population Exposed to Benzene in Outdoor Air: Summary, 2006. Available at http://www.carexcanada.ca/en/benzene/environmental_exposure_estimates/outdoor_air/phase_1/ [accessed February 3, 2012].

CAREX Canada. 2011b. Surveillance of Environmental and Occupational Exposures for Cancer Prevention: Environmental Exposure Estimates. Available at http://www.carexcanada.ca/en/benzene/environmental_exposure_estimates/ [accessed February 3, 2012].

CAREX Canada. 2011c. Occupational Exposure Estimates: Styrene and Styrene-7,8-oxide. Available at http://www.carexcanada.ca/cen/styrene/occupational_exposure_estimates/phase_2/ [accessed February 22, 2012].

CAREX Canada. 2011d. Styrene and Styrene-7,8-oxide. Available at http://www.carexcanada.ca/en/styrene/ [accessed March 2, 2012].

CAREX Canada. 2011e. Occupational Exposure Estimates: Benzene. Available at http://www.carexcanada.ca/en/benzene/occupational_exposure_estimates/phase_2/ [accessed February 3, 2012].

CAREX Canada. 2012. Trichloroethylene. Available at http://www.carexcanada.ca/en/trichloroethylene/ [accessed May 20, 2012].

CDC (Centers for Disease Control and Prevention). 2009. Fourth National Report on Human Exposure to Environmental Chemicals. Department of Health and Human Services. Atlanta, GA: CDC.

Chan JK, Xing GH, Xu Y, Liang Y, Chen LX, Wu SC, et al. 2007. Body loadings and health risk assessment of polychlorinated dibenzo-*p*-dioxins and dibenzofurans at an intensive electronic waste recycling site in China. Environ Sci Technol 41: 7668–7674.

Clapp RW, Howe GK, Jacobs MM. 2005. Environmental and Occupational Causes of Cancer: A Review of Recent Scientific Literature. Lowell, MA: Lowell Center for Sustainable Production, University of Massachusetts Lowell.

Clapp R, Howe G, Jacobs M. 2006. Environmental and occupational causes of cancer revisited. J Public Health Policy 27:61–76.

Clapp RC, Jacobs M, Loechler EL. 2008. Environmental and occupational causes of cancer: New evidence 2005–2007. Rev Environ Health 23:1–37.

Cogliano VJ, Baan R, Straif K, Grosse Y, Lauby-Secretan B, El Ghissassi F, et al. 2011. Preventable exposures associated with human cancers. J Natl Cancer Inst 103: 1827–1839.

Cohen AJ. 2000. Outdoor air pollution and lung cancer. Environ Health Perspect 108(Suppl 4):743–750.

Cohen AJ, Pope AC, III. 1995. Lung cancer and air pollution. Environ Health Perspect 103(Suppl 8):219–224.

Cohn BA, Wolff MS, Cirillo PM, Sholtz RI. 2007. DDT and breast cancer in young women: New data on the significance of age at exposure. Environ Health Perspect 115:1406–1414.

Colt JS, Blair A. 1998. Parental occupational exposures and risk of childhood cancer. Environ Health Perspect 106(Suppl 3):909–925.

Cordier S, Mandereau L, Preston-Martin S, Little J, Lubin F, Mueller B, et al. 2001. Parental occupations and childhood brain tumors: Results of an international case-control study. Cancer Causes Control 12:865–874.

Creech JL, Jr., Johnson MN. 1974. Angiosarcoma of liver in the manufacture of polyvinyl chloride. J Occup Med 16:150–151.

Croplife International. 2010. Facts and Figures: The Status of Global Agriculture. Available at http://www.croplife.org/view_document.aspx?docId=2877 [accessed May 23, 2013].

Davis DL. 1990. Trends in cancer mortality in industrial countries: Report of an international workshop, Carpi, Italy, October 21–22, 1989. Ann N Y Acad Sci 609:4.

Davis S. 2009. Chemical Economics Handbook Marketing Research Report: Styrene. Menlo Park, CA: SRI Consulting.

Davis S. 2011. Chemical Economics Handbook Marketing Research Report: Benzene. Menlo Park, CA: SRI Consulting.

Dich J, Zahm SH, Hanberg A, Adami HO. 1997. Pesticides and cancer. Cancer Causes Control 8:420–443.

Doll R. 1998. Epidemiological evidence of the effects of behaviour and the environment on the risk of human cancer. Recent Results Cancer Res 154:3–21.

Doll R, Peto R. 1981. The causes of cancer: Quantitative estimates of avoidable risks of cancer in the United States today. J Natl Cancer Inst 66:1191–1308.

Durando M, Kass L, Piva J, Sonnenschein C, Soto AM, Luque EH, et al. 2007. Prenatal bisphenol A exposure induces preneoplastic lesions in the mammary gland in Wistar rats. Environ Health Perspect 115:80–86.

European Food Safety Authority. 2010. Results of the monitoring of dioxin levels in food and feed. EFSA J 8:1385.

Feychting M, Plato N, Nise G, Ahlbom A. 2001. Paternal occupational exposures and childhood cancer. Environ Health Perspect 109:193–196.

Freedonia Group. 2010. World Pesticides to 2014—Demand and Sales Forecasts, Market Share, Market Size, Market Leaders. Study #: 2664. Cleveland, OH: Freedonia Group.

Glauser J, Ishikawa Y. 2008. Chemical Economics Handbook Marketing Research Report: C2 Chlorinated Solvents. Menlo Park, CA: SRI Consulting.

Greiner EOC, Funada C. 2010. Chemical Economics Handbook Marketing Research Report: Bisphenol A. Menlo Park, CA: SRI Consulting.

Grosse Y, Baan R, Secretan-Lauby B, El Ghissassi F, Bouvard V, Benbrahim-Tallaa L, et al. 2011. Carcinogenicity of chemicals in industrial and consumer products, food contaminants and flavourings, and water chlorination by-products. Lancet Oncol 12:328–329.

Hanahan D, Weinberg RA. 2011. Hallmarks of Cancer: The Next Generation. Cell 144:646–674.

Hardell L, Eriksson M. 2003. Is the decline of the increasing incidence of non-Hodgkin lymphoma in Sweden and other countries a result of cancer preventive measures? Environ Health Perspect 111:1704–1706.

Harling M, Schablon A, Schedlbauer G, Dulon M, Nienhaus A. 2010. Bladder cancer among hairdressers: A meta-analysis. Occup Environ Med 67:351–358.

Hayes HM, Tarone RE, Cantor KP, Jessen CR, McCurnin DM, Richardson RC. 1991. Case-control study of canine malignant lymphoma: Positive association with dog owner's use of 2,4-dichlorophenoxyacetic acid herbicides. J Natl Cancer Inst 83: 1226–1231.

Ho SM, Tang WY, Belmonte de Frausto J, Prins GS. 2006. Developmental exposure to estradiol and bisphenol A increases susceptibility to prostate carcinogenesis and epigenetically regulates phosphodiesterase type 4 variant 4. Cancer Res 66: 5624–5632.

Hoar SK, Blair A, Holmes FF, Boysen CD, Robel RJ, Hoover R, et al. 1986. Agricultural herbicide use and risk of lymphoma and soft-tissue sarcoma. JAMA 256: 1141–1147.

Hutchings S, Rushton L. 2011. Toward risk reduction: Predicting the future burden of occupational cancer. Am J Epidemiol 173:1069–1077.

IARC (International Agency for Research on Cancer). 1987. IARC Monographs Supplement 7: Benzene.

IARC (International Agency for Research on Cancer). 2002. IARC Monographs Volume 82: Styrene. Lyon, France: IARC.

IARC (International Agency for Research on Cancer). 2008. World Cancer Report. Lyon, France: IARC.

IARC (International Agency for Research on Cancer). 2012a. Agents Classified by the IARC Monographs, Volumes 1–104. Available at http://monographs.iarc.fr/ENG/Classification/ClassificationsAlphaOrder.pdf [accessed May 29, 2012].

IARC (International Agency for Research on Cancer). 2012b. GLOBOCAN 2008. Available at http://globocan.iarc.fr/ [accessed May 29, 2012].

Ibrahim MA, Bond GG, Burke TA, Cole P, Dost FN, Enterline PE, et al. 1991. Weight of evidence on the human carcinogenicity of 2,4-D. Environ Health Perspect 96:213–222.

Industry Task Force II on 2,4-D Research Data. 2012. Available at http://www.24d.org [accessed July 25, 2012].

Iso T, Watanabe T, Iwamoto T, Shimamoto A, Furuichi Y. 2006. DNA damage caused by bisphenol A and estradiol through estrogenic activity. Biol Pharm Bull 29:206–210.

Jacobs M, Tickner J, Kriebel D. 2011. Regulating methylene chloride: A cautionary tale about setting health standards one chemical at a time. In: Lessons Learned: Solutions for Workplace Safety and Health (Kriebel D, Jacobs M, Markkanen P, Tickner J, eds.). Lowell, MA: Lowell Center for Sustainable Production, University of Massachusetts Lowell, pp. 81–97.

Jenkins S, Raghuraman N, Eltoum I, Carpenter M, Russo J, Lamartiniere CA. 2009. Oral exposure to bisphenol A increases dimethylbenzanthracene-induced mammary cancer in rats. Environ Health Perspect 117:910–915.

Kaatsch P, Steliarova-Foucher E, Crocetti E, Magnani C, Spix C, Zambon P. 2006. Time trends of cancer incidence in European children (1978–1997): Report from the automated childhood cancer information system project. Eur J Cancer 42: 1961–1971.

Kielhorn J, Melber C, Wahnschaffe U, Aitio A, Mangelsdorf I. 2000. Vinyl chloride: Still a cause for concern. Environ Health Perspect 108:579–588.

Kluwe WM, McConnell EE, Huff JE, Haseman JK, Douglas JF, Hartwell WV. 1982. Carcinogenicity testing of phthalate esters and related compounds by the National Toxicology Program and the National Cancer Institute. Environ Health Perspect 45:129–133.

Landrigan PJ, Baker DB. 1995. Clinical recognition of occupational and environmental disease. Mt Sinai J Med 62:406–411.

Landrigan P, Markowitz S, Nicholson W, Baker D. 1995. Cancer prevention in the workplace. In: Cancer Prevention and Control (Greenwald P, Kramer B, Weed D, eds.). New York: Marcel Dekker, Inc., pp. 393–410.

Lauby-Secretan B, Loomis D, Grosse Y, El Ghissassi F, Bouvard V, Benbrahim-Tallaa L, et al. 2013. Carcinogenicity of polychlorinated biphenyls and polybrominated biphenyls. International Agency for Research on Cancer Monograph Working Group IARC, Lyon, France. Lancet Oncol 14:287–288.

Linabery AM, Ross JA. 2008. Trends in childhood cancer incidence in the U.S. (1992–2004). Cancer 112:416–432.

Lunder S, Andrews D, Houlihan J. 2010. Synthetic Estrogen BPA Coats Cash Register Receipts. Washington, DC: Environmental Working Group.

Ma X, Buffler PA, Gunier RB, Dahl G, Smith MT, Reinier K, et al. 2002. Critical windows of exposure to household pesticides and risk of childhood leukemia. Environ Health Perspect 110:955–960.

Maffini MV, Rubin BS, Sonnenschein C, Soto AM. 2006. Endocrine disruptors and reproductive health: The case of bisphenol A. Mol Cell Endocrinol 254–255: 179–186.

Markey CM, Luque EH, Munoz De Toro M, Sonnenschein C, Soto AM. 2001. In utero exposure to bisphenol A alters the development and tissue organization of the mouse mammary gland. Biol Reprod 65:1215–1223.

Meinert R, Schuz J, Kaletsch U, Kaatsch P, Michaelis J. 2000. Leukemia and non-Hodgkin's lymphoma in childhood and exposure to pesticides: Results of a register-based case-control study in Germany. Am J Epidemiol 151:639–646.

Menegaux F, Baruchel A, Bertrand Y, Lescoeur B, Leverger G, Nelken B, et al. 2006. Household exposure to pesticides and risk of childhood acute leukaemia. Occup Environ Med 63:131–134.

Mills PK, Yang R. 2005. Breast cancer risk in Hispanic agricultural workers in California. Int J Occup Environ Health 11:123–131.

Monge P, Wesseling C, Guardado J, Lundberg I, Ahlbom A, Cantor KP, et al. 2007. Parental occupational exposure to pesticides and the risk of childhood leukemia in Costa Rica. Scand J Work Environ Health 33:293–303.

Muñoz-de-Toro M, Markey CM, Wadia PR, Luque EH, Rubin BS, Sonnenschein C, et al. 2005. Perinatal exposure to bisphenol-A alters peripubertal mammary gland development in mice. Endocrinology 146:4138–4147.

Murray TJ, Maffini MV, Ucci AA, Sonnenschein C, Soto AM. 2007. Induction of mammary gland ductal hyperplasias and carcinoma in situ following fetal bisphenol A exposure. Reprod Toxicol 23:383–390.

Noonan GO, Ackerman L, Begley TH. 2011. Concentration of bisphenol A in highly consumed canned foods on the U.S. market. J Agric Food Chem 59: 7178–7185.

NRC (National Research Council). 2009. Veterans and Agent Orange—Update 2008. Washington, DC: The National Academies Press.

NTP (National Toxicology Program). 2011a. Toxicology and Carcinogenesis Studies of 1-Bromopropane (CAS No. 106-94-5) in F344/N Rats and B6C3F1 Mice (Inhalation

Studies). National Institutes of Health Pub No. 11-5906. Available at http://ntp.niehs.nih.gov/ntp/htdocs/LT_rpts/TR564.pdf [accessed October 10, 2011].

NTP (National Toxicology Program). 2011b. Report on Carcinogens, 12th edition. Research Triangle Park, NC: National Toxicology Program.

OECD (Organization for Economic Cooperation and Development). 2001. OECD Environmental Outlook for the Chemicals Industry. Paris: OECD.

Pavuk M, Michalek JE, Ketchum NS. 2006. Prostate cancer in US air force veterans of the Vietnam War. J Expo Sci Environ Epidemiol 16:184–190.

Pesatori AC, Consonni D, Rubagotti M, Grillo P, Bertazzi PA. 2009. Cancer incidence in the population exposed to dioxin after the "Seveso accident": Twenty years of follow-up. Environ Health 8:39.

Pope CA, 3rd, Dockery DW. 2006. Health effects of fine particulate air pollution: Lines that connect. J Air Waste Manag Assoc 56:709–742.

Prüss-Üstün A, Vickers C, Haefliger P, Bertollini R. 2011. Knowns and unknowns on burden of disease due to chemicals: A systematic review. Environ Health 10:9.

Reese CD. 2008. Industrial Safety and Health for Infrastructure Services. Appendix B: The 50 Most Common Chemicals Found on Hazardous Waste Sites by Frequency of Occurrence. Boca Raton, FL: CRC Press.

Robinson BH. 2009. E-waste: An assessment of global production and environmental impacts. Sci Total Environ 408:183–191.

Rothman KJ. 2002. Epidemiology: An Introduction. Oxford: Oxford University Press.

Rudel RA, Attfield KR, Schifano JN, Brody JG. 2007. Chemicals causing mammary gland tumors in animals signal new directions for epidemiology, chemicals testing, and risk assessment for breast cancer prevention. Cancer 109(12 Suppl): 2635–2666.

Rushton L, Bagga S, Bevan R, Brown TP, Cherrie JW, Holmes P, et al. 2010. The Burden of Occupational Cancer in Great Britain: Overview Report. London: Health and Safety Executive (Prepared by Imperial College London, the Institute of Environment and Health, the Health and Safety Laboratory and the Institute of Occupational Medicine for the Health and Safety Executive).

Sandin S, Hjalgrim H, Glimellus B, Rostgaard K. 2006. Incidence of non-Hodgkin's lymphoma in Sweden, Denmark, and Finland from 1960 through 2003: An epidemic that was. Cancer Epidemiol Biomarkers Prev 15:1295–1300.

Sasco AJ. 2007. Cancer, environnement et populations à l'heure de la mondialisation. Oncologie 9:380–391.

Sasco AJ. 2008. Cancer and globalization. Biomed Pharmacother 62:110–121.

Schecter A, Pavuk M, Constable JD, Dai le C, Papke O. 2002. A follow-up: High level of dioxin contamination in Vietnamese from Agent Orange, three decades after the end of spraying. J Occup Environ Med 44:218–220.

Schecter A, Quynh HT, Papke O, Tung KC, Constable JD. 2006. Agent orange, dioxins and other chemicals of concern in Vietnam: Update 2006. J Occup Environ Med 48:408–413.

Schecter A, Malik N, Haffner D, Smith S, Harris TR. 2010. Bisphenol A (BPA) in U.S. Food Sci Technol 44:9425–9430.

Siemiatycki J, Richardson L, Straif K, Latreille B, Lakhani R, Campbell S, et al. 2004. Listing occupational carcinogens. Environ Health Perspect 112:1447–1459.

Silverman DT, Samanic CM, Lubin JH, Blair AE, Stewart PA, Vermeulen R, et al. 2012. The diesel exhaust in miners study: A nested case-control study of lung cancer and diesel exhaust. J Natl Cancer Inst 104:855–868.

Ulaszewska MM, Zuccato E, Davoli E. 2011. PCDD/Fs and dioxin-like PCBs in human milk and estimation of infants' daily intake: A review. Chemosphere 83: 774–782.

United Nations Environment Programme, Ozone Secretariat. 2010. Meeting of the Parties to the Montreal Protocol. Decision XIII/7; n-Propyl Bromide. Available at http://ozone.unmfs.org/new_site/fr/Treaties/decisions_text.php?dec_id=568 [accessed May 29, 2012].

U.S. EPA (U.S. Environmental Protection Agency). 2005. Reregistration Eligibility Decision for 2,4-D. EPA 738-R-05-002. Washington, DC: U.S. EPA.

U.S. EPA (U.S. Environmental Protection Agency). 2006. An Inventory of Sources and Environmental Releases of Dioxin-Like Compounds in the U.S. for the Years 1987, 1995, and 2000. EPA/600/P-03/002F. Washington, DC: U.S. EPA.

U.S. EPA (U.S. Environmental Protection Agency). 2011a. Technology Transfer Network Air Toxics: 2005 National-Scale Air Toxics Assessment (NATA): 2005 Assessment Results. Available at http://www.epa.gov/ttn/atw/nata2005/tables.html [accessed February 23, 2012].

U.S. EPA (U.S. Environmental Protection Agency). 2011b. Toxicological Review of Trichloroethylene. EPA/635/R-09/011F. Washington, DC: U.S. EPA.

U.S. President's Cancer Panel. 2010. Reducing Environmental Cancer Risk: What We Can Do Now. President's Cancer Panel 2008–2009 Annual Report. Bethesda, MD: US Department of Health and Human Services, President's Cancer Panel. Available at http://deainfo.nci.nih.gov/advisory/pcp/annualreports/pcp08-09rpt/PCP_Report_08-09_508.pdf [accessed May 29, 2012].

Van Leeuwen FXR, Malisch R. 2002. Results of the third round of the WHO coordinated exposure study on the level of PCBs, PCDDs, and PCFDs in human milk. Organohalogen Compds 56:311–316.

Viel JF, Fournier E, Danzon A. 2010. Age-period-cohort modeling of non-Hodgkin's lymphoma incidence in a French region: A period effect compatible with an environmental exposure. Environ Health 9:47–53.

Voss C, Zerban H, Bannasch P, Berger MR. 2005. Lifelong exposure to di-(2-ethylhexyl)-phthalate induces tumors in liver and testes of Sprague-Dawley rats. Toxicology 206:359–371.

Ward EM, Thun MJ, Hannan LM, Jemal A. 2006. Interpreting cancer trends. Ann N Y Acad Sci 1076:29–53.

WHO (World Health Organization). 2010. Dioxins and their effects on human health. Fact sheet no. 225. Available at http://www.who.int/mediacentre/factsheets/fs225/en/index.html [accessed July 27, 2012].

WHO (World Health Organization). 2011. Primary prevention of cancer through m itigation of environmental and occupational determinants. International Conference on Environmental and Occupational Determinants of Cancer: Interventions for

Primary Prevention. Asturias, Spain. Available at http://www.who.int/phe/news/events/international_conference/Background_interventions.pdf [accessed June 5, 2012].

Zahm SH. 1997. Mortality study of pesticide applicators and other employees of a lawn care service company. J Occup Environ Med 39:1055–1067.

Zahm SH, Ward MH. 1998. Pesticides and childhood cancer. Environ Health Perspect 106(Suppl 3):893–908.

Zahm SH, Weisenburger DD, Babbitt PA, Saal RC, Vaught JB, Cantor KP, et al. 1990. A case-control study of non-Hodgkin's lymphoma and the herbicide 2,4-dichlorophenoxyacetic acid (2,4-D) in eastern Nebraska. Epidemiology 1:349–356.

Zahm SH, Weisenburger DD, Babbitt PA, Saal RC, Vaught JB, Blair A. 1992. Use of hair coloring products and the risk of lymphoma, multiple myeloma, and chronic lymphocytic leukemia. Am J Public Health 82:990–997.

Zhang Y, Holford TR, Leaderer B, Boyle P, Zahm SH, Flynn S, et al. 2004. Hair-coloring product use and risk of non-Hodgkin's lymphoma: A population-based case-control study in Connecticut. Am J Epidemiol 159:148–154.

CHAPTER 4

Carcinogenicity and Mechanisms of Persistent Organic Pollutants

GABRIELE LUDEWIG, LARRY W. ROBERTSON, and HOWARD P. GLAUERT

ABSTRACT

Background: Persistent organic pollutants (POPs) are a group of halogenated organic compounds that were produced in large amounts as pesticides, industrial compounds, and production by-products during the 1950s–1980s. Examples include the pesticides DDT and dieldrin, the polychlorinated biphenyls (PCBs) and the notorious dioxin 2,3,7,8-tetrachlorodibenzo-*p*-dioxin (TCDD). Some of these compounds are still in use and/or produced today. However, because of their lipophilicity and resistance to destruction, they tend to biomagnify in the food chain and bioaccumulate in our bodies where they may produce a host of adverse health effects.

Objectives: Most of these compounds are now classified as probable human carcinogens, some even as proven human carcinogens (TCDD and most recently PCBs as a group). This chapter will provide an insight into the stages and mechanisms of chemical carcinogenesis and the activity of these POPS in them.

Discussion: The stages of chemical carcinogenesis are initiation, promotion, and progression, and each has different mechanisms and events that trigger these steps. Most POPs are believed to be promoters by changing gene transcription, acting as endocrine disruptors, producing reactive oxygen species and other mechanisms. Some members of these POPs can be regarded as initiators since they produce DNA damage either directly, after being metabolically bioactivated, or indirectly through the production of other genotoxic compounds. Some POPs may enhance progression by inducing genome instability, for example through interference with cellular proteins like telomerase or the cytoskeleton.
(*Continued*)

Effects of Persistent and Bioactive Organic Pollutants on Human Health, First Edition.
Edited by David O. Carpenter.

The requirements for these stages of carcinogenesis regarding chemical structure and metabolic activation pathways are described in more detail using the large and diverse group of PCBs as an example.

Conclusions: This review shows that our knowledge about the activators and mechanisms of chemical carcinogensis is incomplete. In our world of increasing exposure to mixtures of diverse chemicals a better understanding of the forces in chemical carcinogenesis is needed, to provide precautionary risk assessment and public health protective regulation.

Abbreviations: AhR, aryl hydrocarbon receptor; ANF, α-naphthoflavone; CYP, cytochrome P-450; DDT, dichlorodiphenyltrichloroethane; GJC, gap junctional communication; HCB, hexachlorobenzene; PBBs, polybrominated biphenyls; PCBs, polychlorinated biphenyls; POPs, persistent organic pollutants; ROS, reactive oxygen species; SCEs, sister chromatid exchanges; TCDD, 2,3,7,8-tetrachlorodibenzo-*p*-dioxin.

INTRODUCTION

Cancer is a major cause of death worldwide. Even though genes play an important role in this process, few cancers are known to be caused by specific inherited gene defects alone. If exposure to endogenous and exogenous factors is driving the development of cancer, then intervention in the exposure can prevent its occurrence. For this, two steps are needed: the identification of carcinogenic compounds and an understanding of their mechanisms of action. The latter will help to identify other, similar carcinogenic compounds, essential for the assessment of risk (toxic dose level, additive or even synergistic interaction with other compounds), and help in the development of treatment options. This review will provide an overview of our current knowledge of the carcinogenic potential and proposed mechanisms of some persistent organic pollutants (POPs), focusing on the "dirty dozen" of the Stockholm treaty (Kaiser and Enserink 2000). These halogenated organic compounds were produced in the last century in large amounts as pesticides, industrial mixtures or production by-products, and have become ubiquitous contaminants in our environment, our foods, and our bodies, and may exert their toxic action alone or in combination with other compounds for many decades to come.

MANY POPs ARE COMPLETE CARCINOGENS IN ANIMAL STUDIES

The compounds that are described in this chapter have some characteristics in common: they all have halogens, chlorine or bromine atoms, in their structures (Figure 4.1; chemical structures of some POPs), which confers upon them

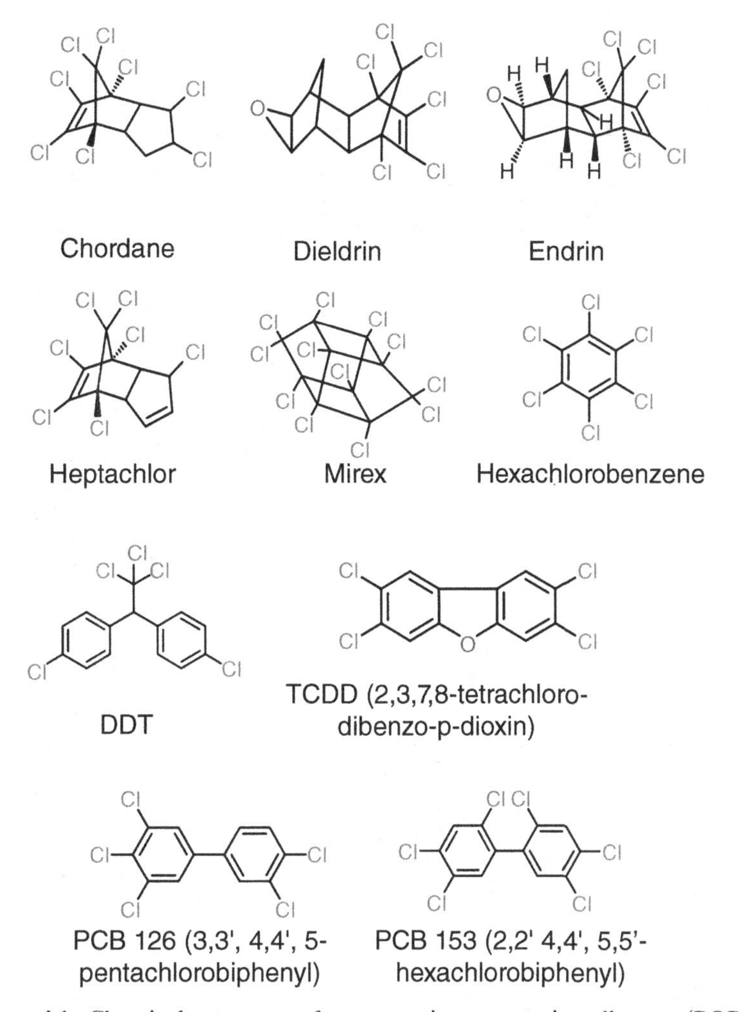

Figure 4.1. Chemical structures of some persistent organic pollutants (POPs).

chemical stability and resistance to metabolic degradation. The carcinogenic activities of many environmental pollutants have been examined in laboratory animal models (Table 4.1; carcinogenic and tumor-promoting activity of some POPs), usually chronic 2-year feeding studies with rats or mice. Most of these agents are active as liver carcinogens, but other tissues are also affected by some of the agents. Carcinogenicity in rodents is often the first indication that a compound may be a human carcinogen.

Aldrin and dieldrin are pesticides that were used until 1987. Both aldrin and dieldrin have been examined for their carcinogenic activity *in vivo* (Stevenson et al. 1999). Both agents have undergone analysis in long-term carcinogenesis studies, with the only positive response being observed in mouse liver, for both agents.

TABLE 4.1. Hepatic Carcinogenic and Tumor-Promoting Activities of Environmental Pollutants

Chemical	EPA[a] (Year)	IARC[a] (Year)	Carcinogenic Activity	Tumor-Promoting Activity	References
Aldrin	B2 (1993)	3 (1987)	Mouse liver only	Not studied	Stevenson et al. (1999)
Chlordane	B2 (1998)	2B (2001)	Mouse liver only	Tumor-promoting activity in mice	Khasawinah and Grutsch (1989), Malarkey et al. (1995), Reuber and Ward (1979), Williams and Numoto (1984)
DDT	B2 (1991)	2B (1991)	Tumors induced in livers of mice and rats, lung tumors and lymphomas induced in mice, adrenal tumors induced in hamsters	Tumor-promoting activity in rats and mice	Angsubhakorn et al. (2002), Flodstrom et al. (1990), Fukushima et al. (2005), Harada et al. (2003), Numoto et al. (1985), Oesterle and Deml (1981), Turusov et al. (2002), Williams and Numoto (1984)
Dieldrin	B2 (1993)	3 (1987)	Mouse liver only	Tumor-promoting activity in mice but not rats	Stevenson et al. (1999)
Endrin	D (1993)	3 (1987)	Tumors induced in several tissues	Not studied	Reuber and Ward (1979)
Heptachlor	B2 (1993)	2B (2001)	Tumors induced in liver and other tissues in rats and mice	Tumor-promoting activity in mice	Numoto et al. (1985), Reuber (1987), Williams and Numoto (1984)
Hexachlorobenzene (HCB)	B2 (1996)	2B (2001)	Tumors induced in livers of rats and mice, in the thyroid of hamsters, and in the kidneys of rats	HCB had promoting activity in female rats; in males, HCB had promoting activity in energy-restricted rats but not in control rats	Agency for Toxic Substances and Disease Registry (2002), Kishima et al. (2000), Pereira et al. (1982), Reed et al. (2007)

Mirex	N/C	2B(1987)	In rats, tumors induced in liver, adrenal gland, and kidney	Not studied	Moser et al. (1992, 1993), NTP (1990), Porter et al. (2002), Ulland et al. (1977)
Toxaphene	B2 (1991)	2B (2001)	Tumors induced in the liver in mice and in the thyroid in rats	No tumor-promoting activity in rats	Besselink et al. (2008), Goodman et al. (2000)
PBBs	N/C	2A (2013)[b]	PBB mixtures induced liver tumors	PBB mixtures and individual PBB congeners have promoting activity	Silberhorn et al. (1990)
PCBs	B2 (1997)	1 (2013)[b] PCB 126 = 1 (2012)[c]	PCB mixtures induced liver tumors	PCB mixtures have promoting activity; for individual congeners, both Ah and constitutive androstane/active receptor activators have promoting activity	Glauert et al. (2001); Lauby-Secretan et al. (2013); Silberhorn et al. (1990)
TCDD	Under review (2012)	1 (2012)	Tumors induced in the liver, thyroid, oral cavity, and lung in rats, and in the liver, thymus, and skin in mice.	TCDD has promoting activity in rats and in one mouse strain but not in two others	Knerr and Schrenk (2006)
Peroxisome proliferators	N/C	N/C	Liver tumors induced in rats and mice but not other species	Several peroxisome proliferators have hepatic tumor-promoting activity	Cattley et al. (1998), Rao and Reddy (1987)

[a]The numbers in brackets indicate the most recent year of classification; EPA–International Agency for Research on Cancer (IARC) groups: A—1 (human carcinogen), B1/2—2A (probable human carcinogen, limited or inadequate evidence in humans so far), C—2B (possible human carcinogen), D—3 (not classifiable), E—4 (evidence of noncarcinogenicity for humans.

[b]Lauby-Secretan et al. (2013).

[c]Cogliano et al (2012).

N/C, not classified.

Chlordane is a pesticide that was banned in 1988. Chlordane has been examined in three long-term carcinogenesis studies and has been found only to induce liver neoplasms (Khasawinah and Grutsch 1989; Malarkey et al. 1995; Reuber and Ward 1979).

Endrin is a pesticide that has not been produced since 1986. There are very few studies available on endrin. Most indicate that endrin does not induce cancer, but in others, endrin has been found to induce tumors in a number of tissues, depending on the animal model studied (Reuber 1979). The Environmental Protection Agency (EPA) classifies endrin as Group D and IARC in 1987 as Group 3 compound, that is, as not classifiable due to insufficient data.

Heptachlor is an insecticide that was used until 1988. Heptachlor has been found to induce liver tumors in both rats and mice, as well as tumors in several other tissues (Reuber 1987).

Mirex is an insecticide and flame retardant that has not been produced since 1978. It was found to induce tumors in the livers of rats, as well as in their adrenal gland and kidney (NTP 1990; Ulland et al. 1977). IARC classified mirex as Class 2B, possibly carcinogenic to humans (1987).

Toxaphene is an insecticide that was banned in 1990. It has been found to be carcinogenic in the livers of mice in two studies and in the thyroid gland of rats (Goodman et al. 2000).

DDT is a pesticide that was banned in the United States in 1972. It has been found to be carcinogenic in the livers of both rats and mice (Turusov et al. 2002). In addition, it induced lung tumors and lymphomas in mice as well as adrenal adenomas in hamsters.

Hexachlorobenzene (HCB) is a pesticide and fungicide and is also produced during the manufacture of other agents. It induces liver tumors in rats and mice as well as tumors in other tissues (Agency for Toxic Substances and Disease Registry 2002; Reed et al. 2007).

2,3,7,8-Tetrachlorodibenzo-*p*-dioxin (TCDD) is a contaminant that is found in chlorophenoxy acid herbicides. It has been reported to induce tumors in several tissues in mice and rats, including the liver of both species, the thyroid, oral cavity, and lung of rats, and the thymus and skin of mice (Knerr and Schrenk 2006). In a recent reanalysis, IARC grouped TCDD as a Class 1, proven human carcinogen (Cogliano et al. 2012).

Polychlorinated biphenyls (PCBs) had a wide variety of uses, including in electrical equipment, as plasticizers, and as flame retardants; their production was banned in the United States in 1979. PCBs are used as mixtures, with a total of 209 possible congeners. PCB mixtures have complete carcinogenic activities in the liver (Glauert et al. 2001; Silberhorn et al. 1990). One of the congeners, 3,3′,4,4′,5-pentachlorobiphenyl (PCB 126), was classified by IARC as a human carcinogen, Class 1 (Cogliano et al. 2012). All other PCB congeners were upgraded to Class 1 in 2013 (Lauby-Secretan et al. 2013).

Polybrominated biphenyls (PBBs) were formerly used as flame retardants. Like PCBs, they were produced as mixtures of the 209 possible congeners. PBB mixtures induced liver tumors (Silberhorn et al. 1990). Due to insufficient evidence in human studies, PBBs are now classified by IARC as Group 2A, probable human carcinogens (Lauby-Secretan et al. 2013).

Several environmental agents, including phthalate plasticizers, certain herbicides, and trichloroacetic acid, activate the peroxisome proliferator-activated receptor α and induce peroxisome proliferation (Reddy and Lalwani 1983; Roberts 1999). Several peroxisome proliferators are hepatocarcinogenic (Cattley et al. 1998; Rao and Reddy 1987; Reddy and Lalwani 1983; Roberts 1999).

ASPECTS OF CANCER INDUCTION

Carcinogenesis Is a Multistep Process

From animal studies it was concluded that the process of carcinogenesis involves multiple steps that occur in a certain order (Figure 4.2, multistep process of carcinogenesis) (Pitot and Dragan 1994). The first step, initiation, is a genotoxic event. The DNA sequence or amount in a cell is changed irreversibly and in a heritable way. A single exposure to an initiating compound may be enough if such mutagenic changes occur in oncogenes or tumor suppressor genes, which lead to an initiated cell. The second step, promotion, has to occur after initiation and gives such initiated cells a growth advantage, for example, by stimulating their proliferation or inhibiting their death. Multiple and prolonged exposure to a promoting agent is required, during which the initiated cells multiply and develop into a focal lesion. Unlike initiation, promotion is nongenotoxic; exposure to the promoting agent has to occur above a specific threshold dose/concentration and for a long time interval, below which promotion is reversible. The final step is called progression. Here, significant DNA modifications and changes in the karyotype are observed and the cells change from preneolastic to benign neoplastic and later malignant cell types. Progression is irreversible and usually occurs over an extended period of time, although it is not known whether repeated exposures are needed. A carcinogenic compound may be a complete carcinogen, able to accomplish the change of a normal cell to a malignant neoplastic cell. Many compounds are either genotoxic (initiators) or promoters. Progression may occur spontaneously or may be driven by compounds, many of which have

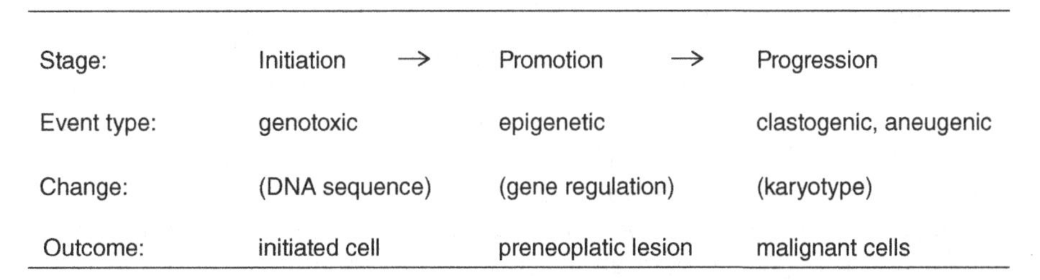

Stage:	Initiation →	Promotion →	Progression
Event type:	genotoxic	epigenetic	clastogenic, aneugenic
Change:	(DNA sequence)	(gene regulation)	(karyotype)
Outcome:	initiated cell	preneoplatic lesion	malignant cells

Figure 4.2. Stages of chemical carcinogenesis (simplified scheme).

clastogenic (chromosome breaking) or aneugenic (producing chromosome mis-segregation) activity. In the case of such incomplete carcinogens, an exposure to several different compounds or, as is most likely the case for PCBs, to different congeners or metabolites (Ludewig 2001), has to occur for cancer induction. In this chapter, we will describe the initiating and promoting activity of some POPs and discuss some, but not all, possible mechanisms using the large family of PCBs as example.

Compounds May Require Metabolic Bioactivation to Reactive Intermediates and/or By-Products to Act as Carcinogens: Example Lower-Chlorinated Biphenyls

Often a compound itself has no carcinogenic activities, but a metabolite or degradation intermediate or by-product may. As mentioned earlier, the prime characteristic of POPs is that they are extremely chemically stable. This usually desirable feature is the reason for their accumulation and biomagnifications in our bodies, the environment, and the food chain. It also means, however, that these compounds are not very reactive themselves and are not very prone to be changed by environmental or biological factors to more reactive intermediates. As a consequence of this, POPs are generally seen as being nongenotoxic and therefore not initiators or involved in progression. Recent results with lower-chlorinated biphenyls have changed this perception. In addition, changes in the chemical structure of the compound may change or increase the binding of such an intermediate compound with receptors, transcription factors, or specific transport proteins, thereby increasing their tumor-promoting activities. Of the POPs mentioned earlier, the large group of PCBs is a good example for very different biochemical behavior due to minor differences in chemical structure.

The 209 PCB congeners vary greatly in their propensity for metabolic attack, the first line of which is monooxygenation by members of the cytochrome P-450 (CYP) superfamily. There are 837 possible monofunctional hydroxylated products (Rayne and Forest 2010). The enzymic hydroxylation mechanisms may occur via a "direct insertion" mechanism or via the intermediacy of an arene oxide (Darbyshire et al. 1996; Guengerich 2001; Jerina and Daly 1974). The latter may be a highly reactive, electrophilic species, depending on the number of chlorines present on biphenyl (the greater the number of chlorines, the more stable the arene oxide intermediates).

Monohydroxylated PCBs may undergo an additional hydroxylation producing a dihydroxylated PCB derivative (Figure 4.3; PCB metabolism). Depending on the relative positions of the hydroxyl groups, catechols (ortho to each other) or hydroquinones (para) may be formed (McLean et al. 1996a). The biosynthesis of dihydroxylated PCBs is catalyzed primarily by CYPs, as far as is known, while the oxidation of PCB catechols and hydroquinones may be catalyzed by peroxidases (Amaro et al. 1996; Oakley et al. 1996), prostaglandin synthase (Wangpradit et al. 2009), and likely other enzymes as well,

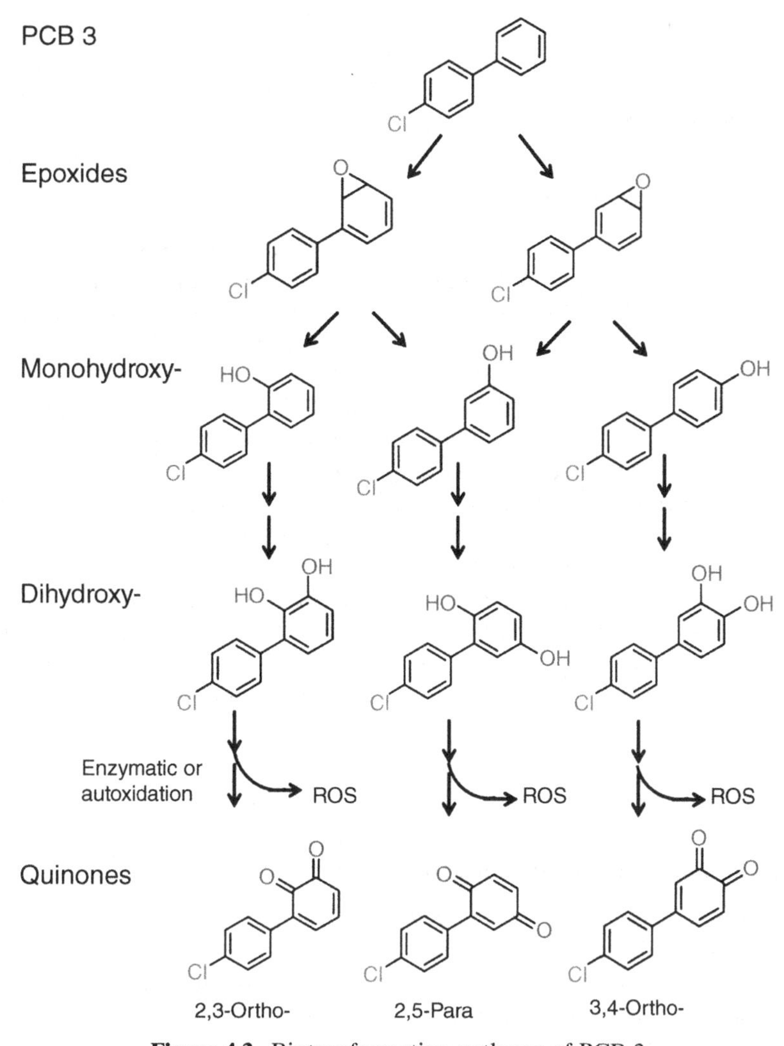

Figure 4.3. Biotranformation pathway of PCB 3.

yielding highly reactive electrophilic PCB quinones (Amaro et al. 1996; Wangpradit et al. 2009).

Presumably the oxygenated PCB intermediates/metabolites, the arene oxides and quinones, are the most relevant to PCB carcinogenesis, but it should be remembered that in addition to the hydroxylated PCBs, with their myriad biological effects, many other metabolites are formed from PCBs. Recently, it was shown that the hydroxylated PCBs are substrates for glucuronidation (Tampal et al. 2002) and sulfation (Ekuase et al. 2011; Liu et al. 2006, 2009). Very little is known about the consequences of hydroxylation and/or conjugation with respect to the promoting activity, but first reports indicate that these metabolites have equal or higher endocrine-disrupting activities as their parent compound, which may increase their tumor-promoting activities. All metabolites with higher chemical reactivity, however, have to be regarded as more likely cancer initiators.

Direct versus Indirect Action

If the interaction of a compound or its metabolite with DNA or proteins is responsible for carcinogenic changes, then this compound is a direct-acting carcinogen. But sometimes it is not the compound itself that is producing the carcinogenic damage. For example, the autoxidation of a PCB hydroquinone may produce reactive oxygen species (ROS) (Amaro et al. 1996; Oakley et al. 1996). ROS are considered to be active in initiation, promotion, and progression (Klaunig et al. 2011). Also, many POPs induce metabolizing enzymes and may change the metabolic pathway of endogenous or other exogenous compounds. For example, PCBs induce CYPs in the liver, which may change the metabolism of endogenous estrogen to more harmful estrogen catechols (Ho et al. 2008) or ROS producing estrogen quinones (Brown et al. 2007). In addition, an overabundance of CYP may produce ROS (see section on "Mechanisms of Tumor Promotion by Environmental Pollutants" for more on ROS and estrogen metabolism). In all these cases, PCBs are acting as an indirect agent, generating another compound that acts as initiator, promoter, or progressor.

CANCER-INITIATING ACTIVITY OF POPs

Cancer Initiation Is Based on Genotoxic Events

Initiation of carcinogenesis is believed to be the first step in cancer development. Initiation involves a heritable, irreversible alteration of the cellular DNA. Such changes include point mutations, that is, change, deletion, or addition of one to a few hundred base pairs, as well as larger changes due to chromosome breaks and rearrangements resulting in the duplication or loss of genes or changes in their transcriptional control. Mechanisms that produce such changes include direct adduction (covalent binding) or steric interference with the DNA or DNA-related proteins by the compound or one of its metabolites or by-products. The genotoxicity of a compound can be assessed with a large number of *in vitro* and *in vivo* assays. Since no test method exists that is able to detect all different types of genotoxic damage, a test battery of several assays that measure different end points is needed to fully evaluate the cancer-initiating potential of a compound. Understanding the environmental and endogenous transformation pathway of a compound and the chemical reactivity of the compound and (bio)degradation/activation products is essential for the selection of the best type of assay and assay conditions to avoid false negative results due to the choice of inappropriate tests and test conditions.

Initiating and Genotoxic Activity of POPs *In Vitro* and *In Vivo*

The multitude of genotoxic assays and the complexity of effects can best be illustrated using the example of the PCBs. This group of POPs is composed of 209 different individual congeners that display differing affinities for various receptors and other macromolecules and differing likelihoods of being metabolically altered to reactive intermediates. Therefore, the different assays will be described next with a description of the test results with PCBs, followed by a brief summary of genotoxicity results with other POPs. The list of references for each assay and compound is not meant to be complete, but only to provide some examples of findings with each compound.

In Vivo *Initiation (Solt–Farber)* Very few assays exist that examine the initiating activity of a compound *in vivo*. In the mouse two-stage skin carcinogenesis model, the test compound is applied as a possible initiator, followed by a longer application of a known promoter. The commercial PCB mixture Arochlor 1254, with predominantly tetra- and pentachlorainted biphenyls, acted as a weak tumor initiator in this assay (DiGiovanni et al. 1977). In the Solt–Farber protocol (Laconi et al. 1995; Semple-Roberts et al. 1987; Solt et al. 1977; Tsuda et al. 1980), initiation is brought about by induction of cell proliferation (necrogenic dose, partial hepatectomy or fasting/refeeding) followed by the application of the test compound, and 2 weeks later by a selection procedure with 2-acetylaminofluorene and carbon tetrachloride and analysis of the resulting preneoplastic foci or nodules. Using this rat liver initiation assay in male F344 rats, Hayes and coworkers analyzed Aroclor 1254 and the congeners PCB 153, PCB 52, and PCB 47 and obtained only negative results (Hayes et al. 1985). One possible explanation for this could be that these PCB congeners contain four or more chlorines and are not very prone to metabolic activation. Therefore, lower-chlorinated biphenyl congeners that are easily biotranformed to potential reactive and genotoxic metabolites were tested for their ability to induce preneoplastic foci, including PCB 3 (4-Cl-,), PCB 12 (3,4-diCl-), and PCB 38 (3,4,5-triCl-BP) (Espandiari et al. 2003). In this study, a single dose of 100 mg/kg of the initiator diethylnitrosamine (DEN) was used as a positive control. All animals in the positive control group showed visible nodules, while no nodules were apparent in animals in the negative vehicle control group, or groups receiving the dichloro- or trichloro-PCBs as initiators. However, the monochloro congener (PCB 3) induced grossly visible nodules in 50% of the rats treated. Histological examination of liver sections further revealed the presence of altered hepatic foci and nodules in all DEN-treated animal (5500 foci/liver) and in 80% of PCB 3-treated animal (1500 foci/liver). In a second experiment, PCB 15 (4,4′-diCl-), PCB 77 (3,3′,4,4′-tetraCl-), PCB 52 (2,2′,5,5′-tetraCl-BP), and a combination of PCB 77 and PCB 52 were tested (Espandiari et al. 2003). All four treatment groups increased the number of foci per cubic centimeter in the liver, with DEN > PCB 15 >> PCB 52 > PCB 77 > PCB 52/77. The conclusion from these experiments is that lower-chlorinated

PCBs with one to four chlorines are able to initiate hepatocarcinogenesis in the rats, but from these few congeners tested, a structure–activity relationship could not be assigned.

A series of experiments with synthetic PCB 3 metabolites were tested in an effort to determine the metabolic activation pathway and the ultimate initiating agent (Espandiari et al. 2005). Test compounds included the 2-OH-, 3-OH-, 4-OH-, 2,3-diOH-, 3,4-diOH-, 2,5-diOH-, 2,3-quinone, 3,4-quinone, and 2,5-quinone metabolite of PCB 3. The 4-OH- and 3,4-quinone metabolites of PCB 3 significantly increased the number of foci per cubic centimeter, the number of foci per liver and the focal volume (% of liver). In fact, 100 µmol/kg 3,4-quinone of PCB 3 was more active than 20 mg/kg DEN with respect to foci number. None of the other PCB 3 metabolites had a significant effect on either foci number or foci volume (Espandiari et al. 2004). The conclusion is that the 3,4-ortho-quinone of PCB 3 is the ultimate initiating metabolite and that PCB 3 is metabolized *in vivo* in the rat liver to this ultimate carcinogen.

We are mostly exposed to such lower halogenated and therefore semivolatile PCB congeners through the indoor and outdoor air (Fitzgerald et al. 2011; Norstrom et al. 2010). The implication of the studies mentioned earlier is that PCB congeners like PCBs 5, 8, 15, 17, 18, 28, 31, 33, 44, 52, 66, 84, 92, 95, 101, and 110, all major congeners found in the air of cities and contaminated buildings, may be activated through the same pathway and may be possibly participating in liver and lung cancer initation.

Role of DNA and Protein Adduction and Production of ROS Adduct formation with DNA is a prerequisite for the induction of gene mutations by direct-acting compounds. A very sensitive method to measure protein and DNA binding employs radioactively labeled test compounds. Using this technique, Morales and Matthews (1979) had shown that PCB 153 (2,2′,4,4′,5,5′-hexachlorobiphenyl) strongly bioaccumulates in mouse tissues, but that PCB 136 (2,2′,3,3′,6,6′-hexachlorobiphenyl) has an at least 10 times higher covalent binding rate to macromolecules, with RNA > protein > DNA. PCB 136 is much more rapidly metabolized than PCB 153, pointing toward biotransformation intermediates as binding species. Moreover, focusing on tissue levels to evaluate the risks for cancer initiation may be very misleading since compounds that are prone to biotransformation, and therefore fast disappearing, are obviously the more dangerous ones in the case of PCBs. Using ^{3}H- or ^{14}C-labeling and microsomal bioactivation systems with various modifications by inhibitors and inducers *in vitro* to analyze the activation pathway, PCB 3 and the tetrachlorinated congeners PCB 47, PCB 49, PCB 52, and PCB 77 were found to bind efficiently to proteins and DNA (Hesse et al. 1978; Shimada and Sato 1978, 1980; Shimada and Sawabe 1984; Wyndham et al. 1976). Higher-chlorinated congeners were usually not binding to macromolecules. Based on the results with the metabolism modifiers, the authors suggested that CYP-mediated metabolic activation was required and that an arene oxide intermediate and possibly a semiquinone or quinone were the binding species.

Another very sensitive analytical method specifically to study DNA adducts is ^{32}P-postlabeling. Using this technique in a environmental pollution biomonitoring study of the Meuse river with crayfish, a Dutch group had found a strong positive correlation between DNA adducts in hepatopancreatic tissue and body burdens of PCB 28, PCB 52, and PCB 101 (Schilderman et al. 1999). No correlation was observed for higher-chlorinated PCBs like PCB 118, PCB 138, PCB 153, and PCB 180, or other polycyclic aromatic hydrocarbons. To evaluate the metabolic activation pathway and requirements, a series of 15 mono- and dichlorinated PCBs were incubated with DNA *in vitro* in the presence of microsomes and horseradish peroxidase (HRP) and hydrogen peroxide (H_2O_2). Significant DNA adduct formation was observed, which was not seen when HRP/H_2O_2 was omitted and/or ascorbic acid was added to the incubation mixture (McLean et al. 1996b). This suggested that quinones were the ultimate genotoxic agent. The majority of these adducts were with guanine, and increasing chlorination in one ring increased adduct formation (Oakley et al. 1996). In fact, incubation of the synthetic para-quinone of PCB 3, PCB 12, and PCB 38 with DNA resulted in two major and two to four minor DNA adducts with each quinone. Quantitation of adducts showed that at equimolar concentrations, PCB 38 was most reactive (298 adducts/10^9 nucleotides), followed by PCB 3 (140 adducts/10^9 nucleotides). PCB 12 was least reactive (56 adducts/10^9 nucleotides).

Zhao and colleagues reported on the characterization and quantitative studies of DNA adducts formed from lower-chlorinated PCB-derived quinones (Zhao et al. 2004). Their quantitation (by HPLC/ESI-MS/MS and ^{32}P-postlabeling) revealed adduct levels in the range of 3–1200 adducts per 10^8 nucleotides, the highest level for the metabolite of PCB 3 and for all congeners mostly to guanosine. Increasing chlorine content diminished adduct yield. Interestingly, the ^{32}P-postlabeling method, while much more sensitive, "severely underestimated" the number of adducts, compared to the MS method, indicating that our "tools" to detect DNA adduction are still inadequate and too blunt to fully measure the extent of this risk factor for cancer initiation.

The observation that PCB 52, a major congener in inner city air, formed quinonoid protein adducts in the liver and brain of exposed rats indicates that a major bioactivation pathway for PCBs leads indeed to reactive quinones and that these metabolites can reach and bind to proteins and probably DNA in such distant and protected tissues as our brain (Lin et al. 2000). A study with radiolabeled PCB 3 and PCB 77 in mice discovered significant amounts of both PCBs in various tissues, with liver > kidneys > lung (Pereg et al. 2001). Bound PCBs were mostly localized in the cytosol or organelles, less in the microsomes, and significant amounts of PCB 3 and PCB 77 were covalently bound to proteins in the nucleus of the liver of these animals, with PCB 3 >> PCB 77. There was also a substantial covalent binding of these PCBs to liver DNA, but it was below the significance level. Protein binding of PCB quinones occurs preferably to cysteines, but also to arginine and histidine (Amaro et al. 1996), and binding of PCB 3 quinone to topoisomerase II was observed, a nuclear protein that is essential for proper DNA maintenance and replication (Bender et al. 2006; Srinivasan et al. 2002). Thus, besides binding to DNA, PCB quinones bind to cellular proteins, a process that may also lead to genotoxicity and cancer initiation.

Aside from direct DNA and protein adduction, dihyroxylated PCBs and their corresponding PCB quinones are able to redox cycle (McLean et al. 2000), with the production of ROS and the formation of oxidized nucleotides like 8-oxodG (Oakley et al. 1996), a DNA modification known to produce gene mutations. Generated ROS produce DNA strand breaks (Srinivasan et al. 2001). Thus, there are at least three mechanisms, DNA adduction, protein binding, and ROS generation that can lead to genotoxic damage.

Gene Mutation Assays A fast, easy, and inexpensive method to test the point mutation activity of a compound is the Ames test. This assay uses specific strains of the bacterium *Salmonella* which were created by introducing mutations into genes of the histidine synthesis pathway. These cells called histidine auxotrophs require histidine in the medium for growth. In the Ames test, the ability of compounds to reverse this mutation is analyzed. Different strains have different known types of mutations and therefore indicate whether the test compound induced a base pair substitution, deletion, or other type of point mutation. To mimic the mammalian biotransformation system, a microsomal system (S9) is often added. Most PCB mixtures and the few congeners that were tested were negative in the Ames test with and without S9 (reviewed in Ludewig 2001). A negative result in the Ames test is not uncommon for compounds with complicated and multistep activation pathways as the one proposed for lower-chlorinated PCBs, that is, the metabolic activation to quinones. Thus, a bacterial mutagenicity test is most likely not an appropriate assay to evaluate the genotoxicity of PCBs.

Other *in vitro* point mutation assays use mammalian cells, but PCB mixtures or congeners were hardly tested in these assays and were always negative (Hattula 1985). The cell lines used for such mutation studies like the Chinese hamster lung fibroblast (V79), Chinese hamster ovary fibroblast, and mouse lymphoma (L5178Y) cell lines have only very limited or no biotransformation capability, a crucial requisite if the test compound is to be metabolically activated. To overcome this problem, a series of synthetic PCB 3 metabolites was tested in the V79 gene mutation assay (Zettner et al. 2007). Both the ortho- (3,4-) and para- (2,5-) quinone efficiently induced mutations at the hypoxanthine-guanine phosphoribosyltransferase locus, while none of the tested mono- or dihydroxylated metabolites or PCB 3 itself were mutagenic. This is in agreement with the hypothesis that the PCB quinones are genotoxic and potentially initiating carcinogens.

To analyze the mutagenic activity of PCB 3, a transgenic *in vivo* assay was employed. The BigBlue rat (Stratagene, La Jolla, CA) is a transgenic rat that carries multiple copies of an indicator gene that can be isolated, packed in phages, and selected on agar plates after transfection into bacteria. Male rats were treated weekly for 4 weeks with intraperitoneal (i.p.) injections of 3-methylcholanthrene, PCB 3, 4-OH-PCB 3, or vehicle alone (corn oil) and were killed after a period of an additional 10 days. Liver tissue was collected, the target DNA isolated and packed into phages, transfected into bacteria, and mutated (blue) colonies identified. The mutation frequency per 100,000 pfu was 1.7 in negative control, and 8.8 (3-methylcholanthrene), 4.8 (PCB 3), and 4.0 (4-OH-PCB 3) (Lehmann et al. 2007). This mutation frequency of 3-methylcholanthrene and PCB 3 in the liver was statistically significant. Moreover, in these treatment groups the mutation spectrum was altered from predominantly transitions to predominantly GC→TA transversions. 4-OH-PCB 3 had a similar but smaller effect, causing a doubling of the mutation rate that was below the level of statistical significance (Lehmann et al. 2007). This demonstrates that this PCB congener is mutagenic *in vivo* in the target organ, the liver. Unfortunately, these results do not definitely explain the mechanism of genotoxicity (DNA adduction or ROS), or the ultimate mutagenic metabolite (ortho- or para-quinone, or epoxide or other metabolite), but all indications from other assays point toward adduct formation by a quinone, most likely the ortho (3,4-) quinone as one highly likely mechanism.

Chromosome Aberrations, Micronuclei, Comet, Aneuploidy, Polyploidy Besides gene mutations, DNA strand breaks and anomalous segregation of chromosomes represent a major form of genotoxicity. Compounds that induce strand breaks are called clastogens, while compounds that induce an anomalous distribution of chromosomes are called aneugens. Assays that measure these types of genotoxicity include the chromosome aberration assay, a visual inspection of the number and shape of the chromosomes of a cell, the micronucleus assay, a counting of cells with small "micro"nuclei that contain pieces or a whole chromosome, and the Comet assay, in which cells are embedded in agarose and subjected to a low electric current that pulls small pieces of broken DNA out of the nucleus where it forms a kind of a comet tail. All three test systems can be used with cells in culture or from animal or human biomonitoring studies. Different mechanisms can result in these types of damage, including DNA adduction, adduction to proteins like topoisomerase or the cytoskeleton during chromosome distribution, and also damage due to production of ROS.

The Comet assay, which measures DNA stand breaks, showed a positive response with PCB 52 and PCB 77 in human lymphocytes (Sandal et al. 2008), and PCB 101 and PCB 118 in fish cells *in vitro* (Marabini et al. 2011). The Comet assay can be modified to examine the percent of strand breaks due to ROS, which indicated that ROS were the clastogenic species in the fish cells. In human HL60 promyelocytic leukemia cells, the redox cycling pair PCB 3-hydroquinone and PCB 3-p-quinone both induced Comets (Xie et al. 2010). A closer analysis of the Comet induction by these PCB3 metabolites revealed that the hydroquinone activity required cellular myeloperoxidase activity, while the quinone activity was independent of the peroxidase activity.

With both compounds, about half of the breaks were due to ROS, while the other half was most likely produced by a direct interaction of the quinone with the DNA or DNA-related proteins.

Increased levels of micronuclei were observed in several fish species after exposure to commercial PCB mixtures or to PCBs and DDT in contaminated areas (reviewed in Ludewig 2001). PCBs 153, 138, 101, and 118 also induced micronuclei in fish cells *in vitro* (Marabini et al. 2011). In addition, contaminated soil from industrial or irrigation sites induced micronuclei in plant species (Cotelle et al. 1999; Song et al. 2006). No increase was seen after exposure of human lymphocytes and human keratinocytes to PCB mixtures *in vitro* (Belpaeme et al. 1996; van Pelt et al. 1991). Chromosome breaks and chromosome loss most likely require the action of reactive intermediates. To analyze whether and which metabolites are inducing micronuclei, V79 Chinese hamster lung fibroblasts were exposed to the three mono-, two dihydroxy-, and two quinone metabolites of PCB 3 (Zettner et al. 2007). PCB 3 was negative in the assay, but the mono- and, more efficiently, the dihydroxylated and quinonoid metabolites induced micronuclei formation. Interestingly, with the para-quinone mainly chromosome breaks were seen, while the monohydroxylated and to a lesser extent the other metabolites produced micronuclei predominantly by chromosome loss. It can be speculated that the highly reactive quinone interacts with the DNA or DNA maintenance proteins like topoisomerase II leading to chromosome breaks, while the other compounds may have another mechanism to produce the observed chromosome loss.

Some epidemiological studies observed abnormal karyotypes in workers exposed to PCBs (Kalina et al. 1991; Tretjak et al. 1990), while others reported negative results (Elo et al. 1985). Most studies with rats and commercial PCB mixtures did not find chromosome aberrations in bone marrow or spermatogonia, while studies with fish were positive (reviewed in Ludewig 2001). Sargent observed chromosome breaks and rearrangements in human lymphocytes exposed to Arochlor 1254, PCB 77, or PCB 153 *in vitro* (Sargent et al. 1989, 1991). PCB 52 was negative, but a combination of PCB 52 with PCB 77 acted synergistically, producing more aberrations than the sum. Similar results were obtained in hepatocytes of rats treated *in vivo* (Meisner et al. 1992; Sargent et al. 1992, 2001). Besides chromosome aberrations, an increase in polyploid cells was observed. This was also seen in V79 hamster cells *in vitro*, where PCB 52 and PCB 105 were strong inducers of aberrant mitosis, while PCB 77 and PCB 153 were inactive (Jensen et al. 2000). This points toward metabolic activation as a required factor. In addition, a strong synergistic action was seen when PCB 105 (2,3,3′,4,4′-pentachlorobiphenyl) was combined with the spindle poison triphenyltin, indicating an interaction with the cytoskeleton as a mechanism of this effect (Jensen et al. 2000). Interestingly, hydroquinones of PCB 2 and PCB 3, but not the PCB 3 3,4-catechol or the PCB1 hydroquinone, induced up to 90% polyploidization of V79 cells (Flor and Ludewig 2010). Further studies indicate an interaction of these metabolites with the cytoskeleton. These results also show how very structure dependent these genotoxic effects are.

Sister Chromatid Exchanges (SCEs) SCEs are formed only during the S-phase of the cell cycle at/near the replication fork by a double-strand break and a reciprocal rejoining of the strands. Although this event seems to be very exact, it was hypothesized that sometimes a few base pairs may be lost or gained. Alternatively, this event may disturb the methylation of the newly synthesized strand. Because the mechanism and consequences of SCE are not known, this assay is not used as frequently anymore and most data are derived from the 1990s. However, the fact that a very high correlation (76%, $n = 379$) was observed between compounds that are carcinogens and those that induce SCE supports the relevance of this assay in carcinogen screening (Abe and Sasaki 1982).

One advantage of the SCE assay is that it can be used under controlled conditions with cultured cells or animals, but also for biomonitoring of potentially exposed populations. For example, a human population study found that the frequency of SCEs in peripheral lymphocytes was significantly higher among pesticide applicators (DDT, but also other compounds) when compared with controls (Rupa et al. 1991). Lymphocytes of 10-month-old Japanese infants showed an increasing trend with the increasing exposure to DDT and also hexachlorohexanes (HCHs) through the breast milk (Nagayama et al. 2003). However, SCE frequencies were not elevated in workers handling aldrin (Edwards and Priestly 1994). Experimental exposure to endrin and chlordane caused significant increases in the frequencies of SCE in fish (Central mudminnows) (Vigfusson et al. 1983). Similarly, human lymphoid cells exposed *in vitro* to chlordane or toxaphene showed statistically significant higher SCE levels, which were further increased on metabolic activation with rat liver microsomal S-9 enzymes for chlordane (Sobti et al. 1983).

Vietnam war veterans that had sprayed TCDD-contaminated phenoxylic herbicides during 1965–1971 had highly significantly ($p < 0.001$) more SCE than a control group and an exceptionally high proportion of cells with high SCE frequencies (Rowland et al. 2007), but no statistically significant increases were seen in TCDD-exposed workers (Zober et al. 1997) and patients (Valic et al. 2004), while TCDD levels in mothers' milk correlated with SCE in lymphocytes from babies in Japan (Nagayama et al. 2001b). Also, in animal studies, TCDD failed to induce SCE in rhesus monkeys (Lim et al. 1987), in bone marrow cells of mice (Meyne et al. 1985), in lymphocytes of rodents (Huff et al. 1991), or in mice exposed to various mixtures of 2,4-dichlorophenoxyacetic acid, 2,4,5-trichlorophenoxyacetic acid, and TCDD (Lamb et al. 1981). However, sheep exposed to TCDD (Iannuzzi et al. 2004) or dioxins (17 different congeners) had elevated SCE levels (Perucatti et al. 2006) and a small but significant increase of SCE in lymphocytes from TCDD-exposed Han/Wistar rats was reported (Mustonen et al. 1989). Moreover, lymphocytes from TCDD-treated rats had higher SCE levels if exposed to α-naphthoflavone (ANF) *in vitro* than lymphocytes from control rats, indicating a sensitizing effect of TCDD (Lundgren et al. 1986).

Human biomonitoring studies from the 1980s and 1990s reported higher SCE levels in peripheral lymphocytes of PCB-exposed workers or victims of the 1978 (Yu-Cheng) rice oil PCB poisoning in Taiwan (Kalina et al. 1991; Lundgren et al. 1988). Often, a long occupational exposure time or coexposure to cigarette smoke was needed, or the cells were shown to be sensitized to exposure to other compounds like ANF. Organochlorine mixtures (PCBs, polychlorinated dibenzo-*p*-dioxin [PCDD], polychlorinated dibenzofuran [PCDF], DDT, and others) as those found in human blood around 1990 induced SCE and enhanced SCE induction by ANF in cultured lymphocytes (Nagayama et al. 1991). The EC_{50} for SCE induction was only three times higher than the average concentration in healthy humans at that time. Recent biomonitoring reports show no increase of SCE in lymphocytes of teachers from a PCB-contaminated school (Wiesner et al. 2000), or of current Japanese Yusho patients (rice oil PCB poisoning in 1968), even though they still had a seven times higher mean total toxic equivalent (TEQ) levels of PCDDs, PCDFs, and dioxin-like PCBs in the blood, than did healthy Japanese controls (Nagayama et al. 2001a).

The challenge with human studies is that workers and other contaminated population groups usually were exposed to multiple compounds. Only three individual PCB congeners were tested *in vitro* for SCE induction and none of them (4-monochloro, 2,2′,5,5′-, and 3,3′,4,4′-tetrachlorobiphenyl, i.e., PCB 3, PCB 52, and PCB 77, respectively) were positive (Flor and Ludewig 2010; Sargent et al. 1989). If PCB congeners themselves are not SCE inducing, the metabolites may. Indeed, two of five methylsulfonate metabolites of PCBs (MSF-PCBs) tested, namely, 3-MSF-2,5,2′,4′,5′- and 4-MSF-2,5,2′,3′,4′-pentachlorobiphenyl significantly enhanced the frequency of SCEs. However, doses (>5 ppm) that were about 35,000 times higher than the concentrations in the lungs and adipose tissue of healthy Japanese people were needed to produce significant effects (Nagayama et al. 1999). Interestingly, of several monochlorinated biphenyl metabolites tested, only the PCB 3-3,4-catechol induced SCE in V79 cells, not the structurally closely related PCB 1, PCB 2, or PCB 3-hydroquinones, pointing toward a very strict structure–activity relationship for SCE induction (Flor and Ludewig 2010). Taken together, these results indicate that PCBs themselves may not be efficient inducers of SCE, but that even low environmental exposure to PCBs is sufficient to make human lymphocytes more sensitive to other procarcinogens, possibly by inducing drug-metabolizing enzymes.

The mechanism and implications of SCE induction are not clearly understood. It is thus of interest that in human lymphocytes, TCDD, 2,3,4,7,8-pentachlorodibenzofuran (PenCDF), and PCB 126, three potent aryl hydrocarbon receptor (AhR) agonists, significantly increased the frequency of SCEs with almost the same dose relationship with respect to their TEQ (Nagayama et al. 1995). The authors also reported an enhancement of SCE induction by ANF and an additive effect of these compounds. They were regarded as very efficient SCE inducers since their EC_{50} for SCE enhancement was only 5–10 times higher than the level in adipose tissue in healthy Japanese, namely, 70 ppt as TCDD equivalent (Nagayama et al. 1995). All this could simply indicate that they act by increasing the metabolic capacity of the cells resulting in the more efficient bioactivation of exogenous and/or endogenous compounds. But AhR agonists are considered to be promoters of carcinogenicity as well. As mentioned earlier, one hypothesis for the correlation of SCE and carcinogenesis is that SCE may disturb the methylation of newly synthesized DNA. It was recently reported that the total DNA methylation levels in Greenland Inuits were inversely correlated with total organochlorine contamination and also with chlordanes, DDT, and some PCB congeners (Rusiecki et al. 2008). Changes in DNA methylation is seen as an epigenetic event that alters gene expression levels and thus could be involved in cancer promotion. Thus, alternatively, it could be hypothesized that SCE induction is not primarily a genotoxic event and an indicator for cancer initiation, but rather an indicator for cancer-promoting activity.

Genotoxicity of Individual Pesticides

The existing literature about genotoxicity of pesticides is by far smaller than that on PCBs. This can in part be explained by the fact that they are single compounds or a relatively undefined mixture and not 209 well-classified individual compounds as in the case of PCBs. Next is a concise summary of the reported findings using a selection of the existing publications.

Aldrin Since aldrin is very fast and efficiently biotransformed to dieldrin and is not very water soluble, making *in vitro* studies more difficult and limited in concentration, most genotoxicity studies used dieldrin. Floriculturists, which were exposed to aldrin and several other pesticides, had increased SCE rates but not more chromosome aberrations than the control group (Dulout et al. 1985). Aldrin was not mutagenic in bacterial assays (Moriya et al. 1983; Zeiger et al. 1992).

Chlordane This pesticide was used in the United States for four decades (1948–1988) before the EPA banned its use due to environmental and human health concerns. *cis*-Chlordane is metabolized by CYPs via hydroxylation and dehydrochlorination to reactive intermediates that bind irreversibly to cellular macromolecules (Brimfield and Street 1981). Metabolites include mono- and dihydroxy- as well as epoxy-intermediates, one of them oxychlordane. *trans*-Chlordane is metabolized to heptachlor. Chlordane may also increase cellular levels of superoxide (Suzaki et al. 1988). Nevertheless, bacterial mutagenicity tests were mostly negative and mammalian gene mutation test equivocal

(McGregor et al. 1988; Probst et al. 1981), and no DNA adduct formation was observed in mouse liver using the ^{32}P-postlabeling assay (Whysner et al. 1998). The finding that only chlordane-exposed mice and not rats developed cancer indicate that specific metabolic tissue requirement seemed to be necessary for chlordane bioactivation. Despite these mostly negative or conflicting results, micronucleus induction was seen in whale skin fibroblasts by chlordane (Gauthier et al. 1999), and an increase in SCE in fish and human lymphocytes was reported (Sobti et al. 1983; Vigfusson et al. 1983).

Endrin Very few studies analyzed the genotoxicity of endrin with and without microsomal activation and all of them showed negative results (McGregor et al. 1991; Zeiger et al. 1992). Although single-strand breaks were observed in hepatocytes, it was hypothesized that these were due to oxidative damage and that endrin was not a direct-acting genotoxin (Hassoun and Stohs 1996; Hassoun et al. 1993).

Dieldrin Dieldrin, chemically an epoxide of aldrin, is a pesticide and the metabolic and environmental breakdown product of aldrin. The results from the limited number of genotoxicity assays are mixed, showing negative as well as positive results with bacterial assays, mostly positive results with mammalian mutation, chromosome aberration, and SCE assays, and abnormal metaphases, chromosome aberrations and ploidy changes in mice, the more sensitive species, but not in rats (Galloway et al. 1987; Kamendulis et al. 2001; Majumdar et al. 1976; McGregor et al. 1991). Dieldrin was also positive in an *in vitro* transformation assay using focus formation in a stable, bovine papillomavirus type 1 DNA carrying C3H/10T(1/2) mouse embryo fibroblast cell line (T1) (Kowalski et al. 2000). Thus, dieldrin may have some initiating activity, even though its promoting activity through production of oxidative stress and inhibition of gap junctional communication (GJC) (see further) may be the major mechanism in its carcinogenicity.

Heptachlor Heptachlor is metabolized to an epoxide, which is the more common form found in the environment. Several studies with bacterial and mammalian cells, with and without metabolic activation, showed mostly negative results, but heptachlor induced gene mutations in mouse lymphoma cells and SCE with metabolic activation and also chromosome aberrations in CHO cells (McGregor et al. 1988). Interestingly, several studies showed clastogenic activity of heptachlor in plants (Sandhu et al. 1989).

HCB HCB was not genotoxic in short-term tests (Gorski et al. 1986) and did not increase SCE levels in a human cell line *in vitro* (Brusick 1986). However, a weak positive result was observed in the micronucleus assay, and some adduct formation with DNA and protein was also observed (Gopalaswamy and Nair 1992; McGregor et al. 1988). HCB is metabolized to a radical or arene oxide intermediate, both of which can be expected to interact with the DNA. Inadequate metabolic activation systems in *in vitro* assays could be the reason for the weak response.

Mirex Very few genotoxicity studies with exposed humans, or with animal models or cells in culture, were performed. No gene mutations in bacteria or rodent cells, or chromosomal aberrations or SCE in CHO cells in the presence or absence of S9, were observed (NTP 1990). One unusual observation was that mirex strongly reduced the number of polyploidy hepatocytes in the livers of adult rats (Abraham et al. 1983). The number of polyploid hepatocytes is reduced during hepatocarcinogenesis in rats (Saeter et al. 1988). Based on these few genotoxicity studies, the fact that mirex is not metabolized, and observations that it inhibits GJC (Tsushimoto et al. 1982b), it is assumed that this pesticide is nongenotoxic and is instead a cancer promoter.

Toxaphene Toxaphene is a mixture of hundreds of different chlorinated camphenes and related compounds. Lymphocytes from women exposed to the technical mixture had an increased rate of chromosome aberrations (Samosh 1974). No DNA adduction (Hedli et al. 1998) or DNA damage (Kitchin and Brown 1994) was observed in exposed rats. Toxaphene significantly and dose dependently induced SCE in Chinese hamster lung (don) cells (Steinel et al. 1990), but no significant elevation of SCE was seen in Chinese hamster V79 cells (Schrader et al. 1998). Mutagenicity assays with mammalian cells showed mostly negative results (Schrader et al. 1998). However, bacterial mutagenicity tests without metabolic activation system were almost always clearly positive, while those with activation systems showed a reduced mutagenicity (Mortelmans et al. 1986; Schrader et al. 1998). These results indicate that the technical toxaphene mixture is genotoxic. However, since toxaphene is such a complex chemical mixture, the chemical structure of the direct-acting genotoxin remains unknown.

DDT DDT is still in use to fight malaria in areas where the disease is endemic by reducing the prevalence of the vector, mosquitoes. Its use was stopped wherever possible due to its bioaccumulation and environmental effects. There are also enough data available to classify it as a probable human carcinogen. Several epidemiological studies with exposed factory workers and pesticide applicators found increased chromosome aberrations and SCE in peripheral blood lymphocytes, but it has to be kept in mind that these cohorts are usually exposed to a large mixture of pesticides (Rupa et al. 1988). *In vivo* experiments with rodents were positive or negative for chromosome aberrations, positive for SCE, and showed an effect on chromosome number and ploidy (Amer et al. 1996; Clark 1974; Larsen and Jalal 1974; Uppala et al. 2005), indicating that this type of DNA damage may deserve further attention. Dichlorodiphenyldichloroethylene, a metabolite of DDT, was also positive in the micronucleus assay, which measures clastogenic and aneugenic activity (Ennaceur et al. 2008). DDT did not, however, induce mutations in bacteria test systems and microsomal activation systems (Zeiger et al. 1992) and was negative in most mammalian *in vitro* gene mutation tests (Amacher and Zelljadt 1984; Tsushimoto et al. 1983), indicating that prokaryotic mutation assays and even mammalian mutagenicity tests *in vitro* may not be appropriate or sufficient systems to capture the genotoxicity of refractile compounds, like POPs (Ashby 1986), that may require a complicated multistep activation system.

TUMOR-PROMOTING ACTIVITIES OF ENVIRONMENTAL POLLUTANTS

Promoting Activity of Individual POPs

Cancer promotion is a nongenotoxic event and only a few assays exist to determine the promoting activity of a compound. The standard assays use rodents and an exposure regimen that starts with a single application of an initiator followed by repeated, long-term exposure to the suspected promoting compound. The two major models use either mouse skin or rat liver as the target organ of carcinogenesis. The promoting activities of many environmental pollutants have been examined in laboratory animal models (Table 4.1).

Several, but not all, pesticides have been tested in rodent tumor promotion assays. The promoting activity of aldrin has not been studied, but dieldrin has been found to have promoting activity in mouse liver but not in rat liver (Stevenson et al. 1999). Chlordane has been found to have tumor-promoting activity in mouse liver (Williams and Numoto 1984). No studies have examined the tumor-promoting activity of endrin. The hepatic tumor-promoting activity of heptachlor has been examined in mice, and it has been found to be positive (Numoto et al. 1985; Williams and Numoto 1984). Mirex was found to have tumor-promoting activity in the skin but has not been examined in the liver (Moser et al. 1992, 1993; Porter et al. 2002). Toxaphene, in the form of technical toxaphene, UV-irradiated toxaphene, or cod liver oil from cod exposed to toxaphene, was not found to have promoting activity in rats (Besselink et al. 2008). However, all three forms were found to inhibit gap junctional intercellular communication, which frequently is correlated with tumor-promoting activity (Besselink et al. 2008). DDT has promoting activity in liver in both rats and mice (Angsubhakorn et al. 2002; Flodstrom et al. 1990; Harada et al. 2003; Numoto et al. 1985; Oesterle and Deml 1981; Williams and Numoto 1984), although one study observed an inhibitory (hormetic) effect at a low dose (Fukushima et al. 2005). HCB has promoting activity in the liver of female rats; in male rats, promoting activity was only seen in caloric-restricted animals (Agency for Toxic Substances and Disease Registry 2002; Reed et al. 2007).

TCDD, a by-product of the production of chlorophenoxy acid herbicides and of waste combustion at inadequate temperatures, has promoting activity in the liver of rats and in the liver of the B6D2F1 strain of mice, with no effect in two other strains. Additionally, the skin of mice and the lung of mice and rats have been shown to be sensitive to the promoting effects of TCDD in some studies (Knerr and Schrenk 2006).

PCBs are ubiquitous environmental contaminants. PCB mixtures have tumor-promoting activities in the liver (Glauert et al. 2001; Silberhorn et al. 1990). Many of the 209 individual PCB congeners also have promoting activity, including congeners that activate the AhR and constitutive androstane/active receptor (Glauert et al. 2001). Additionally, promoting activity in the lung and skin has been observed. Commercial mixtures of the PBBs and individual PBB congeners have tumor-promoting activity in the liver (Silberhorn et al. 1990).

Several of the peroxisome proliferators, which include diverse compounds like phthalate plasticizers, certain herbicides, and trichloroacetic acid, have hepatic tumor-promoting activity (Cattley et al. 1998; Rao and Reddy 1987; Reddy and Lalwani 1983; Roberts 1999).

Mechanisms of Tumor Promotion by Environmental Pollutants

Effects on Cell Kinetics Environmental pollutants can influence both cell proliferation and apoptosis, both of which can affect the growth of preneoplastic lesions or tumors. The induction of cell proliferation can increase the likelihood of an initiating event and can also expand the population of initiated cells. Several environmental pollutants have been found to increase hepatic cell proliferation, including chlordane (Barrass et al. 1993; Ross et al. 2010), DDT (Harada et al. 2003; Kostka et al. 1996; Schulte-Hermann et al. 1983), dieldrin (mouse, but not rat) (Kamendulis et al. 2001), heptachlor (Okoumassoun et al. 2003), HCB (Giribaldi et al. 2011; Ou et al. 2001, 2003), mirex (Yarbrough et al. 1991), PBBs (Mirsalis et al. 1989), PCBs (Lu et al. 2003; National Toxicology Program 2006; Tharappel et al. 2002; Whysner and Wang 2001), peroxisome proliferators (Cattley 2003), and TCDD (Dragan and Schrenk 2000; Tritscher et al. 1995). Several of these agents have also been shown to increase cell proliferation in other cell types, including the thyroid and mammary gland (Brown and Lamartiniere 1995; Cossette et al. 2002; Garcia et al. 2010; Uppala et al. 2005). Mixtures of these agents are usually more efficacious at inducing cell proliferation than a single agent alone (Aube et al. 2011; Payne et al. 2001; Wade et al. 2002).

Alterations in the rate of apoptosis can also play a major role in the expansion of the population of initiated cells. Environmental pollutants have been found in most studies to inhibit hepatic cell apoptosis, including DDT (Buchmann et al. 1999), dieldrin (Buchmann et al. 1999; Buenemann et al. 2001), heptachlor (Okoumassoun et al. 2003), TCDD (Chopra et al. 2009; Stinchcombe et al. 1995; Worner and Schrenk 1996, 1998), PCBs, with the most consistent inhibition observed with *ortho*-substituted PCBs (Al-Anati et al. 2010; Bohnenberger et al. 2001; Glauert et al. 2008; Kolaja et al. 2000; Lu et al. 2004; Tharappel et al. 2002), and peroxisome proliferators (Bayly et al. 1994; Hasmall et al. 2000; Roberts et al. 1998). Other studies have found environmental pollutants to induce (HCB, TCDD) or not influence (dieldrin) hepatic cell apoptosis (Giribaldi et al. 2011; Kamendulis et al. 2001; Kolaja et al. 1996; Patterson et al. 2003). In other cell types, some environmental agents induce apoptosis in most studies, including dieldrin (Hallegue et al. 2002; Kanthasamy et al. 2008; Kitazawa et al. 2003), HCB (Chiappini et al. 2009), and toxaphene (Gauthier et al. 2001; Lavastre et al. 2002). Other pollutants have variable effects (enhancement, no effect, or inhibition of apoptosis) that depend on the cell type and experimental design, including DDT (Burow et al. 1999; Perez-Maldonado et al. 2004; Tebourbi et al. 1998), heptachlor (Rought et al. 2000), TCDD (Davis et al. 2000, 2001; Kamath et al. 1997; Pryputniewicz et al. 1998; Puebla-Osorio et al. 2004; Sanchez-Martin et al. 2010; Yang and Lee 2010), and PCBs (Jeon et al. 2002; Lee et al. 2003; Sanchez-Alonso et al. 2004). Overall, most environmental pollutants inhibit apoptosis in the liver, which would lead to an expansion in initiated cell populations. In other tissues, however, most studies find that these agents induce apoptosis, which may be inhibitory to cancer development, but which could compromise tissue function.

ROS Another mechanism by which environmental pollutants could exert their tumor-promoting activity is by increasing oxidative stress, which could result in lipid peroxidation, oxidative DNA damage, or alterations in gene expression. Oxidative stress could be increased by several mechanisms (Figure 4.4; ROS production). Mitochondria release electrons as a by-product of the electron transport pathway; these electrons can react with oxygen to form superoxide (Fariss et al. 2005). Environmental pollutants can induce specific CYP isozymes. CYP can release active oxygen in the form of superoxide or hydrogen peroxide as a by-product (Bondy and Naderi 1994; Zangar et al. 2004). Environmental agents that activate the PPARα increase the β-oxidation pathway in peroxisomes, which releases hydrogen peroxide as a by-product, as do other enzymes in peroxisomes (O'Brien et al. 2005; Rao and Reddy 1987). Superoxide or hydrogen peroxide produced via these mechanisms can be metabolized to the hydroxyl radical, which then reacts instantaneously with other cellular components, including lipids, protein, and DNA. In addition, some of these agents can decrease the expression of antioxidant enzymes, such

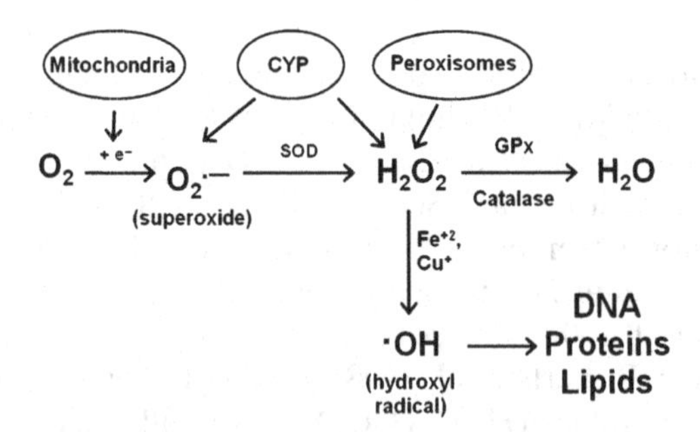

Figure 4.4. Sources and fate of reactive oxygen species in the cell.

as glutathione peroxidase, or the concentrations of cellular antioxidants, such as vitamin E (Glauert et al. 2001; O'Brien et al. 2005; Valavanidis et al. 2006; Yoshida and Ogawa 2000), thus further increasing oxidative stress in the cell.

This increased oxidative stress can lead to several changes in the cell, including lipid peroxidation, oxidative DNA damage, and changes in gene expression. Lipid peroxidation can cause several deleterious changes, including toxicity, DNA damage, and changes in gene expression (Niki 2009). Several environmental pollutants, including chlordane, DDT, dieldrin, endrin, heptachlor, HCBs, PCBs, and TCDD, induce lipid peroxidation in the liver (Glauert et al. 2001; Goel et al. 1988; Hassoun et al. 1993; Horvath et al. 2001; Izushi and Ogata 1990; Stevenson et al. 1999; Stohs et al. 1986). Oxidative DNA damage can be produced either from the production of malonaldehyde, which can form DNA adducts (Marnett 2002), or from the hydroxyl radical, produced from hydrogen peroxide, which can form oxidative products such as 8-hydroxydeoxyguanosine (Cadet et al. 2010). DDT, HCB, PCBs, peroxisome proliferators, and TCDD have been found to increase oxidative DNA damage (Harada et al. 2003; Horvath et al. 2001; Oakley et al. 1996; O'Brien et al. 2005; Schilderman et al. 2000; Tritscher et al. 1996), but other studies have observed that oxidative DNA damage is not affected by these agents (Bachowski et al. 1998; Kushida et al. 2005; O'Brien et al. 2005). Finally, oxidative stress can lead to alterations in gene expression. As mentioned earlier, some products of lipid peroxidation can act on signal transduction pathways (Uchida et al. 1999; Welsch 1995). Oxidative stress can also activate specific transcription factors, such as nuclear factor-kappa β (NF-κB) (Gloire et al. 2006; Morgan and Liu 2011). NF-κB has been found to be activated by several environmental agents, including PCBs and peroxisome proliferators (Li et al. 1996; Lu et al. 2003).

Estrogen Metabolism Several environmental pollutants may influence hepatic tumor-promoting activity by changing estrogen metabolism; in addition, some agents have estrogenic activity, whereas others are antiestrogenic. Estrogen influences both human liver carcinogenesis and experimental animal carcinogenesis. In animals, rats but not mice are sensitive to the hepatocarcinogenic effects of estrogens (Holsapple et al. 2006). In humans, several epidemiological studies have observed that women taking oral contraceptives have a higher risk of developing liver cancer (Holsapple et al. 2006). Men, however, have a much higher risk of developing liver cancer than women (Bosch et al. 2004). Estrogen inhibits interleukin (IL)-6 production by hepatic Kupffer cells, which could decrease the risk of liver cancer in women (Naugler et al. 2007).

Several environmental pollutants have estrogenic activity, including DDT and its analogs, aldrin, dieldrin, HCB, toxaphene, and some PCB congeners (Aube et al. 2011; Cooke et al. 2001; Nelson et al. 1978; Shekhar et al. 1997). The primary mechanism is by activation of estrogen receptors (ERα and/or ERβ). Other environmental agents are antagonistic to estrogen receptors, including chlordane, endrin, TCDD, toxaphene, and PCB congeners that activate the AhR (Arcaro et al. 2000; Ariazi and Jordan 2006; Cooke et al. 2001; Lemaire et al. 2006; Safe et al. 1991; Yang and Chen 1999). The primary mechanism is by the activation of the AhR, which then can cross-talk with estrogen receptors, resulting in the ubiquitination and degradation of the receptors (Ohtake et al. 2011).

Inhibition of GJC One mechanism for promotion is believed to be through suppression of exchange of negative proliferation control signals by inhibition of intracellular communication between cells through gap junctions (Trosko et al. 1983). Promoting agents that were found to inhibit GJC include the nongenotoxic carcinogen mirex and PBBs (Tsushimoto et al. 1982a,b). Inhibition of GJC with dieldrin was seen in a variety of cell types and is hypothesized to be a major factor in its carcinogenicity (Klaunig and Ruch 1987; Trosko et al. 1987). Chlordane inhibited gap junctional intercellular communication in Syrian hamster embryo cells (Rivedal et al. 2000). Many PCB congeners (Bager et al. 1994, 1997a,b; Brevini et al. 2004; Kang et al. 1996; Swierenga et al. 1990) and their hydroxylated, quinonoid, and methylsulfonyl metabolites (Kato et al. 1998; Machala et al. 2004) were found to be efficient inhibitors of GJC, and the mechanism is believed to involve enhanced degradation or misplacement of connexins (Heikaus et al. 2002; Kang et al. 1996; Krutovskikh et al. 1995). Overall, strong evidence is accumulating that inhibition of metabolic cooperation between cells may be a major mechanism of tumor promotion by persistent halogenated organic compounds.

PROGRESSION IN CARCINOGENESIS

Very little is known about the mechanisms of progression in carcinogenesis. One aspect is the increase in karyotype changes during this phase. As discussed in the section "Chromosome Aberrations, Micronuclei, Comet, Aneuploidy, Polyploidy," several POPs, but particularly PCBs, were observed to produce changes in the karyotype. Recent results also indicate that severely shortened telomeres, the protective "caps" at the end of chromosomes, cause a cellular crisis and, after this, the emergence of immortal cells with aberrant karyotypes. This is believed to be created by stickiness of chromosome ends with short telomeres, resulting in fusion of chromosomes and missegregation during cell divisions (Greenberg 2005). If this is a mechanism of cancer progression, then PCBs may be involved in this process via this mechanism. Several PCB congeners and PCB 3 p-quinone were observed to induce telomere shortening, the congeners most likely by reducing telomerase activity, the PCB 3-quinone most likely via ROS generation (Jacobus et al. 2008; Senthilkumar et al. 2011, 2012). This area of research is only at its beginning, and if confirmed, it would add another mechanism of carcinogenesis to the already very large list of mechanisms of POPs.

SUMMARY AND CONCLUSION

Most POPs of the Stockholm convention are probably animal and human carcinogens. Even though they share some structural features—they are all halogenated organic compounds—they show differing mechanisms of carcinogenicity. This is most obvious in the 209-member family of PCBs. The higher-chlorinated congeners interact with different receptors, often resulting in the generation of ROS, two potential mechanisms of promotion. The lower-chlorinated congeners can be metabolically activated to reactive species that bind to DNA and cellular proteins, thereby inducing DNA damage and anomalous chromosome segregation. These are just a few of the observed mechanisms. Taken together, we have to remember five aspects, if we want to evaluate the risk from exposure to these compounds: (1) We are always exposed to mixtures of these compounds; in fact, most of us probably have all or most of these compounds in our tissues; (2) different compounds may act through the same mechanisms, possibly in an unpredictable additive or synergistic way; (3) some members of the mixtures we are exposed to are initiators, while others are promoters or progressors; thus, they have the potential to complement each other and induce every step of chemical carcinogenesis; (4) focusing on tissue levels alone may be misleading since the metabolites may be more dangerous than the original compound; (5) we are constantly adding new compounds to the old POPs, but we are still unable to predict mixture effects and possible consequences. The last decade has brought substantial, new knowledge about the mechanisms and roles of these compounds in carcinogenesis, but significantly more research is needed before we can predict the effects of compounds and reduce the risk of cancer induction.

ACKNOWLEDGMENTS

The authors thankfully acknowledge the grants and support that made some of the research that was described in this chapter possible, particularly the KY Superfund grant P42 ES07380 and the IA Superfund grant P42 ES013661 from NIEHS, DAMD 17-96-1-6162 and DAMD17-02-1-0241 from the DOD, EPA STAR R-82902102-0, the Kentucky Agricultural Experiment Station, and the Iowa Center for Health Effects of Environmental Contaminants.

REFERENCES

Abe S, Sasaki M. 1982. SCE as an index of mutagenicity and/or carcinogenicity. In: Sister Chromatid Exchange (Sandberg AA, ed.). New York: Alan R. Liss, pp. 461–514.

Abraham R, Benitz KF, Mankes R. 1983. Ploidy patterns in hepatic tumors induced by Mirex. Exp Mol Pathol 38:271–282.

Agency for Toxic Substances and Disease Registry. 2002. Toxicological Profile for Hexachlorobenzene. Atlanta, GA: Agency for Toxic Substances and Disease Registry, Division of Toxicology/Toxicology Information Branch.

Al-Anati L, Hogberg J, Stenius U. 2010. Non-dioxin-like PCBs interact with benzo[a]pyrene-induced p53-responses and inhibit apoptosis. Toxicol Appl Pharmacol 249:166–177.

Amacher DE, Zelljadt I. 1984. Mutagenic activity of some clastogenic chemicals at the hypoxanthine guanine phosphoribosyl transferase locus of Chinese hamster ovary cells. Mutat Res 136:137–145.

Amaro AR, Oakley GG, Bauer U, Spielmann HP, Robertson LW. 1996. Metabolic activation of PCBs to quinones: Reactivity toward nitrogen and sulfur nucleophiles and influence of superoxide dismutase. Chem Res Toxicol 9:623–629.

Amer SM, Fahmy MA, Donya SM. 1996. Cytogenetic effect of some insecticides in mouse spleen. J Appl Toxicol 16:1–3.

Angsubhakorn S, Pradermwong A, Phanwichien K, Nguansangiam S. 2002. Promotion of aflatoxin B1-induced hepatocarcinogenesis by dichlorodiphenyl trichloroethane (DDT). Southeast Asian J Trop Med Public Health 33:613–623.

Arcaro KF, Yang Y, Vakharia DD, Gierthy JF. 2000. Toxaphene is antiestrogenic in a human breast-cancer cell assay. J Toxicol Environ Health A 59:197–210.

Ariazi EA, Jordan VC. 2006. Estrogen-related receptors as emerging targets in cancer and metabolic disorders. Curr Top Med Chem 6:203–215.

Ashby J. 1986. The value and limitations of short-term genotoxicity assays and the inadequacy of current criteria for selecting chemicals for cancer bioassays. Food Chem Toxicol 24:663–666.

Aube M, Larochelle C, Ayotte P. 2011. Differential effects of a complex organochlorine mixture on the proliferation of breast cancer cell lines. Environ Res 111:337–347.

Bachowski S, Xu Y, Stevenson DE, Walborg EF, Jr., Klaunig JE. 1998. Role of oxidative stress in the selective toxicity of dieldrin in the mouse liver. Toxicol Appl Pharmacol 150:301–309.

Bager Y, Kenne K, Krutovskikh V, Mesnil M, Traub O, Warngard L. 1994. Alteration in expression of gap junction proteins in rat liver after treatment with the tumour promoter 3,4,5,3′,4′-pentachlorobiphenyl. Carcinogenesis 15:2439–2443.

Bager Y, Kato Y, Kenne K, Warngard L. 1997a. The ability to alter the gap junction protein expression outside GST-P positive foci in liver of rats was associated to the tumour promotion potency of different polychlorinated biphenyls. Chem Biol Interact 103:199–212.

Bager Y, Lindebro MC, Martel P, Chaumontet C, Warngard L. 1997b. Altered function, localization and phosphorylation of gap junctions in rat liver epithelial, IAR 20, cells after treatment with PCBs or TCDD. Environ Toxicol Pharmacol 3: 257–266.

Barrass N, Stewart M, Warburton S, Aitchison J, Jackson D, Wadsworth P, et al. 1993. Cell proliferation in the liver and thyroid of C57Bl/10J mice after dietary administration of chlordane. Environ Health Perspect 101(Suppl 5):219–223.

Bayly AC, Roberts RA, Dive C. 1994. Suppression of liver cell apoptosis in vitro by the non-genotoxic hepatocarcinogen and peroxisome proliferator nafenopin. J Cell Biol 125:197–203.

Belpaeme K, Delbeke K, Zhu L, Kirsch-Volders M. 1996. PCBs do not induce DNA breakage in vitro in human lymphocytes. Mutagenesis 11:383–389.

Bender RP, Lehmler HJ, Robertson LW, Ludewig G, Osheroff N. 2006. Polychlorinated biphenyl quinone metabolites poison human topoisomerase IIalpha: Altering enzyme function by blocking the N-terminal protein gate. Biochemistry 45: 10140–10152.

Besselink H, Nixon E, McHugh B, Rimkus G, Klungsoyr J, Leonards P, et al. 2008. Evaluation of tumour promoting potency of fish borne toxaphene residues, as compared to technical toxaphene and UV-irradiated toxaphene. Food Chem Toxicol 46:2629–2638.

Bohnenberger S, Wagner B, Schmitz HJ, Schrenk D. 2001. Inhibition of apoptosis in rat hepatocytes treated with "non-dioxin-like" polychlorinated biphenyls. Carcinogenesis 22:1601–1606.

Bondy SC, Naderi S. 1994. Contribution of hepatic cytochrome P450 systems to the generation of reactive oxygen species. Biochem Pharmacol 48:155–159.

Bosch FX, Ribes J, Diaz M, Cleries R. 2004. Primary liver cancer: Worldwide incidence and trends. Gastroenterology 127(Suppl 1):S5–S16.

Brevini TA, Vassena R, Paffoni A, Francisci C, Fascio U, Gandolfi F. 2004. Exposure of pig oocytes to PCBs during in vitro maturation: Effects on developmental competence, cytoplasmic remodelling and communications with cumulus cells. Eur J Histochem 48:347–356.

Brimfield AA, Street JC. 1981. Microsomal activation of chlordane isomers to derivatives that irreversibly interact with cellular macromolecules. J Toxicol Environ Health 7:193–206.

Brown NM, Lamartiniere CA. 1995. Xenoestrogens alter mammary gland differentiation and cell proliferation in the rat. Environ Health Perspect 103:708–713.

Brown JF, Jr., Mayes BA, Silkworth JB, Hamilton SB. 2007. Polychlorinated biphenyls-modulated tumorigenesis in Sprague Dawley rats: Correlation with mixed function oxidase activities and superoxide ($O_2^{\bullet-}$) formation potentials and implied mode of action. Toxicol Sci 98:375–394.

Brusick DJ. 1986. Genotoxicity of hexachlorobenzene and other chlorinated benzenes. IARC Sci Publ 77:393–397.

Buchmann A, Willy C, Buenemann CL, Stroh C, Schmiechen A, Schwarz M. 1999. Inhibition of transforming growth factor beta1-induced hepatoma cell apoptosis by liver tumor promoters: Characterization of primary signaling events and effects on CPP32-like caspase activity. Cell Death Differ 6:190–200.

Buenemann CL, Willy C, Buchmann A, Schmiechen A, Schwarz M. 2001. Transforming growth factor-beta 1-induced Smad signaling, cell-cycle arrest and apoptosis in hepatoma cells. Carcinogenesis 22:447–452.

Burow ME, Tang Y, Collins-Burow BM, Krajewski S, Reed JC, McLachlan JA, et al. 1999. Effects of environmental estrogens on tumor necrosis factor alpha-mediated apoptosis in MCF-7 cells. Carcinogenesis 20:2057–2061.

Cadet J, Douki T, Ravanat JL. 2010. Oxidatively generated base damage to cellular DNA. Free Radic Biol Med 49:9–21.

Cattley RC. 2003. Regulation of cell proliferation and cell death by peroxisome proliferators. Microsc Res Tech 61:179–184.

Cattley RC, DeLuca J, Elcombe C, FennerCrisp P, Lake BG, Marsman DS, et al. 1998. Do peroxisome proliferating compounds pose a hepatocarcinogenic hazard to humans? Regul Toxicol Pharmacol 27:47–60.

Chiappini F, Alvarez L, Lux-Lantos V, Randi AS, Kleiman de Pisarev DL. 2009. Hexachlorobenzene triggers apoptosis in rat thyroid follicular cells. Toxicol Sci 108: 301–310.

Chopra M, Dharmarajan AM, Meiss G, Schrenk D. 2009. Inhibition of UV-C light-induced apoptosis in liver cells by 2,3,7,8-tetrachlorodibenzo-p-dioxin. Toxicol Sci 111:49–63.

Clark JM. 1974. Mutagenicity of DDT in mice, *Drosophila melanogaster* and *Neurospora crassa*. Aust J Biol Sci 27:427–440.

Cogliano VJ, Baan R, Straif K, Grosse Y, Lauby-Secretan B, El Ghissassi F, et al. 2012. Preventable exposures associated with human cancers. J Natl Cancer Inst 103: 1827–1839.

Cooke PS, Sato T, Buchanan DL. 2001. Disruption of steroid hormone signaling by PCBs. In: PCBs: Recent Advances in Environmental Toxicology and Health Effects (Robertson LW, Hansen LG, eds.). Lexington, KY: University Press of Kentucky, pp. 257–263.

Cossette LJ, Gaumond I, Martinoli MG. 2002. Combined effect of xenoestrogens and growth factors in two estrogen-responsive cell lines. Endocrine 18:303–308.

Cotelle S, Masfaraud JF, Ferard JF. 1999. Assessment of the genotoxicity of contaminated soil with the Allium/Vicia-micronucleus and the Tradescantia-micronucleus assays. Mutat Res 426:167–171.

Darbyshire JF, Iyer KR, Grogan J, Korzekwa KR, Trager WF. 1996. Substrate probe for the mechanism of aromatic hydroxylation catalyzed by cytochrome P450. Drug Metab Dispos 24:1038–1045.

Davis JW, 2nd, Melendez K, Salas VM, Lauer FT, Burchiel SW. 2000. 2,3,7,8-Tetrachlorodibenzo-p-dioxin (TCDD) inhibits growth factor withdrawal-induced apoptosis in the human mammary epithelial cell line, MCF-10A. Carcinogenesis 21:881–886.

Davis JW, 2nd, Lauer FT, Burdick AD, Hudson LG, Burchiel SW. 2001. Prevention of apoptosis by 2,3,7,8-tetrachlorodibenzo-p-dioxin (TCDD) in the MCF-10A cell line: Correlation with increased transforming growth factor alpha production. Cancer Res 61:3314–3320.

DiGiovanni J, Viaje A, Berry DL, Slaga TJ, Juchau MR. 1977. Tumor-initiating ability of 2,3,7,8-tetrachlorodibenzo-p-dioxin (TCDD) and Arochlor 1254 in the two-stage system of mouse skin carcinogenesis. Bull Environ Contam Toxicol 18:552–557.

Dragan YP, Schrenk D. 2000. Animal studies addressing the carcinogenicity of TCDD (or related compounds) with an emphasis on tumour promotion. Food Addit Contam 17:289–302.

Dulout FN, Pastori MC, Olivero OA, Gonzalez Cid M, Loria D, Matos E, et al. 1985. Sister-chromatid exchanges and chromosomal aberrations in a population exposed to pesticides. Mutat Res 143:237–244.

Edwards JW, Priestly BG. 1994. Effect of occupational exposure to aldrin on urinary D-glucaric acid, plasma dieldrin, and lymphocyte sister chromatid exchange. Int Arch Occup Environ Health 66:229–234.

Ekuase EJ, Liu Y, Lehmler HJ, Robertson LW, Duffel MW. 2011. Structure-activity relationships for hydroxylated polychlorinated biphenyls as inhibitors of the sulfation of dehydroepiandrosterone catalyzed by human hydroxysteroid sulfotransferase SULT2A1. Chem Res Toxicol 24:1720–1728.

Elo O, Vuojolahti P, Janhunen H, Rantanen J. 1985. Recent PCB accidents in Finland. Environ Health Perspect 60:315–319.

Ennaceur S, Ridha D, Marcos R. 2008. Genotoxicity of the organochlorine pesticides 1,1-dichloro-2,2- bis(p-chlorophenyl)ethylene (DDE) and hexachlorobenzene (HCB) in cultured human lymphocytes. Chemosphere 71:1335–1339.

Espandiari P, Glauert HP, Lehmler HJ, Lee EY, Srinivasan C, Robertson LW. 2003. Polychlorinated biphenyls as initiators in liver carcinogenesis: Resistant hepatocyte model. Toxicol Appl Pharmacol 186:55–62.

Espandiari P, Glauert HP, Lehmler HJ, Lee EY, Srinivasan C, Robertson LW. 2004. Initiating activity of 4-chlorobiphenyl metabolites in the resistant hepatocyte model. Toxicol Sci 79:41–46.

Espandiari P, Robertson LW, Srinivasan C, Glauert HP. 2005. Comparison of different initiation protocols in the resistant hepatocyte model. Toxicology 206:373–381.

Fariss MW, Chan CB, Patel M, Van Houten B, Orrenius S. 2005. Role of mitochondria in toxic oxidative stress. Mol Interv 5:94–111.

Fitzgerald EF, Shrestha S, Palmer PM, Wilson LR, Belanger EE, Gomez MI, et al. 2011. Polychlorinated biphenyls (PCBs) in indoor air and in serum among older residents of upper Hudson River communities. Chemosphere 85:225–231.

Flodstrom S, Hemming H, Warngard L, Ahlborg UG. 1990. Promotion of altered hepatic foci development in rat liver, cytochrome P450 enzyme induction and inhibition of cell-cell communication by DDT and some structurally related organohalogen pesticides. Carcinogenesis 11:1413–1417.

Flor S, Ludewig G. 2010. Polyploidy-induction by dihydroxylated monochlorobiphenyls: Structure-activity-relationships. Environ Int 36:962–969.

Fukushima S, Kinoshita A, Puatanachokchai R, Kushida M, Wanibuchi H, Morimura K. 2005. Hormesis and dose-response-mediated mechanisms in carcinogenesis: Evidence for a threshold in carcinogenicity of non-genotoxic carcinogens. Carcinogenesis 26:1835–1845.

Galloway SM, Armstrong MJ, Reuben C, Colman S, Brown B, Cannon C, et al. 1987. Chromosome aberrations and sister-chromatid exchanges in Chinese hamster ovary cells: Evaluations of 108 chemicals. Environ Mol Mutagen 10(Suppl 10):1–175.

Garcia MA, Pena D, Alvarez L, Cocca C, Pontillo C, Bergoc R, et al. 2010. Hexachlorobenzene induces cell proliferation and IGF-I signaling pathway in an estrogen receptor alpha-dependent manner in MCF-7 breast cancer cell line. Toxicol Lett 192:195–205.

Gauthier JM, Dubeau H, Rassart E. 1999. Induction of micronuclei in vitro by organochlorine compounds in beluga whale skin fibroblasts. Mutat Res 439:87–95.

Gauthier M, Roberge CJ, Pelletier M, Tessier PA, Girard D. 2001. Activation of human neutrophils by technical toxaphene. Clin Immunol 98:46–53.

Giribaldi L, Chiappini F, Pontillo C, Randi AS, Kleiman de Pisarev DL, Alvarez L. 2011. Hexachlorobenzene induces deregulation of cellular growth in rat liver. Toxicology 289:19–27.

Glauert HP, Robertson LW, Silberhorn EM. 2001. PCBs and tumor promotion. In: PCBs: Recent Advances in Environmental Toxicology and Health Effects (Robertson LW, Hansen LG, eds.). Lexington, KY: University Press of Kentucky, pp. 355–371.

Glauert HP, Tharappel JC, Banerjee S, Chan LS, Kania-Korwel I, Lehmler HJ, et al. 2008. Inhibition of the promotion of hepatocarcinogenesis by 2,2′,4,4′,5,5′-hexachlorobiphenyl (PCB-153) by the deletion of the p50 subunit of NF-kB in mice. Toxicol Appl Pharmacol 232:302–308.

Gloire G, Legrand-Poels S, Piette J. 2006. NF-kappaB activation by reactive oxygen species: Fifteen years later. Biochem Pharmacol 72:1493–1505.

Goel MR, Shara MA, Stohs SJ. 1988. Induction of lipid peroxidation by hexachlorocyclohexane, dieldrin, TCDD, carbon tetrachloride, and hexachlorobenzene in rats. Bull Environ Contam Toxicol 40:255–562.

Goodman JI, Brusick DJ, Busey WM, Cohen SM, Lamb JC, Starr TB. 2000. Reevaluation of the cancer potency factor of toxaphene: Recommendations from a peer review panel. Toxicol Sci 55:3–16.

Gopalaswamy UV, Nair CK. 1992. DNA binding and mutagenicity of lindane and its metabolites. Bull Environ Contam Toxicol 49:300–305.

Gorski T, Gorska E, Gorecka D, Sikora M. 1986. Hexachlorobenzene is non-genotoxic in short-term tests. IARC Sci Publ 77:399–401.

Greenberg RA. 2005. Telomeres, crisis and cancer. Curr Mol Med 5(2):213–218.

Guengerich FP. 2001. Common and uncommon cytochrome P450 reactions related to metabolism and chemical toxicity. Chem Res Toxicol 14:611–650.

Hallegue D, Ben Rhouma K, Krichah R, Sakly M. 2002. Dieldrin initiates apoptosis in rat thymocytes. Indian J Exp Biol 40:1147–1150.

Harada T, Yamaguchi S, Ohtsuka R, Takeda M, Fujisawa H, Yoshida T, et al. 2003. Mechanisms of promotion and progression of preneoplastic lesions in hepatocarcinogenesis by DDT in F344 rats. Toxicol Pathol 31:87–98.

Hasmall SC, James NH, Macdonald N, Gonzalez FJ, Peters JM, Roberts RA. 2000. Suppression of mouse hepatocyte apoptosis by peroxisome proliferators: Role of PPAR alpha and TNF alpha. Mutat Res Fundam Mol Mech Mutagen 448:193–200.

Hassoun EA, Stohs SJ. 1996. TCDD, endrin and lindane induced oxidative stress in fetal and placental tissues of C57BL/6J and DBA/2J mice. Comp Biochem Physiol C Pharmacol Toxicol Endocrinol 115:11–18.

Hassoun E, Bagchi M, Bagchi D, Stohs SJ. 1993. Comparative studies on lipid peroxidation and DNA-single strand breaks induced by lindane, DDT, chlordane and endrin in rats. Comp Biochem Physiol C 104:427–431.

Hattula ML. 1985. Mutagenicity of PCBs and their pyrosynthetic derivatives in cell-mediated assay. Environ Health Perspect 60:255–257.

Hayes MA, Safe SH, Armstrong D, Cameron RG. 1985. Influence of cell proliferation on initiating activity of pure polychlorinated biphenyls and complex mixtures in resistant hepatocyte in vivo assays for carcinogenicity. J Natl Cancer Inst 74: 1037–1041.

Hedli CC, Snyder R, Kinoshita FK, Steinberg M. 1998. Investigation of hepatic cytochrome P-450 enzyme induction and DNA adduct formation in male CD/1 mice following oral administration of toxaphene. J Appl Toxicol 18:173–178.

Heikaus S, Winterhager E, Traub O, Grummer R. 2002. Responsiveness of endometrial genes Connexin26, Connexin43, C3 and clusterin to primary estrogen, selective estrogen receptor modulators, phyto- and xenoestrogens. J Mol Endocrinol 29: 239–249.

Hesse S, Mezger M, Wolff T. 1978. Activation of [14C] chlorobiphenyls to protein-binding metabolites by rat liver microsomes. Chem Biol Interact 20:355–365.

Ho PW, Garner CE, Ho JW, Leung KC, Chu AC, Kwok KH, et al. 2008. Estrogenic phenol and catechol metabolites of PCBs modulate catechol-O-methyltransferase expression via the estrogen receptor: Potential contribution to cancer risk. Curr Drug Metab 9:304–309.

Holsapple MP, Pitot HC, Cohen SM, Boobis AR, Klaunig JE, Pastoor T, et al. 2006. Mode of action in relevance of rodent liver tumors to human cancer risk. Toxicol Sci 89:51–56.

Horvath ME, Faux SP, Blazovics A, Feher J. 2001. Lipid and DNA oxidative damage in experimentally induced hepatic porphyria in C57BL/10ScSn mice. Z Gastroenterol 39:453–455, 458.

Huff JE, Salmon AG, Hooper NK, Zeise L. 1991. Long-term carcinogenesis studies on 2,3,7,8-tetrachlorodibenzo-p-dioxin and hexachlorodibenzo-p-dioxins. Cell Biol Toxicol 7:67–94.

Iannuzzi L, Perucatti A, Di Meo GP, Polimeno F, Ciotola F, Incarnato D, et al. 2004. Chromosome fragility in two sheep flocks exposed to dioxins during pasturage. Mutagenesis 19:355–359.

Izushi F, Ogata M. 1990. Hepatic and muscle injuries in mice treated with heptachlor. Toxicol Lett 54:47–54.

Jacobus JA, Flor S, Klingelhutz A, Robertson LW, Ludewig G. 2008. 2-(4′-Chlorophenyl)-1,4-benzoquinone increases the frequency of micronuclei and shortens telomeres. Environ Toxicol Pharmacol 25:267–272.

Jensen KG, Wiberg K, Klasson-Wehler E, Onfelt A. 2000. Induction of aberrant mitosis with PCBs: Particular efficiency of 2, 3,3′,4,4′-pentachlorobiphenyl and synergism with triphenyltin. Mutagenesis 15:9–15.

Jeon YJ, Youk ES, Lee SH, Suh J, Na YJ, Kim HM. 2002. Polychlorinated biphenyl-induced apoptosis of murine spleen cells is aryl hydrocarbon receptor independent but caspases dependent. Toxicol Appl Pharmacol 181:69–78.

Jerina DM, Daly JW. 1974. Arene oxides: A new aspect of drug metabolism. Science 185(4151):573–582.

Kaiser J, Enserink M. 2000. Environmental toxicology. Treaty takes a POP at the dirty dozen. Science 290:2053.

Kalina I, Sram RJ, Konecna H, Ondrussekova A. 1991. Cytogenetic analysis of peripheral blood lymphocytes in workers occupationally exposed to polychlorinated biphenyls. Teratog Carcinog Mutagen 11:77–82.

Kamath AB, Xu H, Nagarkatti PS, Nagarkatti M. 1997. Evidence for the induction of apoptosis in thymocytes by 2,3,7,8-tetrachlorodibenzo-p-dioxin in vivo. Toxicol Appl Pharmacol 142:367–377.

Kamendulis LM, Kolaja KL, Stevenson DE, Walborg EF, Jr., Klaunig JE. 2001. Comparative effects of dieldrin on hepatic ploidy, cell proliferation, and apoptosis in rodent liver. J Toxicol Environ Health A 62:127–141.

Kang KS, Wilson MR, Hayashi T, Chang CC, Trosko JE. 1996. Inhibition of gap junctional intercellular communication in normal human breast epithelial cells after treatment with pesticides, PCBs, and PBBs, alone or in mixtures. Environ Health Perspect 104:192–200.

Kanthasamy AG, Kitazawa M, Yang Y, Anantharam V, Kanthasamy A. 2008. Environmental neurotoxin dieldrin induces apoptosis via caspase-3-dependent proteolytic activation of protein kinase C delta (PKCdelta): Implications for neurodegeneration in Parkinson's disease. Mol Brain 1:12.

Kato Y, Kenne K, Haraguchi K, Masuda Y, Kimura R, Warngard L. 1998. Inhibition of cell-cell communication by methylsulfonyl metabolites of polychlorinated biphenyl congeners in rat liver epithelial IAR 20 cells. Arch Toxicol 72:178–182.

Khasawinah AM, Grutsch JF. 1989. Chlordane: Thirty-month tumorigenicity and chronic toxicity test in rats. Regul Toxicol Pharmacol 10:95–109.

Kishima MO, Barbisan LF, Estevao D, Rodrigues MAM, Decamargo JLV. 2000. Promotion of hepatocarcinogenesis by hexachlorobenzene in energy-restricted rats. Cancer Lett 152:37–44.

Kitazawa M, Anantharam V, Kanthasamy AG. 2003. Dieldrin induces apoptosis by promoting caspase-3-dependent proteolytic cleavage of protein kinase Cdelta in dopaminergic cells: Relevance to oxidative stress and dopaminergic degeneration. Neuroscience 119:945–964.

Kitchin KT, Brown JL. 1994. Dose-response relationship for rat liver DNA damage caused by 49 rodent carcinogens. Toxicology 88:31–49.

Klaunig JE, Ruch RJ. 1987. Strain and species effects on the inhibition of hepatocyte intercellular communication by liver tumor promoters. Cancer Lett 36:161–168.

Klaunig JE, Wang Z, Pu X, Zhou S. 2011. Oxidative stress and oxidative damage in chemical carcinogenesis. Toxicol Appl Pharmacol 254:86–99.

Knerr S, Schrenk D. 2006. Carcinogenicity of 2,3,7,8-tetrachlorodibenzo-p-dioxin in experimental models. Mol Nutr Food Res 50:897–907.

Kolaja KL, Stevenson DE, Walborg EF, Klaunig JE. 1996. Selective dieldrin promotion of hepatic focal lesions in mice. Carcinogenesis 17:1243–1250.

Kolaja KL, Engelken DT, Klaassen CD. 2000. Inhibition of gap-junctional-intercellular communication in intact rat liver by nongenotoxic hepatocarcinogens. Toxicology 146:15–22.

Kostka G, Kopec-Szlezak J, Palut D. 1996. Early hepatic changes induced in rats by two hepatocarcinogenic organohalogen pesticides: Bromopropylate and DDT. Carcinogenesis 17:407–412.

Kowalski LA, Laitinen AM, Mortazavi-Asl B, Wee RK, Erb HE, Assi KP, et al. 2000. In vitro determination of carcinogenicity of sixty-four compounds using a bovine papillomavirus DNA-carrying C3H/10T(1/2) cell line. Environ Mol Mutagen 35:300–311.

Krutovskikh VA, Mesnil M, Mazzoleni G, Yamasaki H. 1995. Inhibition of rat liver gap junction intercellular communication by tumor-promoting agents in vivo. Association with aberrant localization of connexin proteins. Lab Invest 72:571–577.

Kushida M, Sukata T, Uwagawa S, Ozaki K, Kinoshita A, Wanibuchi H, et al. 2005. Low dose DDT inhibition of hepatocarcinogenesis initiated by diethylnitrosamine in male rats: Possible mechanisms. Toxicol Appl Pharmacol 208:285–294.

Laconi E, Tessitore L, Milia G, Yusuf A, Sarma DSR, Todde P, Pani P. 1995. The enhancing effect of fasting/refeeding on the growth of nodules selectable by the resistant hepatocyte model in rat liver. Carcinogenesis 16:1865–1869.

Lamb JCT, Marks TA, Gladen BC, Allen JW, Moore JA. 1981. Male fertility, sister chromatid exchange, and germ cell toxicity following exposure to mixtures of chlorinated phenoxy acids containing 2,3,7,8-tetrachlorodibenzo-p-dioxin. J Toxicol Environ Health 8:825–834.

Larsen KD, Jalal SM. 1974. DDT induced chromosome mutations in mice—further testing. Can J Genet Cytol 16:491–497.

Lauby-Secretan B, Loomis D, Grosse Y, El Ghissassi F, Bouvard V, Benbrahim-Tallaa L, et al. 2013. Carcinogenicity of polychlorinated biphenyls and polybrominated biphenyls. International Agency for Research on Cancer Monograph Working Group IARC, Lyon, France. Lancet Oncol 14:287–288.

Lavastre V, Roberge CJ, Pelletier M, Gauthier M, Girard D. 2002. Toxaphene, but not beryllium, induces human neutrophil chemotaxis and apoptosis via reactive oxygen species (ROS): Involvement of caspases and ROS in the degradation of cytoskeletal proteins. Clin Immunol 104:40–48.

Lee YW, Park HJ, Son KW, Hennig B, Robertson LW, Toborek M. 2003. 2,2′,4,6,6′-Pentachlorobiphenyl (PCB 104) induces apoptosis of human microvascular endothelial cells through the caspase-dependent activation of CREB. Toxicol Appl Pharmacol 189:1–10.

Lehmann L, Esch H, Kirby P, Robertson LW, Ludewig G. 2007. 4-Monochlorobiphenyl (PCB3) induces mutations in the livers of transgenic Fisher 344 rats. Carcinogenesis 28:471–478.

Lemaire G, Mnif W, Mauvais P, Balaguer P, Rahmani R. 2006. Activation of alpha- and beta-estrogen receptors by persistent pesticides in reporter cell lines. Life Sci 79:1160–1169.

Li Y, Leung LK, Glauert HP, Spear BT. 1996. Treatment of rats with the peroxisome proliferator ciprofibrate results in increased liver NF-kappa B activity. Carcinogenesis 17:2305–2309.

Lim M, Jacobson-Kram D, Bowman RE, Williams JR. 1987. Effect of chronic exposure to 2,3,7,8-tetrachlorodibenzo-p-dioxin on sister chromatid exchange levels in peripheral lymphocytes of the rhesus monkey. Cell Biol Toxicol 3:279–284.

Lin PH, Sangaiah R, Ranasinghe A, Upton PB, La DK, Gold A, et al. 2000. Formation of quinonoid-derived protein adducts in the liver and brain of Sprague-Dawley rats treated with 2,2′,5,5′-tetrachlorobiphenyl. Chem Res Toxicol 13:710–718.

Liu Y, Apak TI, Lehmler HJ, Robertson LW, Duffel MW. 2006. Hydroxylated polychlorinated biphenyls are substrates and inhibitors of human hydroxysteroid sulfotransferase SULT2A1. Chem Res Toxicol 19:1420–1425.

Liu Y, Smart JT, Song Y, Lehmler HJ, Robertson LW, Duffel MW. 2009. Structure-activity relationships for hydroxylated polychlorinated biphenyls as substrates and inhibitors of rat sulfotransferases and modification of these relationships by changes in thiol status. Drug Metab Dispos 37:1065–1072.

Lu Z, Tharappel JC, Lee EY, Robertson LW, Spear BT, Glauert HP. 2003. Effect of a single dose of polychlorinated biphenyls on hepatic cell proliferation and the DNA binding activity of NF-kappaB and AP-1 in rats. Mol Carcinog 37:171–180.

Lu Z, Lee EY, Robertson LW, Glauert HP, Spear BT. 2004. Effect of 2,2′,4,4′,5,5′-hexachlorobiphenyl (PCB-153) on hepatocyte proliferation and apoptosis in mice deficient in the p50 subunit of the transcription factor NF-kB. Toxicol Sci 81: 35–42.

Ludewig G. 2001. Cancer initiation by PCBs. In: PCBs: Recent Advances in Environmental Toxicology and Health Effects (Robertson LW, Hansen LG, eds.). Lexington: University Press of Kentucky, pp. 337–354.

Lundgren K, Andries M, Thompson C, Lucier GW. 1986. Dioxin treatment of rats results in increased in vitro induction of sister chromatid exchanges by alpha-naphthoflavone: An animal model for human exposure to halogenated aromatics. Toxicol Appl Pharmacol 85:189–195.

Lundgren K, Collman GW, Wang-Wuu S, Tiernan T, Taylor M, Thompson CL, et al. 1988. Cytogenetic and chemical detection of human exposure to polyhalogenated aromatic hydrocarbons. Environ Mol Mutagen 11:1–11.

Machala M, Blaha L, Lehmler HJ, Pliskova M, Majkova Z, Kapplova P, et al. 2004. Toxicity of hydroxylated and quinoid PCB metabolites: Inhibition of gap junctional intercellular communication and activation of aryl hydrocarbon and estrogen receptors in hepatic and mammary cells. Chem Res Toxicol 17:340–347.

Majumdar SK, Kopelman HA, Schnitman MJ. 1976. Dieldrin-induced chromosome damage in mouse bone-marrow and WI-38 human lung cells. J Hered 67:303–307.

Malarkey DE, Devereux TR, Dinse GE, Mann PC, Maronpot RR. 1995. Hepatocarcinogenicity of chlordane in B6C3F1 and B6D2F1 male mice: Evidence for regression in B6C3F1 mice and carcinogenesis independent of ras proto-oncogene activation. Carcinogenesis 16:2617–2625.

Marabini L, Calo R, Fucile S. 2011. Genotoxic effects of polychlorinated biphenyls (PCB 153, 138, 101, 118) in a fish cell line (RTG-2). Toxicol in Vitro 25:1045–1052.

Marnett LJ. 2002. Oxy radicals, lipid peroxidation and DNA damage. Toxicology 181-182:219–222.

McGregor DB, Brown A, Cattanach P, Edwards I, McBride D, Riach C, et al. 1988. Responses of the L5178Y tk+/tk- mouse lymphoma cell forward mutation assay: III. 72 coded chemicals. Environ Mol Mutagen 12:85–154.

McGregor DB, Brown AG, Howgate S, McBride D, Riach C, Caspary WJ. 1991. Responses of the L5178Y mouse Lymphoma cell forward mutation assay. V: 27 coded chemicals. Environ Mol Mutagen 17:196–219.

McLean MR, Bauer U, Amaro AR, Robertson LW. 1996a. Identification of catechol and hydroquinone metabolites of 4-monochlorobiphenyl. Chem Res Toxicol 9:158–164.

McLean MR, Robertson LW, Gupta RC. 1996b. Detection of PCB adducts by the 32P-postlabeling technique. Chem Res Toxicol 9:165–171.

McLean MR, Twaroski TP, Robertson LW. 2000. Redox cycling of 2-(x′-mono, -di, -trichlorophenyl)- 1,4-benzoquinones, oxidation products of polychlorinated biphenyls. Arch Biochem Biophys 376:449–455.

Meisner LF, Roloff B, Sargent L, Pitot H. 1992. Interactive cytogenetic effects on rat bone-marrow due to chronic ingestion of 2,5,2′,5′ and 3,4,3′,4′ PCBs. Mutat Res 283:179–183.

Meyne J, Allison DC, Bose K, Jordan SW, Ridolpho PF, Smith J. 1985. Hepatotoxic doses of dioxin do not damage mouse bone marrow chromosomes. Mutat Res 157:63–69.

Mirsalis JC, Tyson CK, Steinmetz KL, Loh EK, Hamilton CM, Bakke JP, et al. 1989. Measurement of unscheduled DNA synthesis and S-phase synthesis in rodent hepatocytes following in vivo treatment: Testing of 24 compounds. Environ Mol Mutagen 14:155–164.

Morales NM, Matthews HB. 1979. In vivo binding of 2,3,6,2′,3′,6′-hexachlorobiphenyl and 2,4,5,2′,4′,5′-hexachlorobiphenyl to mouse liver macromolecules. Chem Biol Interact 27:99–110.

Morgan MJ, Liu ZG. 2011. Crosstalk of reactive oxygen species and NF-kappaB signaling. Cell Res 21:103–115.

Moriya M, Ohta T, Watanabe K, Miyazawa T, Kato K, Shirasu Y. 1983. Further mutagenicity studies on pesticides in bacterial reversion assay systems. Mutat Res 116:185–216.

Mortelmans K, Haworth S, Lawlor T, Speck W, Tainer B, Zeiger E. 1986. Salmonella mutagenicity tests: II. Results from the testing of 270 chemicals. Environ Mutagen 8(Suppl 7):1–119.

Moser GJ, Meyer SA, Smart RC. 1992. The chlorinated pesticide mirex is a novel non-phorbol ester-type tumor promoter in mouse skin. Cancer Res 52:631–636.

Moser GJ, Robinette CL, Smart RC. 1993. Characterization of skin tumor promotion by mirex: Structure-activity relationships, sexual dimorphism and presence of Ha-ras mutation. Carcinogenesis 14:1155–1160.

Mustonen R, Elovaara E, Zitting A, Linnainmaa K, Vainio H. 1989. Effects of commercial chlorophenolate, 2,3,7,8-TCDD, and pure phenoxyacetic acids on hepatic peroxisome proliferation, xenobiotic metabolism and sister chromatid exchange in the rat. Arch Toxicol 63:203–208.

Nagayama J, Nagayama M, Wada K, Iida T, Hirakawa H, Matsueda T, et al. 1991. The effect of organochlorine compounds on the induction of sister chromatid exchanges in cultured human lymphocytes. Fukuoka Igaku Zasshi 82:221–227.

Nagayama J, Nagayama M, Haraguchi K, Kuroki H, Masuda Y. 1995. Effect of 2, 3, 4, 7, 8-pentachlorodibenzofuran and its analogues on induction of sister chromatid exchanges in cultured human lymphocytes. Fukuoka Igaku Zasshi 86:184–189.

Nagayama J, Nagayama M, Haraguchi K, Kuroki H, Masuda Y. 1999. Induction of sister chromatid exchanges in cultured human lymphocytes with methylsulphonyl PCB congeners. Fukuoka Igaku Zasshi 90:238–245.

Nagayama J, Nagayama M, Iida T, Hirakawa H, Matsueda T, Ohki M, et al. 2001a. Comparison between "Yusho" patients and healthy Japanese in contamination level of dioxins and related chemicals and frequency of sister chromatid exchanges. Chemosphere 43:931–936.

Nagayama J, Nagayama M, Iida T, Hirakawa H, Matsueda T, Yanagawa T, et al. 2001b. Effect of dioxins in mother's milk on sister chromatid exchange frequency in infant lymphocytes. Fukuoka Igaku Zasshi 92:177–183.

Nagayama J, Nagayama M, Nakagawa R, Hirakawa H, Matsueda T, Iida T, et al. 2003. Frequency of SCEs in Japanese infants lactationally exposed to organochlorine pesticides. Fukuoka Igaku Zasshi 94:166–173.

National Toxicology Program. 2006. NTP technical report on the toxicology and carcinogenesis studies of 2,2′,4,4′,5,5′-hexachlorobiphenyl (PCB 153) (CAS No. 35065-27-1) in female Harlan Sprague-Dawley rats (Gavage studies). Natl Toxicol Program Tech Rep Ser 529:4–168.

Naugler WE, Sakurai T, Kim S, Maeda S, Kim K, Elsharkawy AM, et al. 2007. Gender disparity in liver cancer due to sex differences in MyD88-dependent IL-6 production. Science 317:121–124.

Nelson JA, Struck RF, James R. 1978. Estrogenic activities of chlorinated hydrocarbons. J Toxicol Environ Health 4:325–339.

Niki E. 2009. Lipid peroxidation: Physiological levels and dual biological effects. Free Radic Biol Med 47:469–484.

Norstrom K, Czub G, McLachlan MS, Hu D, Thorne PS, Hornbuckle KC. 2010. External exposure and bioaccumulation of PCBs in humans living in a contaminated urban environment. Environ Int 36:855–861.

NTP. 1990. NTP Toxicology and Carcinogenesis Studies of Mirex (1,1a,2,2,3,3a,4,5,5,5a,5b,6-Dodecachlorooctahydro-1,3,4- metheno-1H-cycloutа[cd]pentalene) (CAS No. 2385-85-5) in F344/N Rats (Feed Studies). Natl Toxicol Program Tech Rep Ser 313:1–140.

Numoto S, Tanaka T, Williams GM. 1985. Morphologic and cytochemical properties of mouse liver neoplasms induced by diethylnitrosamine and promoted by 4,4′-dichlorodiphenyltrichloroethane, chlordane, or heptachlor. Toxicol Pathol 13:325–334.

Oakley GG, Devanaboyina U, Robertson LW, Gupta RC. 1996. Oxidative DNA damage induced by activation of polychlorinated biphenyls (PCBs): Implications for PCB-induced oxidative stress in breast cancer. Chem Res Toxicol 9:1285–1292.

O'Brien ML, Spear BT, Glauert HP. 2005. Role of oxidative stress in peroxisome proliferator-mediated carcinogenesis. Crit Rev Toxicol 35:61–88.

Oesterle D, Deml E. 1981. Promoting effect of various PCBs and DDT on enzyme-altered islands in rat liver. Naunyn Schmiedebergs Arch Pharmacol 316:R16.

Ohtake F, Fujii-Kuriyama Y, Kawajiri K, Kato S. 2011. Cross-talk of dioxin and estrogen receptor signals through the ubiquitin system. J Steroid Biochem Mol Biol 127: 102–107.

Okoumassoun LE, Averill-Bates D, Marion M, Denizeau F. 2003. Possible mechanisms underlying the mitogenic action of heptachlor in rat hepatocytes. Toxicol Appl Pharmacol 193:356–369.

Ou YC, Conolly RB, Thomas RS, Xu Y, Andersen ME, Chubb LS, et al. 2001. A clonal growth model: Time-course simulations of liver foci growth following penta- or hexachlorobenzene treatment in a medium-term bioassay. Cancer Res 61:1879–1889.

Ou YC, Conolly RB, Thomas RS, Gustafson DL, Long ME, Dobrev ID, et al. 2003. Stochastic simulation of hepatic preneoplastic foci development for four chlorobenzene congeners in a medium-term bioassay. Toxicol Sci 73:301–314.

Patterson RM, Stachlewitz R, Germolec D. 2003. Induction of apoptosis by 2,3,7,8-tetrachlorodibenzo-p-dioxin following endotoxin exposure. Toxicol Appl Pharmacol 190:120–134.

Payne J, Scholze M, Kortenkamp A. 2001. Mixtures of four organochlorines enhance human breast cancer cell proliferation. Environ Health Perspect 109:391–397.

van Pelt FN, Haring RM, Weterings PJ. 1991. Micronucleus formation in cultured human keratinocytes: Involvement of intercellular bioactivation. Toxicol in Vitro 5:515–518.

Pereg D, Tampal N, Espandiari P, Robertson LW. 2001. Distribution and macromolecular binding of benzo[a]pyrene and two polychlorinated biphenyl congeners in female mice. Chem Biol Interact 137:243–258.

Pereira MA, Herren SL, Britt AL, Khoury MM. 1982. Sex difference in enhancement of GGTase-positive foci by hexachlorobenzene and lindane in rat liver. Cancer Lett 15:95–101.

Perez-Maldonado IN, Diaz-Barriga F, de la Fuente H, Gonzalez-Amaro R, Calderon J, Yanez L. 2004. DDT induces apoptosis in human mononuclear cells in vitro and is associated with increased apoptosis in exposed children. Environ Res 94:38–46.

Perucatti A, Di Meo GP, Albarella S, Ciotola F, Incarnato D, Jambrenghi AC, et al. 2006. Increased frequencies of both chromosome abnormalities and SCEs in two sheep flocks exposed to high dioxin levels during pasturage. Mutagenesis 21:67–75.

Pitot HC, Dragan YP. 1994. Chemical induction of hepatic neoplasia. In: The Liver: Biology and Pathobiology, 3rd edition (Arias IM et al., eds.). New York: Raven, pp. 1467–1495.

Porter KL, Chanda S, Wang HQ, Gaido KW, Smart RC, Robinette CL. 2002. 17Beta-estradiol is a hormonal regulator of mirex tumor promotion sensitivity in mice. Toxicol Sci 69:42–48.

Probst GS, McMahon RE, Hill LE, Thompson CZ, Epp JK, Neal SB. 1981. Chemically-induced unscheduled DNA synthesis in primary rat hepatocyte cultures: A comparison with bacterial mutagenicity using 218 compounds. Environ Mutagen 3:11–32.

Pryputniewicz SJ, Nagarkatti M, Nagarkatti PS. 1998. Differential induction of apoptosis in activated and resting T cells by 2,3,7,8-tetrachlorodibenzo-p-dioxin (TCDD) and its repercussion on T cell responsiveness. Toxicology 129:211–226.

Puebla-Osorio N, Ramos KS, Falahatpisheh MH, Smith R, 3rd, Berghman LR. 2004. 2,3,7,8-Tetrachlorodibenzo-p-dioxin elicits aryl hydrocarbon receptor-mediated apoptosis in the avian DT40 pre-B-cell line through activation of caspases 9 and 3. Comp Biochem Physiol C Toxicol Pharmacol 138:461–468.

Rao MS, Reddy JK. 1987. Peroxisome proliferation and hepatocarcinogenesis. Carcinogenesis 8:631–636.

Rayne S, Forest K. 2010. pK(a) values of the monohydroxylated polychlorinated biphenyls (OH-PCBs), polybrominated biphenyls (OH-PBBs), polychlorinated diphenyl ethers (OH-PCDEs), and polybrominated diphenyl ethers (OH-PBDEs). J Environ Sci Health A Tox Hazard Subst Environ Eng 45:1322–1346.

Reddy JK, Lalwani ND. 1983. Carcinogenesis by hepatic peroxisome proliferators: Evaluation of the risk of hypolipidemic drugs and industrial plasticizers to humans. Crit Rev Toxicol 12:1–58.

Reed L, Buchner V, Tchounwou PB. 2007. Environmental toxicology and health effects associated with hexachlorobenzene exposure. Rev Environ Health 22:213–243.

Reuber MD. 1979. Carcinogenicity of endrin. Sci Total Environ 72:101–135.

Reuber MD. 1987. Carcinogenicity of heptachlor and heptachlor epoxide. J Environ Pathol Toxicol Oncol 7:85–114.

Reuber MD, Ward JM. 1979. Histopathology of liver carcinomas in (C57BL/6N X C3H/HeN)F1 mice ingesting chlordane. J Natl Cancer Inst 63:89–92.

Rivedal E, Mikalsen SO, Sanner T. 2000. Morphological transformation and effect on gap junction intercellular communication in Syrian hamster embryo cells as screening tests for carcinogens devoid of mutagenic activity. Toxicol in Vitro 14:185–192.

Roberts RA. 1999. Peroxisome proliferators: Mechanisms of adverse effects in rodents and molecular basis for species differences. Arch Toxicol 73:413–418.

Roberts RA, James NH, Woodyatt NJ, Macdonald N, Tugwood JD. 1998. Evidence for the suppression of apoptosis by the peroxisome proliferator activated receptor alpha (PPAR alpha). Carcinogenesis 19:43–48.

Ross J, Plummer SM, Rode A, Scheer N, Bower CC, Vogel O, et al. 2010. Human constitutive androstane receptor (CAR) and pregnane X receptor (PXR) support the hypertrophic but not the hyperplastic response to the murine nongenotoxic hepatocarcinogens phenobarbital and chlordane in vivo. Toxicol Sci 116:452–466.

Rought SE, Yau PM, Guo XW, Chuang LF, Doi RH, Chuang RY. 2000. Modulation of CPP32 activity and induction of apoptosis in Human CEM X 174 lymphocytes by heptachlor, a chlorinated hydrocarbon insecticide. J Biochem Mol Toxicol 14:42–50.

Rowland RE, Edwards LA, Podd JV. 2007. Elevated sister chromatid exchange frequencies in New Zealand Vietnam War veterans. Cytogenet Genome Res 116:248–251.

Rupa DS, Rita P, Reddy PP, Reddi OS. 1988. Screening of chromosomal aberrations and sister chromatid exchanges in peripheral lymphocytes of vegetable garden workers. Hum Toxicol 7:333–336.

Rupa DS, Reddy PP, Sreemannarayana K, Reddi OS. 1991. Frequency of sister chromatid exchange in peripheral lymphocytes of male pesticide applicators. Environ Mol Mutagen 18:136–138.

Rusiecki JA, Baccarelli A, Bollati V, Tarantini L, Moore LE, Bonefeld-Jorgensen EC. 2008. Global DNA hypomethylation is associated with high serum-persistent organic pollutants in Greenlandic Inuit. Environ Health Perspect 116:1547–1552.

Saeter G, Schwarze E, Seglen O. 1988. Shift from polyploidizing to nonpolyploidizing growth in carcinogen-treated rat liver. J Natl Cancer Inst 80:950–958.

Safe S, Astroff B, Harris M, Zacharewski T, Dickerson R, Romkes M, et al. 1991. 2,3,7,8-Tetrachlorodibenzo-p-dioxin (TCDD) and related compounds as antioestrogens: Characterization and mechanism of action. Pharmacol Toxicol 69:400–409.

Samosh LV. 1974. Chromosome aberrations and the character of satellite associations following accidental exposure of the human body to polychlorcamphene. Tsitol Genet 8:24–27.

Sanchez-Alonso JA, Lopez-Aparicio P, Recio MN, Perez-Albarsanz MA. 2004. Polychlorinated biphenyl mixtures (Aroclors) induce apoptosis via Bcl-2, Bax and caspase-3 proteins in neuronal cell cultures. Toxicol Lett 153:311–326.

Sanchez-Martin FJ, Fernandez-Salguero PM, Merino JM. 2010. 2,3,7,8-Tetrachlorodibenzo-p-dioxin induces apoptosis in neural growth factor (NGF)-differentiated pheochromocytoma PC12 cells. Neurotoxicology 31:267–276.

Sandal S, Yilmaz B, Carpenter DO. 2008. Genotoxic effects of PCB 52 and PCB 77 on cultured human peripheral lymphocytes. Mutat Res 654:88–92.

Sandhu SS, Ma TH, Peng Y, Zhou XD. 1989. Clastogenicity evaluation of seven chemicals commonly found at hazardous industrial waste sites. Mutat Res 224:437–445.

Sargent L, Roloff B, Meisner L. 1989. In vitro chromosome damage due to PCB interactions. Mutat Res 224:79–88.

Sargent L, Dragan YP, Erickson C, Laufer CJ, Pitot HC. 1991. Study of the separate and combined effects of the non-planar 2,5,2′,5′- and the planar 3,4,3′,4′-tetrachlorobiphenyl in liver and lymphocytes in vivo. Carcinogenesis 12:793–800.

Sargent LM, Sattler GL, Roloff B, Xu YH, Sattler CA, Meisner L, et al. 1992. Ploidy and specific karyotypic changes during promotion with phenobarbital, 2,5,2′,5′-

tetrachlorobiphenyl, and/or 3,4,3′4′-tetrachlorobiphenyl in rat liver. Cancer Res 52:955–962.

Sargent LM, Nelson MA, Lowry DT, Senft JR, Jefferson AM, Ariza ME, et al. 2001. Detection of three novel translocations and specific common chromosomal break sites in malignant melanoma by spectral karyotyping. Genes Chromosomes Cancer 32:18–25.

Schilderman PA, Moonen EJ, Maas LM, Welle I, Kleinjans JC. 1999. Use of crayfish in biomonitoring studies of environmental pollution of the river Meuse. Ecotoxicol Environ Saf 44:241–252.

Schilderman PA, Maas LM, Pachen DM, de Kok TM, Kleinjans JC, van Schooten FJ. 2000. Induction of DNA adducts by several polychlorinated biphenyls. Environ Mol Mutagen 36:79–86.

Schrader TJ, Boyes BG, Matula TI, Heroux-Metcalf C, Langlois I, Downie RH. 1998. In vitro investigation of toxaphene genotoxicity in *S. typhimurium* and Chinese hamster V79 lung fibroblasts. Mutat Res 413:159–168.

Schulte-Hermann R, Schuppler J, Timmermann-Trosiener I, Ohde G, Bursch W, Berger H. 1983. The role of growth of normal and preneoplastic cell populations for tumor promotion in rat liver. Environ Health Perspect 50:185–194.

Semple-Roberts E, Hayes MA, Armstrong D, Becker RA, Racz WJ, Farber E. 1987. Alternative methods of selecting rat hepatocellular nodules resistant to 2-acetylaminofluorene. Int J Cancer 40:643–645.

Senthilkumar PK, Klingelhutz AJ, Jacobus JA, Lehmler H, Robertson LW, Ludewig G. 2011. Airborne polychlorinated biphenyls (PCBs) reduce telomerase activity and shorten telomere length in immortal human skin keratinocytes (HaCat). Toxicol Lett 204:64–70.

Senthilkumar PK, Robertson LW, Ludewig G. 2012. PCB153 reduces telomerase activity and telomere length in immortalized human skin keratinocytes (HaCaT) but not in human foreskin keratinocytes (NFK). Toxicol Appl Pharmacol 259:115–123.

Shekhar PV, Werdell J, Basrur VS. 1997. Environmental estrogen stimulation of growth and estrogen receptor function in preneoplastic and cancerous human breast cell lines. J Natl Cancer Inst 89:1774–1782.

Shimada T, Sato R. 1978. Covalent binding in vitro of polychlorinated biphenyls to microsomal macromolecules. Involvement of metabolic activation by a cytochrome P-450-linked mono-oxygenase system. Biochem Pharmacol 27:585–593.

Shimada T, Sato R. 1980. Covalent binding of polychlorinated biphenyls to rat liver microsomes in vitro: Nature of reactive metabolites and target macromolecules. Toxicol Appl Pharmacol 55:490–500.

Shimada T, Sawabe Y. 1984. Comparative studies on distribution and covalent tissue binding of 2,4,2′,4′- and 3,4,3′,4′-tetrachlorobiphenyl isomers in the rat. Arch Toxicol 55:182–185.

Silberhorn EM, Glauert HP, Robertson LW. 1990. Carcinogenicity of polyhalogenated biphenyls: PCBs and PBBs. Crit Rev Toxicol 20:439–496.

Sobti RC, Krishan A, Davies J. 1983. Cytokinetic and cytogenetic effect of agricultural chemicals on human lymphoid cells in vitro. II. Organochlorine pesticides. Arch Toxicol 52:221–231.

Solt DB, Medline A, Farber E. 1977. Rapid emergence of carcinogen-induced hyperplastic lesions in a new model for the sequential analysis of liver carcinogenesis. Am J Pathol 88:595–618.

Song YF, Wilke BM, Song XY, Gong P, Zhou QX, Yang GF. 2006. Polycyclic aromatic hydrocarbons (PAHs), polychlorinated biphenyls (PCBs) and heavy metals (HMs) as well as their genotoxicity in soil after long-term wastewater irrigation. Chemosphere 65:1859–1868.

Srinivasan A, Lehmler HJ, Robertson LW, Ludewig G. 2001. Production of DNA strand breaks in vitro and reactive oxygen species in vitro and in HL-60 cells by PCB metabolites. Toxicol Sci 60:92–102.

Srinivasan A, Robertson LW, Ludewig G. 2002. Sulfhydryl binding and topoisomerase inhibition by PCB metabolites. Chem Res Toxicol 15:497–505.

Steinel HH, Arlauskas A, Baker RS. 1990. SCE induction and cell-cycle delay by toxaphene. Mutat Res 230:29–33.

Stevenson DE, Walborg EF, Jr., North DW, Sielken RL, Jr., Ross CE, Wright AS, et al. 1999. Monograph: Reassessment of human cancer risk of aldrin/dieldrin. Toxicol Lett 109:123–186.

Stinchcombe S, Buchmann A, Bock KW, Schwarz M. 1995. Inhibition of apoptosis during 2,3,7,8-tetrachlorodibenzo-p-dioxin-mediated tumour promotion in rat liver. Carcinogenesis 16:1271–1275.

Stohs SJ, Al-Bayati ZF, Hassan MQ, Murray WJ, Mohammadpour HA. 1986. Glutathione peroxidase and reactive oxygen species in TCDD-induced lipid peroxidation. Adv Exp Med Biol 197:357–365.

Suzaki E, Inoue B, Okimasu E, Ogata M, Utsumi K. 1988. Stimulative effect of chlordane on the various functions of the guinea pig leukocytes. Toxicol Appl Pharmacol 93:137–145.

Swierenga SH, Yamasaki H, Piccoli C, Robertson L, Bourgon L, Marceau N, et al. 1990. Effects on intercellular communication in human keratinocytes and liver-derived cells of polychlorinated biphenyl congeners with differing in vivo promotion activities. Carcinogenesis 11:921–926.

Tampal N, Lehmler HJ, Espandiari P, Malmberg T, Robertson LW. 2002. Glucuronidation of hydroxylated polychlorinated biphenyls (PCBs). Chem Res Toxicol 15: 1259–1266.

Tebourbi O, Rhouma KB, Sakly M. 1998. DDT induces apoptosis in rat thymocytes. Bull Environ Contam Toxicol 61:216–223.

Tharappel JC, Lee EY, Robertson LW, Spear BT, Glauert HP. 2002. Regulation of cell proliferation, apoptosis, and transcription factor activities during the promotion of liver carcinogenesis by polychlorinated biphenyls. Toxicol Appl Pharmacol 179: 172–184.

Tretjak Z, Volavsek C, Beckmann SL. 1990. Structural chromosome aberrations and industrial waste. Lancet 335:1288.

Tritscher AM, Clark GC, Sewall C, Sills RC, Maronpot R, Lucier GW. 1995. Persistence of TCDD-induced hepatic cell proliferation and growth of enzyme altered foci after chronic exposure followed by cessation of treatment in DEN initiated female rats. Carcinogenesis 16:2807–2811.

Tritscher AM, Seacat AM, Yager JD, Groopman JD, Miller BD, Bell D, et al. 1996. Increased oxidative DNA damage in livers of 2,3,7,8-tetrachlorodibenzo-p-dioxin treated intact but not ovariectomized rats. Cancer Lett 98:219–225.

Trosko JE, Jone C, Chang CC. 1983. Oncogenes, inhibited intercellular communication and tumor promotion. Princess Takamatsu Symp 14:101–113.

Trosko JE, Jone C, Chang CC. 1987. Inhibition of gap junctional-mediated intercellular communication in vitro by aldrin, dieldrin, and toxaphene: A possible cellular mechanism for their tumor-promoting and neurotoxic effects. Mol Toxicol 1:83–93.

Tsuda H, Lee G, Farber E. 1980. Induction of resistant hepatocytes as a new principle for a possible short-term in vivo test for carcinogens. Cancer Res 40:1157–1164.

Tsushimoto G, Trosko JE, Chang CC, Aust SD. 1982a. Inhibition of metabolic cooperation in Chinese hamster V79 cells in culture by various polybrominated biphenyl (PBB) congeners. Carcinogenesis 3:181–185.

Tsushimoto G, Trosko JE, Chang CC, Matsumura F. 1982b. Inhibition of intercellular communication by chlordecone (kepone) and mirex in Chinese hamster v79 cells in vitro. Toxicol Appl Pharmacol 64:550–556.

Tsushimoto G, Chang CC, Trosko JE, Matsumura F. 1983. Cytotoxic, mutagenic, and cell-cell communication inhibitory properties of DDT, lindane, and chlordane on Chinese hamster cells in vitro. Arch Environ Contam Toxicol 12:721–729.

Turusov V, Rakitsky V, Tomatis L. 2002. Dichlorodiphenyltrichloroethane (DDT): Ubiquity, persistence, and risks. Environ Health Perspect 110:125–128.

Uchida K, Shiraishi M, Naito Y, Torii Y, Nakamura Y, Osawa, T. 1999. Activation of stress signaling pathways by the end product of lipid peroxidation. 4-hydroxy-2-nonenal is a potential inducer of intracellular peroxide production. J Biol Chem 274:2234–2242.

Ulland BM, Page NP, Squire RA, Weisburger EK, Cypher RL. 1977. A carcinogenicity assay of Mirex in Charles River CD rats. J Natl Cancer Inst 58:133–140.

Uppala PT, Roy SK, Tousson A, Barnes S, Uppala GR, Eastmond DA. 2005. Induction of cell proliferation, micronuclei and hyperdiploidy/polyploidy in the mammary cells of DDT- and DMBA-treated pubertal rats. Environ Mol Mutagen 46:43–52.

Valavanidis A, Vlahogianni T, Dassenakis M, Scoullos M. 2006. Molecular biomarkers of oxidative stress in aquatic organisms in relation to toxic environmental pollutants. Ecotoxicol Environ Saf 64:178–189.

Valic E, Jahn O, Papke O, Winker R, Wolf C, Rudiger WH. 2004. Transient increase in micronucleus frequency and DNA effects in the comet assay in two patients after intoxication with 2,3,7,8-tetrachlorodibenzo- p-dioxin. Int Arch Occup Environ Health 77:301–306.

Vigfusson NV, Vyse ER, Pernsteiner CA, Dawson RJ. 1983. In vivo induction of sister-chromatid exchange in Umbra limi by the insecticides endrin, chlordane, diazinon and guthion. Mutat Res 118:61–68.

Wade MG, Foster WG, Younglai EV, McMahon A, Leingartner K, Yagminas A, et al. 2002. Effects of subchronic exposure to a complex mixture of persistent contaminants in male rats: Systemic, immune, and reproductive effects. Toxicol Sci 67:131–143.

Wangpradit O, Mariappan SV, Teesch LM, Duffel MW, Norstrom K, Robertson LW, et al. 2009. Oxidation of 4-chlorobiphenyl metabolites to electrophilic species by prostaglandin H synthase. Chem Res Toxicol 22:64–71.

Welsch CW. 1995. Review of the effects of dietary fat on experimental mammary gland tumorigenesis: Role of lipid peroxidation. Free Radic Biol Med 18:757–773.

Whysner J, Wang CX. 2001. Hepatocellular iron accumulation and increased cell proliferation in polychlorinated biphenyl-exposed Sprague-Dawley rats and the development of hepatocarcinogenesis. Toxicol Sci 62:36–45.

Whysner J, Montandon F, McClain RM, Downing J, Verna LK, Steward RE, 3rd, et al. 1998. Absence of DNA adduct formation by phenobarbital, polychlorinated biphenyls, and chlordane in mouse liver using the 32P-postlabeling assay. Toxicol Appl Pharmacol 148:14–23.

Wiesner G, Wild KJ, Gruber M, Lindner R, Taeger K. 2000. A cytogenetic study on the teaching staff of a polluted school with a questionable increased incidence of malignancies. Int J Hyg Environ Health 203:141–146.

Williams GM, Numoto S. 1984. Promotion of mouse liver neoplasms by the organochlorine pesticides chlordane and heptachlor in comparison to dichlorodiphenyltrichloroethane. Carcinogenesis 5:1689–1696.

Worner W, Schrenk D. 1996. Influence of liver tumor promoters on apoptosis in rat hepatocytes induced by 2-acetylaminofluorene, ultraviolet light, or transforming growth factor beta 1. Cancer Res 56:1272–1278.

Worner W, Schrenk D. 1998. 2,3,7,8-Tetrachlorodibenzo-p-dioxin suppresses apoptosis and leads to hyperphosphorylation of p53 in rat hepatocytes. Environ Toxicol Pharmacol 6:239–247.

Wyndham C, Devenish J, Safe S. 1976. The in vitro metabolism, macromolecular binding and bacterial mutagenicity of 4-chloribiphenyl, a model PCB substrate. Res Commun Chem Pathol Pharmacol 15:563–570.

Xie W, Wang K, Robertson LW, Ludewig G. 2010. Investigation of mechanism(s) of DNA damage induced by 4-monochlorobiphenyl (PCB3) metabolites. Environ Int 36:950–961.

Yang C, Chen S. 1999. Two organochlorine pesticides, toxaphene and chlordane, are antagonists for estrogen-related receptor alpha-1 orphan receptor. Cancer Res 59:4519–4524.

Yang JH, Lee HG. 2010. 2,3,7,8-Tetrachlorodibenzo-p-dioxin induces apoptosis of articular chondrocytes in culture. Chemosphere 79:278–284.

Yarbrough J, Cunningham M, Yamanaka H, Thurman R, Badr M. 1991. Carbohydrate and oxygen metabolism during hepatocellular proliferation: A study in perfused livers from mirex-treated rats. Hepatology 13:1229–1234.

Yoshida R, Ogawa Y. 2000. Oxidative stress induced by 2,3,7,8-tetrachlorodibenzo-p-dioxin: An application of oxidative stress markers to cancer risk assessment of dioxins. Ind Health 38(1):5–14.

Zangar RC, Davydov DR, Verma S. 2004. Mechanisms that regulate production of reactive oxygen species by cytochrome P450. Toxicol Appl Pharmacol 199:316–331.

Zeiger E, Anderson B, Haworth S, Lawlor T, Mortelmans K. 1992. Salmonella mutagenicity tests: V. Results from the testing of 311 chemicals. Environ Mol Mutagen 19(Suppl 21):2–141.

Zettner MA, Flor S, Ludewig G, Wagner J, Robertson LW, Lehmann L. 2007. Quinoid metabolites of 4-monochlorobiphenyl induce gene mutations in cultured Chinese hamster v79 cells. Toxicol Sci 100:88–98.

Zhao S, Narang A, Ding X, Eadon G. 2004. Characterization and quantitative analysis of DNA adducts formed from lower chlorinated PCB-derived quinones. Chem Res Toxicol 17:502–511.

Zober A, Messerer P, Ott MG. 1997. BASF studies: Epidemiological and clinical investigations on dioxin-exposed chemical workers. Teratog Carcinog Mutagen 17:249–256.

CHAPTER 5

Diabetes and the Metabolic Syndrome

DUK-HEE LEE and DAVID R. JACOBS, JR.

ABSTRACT

Background: Type 2 diabetes (T2D) is a chronic disease that has increased at an alarming rate worldwide. Even though there is a common belief that both an excess of dietary calories and a lack of exercise have led to obesity and finally the diabetes epidemic, compelling experimental and epidemiological evidence has recently emerged that links environmental chemicals with T2D.

Objectives: Among various possible chemicals, persistent organic pollutants (POPs) have recently become a focus due to their strong epidemiological evidence. We summarize the currently available evidence, focusing on key epidemiological studies, and discuss implications for preventing and controlling T2D.

Discussion: The earliest evidence linking exposure to POPs with diabetes came from a series of studies on 2,3,7,8-tetrachlorodibenzo-*p*-dioxin (TCDD) among U.S. Air Force veterans involved in spraying defoliants during the Vietnam War. However, in the general population, organochlorine pesticides or polychlorinated biphenyls (PCBs) have shown much stronger associations in many cross-sectional studies. Recent prospective studies mostly confirmed cross-sectional findings, although the specific kinds of POPs predicting T2D and the shapes of the dose–response curves varied across studies. Interestingly, in one cross-sectional study, obesity was not associated with T2D among persons with very low levels of POPs, suggesting that the POPs that have accumulated in adipose tissue, rather than the adiposity itself, play a critical role in the pathogenesis of T2D.

Conclusions: Based on current available evidence, the background exposure to POPs in the general population appears to be causally linked to the development of T2D. Any efforts to reduce the exposure to POPs may be needed to decrease the burden of T2D.

Effects of Persistent and Bioactive Organic Pollutants on Human Health, First Edition.
Edited by David O. Carpenter.

INTRODUCTION

Diabetes is a metabolic disease in which glucose homeostasis is lost. Often, diabetes results when the body does not adequately produce or properly use insulin, eventually leading to increased circulating glucose concentrations. Insulin is a hormone produced by pancreatic beta cells that is critical in glucose homeostasis. It is necessary for glucose to move into cells. The pancreas sensitively controls the amount of insulin depending on the level of glucose in the blood, thereby maintaining blood glucose levels within the physiological range. When this process breaks down, blood glucose levels become high and diabetes develops.

There are two major types of diabetes. In type 1 diabetes (T1D), the pancreas produces very little or no insulin, often due to autoimmune destruction of pancreatic beta cells. T1D is common during childhood and adolescence. In type 2 diabetes (T2D), the pancreas can produce insulin but not as much as required. T2D, particularly when preceded by obesity or metabolic syndrome, appears to be closely linked to insulin resistance, meaning that insulin-sensitive tissues do not respond well to insulin. Thus, such tissue needs more insulin to maintain proper glucose levels. When insulin sensitivity becomes low, pancreatic beta-cell mass and function adapt to this situation to maintain glucose homeostasis by overproducing insulin, which may avoid the development of T2D. However, T2D develops when the capacity for producing enough insulin to counteract this insulin-resistant state is impaired. A small proportion of T2D occurs without obesity or metabolic syndrome, suggesting a different etiology. The majority of diabetes is T2D in association with obesity or the metabolic syndrome. Even though T2D was once considered a disease of older adults, its prevalence is now increasing in children. Another type is gestational diabetes (GD), which means loss of glucose homeostasis during pregnancy in a woman without T1D or T2D. Because pregnancy is, in general, an insulin-resistant state, GD likely has similar etiology to T2D, usually disappearing after the birth of the baby when relative insulin sensitivity returns. However, GD increases the risk for developing T2D later in life (Gunderson et al. 2007).

In all types of diabetes, the glucose cannot move into the cells well and blood glucose levels become high. As the current evidence on persistent organic pollutants (POPs) and diabetes has been mostly in relation to T2D, often with concurrent study of insulin resistance, we focus mainly on T2D in this chapter.

WHY IS THERE AN EPIDEMIC OF T2D?

T2D is a chronic disease that has increased at an alarming rate worldwide. As reviewed by Crinnion (2011), a study in 2004 estimated that 171 million persons (2.8%) for all age groups had diabetes worldwide and predicted that number as 366 million (4.4%) in 2030 (Wild et al. 2004). However, the situation seems to be even more serious than what was expected. A more recent study in 2010 estimated the prevalence of T2D as 6.4% in adults aged 20–79 and predicted the number affected as 439 million (7.7%) in 2030 (Shaw et al. 2010). Furthermore, the prevalence of T2D is expected to increase more rapidly in developing countries (69% increase) compared with developed countries (20% increase) during the following two decades (Shaw et al. 2010). In addition, T2D has traditionally been rare in childhood and in adolescence but has increased alarmingly in this age group.

The pathogenesis of T2D involves multitudes of both genetic and environmental factors that can adversely affect response to insulin and insulin action. Although several genes have been identified on some chromosomes that predispose to T2D development, genetic factors cannot explain the rapid increases of T2D during recent decades. The human gene pool has remained stable during this period, indicating that environmental factors may play a significant role in the development of T2D.

There is a common belief that lifestyle changes, in particular, an excess of dietary calories and fat and a lack of exercise, characteristic of the second half of the 20th century, have led to obesity and finally the diabetes epidemic. Despite a strong dogma on the role of obesity in the development of T2D, trends of obesity versus T2D are discrepant (Table 5.1, Figure 5.1). For example, the prevalence of obesity in the United States is about 10 times higher than in Asian countries, but the prevalence of T2D in the United States is not higher than those in most Asian countries (Yoon et al. 2006). At present, rapid change of lifestyles and a strong genetic susceptibility to T2D in Asians, characterized by early pancreatic beta-cell failure and prominent central obesity, are considered to be the main causes of this discrepancy (Yoon et al. 2006). However, recent evidence on various environmental chemicals in the development of obesity, insulin resistance, and T2D suggests a role of other environmental factors, which may explain the discrepancy of trends between obesity and T2D.

Notably, not every individual with obesity has insulin resistance, a phenotype usually named metabolically healthy obese. By contrast, humans with normal weight that have serious metabolic problems, including insulin resistance and T2D, have a phenotype referred to as metabolically obese normal weight. These two opposite groups are common: 31.7% of U.S. obese adults are metabolically healthy obese and 23.5% of normal-weight adults are metabolically abnormal normal weight (Wildman et al. 2008). These findings also suggest the possible presence of other T2D risk factors in addition to obesity.

HUMAN EVIDENCE LINKING POPs AND T2D

During the last decade, compelling experimental and epidemiological evidence has emerged that links environmental chemicals with obesity, insulin resistance, and T2D (Thayer et al. 2012). As obesity will be dealt with in another chapter, we focus on T2D here, except when the discussion on obesity

TABLE 5.1. Prevalence of Obesity and Type 2 Diabetes in 30 Countries

Countries	Prevalence[a] of Obesity (BMI ≥ 30 kg/m^2) (%)	Prevalence[b] of Type 2 Diabetes (%)
United States	33.8	10.3
Mexico	30.0	10.8
New Zealand	26.5	5.2
Ireland	25.0	5.2
Australia	24.5	5.7
Canada	24.2	9.2
United Kingdom	23.5	3.6
Iceland	20.1	1.6
Luxembourg	20.0	5.3
Hungary	18.8	6.4
Greece	18.1	6.0
Spain	17.1	6.6
Czech Republic	17.0	6.4
Slovak Republic	16.9	6.4
Germany	16.0	8.9
Finland	15.7	5.7
Portugal	15.4	7.7
Turkey	15.2	8.0
Belgium	13.8	5.3
Poland	12.5	7.6
Austria	12.4	8.9
The Netherlands	11.8	5.3
Denmark	11.4	5.6
France	11.2	6.7
Sweden	10.2	5.2
Norway	10.0	3.6
Italy	9.9	5.9
Switzerland	8.1	8.9
Korea	3.8	7.9
Japan	3.4	5.0

Estimates of the number of persons aged 20–79 years old living with diabetes and crude and age-standardized rates (World Standard Population) of diabetes prevalence in OECD countries in 2010, as compiled by the International Diabetes Federation.

[a]*Source*: http://www.noo.org.uk/uploads/doc799_2_International_Comparisons_Obesity_Prevalence2.pdf.

[b]*Source*: http://www.ecosante.org/OCDEENG/68.html.

is necessary for this chapter. Chemicals that are related to T2D can be classified into obesogens and diabetogens, recognizing, however, that some chemicals may be in both groups and that the effect of chemicals may be modified in mixtures. First, obesity-inducing chemicals (obesogens) can theoretically be diabetogens because obesity leads to insulin resistance and many persons with insulin resistance develop T2D with the exhaustion of pancreatic beta cells.

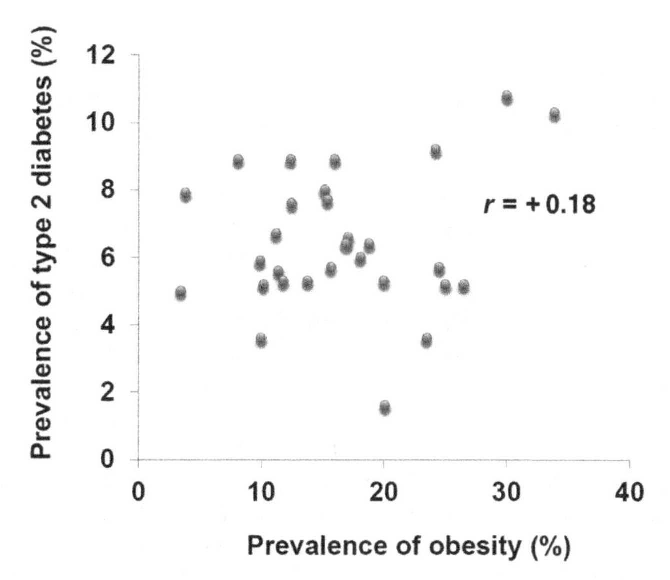

Figure 5.1. Ecologic correlation coefficient between prevalence of obesity (BMI $> 30\,kg/m^2$) and type 2 diabetes among 30 Organization for Economic Cooperation and Development (OECD) countries.

Second, even without clear obesogenic effects, certain chemicals can be diabetogenic if they directly affect pancreatic beta cells.

Among various possible chemicals, background exposure to POPs has recently become a focus in the context of their possible connection with T2D due to their strong epidemiological evidence. This book chapter is not a systemic review on POPs and T2D. We generally summarize the currently available evidence, focusing on key epidemiological studies, and discuss implications for preventing and controlling T2D. Molecular mechanisms in relation to these associations are dealt with in the following chapter.

We classify human exposure to POPs into two types. The first one is accidental or occupational exposure to POPs, while the second one is environmental exposure to POPs in the general population. Compared with high-dose exposure during adulthood to one or several selected POPs in occupational settings, background exposure in the general population is both low dose and long term, mostly throughout the lifetime, and exposure is to various POP mixtures.

EARLIER EVIDENCE

The earliest evidence on POPs and T2D came from 2,3,7,8-tetrachlorodibenzo-*p*-dioxin (TCDD), the most toxic POP from the viewpoint of traditional toxicology. Exposure to TCDD in occupational or accidental settings has been associated with an increased risk of T2D, modified glucose metabolism, or insulin resistance in some studies, but evidence is not completely consistent (Bertazzi et al. 1998; Henriksen et al. 1997; Longnecker and Michalek 2000; Remillard and Bunce 2002; Steenland et al. 1999; Sweeney et al. 1997; Vena et al. 1998; Zober et al. 1994).

The most extensive study was performed among U.S. Air Force veterans of Operation Ranch Hand, the unit responsible for the aerial spraying of herbicides from 1962 to 1971 during the Vietnam War. Among the herbicides was Agent Orange, which included 2,3,7,8-TCDD as a constituent. Ranch Hand veterans were exposed to herbicides during flight operations and maintenance of the aircraft and herbicide spray equipment. The study compared the current health and cumulative mortality experience of Ranch Hand veterans with that of a comparison group of other Air Force veterans who served in Southeast Asia during the same period that the Ranch Hand unit was active and who were not involved with spraying herbicides. The study included periodic analyses of noncombat mortality, in-person interviews, and physical examinations. In 1987, blood from willing participants was collected and assayed for dioxin. The initial dioxin dose at the end of the duty in Vietnam in Ranch Hand was estimated extrapolating backward using a constant half-life of 8.7 years.

Compared with other U.S. Air Force veterans without exposure to Agent Orange, exposed veterans had about three times higher serum dioxin level in 1987, with 40% higher risk of fasting or 2-hour postprandial glucose abnormality and 50% higher risk of T2D, suggesting an adverse relation between dioxin exposure and T2D (Henriksen et al. 1997). Based on the combined evidence, the Department of Veterans Affairs in the United States added T2D to the list of diseases presumptively associated with exposure to dioxin-containing Agent Orange in Vietnam (IOM 2000).

On the other hand, the relation between dioxin and T2D was also examined only among the comparison group to explore if background-level exposure might have adverse effects on insulin–glucose metabolism considering the high toxicity of dioxin (Longnecker and Michalek 2000). These study subjects never had contact with dioxin-contaminated herbicides and had serum dioxin levels within the range of background exposure typically seen in the U.S. general population. Interestingly, dose–response relations between dioxin and T2D tended to be clearer when study subjects were restricted to the comparison group (Longnecker and Michalek 2000) than in earlier studies including Ranch Hand veterans with high levels of TCDD, which used the experience in the comparison group without differentiation of within comparison group gradients in diabetes risk by dioxin levels as the reference (Henriksen et al. 1997).

A molecular epidemiological study observed relations between the GLUT4:NFkB ratio, a biomarker for the diabetogenic action of dioxin, and serum dioxin levels (Fujiyoshi et al. 2006). In this study, the relations were clearly observed in the comparison group without the exposure to Agent Orange rather than exposed group. The correlation in the comparison group was particularly significant among those with known risk factors such as obesity and family history of diabetes (Fujiyoshi et al. 2006). All these findings suggested the possibility of a diabetogenic effect at low dioxin dose and, therefore, indicated the necessity for epidemiological studies in the general population, in particular, in relation to known risk factors for T2D.

The accident that occurred in a chemical plant near the town of Seveso, Italy, in 1976 also provided an opportunity to examine the association between exposure to 2,3,7,8-TCDD and T2D (Bertazzi et al. 1998). Although the interpretation was limited because they used death due to T2D as the study outcome, one interesting finding was that residents living in the medium-exposure area showed a higher risk of T2D mortality than those living in the high-exposure area (Bertazzi et al. 1998). In fact, this evidence for a nonmonotonic dose–response relation seems to be consistent with the clearer dose–response relation observed in U.S. Air Force veterans with lower than with higher exposure to TCDD.

There were also some studies that evaluated the health effects of TCDD among workers with occupational exposures in factories with comparison groups consisting of general populations with background exposure levels (Calvert et al. 1999; Sweeney et al. 1997; Zober et al. 1994). Even though their dioxin levels tended to be higher than those of Ranch Hand veterans, results were inconsistent and there was no clear dose–response relation.

Besides TCDD, there were epidemiological studies among polychlorinated biphenyl (PCB)-exposed workers in occupational settings. However, they revealed no suggestion of an increased level of glucose nor of morbidity or mortality from T2D (Emmett 1985; Kimbrough et al. 2003; Lawton et al. 1985). Therefore, even though one cross-sectional study on PCBs and organochlorine (OC) pesticides observed positive associations with T2D in the general population (Glynn et al. 2003), these findings were mainly interpreted as results of altered toxicokinetics of lipid-soluble POPs by diabetes-related changes in metabolizing capacity and alterations in lipid distribution in the body. There seemed to be a strong belief that it would be impossible to show any causal association of POPs and T2D in the general population with low-dose exposure in the face of occupationally highly exposed workers who did not show any excess propensity for T2D.

RECENT EVIDENCE: CROSS-SECTIONAL STUDIES

Despite some evidence on the link between dioxin and T2D, for a long time, POPs such as OC pesticides or dioxin-unlike PCBs received little consideration as possible risk factors for T2D. It may be that this focus on dioxin was the result of the interpretation that dioxin shows most of its harmful effects, including diabetes, through binding with the aryl hydrocarbon receptor (AhR), an intracellular ligand-dependent transcriptional factor expressed in most tissues of mammals (Remillard and Bunce 2002). As OC pesticides and dioxin-unlike PCBs have no affinity to AhR, they were not suspected to be as harmful as dioxin. In addition, as the strength of association between TCDD and T2D was only weak or modest among persons with high exposure, the traditional toxicologic monotonicity assumption ("the dose makes the poison") implied that any relation with T2D of POPs, like OC pesticides, and PCBs would be smaller than that of TCDD and therefore, would be negligible.

However, over the past several years, several epidemiological studies of associations of diabetes with OC pesticides and PCBs have been carried out. Unlike the early studies of dioxin discussed earlier, most studies were performed in the general population who were not exposed to these chemicals in occupational or accidental settings. As the production and use of most OC pesticides and PCBs were banned several decades ago in most developed countries, their absolute levels in recent epidemiological studies were very low compared with those in earlier times, despite substantial variation among countries reflecting different lifestyles and geographical variation of the environmental contamination by these chemicals. However, we point out that recent populations have a long duration of exposure compared to populations in earlier times.

Evidence on OC pesticides and PCBs has been reported from many countries including the United States, Sweden, Finland, Japan, Korea, Taiwan, and the Slovak Republic (Airaksinen et al. 2011; Lee et al. 2006a; Rignell-Hydbom et al. 2007, 2009; Son et al. 2010; Tanaka et al. 2011; Turyk et al. 2009a,b; Ukropec et al. 2010; Vasiliu et al. 2006; Wang et al. 2008). In this section, we will first discuss the findings from cross-sectional studies. Even though most cross-sectional studies have similarly reported positive associations, details differed, especially in terms of the kinds of POPs that were statistically significant. Later, we will discuss why we have elected not to focus on individual POPs in epidemiological studies of POPs. Therefore, we will focus on several noteworthy findings graphically rather than listing specific findings study by study. However, the lack of association observed in the Greenland Inuit population will be discussed in detail (Jorgensen et al. 2008) because this study provides a unique opportunity to evaluate the epidemiological findings on POPs in a high-exposure general population.

The most compelling evidence that POPs are associated with T2D was found in the U.S. general population using the 1999–2002 dataset from the U.S. National Health and Examination Survey (NHANES) (Lee et al. 2006a). NHANES is designed to be nationally representative of the noninstitutionalized U.S. civilian population and their datasets are open to the public. From the viewpoint of epidemiological research, the quality of this dataset is exemplary.

In this study, out of 49 POPs assayed, 6 kinds of POPs (2 dioxins, 1 PCB, and 3 metabolites of OC pesticides) that were detected in ≥80% of participants in this U.S. general sample were included in analyses. After adjusting for known risk factors including obesity, the prevalence of T2D was strongly positively associated with serum concentrations of all six POPs (Figure 5.2). The adjusted odds ratios of *trans*-nonachlor, oxychlordane, 2,2-bis(*p*-chlorophenyl)-

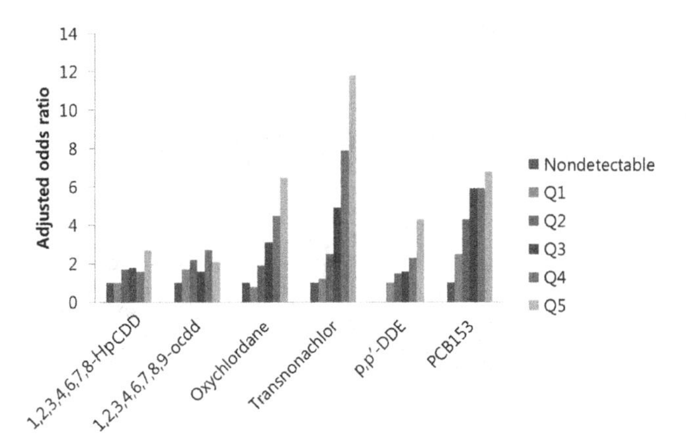

Figure 5.2. Associations of type 2 diabetes with six persistent organic pollutants (POPs), which were most commonly detectable in the U.S. general population. Adjusted for age, race, sex, socioeconomic status, body mass index, and waist circumference (Lee et al. 2006a). Each series of bars goes from nondetectable, farthest to the left, to Q5 farthest to the right. HpCDD, 1,2,3,4,6,7,8-heptachlorodibenzo-p-dioxin; OCDD, 1,2,3,4,6,7,8,9-octachlorodibenzo-p-dioxin.

1,1-dichloroethylene (DDE), and PCB 153 comparing the highest to lowest exposure category were 6.5, 11.8, 4.3, and 6.8, respectively. When the participants were classified by a summary measure (sum of ranks of the six POPs), the risk of T2D was about 10–40 times higher as the summary measure levels increased compared with subjects belonging to the lowest quartile of the summary measure. The associations were stronger among younger persons and Mexican Americans.

Interestingly, in this study, the strengths of association were much stronger with OC pesticides like *trans*-nonachlor, oxychlordane, or p,p′-DDE and PCB 153 rather than dioxins. In addition, toxic equivalents (TEQs), measures of the ability to bind to the AhR, of each POP were not correlated with the strength of association. This suggested that binding to AhR might not be the only critical pathway linking POPs and T2D and that long-term exposure to OC pesticide and PCBs at background levels might be even more harmful than dioxin, at least in relation to T2D. Considering a long-held belief that dioxin is the most dangerous chemical on earth, all these findings were unexpected.

Another important finding was the interaction between POPs and obesity on the risk of T2D (Figure 5.3). The association between POPs and T2D was stronger among the more obese persons. There was a clear positive association between POPs and T2D among normal-weight persons with body mass index (BMI) $< 25\,kg/m^2$. However, obesity was not associated with T2D among persons with very low serum concentrations of POPs in this study, and T2D itself was very rare even among persons with BMI $\geq 30\,kg/m^2$ who had a low POP summary score. Interpreting these findings literally would mean that the

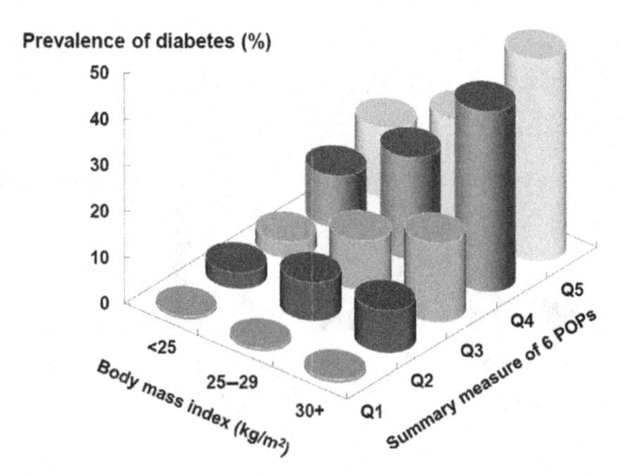

Figure 5.3. Interaction between body mass index (BMI) and persistent organic pollutants (POPs) on the prevalence of type 2 diabetes (Lee et al. 2006a). The summary measure of six POPs was calculated by summing the individual rank of six POPs in Figure 5.2 (1,2,3,4,6,7,8-heptachlorodibenzo-p-dioxin, 1,2,3,4,6,7,8,9-octachlorodibenzo-p-dioxin, p,p′-DDE, oxychlordane, *trans*-nonachlor, and PCB 153). The summary measure was classified into five quintiles from Q1 to Q5. Among persons with the lowest quintile (Q1) of the summary POPs, BMI was not associated with the risk of T2D and T2D itself was very rare even among obese. In addition, the risk of diabetes increased with increasing concentrations of POPs even among lean persons.

POPs accumulated in adipose tissue may play a more critical role in the pathogenesis of T2D than the adipose tissue itself (Lee et al. 2006b; Porta 2006).

In the following study using the same NHANES dataset (Lee et al. 2007a), associations between serum concentrations of POPs and insulin resistance were investigated among nondiabetics. As increased insulin resistance normally precedes the onset of T2D by one to two decades and most patients with T2D are insulin resistant, if the strong associations of POPs with T2D shown in the previous study (Lee et al. 2006a) were causal, the authors thought it would be reasonable to expect similar relations between POPs and insulin resistance even in the prediabetic stage. POPs belonging to OC pesticides or non-dioxin-like PCBs showed significant positive associations with insulin resistance, while POPs with dioxin activity were not clearly associated with insulin resistance.

There had been 20 or more cross-sectional studies on this issue by the end of 2011. The findings from the NHANES dataset were mostly replicated in other cross-sectional studies, although the strengths of association differed somewhat across studies. Also, the particular POPs that had statistically significant associations varied to some extent across studies. However, it is difficult to know which specific POP is really problematic in human studies. These lipophilic chemicals, which move together both in the environment and in the body, and serum concentrations of most POPs are highly correlated with each other in the general population. Therefore, the specific POPs that emerged with significant relationships in epidemiological studies may not identify the real culprit. There is a possibility that a mixture of all POPs or certain kinds of POPs may be the real problem rather than certain individual POPs. Also, the issue that will be discussed in the section "Mismatch of Time Trend: Issues Related to Inverted U-Shaped Associations" will influence specific results among various studies.

Even though most cross-sectional studies were not large enough to test the interaction between POPs and obesity observed in the NHANES dataset, one recent study from Finland, like the first NHANES study (Lee et al. 2006a), demonstrated that the positive associations of T2D with OC pesticides like oxychlordane and *trans*-nonachlor were observed only among the overweight participants, while high POPs were not related to T2D among normal-weight participants (Airaksinen et al. 2011). Additionally, they observed that, among individuals with low POP exposure, overweight and obesity did not seem to increase the prevalence of T2D as much as among individuals with high POP exposure. These findings were in line with the earlier findings from the NHANES dataset (Lee et al. 2006a).

The most provocative aspect of the interaction between POPs and obesity is the absence of association between obesity and T2D among persons with very low serum concentrations of POPs. Taken literally, such interaction would imply a need for a drastic change in the current paradigm on T2D. A definitive study of this issue would be difficult, requiring a large number of study subjects with very low serum concentrations of POPs across a wide range of adiposity levels. Future studies should focus on a less demanding hypothesis, namely, whether obesity has a less profound influence on T2D risk if POPs levels are low than if they are high.

In addition to POPs, other chemicals like arsenic or bisphenol A are also suspected to have links to T2D. Analyses of associations between these other chemicals and T2D have been performed using the NHANES dataset. Although most studies have focused on one or several chemicals by different researchers, one study performed multiple cross-sectional analyses associating 266 unique environmental factors with T2D using an environment-wide association study (EWAS), which is conceptually similar to a genome-wide association study (GWAS) (Patel et al. 2010). They demonstrated that OC pesticides and PCBs showed the most significant associations for various chemicals, suggesting the importance of these chemicals in relation to T2D when examined in the same analyses as other suspect chemicals.

It is worthwhile to review in detail one cross-sectional study that was performed among the Greenland Inuits. They are highly exposed to POPs due to a high intake of marine mammals and have experienced a rapid increase in T2D prevalence over the last 30 years (Jorgensen et al. 2008). In this study, the average body burden of some POPs like PCBs and chlordanes was about 10–12 times higher than those found in the current U.S. general population and DDE concentration was twice as high (Lee et al. 2006a). Compared to the Swedish Baltic Sea fishermen, PCB 153 was five times higher and DDE was twice as high (Rylander et al. 2005). However, no associations were found between POPs and glucose intolerance, markers of insulin resistance, or T2D.

As Greenland Inuits have low levels of POPs with high affinity to AhR, despite having high levels of AhR-unrelated POPs, the authors interpreted that low AhR-associated xenobiotic activity of POPs may explain why they found no association between POPs and glucose intolerance among the Inuit. They also suggested that the potential negative health effect of POPs is counterbalanced by a beneficial dietary factor, especially n-3 fatty acids from fish and marine mammals. They cited that a higher proportion of n-3 fatty acids can improve insulin signaling by increasing membrane fluidity.

However, we suggest another possibility, that the lack of observed association of POPs with glucose intolerance may be related to the low-dose effects of POPs typical of endocrine disruptors, as discussed in detail in the section "Mismatch of Time Trend: Issues Related to Inverted U-Shaped Associations." If the persons with the highest risk of T2D were those who had relatively low levels of exposure among the Inuit population, whose levels were nevertheless much higher than truly low-risk levels in populations such as NHANES, there would be no true reference group, and it would be difficult to observe a monotonically increasing pattern of association at the individual level even though the prevalence of T2D increased at the population level. This perspective on the lack of association in the Inuit population is closely related to our view on the earlier findings about dioxin, which showed a clear dose–response relation in the study of U.S. veterans with background exposure to TCDD rather than those with high exposure to TCDD (Longnecker and Michalek 2000).

On the other hand, even though there was no association between POPs and T2D among the Inuit population, POPs were significantly inversely associated with stimulated insulin concentrations and homeostasis model assessment of beta-cell function, markers related to insulin secretion. This suggests that relatively high-dose POPs may affect insulin secretion more than it affects other aspects of the pathogenesis of insulin resistance. Under the heading "Human Evidence Linking POPs and T2D," we discussed that certain chemicals can be diabetogenic if they directly affect pancreatic beta cells even without any effect on insulin resistance. Thus, there is a possibility that the predominant role of POPs may differ depending on their doses. A low dose of POPs may affect insulin resistance more, while a high dose of POPs may affect pancreatic beta cells relatively more.

RECENT EVIDENCE: PROSPECTIVE STUDIES

Since the publications of strong cross-sectional associations of OC pesticides or PCBs with T2D, several prospective studies have been published (Lee et al. 2010, 2011a; Rignell-Hydbom et al. 2007; Turyk et al. 2009a; Vasiliu et al. 2006). In cross-sectional studies, even those that show strong associations, there is a concern about reverse causality, namely, that T2D might alter the metabolism of POPs as to slow their removal from the body, that is, that T2D might cause high levels of POPs, rather than POPs leading to T2D. This possibility is made less likely by findings in two studies that revealed no change in the rate of elimination of POPs from blood in relation to the status of T2D or the duration of T2D (Michalek et al. 2003; Turyk et al. 2009a). Still, clear evidence from prospective studies would further reduce the possibility of reverse causality.

Broadly speaking, all prospective studies supported the cross-sectional findings; in each study, some OC pesticides and PCBs predicted the future risk of T2D, although the specific POPs that predicted T2D and the shapes of the dose–response curves varied across studies. We briefly summarize the findings from the six prospective studies that were performed in general populations and were published in journals by the end of 2011.

Driven in part by past limitations in analytic technology, many of the cohort studies measured one or several selected POPs as a surrogate marker of all POPs, for example, p,p′-DDE, PCB 153, or total PCBs. With improved technology to detect POPs in relatively small amounts of frozen blood, recent cohort studies have measured a variety of POPs regardless of the starting date of the cohort study itself. As we discussed earlier under the heading "Recent Evidence: Cross-Sectional Studies," there are strong correlations among POPs. Therefore, an association with one particular POP does not necessarily implicate that chemical in a causal pathway; a given chemical may merely mark a different causal compound in both the earlier and more recent cohort studies. Another caveat is that the absolute levels of POPs differed substantially among studies, in particular, depending on the calendar time of measurement.

Three prospective studies were from the United States. The first one was performed in a Michigan cohort aged 20 years and over (two-thirds aged 20–45 years) who were suspected to consume animal products mistakenly contaminated with a fire-retardant product containing polybrominated biphenyls (PBBs) in 1973 (Vasiliu et al. 2006). In this study, in addition to PBB 153 (the most abundant congener among more than 200 PBB congeners), total PCBs were also measured. During the 25 years of follow-up, PBB 153 was not associated with incident T2D, but increased total PCB concentration was associated with diabetes in women, though the somewhat weaker positive trend in men did not reach statistical significance.

Another cohort study consisted of Great Lakes sport fish consumers, average age about 50 years, who were the object of study in part because they consume fish more than the average U.S. resident (Turyk et al. 2009b). In this study, the risk of developing T2D during 10 years (from 1994–1995 to 2005) was associated with p,p′-DDE exposure, but not total PCBs or separately examined PCB 118 exposure. Total PCBs were the sum of congeners 74, 99, 118, 146, 180, 194, 201, 206, 132/153, 138/163, 170/190, 182/187, and 196/200; however, findings for individual PCBs (except PCB 118) were not presented.

The third cohort study was performed among young adults who were aged 20–32 at baseline within the Coronary Artery Risk Development in Young

Adults (CARDIA) cohort (Lee et al. 2010). This study had a nested case-control design. Study subjects were 90 incident cases who were diabetes free in 1987–1988 but developed diabetes by 2005–2006 and 90 controls who remained free of diabetes throughout the follow-up. This study was unique because a variety of POPs (8 OC pesticides, 22 PCBs, and 1 PBB) were measured compared to previous prospective studies. Even though they used serums collected in 1987~1988, the chemicals were recently measured using an advanced technology. Therefore, precise measurements of some POPs with very low concentrations were possible.

In this study, POPs showed nonlinear associations with diabetes risk. The highest risk was observed in the second quartiles of *trans*-nonachlor, oxychlordane, mirex, highly chlorinated PCBs, and PBB 153. An important point was that the findings suggested clear low-dose effects. When serum concentrations of POPs were categorized by quartiles, the highest risk was observed in the second quartile, not the highest quartile, making the inverted U-shaped dose–response curve. The authors formed a summary measure (sum of ranks of these POPs) and categorized study subjects by sextiles, not conventional quartiles, to isolate subjects with very low concentrations of multiple POPs in the lowest sextile of the summary measure. As the authors hypothesized, the strength of association became stronger with the sextile approach than the quartile approach. Also, the association of the summary measure with future diabetes strengthened when study subjects were restricted to obese persons with BMI $\geq 30\,kg/m^2$.

Two prospective studies were performed in Sweden. One was performed among the general female population, aged 50–59 at baseline. The risk of T2D that developed at least 7 years after the baseline was significantly higher among individuals with high baseline exposure to p,p′-DDE and not significantly higher PCB 153, compared with individuals with low exposure. The authors interpreted this to mean that high exposure may predispose individuals to develop T2D later in life (Rignell-Hydbom et al. 2009).

The other prospective study was based among elderly persons who were aged 70 years at baseline with information on a battery of POPs similar to that measured in the CARDIA study. In this study, most PCBs irrespective of dioxin activity and some OC pesticides predicted incident T2D during a 5-year follow-up (Lee et al. 2010). Comparing strengths of association between PCBs and OC pesticides, PCBs showed the stronger associations, while among OC pesticides, *trans*-nonachlor predicted T2D more strongly than p,p′-DDE. However, neither BDE47 nor dioxin was significantly associated with incident diabetes.

A related study was the Yucheng cohort study of people who were accidently exposed to PCBs and PCDFs during the late 1970s through the consumption of rice bran oil laced with PCBs in Taiwan. In fact, the body burden of PCBs was much higher than those of other prospective studies on POPs and T2D. The authors reported that Yucheng women victims, but not men, suffered from an increased incidence of diabetes compared with neighborhood controls. Particularly affected were those who were assumed to retain significant levels of the pollutants, as evident from the diagnosis of chloracne (Wang et al. 2008).

Importantly, while the CARDIA study clearly showed an inverted U-shaped association that featured risk of T2D within the range of higher levels of POPs that was lower than the risk at mid levels of POPs (Lee et al. 2010), other prospective studies that were able to examine dose–response relations (Lee et al. 2011a; Turyk et al. 2009b) did not show a clear inverted U-shaped association. However, even the studies not showing a clear inverted U-shape demonstrated that the risk of future T2D was substantially increased with only a slight increase of concentrations of POPs within the lower-dose range of POPs; that is, they showed a low-dose effect. There was only a slight increase in risk with an increasing dose of POPs within a higher dose of POPs. Therefore, considering all these findings, the biological effects of POPs are induced at very low concentrations of certain POPs; however, the responses at high concentration may differ depending on the study subjects. This phenomenon may be related to the age of study subjects as we discuss further under the heading "Mismatch of Time Trend: Issues Related to Inverted U-Shaped Associations."

HUMAN EVIDENCE LINKING POPs AND METABOLIC SYNDROME

Metabolic syndrome is a condition closely linked to insulin resistance, although metabolic syndrome is not unequivocally related to insulin resistance (McLaughlin et al. 2003). Insulin resistance is considered as the centerpiece of the pathogenesis of T2D. However, insulin resistance is not confined to T2D but is also commonly found among persons with abdominal obesity, hypertension, or dyslipidemia. Importantly, these traits tend to be observed simultaneously within a person (hence the term "syndrome").

The criteria for the diagnosis of metabolic syndrome by the U.S. National Cholesterol Education Program (NCEP) Adult Treatment Panel III (2001) are the presence of at least three of the following: central obesity, impaired fasting glucose, high blood pressure, decreased high-density lipoprotein (HDL), cholesterol, and increased triglycerides. The presence of metabolic syndrome is a major risk factor for the development of T2D and coronary heart disease.

In the NHANES dataset, OC pesticides were most strongly and consistently associated with metabolic syndrome (Lee et al. 2007b). Although PCBs were also positively associated with metabolic syndrome, the shapes of association differed somewhat depending on the kind of PCB. Dioxin-like PCBs were linearly associated with metabolic syndrome, while non-dioxin-like PCBs showed an inverted U-shaped association.

Interestingly, subclasses of POPs were differently related to the five components of metabolic syndrome. OC pesticides and PCBs were mainly associated with four of the components of metabolic syndrome (central obesity, impaired fasting glucose, decreased HDL cholesterol, and increased triglycerides), while dioxins and furans were most strongly associated with high blood pressure but not with other components. In fact, earlier studies that performed factor analyses on the five components of metabolic syndrome have consistently shown that blood pressure elevation clusters less closely with the other metabolic syndrome components (Meigs 2000). The findings on POPs and metabolic syndrome may be related to this pattern.

On the other hand, a study from Japan used TEQs for POPs with dioxin activity. It showed that dioxin TEQs, furan TEQs, and dioxin-like PCB TEQs were all associated with metabolic syndrome among nondiabetic participants (Uemura et al. 2009). Among them, dioxin-like PCB TEQs showed the strongest association. Even though they primarily used TEQ levels in their analyses with the assumption that AhR is the mechanism involved, the analyses using the serum concentration of each POP showed associations similar to those based on TEQs, suggesting that dioxin-like activity may not be critical in these associations.

One prospective study evaluated if low-dose POPs in young adults predicted several traits related to metabolic syndrome (obesity, high triglycerides, low HDL cholesterol, or insulin resistance) measured 18 years later (Lee et al. 2011b), using the controls from the nested case-control study, which examined the associations between POPs and T2D (Lee et al. 2010). As this study adjusted for the baseline values of outcome variables plus potential confounders, findings pertained to the evolution of the study outcomes over 18 years. Statistically significant associations with some dysmetabolic conditions were observed with p,p′-DDE, oxychlordane, *trans*-nonachlor, hexachlorobenzene, or highly chlorinated PCBs.

Also, their associations appeared at low dose, forming inverted U-shaped relations. This dose–response curve was similar to that for T2D (Lee et al. 2010). In the earlier cross-sectional studies, non-dioxin-like PCBs in the U.S. general population using the NHANES data also showed inverted U-shaped associations (Lee et al. 2007b). Even though the study from Japan showed linear trends (Uemura et al. 2009), the risk of metabolic syndrome in the second quartile was already two to three times higher than that in the first quartile, suggesting at least low-dose effects without decreasing risk at higher concentrations of POPs.

In experimental studies, health effects of single POPs might vary by POP. However, simultaneous background exposure to various POPs in humans may partially explain why metabolic abnormalities like obesity, dyslipidemia, insulin resistance, and hypertension tend to occur as a cluster.

MISMATCH OF TIME TREND: ISSUES RELATED TO INVERTED U-SHAPED ASSOCIATIONS

One dilemma in interpreting recent epidemiological findings on POPs is how the strengths of associations of T2D with OC pesticides or PCBs observed in recent general population studies could be so large when the earlier studies of PCBs or TCDD revealed no association or only a modest increase in risk of T2D, despite occupational or accidental high exposure. At first glance, it may be difficult to accept that low-dose exposure in the general population showed huge associations with T2D, while high-dose exposure in occupational or accidental settings showed only modest associations. Another puzzling aspect is that the body burden of chlorinated POPs has declined worldwide over the several recent decades after banning production and usage in most developed countries, while T2D has recently emerged in a worldwide epidemic. If chlorinated POPs are really important in the development of T2D, how are these kinds of mismatches possible?

Understanding how these intuitively contradictory findings are possible is very important. Our explanation of these issues is that a low-dose but persistent exposure might be more harmful than high-dose exposures if chlorinated POPs are involved in the pathogenesis of T2D as endocrine disruptors.

Larger effects from low-dose exposure than high-dose exposure have been proposed as possible biological responses of chemicals with endocrine-disrupting properties (Welshons et al. 2003), unlike traditional toxicology, in which a linear dose–response relation is typically assumed. In fact, POPs are well known to be endocrine disruptors. In biological systems, there is a linearity of effects of dose only up to a dose that occupies about 10% of receptors. At higher doses, the effect of higher hormone receptor occupancy rate does not linearly increase as the dose of hormone increases. Furthermore, a linear biological response is observed at much lower doses than that showing linearity with receptor occupancy. Under the high-dose exposure of hormones, there is even downregulation of receptors as the dose further increases (Medlock et al. 1991). Thus, chemicals with endocrine-disrupting properties can lead to inverted U dose–response curves with certain biological end points, although it remains unclear what kinds of endocrine-disrupting properties of POPs are important in the pathogenesis of T2D.

Under the assumption of an inverted U-shaped relationship between POPs and T2D, results from human studies can be variable; inverted U-shaped, positive, inverse, and null associations are all possible because a common approach to analyze human data is to categorize the concentrations of each chemical into quartile, quintile, and so on, irrespective of their absolute concentrations, and then to compare the risk of disease between persons belonging to higher categories and persons belonging to the lowest category. Thus, the range of absolute concentration of each chemical and the absolute risk of disease in the reference category (i.e., the lowest category) can be substantially different among studies.

For simplicity, let us assume monotonically increasing doses, arbitrarily numbered starting from 0. Suppose a dose of 0~2 entails no significant response; doses 3~9 induce endocrine disruption and a sharp risk increase; and downregulation of receptor occurs from 10 to 19, while toxicity (including strongly adverse effects) starts from dose 20 and higher. As the current general population is exposed to concentrations of chemicals within the presumed safety level, determined from the linear viewpoint of toxicology, we consider the exposure levels from 0 to 19, which are close to environmental exposure.

The only situation in which we could observe an inverted U-shaped relation would be in a population that covered the whole range of nontoxic exposure doses, that is, 0~19 (population A in Figure 5.4). If the population had exposure distribution close to 0~9, we would observe a strongly increasing linear dose–response relation (population B in Figure 5.4). As the lowest concentration in the population increases, the strength of association would weaken. If the population had the distribution of exposure close to 5~15 (population C in Figure 5.4), results might be deemed to be a null association. A more extreme situation would be observation of an inverse association in a population with an exposure dose close to 10~19 (population D in Figure 5.4). An additional feature made clear in Figure 5.4 is attenuation of risk estimation when there is a gradient of risk within the reference category, for example, in a population with exposure doses close to 0~12, in which doses 0~7 were taken as the refer-

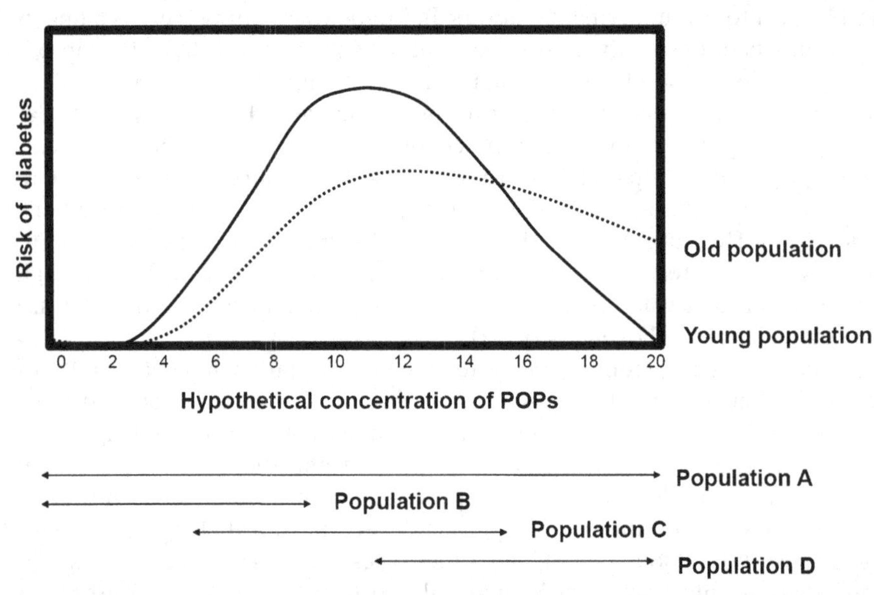

Figure 5.4. Hypothetical dose–response relation between POPs and T2D. The shape of the dose–response curve can differ depending on the different distributions of POPs across populations. Population A shows an inverted U-shaped association; population B shows a strong positive association; population C shows a null association; and population D showed an inverse association. In addition, the shape of the dose–response curve can look different, depending on the sensitivity of the endocrine system. A young population is expected to show a sharper inverted U-shaped association, while an older population would show a blunt shape.

ence group and doses 8~12 were considered the "at-risk" group. Relative risk would be much less than if doses 0~2 were taken as the reference group.

These concepts can help to explain why the strength of association from recent studies in general populations with low-dose exposure to POPs would be much larger than the earlier studies with occupational or accidental high exposure. In fact, most previous epidemiological studies on POPs were performed with subjects who had an exposure to higher concentrations of POPs in occupational or accidental settings taking the "occupationally unexposed or accidentally unexposed" entire general population as the reference group. However, recent epidemiological studies suggest that this kind of approach may not be valid because there may be a much clearer dose–response relation within the lower concentrations of background concentrations of POPs in the general population.

There has been a general belief that the human study of chemicals would best be performed among subjects with high exposure. Therefore, the target population of early studies of chemicals is of occupationally or accidentally high-exposure groups. If chemicals exerted their effects only through toxicity, this approach would be sensible. However, if the purpose of human studies is to evaluate any health effects due to endocrine disruption, the association should be studied among low-exposure populations due to the expected low-dose effects.

Although many human studies showed a low-dose effect, the only study that presented clearly inverted U-shaped associations was the CARDIA study. One defining feature of this prospective study was that the participants had POPs measured as young adults. If POPs are involved in the development of T2D as endocrine disruptors, we can expect that this pattern would be more clearly observed among persons with a more sensitive endocrine system, like the CARDIA subjects (young population in Figure 5.4). However, it is well known that the sensitivity of the endocrine system decreases with aging. Therefore, the inverted U-shaped dose–response curve can be blunted, as in the Prospective Investigation of the Vasculature in Uppsala Seniors (PIVUS) subjects (old population in Figure 5.4). As most chronic diseases including T2D increase with aging, study subjects in most epidemiological studies focusing on chronic diseases are middle-aged or old-aged populations. However, any feature in relation to endocrine disruption may be more sensitively observed among young persons.

Absolute dose presents yet another problem in interpreting the inverted U-shaped associations observed in human studies. In the U.S. general population, serum concentrations of p,p′-DDE range from 100 to 5000 ng/g lipid, while those of *trans*-nonachlor are much lower, ranging from 2 to 10 ng/g lipid. However, when statistical associations were explored using quartile, quintile, or similar approaches, each POP showed a similar association despite the great difference in absolute concentrations. For example, the highest risk for T2D or metabolic dysfunction tended to be observed in the same second quartile of the different POPs.

If these associations were governed by linear dose–response relations, the difference in absolute concentrations among various POPs might not be an issue. Each POP could show significant linear associations within its specific range of concentration. However, under the assumption of nonlinear dose–response relations, the similar shapes of association observed among various POPs with great differences in absolute concentration might be difficult to understand.

There are several possible explanations. First, the important thing may be mixtures of POPs with a complicated exposure matrix rather than individual POPs. Second, some associations might represent true causal relationships, while other associations only reflect the causal associations due to high correlations between them and the actual causal actor. From the viewpoint of regulation, focusing on specific individual POP is most practical and therefore important. When we evaluate possible health effects of POPs in humans, however, the mixtures of POPs may be most salient because it is impossible in the general population to be exposed to some specific POPs while avoiding other specific POPs.

POLYBROMINATED BIPHENYL ETHERS (PBDEs) AND T2D

Unlike chlorinated POPs, which showed declining trends after banning, PBDEs are of special interest because of their recent marked increase in levels in humans as well as in the environment. As they are chemically and toxicologically similar to PCBs, it would be reasonable to suspect that PBDEs may also be related to T2D.

However, epidemiological evidence on PBDEs is not clear. Based on the same NHANES dataset, PBDEs generally showed much weaker associations with T2D compared with associations with chlorinated POPs (Lim et al. 2008). In addition, rather than the strong linear dose–response relation, some PBDEs showed the inverted U-shaped associations with the prevalence of T2D or metabolic syndrome (Lim et al. 2008). However, other studies reported little association (Airaksinen et al. 2011; Turyk et al. 2009b).

Assuming there were a true causal inverted U-shaped association, there is a possibility that the increasing trend of body burden of PBDEs may make it more difficult to detect the association between PBDEs and T2D for the reasons discussed in Figure 5.4. Also, there is an important difference in exposure route of PBDEs compared to chlorinated POPs. Even though PBDEs also bioaccumulate in food chains, the main exposure route of PBDEs in the general population is house dust because it is used as an additive to retard fire and flames in a variety of commercial and household products (Wu et al. 2007). As toxicological studies have reported that different exposure routes of the same chemical can lead to different pharmacokinetics in terms of uptake, distribution, and elimination, the health effects of PBDEs may differ depending on different exposure routes (Sanzgiri et al. 1997).

On the other hand, the worst scenario may be that health effects of PBDEs as endocrine disruptors may clearly come out in epidemiological studies when the body burden of PBDEs decreases to certain levels after banning of these chemicals, as has been the case for PCBs.

THRIFTY GENE HYPOTHESIS AND POPs

The epidemiological links observed between fetal and infant growth and impaired glucose tolerance in adult life led to the formulation of the "thrifty gene hypothesis" (Hales and Barker 1992). The hypothesis postulates that evolutionary pressure from scarcity of food led to the selection of highly efficient insulin secretion and action and that fuel starvation during intrauterine life leads to adaptation to ensure survival but also results in permanent programming of metabolism. However, in the modern world, with enough food, this evolutionarily thrifty metabolism seems to have had maladaptive consequences, including leading to T2D. The mechanism for programming of disease risk may involve inherited non-DNA, epigenetic modification through methylation, and other chemical changes to DNA and histone proteins.

Although low birth weight is generally considered to indicate fetal undernutrition, it is well known that intrauterine exposure to chemicals including POPs also increases the risk of low birth weight (Rylander et al. 2000). There is also evidence that prenatal exposure to DDE may contribute to the obesity epidemic in women (Karmaus et al. 2009). Thus, prenatal exposure to chemicals like POPs might contribute to the associations among low birth weight, future obesity, and T2D.

In fact, this issue is well studied for fetuses that experienced a different toxicant, maternal cigarette smoking rather than POPs. Epidemiological studies of the effects of maternal smoking on infant growth have shown that infants born to smoking mothers weigh less at birth compared with infants born to nonsmoking mothers (Oken et al. 2008). Similar to undernutrition, babies with low birth weight who experience a rapid catch-up period after birth increase the risk of obesity and obesity-related metabolic dysfunction in later life. This phenomenon has similarly been observed with the prenatal exposure to modern nonpersistent pesticides (Wohlfahrt-Veje et al. 2011). Although we think that undernutrition is still an important source of low-birth weight babies in developing countries, chemical-related low birth weight may be more common in modern affluent societies.

POPs AND GLYCEMIC CONTROL AMONG DIABETIC PATIENTS

Hyperglycemia itself not only defines T2D but also is at least a contributing cause of its most characteristic symptoms and long-term complications. If diabetes is not properly controlled, a variety of serious macrovascular and microvascular complications like coronary artery disease, cerebrovascular disease, peripheral vascular disease, retinopathy, nephropathy, and neuropathy arise among patients with T2D. These diseases are a great social and economic burden in developing as well as developed countries. Epidemiological data indicate that the degree and duration of hyperglycemia is associated with the microvascular and macrovascular complications of T2D (Krishnamurti and Steffes 2001).

In addition to the role of POPs in the development of T2D, there is evidence that POPs may increase the risk of complications among patients with T2D. In a cross-sectional study, T2D patients with high levels of POPs, in particular, OC pesticides, had higher HbA1C levels, a biomarker of long-term control of blood glucose, and the prevalence of peripheral neuropathy, a common long-term complication of diabetes (Lee et al. 2008). In fact, OC pesticides were most consistently associated with T2D, insulin resistance, and metabolic syndrome in the NHANES dataset (Lee et al. 2006a, 2007a,b).

The findings from prospective studies on U.S. Air Force veterans of Operation Ranch Hand also supported the possibility that POPs may be involved in poor glycemic control because T2D among veterans with high dioxin levels were clinically more serious, requiring oral medication or insulin therapy compared with patients with low dioxin levels (Henriksen et al. 1997). As serum concentrations of POPs were related to cardiovascular diseases like coronary heart disease or stroke in the general population (Ha et al. 2007), the risk of these complications may also be higher among patients with T2D.

DIETARY INTERVENTIONS IN T2D

Dietary interventions are important to prevent or treat obesity-related metabolic diseases as well as obesity itself. Nutritional guidelines are usually based on energy density and factors such as glycemic index of the diet. However, as POPs have widely contaminated food chains, dietary intervention ignoring the presence of POPs in food may interfere with the expected beneficial effects of dietary recommendations.

As POPs are ubiquitous in the environment, there is no way to completely avoid exposure to them. However, there may be ways to reduce exposure. Currently, exposure mainly happens through two sources: the external source and the internal source. The main external source is diet because POPs have completely contaminated our food chain. In particular, fatty animal food is the main source because animal food has a high position in the food chain. Therefore, avoiding fatty animal food may be one way to reduce external exposure to POPs. This recommendation coincides with many other recommendations to limit meat intake.

It is interesting that some randomized controlled trial reported that vegan diet is more effective to control glycemic levels in patients with T2D than

American Diabetic Association (ADA)-recommended diet (Barnard et al. 2006, 2009). The ADA diet focuses on calorie restriction and glycemic index but does not consider plant or animal food as the source of the diet. Compared to the ADA diet, a vegan diet does not restrict calorie intake, but it is recommended for patients to eat only plant foods. Although there appear to be other advantages of a vegan diet, the possibility that such a diet would avoid POP exposure is worth considering as an additional advantage.

When serum POP levels in adults are serially measured, their levels tend to decrease within persons (Turyk et al. 2009a). This means that, despite ongoing external exposure through diet, internal levels that have bioaccumulated during previous high-exposure periods are gradually excreted, finally leading to a decrease in total POP body burden. One characteristic of the body metabolism of POPs is that they undergo enterohepatic recirculation (Jandacek and Tso 2007). The main excretion route of POPs is bile. The bile is stored in the gallbladder and secreted to digest fatty food during meals. However, some of the bile is reabsorbed at the terminal ileum (enterohepatic recirculation). During this physiological process, POPs mixed with bile are also reabsorbed; this is a known mechanism that lengthens the half-life of POPs. Therefore, anything that would decrease POP absorption in the gut and increase excretion in the feces would prevent reabsorption and therefore reduce the half-life of POPs.

Some dietary supplements or drugs were tested in the 1970s or 1980s for their ability to enhance POP excretion. Among them, dietary fiber was confirmed in animal studies to increase the excretion of POPs (Aozasa et al. 2001; Sera et al. 2005). Plant food has dietary fiber, whereas animal food does not. Therefore, a vegan or other low meat diet may be beneficial for at least two reasons with respect to POPs: avoidance of further external exposure to POPs and increase of excretion rate of POPs from internal exposure.

SUMMARY

Prevalence of metabolic dysfunction such as metabolic syndrome, insulin resistance, and T2D is continuing to increase worldwide. These conditions have become a great medical and economic burden to most societies. Even though obesity is the most well-known risk factor associated with all these conditions, a growing body of evidence links them to background exposure to POPs, in particular, OC pesticides and PCBs. In addition to strong associations observed in many cross-sectional studies that were mainly performed among general populations with only background exposure to these chemicals through the environment, recent prospective studies generally confirmed the cross-sectional findings. Therefore, based on the currently available evidence, the background exposure to POPs seems to play an important role in all the stages of the development of T2D (Figure 5.5). The epidemiological evidence on POPs and T2D suggests that increased adiposity with high levels of POPs puts

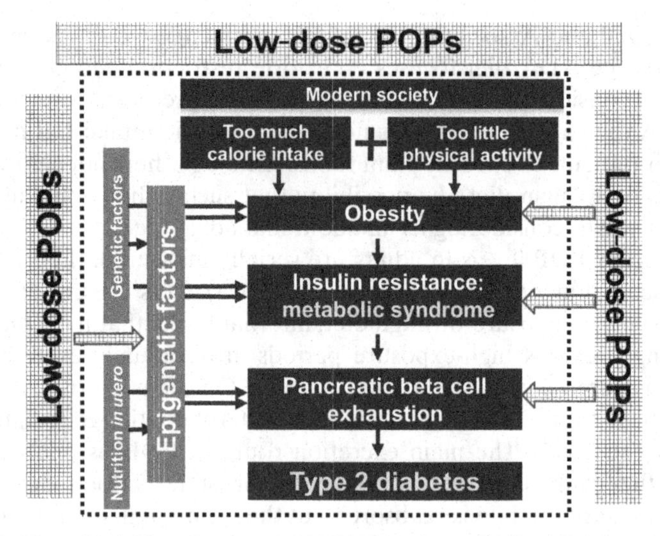

Figure 5.5. Hypothetical pathogenesis of T2D and metabolic syndrome. The dotted box contains the current paradigm of T2D and metabolic syndrome. Checkered boxes and arrows show that low-dose POPs can be involved in the all pathways of T2D and metabolic syndrome including epigenetic modulation of genes, obesity, insulin resistance, and pancreatic beta cell exhaustion.

people at substantial risk, while increased body fatness with low *in vivo* POP levels may not have nearly as strong an effect. Although more work is needed to verify these findings, this evidence suggests that obesity as a risk factor for T2D is of particular concern in persons with high levels of POPs.

However, epidemiological evidence on the shape of dose–response curves has been mixed. Even though more research is required to better understand the dose–response relationship, if POPs are involved in the pathogenesis of these diseases as endocrine disruptors, inconsistent results are possible. Future human studies should carefully consider how the properties of endocrine disruptors can affect their study findings.

Furthermore, some POPs have been consistently linked to risk for T2D, while others have been inconsistently linked. Due to high correlations among serum concentrations of various POPs in humans, it may be difficult to sort out which specific POPs have adverse effects solely through epidemiological studies. Evidence from experimental studies would be very helpful. However, as humans are exposed to a mixture of POPs, distinguishing the role of each POP may not be the most helpful approach since it is far from the reality that humans face. Rather, the most helpful studies would investigate if efforts to reduce further exposure to POPs and to increase the excretion of POPs can be helpful to prevent the development of T2D or complications of T2D in future studies.

REFERENCES

Airaksinen R, Rantakokko P, Eriksson JG, Blomstedt P, Kajantie E, Kiviranta H. 2011 Association between type 2 diabetes and exposure to persistent organic pollutants. Diabetes Care 34:1972–1979.

Aozasa O, Ohta S, Nakao T, Miyata H, Nomura T. 2001. Enhancement in fecal excretion of dioxin isomer in mice by several dietary fibers. Chemosphere 45:195–200.

Barnard ND, Cohen J, Jenkins DJ, Turner-McGrievy G, Gloede L, Jaster B, et al. 2006. A low-fat vegan diet improves glycemic control and cardiovascular risk factors in a randomized clinical trial in individuals with type 2 diabetes. Diabetes Care 29:1777–1783.

Barnard ND, Cohen J, Jenkins DJ, Turner-McGrievy G, Gloede L, Green A, et al. 2009. A low-fat vegan diet and a conventional diabetes diet in the treatment of type 2 diabetes: A randomized, controlled, 74-wk clinical trial. Am J Clin Nutr 89: 1588S–1596S.

Bertazzi PA, Bernucci I, Brambilla G, Consonni D, Pesatori AC. 1998. The Seveso studies on early and long-term effects of dioxin exposure: A review. Environ Health Perspect 106(Suppl 2):625–633.

Calvert GM, Sweeney MH, Deddens J, Wall DK. 1999. Evaluation of diabetes mellitus, serum glucose, and thyroid function among United States workers exposed to 2,3,7,8-tetrachlorodibenzo-p-dioxin. Occup Environ Med 56:270–276.

Crinnion WJ. 2011. The role of persistent organic pollutants in the worldwide epidemic of type 2 diabetes mellitus and the possible connection to farmed Atlantic salmon (*Salmo salar*). Altern Med Rev 16:301–313.

Emmett EA. 1985. Polychlorinated biphenyl exposure and effects in transformer repair workers. Environ Health Perspect 60:185–192.

Fujiyoshi PT, Michalek JE, Matsumura F. 2006. Molecular epidemiologic evidence for diabetogenic effects of dioxin exposure in U.S. Air force veterans of the Vietnam war. Environ Health Perspect 114:1677–1683.

Glynn AW, Granath F, Aune M, Atuma S, Darnerud PO, Bjerselius R, et al. 2003. Organochlorines in Swedish women: Determinants of serum concentrations. Environ Health Perspect 111:349–355.

Gunderson EP, Lewis CE, Tsai AL, Chiang V, Carnethon M, Quesenberry CP, Jr., et al. 2007. A 20-year prospective study of childbearing and incidence of diabetes in young women, controlling for glycemia before conception: The Coronary Artery Risk Development in Young Adults (CARDIA) Study. Diabetes 56:2990–2996.

Ha MH, Lee DH, Jacobs DR. 2007. Association between serum concentrations of persistent organic pollutants and self-reported cardiovascular disease prevalence: Results from the National Health and Nutrition Examination Survey, 1999–2002. Environ Health Perspect 115:1204–1209.

Hales CN, Barker DJ. 1992. Type 2 (non-insulin-dependent) diabetes mellitus: The thrifty phenotype hypothesis. Diabetologia 35:595–601.

Henriksen GL, Ketchum NS, Michalek JE, Swaby JA. 1997. Serum dioxin and diabetes mellitus in veterans of Operation Ranch Hand. Epidemiology 8:252–258.

IOM (Institute of Medicine). 2000. Veterans and Agent Orange: Herbicide/Dioxin Exposure and Type 2 Diabetes. Washinton, DC: The National Academy Press.

Jandacek RJ, Tso P. 2007. Enterohepatic circulation of organochlorine compounds: A site for nutritional intervention. J Nutr Biochem 18:163–167.

Jorgensen ME, Borch-Johnsen K, Bjerregaard P. 2008. A cross-sectional study of the association between persistent organic pollutants and glucose intolerance among Greenland Inuit. Diabetologia 51:1416–1422.

Karmaus W, Osuch JR, Eneli I, Mudd LM, Zhang J, Mikucki D, et al. 2009. Maternal levels of dichlorodiphenyl-dichloroethylene (DDE) may increase weight and body mass index in adult female offspring. Occup Environ Med 66:143–149.

Kimbrough RD, Doemland ML, Mandel JS. 2003. A mortality update of male and female capacitor workers exposed to polychlorinated biphenyls. J Occup Environ Med 45:271–282.

Krishnamurti U, Steffes MW. 2001. Glycohemoglobin: A primary predictor of the development or reversal of complications of diabetes mellitus. Clin Chem 47: 1157–1165.

Lawton RW, Ross MR, Feingold J, Brown JF, Jr. 1985. Effects of PCB exposure on biochemical and hematological findings in capacitor workers. Environ Health Perspect 60:165–184.

Lee DH, Lee IK, Song K, Steffes M, Toscano W, Baker BA, et al. 2006a. A strong dose-response relation between serum concentrations of persistent organic pollutants and diabetes: Results from the National Health and Examination Survey 1999–2002. Diabetes Care 29:1638–1644.

Lee DH, Jacobs DR, Jr., Porta M. 2006b. Could low-level background exposure to persistent organic pollutants contribute to the social burden of type 2 diabetes? J Epidemiol Community Health 60:1006–1008.

Lee DH, Lee IK, Jin SH, Steffes M, Jacobs DR, Jr. 2007a. Association between serum concentrations of persistent organic pollutants and insulin resistance among nondiabetic adults: Results from the National Health and Nutrition Examination Survey 1999–2002. Diabetes Care 30:622–628.

Lee DH, Lee IK, Porta M, Steffes M, Jacobs DR, Jr. 2007b. Relationship between serum concentrations of persistent organic pollutants and the prevalence of metabolic syndrome among non-diabetic adults: Results from the National Health and Nutrition Examination Survey 1999–2002. Diabetologia 50:1841–1851.

Lee DH, Jacobs DR, Jr., Steffes M. 2008. Association of organochlorine pesticides with peripheral neuropathy in patients with diabetes or impaired fasting glucose. Diabetes 57:3108–3111.

Lee DH, Steffes MW, Sjodin A, Jones RS, Needham LL, Jacobs DR, Jr. 2010. Low dose of some persistent organic pollutants predicts type 2 diabetes: A nested case-control study. Environ Health Perspect 118:1235–1242.

Lee DH, Lind PM, Jacobs DR, Jr., Salihovic S, van Bavel B, Lind L. 2011a. Polychlorinated biphenyls and organochlorine pesticides in plasma predict development of type 2 diabetes in the elderly: The Prospective Investigation of the Vasculature in Uppsala Seniors (PIVUS) study. Diabetes Care 34:1778–1784.

Lee DH, Steffes MW, Sjodin A, Jones RS, Needham LL, Jacobs DR, Jr. 2011b. Low dose organochlorine pesticides and polychlorinated biphenyls predict obesity, dyslipidemia, and insulin resistance among people free of diabetes. PLoS ONE 6: e15977.

Lim JS, Lee DH, Jacobs DR, Jr. 2008. Association of brominated flame retardants with diabetes and metabolic syndrome in the U.S. population, 2003–2004. Diabetes Care 31:1802–1807.

Longnecker MP, Michalek JE. 2000. Serum dioxin level in relation to diabetes mellitus among Air Force veterans with background levels of exposure. Epidemiology 11:44–48.

McLaughlin T, Abbasi F, Cheal K, Chu J, Lamendola C, Reaven G. 2003. Use of metabolic markers to identify overweight individuals who are insulin resistant. Ann Intern Med 139:802–809.

Medlock KL, Lyttle CR, Kelepouris N, Newman ED, Sheehan DM. 1991. Estradiol down-regulation of the rat uterine estrogen receptor. Proc Soc Exp Biol Med 196:293–300.

Meigs JB. 2000. Invited commentary: Insulin resistance syndrome? Syndrome X? Multiple metabolic syndrome? A syndrome at all? Factor analysis reveals patterns in the fabric of correlated metabolic risk factors. Am J Epidemiol 152:908–911; discussion 912.

Michalek JE, Ketchum NS, Tripathi RC. 2003. Diabetes mellitus and 2,3,7,8-tetrachlorodibenzo-p-dioxin elimination in veterans of Operation Ranch Hand. J Toxicol Environ Health A 66:211–221.

Oken E, Levitan EB, Gillman MW. 2008. Maternal smoking during pregnancy and child overweight: Systematic review and meta-analysis. Int J Obes (Lond) 32:201–210.

Patel CJ, Bhattacharya J, Butte AJ. 2010. An environment-wide association study (EWAS) on type 2 diabetes mellitus. PLoS ONE 5:e10746.

Porta M. 2006. Persistent organic pollutants and the burden of diabetes. Lancet 368:558–559.

Remillard RB, Bunce NJ. 2002. Linking dioxins to diabetes: Epidemiology and biologic plausibility. Environ Health Perspect 110:853–858.

Rignell-Hydbom A, Rylander L, Hagmar L. 2007. Exposure to persistent organochlorine pollutants and type 2 diabetes mellitus. Hum Exp Toxicol 26:447–452.

Rignell-Hydbom A, Lidfeldt J, Kiviranta H, Rantakokko P, Samsioe G, Agardh CD, et al. 2009. Exposure to p,p′-DDE: A risk factor for type 2 diabetes. PLoS ONE 4:e7503.

Rylander L, Stromberg U, Hagmar L. 2000. Lowered birth weight among infants born to women with a high intake of fish contaminated with persistent organochlorine compounds. Chemosphere 40:1255–1262.

Rylander L, Rignell-Hydbom A, Hagmar L. 2005. A cross-sectional study of the association between persistent organochlorine pollutants and diabetes. Environ Health 4:28.

Sanzgiri UY, Srivatsan V, Muralidhara S, Dallas CE, Bruckner JV. 1997. Uptake, distribution, and elimination of carbon tetrachloride in rat tissues following inhalation and ingestion exposures. Toxicol Appl Pharmacol 143:120–129.

Sera N, Morita K, Nagasoe M, Tokieda H, Kitaura T, Tokiwa H. 2005. Binding effect of polychlorinated compounds and environmental carcinogens on rice bran fiber. J Nutr Biochem 16:50–58.

Shaw JE, Sicree RA, Zimmet PZ. 2010. Global estimates of the prevalence of diabetes for 2010 and 2030. Diabetes Res Clin Pract 87:4–14.

Son HK, Kim SA, Kang JH, Chang YS, Park SK, Lee SK, et al. 2010. Strong associations between low-dose organochlorine pesticides and type 2 diabetes in Korea. Environ Int 36:410–414.

Steenland K, Piacitelli L, Deddens J, Fingerhut M, Chang LI. 1999. Cancer, heart disease, and diabetes in workers exposed to 2,3,7,8-tetrachlorodibenzo-p-dioxin. J Natl Cancer Inst 91:779–786.

Sweeney MH, Calvert GM, Egeland GA, Fingerhut MA, Halperin WE, Piacitelli LA. 1997. Review and update of the results of the NIOSH medical study of workers exposed to chemicals contaminated with 2,3,7,8-tetrachlorodibenzodioxin. Teratog Carcinog Mutagen 17:241–247.

Tanaka T, Morita A, Kato M, Hirai T, Mizoue T, Terauchi Y, et al. 2011. Congener-specific polychlorinated biphenyls and the prevalence of diabetes in the Saku Control Obesity Program (SCOP). Endocr J 58:589–596.

Thayer KA, Heindel JJ, Bucher JR, Gallo MA. 2012. Role of environmental chemicals in diabetes and obesity: A national toxicology program workshop report. Environ Health Perspect 120:779–789.

Turyk M, Anderson H, Knobeloch L, Imm P, Persky V. 2009a. Organochlorine exposure and incidence of diabetes in a cohort of Great Lakes sport fish consumers. Environ Health Perspect 117:1076–1082.

Turyk M, Anderson HA, Knobeloch L, Imm P, Persky VW. 2009b. Prevalence of diabetes and body burdens of polychlorinated biphenyls, polybrominated diphenyl ethers, and p,p′-diphenyldichloroethene in Great Lakes sport fish consumers. Chemosphere 75:674–679.

Uemura H, Arisawa K, Hiyoshi M, Kitayama A, Takami H, Sawachika F, et al. 2009. Prevalence of metabolic syndrome associated with body burden levels of dioxin and related compounds among Japan's general population. Environ Health Perspect 117:568–573.

Ukropec J, Radikova Z, Huckova M, Koska J, Kocan A, Sebokova E, et al. 2010. High prevalence of prediabetes and diabetes in a population exposed to high levels of an organochlorine cocktail. Diabetologia 53:899–906.

Vasiliu O, Cameron L, Gardiner J, Deguire P, Karmaus W. 2006. Polybrominated biphenyls, polychlorinated biphenyls, body weight, and incidence of adult-onset diabetes mellitus. Epidemiology 17:352–359.

Vena J, Boffetta P, Becher H, Benn T, Bueno-de-Mesquita HB, Coggon D, et al. 1998. Exposure to dioxin and nonneoplastic mortality in the expanded IARC international cohort study of phenoxy herbicide and chlorophenol production workers and sprayers. Environ Health Perspect 106(Suppl 2):645–653.

Wang SL, Tsai PC, Yang CY, Leon Guo Y. 2008. Increased risk of diabetes and polychlorinated biphenyls and dioxins: A 24-year follow-up study of the Yucheng cohort. Diabetes Care 31:1574–1579.

Welshons WV, Thayer KA, Judy BM, Taylor JA, Curran EM, vom Saal FS. 2003. Large effects from small exposures. I. Mechanisms for endocrine-disrupting chemicals with estrogenic activity. Environ Health Perspect 111:994–1006.

Wild S, Roglic G, Green A, Sicree R, King H. 2004. Global prevalence of diabetes: Estimates for the year 2000 and projections for 2030. Diabetes Care 27:1047–1053.

Wildman RP, Muntner P, Reynolds K, McGinn AP, Rajpathak S, Wylie-Rosett J, et al. 2008. The obese without cardiometabolic risk factor clustering and the normal weight with cardiometabolic risk factor clustering: Prevalence and correlates of 2 phenotypes among the US population (NHANES 1999–2004). Arch Intern Med 168:1617–1624.

Wohlfahrt-Veje C, Main KM, Schmidt IM, Boas M, Jensen TK, Grandjean P, et al. 2011. Lower birth weight and increased body fat at school age in children prenatally exposed to modern pesticides: A prospective study. Environ Health 10:79.

Wu N, Herrmann T, Paepke O, Tickner J, Hale R, Harvey LE, et al. 2007. Human exposure to PBDEs: Associations of PBDE body burdens with food consumption and house dust concentrations. Environ Sci Technol 41:1584–1589.

Yoon KH, Lee JH, Kim JW, Cho JH, Choi YH, Ko SH, et al. 2006. Epidemic obesity and type 2 diabetes in Asia. Lancet 368:1681–1688.

Zober A, Ott MG, Messerer P. 1994. Morbidity follow up study of BASF employees exposed to 2,3,7, 8-tetrachlorodibenzo-p-dioxin (TCDD) after a 1953 chemical reactor incident. Occup Environ Med 51:479–486.

CHAPTER 6

Mechanistic Basis for Elevation in Risk of Diabetes Caused by Persistent Organic Pollutants

JÉRÔME RUZZIN

ABSTRACT

Background: Persistent organic pollutants (POPs) are potent endocrine and metabolic disruptors. However, the modes of action of POPs remain mostly unknown.

Objective: The objective of this chapter was to explore potential molecular mechanisms by which POPs can interact with insulin as well as glucose and lipid homeostasis.

Discussion: Relevant studies were identified by searching MEDLINE and Web of Knowledge. The search was restricted to studies focusing on the aryl hydrocarbon receptor (AhR), constitutive androstane receptor (CAR), steroid and xenobiotic receptor (SXR, also known as pregnane X receptor [PXR]), and metabolic disorders, with a special emphasis on diabetes.

Conclusions: Activation of AhR, CAR, and SXR can deregulate endocrine and metabolic homeostasis. However, the exact biochemical events involved are still poorly identified.

INTRODUCTION

The global prevalence of diabetes has dramatically increased during the last three decades. Estimates indicate that one in four U.S. adults has diabetes and that over 6 million people worldwide develop diabetes each year. These fright-

Effects of Persistent and Bioactive Organic Pollutants on Human Health, First Edition.
Edited by David O. Carpenter.

ening statistics highlight the reality that the origins of the disease remain mostly unknown and that some etiological factors have likely been overlooked. In this chapter, the mechanistic contribution of persistent organic pollutants (POPs) to metabolic disorders linked to diabetes is discussed.

DIABETES

At an early stage, high blood glucose level is a major characteristic of persons with diabetes either because the pancreatic β cells do not produce enough insulin (type 1 diabetes) or because insulin-insensitive cells fail to use insulin properly (type 2 diabetes). Over time, when glucose accumulates in the blood instead of going into cells, the risk of developing metabolic complications significantly increases, and impaired lipid and glucose metabolism becomes a common characteristic of diabetes. As a consequence, diabetes is often associated with obesity, high blood pressure, cardiovascular disease, stroke, and/or nephropathy. This cluster of metabolic disturbances is dramatic. According to the Centers for Disease Control (CDC), two out of three people with diabetes are known to die from stroke or heart disease; 44% of new cases of kidney failure are associated with diabetes; more than 60% of nontraumatic lower-limb amputations occur in people with diabetes; and about 60–70% of people with diabetes have mild to severe forms of nervous system damage (CDC 2011).

Type 2 diabetes is by far the most common form of diabetes, representing 90–95% of all cases. Type 1 diabetes results from an autoimmune attack of the pancreatic β cells leading to a loss of insulin production. On the other hand, the origins of type 2 diabetes have classically focused on lifestyle factors like physical inactivity and excess energy intake. However, these two conventional risk factors alone are unlikely to fully explain the worldwide explosive and uncontrolled rise of metabolic diseases (Ruzzin et al. 2012; Thayer et al. 2012).

POPs: NOVEL DIABETOGENIC FACTORS

POPs are toxic chemicals mostly created by humans in industrial processes such as organochlorine pesticides (OPs), polychlorinated biphenyls (PCBs), and dioxins. These chemicals are omnipresent in our environment because of their high resistance to degradation. Humans are particularly exposed to POPs through consumption of food of animal origin, such as fatty fish, meat, and dairy products (Fisher 1999; Schecter et al. 2010). Initially described for their ability to induce immunotoxicity, dermal toxicity (chloracne), neurotoxicity, hepatotoxicity, and reproductive dysfunction, and to modulate cell and tumor growth (Skene et al. 1989), POPs are now recognized as potent endocrine disruptors (Alonso-Magdalena et al. 2011; Neel and Sargis 2011).

In humans, several cross-sectional and prospective studies have reported an association between diabetes and POP exposure (see Chapter 5). To determine the causal role of POPs on diabetes, we investigated the metabolic impacts associated with the consumption of fish oil obtained from farmed Atlantic salmon (*Salmo salar*), which contains environmental levels of POPs, using both *in vivo* and *in vitro* models. After 28 days, rats exposed to POPs developed insulin resistance, glucose intolerance, visceral obesity, and nonalcoholic fatty liver (Ruzzin et al. 2010). Importantly, the concentrations of POPs in adipose tissue in these animals were similar to those observed in humans, thereby indicating that environmental concentrations of POPs can be sufficient to induce metabolic disturbances. In line with these findings, POP mixtures extracted from burbot (*Lota lota*) liver were found to impair insulin signaling and to stimulate body weight gain in zebrafish (Lyche et al. 2010).

To further demonstrate the ability of POPs to interact with insulin action, we exposed 3T3L1 cells (differentiated adipocytes) to different POP mixtures, mimicking those present in the salmon oil. Interestingly, the ability of insulin to stimulate glucose uptake was robustly impaired after exposure to OPs, PCBs, and dioxin-like polychlorinated biphenyls (dl-PCBs), whereas insulin action was normal in cells treated with polychlorinated dibenzo-*p*-dioxins (PCDDs) and polychlorinated dibenzofurans (PCDFs) (Ruzzin et al. 2010). Furthermore, our *in vitro* investigations revealed that impaired insulin action was independent of the total toxic equivalent (TEQ) concentration (Van den Berg et al. 2006) of the POP mixtures, thereby demonstrating that current risk assessment based on TEQ is unlikely to predict the risk of insulin resistance (Ruzzin 2012).

More recently, we further studied the metabolic impacts of fatty fish, which is usually the most contaminated food product that humans can be exposed to (Darnerud et al. 2006; Fromberg et al. 2011; Schafer and Kegley 2002; Schecter et al. 2010). We demonstrated that, after 8 weeks, mice fed with farmed Atlantic salmon fillet were characterized by many metabolic defects linked to type 2 diabetes (Ibrahim et al. 2011). Interestingly, these metabolic disturbances were reduced in mice fed with farmed Atlantic salmon fillet containing lower POP levels. These findings indicate that POPs can directly affect the health outcomes of food products and may, in part, explain why recent prospective studies reported a positive association between fatty fish intake and the risk of type 2 diabetes (Djousse et al. 2011; Kaushik et al. 2009; Ruzzin and Jacobs 2012). Taken together, studies performed in both humans and animals, as well as cell models, have demonstrated that exposure to POPs can contribute to metabolic disorders linked to type 2 diabetes. Yet, the mechanistic basis for elevation in the risk of diabetes caused by POPs remains largely unclear.

POPs AND DIABETES: MECHANISTIC BASIS

Once absorbed, POPs can activate different xenobiotic receptors, including aryl hydrocarbon receptor (AhR), constitutive androstane receptor (CAR), and steroid and xenobiotic receptor (SXR, also known as pregnane X receptor [PXR]). The actual spectrum of POPs that can bind AhR, CAR, or SXR is complex and still poorly understood. The primary function of AhR, CAR, and SXR is to act as intracellular sensors for toxic substances including foreign chemicals, that is, xenobiotics, and to regulate pollutant and drug elimination pathways. However, recent studies have highlighted that these receptors can also interact with metabolic pathways.

AhR

AhR is a basic helix–loop–helix (bHLH)- and Per–Arnt–Sim (PAS)-containing transcription factor that regulates the expression of genes in a ligand-dependent manner. AhR is ubiquitously expressed and regulates the biological response to multiple POPs, among which 2,3,7,8-tetrachlorodibenzo-*p*-dioxin (TCDD) is commonly recognized as the most potent AhR activator. The traditional view of AhR ligands suggests that only planar, aromatic and hydrophobic ligands with maximal dimensions of $14\,\text{Å} \times 12\,\text{Å} \times 5\,\text{Å}$ like halogenated aromatic hydrocarbon (PCDDs, PCDFs, and biphenyls) and polycyclic aromatic hydrocarbons (benzo(*a*)pyrene, aromatic amines, and benzoflavones) can fit into the ligand-binding pocket of the AhR (Machala et al. 2001; Waller and McKinney 1995). However, "nontraditional" chemicals such as pesticides were also found to interact with AhR (Randi et al. 2008; Takeuchi et al. 2008). AhR resides in the cytosol as a multiprotein complex containing two molecules of heat shock protein of 90 kDa (Hsp 90), the X-associated protein 2 (XAP2), and the 23 kDa cochaperone protein (p23) (Kazlauskas et al. 1999; Meyer et al. 1998). Ligand binding induces dissociation of the multiprotein complex allowing the translocation of AhR to the nucleus (Figure 6.1). Once into the nucleus, AhR interacts with the AhR nuclear translocator (Arnt) and binds gene promoters containing a core enhancer regulatory element known as the dioxin response element (DRE), which increases transcription of target genes like those regulating drug-metabolizing enzymes, including the phase I enzyme cytochrome P450 isoform 1A1 (CYP1A1) and the phase II enzymes uridine diphosphate-5′-glucuronosyltransferase 1A1 (UGT1A1) and UGT1A6 (Ashida et al. 2008).

Interestingly, gene expression profiles of pancreatic islets by microarray and real-time polymerase chain reaction (PCR) analysis revealed that Arnt was strongly decreased in subjects with type 2 diabetes compared with healthy subjects (Gunton et al. 2005). Furthermore, reduction of Arnt or AhR protein expression by small interfering RNA (siRNA) in Min6 cells, a glucose-responsive β cell, impaired glucose-stimulated insulin secretion (Gunton et al. 2005). In line with these findings, islets from specific β-Arnt knockout mice exhibited defective insulin release upon glucose stimulation (Gunton et al. 2005).

A reduction in Arnt expression in the liver of patients with type 2 diabetes was later reported, and mice lacking hepatic Arnt were found to exhibit

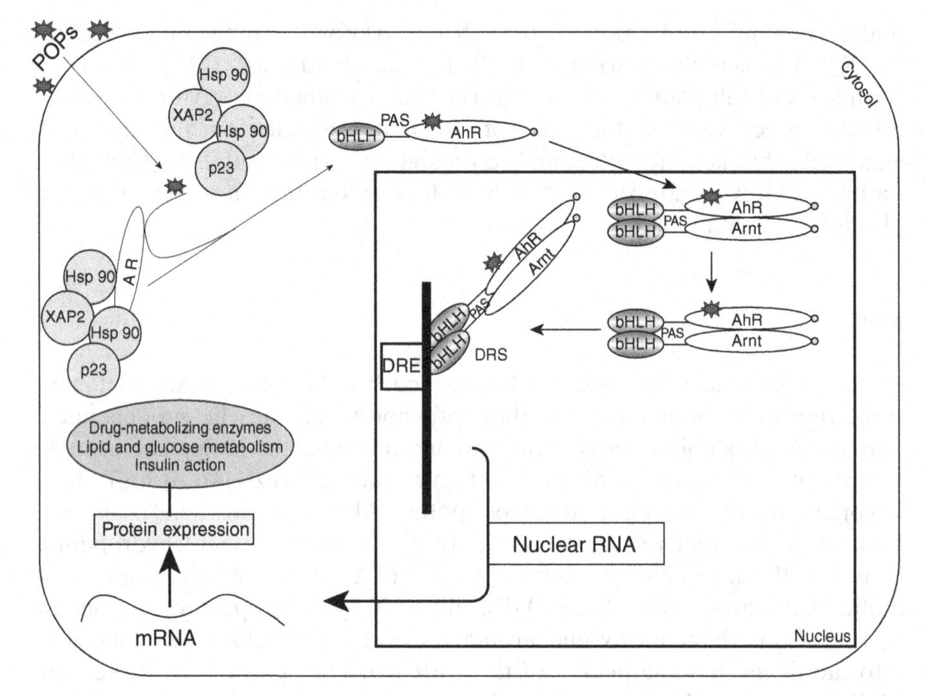

Figure 6.1. Signal transduction mechanisms of AhR. POPs can diffuse across the cell membrane and bind to AhR. Formation of the AhR:Arnt heterodimer induces the transcription of target genes. Hsp 90, heat shock protein of 90 kDa; XAP2, X-associated protein 2; p23, 23 kDa cochaperone protein; Arnt, AhR nuclear translocator; DRE, dioxin response element.

disorders associated with type 2 diabetes, including an increase in gluconeogenesis and lipogenesis, and hyperinsulinemia (Wang et al. 2009). On the other hand, mice lacking AhR were characterized by increased insulin sensitivity and improved glucose tolerance and showed decreased gene expression of peroxisome proliferator-activated receptor-α (*PPAR-α*), acyl-CoA oxidase (*ACO*), carnitine palmitoyl transferase (*CPT*), glucose 6 phosphatase (G6Pase), and phosphoenolpyruvate carboxykinase (*PEPCK*), which are key genes involved in lipid and glucose homeostasis (Wang et al. 2011). In transgenic mice expressing active AhR in the liver, hepatic stenosis developed spontaneously and these animals had decreased body weight and fat mass (Lee et al. 2010), a phenotype that closely mimics the wasting syndrome observed in animals treated with high TCDD concentrations (Croutch et al. 2005).

Exploration of gene expression by Affymetrix microarray analysis revealed that cluster of differentiation 36 (CD36), a fatty acid translocase responsible for the uptake of long-chain fatty acids, but not sterol regulatory element-binding protein 1 (SREBP1), fatty acid synthase (FAS), acetyl-CoA-carboxylase (ACC), and stearoyl-CoA desaturase-1 (SCD-1), was increased in the liver of transgenic mice, and that both mouse and human CD36 gene promoters are transcriptional targets of AhR (Lee et al. 2010).

Additional studies have demonstrated that the deleterious metabolic effects of AhR may occur at a very low dose of exposure. For example, while TCDD treatment at high doses induced weight loss in rats, animals treated with lower doses (<1.0 μg/kg body weight) exhibited significantly greater body weight gain (Croutch et al. 2005). In line with this finding, exposure to low concentrations of TCDD was reported to accelerate differentiation and lipid accumulation in 3T3L1 adipocytes, whereas higher doses of TCDD induced inhibitory effects (Arsenescu et al. 2008).

CAR AND SXR

In contrast to AhR, CAR (NR1I3) and SXR (NR1I2) are members of the nuclear receptor superfamily and are mainly expressed in the liver and intestine (Chang and Waxman 2006; di Masi et al. 2009). Both receptors have the ability to respond to a diverse set of low-affinity ligands, especially those that belong to the pesticide family (Casabar et al. 2010; Coumoul et al. 2002; Kretschmer and Baldwin 2005). Once activated, CAR and SXR form heterodimers with the retinoid X receptor (RXR) and regulate the expression of genes involved in the inactivation and elimination of POPs and xenobiotics. While these two receptors share common ligands and regulate overlapping sets of target genes, their modes of action are different. SXR has a low basal activity, is highly activated upon ligand binding, and is located in the nucleus (Figure 6.2). In contrast, CAR is retained in the cytosol and its localization and activity are regulated by multiple protein phosphorylation cascades. Similar to AhR, both CAR and SXR were recently reported to activate signal transduction pathways linked to metabolic and energy functions.

Activation of CAR by 1,4-bis[2-(3,5-dichloropyridyloxy)]benzene (TCPOBOP) [intraperitoneal (i.p.) injection of 0.5 mg/kg once a week for 5 weeks] was found to improve insulin action as well as glucose and lipid metabolic pathways through the regulation of PEPCK, G6Pase, SREBP1, and SCD-1 (Dong et al. 2009; Gao et al. 2009). Earlier studies reported that 10 hours after TCPOBOP injection (i.p. injection of 10 mg/kg), hepatic triglyceride levels significantly decreased and were paralleled by a robust increase in insulin-induced gene-1 (*Insig-1*) mRNA expression (Roth et al. 2008). Importantly, these authors reported a functional binding site for CAR (as well as SXR) in the upstream promoter region of *Insig-1*, thereby suggesting that CAR and SXR can directly account for the induction of *Insig-1*. In contrast, other studies reported that TCPOBOP treatment (i.p. injection of 0.3 mg/kg once daily for 14 days) decreased hepatic PPAR-α, *Lpin1*, and elongation of very long-chain fatty acid protein 5 (ELOVL5) expression, and significantly increased serum triglycerides (Maglich et al. 2009). Interestingly, we found that *Lpin1* and *Insig-1* mRNA expression was reduced in both liver and adipose

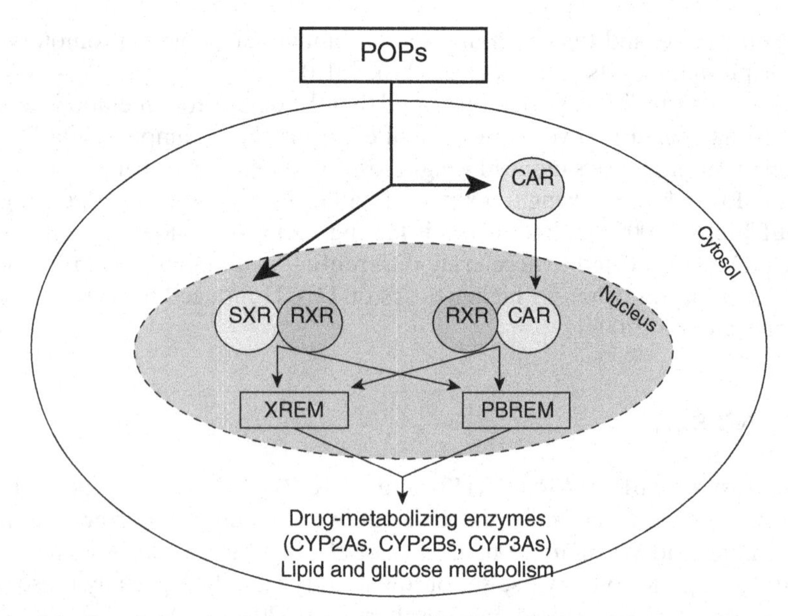

Figure 6.2. Signal transduction mechanisms of SXR and CAR. After entering the cell, POPs can bind to CAR or SXR. Once activated, CAR and SXR heterodimerize with RXR and bind to response elements present in the promoters of target genes. In the flanking regions of several genes, response elements can be activated by both PXR and CAR allowing direct cross talk of these receptors. RXR, retinoid X receptor; XREM, xenobiotic-responsive enhancer module; PBREM, phenobarbital-responsive enhancer module.

tissue of rats exposed to POPs through the intake of farmed Atlantic salmon oil (Ruzzin et al. 2010). Similar findings were also observed in 3T3L1 exposed to POP mixtures (Ruzzin et al. 2010).

The regulation of the lipogenic thyroid hormone-responsive spot 14 protein (THRSP) expression by CAR in human and murine hepatocytes has been suggested as a potential mechanism in the development of hepatic stenosis (Breuker et al. 2010). In humans, chronic treatment with the antiepileptic drug, phenobarbital (PB) (a CAR activator), induced considerable changes in hepatic and plasma lipid profiles (Calandre et al. 1991; Luoma 1988), and patients with type 2 diabetes treated with PB showed decreased plasma glucose (Lahtela et al. 1985).

Acute activation of SXR by pregnenolone-16α-carbonitrile (PCN; i.p. injection of 40 mg/kg) was found to decrease hepatic triglyceride and cholesterol levels and to reduce the expression of the lipogenic transcription factor SREBP1 (Roth et al. 2008). On the other hand, acute treatment with PCN at low doses (i.p. injection of 20 mg/kg) was reported to increase hepatic triglycerides in mice (Nakamura et al. 2007). Interestingly, transgenic mice expressing activated SXR in liver (chronic SXR activation) or humanized SXR (hSXR) mice treated with rifampicin (agonist of hSXR) for 5 weeks (i.p. injection of 10 mg/kg) exhibited nonalcoholic fatty liver (Zhou et al. 2006).

These hepatic lipid disturbances were mainly explained by the ability of activated SXR to stimulate fatty acid uptake by directly binding the promoter of *CD36*. No activation of the lipogenic transcriptional factor SREBP-1α and its primary lipogenic targets, *FAS* and *ACC*, was found (Zhou et al. 2006). In addition, SXR activation decreased hepatic fatty acid oxidation through the reduction of mRNA levels of CPT1A (Nakamura et al. 2007), which controls the entry of long-chain fatty acids into the mitochondria, through repression of insulin response forkhead factor FoxA2, hepatocyte nuclear factor 4α (HNF4α), and PPAR-γ coactivator-1α (PGC1α) (Bhalla et al. 2004; Li and Chiang 2005). Also, the constitutively activated mutant of SXR or the pharmacological activation of SXR induced the mRNA regulation of the nuclear receptor PPAR-γ and two lipogenic enzymes, SCD1 and fatty acid elongase (Nakamura et al. 2007; Zhou et al. 2006).

More recently, it was demonstrated that THRSP homolog (S14), a protein that transduces hormone- and nutrient-related signals to genes involved in lipid metabolism, is an SXR target gene in human hepatocytes (Moreau et al. 2009). Consistent with a metabolic role of SXR, hepatosteatosis and impaired lipid metabolism have been observed in patients treated with rifampicin (Morere et al. 1975), carbapazepin (Grieco et al. 2005), or nifedipin (Babany et al. 1989), which are pharmaceuticals known to activate SXR.

There is also evidence that SXR regulates hepatic gluconeogenic enzymes and genes (Argaud et al. 1991; Kodama et al. 2004; Ueda et al. 2002). For instance, *PEPCK1* and *G6Pase* genes were found downregulated in transgenic mice expressing constitutively activated SXR (Zhou et al. 2006) or in mice treated with PCN (Nakamura et al. 2007). In the fasting state, where the liver increases production of glucose by enhancing gluconeogenesis and glycogenolysis, SXR activation by PCN treatment (i.p. 40 mg/kg) caused an SXR-dependent decreased binding of cAMP-response element-binding protein (CREB) to the G6Pase promoter (Kodama et al. 2007), a mechanism explaining how SXR can inhibit the *G6Pase* gene expression.

Altogether, there is evidence that AhR and Arnt can affect β-cell function and that lipid and glucose homeostasis can be deregulated through the activation of AhR, CAR, and SXR. However, contrasting results regarding metabolic outcomes associated with AhR, CAR, and SXR activation exist. This may result from the dose and time of exposure to xenobiotics, as well as the route of administration. In addition, recent data have shown that some CAR and SXR ligands, like PB, may not be highly specific as previously thought. For instance, PB was unexpectedly found to induce mRNA expression of PPAR-α, HNF4α, CYP4A10, and CYP4A14 in cells lacking CAR and SXR (Tamasi et al. 2009). More research studies are warranted to further characterize the metabolic functions of AhR, CAR, and SXR.

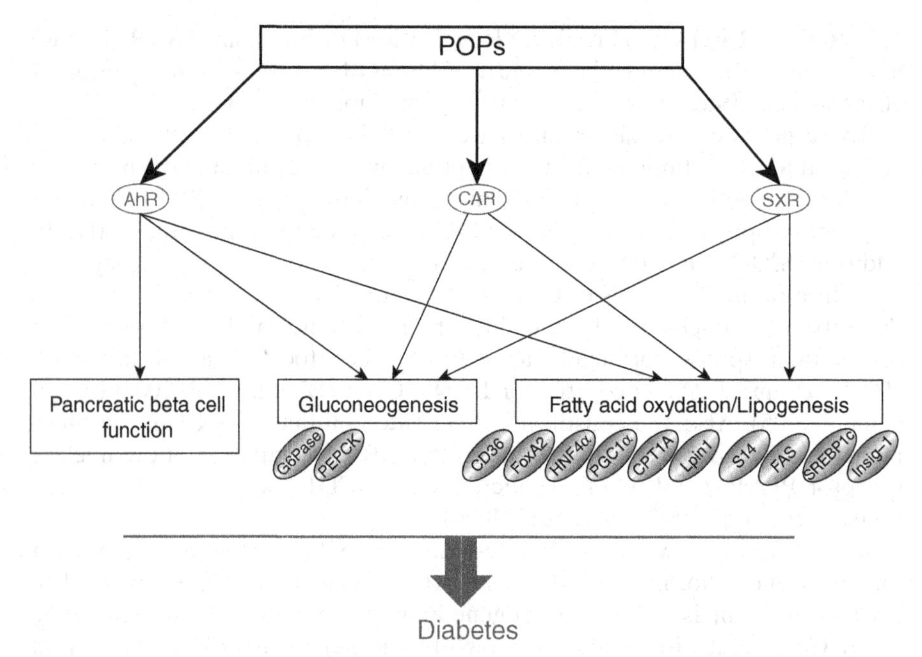

Figure 6.3. Mechanistic basis for elevation in risk of diabetes caused by persistent organic pollutants (POPs). Chronic and daily exposure to POPs may disrupt metabolic homeostasis and increase the risk of diabetes. Activation of AhR, CAR, and SXR may in concert regulate the expression of critical genes involved in lipid and glucose homeostasis.

CHALLENGES AND PERSPECTIVES

Although significant advances in the field have clearly defined a causal role of POPs in the development of diabetes and metabolic disorders, the exact biochemical events involved are still poorly identified. Activation of AhR, CAR, or SXR is likely to affect insulin action as well as glucose and lipid metabolism, thereby increasing the risk to develop diabetes (Figure 6.3). POPs may also induce their deleterious metabolic effects through additional mechanisms including competitive binding to receptors and regulation of epigenetic programming. Better understanding of these mechanisms will help to establish novel strategies aimed at preventing and treating diabetes.

REFERENCES

Alonso-Magdalena P, Quesada I, Nadal A. 2011. Endocrine disruptors in the etiology of type 2 diabetes mellitus. Nat Rev Endocrinol 7:346–353.

Argaud D, Halimi S, Catelloni F, Leverve XM. 1991. Inhibition of gluconeogenesis in isolated rat hepatocytes after chronic treatment with phenobarbital. Biochem J 280:663–669.

Arsenescu V, Arsenescu RI, King V, Swanson H, Cassis LA. 2008. Polychlorinated biphenyl-77 induces adipocyte differentiation and proinflammatory adipokines and promotes obesity and atherosclerosis. Environ Health Perspect 116:761–768.

Ashida H, Nishiumi S, Fukuda I. 2008. An update on the dietary ligands of the AhR. Expert Opin Drug Metab Toxicol 4:1429–1447.

Babany G, Uzzan F, Larrey D, Degott C, Bourgeois P, Rene E, et al. 1989. Alcoholic-like liver-lesions induced by nifedipine. J Hepatol 9:252–255.

Bhalla S, Ozalp C, Fang S, Xiang L, Kemper JK. 2004. Ligand-activated pregnane X receptor interferes with HNF-4 signaling by targeting a common coactivator PGC-1alpha. Functional implications in hepatic cholesterol and glucose metabolism. J Biol Chem 279:45139–45147.

Breuker C, Moreau A, Lakhal L, Tamasi V, Parmentier Y, Meyer U, et al. 2010. Hepatic expression of thyroid hormone-responsive spot 14 protein is regulated by constitutive androstane receptor (NR1I3). Endocrinology 151:1653–1661.

Calandre EP, Rodriquez-Lopez C, Blazquez A, Cano D. 1991. Serum lipids, lipoproteins and apolipoproteins A and B in epileptic patients treated with valproic acid, carbamazepine or phenobarbital. Acta Neurol Scand 83:250–253.

Casabar RCT, Das PC, DeKrey GK, Gardiner CS, Cao Y, Rose RL, et al. 2010. Endosulfan induces CYP2B6 and CYP3A4 by activating the pregnane X receptor. Toxicol Appl Pharmacol 245:335–343.

CDC. 2011. National Diabetes Fact Sheet, 2011. Available: http://www.cdc.gov/diabetes/pubs/pdf/ndfs_2011.pdf [accessed April 11, 2013].

Chang TK, Waxman DJ. 2006. Synthetic drugs and natural products as modulators of constitutive androstane receptor (CAR) and pregnane X receptor (PXR). Drug Metab Rev 38:51–73.

Coumoul X, Diry M, Barouki R. 2002. PXR-dependent induction of human CYP3A4 gene expression by organochlorine pesticides. Biochem Pharmacol 64:1513–1519.

Croutch CR, Lebofsky M, Schramm KW, Terranova PF, Rozman KK. 2005. 2,3,7,8-Tetrachlorodibenzo-p-dioxin (TCDD) and 1,2,3,4,7,8-hexachlorodibenzo-p-dioxin (HxCDD) alter body weight by decreasing insulin-like growth factor I (IGF-I) signaling. Toxicol Sci 85:560–571.

Darnerud PO, Atuma S, Aune M, Bjerselius R, Glynn A, Grawe KP, et al. 2006. Dietary intake estimations of organohalogen contaminants (dioxins, PCB, PBDE and chlorinated pesticides, e.g. DDT) based on Swedish market basket data. Food Chem Toxicol 44:1597–1606.

Djousse L, Gaziano JM, Buring JE, Lee IM. 2011. Dietary omega-3 fatty acids and fish consumption and risk of type 2 diabetes. Am J Clin Nutr 93:143–150.

Dong B, Saha PK, Huang WD, Chen WL, Abu-Elheiga LA, Wakil SJ, et al. 2009. Activation of nuclear receptor CAR ameliorates diabetes and fatty liver disease. PNAS 106:18831–18836.

Fisher BE. 1999. Most unwanted. Environ Health Perspect 107:A18–A23.

Fromberg A, Granb K, Hojgard A, Fagt S, Larsen JC. 2011. Estimation of dietary intake of PCB and organochlorine pesticides for children and adults. Food Chem 125: 1179–1187.

Gao J, He JH, Zhai YG, Wada TR, Xie W. 2009. The constitutive androstane receptor is an anti-obesity nuclear receptor that improves insulin sensitivity. J Biol Chem 284:25984–25992.

Grieco A, Forgione A, Miele L, Vero V, Greco AV, Gasbarrini A, et al. 2005. Fatty liver and drugs. Eur Rev Med Pharmacol Sci 9:261–263.

Gunton JE, Kulkarni RN, Yim SH, Okada T, Hawthorne WJ, Tseng YH, et al. 2005. Loss of ARNT/HIF1 beta mediates altered gene expression and pancreatic-islet dysfunction in human type 2 diabetes. Cell 122:337–349.

Ibrahim MM, Fjaere E, Lock EJ, Naville D, Amlund H, Meugnier E, et al. 2011. Chronic consumption of farmed salmon containing persistent organic pollutants causes insulin resistance and obesity in mice. PLoS ONE 6:e25170.

Kaushik M, Mozaffarian D, Spiegelman D, Manson JE, Willett WC, Hu FB. 2009. Long-chain omega-3 fatty acids, fish intake, and the risk of type 2 diabetes mellitus. Am J Clin Nutr 90:613–620.

Kazlauskas A, Poellinger L, Pongratz I. 1999. Evidence that the co-chaperone p23 regulates ligand responsiveness of the dioxin (aryl hydrocarbon) receptor. J Biol Chem 274:13519–13524.

Kodama S, Koike C, Negishi M, Yamamoto Y. 2004. Nuclear receptors CAR and PXR cross talk with FOXO1 to regulate genes that encode drug-metabolizing and gluconeogenic enzymes. Mol Cell Biol 24:7931–7940.

Kodama S, Moore R, Yamamoto Y, Negishi M. 2007. Human nuclear pregnane X receptor cross-talk with CREB to repress cAMP activation of the glucose-6-phosphatase gene. Biochem J 407:373–381.

Kretschmer XC, Baldwin WS. 2005. CAR and PXR: Xenosensors of endocrine disrupters? Chem Biol Interact 155:111–128.

Lahtela JT, Arranto AJ, Sotaniemi EA. 1985. Enzyme inducers improve insulin sensitivity in non-insulin-dependent diabetic subjects. Diabetes 34:911–916.

Lee JH, Wada T, Febbraio M, He J, Matsubara T, Lee MJ, et al. 2010. A novel role for the dioxin receptor in fatty acid metabolism and hepatic steatosis. Gastroenterology 139:653–663.

Li T, Chiang JY. 2005. Mechanism of rifampicin and pregnane X receptor inhibition of human cholesterol 7 alpha-hydroxylase gene transcription. Am J Physiol Gastrointest Liver Physiol 288:G74–G84.

Luoma PV. 1988. Microsomal enzyme induction, lipoproteins and atherosclerosis. Pharmacol Toxicol 62:243–249.

Lyche JL, Nourizadeh-Lillabadi R, Almaas C, Stavik B, Berg V, Skare JU, et al. 2010. Natural mixtures of persistent organic pollutants (Pop) increase weight gain, advance puberty, and induce changes in gene expression associated with steroid hormones and obesity in female zebrafish. J Toxicol Environ Health A 73: 1032–1057.

Machala M, Vondracek J, Blaha L, Ciganek M, Neca J. 2001. Aryl hydrocarbon receptor-mediated activity of mutagenic polycyclic aromatic hydrocarbons determined using in vitro reporter gene assay. Mutat Res-Genet Toxicol Environ Mutagen 497:49–62.

Maglich JM, Lobe DC, Moore JT. 2009. The nuclear receptor CAR (NR1I3) regulates serum triglyceride levels under conditions of metabolic stress. J Lipid Res 50: 439–445.

di Masi A, De Marinis E, Ascenzi P, Marino M. 2009. Nuclear receptors CAR and PXR: Molecular, functional, and biomedical aspects. Mol Aspects Med 30:297–343.

Meyer BK, Pray-Grant MG, Vanden Heuvel JP, Perdew GH. 1998. Hepatitis B virus X-associated protein 2 is a subunit of the unliganded aryl hydrocarbon receptor core complex and exhibits transcriptional enhancer activity. Mol Cell Biol 18: 978–988.

Moreau A, Teruel C, Beylot M, Albalea V, Tamasi V, Umbdenstock T, et al. 2009. A novel pregnane X receptor and S14-mediated lipogenic pathway in human hepatocyte. Hepatology 49:2068–2079.

Morere P, Nouvet G, Stain JP, Paillot B, Metayer J, Hemet J. 1975. Information supplied by liver-biopsy in 100 patients with tuberculosis. Sem Hop 51:2095–2102.

Nakamura K, Moore R, Negishi M, Sueyoshi T. 2007. Nuclear pregnane X receptor cross-talk with FoxA2 to mediate drug-induced regulation of lipid metabolism in fasting mouse liver. J Biol Chem 282:9768–9776.

Neel BA, Sargis RM. 2011. The paradox of progress: Environmental disruption of metabolism and the diabetes epidemic. Diabetes 60:1838–1848.

Randi AS, Sanchez MS, Alvarez L, Cardozo J, Pontillo C, de Pisarev DLK. 2008. Hexachlorobenzene triggers AhR translocation to the nucleus, c-Src activation and EGFR transactivation in rat liver. Toxicol Lett 177:116–122.

Roth A, Looser R, Kaufmann M, Blattler SM, Rencurel F, Huang WD, et al. 2008. Regulatory cross-talk between drug metabolism and lipid homeostasis: Constitutive androstane receptor and pregnane X receptor increase Insig-1 expression. Mol Pharm 73:1282–1289.

Ruzzin J. 2012. Public health concern behind the exposure to persistent organic pollutants and the risk of metabolic diseases. BMC Public Health 12:298.

Ruzzin J, Jacobs DR, Jr. 2012. The secret story of fish: Decreasing nutritional value due to pollution? Br J Nutr 108:397–399.

Ruzzin J, Petersen R, Meugnier E, Madsen L, Lock EJ, Lillefosse H, et al. 2010. Persistent organic pollutant exposure leads to insulin resistance syndrome. Environ Health Perspect 118:465–471.

Ruzzin J, Lee DH, Carpenter DO, Jacobs D, Jr. 2012. Reconsidering metabolic diseases: The impacts of persistent organic pollutants. Atherosclerosis 224:1–3.

Schafer KS, Kegley SE. 2002. Persistent toxic chemicals in the US food supply. J Epidemiol Community Health 56:813–817.

Schecter A, Colacino J, Haffner D, Patel K, Opel M, Papke O, et al. 2010. Perfluorinated compounds, polychlorinated biphenyls, and organochlorine pesticide contamination in composite food samples from Dallas, Texas, USA. Environ Health Perspect 118:796–802.

Skene SA, Dewhurst IC, Greenberg M. 1989. Polychlorinated dibenzo-para-dioxins and polychlorinated dibenzofurans—The risks to human health—A review. Hum Toxicol 8:173–203.

Takeuchi S, Iida M, Yabushita H, Matsuda T, Kojima H. 2008. In vitro screening for aryl hydrocarbon receptor agonistic activity in 200 pesticides using a highly sensitive reporter cell line, DR-EcoScreen cells, and in vivo mouse liver cytochrome P450-1A induction by propanil, diuron and linuron. Chemosphere 74:155–165.

Tamasi V, Juvan P, Beer M, Rozman D, Meyer UA. 2009. Transcriptional activation of PPARalpha by phenobarbital in the absence of CAR and PXR. Mol Pharm 6:1573–1581.

Thayer KA, Heindel JJ, Bucher JR, Gallo MA. 2012. Role of environmental chemicals in diabetes and obesity: A national toxicology program workshop report. Environ Health Perspect 120:779–789.

Ueda A, Hamadeh HK, Webb HK, Yamamoto Y, Sueyoshi T, Afshari CA, et al. 2002. Diverse roles of the nuclear orphan receptor CAR in regulating hepatic genes in response to phenobarbital. Mol Pharmacol 61:1–6.

Van den Berg M, Birnbaum LS, Denison M, De Vito M, Farland W, Feeley M, et al. 2006. The 2005 World Health Organization reevaluation of human and mammalian toxic equivalency factors for dioxins and dioxin-like compounds. Toxicol Sci 93:223–241.

Waller CL, McKinney JD. 1995. 3-Dimensional quantitative structure-activity-relationships of dioxins and dioxin-like compounds—Model validation and Ah receptor characterization. Chem Res Toxicol 8:847–858.

Wang XL, Suzuki R, Lee K, Tran T, Gunton JE, Saha AK, et al. 2009. Ablation of ARNT/HIF1 beta in liver alters gluconeogenesis, lipogenic gene expression, and serum ketones. Cell Metab 9:428–439.

Wang C, Xu CX, Krager SL, Bottum KM, Liao DF, Tischkau SA. 2011. Aryl hydrocarbon receptor deficiency enhances insulin sensitivity and reduces PPAR-alpha pathway activity in mice. Environ Health Perspect 119:1739–1744.

Zhou J, Zhai YG, Mu Y, Gong HB, Uppal H, Toma D, et al. 2006. A novel pregnane X receptor-mediated and sterol regulatory element-binding protein-independent lipogenic pathway. J Biol Chem 281:15013–15020.

CHAPTER 7

Cardiovascular Disease and Hypertension

MARIAN PAVUK and NINA DUTTON

ABSTRACT

Background: Persistent organic pollutants (POPs) have been implicated in altering a variety of physiological processes related to cardiovascular function. Occupationally and accidentally exposed cohorts have been studied for mortality from cardiovascular disease (CVD) and exposure to dioxins since the 1970s. Recently, relationships between hypertension and low-level exposure to polychlorinated biphenyls (PCBs) and other POPs have been intensely studied.

Objectives: This chapter reviews evidence for relationships between dioxin exposure and morbidity from CVD and ischemic heart disease (IHD), focusing on recent cohort updates. Associations between hypertension and exposure to PCBs and other POPs are evaluated in studies of the general U.S. population and populations near contaminated sites.

Discussion: Results of CVD mortality studies in dioxin-exposed cohorts using external comparisons found little evidence of association. In internal comparison studies with detailed exposure assessments, more evidence of increased IHD and CVD mortality was reported. Two cohort studies with proper adjustment for major CVD risk factors showed little or no association with CVD mortality. In hypertension studies, effects of dioxin-like or non-dioxin-like PCB congeners could not be separated clearly.

Conclusions: Mixed results of studies in dioxin-exposed cohorts do not support a strong association between dioxin exposure and cardiovascular mortality in humans. Suggestive findings on PCBs and other POPs and hypertension require confirmation in follow-up studies.

Effects of Persistent and Bioactive Organic Pollutants on Human Health, First Edition.
Edited by David O. Carpenter.

INTRODUCTION

In the first part, this chapter will review evidence on relationships between mortality from cardiovascular disease (CVD) and ischemic heart disease (IHD) and exposure to 2,3,7,8-tetrachlorodibenzo-p-dioxin (TCDD) and other dioxin-like compounds. Information on morbidity from CVD and IHD will also be reviewed where available. A number of occupationally and accidentally exposed cohorts were studied for these associations; we will focus on recent updates for these cohorts and compare the results with earlier data. In the second part, associations between hypertension and exposure to polychlorinated biphenyls (PCBs) and various other persistent organic pollutants (POPs) will be evaluated in a smaller number of studies of the general U.S. population and specific populations residing near contaminated sites.

TCDD (also often referred to as dioxin) was an unintended contaminant in the production of trichlorophenol and phenoxy herbicides, specifically that of 2,4,5-trichlorophenoxyacetic acid (2,4,5-T). It is the most studied and most toxic of the group of chlorinated dibenzo-p-dioxins and dibenzofurans. Different products at different manufacturing facilities were contaminated by different amounts of TCDD (from ranges of parts per trillion to parts per million) with no or very little contamination from other dioxins (Piacitelli et al. 2000). The production of pentachlorophenol and related products led to contamination and exposure of workers to higher chlorinated dioxins (Collins et al. 2009b). By contrast, PCBs were produced in large quantities and widely used in numerous industrial applications; some PCB congeners are dioxin-like, while others are non-dioxin-like and act through other mechanisms (ATSDR 2000).

Historically, human health studies of exposure to dioxins can be divided into cohorts of occupationally exposed chemical workers, environmentally exposed cohorts, Vietnam veteran cohorts, and food poisoning cohorts. The occupationally exposed chemical worker studies include the BASF cohort (Ott and Zober 1996), Dutch cohort (Boers et al. 2010, 2012; Hooiveld et al. 1998), Hamburg cohort (Flesch-Janys et al. 1995, 1998; Manuwald et al. 2012), Midland cohort (Collins et al. 2009a,b), New Zealand cohort (McBride et al. 2009a,b; 't Mannetje et al. 2005), National Institute for Occupational Safety and Health (NIOSH) cohort (Fingerhut et al. 1991; Steenland et al. 1999), and the International Agency for Research on Cancer (IARC) cohort (Vena et al. 1998), which is a combination of 36 different cohorts of chemical workers including some of the same workers from the NIOSH, Dutch, Midland, New Zealand, and Hamburg cohorts. Several other occupational cohorts were also included such as Finnish herbicide sprayers (Asp et al. 1994) and Canadian sawmill workers (Demers et al. 2006; Hertzman et al. 1997) because they included long-term follow-up with detailed mortality information. The largest environmentally TCDD-exposed population is the Seveso population, which was exposed during the industrial accident in Seveso, Italy (Bertazzi et al. 2001; Consonni et al. 2008). Vietnam veterans involved in spraying of herbicides include the cohort of veterans involved in Operation Ranch Hand (RH) (Ketchum and

Michalek 2005; Michalek et al. 1998) and the Army's Chemical Corps veterans (Dalager and Kang 1997). Food poisoning cohorts include Yucheng (Hsu et al. 1985; Tsai et al. 2007) and Yusho (Kashima et al. 2011; Kuratsune et al. 1972), both of which were rice oil contamination incidents where most of the exposure was to dibenzofurans (PCDFs) and PCBs (Ryan et al. 1990).

Because of the lack of individual TCDD or dioxin exposure assessments, this chapter does not include studies of workers from the paper and pulp industry (McLean et al. 2006) and leather tanning and processing (Mikoczy et al. 1994), or of fishermen (Mikoczy and Rylander 2009).

Multiple end points, with a focus on cancer mortality, were usually examined for most cohorts and multiple references are available for each cohort (Boffetta et al. 2011; IARC 1997; National Academy of Sciences 2007). CVD mortality was usually included in the overall assessment of mortality along with other underlying causes of deaths. In this chapter, we will provide and contrast, where available, results for the most recent mortality update with earlier follow-up results and/or alternative statistical analyses of the same cohort.

Total CVD and IHD mortality are reviewed for the cohorts noted earlier. Other subtypes of CVD mortality (e.g., myocardial infarction, hypertension, cardiopathy, chronic ischemic disease) were inconsistently reported and are not examined in this chapter. Hypertension not related to CVD mortality is examined in the second part of this chapter. CVD mortality is generally accurately coded on death certificates. The *International Classification of Diseases*, 9th Revision (WHO 1978) or 10th Revision (WHO 2010), was used to code causes of death attributed to IHD and all CVDs in most studies.

The majority of occupational studies, as well as of the Seveso and Vietnam veteran cohorts, focused on mortality associated with exposure to TCDD. The term "dioxin," which is sometimes used to refer to this congener only, also refers to a diverse group of structurally related, environmentally persistent chemicals that exert toxic effects through a common pathway mediated by the aryl hydrocarbon receptor (AhR) (Van den Berg et al. 1998, 2006). We will use TCDD to refer to 2,3,7,8-tetrachlorodibenzo-*p*-dioxin and "dioxins" for the broader definition of dioxin-like congeners.

ANIMAL STUDIES AND LABORATORY EVIDENCE

Dioxins, including dioxin-like coplanar PCBs, have been shown to alter cardiovascular end points in different groups of laboratory animals (Dalton et al. 2001; Hermansky et al. 1988; Jokinen et al. 2003; Kopf et al. 2010; Lind et al. 2004; NTP 2006).

In mice exposed to acute high doses of TCDD, high triglyceride levels and high blood pressure were observed (Dalton et al. 2001). Earlier and more severe atherosclerotic lesions were also seen in *ApoE−/−* mice exposed to subchronic doses of TCDD (Dalton et al. 2001). TCDD-exposed adult mice developed elevated blood pressure, markers of oxidative stress, and higher heart weight (Kopf et al. 2008).

In rats exposed chronically to TCDD or PCB 126, increased incidence of degenerative cardiovascular lesions, including cardiomyopathy and chronic active arthritis, was reported (Jokinen et al. 2003). The dioxin-like PCB 126 also increased serum cholesterol levels, blood pressure, and heart weight (Lind et al. 2004). As with other toxic outcomes, fetal TCDD exposure may result in more tissue damage and TCDD toxicity than seen in adult animals (Kopf and Walker 2009; Thackaberry et al. 2005).

Xie et al. 2006 reported that some direct effects in heart muscle cells were observed, including abnormal depolarization of key calcium signaling pathways (Xie et al. 2006) altering inflammation, oxidative stress, or changes in gene activation (Arzuaga et al. 2007; Lund et al. 2005) following exposure to TCDD. Mitochondrial dysfunction reported in cell culture (Biswas et al. 2008) may suggest an additional mechanism for the effect of dioxin on the cardiovascular system.

It was suggested by Lind et al. (2004) that gender and estrogen levels may be factors of concern in the cardiovascular effects of dioxins. Molecular interactions between the AhR and the estrogen receptors (ERα and ERβ) are well documented, but the interactive effects between AhR agonists and ER agonists in human endothelial cells have not been previously investigated (Beischlag et al. 2008; Ohtake et al. 2003). Studies investigating the interactions between AhR and ER signaling mostly refer to cells and tissues predominately expressing ERα (Chang et al. 2007; Spink et al. 2009); however, the gene expression of PCB 126-induced CYP1A1 and CYP1B1 was unaffected by the addition of estradiol in another study in breast cancer cells (Matthews et al. 2007).

Heart disease in humans is complex, chronic, and involves a multistage process unfolding over decades of life with the interplay of numerous inherited and modifiable factors. Laboratory studies, while valuable in uncovering underlying mechanisms and formulating hypotheses, are no substitute for studies in humans designed to identify and analyze life experiences and exposures unique to humans.

STANDARDIZED MORTALITY RATIOS (SMRs) FOR HEART DISEASE

External Comparisons of Exposed versus Unexposed Groups

The measure of risk for analyses using external comparisons is the SMRs. The SMR compares the number of deaths observed in the exposed group (e.g., dioxin-exposed factory workers) to the number of deaths expected in the unexposed group, usually the general population. The SMR is standardized for age and sex and sometimes calendar year or country. In occupational studies, the exposed group—an employed population—is on average healthier than the general population. This is known as the healthy worker effect and results in a bias toward the null (no effect of exposure on mortality). The SMRs for CVD may be biased downward (generally below 1) (Burns et al. 2011;

McMichael 1976) because of the healthy worker effect. As such, observed elevated SMRs are thus useful indicators of excess mortality in worker populations and warrant further investigation using more robust study designs and statistical analyses to identify occupational exposures and risk factors potentially associated with an observed excess mortality.

Most SMR analyses presented in Table 7.1 found no elevated risk for either overall CVD or IHD. This included the Seveso (Bertazzi et al. 2001; Consonni et al. 2008) and Yucheng (Tsai et al. 2007) studies. The most recent update of the Yusho study found an increase in CVD but a decrease in IHD mortality (Kashima et al. 2011). Two occupational studies found elevated SMRs for CVD or IHD; the follow-up of the NIOSH cohort reported elevated SMR for IHD (1.09, 95% confidence interval [CI] = 1.00–1.20, IHD) (Steenland et al. 1999), and the update of the Hamburg cohort (Manuwald et al. 2012) reported a statistically significant SMR for all CVD in men (SMR = 1.16, 95% CI = 1.02–1.31) and a statistically significantly lower risk in women (SMR = 0.74, 95% CI = 0.56–0.94). Earlier results from the NIOSH cohort by Fingerhut et al. (1991), earlier IHD mortality in Hamburg cohort (Flesch-Janys et al. 1998), and the results from IARC cohort for CVD and IHD mortality reported SMRs below 1 (Vena et al. 1998).

Several studies used internal comparisons (unexposed study participants) to analyze mortality. The RH study (Ketchum and Michalek 2005) showed a modestly elevated nonsignificant relative risk (RR) for all CVDs (RR = 1.3; $p = 0.07$) at the 20-year follow-up in contrast to an RR of 1.00 at the 15-year mortality follow-up. The enlisted ground crew, the highest exposed occupational category of RH, had an elevated risk of IHD mortality at 20-year follow-up, but not at the 15-year mark. In the study of the Army Chemical Corps Vietnam veterans, Dalager and Kang (1997) did not report a statistically significant increased risk for IHD (RR = 1.06, 95% CI = 0.62–1.82). No significant elevations in hazard ratios (HRs) were observed in Dutch workers from factory A or B with HRs marginally higher for IHD (Boers et al. 2010). The only statistically significant increases in CVD or IHD were reported for the comparison of exposed to unexposed workers in the IARC study (RR = 1.51, 95% CI = 1.17–1.96 for all CVD; and RR = 1.67, 95% CI = 1.23–2.26 for IHD) (Vena et al. 1998).

External Comparisons with More Detailed Exposure Assessment

Four occupational cohorts and the Seveso population were studied using stratified TCDD exposure indicators in the SMR analyses (Collins et al. 2009b; McBride et al. 2009a; Ott and Zober 1996; Steenland et al. 1999). In the BASF study (Ott and Zober 1996), the IHD SMRs (95% CIs) corresponding to the exposure categories of <0.1, 0.1–0.99, and ≥1 µg/kg TCDD body weight were 0.9 (0.3–1.8), 0.7 (0.2–1.7), and 0.6 (0.2–1.3), respectively. The exposure categories for the Midland, New Zealand, and NIOSH study are shown in Table 7.2 and Table 7.5. The SMRs (95% CIs) for IHD corresponding to those exposure categories in the cohort of Midland pentachlorophenol (PCP) workers were 1.2 (0.9–1.6), 1.1 (0.78–1.4), and 1.0 (0.7–1.4) (Collins et al. 2009b); in the New Zealand cohort of PCP workers, the SMRs (95% CI) were 1.4 (0.8–2.3), 1.3

TABLE 7.1. SMRs for Mortality from All CVD and IHD for External Comparisons and Relative Risks for Internal Comparisons by Dioxin Exposure Status (Exposed/Nonexposed)

Cohort	No.	Obs. Deaths	All CVD	Obs. Deaths	IHD	Reference
External comparisons						
BASF (1953–1992)	243	37	0.70 (0.4–1.1)	16	0.80 (0.6–1.2)	Ott and Zober (1996)
Canadian sawmills (1950–1990)	23,829	2,013	0.74 (0.71–0.76)		NA	Hertzman et al. (1997)
Canadian sawmills (1950–1995)	27,464	2,510	0.98 (0.94–1.02)		NA	Demers et al. (2006)
Dutch: factory A (1955–1991)	549	45	1.0 (0.8–1.4)	33	1.2 (0.8–1.6)	Hooiveld et al. (1998)
Finnish sprayers (1972–1989)	1,909		NA	148	0.94 (0.80–1.10)	Asp et al. (1994)
Hamburg (1952–1989)	1,177	156	1.06 (0.90–1.24)	76	0.97 (0.77–1.22)	Flesch-Janys et al. (1998)
Hamburg (1955–2007)—men	1,132	251	1.16 (1.02–1.31)		NA	Manuwald et al. (2012)
Hamburg (1955–2007)—women	387	58	0.74 (0.56–0.94)		NA	Manuwald et al. (2012)
IARC (1939–1992), exposed to TCDD/HCD	13,831	1,170	0.94 (0.88–0.99)	789	0.97 (0.90–1.04)	Vena et al. (1998)
IARC (1939–1992), not exposed to TCDD/HCD	7,553	588	0.86 (0.79–0.93)	494	0.85 (0.77–0.94)	Vena et al. (1998)
Midland, TCP workers (1942–2003)	1,615		NA	218	1.1 (0.9–1.2)	Collins et al. (2009a)
Midland, PCP workers (1940–2003)	773		NA	131	1.1 (0.9–1.3)	Collins et al. (2009b)
New Zealand (1969–2000)—Workers	813	51	0.96 (0.72–1.27)	38	1.04 (0.74–1.43)	‘t Mannetje et al. (2005)
New Zealand (1969–2000)—Sprayers	699	33	0.52 (0.36–0.72)	22	0.49 (0.31–0.75)	

New Zealand—TCP (1969–2004)	1,599	66	1.1 (0.9–1.5)	61	1.1 (0.9–1.5)	McBride et al. (2009a)
New Zealand—New Plymouth (1969–2004)	1,754	82	1.02 (0.81–1.27)	75	1.09 (0.86–1.36)	McBride et al. (2009b)
NIOSH (1942–1987)	5,172		NA	393	0.96 (0.87–1.06)	Fingerhut et al. (1991)
NIOSH (1942–1993)	5,132		NA	456	1.09 (1.00–1.20)	Steenland et al. (1999)
Seveso (1976–1996)	6,745	265	1.00 (0.8–1.1)	97	1.0 (0.8–1.2)	Bertazzi et al. (2001)
Seveso: zone A (1976–2001)	804	45	1.06 (0.79, 1.42)	13	0.83 (0.48–1.43)	Consonni et al. (2008)
Seveso: zone B (1976–2001)	5,941	289	0.99 (0.88, 1.11)	102	0.95 (0.78, 1.15)	Consonni et al. (2008)
Yucheng (1980–2003)	1,823	54	1.00 (0.8–1.3)		NA	Tsai et al. (2007)
Yusho (1968–2009)	2,000	180	1.17 (1.01, 1.35)	56	0.85 (0.65, 1.10)	Kashima et al. (2011)
Internal comparisons						
Army Chemical Corps (1965–1991)	2,872/2,737	50/20	1.06 (0.62–1.82)		NA	Dalager and Kang (1997)
Ranch Hand (1962–1993)	1,261/19,080	39	1.00 (0.7–1.3)	16	1.4 (0.8–2.1)[a]	Michalek et al. (1998)
Ranch Hand (1962–1999)	20,340	66/745	1.30 (1.0–1.6)	28/281	1.7 (1.1–2.5)[a]	Ketchum and Michalek (2005)
IARC (1939–1993)	13,831/7553	1151/582	1.51 (1.17–1.96)	775/391	1.67 (1.23–2.26)	Vena et al. (1998)
Dutch: factory A (1955–2006)	539/482	77/37	1.07 (0.70–1.62)	43/18	1.15 (0.66–1.98)	Boers et al. 2010
Dutch: factory B (1955–2006)	411/622	31/37	1.06 (0.66–1.70)	18/15	1.56 (0.79–3.11)	Boers et al. (2010)

[a]In Ranch Hand enlisted ground personnel only.
NA, not available.

TABLE 7.2. Mortality from IHD in 773 Midland and 1599 New Zealand Pentachlorophenol (PCP) Workers

Author/Study/ Exposure Matrix	No.	CVD	No.	IHD	Risk Factors Adjusted for
Collins et al. (2009b) Midland—PCP (cumulative TEQ ppb-years)			773	RR	Birth year, hire year
0.01–0.69		NA	46	1.0 (Ref.)	
0.70–3.99		NA	46	1.8 (0.4–8.8)	
4.00–113.37		NA	39	1.6 (0.4–7.0)	p-value for trend = 0.74
McBride et al. (2009a) New Zealand—PCP (cumulative TCDD ppt)				RR	Age, sex, hire year, and birth year
0–68.3		NA	–	1.00 (Ref.)	
68.4–475		NA	–	1.24 (0.58–2.64)	
475.1–2085.7		NA	–	1.32 (0.60–2.90)	
2085.8+		NA	–	0.93 (0.37–2.35)	p-value for trend = 0.50

(0.7–2.0), 1.1 (0.6–1.8), and 0.9 (0.5–1.6) (McBride et al. 2009a); and for each category of the cumulative exposure score in the NIOSH study, the SMRs were 0.93, 1.00, 1.05, 0.97, 1.10, 1.20, and 1.28 (Steenland et al. 1999; no 95% CIs were shown, p-trend = 0.14). Results for the BASF, New-Zealand, and Midland cohorts are in agreement with those shown for internal comparisons analyses (Table 7.2 and Table 7.7). In the NIOSH cohort, there was no dose–response trend (p = 0.14) using the external comparison of the U.S. general population, but a significant trend was observed using the same exposure categories and the internal comparison (p = 0.05; Table 7.5). SMRs for the two highest-exposure zones in the Seveso study are shown in Table 7.1, neither showing a significant increase in CVD or IHD SMRs.

An update to the Hamburg cohort (Manuwald et al. 2012) also includes SMR analyses stratified by TCDD levels in chemical workers; results support the finding from the overall SMR analyses of higher risk of CVD mortality. These data are shown in Table 7.4 and are discussed in the section on the Hamburg cohort mortality using internal comparisons because of new results on female workers. However, regression analyses with the previously used gas company worker or internal chemical worker comparisons (Flesch-Janys et al. 1995) based on TCDD levels were not presented by Manuwald et al. (2012).

INTERNAL COMPARISONS

New Zealand and Midland Cohorts

Since the last mortality review by Humblet et al. (2008), results on these two new cohorts were published using internal comparisons with dioxin exposure categories based on individual measurements. The New Zealand cohort measured TCDD in 346 of 1599 workers (McBride et al. 2009b). Seventy percent (241/346) of the serum sample participants were exposed workers. The current TCDD levels were used to calculate cumulative TCDD estimates in all workers included in the statistical analyses using detailed information on duration of employment and contamination of workplace during production. No increase in the IHD mortality was observed and the RR in the highest-exposure category was below 1.

In the Midland cohort, total dioxin toxic equivalent (TEQ) based on five dioxins was calculated to estimate the total dioxin cumulative exposure (Collins et al. 2009b). Serum measurements of dioxins were completed in 128 out of 773 PCP workers. No elevation in IHD mortality was found (p-value for trend = 0.74). In the tricholorophenol (TCP) workers, 17% (280 of 1615) of workers had TCDD measured and historical exposures estimated (Aylward et al. 2010). The results for continuous cumulative TCDD and IHD mortality are shown in Table 7.7 (Collins et al. 2009a).

Dutch Cohort Mortality

The Boers et al. update of the Dutch cohort extended the follow-up through 2006 and utilized revised exposure categories (Boers et al. 2010). The authors stepped back from using the cumulative TCDD categories estimated for all workers (Hooiveld et al. 1998), citing small numbers of nonrandomly selected workers, and chose to use broader exposure categories. Categories of occasionally exposed main production workers and 1963 accident workers comprised factory A. Workers in factory B were not exposed to TCDD (Boers et al. 2010). Previously used estimated cumulative TCDD categories from Hooiveld et al. (1998) are shown in Table 7.3 for comparison. Extending the follow-up to 2006 and using the new exposure categories, the authors report no association with CVD or IHD for factory A or B. The increased risk of mortality from IHD in the high-exposure group from factory A reported by Hooiveld et al. 1998 was no longer apparent.

In yet another reanalysis of the same mortality data by Boers et al. (2012), the risk for IHD in factory A was again significantly elevated in the high-exposure category (low HR = 1.17 [95% CI 0.7–2.1]; medium HR = 1.00 [95% CI 0.5–1.9]; high HR = 2.60 [95% CI = 1.6–4.3]). Serum samples from 101 workers from factory A and 86 workers from factory B were analyzed for TCDD. Using a TCDD half-life of 7.1 years, serum TCDD levels (ppt) were modeled and back extrapolated for all workers without individual serum TCDD measurements. Relative rankings and absolute estimated exposure levels changed (background ≤0.4, low 0.4–1.9, medium 1.9–9.9, high ≥9.9 TCDD ppt) in comparison to Hooiveld et al. (1998), with the number of workers in the high category increasing from 19 to 28. However, the overall (rank) correlation between the two analyses was high ($r = 0.79$) with no qualitative difference in the results of IHD mortality. Results for CVD mortality were not presented. CVD mortality was not associated with continuous

TABLE 7.3. Dutch Cohort Mortality

Study/Exposure Matrix	No.	CVD	No.	IHD	Risk Factors Adjusted for
Hooiveld et al. (1998)	1031	RR		RR	Age
Dutch factory A (TCDD ppt lipid)					
Low (7.1)	530	1.00 (Ref.)		1.00	
Medium (7.7–124.1)	259	1.50 (0.88–2.8)		1.50 (0.7–3.6)	
High (124.2–7307.5)	242	1.50 (0.7–1.24)		2.3 (1.0–5.0)	
Boers et al. (2010)		HR		HR	Age and year at first employment
Dutch factory A (TCDD ppt lipid)					
Nonexposed (7.6)	37/482	Ref.	18/482	Ref.	
Occasionally exposed	39/286	1.22 (0.77–1.93)	9/121	1.1 (0.57–2.14)	
Main production (608.2)	17/121	0.94 (0.53–1.67)	17/286	1.02 (0.36–4.70)	
Accident 1963 (1841.8)	21/139	0.88 (0.48–1.60)	17/139	1.60 (0.72–3.55)	
Boers et al. (2010)		HR		HR	
Dutch factory B					
Nonexposed	37/482	Ref.	15/626	Ref.	
Occasionally exposed	25/276	1.12 (0.67–1.88)	13/276	1.55 (0.73–3.29)	
Main production	6/137	0.90 (0.38–2.13)	5/137	1.67 (0.61–4.56)	

log-transformed TCDD in all workers or in factory A, while the HRs for IHD were elevated (shown in Table 7.7; Boers et al. 2012).

Hamburg Cohort Mortality

This is one of two cohorts where measurements of individual dioxins other than TCDD were analyzed and estimates of total TEQ were used in statistical analyses (Flesch-Janys et al. 1995). The Hamburg cohort used a subsample of 190 workers (out of 1173) to measure polychlorinated dibenzo-*p*-dioxins

(PCDDs), polychlorinated dibenzofurans (PCDFs), and coplanar PCBs to estimate the cumulative TCDD and cumulative I-TEQ using 1998 North Atlantic Treaty Organization (NATO) Toxicity Equivalency Factors (TEFs) (NATO 1988). Forty-eight workers had repeated TCDD measurements to estimate half-lives, and the duration of employment in 14 different departments was used to calculate estimated totals for each member of the cohort.

No association with CVD or IHD was found for the cumulative TCDD (*p*-value for trend = 0.18 and 0.14) using chemical workers as reference group based on the estimated TCDD. A dose–response trend and an increased risk in the highest-exposure category of total I-TEQ were found for mortality from IHD (*p*-value for trend = 0.03) and, to lesser extent, for CVD ($p = 0.05$) with the same reference group (Table 7.4). When the cohort of gas workers was used as a reference, both TCDD and total I-TEQ analyses showed significant association with CVD and IHD (all $p \leq 0.01$). No information on major risk factors of heart disease was available. Using gas workers as the reference illustrates potential lack of control for smoking, a major risk factor for CVD.

An update of mortality follow-up for the Hamburg cohort for the period from 1952 to 2007 was recently completed (Manuwald et al. 2012). The vital status could be determined for 1145 men and 389 women and all CVD mortality was assessed. The "intensity" of exposure to TCDD has been estimated retrospectively for the different workplaces in the plant in a previous analysis, based on dioxin analyses in blood or fat tissue samples (Becher et al. 1998). Total cumulative exposure for each individual was calculated as the sum of the cumulative exposures in each of the workplaces where the worker had been employed.

For men, there was a nonmonotonic increase in SMRs across the cumulative TCDD categories (p = close to 0 for Cochran–Armitage trend test), with the lowest SRM = 1.08 (95% CI = 0.83–1.39) in the fourth quartile of the cumulative TCDD category (≥334.5) and the highest SMR = 1.20 in the second and third quartile of the cumulative TCDD (95% CI = 0.92–1.53 and 95% CI = 0.94–1.52, respectively). For women, all SMRs were below 1 ranging from 0.65 to 0.79 (p = close to 0 for Cochran–Armitage trend test) (Manuwald et al. 2012). No analysis with total dioxin TEQ was presented, and only all CVD mortality results were included in the update. We should reiterate that no internal comparison analyses were performed in this update and the SMR analyses presented in Table 7.4 were stratified by cumulative TCDD using external comparisons. As such, direct comparisons with Flesh-Janys et al. (1995) regression analysis results are not possible.

Significant increases in SMRs for most smoking-related cancers (lip, oral cavity, and pharynx; oesophagus; larynx; trachea, bronchus and lung; bladder; and kidney) but not for non-smoking-related cancers in this group of male chemical workers suggest that smoking rates may have been higher than in the general population of Hamburg men (Manuwald et al. 2012). Differences in smoking rates may have also partly contributed to the observed CVD mortality results.

TABLE 7.4. Hamburg Cohort Results

Study/ Exposure Matrix	No.	CVD	No.	IHD	Risk Factors Adjusted for
Flesch-Janys et al. (1995) Hamburg Germany (cumulative TCDD ng/g lipid)[a]		RR		RR	None
0–14.4		1.00 (Ref.)		1.00 (Ref.)	
14.5–49.2		1.34 (0.85–2.10)		1.12 (0.58–2.16)	
49.3–156.7		1.25 (0.78–1.99)		0.56 (0.31–1.41)	
156.8–344.6		1.10–(0.63–1.92)		1.18 (0.56–2.49)	
344.7–3890.2		1.28 (0.67–2.46)		1.59 (0.71–3.59)	
		p for trend = 0.14		p for trend = 0.18	
Hamburg (total I-TEQ ng/g lipid)[a]	157/1177	RR	76/1177	RR	
1.19–39.5	471[b]	1.00 (Ref.)		1.00 (Ref.)	
39.6–98.9	235	1.34 (0.85–2.13)		0.85 (0.41–1.75)	
99.0–278.5	235	1.18 (0.71–1.95)		0.86 (0.41–1.83)	
278.6–545.2	118	1.21 (0.66–2.25)		1.31 (0.57–3.00)	
545.3–4361.9	118	1.40 (0.71–2.76)		1.89 (0.79–4.51)	
		p for trend = 0.05		p for trend = 0.03	
Manuwald et al. (2012) Cumulative TCDD (pg/g lipid)		SMR			
Men					
<13.1	287[c]	1.14 (0.86–1.49)			
13.1–77.3	287	1.20 (0.92–1.53)			
77.4–334.4	284	1.20 (0.94–1.53)			
>334.5	287	1.08 (0.83–1.39)			
		p for trend = close to 0			
Women		SMR			
0	102[c]	0.65 (0.35–1.20)			
>0–19.5	91	0.79 (0.39–1.41)			
19.6–78.3	99	0.78 (0.46–1.23)			
>78.3	97	0.73 (0.44–1.15)			
		p for trend = close to 0			

[a]Serum dioxin extrapolated back to the end of occupational exposure.
[b]Number for each quintile estimated as done by combining the two lowest quintiles and halving the highest quintile as done by Flesch-Janys et al. (1995).
[c]Number of observed and expected diseases of the circulatory system (all CVD) for cumulative job exposure categories were 55/48.1, 63/52.7, 70/58.3, and 63/58.1 for men; and 10/15.3, 11/14.0, 18/23.2, and 19/25.9 for women.

Mortality in NIOSH Cohort

The NIOSH study included workers from 12 different plants in the United States to evaluate overall and cause-specific mortality from exposure to dioxins. Data on 3538 workers from eight plants with detailed worker histories were used to develop a job exposure matrix using extensive plant-specific information on production processes, work place environment, and product contamination (Piacitelli et al. 2000). A cumulative dioxin exposure score was created using TCDD measurements from process materials, fraction of the day exposed, and contact level related to job category as described in Steenland et al. (1999). The validity of the exposure score was supported by serum TCDD measures determined in a nonrandom subsample of 193 workers from one plant. The exposure scores represented a quantitative exposure ranking of workers across different jobs and plants rather than an assignment of a specific dose of TCDD. Nonetheless, the scores were thought to reflect the relative exposure level to TCDD among workers.

No adjustment for major risk factors of CVD was available. Limited smoking data were available for 223 workers from two of the plants, so generalization to other plants was problematic. An apparent dose–response trend ($p = 0.05$) was observed with RR estimates ranging from 1.23 to 1.75, with 95% CI statistically significant in the highest septile. Analyses included 290 deaths from IHD (Table 7.5).

Mortality in RH Cohort

The RH cohort examined mortality in U.S. Air Force veterans involved in spraying Agent Orange and other herbicides during the Vietnam War. It is unique among all cohorts in this review because all veterans (RH and comparisons) included in the internal comparison analyses had serum TCDD measures (Ketchum and Michalek 2005). Serum samples were collected between 17 and 35 years after the last potential exposure in Vietnam. A total of 2452 veterans were included in the analyses of CVD mortality (Table 7.6). Multiple TCDD measurements were available for 343 RH veterans to calculate TCDD half-life and to estimate the TCDD exposure extrapolated back to the end of service in Vietnam for low and high RH groups but not for comparison and background categories (Michalek and Tripathi 1999).

Major risk factors for CVD such as age, military occupation, smoking, and family history of heart disease were included in the final analytic model. Additional risk factors were also examined but were not found to confound the association with CVD mortality. CVD mortality was not significantly elevated in RH veterans; the RR was lower in the high log TCDD RH category (RR = 1.50) than in the low log TCDD RH category (RR = 1.80) (Table 7.6).

In addition, the study collected and periodically analyzed prevalence of heart disease with information on additional risk factors for CVD including total cholesterol, body mass index (BMI), alcohol consumption, and physical activity as detailed in the "Discussion" section (Michalek et al. 2005).

TABLE 7.5. NIOSH Cohort Results for IHD Mortality in 3538 Workers with Cumulative Dioxin Exposure Scores

Study/Exposure Variable	No.	CVD	No. of Deaths	IHD	Risk Factors Adjusted for
Steenland et al. (1999)				Rate ratio	
NIOSH (cumulative dioxin exposure score)					
0–19		NA	29	1.00 (Ref.)	Age and year of birth
19–139		NA	39	1.23 (0.75–2.00)	
139–581		NA	45	1.34 (0.83–2.18)	
581–1,650		NA	42	1.30 (0.79–2.13)	
1,650–5,740		NA	48	1.39 (0.86–2.24)	
5,740–20,200		NA	44	1.57 (0.96–2.56)	
>20,200		NA	43	1.75 (1.07–2.87)	*p*-value for trend = 0.05

TABLE 7.6. CVD Mortality for 2452 U.S. Air Force Veterans with Individual TCDD Measurements

Study/Exposure Matrix	No.	CVD	No.	IHD	Risk Factors Adjusted for
Ketchum and Michalek (2005)		RR			
Ranch Hand (log TCCD pg/g lipid)					
Comparison	31/1436	1.00 (Ref.)		NA	Military occupation, birth year, smoking, and family history of heart disease
Background	8/442	0.80 (0.4–1.8)		NA	
Low (32.2–117.4)	12/287	1.80 (0.9–3.5)		NA	
High (117.9–4221.9)	9/287	1.50 (0.7–3.3)		NA	*p*-value for trend = 0.07

We should note that the number of deaths from IHD in the most exposed occupational group of RH veterans—the enlisted ground crew, responsible for loading and cleaning the planes, was elevated when compared to the comparison enlisted ground crew (28 [4.6%] vs. 281 [2.6%], IHD deaths, shown in Table 7.1). The adjusted RR for mortality from IHD was 1.7 (95% CI 1.1–2.5) using log-transformed serum TCDD measures (Ketchum and Michalek 2005). The risk of mortality from IHD was not significantly elevated in the enlisted ground crew in the earlier mortality follow-up (RR = 1.4, 95% CI 0.8–2.1) (Michalek et al. 1998).

Statistical Analyses Using Continuous TCDD/TEQ Estimates

Statistical analyses using the continuous TCDD estimates were presented for the BASF and Midland cohort analyses of PCP and trichlorophenol workers. The Midland cohort also used cumulative 1 ppb-year change in total dioxin TEQ as an exposure measure. In the BASF cohort, risk estimates were below 1 for CVD when adjusted for age, BMI, and smoking and were also below 1 for IHD mortality in the Midland cohort after adjusting for year of birth and year of hire (Collins et al. 2009a,b; Ott and Zober 1996). In the second reanalyses of the Dutch cohort mortality data (1955–2006), no increased risk in mortality from CVD was found, but an elevation was reported for IHD mortality in factory A and in the overall cohort including nonexposed workers (Boers et al. 2012) (Table 7.7).

SUMMARY

Of the occupationally TCDD-exposed cohort studies with sufficient serum dioxin data that were conducted using internal comparisons, three showed no increased risk of mortality from heart disease overall or from IHD (BASF,

TABLE 7.7. Evaluation of CVD and IHD Mortality Using Continuous Measure of Serum TCDD

Study/Exposure Variable	No.	CVD	No.	IHD	Risk Factors Adjusted for
Ott and Zober (1996) BASF (1 μg/kg increase in TCDD)	243	0.93 (0.70–1.24)		NA	Age, smoking, and BMI
Collins et al. (2009a) Midland, TCP workers (1 ppb-year increase in cumulative TCDD)		NA	1615	0.999 (0.989–1.009)	Birth year and hire year
Collins et al. (2009b) Midland, PCP workers (1 ppb-year increase in cumulative TEQ)		NA	773	0.997 (0.983–1.011)	Birth year and hire year
Boers et al. (2012) Dutch factory A (log-transformed TCDD ppt)	1020	1.04 (0.94–1.16)		1.24 (1.09–1.43)	Age
Boers et al. (2012) Dutch total cohort (log-transformed TCDD ppt)	2056	1.07 (0.98–1.16)		1.19 (1.08–1.32)	Age

Midland, and New Zealand cohorts; Collins et al. 2009a,b; McBride et al. 2009a,b; Ott and Zober 1996) and two found increased risk across exposure categories (Hamburg and NIOSH cohort; Flesch-Janys et al. 1995; Manuwald et al. 2012; Steenland et al. 1999). Serum dioxin measures in a large number of New Zealand and Midland workers provided validation of the exposure categories used in those studies and supported the findings of no associations with heart disease mortality from the earlier BASF cohort. Mixed results were reported from the Dutch cohort of workers. One of the two reanalyses of the Dutch cohort with longer follow-up using revised exposure categories did not find elevated risk of CVD or IHD mortality, while the other found increases with IHD mortality in both exposed workers and the total cohort including nonexposed workers (Boers et al. 2010, 2012).

The Hamburg cohort reported positive associations with IHD and to a lesser extent with CVD and showed a dose–response trend with an estimated total I-TEQ (0.03 and 0.05, respectively), but not with estimated cumulative serum TCDD (0.18 and 0.14, respectively; Flesch-Janys et al. 1995). In contrast, extended mortality follow-up of the Hamburg cohort reported an increased SMR for all CVDs with TCDD in men but not in women and did not report associations with total dioxin TEQ or with IHD (Manuwald et al. 2012). Significant tests for trend but no significant SMR increase across categories of cumulative job exposure based on TCDD measurements were found, with the direction and magnitude of association similar to external comparisons analyses (Manuwald et al. 2012). A borderline significant trend and an apparent dose–response across septiles of exposure scores were also reported in the NIOSH cohort, with the RR statistically significant in the highest septile (RR = 1.75, 95% CI 1.07–2.87; Steenland et al. 1999).

No dose–response trend was observed for mortality from CVD in RH veterans across exposure categories based on log-transformed serum TCDD measures (Ketchum and Michalek 2005).

In agreement with the majority of occupational studies, the study of the Seveso accident (Bertazzi et al. 2001; Consonni et al. 2008) reported no increase in CVD mortality, overall or when exposure was stratified into more refined exposure categories, both of which were compared to an external referent group. Small differences were observed in CVD or IHD mortality after 20 or 25 years of follow-up, and the exposure categories were further validated by additional serum TCDD measurements in stored samples from 1976 to 1977 and from 1992 to 1996. Results from food poisoning studies where most of the toxic exposure was ascribed mainly to PCDFs showed statistically significant increases in risk of mortality from CVD in the Yusho (Kashima et al. 2011) but not in the Yucheng cohort (Tsai et al. 2007), with no increase in IHD mortality.

In conclusion, the data presented provide limited evidence of a positive association between exposure to dioxins and heart disease mortality, with mixed support of an association in the majority of studies using internal comparisons and individual exposure assessment.

DISCUSSION

Results of analyses using external comparisons and dichotomized exposure categories (exposed, nonexposed) found no association between dioxin exposure and CVD mortality. The bias from the healthy worker effect (Monson 1986; Weed et al. 1987) can occur when the rates in the general population are used to contrast rates observed in worker cohorts. As noted by Burns et al. (2011), up to a 20% decrease in the overall mortality, and in heart disease mortality specifically, has been observed historically and was recently reported in the large cohort of U.S. chemical workers employed between 1960 and 2005. Analyses using an internal group for comparison avoid this bias, but the findings regarding TCDD exposure and increased mortality from IHD and/or CVD were similar to the results reported from analyses using external comparisons. The lack of association with CVD mortality in studies using an external comparison from the general population (Seveso, Yucheng) where the healthy worker effect was not a factor similarly argues against an association between dioxins and CVD mortality following the occupational or environmental (accidental) exposures.

A major concern in the reviewed studies was potential confounding by the major risk factors for CVD (e.g., diet, smoking, physical activity). If these risk factors were strongly associated with dioxin exposure—or some of its surrogates used in statistical analyses, such as the duration of employment—they could confound the association between dioxins and CVD, biasing it either upward or downward. Of the included studies, only the RH (Ketchum and Michalek 2005) and BASF (Ott and Zober 1996) studies adjusted for possible confounding by the major risk factors for CVD mortality in the statistical analysis (smoking and family history of heart disease in RH, alcohol consumption in some analyses, smoking and BMI in BASF). The RH study reported some increases in adjusted RRs, but estimates were higher for low exposed RH than for high exposed RH for all CVD mortality resulting in a nonsignificant dose–response trend ($p = 0.07$). The BASF study found no elevated RRs for all CVDs in the adjusted analysis.

The RH study was the only study that evaluated in detail other major risk factors for heart disease, but those were not found to confound the association in the multivariate Cox proportional hazards models of mortality outcomes and were not included in the final model (not described in the published article; J.E. Michalek, pers. comm.). The Air Force Health Study (AFHS, acronym used by the U.S. Air Force for Operation RH) reports looked at the prevalence of heart disease (excluding essential hypertension) and IHD and numerous other outcomes at each of the six follow-up examinations as a part of a comprehensive cardiovascular assessment. No elevation in RRs was reported over the course of the study for CVD or IHD prevalence when including all major risk factors in the statistical analyses (Michalek et al. 2005).

The covariates included in AFHS cardiovascular assessment were age, race, military occupation, lifetime cigarette smoking, current cigarette smoking, lifetime and current alcohol use, uric acid levels, BMI, waist-to-hip ratio, (total) cholesterol, high-density lipoprotein (HDL), cholesterol:HDL ratio, family history of heart disease, family history of heart disease before age 45, taking blood pressure medication, and diabetic class (Michalek et al. 2005). Unadjusted RR at the 2002 examination were 1.50 (95% CI = 1.07–2.11) for background RH, 1.13 (95% CI = 0.76–1.67) for low RH, and 0.99 (95% CI = 0.68–1.45) for high RH. Adjusted RRs were 1.33 (95% CI = 0.94–1.89), 1.03 (95% CI = 0.68–1.54), and 1.21 (95% CI = 0.81–1.52), respectively. The effect of confounding, if any, is not obvious when there is no apparent positive association present in the unadjusted analyses. TCDD categories used in the analysis of heart disease prevalence were similar to those used in mortality analyses (comparison veterans—referents, 1987 TCDD ≤ 10 ppt for background RH, 1987 TCDD > 10 ppt and initial TCDD ≤ 118 ppt for low RH, 1987 TCDD > 10 ppt and initial TCDD > 118 ppt for high RH [initial TCDD refers to the TCDD recalculated to the end of exposure in Vietnam using 7.6 years' half-life for TCDD and one-compartment linear decay model]).

Two studies estimated the exposure to polychlorinated dibenzo-*p*-dioxin and dibenzofuran (PCDD/F) congeners other than TCDD based on individual serum measurements. The total dioxin TEQ (combined dioxin toxic equivalencies of each dioxin-like congener relative to TCDD) may represent the cumulative potency of the multiple dioxin congeners and could be considered more biologically relevant than using TCDD alone. The Hamburg study (Flesch-Janys et al. 1995) reported stronger trends using estimated cumulative total I-TEQ (NATO 1988) based on measurements of PCDD/F congeners in 190 out of 1077 workers included in the analyses. Using estimated cumulative TCDD concentration, no association was observed with IHD or CVD in the same internal cohort comparison analyses (Table 7.4). The Midland workers also had other PCDD, PCDF, and PCB congeners measured in 128 out of 773 workers and reported no association with IHD mortality with either cumulative TCDD or cumulative TEQ as the exposure matrix (Table 7.2, Collins et al. 2009b).

Dioxin exposure was frequently accompanied by coexposure to other chemicals, with wide variation between occupational settings and particular plants. Attempts were made in several studies to separate the statistical analyses for those workers most likely exposed to TCDD only (i.e., TCP workers, 2,4,5-T production workers) and those that included workers exposed to higher-chlorinated dioxins that contaminated PCP-based products. The Dutch, New Zealand, and Midland cohorts conducted separate analyses of workers exposed to TCP/PCP with no elevation in CVD or IHD mortality in any of the cohorts (Boers et al. 2010, 2012; Collins et al. 2009a,b; McBride et al. 2009a,b). The workers potentially exposed to PCP and higher-chlorinated dioxins were also excluded from the statistical analyses in the internal analyses of the NIOSH study (Steenland et al. 1999). The results of the internal comparison analyses including PCP workers were not shown. Ruder and Yiin (2011) examined mortality in the PCP-exposed workers of the NIOSH cohort (*n* = 2012); the SMR for IHD was 1.00 (95% CI = 0.88–1.14) in 1402 workers exposed to PCP but not TCP.

Chlorophenols and their derivatives, which were the primary coexposure in the occupational studies, were thought not to be a factor in mortality study analyses as the available toxicological data did not suggest that they had an adverse toxic effect on the cardiovascular system (ATSDR 1999).

There are challenges in accurately assessing personal exposure retrospectively. TCDD or other dioxin measurements were usually performed years or decades after the last (occupational) exposure. Bioaccumulation, retention, and long half-lives of many dioxin congeners make that task more feasible and make the individual measurements of these chemical compounds the gold standard of exposure assessment (Schecter et al. 2009). The exposure misclassification can further be reduced when a random sample or a high proportion of study subjects are sampled. The Seveso study had the benefit of having a large number of samples collected within a year of the industrial accident, thus accurately establishing the exposure categories close to the end of exposure (Consonni et al. 2008).

Measuring the dioxins in all members of the study population is impractical, costly, and often not possible (e.g., in deceased workers). The NIOSH study illustrates challenges of a comprehensive approach in exposure assessment in the multiplant chemical worker cohort. Having performed one of the most detailed and extensive industrial hygiene assessments with measurements of TCDD contamination in final and intermediate products and working environment in most of the plants, the investigators developed categories of exposure based on the job classification, fraction of day exposed, length of employment, product contamination, and other factors (Piacitelli et al. 2000; Steenland et al. 1999). Additionally, serum samples were collected from a group of workers from only one of eight plants included in the final mortality analyses. With a low proportion of workers with serum TCDD measurements coming from mostly one of the eight plants, several sources of error and bias must be considered. One is the apparent lack of random selection for the subgroup with TCCD measurement. In addition, bias in the estimated TCDD from regression models based on duration of employment in a particular department and contamination factor could arise if workers with long durations (of employment) and low TCDD levels were systematically underrepresented in the validation study of the exposure scores. The correlation with the length of employment and measured TCDD was 0.70 for workers from two plants (Fingerhut et al. 1991) but was only 0.42 using the cumulative TCDD exposure scores (Steenland et al. 1999).

While the products and processes across the plants were similar, the exposure certainly varied between and within the plants and may have not been fully captured using the dioxin exposure scores. Using data from Midland workers, modeled occupational dose rates for 1,2,3,4,6,7,8-heptachlorodibenzo-*p*-dioxin (HpCDD), and octachlorodibenzo-*p*-dioxin (OCDD) showed an inverse pattern, with higher dose rates associated with lower exposure scores, and a similar pattern was observed for the middle dose scores for TCDD (Aylward et al. 2007).

The fact that an increased risk of IHD mortality was observed specifically in two multiplant cohorts (IARC is the other cohort) that used a form of exposure scores for TCDD estimates and assigned those scores to the majority of workers without TCDD measurements may indirectly illustrate bias resulting from misclassification. It is noteworthy, however, that no elevated risk was found in the analysis of the prevalence of IHD (Calvert et al. 1998) or in the earlier evaluation of IHD mortality in the NIOSH cohort (Fingerhut et al. 1991) that did not use these exposure scores.

Results of occupational or veterans studies were mostly not able to address the risks of dioxin exposure in females because none of those studies included substantial numbers of women. The association between dioxin exposure and mortality from all circulatory diseases was somewhat stronger in men in the Seveso study in several smaller subcategories of heart disease mortality, but overall results were negative (Consonni et al. 2008). The Hamburg cohort included women and Manuwald et al. (2012) provides SMRs for females, reporting a reduction in risk for women and CVD mortality (Table 7.4).

In agreement with earlier mortality results (Fingerhut et al. 1991), the retrospective examination of cardiovascular morbidity was not significantly associated with dioxin exposure in the NIOSH cohort (Calvert et al. 1998). Similarly, no increase in RRs with serum TCDD levels was reported for the prevalence of heart disease in the RH cohort (Michalek et al. 2005). Self-reported spraying of phenoxy herbicides was associated with an odds ratio (OR) of 1.41 (95% CI = 1.06–1.89) for heart disease adjusting for age, BMI, and regular smoking in another study of U.S. Vietnam veterans (Kang et al. 2006). The mean current serum TCDD in those who reported spraying herbicides was 4.3 ppt versus 2.7 ppt in those not reporting spraying. Exposure assessment in a Korean veterans study was insufficient to make a valid distinction between those exposed and nonexposed to TCDD (Kim et al. 2003).

In the analyses of 1999–2002 data from the National Health and Nutrition Examination Survey (NHANES) representing the general U.S. population, Ha et al. (2007) reported an association between the sum of PCDDs (but not PCDFs) and self-reported CVD. After adjustment for age, race, BMI, smoking, alcohol consumption, exercise, cholesterol, hypertension, and C-reactive protein, the ORs were 1.0 (reference category), 1.4, 1.7, and 1.9 (p for trend = 0.07) for PCDD exposure quartile. The study used study participants with dioxins levels below the limit of detection as the reference category. A selection bias may be introduced using such an approach resulting in the reference category with little or no disease and without the presence of established risk factors for CVD. Irrespective of actual exposure, higher exposed groups may show apparent increased risk of disease in comparison with such a reference group. There are numerous quality control and other non-exposure-related reasons why a sample from a particular person may be under the limit of detection, some of those related to insufficient volume of the serum sample, which was not uncommon in NHANES data (Patterson et al. 2009).

The present review incorporated results from additional follow-up available in several studies, and we tried to present results for both shorter and longer duration of follow-up. Longer follow-up time seemed to attenuate risk estimates for CVD, but not for IHD in the Dutch cohort (in one analysis), did not materially change the estimates in Seveso, and slightly increased the risk estimates in RH and Hamburg cohorts over time in the external comparison analyses. Only CVD mortality was assessed in the extended Hamburg follow-up (Manuwald et al. 2012) with no analyses presented for total dioxin TEQ. The addition of four newly updated worker studies (Dutch, Hamburg, Midland, and New Zealand), updates of Seveso, Yucheng, and Yusho results, and a more detailed review of CVD morbidity in RH and NIOSH cohorts may have contributed to differences with the earlier review by Humblet et al. (2008).

CONCLUSIONS

A review of CVD mortality in dioxin-exposed cohorts, using both external and internal comparison populations with detailed exposure assessments, indicates that there is mixed evidence of increased mortality from both IHD and CVD (Table 7.8). The control of confounding by major risk factors for CVD was examined in some detail in two cohorts that found little or no association with mortality from CVD. The inferences that can be made are limited by the lack of adjustment for risk factors of CVD in most of the earlier and recent occupational cohorts studying dioxin exposure and mortality from CVD and IHD.

Because the production of dioxin-contaminated products ceased mostly in the mid-1980s, further follow-up and updates of the remaining occupational cohorts may be few. The length of follow-up in some of the cohorts has reached up to 64 years, and major changes or increases in precision of the reported results are unlikely. Future studies in both animals and humans should assess whether cardiovascular effects are present at current environmentally relevant doses. Incidence in contrast to the prevalence or mortality of CVD may be a more sensitive end point that should be investigated in cohort studies of environmental exposure to dioxins. The mixed results of mortality studies in TCDD-exposed workers and veterans, as well as limited findings from cardiovascular morbidity studies, do not support the conclusion of strong association between dioxin exposure and CVD or IHD mortality in humans.

HYPERTENSION AND EXPOSURE TO PCBs

Exposure to persistent organic pollutants (POPs), particularly PCBs, have been examined in relation to hypertension as diagnosed by a physician (self-reported) or based on elevated blood pressure measurements in several large cross-sectional studies of the general population (Christensen and White 2011;

TABLE 7.8. Association between Hypertension and Elevated Blood Pressure and PCBs

Study Name/ Location	Congeners Studied	No. of Hypertensives	Congeners Associated with Hypertension/ Blood Pressure	Reference
Triana, AL	Total PCBs (Aroclor 1260)	191/458		Kreiss et al. (1981)
NHANES 1999–2002	dx-like PCBs 74,118, 126,156, 169	890/2,074	74, 118,126	Everett et al. (2008a)
	Non-dx like PCBs 99,138/158, 153, 170, 180, 187		99, 138/158,170, 187	
NHANES 1999–2004	dx-like PCBs 74,118, 126,156, 169	1,377/3,326	74, 118, 126	Everett et al. (2008b)
	Non-dx like PCBs 99,138/158, 153, 170, 180, 187			
NHANES 1999–2002	dx-like PCBs 74,118, 126,156, 169	524	156 (in males only)	Ha et al. (2009)
	Non-dx like PCBs 99,138/158, 153, 170, 180, 187			
Anniston, AL, 2005–2007	35 *ortho*-substituted PCBs	394	156, 157, 189 138/158, 146, 153, 170, 172, 180, 194 177, 178, 183, 187, 195, 196/203, 199, 206, 209	Goncharov et al. (2011)
NHANES 1999–2004	21 PCBs measured over three cycles of NHANES	1,808/4,119	66, 101, 118, 128, 187	Christensen and White (2011)

Everett et al. 2008a,b; Ha et al. 2009) and in two studies of the general population exposed to higher than background levels of PCBs (Goncharov et al. 2010, 2011; Kreiss et al. 1981). Most of these studies had an advantage of detailed individual exposure assessment that focused on PCBs or more broadly on dioxins, dibenzofurans, PCBs, and chlorinated pesticides. These measurements were available for all study subjects included in the statistical analyses. For the most part, a good control of confounding and adjustment for major risk factors of hypertension or heart disease was present.

Laboratory Evidence

Endothelial dysfunction, which is defined as impaired vasodilation, is a major attribute of elevated blood pressure and a risk factor for adverse cardiovascular outcomes (Versari et al. 2009). Endothelial dysfunction in essential hypertension in humans is characterized by increased production of vasoconstriction factors, such as cyclooxygenase-derived prostaglandins and reactive oxygen species (ROS) in combination with decreased bioavailability of nitric oxide (NO) (Bian et al. 2008; Taddei et al. 1997, 1998; Vanhoutte and Tang 2008).

The effects of coplanar PCBs and TCDD on the endothelium have been associated with AhR-dependent induction of CYP1A1 and the accompanying formation of ROS (Beischlag et al. 2008; Kopf et al. 2010; Slim et al. 1999). Induction of cytochrome P450 CYP1A1 has proinflammatory effects and can cause oxidative stress in endothelial cells of various origins, suggesting that the endothelium may be a target for these pollutants (Arzuaga et al. 2007; Hennig et al. 2002). Human umbilical vein endothelial cells (HUVECs) were used to test the effects of PCB 126 and estradiol (E_2) by Andersson et al. (2011). PCB 126 slightly increased ROS production and decreased NO production in HUVEC. The addition of E_2 enhanced PCB 126-induced transcription of CYP1A1, CYP1B1, and COX-2 in HUVEC, whereas an increased transcription of eNOS only occurred following combined treatment with E_2 and PCB 126. These findings suggest that PCB 126 induced changes in human endothelial cells that are characteristic for endothelial dysfunction and that these changes could be modified by increased estrogen levels.

Results from Reviewed Studies

In the seminal early PCB study, Kreiss et al. (1981) investigated the association of PCBs with hypertension and heart disease among 458 residents of Triana, Alabama. Measurements of total PCB concentration were based on the Aroclor 1260 standard. The multiple regression analyses found no association with systolic blood pressure but increased odds of elevated diastolic blood pressure. The analyses were adjusted for age, gender, BMI, social class, total cholesterol, triglyceride, smoking, and race.

Data from participants in 1999–2002 NHANES were used to examine the relationship among 11 PCB congeners and hypertension (self-reported physician diagnosed and newly diagnosed [≥140/90 mmHg]) in logistic regression models adjusted for age, gender, race, smoking status, BMI, exercise, total cholesterol, and family history of coronary heart disease (Everett et al. 2008a). Associations with dioxin-like PCB congeners 74, 118, 126, 156, and 169 and with non-dioxin-like PCBs 99, 138/158, 153, 170, 180, and 187 were assessed. Significant associations with hypertension were reported for PCBs 126, 74, 118, 99, 138/158, 170, and 187. Using additional 2 years of data, the associations among the same 11 PCB congeners and the prevalent hypertension were examined again by Everett et al. (2008b). In this expanded analysis, only the associations with the dioxin-like PCB congeners 74, 118, and 126 were statistically significant.

In another analysis of the 1999–2002 NHANES data, newly diagnosed hypertension was evaluated among 524 participants above 40 years of age (Ha et al. 2009). Subjects with previously diagnosed hypertension and diabetes were excluded from the statistical analyses. Men and women were evaluated separately and logistic regressions were adjusted for age, race, poverty:income ratio, BMI, smoking status, serum cotinine, alcohol consumption and exercise. Dioxin-like PCB congeners 74, 118, 126, 156, and 169 and non-dioxin-like PCB congeners 99, 138, 153, 170, 180, and 187 were assessed in the statistical analyses. A significant association with undiagnosed hypertension was reported only for the mono-*ortho*-substituted PCB 156 among men (OR = 3.3, 95% CI 1.2–9.1 for ≥75th percentile). No significant association was reported for women. We should note that the study also reported statistically significant ORs for both dioxins and furans, although the trend was only significant for furans. It was not clear whether any analyses were conducted to test for the effect of the modification of the dioxins and furans on the association with PCBs.

Residents of Anniston, Alabama, living in close proximity to the former PCB production facility were studied by Goncharov et al. (2010). A total of 35 PCB congeners were measured in 765 participants with levels on average several times higher than those reported in the U.S. general population. Overall, ORs for clinical hypertension from logistic regression models were significantly elevated in the third and fifth quintiles of total PCBs (wet weight) adjusted for age, BMI, total serum lipids, gender, race, smoking status, and physical activity. There was no association of total PCBs with hypertension among those on antihypertensive medication. Among those not on medication, the third tertile of total PCBs had ORs of 3.87 (95% CI 1.13–13.17) for systolic hypertension and 4.49 (95% CI 1.34–14.99) for diastolic hypertension, compared to the first tertile. A follow-up publication examined associations between individual PCB congeners and diastolic and systolic blood pressure in individuals not taking antihypertensive medication (Goncharov et al. 2011). Significant associations were found for both dioxin-like (mono-*ortho*) and non-dioxin-like di-*ortho* and higher *ortho*-substituted PCB congeners. Association with PCB congeners was seen even among those in the normotensive range.

A novel methodological approach to the analyses of the hypertension data among participants of the 1999–2004 NHANES was presented by Christensen and White (2011). Initially, unconditional multivariate logistic regression was used to estimate ORs and associated 95% CIs followed by evaluation of correlation and multicollinearity among PCB congeners. Logistic regression results showed mixed findings by congener and class. The clustering analyses performed to determine groups of related congeners (four clusters identified), discriminant analyses (12 informative congeners found), and principal component analyses (four components with eigenvalues > 1) were also conducted. Finally, a weighted sum was constructed to represent the relative importance of each congener in relation to hypertension risk. PCBs 66, 101, 118, 128, and

187 were significantly associated with increased risk of hypertension using an optimization of weighted sum approach to equalize different ranges and potencies. The study illustrated methodological approaches to analyzing complex data with a large number of exposures correlated and with varying exposure ranges and potencies. The most informative congeners identified by the weighted sum approach do not share a common structure or activity group. PCBs 66 and 118 are mono-*ortho*-substituted; PCBs 101 and 128 are di-*ortho*-substituted; and PCB 187 is tri-*ortho*-substituted. PCBs 66 and 128 are also considered estrogenic. Authors stated that the presented analyses were to illustrate a methodological approach rather than to test a hypothesis regarding the true association between PCB exposure and hypertension in the general population, and they did not account for the complex survey design of the NHANES in these analyses nor did they adjust for the testing of multiple hypotheses. However, the introduction of flexible analytical techniques and analyses of multiple congeners at the same time provides an example for the future evaluation of mixtures in the study of human health outcomes.

Among the studies reviewed, none were longitudinal in design. Similar to the prevalence of CVD, results from longitudinal studies that are required to properly evaluate the relationship between PCB congeners and hypertension in the general population are not available. It would also be useful to (1) obtain repeated measurements of organochlorines for proper estimates of exposure, (2) study nonmonotonic associations observed, and (3) evaluate the presence or absence of reverse causation.

In general, no association of hypertension with chlorinated pesticides was reported in any of the studies that measured and analyzed these compounds (Table 7.8).

Conclusion

Currently, limited information exists on investigations of PCBs and hypertension. Everett et al. (2008b) reported associations between dioxin-like PCB congeners 74, 118, and 126 and clinical hypertension in a representative sample of the U.S. population from the NHANES, while Christensen and White (2011) showed associations with both mono-*ortho* and di-*ortho* (non-dioxin-like) congeners. Associations with both dioxin-like and nondioxin congeners were also observed with elevated systolic and diastolic blood pressures in residents of Anniston, but non-*ortho* dioxin-like PCBs were not measured (Goncharov et al. 2011).

There is a lack of consistency in the reported observed associations for specific PCB congeners and hypertension, but ranges and timing of exposure varied widely in the studies reviewed. Longitudinal epidemiological studies are warranted to assess the validity of the suggestive relationships of PCBs with hypertension because cross-sectional designs are too limited by design to provide for the more advanced analyses required for these complex associations.

ACKNOWLEDGMENT

This research was supported in part by an appointment to the Research Participation Program at the Centers for Disease Control and Prevention administered by the Oak Ridge Institute for Science and Education through an interagency agreement between the U.S. Department of Energy and CDC/ATSDR.

REFERENCES

Andersson H, Garscha U, Brittebo E. 2011. Effects of PCB126 and 17β-oestradiol on endothelium-derived vasoactive factors in human endothelial cells. Toxicology 285:46–56.

Arzuaga X, Reiterer G, Majkova Z, Kilgore MW, Toborek M, Hennig B. 2007. PPARalpha ligands reduce PCB-induced endothelial activation: Possible interactions in inflammation and atherosclerosis. Cardiovasc Toxicol 7:264–272.

Asp S, Riihimaki V, Hernberg S, Pukkala E. 1994. Mortality and cancer morbidity of Finnish chlorophenoxy herbicide applicators: An 18-year prospective follow-up. Am J Ind Med 26:243–253.

ATSDR. (Agency for Toxic Substances and Disease Registry). 1999. Toxicological Profile for Chlorophenols. Atlanta, GA: Agency for Toxic Substances and Disease Registry.

ATSDR (Agency for Toxic Substances and Disease Registry). 2000. Toxicological Profile for Polychlorinated Biphenyls (PCBs). U.S. Department of Health and Human Services, Atlanta.

Aylward LL, Bodner KM, Collins JJ, Hays SM. 2007. Exposure reconstruction for a dioxin-exposed cohort: Integration of serum sampling data and work histories. Organohal Comp 69:2063–2066.

Aylward LL, Bodner KM, Collins JJ, Wilken M, McBride D, Burns CJ, et al. 2010. TCDD exposure estimation for workers at a New Zealand 2,4,5-T manufacturing facility based on serum sampling data. J Expo Sci Environ Epidemiol 20:417–426.

Becher H, Steindorf K, Flesch-Janys D. 1998. Quantitative cancer risk assessment for dioxins using an occupational cohort. Environ Health Perspect 106(Suppl 2): 663–670.

Beischlag TV, Luis Morales J, Hollingshead BD, Perdew GH. 2008. The aryl hydrocarbon receptor complex and the control of gene expression. Crit Rev Eukaryot Gene Expr 18:207–250.

Bertazzi PA, Consonni D, Bachetti S, Rubagotti M, Baccarelli A, Zocchetti C, et al. 2001. Health effects of dioxin exposure: A 20-year mortality study. Am J Epidemiol 153:1031–1044.

Bian K, Doursout MF, Murad F. 2008. Vascular system: Role of nitric oxide in cardiovascular diseases. J Clin Hypertens (Greenwich) 10:304–310.

Biswas G, Srinivasan S, Anandatheerthavarada HK, Avadhani NG. 2008. Dioxin-mediated tumor progression through activation of mitochondria-to-nucleus stress signaling. Proc Natl Acad Sci U S A 105:186–191.

Boers D, Portengen L, Bueno-de-Mesquita HB, Heederik D, Vermeulen R. 2010. Cause-specific mortality of Dutch chlorophenoxy herbicide manufacturing workers. Occup Environ Med 67:24–31.

Boers D, Portengen L, Turner WE, Bueno-de-Mesquita HB, Heederik D, Vermeulen R. 2012. Plasma dioxin levels and cause-specific mortality in an occupational cohort of workers exposed to chlorophenoxy herbicides, chlorophenols and contaminants. Occup Environ Med 69:113–118.

Boffetta P, Mundt KA, Adami HO, Cole P, Mandel JS. 2011. TCDD and cancer: A critical review of epidemiologic studies. Crit Rev Toxicol 41:622–636.

Burns CJ, Bodner KM, Jammer BL, Collins JJ, Swaen GM. 2011. The healthy worker effect in US chemical industry workers. Occup Med (Lond) 61:40–44.

Calvert GM, Wall DK, Sweeney MH, Fingerhut MA. 1998. Evaluation of cardiovascular outcomes among U.S. workers exposed to 2,3,7,8-tetrachlorodibenzo-p-dioxin. Environ Health Perspect 106(Suppl 2):635–643.

Chang LW, Chang YC, Ho CC, Tsai MD, Lin P. 2007. Increase of carcinogenic risk via enhancement of cyclooxygenase-2 expression and hydroxyestradiol accumulation in human lung cells as a result of interaction between BaP and 17-beta estradiol. Carcinogenesis 28:1606–1612.

Christensen KL, White P. 2011. A methodological approach to assessing the health impact of environmental chemical mixtures: PCBs and hypertension in the National Health and Nutrition Examination Survey. Int J Environ Res Public Health 8:4220–4237.

Collins JJ, Bodner K, Aylward LL, Wilken M, Bodnar CM. 2009a. Mortality rates among trichlorophenol workers with exposure to 2,3,7,8-tetrachlorodibenzo-p-dioxin. Am J Epidemiol 170:501–506.

Collins JJ, Bodner K, Aylward LL, Wilken M, Swaen G, Budinsky R, et al. 2009b. Mortality rates among workers exposed to dioxins in the manufacture of pentachlorophenol. J Occup Environ Med 51:1212–1219.

Consonni D, Pesatori AC, Zocchetti C, Sindaco R, D'Oro LC, Rubagotti M, et al. 2008. Mortality in a population exposed to dioxin after the Seveso, Italy, accident in 1976: 25 years of follow-up. Am J Epidemiol 167:847–858.

Dalager NA, Kang HK. 1997. Mortality among Army Chemical Corps Vietnam veterans. Am J Ind Med 31:719–726.

Dalton TP, Kerzee JK, Wang B, Miller M, Dieter MZ, Lorenz JN, et al. 2001. Dioxin exposure is an environmental risk factor for ischemic heart disease. Cardiovasc Toxicol 1:285–298.

Demers PA, Davies HW, Friesen MC, Hertzman C, Ostry A, Hershler R, et al. 2006. Cancer and occupational exposure to pentachlorophenol and tetrachlorophenol (Canada). Cancer Causes Control 17:749–758.

Everett CJ, Mainous AG, 3rd, Frithsen IL, Player MS, Matheson EM. 2008a. Association of polychlorinated biphenyls with hypertension in the 1999–2002 National Health and Nutrition Examination Survey. Environ Res 108:94–97.

Everett CJ, Mainous AG, 3rd, Frithsen IL, Player MS, Matheson EM. 2008b. Commentary on the association of polychlorinated biphenyls with hypertension. Environ Res 108:428–429.

Fingerhut MA, Halperin WE, Marlow DA, Piacitelli LA, Honchar PA, Sweeney MH, et al. 1991. Cancer mortality in workers exposed to 2,3,7,8-tetrachlorodibenzo-p-dioxin. N Engl J Med 324:212–218.

Flesch-Janys D, Berger J, Gurn P, Manz A, Nagel S, Waltsgott H, et al. 1995. Exposure to polychlorinated dioxins and furans (PCDD/F) and mortality in a cohort of workers from a herbicide-producing plant in Hamburg, Federal Republic of Germany. Am J Epidemiol 142:1165–1175.

Flesch-Janys D, Steindorf K, Gurn P, Becher H. 1998. Estimation of the cumulated exposure to polychlorinated dibenzo-p-dioxins/furans and standardized mortality ratio analysis of cancer mortality by dose in an occupationally exposed cohort. Environ Health Perspect 106(Suppl 2):655–662.

Goncharov A, Bloom M, Pavuk M, Birman I, Carpenter DO. 2010. Blood pressure and hypertension in relation to levels of serum polychlorinated biphenyls in residents of Anniston, Alabama. J Hypertens 28:2053–2060.

Goncharov A, Pavuk M, Foushee HR, Carpenter DO. 2011. Blood pressure in relation to concentrations of PCB congeners and chlorinated pesticides. Environ Health Perspect 119:319–325.

Ha MH, Lee DH, Jacobs DR. 2007. Association between serum concentrations of persistent organic pollutants and self-reported cardiovascular disease prevalence: Results from the National Health and Nutrition Examination Survey, 1999–2002. Environ Health Perspect 115:1204–1209.

Ha MH, Lee DH, Son HK, Park SK, Jacobs DR, Jr. 2009. Association between serum concentrations of persistent organic pollutants and prevalence of newly diagnosed hypertension: Results from the National Health and Nutrition Examination Survey 1999–2002. J Hum Hypertens 23:274–286.

Hennig B, Meerarani P, Slim R, Toborek M, Daugherty A, Silverstone AE, et al. 2002. Proinflammatory properties of coplanar PCBs: In vitro and in vivo evidence. Toxicol Appl Pharmacol 181:174–183.

Hertzman C, Teschke K, Ostry A, Hershler R, Dimich-Ward H, Kelly S, et al. 1997. Mortality and cancer incidence among sawmill workers exposed to chlorophenate wood preservatives. Am J Public Health 87:71–79.

Hooiveld M, Heederik DJ, Kogevinas M, Boffetta P, Needham LL, Patterson DG, Jr., et al. 1998. Second follow-up of a Dutch cohort occupationally exposed to phenoxy herbicides, chlorophenols, and contaminants. Am J Epidemiol 147:891–901.

Hsu ST, Ma CI, Hsu SK, Wu SS, Hsu NH, Yeh CC, Wu SB. 1985. Discovery and epidemiology of PCB poisoning in Taiwan: a four-year followup. Environ Health Perspect 59:5–10.

Humblet O, Birnbaum L, Rimm E, Mittleman MA, Hauser R. 2008. Dioxins and cardiovascular disease mortality. Environ Health Perspect 116:1443–1448.

IARC (International Agency for Research on Cancer). 1997. Polychlorinated dibenzo-para-dioxins and polychlorinated dibenzofurans. IARC Monogr Eval Carcinog Risks Hum 69:33–343.

Jokinen MP, Walker NJ, Brix AE, Sells DM, Haseman JK, Nyska A. 2003. Increase in cardiovascular pathology in female Sprague-Dawley rats following chronic treatment with 2,3,7,8-tetrachlorodibenzo-p-dioxin and 3,3′,4,4′,5-pentachlorobiphenyl. Cardiovasc Toxicol 3:299–310.

Kang HK, Dalager NA, Needham LL, Patterson DG, Jr., Lees PS, Yates K, et al. 2006. Health status of Army Chemical Corps Vietnam veterans who sprayed defoliant in Vietnam. Am J Ind Med 49:875–884.

Kashima S, Yorifuji T, Tsuda T. 2011. Acute non-cancer mortality excess after polychlorinated biphenyls and polychlorinated dibenzofurans mixed exposure from contaminated rice oil: Yusho. Sci Total Environ 409:3288–3294.

Ketchum NS, Michalek JE. 2005. Postservice mortality of Air Force veterans occupationally exposed to herbicides during the Vietnam War: 20-year follow-up results. Mil Med 170:406–413.

Kim JS, Lim HS, Cho SI, Cheong HK, Lim MK. 2003. Impact of Agent Orange exposure among Korean Vietnam veterans. Ind Health 41:149–157.

Kopf PG, Walker MK. 2009. Overview of developmental heart defects by dioxins, PCBs, and pesticides. J Environ Sci Health C Environ Carcinog Ecotoxicol Rev 27:276–285.

Kopf PG, Huwe JK, Walker MK. 2008. Hypertension, cardiac hypertrophy, and impaired vascular relaxation induced by 2,3,7,8-tetrachlorodibenzo-p-dioxin are associated with increased superoxide. Cardiovasc Toxicol 8:181–193.

Kopf PG, Scott JA, Agbor LN, Boberg JR, Elased KM, Huwe JK, et al. 2010. Cytochrome P4501A1 is required for vascular dysfunction and hypertension induced by 2,3,7,8-tetrachlorodibenzo-p-dioxin. Toxicol Sci 117:537–546.

Kreiss K, Zach MM, Kimbrough RD, Needham LL, Smrek AL, Jones BT. 1981. Association of blood pressure and polychlorinated biphenyl levels. JAMA 245:2505–2509.

Kuratsune M, Yoshimura T, Matsuzaka J, Yamaguchi A. 1972. Epidemiologic study on Yusho, a poisoning caused by ingestion of rice oil contaminated with a commercial brand of polychlorinated biphenyls. Environ Health Perspect 1:119–128.

Lind PM, Orberg J, Edlund UB, Sjoblom L, Lind L. 2004. The dioxin-like pollutant PCB 126 (3,3′,4,4′,5-pentachloro-biphenyl) affects risk factors for cardiovascular disease in female rats. Toxicol Lett 150:293–299.

Lund AK, Peterson SL, Timmins GS, Walker MK. 2005. Endothelin-1-mediated increase in reactive oxygen species and NADPH oxidase activity in hearts of aryl hydrocarbon receptor (AhR) null mice. Toxicol Sci 88:265–273.

Manuwald U, Velasco Garrido M, Berger J, Manz A, Baur X. 2012. Mortality study of chemical workers exposed to dioxins: follow-up 23 years after chemical plant closure. Occup Environ Med 69(9):636–642.

Matthews J, Wihlén B, Heldring N, MacPherson L, Helguero L, Treuter E, et al. 2007. Co-planar 3,3′,4,4′,5-pentachlorinated biphenyl and non-co-planar 2,2′,4,6,6′-pentachlorinated biphenyl differentially induce recruitment of oestrogen receptor alpha to aryl hydrocarbon receptor target genes. Biochem J 406:343–353.

McBride DI, Burns CJ, Herbison GP, Humphry NF, Bodner K, Collins JJ. 2009a. Mortality in employees at a New Zealand agrochemical manufacturing site. Occup Med (Lond) 59:255–263.

McBride DI, Collins JJ, Humphry NF, Herbison P, Bodner KM, Aylward LL, et al. 2009b. Mortality in workers exposed to 2,3,7,8-tetrachlorodibenzo-p-dioxin at a trichlorophenol plant in New Zealand. J Occup Environ Med 51:1049–1056.

McLean D, Pearce N, Langseth H, Jäppinen P, Szadkowska-Stanczyk I, Persson B, et al. 2006. Cancer mortality in workers exposed to organochlorine compounds in the pulp and paper industry: An international collaborative study. Environ Health Perspect 114:1007–1012.

McMichael AJ. 1976. Standardized mortality ratios and the "healthy worker effect": Scratching beneath the surface. J Occup Med 18:165–168.

Michalek JE, Tripathi RC. 1999. Pharmacokinetics of TCDD in veterans of Operation Ranch Hand: 15-year follow-up. J Toxicol Environ Health A 57:369–378.

Michalek JE, Ketchum NS, Akhtar FZ. 1998. Postservice mortality of US Air Force veterans occupationally exposed to herbicides in Vietnam: 15-year follow-up. Am J Epidemiol 148:786–792.

Michalek JE, Robinson JN, Fox KA, Pavuk M, Grubbs W. 2005. Air Force Health Study: An epidemiologic investigation of health effects in Air Force personnel following exposure to herbicides—2002 follow-up examination results. Air Force Research Laboratory, Brooks City-Base, TX, F41624-96-C-1012.

Mikoczy Z, Rylander L. 2009. Mortality and cancer incidence in cohorts of Swedish fishermen and fishermen's wives: Updated findings. Chemosphere 74:938–943.

Mikoczy Z, Schutz A, Hagmar L. 1994. Cancer incidence and mortality among Swedish leather tanners. Occup Environ Med 51:530–535.

Monson RR. 1986. Observations on the healthy worker effect. J Occup Med 28:425–433.

National Academy of Sciences. 2007. Veterans and Agent Orange: Update 2006. Washington, DC: National Academies Press.

NATO (North Atlantic Treaty Organization, Committee on the Challenges of Modern Society). 1988. Pilot Study on International Information Exchange on Dioxin and Related Compounds. Scientific Basis for the Development of the International Toxicity Equivalency Factor (I-TEF): Method of Risk Assessment for Complex Mixtures of Dioxins and Related Compounds. No. 178. Brussels: North Atlantic Treaty Organization, Committee on the Challenges of Modern Society.

Ohtake F, Takeyama K, Matsumoto T, Kitagawa H, Yamamoto Y, Nohara K, et al. 2003. Modulation of oestrogen receptor signalling by association with the activated dioxin receptor. Nature 423:545–550.

Ott MG, Zober A. 1996. Cause specific mortality and cancer incidence among employees exposed to 2,3,7,8-TCDD after a 1953 reactor accident. Occup Environ Med 53:606–612.

Patterson DG, Jr., Wong LY, Turner WE, Caudill SP, Dipietro ES, McClure PC, et al. 2009. Levels in the U.S. population of those persistent organic pollutants (2003–2004) included in the Stockholm Convention or in other long range transboundary air pollution agreements. Environ Sci Technol 43:1211–1218.

Piacitelli L, Marlow D, Fingerhut M, Steenland K, Sweeney MH. 2000. A retrospective job exposure matrix for estimating exposure to 2,3,7, 8-tetrachlorodibenzo-p-dioxin. Am J Ind Med 38:28–39.

Ruder AM, Yiin JH. 2011. Mortality of US pentachlorophenol production workers through 2005. Chemosphere 83:851–861.

Ryan JJ, Gasiewicz TA, Brown JF, Jr. 1990. Human body burden of polychlorinated dibenzofurans associated with toxicity based on the Yusho and Yucheng incidents. Fundam Appl Toxicol 15:722–731.

Schecter A, Needham L, Pavuk M, Michalek J, Colacino J, Ryan J, et al. 2009. Agent Orange exposure, Vietnam war veterans, and the risk of prostate cancer. Cancer 115:3369–3371.

Slim R, Toborek M, Robertson LW, Hennig B. 1999. Antioxidant protection against PCB-mediated endothelial cell activation. Toxicol Sci 52:232–239.

Spink BC, Bennett JA, Pentecost BT, Lostritto N, Englert NA, Benn GK, Goodenough AK, Turesky RJ, Spink DC. 2009. Long-term estrogen exposure promotes carcinogen bioactivation, induces persistent changes in gene expression, and enhances the tumorigenicity of MCF-7 human breast cancer cells. Toxicol Appl Pharmacol 240(3):355–366.

Steenland K, Piacitelli L, Deddens J, Fingerhut M, Chang LI. 1999. Cancer, heart disease, and diabetes in workers exposed to 2,3,7,8-tetrachlorodibenzo-p-dioxin. J Natl Cancer Inst 91:779–786.

Taddei S, Virdis A, Mattei P, Ghiadoni L, Fasolo CB, Sudano I, et al. 1997. Hypertension causes premature aging of endothelial function in humans. Hypertension 29:736–743.

Taddei S, Virdis A, Ghiadoni L, Salvetti A. 1998. The role of endothelium in human hypertension. Curr Opin Nephrol Hypertens 7:203–209.

Thackaberry EA, Nunez BA, Ivnitski-Steele ID, Friggins M, Walker MK. 2005. Effect of 2,3,7,8-tetrachlorodibenzo-p-dioxin on murine heart development: Alteration in fetal and postnatal cardiac growth, and postnatal cardiac chronotropy. Toxicol Sci 88:242–249.

't Mannetje A, McLean D, Cheng S, Boffetta P, Colin D, Pearce N. 2005. Mortality in New Zealand workers exposed to phenoxy herbicides and dioxins. Occup Environ Med 62:34–40.

Tsai PC, Ko YC, Huang W, Liu HS, Guo YL. 2007. Increased liver and lupus mortalities in 24-year follow-up of the Taiwanese people highly exposed to polychlorinated biphenyls and dibenzofurans. Sci Total Environ 374:216–222.

Van den Berg M, Birnbaum L, Bosveld AT, Brunstrom B, Cook P, Feeley M, et al. 1998. Toxic equivalency factors (TEFs) for PCBs, PCDDs, PCDFs for humans and wildlife. Environ Health Perspect 106:775–792.

Van den Berg M, Birnbaum LS, Denison M, De Vito M, Farland W, Feeley M, et al. 2006. The 2005 World Health Organization reevaluation of human and mammalian toxic equivalency factors for dioxins and dioxin-like compounds. Toxicol Sci 93: 223–241.

Vanhoutte PM, Tang EH. 2008. Endothelium-dependent contractions: When a good guy turns bad! J Physiol 586:5295–5304.

Vena J, Boffetta P, Becher H, Benn T, Bueno-de-Mesquita HB, Coggon D, et al. 1998. Exposure to dioxin and nonneoplastic mortality in the expanded IARC international cohort study of phenoxy herbicide and chlorophenol production workers and sprayers. Environ Health Perspect 106(Suppl 2):645–653.

Versari D, Daghini E, Virdis A, Ghiadoni L, Taddei S. 2009. Endothelium-dependent contractions and endothelial dysfunction in human hypertension. Br J Pharmacol 157:527–536.

Weed DL, Tyroler HA, Shy C. 1987. The healthy worker effect in actively working communications workers. J Occup Med 29:335–339.

WHO. 1978. International Classification of Diseases. Ninth Revision. Geneva: World Health Organization.

WHO. 2010. International Classification of Diseases. Tenth Revision. Geneva: World Health Organization. Available: http://apps.who.int/classifications/icd10/browse/2010/en [accessed April 8, 2012].

Xie A, Walker NJ, Wang D. 2006. Dioxin (2,3,7,8-tetra-chlorodibenzo-p-dioxin) enhances triggered after depolarizations in rat ventricular myocytes. Cardiovasc Toxicol 6:99–110.

CHAPTER 8

Obesity

EVELINE DIRINCK, ADRIAN COVACI, LUC VAN GAAL, and STIJN VERHULST

ABSTRACT

Background: Recent studies have shown that exposure to a variety or organic chemicals increase risk of development of obesity.

Objectives: This chapter reviews the basic biology of adipocytes in the human body and how they are regulated by a variety of hormones. Then we will review current information on how these mechanisms are altered by organic chemicals.

Discussion: Organic chemicals that have endocrine disruptive activities affect risk of development of obesity. This is true especially for organics that have estrogenic activity, but also those that alter androgenic, thyroid and glucocorticoid function and those that act via adipoctokines an peroxisome proliferator-activated receptor γ.

Conclusions: Recent investigations show clearly that obesity is not solely a consequence of overeating and lack of exercise, but that exposure to some endocrine-disrupting organic chemicals significantly increases risk of development of obesity.

INTRODUCTION

Obesity is quickly becoming one of the most significant human health threats worldwide. At the end of the 20th century, more than half a billion people worldwide were estimated to be overweight or obese (Mokdad et al. 2001). Traditionally, the increase in obesity is attributed to an increased caloric intake and a concomitant significant reduction in physical activity and energy expen-

Effects of Persistent and Bioactive Organic Pollutants on Human Health, First Edition.
Edited by David O. Carpenter.

diture (Prentice 2001). However, this does not fully explain the extent of the current epidemic. Concurrent with the obesity epidemic is the exponential increase of human exposure to synthetic chemicals worldwide. It has been postulated that certain chemicals potentially play a causative role in the development of obesity, by affecting or causing changes in fat mass (Baillie-Hamilton 2002).

Several chemicals and consumer products are known to mimic, enhance, or inhibit the action of hormones and are therefore called endocrine-disrupting chemicals (EDCs). Endocrine disruption is the inappropriate modulation of the endocrine system by dietary and environmental chemicals; these can interfere with the synthesis, secretion, transport, binding, or action of natural hormones in the body that are responsible for the maintenance of homeostasis, reproduction, development, and behavior (Vos et al. 2000). Today, we are dealing with thousands of these chemicals, either as individual compounds or as part of a potentially dangerous cocktail. EDCs have been shown to alter the biological actions of hormones such as estrogen, testosterone, and thyroxin and have already been linked to infertility (Foster et al. 2008), developmental changes (Patisaul and Adewale 2009), and cancer (Soto and Sonnenschein 2010). It is provocative, but plausible, to associate the recent obesity epidemic with the exponentially increased use of industrially produced EDCs over the past 40 years.

WHY SUCH AN ENDOCRINE HYPOTHESIS?

Obesity and endocrinology are strongly linked, as hormones and growth factors control adipose tissue metabolism (lipolysis), adipogenesis, and centrally mediated food intake and appetite (Pasquali and Vicennati 2000). In rodents, estrogen has a strong influence on body fat stores: various estrogen analogues decrease weight, fat stores, and food intake in mice and rats (Wei et al. 2011). Although the mechanisms responsible for estrogen's regulation of adipogenesis remain unclear, they may involve suppression of lipoprotein lipase, an enzyme that regulates the metabolism of plasma triglycerides to free fatty acids and increases lipid storage by adipocytes (Deroo and Korach 2006). Sex hormones such as estrogen and testosterone also cause permanent changes in the architecture of the limbic-hypothalamic circuits, responsible for appetite and food intake (Hirschberg 2012). It has long been recognized that thyroid hormone regulates energy expenditure in humans (Bianco 2011). What is less recognized is that thyroid hormone also stimulates appetite (Amin et al. 2011) and lipogenesis (Yen 2001). Glucocorticoids (GCs) are potent inducers of adipogenesis *in vivo*, and hypercortisolism is associated with accumulation of abdominal, visceral adiposity. Adipocytes are uniquely equipped to function in energy storage and balance under tight hormonal and neuronal control. With the realization that adipocytes secrete factors, the so-called adipocytokines, involved in the regulation of food intake and energy homeostasis, the

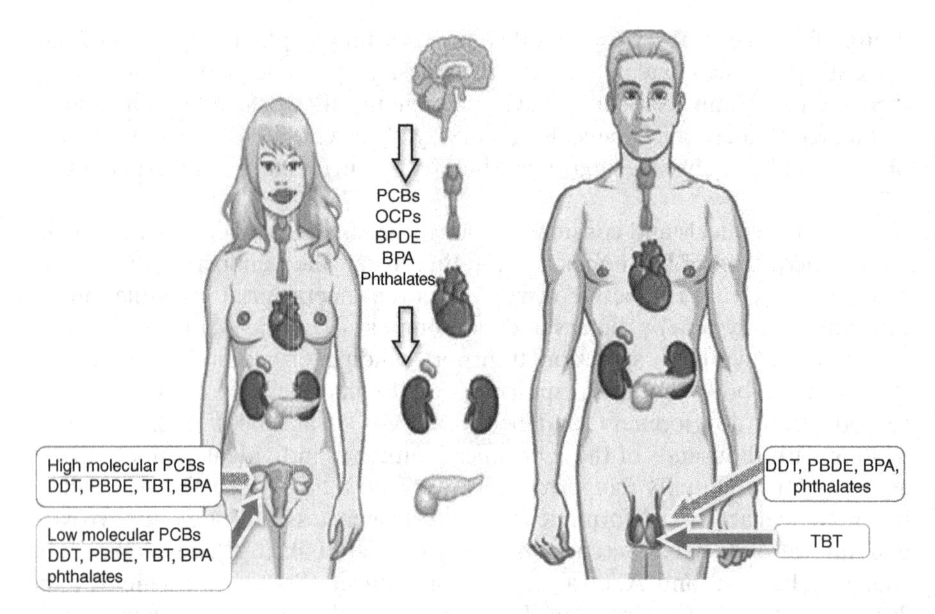

Figure 8.1. Overview of the various influences of EDCs on the thyroid, glucocorticoid, and sex hormone axes. Checkered arrows indicate antagonistic activity; filled arrows indicate agonistic activity; and unfilled arrows indicate mixed and/or an alternative influence.

dynamic role of adipocytes in energy metabolism has emerged (Van Gaal et al. 2003) (Figure 8.1).

WHICH CHEMICALS MIGHT PLAY A ROLE?

Several groups of chemicals have been studied to understand their effect on energy homeostasis and the development of obesity. In general, most EDCs are chemically stable compounds, resistant to degradation and, thus, subject to a significant bioaccumulation in most compartments of the ecosystem and in humans. Humans are environmentally exposed through ingestion of contaminated food or water, inhalation of polluted air, or dermal exposure.

Polychlorinated biphenyls (PCBs) are man-made chemicals, characterized by two benzene rings (see Table 8.1). There are 209 different PCB congeners, and commercially available products always contain a mixture of several congeners.

PCBs are odorless, tasteless, viscous liquids whose chemical properties make them excellent dielectric and coolant fluids. Therefore, they were used worldwide for various industrial purposes as lubricants and coolers in electrical appliances. The production of PCBs was banned in the 1970s. Unfortunately, since PCBs are very resistant to biological degradation, significant

TABLE 8.1. Overview of Chemicals and Their Structures

Chemical
Polychlorinated biphenyls. The possible positions of chlorine atoms on the benzene rings are denoted by numbers assigned to the carbon atoms. 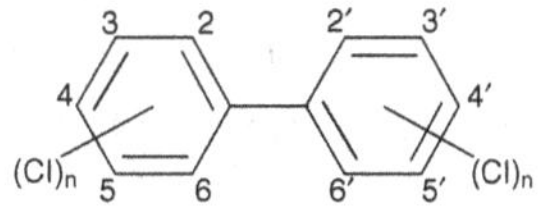
Organochlorine pesticides. The structure of DDT is provided as an exemplary. 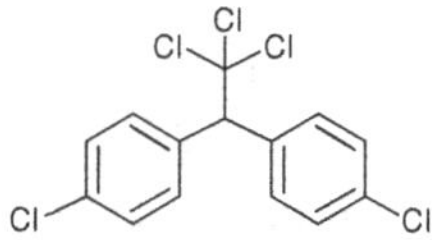
Organotins. The structure of dibutyltin chloride is provided as an exemplary. 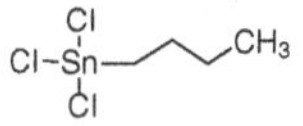
Brominated flame retardants. The general structure of PBDEs is provided. 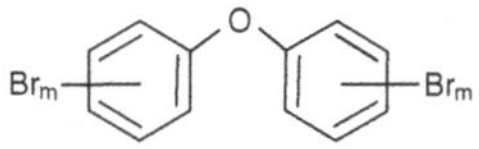
Bisphenol A 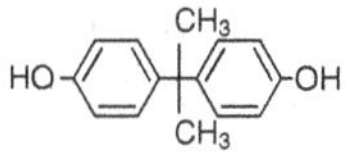
Phthalates 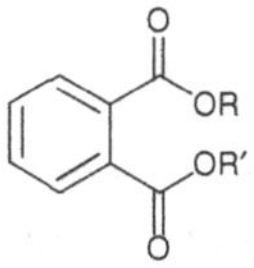

bioaccumulation leads to an ongoing human exposure. PCBs are found in various human tissues around the globe. The U.S. dataset of the National Health and Nutrition Examination Survey (NHANES) 2001–2002 indicated a clear relation between age and blood concentrations for PCB, with median values of 15 ng/g lipid for adolescents, up to 550 ng/g lipid weight for people older than 80 years (Nichols et al. 2007). Our own dataset in an obese Belgian population, with a median age of 40, showed a median PCB concentration of 191 ng/g lipid (Dirinck et al. 2011). Spanish data (Agudo et al. 2009) show a higher median serum concentration of PCBs of 498.8 ng/g lipids, in a population of 55–64 years old. In human milk, European studies detect median levels around 80 ng/g lipid (Colles et al. 2008; Costopoulou et al. 2006; Lignell et al. 2009).

Organochloride pesticides (OCPs) are a group of organic compounds containing at least one chlorine atom (see Table 8.1). The most notorious member of this group is dichlorodiphenyltrichloroethane (DDT). DDT is a man-made crystalline solid. The commercially available DDT is a mixture of closely related isomers, with the p′-p′-isomer as major contributor. DDT is metabolized by the human body into dichlorodiphenyldichloroethylene (DDE).

Other OCPs include aldrin, chlordane, lindane, and mirex. They were used in abundance in the Western world until their ban in the 1960s. Unfortunately, OCPs, and DDT in particular, remain a popular pesticide in the developing world to increase agricultural production and to control various vectors spreading diseases such as malaria. Similar to PCBs, OCPs are very resistant to biological degradation. In the U.S. NHANES 1999–2000 study, mean DDE levels of 260 ng/g lipids were detected in the overall population. There is a distinct difference between ethnicities, with a mean of 217 ng/g lipids in white non-Hispanics, 295 ng/g lipids in non-Hispanic blacks, and up to 674 ng/g lipids in Mexican Americans (Needham et al. 2005). In a recent South African study, mean DDT levels of 34,120 ng/g lipids were reported in an area where DDT is still in use as vector controller (Delport et al. 2011). In Belgium, DDT levels in human milk recently have become undetectably low, whereas the levels of DDE were 124.5 ng/g lipids (Colles et al. 2008).

Organotin compounds are chemicals based on tin with hydrocarbon substituents (see Table 8.1). They are used as fungicides, heat stabilizers in plastics, and antifouling agents in paints. Data on the body burden of organotins in humans are scarce but indicate universal presence. In 1997, butyltin residues were analyzed in nine liver samples of Polish residents of the Southern Baltic Sea area (Kannan and Falandysz 1997), with concentrations of total butyltin compounds varying between 2.4 and 11.0 ng/g wet weight. In 1999, the analysis of four liver samples of Japanese citizens revealed even higher concentrations for butyltin compounds of 59 up to 96 ng/g wet weight (Takahashi et al. 1999). In a Danish study (Nielsen and Strand 2002), concentrations of butyltin compounds between 1.1 and 33 ng/g wet weight were detected in 18 liver samples. Apart from these liver tissue studies, few studies describe the serum burden of organotins. In the United States in Michigan, the mean concentration of butyltin compounds in blood samples of 32 volunteers was found to average 16.4 ng/mL (Kannan et al. 1999). In Germany, triphenyltin was detectable in the serum of eight volunteers at concentrations ranging from 0.17 to 0.67 μg/L (Lo et al. 2003).

Brominated flame retardants are composed of a chemically heterogeneous group of products, with the common property of containing a bromine atom (see Table 8.1). They have an inhibitory effect on the ignition of combustible organic materials and are therefore used in a wide variety of consumer products, from clothing and furniture to electrical appliances. Commercially, the hexabromocyclododecane (HBCD) and polybrominated diphenyl ethers (PBDEs) are the most intensely used. HBCD is detected in both human serum and milk. The concentrations differ greatly between continents. In European human milk, median concentrations of 0.6 ng/g lipid weight in human milk are measured (Roosens et al. 2010). In the United States, similar median concentrations of 0.4 ng/g lipid weight are detected (Schecter et al. 2007). In China, on the other hand, concentrations rise up to 948 ng/g lipid weight (Shi et al. 2009).

The use of PBDEs is in decline in Europe due to legislation bans, but is still on the rise in the United States. This is reflected in the burden of PBDEs in human milk, where there are large differences between the Belgian levels (median 3.0 ng/g lipid weight) and the U.S. levels (median 44 ng/g lipids) (Roosens et al. 2010; Schecter et al. 2007). In serum, the same trend can be detected, with median serum levels of PBDEs of 4.55 ng/g lipid weight in Belgium (Roosens et al. 2010), versus 19.3 ng/g lipid weight in the United States (Stapleton et al. 2008). Analysis of PBDE congeners in human adipose tissue and human liver samples detected average concentrations of 5.3 and 3.6 ng/g lipid weight, respectively (Covaci et al. 2008).

Bisphenol A (BPA) is an organic, solid material, used primarily to make plastics (see Table 8.1). As such, BPA can be found in baby and water bottles, food and beverage cans, sports equipment, medical and dental devices, CDs and DVDs, household electronics, and eyeglass lenses. BPA is still produced on a massive scale, with an estimated annual production of 8 billion pounds yearly.

Given this high production volume, most studies find BPA or its conjugates in more than 90% of tested individuals. An extensive review by Vandenberg et al. (2007) shows human blood levels in the range of 0.4–4.4 ng/mL. Even more troubling is the fact that BPA can be detected in fetal serum, umbilical cord blood, amniotic fluid, and placenta, which indicates that the developing fetus is exposed to BPA in the higher range of 1–3 ng/mL. In urine, BPA conjugate traces are found in the low nanogram per milliliter range (Vandenberg et al. 2007).

Phthalates are another essential compound of plastics (see Table 8.1). Their addition to a plastic makes it more flexible, transparent, and durable. Phthalates can be found in a large variety of products, from enteric coatings of pharmaceutical pills, adhesives and glues, electronics, building materials, personal care products, medical devices, packaging material, children's toys, paints, and printing inks. Phthalates are easily released into the environment because there is no covalent bond between the phthalates and plastics in which they are mixed. As plastics age and break down, the release of phthalates accelerates. Therefore, both Europe and the United States have recently restricted the use of phthalates in children's toys. Similar to BPA, phthalate metabolites are detectable in the vast majority of the population. In the 1999–2002 NHANES, several phthalate metabolite concentrations were measured. Concentrations ranged from $11 \pm 1.3\,\mu g/g$ creatinine for the mono-(2-ethyl)-hexyl phthalate derivate up to $771 \pm 66.7\,\mu g/g$ creatinine for another subtype, mono-ethyl-phthalate (Stahlhut et al. 2007). Similar to PCBs, concentrations varied according to ethnicity. In a European study, mean mono-ethyl-hydroxyhexyl-phthalate (MEHP) concentrations in men were 8.28 ng/mL, and 184 ng/mL for mono-ethyl phthalate (MEP) (Meeker et al. 2007). In children, the exposure to phthalates is known to be higher (Koch et al. 2011).

EPIDEMIOLOGICAL DATA

There are limited epidemiological data linking EDCs with obesity. Furthermore, these studies differ in the timing of exposure, the age of the studied subjects, the number of confounding factors included, as well as the assessment of confounding of other possible EDCs. The epidemiological data will be reviewed according to the EDC.

PCBs

Studies on the effect of prenatal exposure of humans to PCBs provide conflicting results. It has been postulated that these conflicting results depend on the level of exposure at least in prospective studies. Several prospective studies demonstrated that prenatal PCB exposure was associated with increased body mass index (BMI) at relatively low levels of exposure (Hertz-Picciotto et al. 2005; Verhulst et al. 2009) but not at higher levels (Blanck et al. 2002; Rylander et al. 1998). Exposure levels in the medium range showed divergent results (Gladen et al. 2000; Jacobson et al. 1990; Karmaus et al. 2009; Lamb et al. 2006; Patandin et al. 1998). This could indicate that relatively low, but biologically significant, exposure levels could promote obesity, while at higher levels, PCBs mainly exert a toxic effect.

These conflicting results are also reported in cross-sectional studies. In a Belgian adult population, we found a significant negative relation between PCB serum concentration and BMI (Dirinck et al. 2011). This inverse relation between BMI and PCBs was also established in other cohorts (Agudo et al. 2009; Wolff et al. 2005). In the 1999–2002 NHANES survey, a significant relationship was found between serum levels of non-dioxin-like PCBs and an enlarged waist circumference (Lee et al. 2006). In a selected nondiabetic subset of this NHANES population, PCB levels were not associated with BMI (Lee et al. 2007). In a Swedish study, some PCB congeners were positively associated with BMI, others displayed a negative association (Glynn et al. 2003). In Canadian men, PCB levels were not associated with BMI (Hue et al. 2007). In a study that analyzed fat mass by dual-energy X-ray absorptiometry (DXA), some PCBs were positively correlated with fat mass, whereas others were negatively correlated. This effect was more pronounced in women (Rönn et al. 2011). Most importantly, in a prospective 20-year study, a high serum level of PCBs in a U.S. population was associated with increased BMI. This association was congener specific (Lee et al. 2011). The same observation was made in a 5-year follow-up study in a Swedish population, where particularly the less chlorinated PCBs were associated with the development of abdominal obesity. This effect was again more pronounced in women (Lee et al. 2011). This can be explained by the fact that some PCB congeners display estrogenic effects, and others antiestrogenic and/or antiandrogenic effects.

Organochlorine Pesticides

After *in utero* exposure to DDE, studies have revealed a positive relationship with BMI in both female and male offspring (Gladen et al. 2000; Karmaus et al. 2009; Verhulst et al. 2009). In the study by Verhulst et al., there was a significant interaction with DDE and maternal smoking. Smoking could enhance susceptibility to pollutants by interfering with immune function and by promoting oxidative stress, which could sensitize cells to pollutants. Due to power issues, it was not possible to distinguish between mothers who stopped smoking during pregnancy and those who continued smoking during pregnancy. In the first group, weight gain during pregnancy could also modulate cord blood EDC levels and birth weight. In one study, the association between prenatal exposure to DDT and BMI was not significant (Gladen et al. 2004). The former positive studies all showed an association between low levels of DDE and body size, while high levels of exposure were reported in the study by Gladen et al. There are limited data on *in utero* exposure to hexachlorobenzene (HCB) that indicate either a positive association (Smink et al. 2008) or a nonsignificant association (Verhulst et al. 2009).

In the 1999–2002 NHANES survey, a significant relationship was found between serum levels of DDT and an enlarged waist circumference in women, but the inverse was detected in men. Other OCPs in this study, such as oxychlordane, displayed a positive association with BMI in men but not in women (Elobeid et al. 2010). Another analysis of the NHANES cohort showed a positive association with BMI and DDE (Everett et al. 2008). In a nondiabetic subset of the NHANES population, DDE and other OCPs were also positively associated with BMI in both sexes (Lee et al. 2007). In a Swedish study, several OCPs (including DDE and HCB) were positively associated with BMI in women (Glynn et al. 2003). In Canadian men, there seemed to be a positive relation between DDE and BMI, but not so for HCB (Hue et al. 2007). Several OCPs, including DDE, are positively associated with fat mass, assessed by DXA (Rönn et al. 2011). In the aforementioned follow-up study by Lee et al., DDE serum levels were predictive for the development of obesity (Lee et al. 2011). In the Swedish follow-up study, both DDE and octachlordibenzodioxin (OCDD) were associated with the development of obesity (Lee et al. 2011).

Brominated Flame Retardants

Moderate prenatal exposure to polybrominated biphenyls (PBBs) seems to be related to weight, while this relation could not be established after high prenatal exposure (Blanck et al. 2002). Using the dataset of the 2003–2004 NHANES, Lim et al. (2008) demonstrated a significant association between PBDE-153 and both the metabolic syndrome and diabetes.

BPA

In the NHANES 2003–2004 population, no significant difference in urinary BPA concentration was detected across BMI categories (Lang et al. 2008). However, significant associations were found with cardiovascular morbidity, diabetes, and liver disease.

Phthalates

In the 1999–2002 NHANES, concentrations of several prevalent phthalate metabolites showed a statistically significant correlation with abdominal obesity and insulin resistance. The authors provided clinically interpretable data by stating that an increase in phthalate metabolites from the 10th to the 90th percentile equaled an increase in waist circumference from 3.9 to 7.8 cm, depending on the metabolite (Stahlhut et al. 2007). In the same NHANES dataset, some phthalate metabolites are associated with BMI. These results are most consistent in men but less so in women (Hatch et al. 2008).

To our knowledge, there are no epidemiological data in humans on the relationship between organotins and obesity published so far.

MECHANISMS LINKING ENDOCRINE DISRUPTORS WITH OBESITY

Sex Hormones

The influence of sex hormones on our body composition is apparent by the daily observation of the effect of aging in both genders. In women, the loss of circulating estrogen following menopause is associated with an increase in central body fat. As men age, their free testosterone levels fall, and their muscle mass declines and fat mass increases.

Estrogens The estrogen receptor (ER) exists in two main forms, ERα and ERβ, and is expressed in a distinct tissue-specific manner throughout the body. It is important to note that ERs are located both in the nucleus and on the cell membrane (Levin 2009). ERs can be found in human adipose tissue where it is capable of both increasing and inhibiting lipolysis (Palin et al. 2003; Pedersen et al. 2004). Moreover, estrogen has an important role in adipogenesis, adipose deposition, and adipocyte proliferation (Cooke and Naaz 2004), illustrated by the observation that the amount of adipose tissue is from 50% to 100% greater in ERα-knockout mice than in their wild-type counterparts (Heine et al. 2000). At the level of the liver, estrogen suppresses white adipose tissue accumulation by decreasing fatty acid and triglyceride synthesis (D'Eon et al. 2005). Besides the local effects on adipose and liver tissues, estrogen exerts its effect on energy homeostasis through its influence in the central nervous system. Estrogen can act as a "leptinomimetic," enhancing the reducing effect of leptin on food intake (Clegg et al. 2006). Interestingly, aromatase-deficient mice, which do not synthesize estrogen, do not exhibit hyperphagia, but they show a significant reduction in spontaneous physical activity (Jones et al. 2001). It is important to note that all these actions of the ER are present in both sexes.

Androgens It is postulated that testosterone should be considered as a circulating prohormone, locally converted by aromatization to 17β-estradiol, a ligand of ER, or to 5α-dihydrotestosterone, the main ligand of the androgen receptor (AR) (Mauvais-Jarvis 2011). Like the ER, there is extensive evidence for the role of the AR in energy homeostasis. First, men with genetic androgen resistance display elevated levels of visceral fat (Zitzmann et al. 2003). Second, male mice lacking the AR develop late onset visceral obesity with increased lipogenesis in the white adipose tissue and liver (Fan et al. 2005). Third, adiponectin levels are elevated in hypogonadal men and reduced by testosterone therapy (Lanfranco et al. 2004). In concordance with the ER, there are centrally mediated effects of the AR. Leptin signaling in the hypothalamus is enhanced by AR (Fan et al. 2008). Like aromatase-deficient mice, the AR-deficient mice develop obesity as a result of reduced locomotor activity, but without hyperphagia (Fan et al. 2005). The role of the AR in the development of obesity starts during the embryonic phase, as testosterone is capable of promoting the differentiation of pluripotent mesenchymal stem cells into myogenic lineage while inhibiting the adipogenic lineage via an AR-dependent mechanism (Singh et al. 2003).

Influence by EDC

PCBS The toxicological properties of the PCB congeners toward the ER are largely dependent on their structure. Briefly, the lower-molecular-weight PCBs tend to exhibit estrogenic activity, while higher-molecular-weight PCBs tend to be more antiestrogenic (Plísková et al. 2005). Prenatal exposure to PCBs seems to decrease the gene expression of aromatase and 17-α-hydroxylase in adult female offspring (Karmaus et al. 2011). In the same cohort, the gene expression of ERβ was upregulated significantly by a subgroup of dioxin-like PCBs with potential antiestrogenic activity (Warner et al. 2012). In a cohort study of German infants, a significant negative association was found between maternal serum PCB levels and testosterone levels in their female offspring and estradiol levels in their male offspring (Cao et al. 2008). In adult men, serum PCB concentration is inversely correlated with testosterone levels (Goncharov et al. 2009). PCBs exert their effects both through the classical nuclear receptors as through the activation of the membrane-bound G protein-coupled receptor 30 ER (Thomas and Dong 2006).

OCPS DDT and its metabolite DDE are known ERα agonists and AR antagonists (Li et al. 2008). In contrast to PCBs, prenatal DDE concentrations are related to increased aromatase gene expression (Karmaus et al. 2011).

BROMINATED FLAME RETARDANTS The estrogenic effect of PBDEs is suggested by multiple *in vivo* studies in mice, but is hard to corroborate in *in vitro* studies (Legler 2008). In contrast, the antiandrogenic effects of PBDEs in mice were confirmed with several *in vitro* studies (Stoker et al. 2005; van der Ven et al. 2006). Several brominated flame retardants were shown to have AR-antagonistic properties (Hamers et al. 2006). Some PBDE metabolites are aromatase inhibitors *in vitro* (Cantón et al. 2008). In accordance with PCBs, it is suggested that the chemical structure of PBDEs influences the endocrine-disrupting potency (Harju et al. 2007). It is troubling that the newer brominated flame retardants, recently introduced to replace the older, banned products, also exert an antiestrogenic and antiandrogenic activity (Ezechiáš et al. 2012).

ORGANOTINS Initially, the effects of tributyltin (TBT) were investigated in fish, where TBT is capable of downregulating the expression of ERα and upregulating the expression of ERβ and ARβ (Mortensen and Arukwe 2007; Zhang et al. 2009). Recently, TBT was identified as an ERα agonist in preadipocytes and ERβ specific in differentiating adipocytes of mice. Environmentally relevant doses of TBT were shown to increase the fat mass of peripubertal and adult male mice, while this effect was only transient in female mice (Penza et al. 2011).

BPA BPA is a known estrogenic compound, capable of eliciting cellular responses at very low levels (Hugo et al. 2008). In addition, the estrogenic capacity of BPA is exceeded by some of its metabolites (Ben-Jonathan and Steinmetz 1998). BPA binds both the classical ER and the membrane-bound ER (Dong et al. 2011; Thomas and Dong 2006). Moreover, BPA has the capacity to influence the body's response to estrogens long after its exposure period by altering the estrogen sensitivity in a tissue-specific way and by affecting the mRNA levels of ERα and ERβ (Ramos et al. 2003; Wadia et al. 2007). In addition, BPA also exerts an antiandrogenic effect via the AR (Lee et al. 2003).

PHTHALATES Some phthalates have estrogenic activities *in vitro* (Harris et al. 1997). One of the newer phthalates, dibenzyl phthalate, possesses both estrogenic as antiestrogenic properties *in vitro* and *in vivo* (Zhang et al. 2011). Phthalates are better known as antiandrogens, with data from experimental animal models dating back several decades. Interestingly, phthalates do not bind to the AR but interfere with the steroidogenic enzymes in the testosterone biosynthesis (Ye et al. 2011). Exposure to phthalate metabolites alters testicular steroid hormone synthesis, thus affecting the male reproductive system (Parks et al. 2000). In humans, this translates in subtle genital changes in male neonates exposed to high amounts during pregnancy (Swan et al. 2005). In adult males, higher phthalate concentrations are associated with diminished sperm quality and even sperm DNA damage (Hauser et al. 2006, 2007).

Thyroid Hormones

Traditionally, the role of thyroid hormones in energy homeostasis is mainly attributed to thermogenesis. The main action of thyroid hormones is reflected by their ability to stimulate cellular respiration/oxygen consumption while at the same time reducing metabolic efficiency (Silva 2006). Our resting energy expenditure is sensitive to changes in the levels of thyroid hormones, even in euthyroid conditions (al-Adsani et al. 1997). Recent evidence suggests that the hypothalamic–pituitary–thyroid axis may play a direct role in the hypothalamic regulation of appetite, independent of its effects on energy expenditure. This effect could be mediated by the local effects of thyrotropin-releasing hormone (TRH), thyrotropin-stimulating hormone (TSH), and thyroid hormones in the central nervous system, or by influencing the effects of leptin (Amin et al. 2011). At the level of the liver and the adipocytes, thyroid hormones are capable of stimulating both lipogenesis and lipolysis (Brent 1994). Some researchers postulate that the physiological effects of thyroid hormones are poorly reflected by the plasma concentrations. This assumption is based on the observation that intracellular deiodinases are crucial regulators of the effects of T4 entering the cell (Bianco 2011).

Influence by EDCs

PCBs PCBs, and especially their hydroxylated metabolites, bear a structural resemblance to the thyroid hormones (McKinney and Waller 1994). The actions of PCBs on the thyroid receptor are diverse, with reports of both agonistic and antagonistic activity *in vitro*. This diversity seems to be dependent on the type of PCB congener that was used (Zoeller 2007). Moreover, PCBs have an effect on thyroid hormone levels through interference with several enzyme systems involved in its synthesis, intracellular bioactivation, and degradation (Morse et al. 1993; Van Birgelen et al. 1994; Visser et al. 1993). PCBs are also known to displace T4 from its plasma-binding protein transthyretin (TTR) (Lans et al. 1993).

OCPs Organochlorine pesticides decrease thyroid serum levels by an induction of the hepatic enzymes that glucuronidate T4. In order to compensate for this increased T4 metabolism, the thyroid is required to increase thyroid hormone production (Brucker-Davis 1998).

Brominated Flame Retardants PBDEs cause enhanced excretion of T4 (Zhou et al. 2001). PBDE metabolites exhibit competitive binding to the serum transport proteins and the transmembrane transport system (Meerts et al. 2001). Richardson et al. (2008) established a decreased mRNA expression of hepatic TTR after PPDE exposure. Similar to OCPs, PBDE can induce the hepatic glucuronidation enzymes (Hallgren et al. 2001). Recently, BDE-99 was shown to downregulate the gene expression of TRα1 and TRα2 while upregulating the gene expression of TRβ1 (Blanco et al. 2011).

Organotins TBT accelerates the active T3 uptake at the plasma membrane (Shimada and Yamauchi 2004). Moreover, the expression of thyrothropin-releasing hormone is amplified after *in utero* exposure to TBT (Decherf et al. 2010).

BPA BPA is a weak antagonist of the TH receptor (TR) (Freitas et al. 2011; Moriyama et al. 2002).

Phthalates Phthalates are antagonists of the thyroid receptor (Shen et al. 2009). Additionally, phthalates are competitors for the binding of T3 to TTR (Ishihara et al. 2003). Some phthalates interact with the active T3 uptake at the plasma membrane (Shimada and Yamauchi 2004). Phthalates can inhibit the gene expression of the TRβ (Sugiyama et al. 2005). Iodide uptake in thyroid cells is significantly enhanced by phthalates (Wenzel et al. 2005).

GC Hormones

GCs are stress hormones designed to help the body maintain homeostasis during times of increased metabolic demands. The acute effects of stress hormones are mainly catabolic, with a liberation of energy stores such as glucose, amino acids, and fatty acids. Therefore, it is somewhat puzzling that an exogenous or endogenous GC excess such as Cushing's disease causes obesity with a prominent central adiposity. Furthermore, obese individuals do not seem to exhibit an elevation in their morning peak plasma cortisol. Recently, however, an imbalance in the diurnal cortisol rhythm, reflected by an attenuated morning peak and an elevated evening level, has been established in obese individuals (Kumari et al. 2010). What mechanisms explain the adipogenic nature of GCs? First, it is important to note that cortisone, which is not biologically active, can be converted into active cortisol locally, in a tissue-specific manner, by the 11β-hydroxysteroid dehydrogenase (11β-HSD) system. 11β-HSD type 1 is found primarily in the liver, brain, skeletal muscle, and adipose tissue. It is estimated that the GC levels inside adipose tissues are 10–15 times higher compared with circulating levels (Masuzaki et al. 2001). Additionally, 11β-HSD type 1 levels are higher in visceral versus subcutaneous fat (Bujalska et al. 1997). Second, GC alter fatty acid availability by increasing the activity of lipoprotein lipase, thus promoting the uptake and storage of fatty acids (Ottosson et al. 1994). This effect seems to be particularly important in the omental fat of men (Fried et al. 1998). Third, GCs also mediate their effect on energy homeostasis through a central mechanism, influencing neuropeptides such as neuropeptide Y (Epel et al. 2001) and/or the Agouti-related protein (Nakayama et al. 2011). This results in an increased food intake, with a preference for high-caloric foods. Four, GCs are a necessary factor in adipogenesis, inducing the differentiation of preadipocytes into adipocytes (Hauner et al. 1989). And last, next to adipogenesis, GCs also potentiate lipogenesis, an effect that is mainly studied at the level of the liver (Berdanier 1989).

Influence by EDCs

PCBs In a yeast bioassay, PCBs display both agonistic and antagonistic properties (Bovee et al. 2011). In a human cell line, PCB 118 raises cortisol production (Kraugerud et al. 2010). This effect seems to be due to an upregulation of the transcription of different enzymes involved in steroidogenesis (Xu et al. 2006). Animal studies to investigate the result of prenatal exposure have yielded conflicting results, reporting both increased and decreased cortisone levels in exposed offspring (Miller et al. 1993; Zimmer et al. 2011).

Organochlorine Pesticides Increased OCP levels are associated with diminished cortisol levels in wild animals (Ropstad et al. 2006). Endosulfan decreases cortisol secretion *in vitro* in a trout kidney cell. Endrin is a GC agonist and is capable of adipogenesis *in vitro* (Sargis et al. 2010).

Organotins Dibutyltin and monobutyltin were found to upregulate 17β-hydroxysteroid dehydrogenase type 1 (Nakanishi et al. 2006). On the other hand, the enzyme 11β-HSD type 2 was found to be inhibited by TBT, triphenyltin, dibutyltin, and diphenyltin (Ohshima et al. 2005). Dibutyltin can inhibit GR activation through insertion at an allosteric site near the steroid-binding pocket (Gumy et al. 2008).

Brominated Flame Retardants In a human cell line, brominated flame retardants influence the gene expression of enzymes involved in steroidogenesis in various ways, resulting in either up- or downregulation (Ding et al. 2007). In a porcine ovary, BDE-47, but not BDE-100, increases the expression of 17β-hydroxysteroid dehydrogenase type 1 (Karpeta et al. 2011).

BPA BPA is identified as a GC receptor agonist (Prasanth et al. 2010). On its own, however, it does not seem to be capable of inducing adipogenesis. When added to a submaximal stimulus of adipogenesis, though, it facilitates the conversion of a preadipocyte into a mature adipcyte (Sargis et al. 2010).

Phthalates Several phthalates are competitive inhibitors of human 11β-HSD2 enzymatic activity. This activity seems to be dependent on their chemical structure (Zhao et al. 2010).

Adipocytokines

Adipose tissue plays an active role in energy homeostasis and is no longer perceived as a mere lipid storage compartment, but has evolved into an endocrine organ. Adipocytokines, secretagogues from the adipocytes, interact functionally with many peripheral metabolic functions, thereby linking adipose tissue directly to the pathogenesis of obesity-related disorders. Leptin, the main regulator of body fat storage, is released primarily from adipocytes and plays a major role in the control of body fat stores through coordinated regulation of feeding behavior, metabolism, the autonomic nervous system, and body energy balance. It acts on the brain through specific receptors in the hypothalamus to suppress food intake (Wauters et al. 2002) and to induce transmitter signaling involved in energy expenditure and food intake. Circulating leptin levels correlate with BMI and total body fat mass. Many hormones including estrogen, GCs, and thyroid hormones have been shown to modulate leptin secretion by adipocytes *in vitro*. Adiponectin, another recently described adipocyte-specific adipocytokin and a marker of mature adipocytes, is involved in insulin sensitivity and lipid behavior and, as such, is considered as the first adipocytokine being potentially protective against the diabetogenic and atherogenic aspects of (visceral) adipose tissue and its secretagogues.

Influence by EDCs

PCBs Adiponectin levels in obese women are negatively correlated with serum PCB 153 burden, but this relationship could not be established after moderate weight loss or in lean individuals (Mullerova et al. 2008). To our knowledge, there are no studies examining the effect of PCB exposure on leptin levels.

Organochlorine Pesticides *In vitro*, DDE increases the release of leptin and adiponectin from human adipocytes (Howell and Mangum 2011).

Organotins Adipocytes cultured in trimethyltin presence exhibit modified leptin levels (Ravanan et al. 2011).

BPA Perinatal exposure of rats to low doses of BPA predisposes to hyperleptinemia in adulthood. This is accompanied by a severe metabolic syndrome, including obesity, dyslipidemia, hyperglycemia, hyperinsulinemia, and glucose intolerance (Wei et al. 2011). BPA is capable of inhibiting adiponectin release from human adipocytes *in vitro* (Hugo et al. 2008).

Phthalates Prenatal exposure of rats to diisobutyl phthalate reduced leptin levels in both sexes (Boberg et al. 2008).

To our knowledge, there are no studies examining the effect of brominated flame retardants on leptin or adiponectin levels.

Nuclear Receptors

A key factor in the process of adipogenesis is the nuclear peroxisome proliferator-activated receptor gamma (PPAR-γ), whose actions are a necessity in adipocyte differentiation (Tontonoz and Spiegelman 2008). In order to bind to DNA, PPAR-γ forms a heterodimer with the retinoid X receptor (RXR). The identified target genes of PPAR-γ are involved in numerous aspects of lipid metabolism and energy homeostasis. PPAR-γ belongs to the nuclear receptor family of ligand-activated transcription factors. The ligand binding pocket of PPAR-γ is special since it is large and can accommodate a diversity of chemical structures (Maloney and Waxman 1999). The endogenous ligand of PPAR-γ remains largely unknown; however, it is known as the target for the antidiabetic drug thiazolidinedione. PPAR-γ deficiency is embryonically lethal, and PPAR-γ is necessary for the development of adipose tissue *in vivo* (Rosen et al. 1999). Investigating the obesogenicity of chemicals through PPAR-γ is complicated for several reasons. First, some chemicals act as agonists in some cell types, and not in others, due to differential recruitment of coregulators. Second, PPAR-γ expression and action may be influenced in a ligand-independent way, by post-translational modifications (Janesick and Blumberg 2011).

Influence by EDCs So far, disruption of PPAR-γ has only been shown for organotins and phthalates.

Organotins The triorganotins TBT and triphenyltin are potent activators of PPAR-γ and RXR and thus are capable to promote the differentiation of adipocytes *in vitro* and *in vivo* (Grün et al. 2006; Inadera and Shimomura 2005; Kanayama et al. 2005). The organotins are capable of exerting this effect at the very low nanomolar levels, which are environmentally relevant levels. A TBT exposure study demonstrated a significant dose-dependent increase in body weight of young male mice (Zuo et al. 2011). Prenatal exposure to TBT enhances the capacity of stem cells in adult mice white adipose tissue to form adipocytes (Kirchner et al. 2010).

Phthalates Phthalates are known substrates from the PPAR family (Bility et al. 2004). Mono-(2-ethylhexyl) phthalate, in particular, is an agonist for PPAR-γ and selectively activates different PPAR-γ target genes (Feige et al. 2007).

Many More Emerging EDCs

Many other groups of emerging environmental contaminants, such as perfluorinated alkylated chemicals (PFACs), personal care products (musks, parabens, and ultraviolet filters), and other phenolic compounds (halogenated phenols, alkyl phenols) have recently emerged. Their role in endocrine disruption in humans *in vivo* has not been reported. *In vitro*, however, they have also been shown to present a weak to powerful endocrine-disrupting potency even at biologically relevant concentrations. The fact that these groups are high-production volume chemicals, identified as persistent, lipophilic, bioaccumulative in the environment, and potentially harmful for men and biota (Birnbaum and Staskal 2004; Covaci et al. 2002), indicates a potential role in the proposed endocrine disruption theory in obesity.

DISCUSSION

The previous data indicate that exposure to EDCs might play some role in the pandemic of obesity. Indeed, their endocrine-disrupting properties *in vitro* and *in vivo* are compatible with the type of endocrine disturbance observed in obesity. Thus, exposure to EDCs could lead to altered food intake, dysregulation of energy balance, and adipocyte differentiation, therefore playing a role in the etiology of obesity. Despite these multiple data, many particular problems linked to the study of EDCs remain to be elucidated:

1. Unlike the more traditional toxicology, EDCs do not always seem to follow a monotone dose–response relationship. Some EDCs seem to cause a strong effect at low, environmentally relevant doses, but a weakened or no effect at high doses. This inverted U-shape association has been suggested by some experimental studies (Welshons et al. 2006) and has been observed in epidemiological studies (Lee et al. 2011). This is problematic as a linear dose–response relationship is traditionally regarded as a criterion for causality by epidemiologists. Another caveat is the fact that a number of epidemiological studies were performed in a population of professionally exposed individuals. The data retrieved from these studies may not be extrapolated to the general population, which is usually exposed to much lower amounts.
2. Some EDCs metabolize relatively rapidly and bioaccumulate to a lesser extent than others (e.g., BPA and phthalates). As most studies are of cross-sectional design, the level detected during the study may not reflect the real exposure during disease development.
3. Currently, there are very few data on possible pharmacokinetic differences in the metabolism of chemicals between obese and lean individuals. As most endocrine disruptors are lipophylic, it is plausible that their absorption, transport, distribution, and excretion differ between obese and lean people. This is particularly important as most studies are of cross-sectional design and are therefore vulnerable to reverse causality.
4. The time elapse between exposure to the EDC and the development of obesity can extend over many years. In this perspective, exposure of the fetus is of particular interest. As reviewed by Heindel and vom Saal (2009), external effects on the developing fetus might permanently alter metabolic pathways that are crucial for energy homeostasis and thus might alter the "set point" for developing obesity later in life. Determining the exposure that might have occurred in the prenatal period is difficult to assess.
5. Not only our present offspring may be harmed by exposure to EDCs. Some EDCs are capable to cause epigenetic changes, which will be transmitted to progeny (Anway and Skinner 2008). Endocrine disruption might even thus be caused by exposure of an individual's ancestor.
6. People are environmentally exposed to a vast array of different EDCs. These different EDCs might interact, working antagonistically or syn-

ergistically. Coexposure of rats to both a brominated flame retardant and PCBs resulted in an apparent increase in the response observed with either product (Eriksson et al. 2006). Whether such mechanisms play a role in real life has never been studied.

7. For some EDCs demonstrating (anti)estrogenic and/or (anti) androgenic effects, the net effect on obesity might differ between genders as well as among females at various stages of their reproductive lives.
8. Most EDCs are strongly lipophilic. Weight loss, with a substantial decrease of human fat mass, has been proven to increase the serum concentration of some EDCs (Lim et al. 2011).
9. Anthropometric measures are sometimes not included in published studies on EDC levels in humans. This might be the result of a publication bias as positive results have a better chance of being published.
10. Studies on the effect of EDC on obesity in humans should include data on the traditional risk factors, such as a high-fat diet and sedentary lifestyle.

CONCLUSION

Despite all the difficulties mentioned earlier, the myriad of different EDCs poses a vast challenge to the scientific world. However, the cost of the obesity epidemic weighs heavily on patients and health-care systems worldwide. Many preliminary scientific data indeed open up an entirely new front: the hypothesis that disruption of weight homeostasis might, at least partially, be caused by exposure to EDCs. Therefore, there is an urgent need for large, well-designed studies on cellular, animal, and human levels to prove the above-mentioned chemical hypothesis of obesity.

REFERENCES

Agudo A, Goñi F, Etxeandia A, Vives A, Millán E, López R, et al. 2009. Polychlorinated biphenyls in Spanish adults: Determinants of serum concentrations. Environ Res 109:628–628.

al-Adsani H, Hoffer LJ, Silva JE. 1997. Resting energy expenditure is sensitive to small dose changes in patients on chronic thyroid hormone replacement. J Clin Endocrinol Metab 82:1118–1125.

Amin A, Dhillo WS, Murphy KG. 2011. The central effects of thyroid hormones on appetite. J Thyroid Res 2011:306510.

Anway MD, Skinner MK. 2008. Epigenetic programming of the germ line: Effects of endocrine disruptors on the development of transgenerational disease. Reprod Biomed Online 16:23–25.

Baillie-Hamilton PF. 2002. Chemical toxins: A hypothesis to explain the global obesity epidemic. J Altern Complement Med 8:185–192.

Ben-Jonathan N, Steinmetz R. 1998. Xenoestrogens: The emerging story of bisphenol A. Trends Endocrinol Metab 9:124–128.

Berdanier CD. 1989. Role of glucocorticoids in the regulation of lipogenesis. FASEB J 3:2179–2183.

Bianco AC. 2011. Minireview: Cracking the metabolic code for thyroid hormone signaling. Endocrinology 152:3306–3311.

Bility MT, Thompson JT, McKee RH, David RM, Butala JH, Vanden Heuvel JP, et al. 2004. Activation of mouse and human peroxisome proliferator-activated receptors (PPARs) by phthalate monoesters. Toxicol Sci 82:170–182.

Birnbaum LS, Staskal DF. 2004. Brominated flame retardants: Cause for concern? Environ Health Perspect 112:9–17.

Blanck HM, Marcus M, Rubin C, Tolbert PE, Hertzberg VS, Henderson AK, et al. 2002. Growth in girls exposed in utero and postnatally to polybrominated biphenyls and polychlorinated biphenyls. Epidemiology 13:205–210.

Blanco J, Mulero M, López M, Domingo JL, Sánchez DJ. 2011. BDE-99 deregulates BDNF, Bcl-2 and the mRNA expression of thyroid receptor isoforms in rat cerebellar granular neurons. Toxicology 290:305–311.

Boberg J, Metzdorff S, Wortziger R, Axelstad M, Brokken L, Vinggaard AM, et al. 2008. Impact of diisobutyl phthalate and other PPAR agonists on steroidogenesis and plasma insulin and leptin levels in fetal rats. Toxicology 250:75–81.

Bovee TF, Helsdingen RJ, Hamers AR, Brouwer BA, Nielen MW. 2011. Recombinant cell bioassays for the detection of (gluco)corticosteroids and endocrine-disrupting potencies of several environmental PCB contaminants. Anal Bioanal Chem 401:873–882.

Brent GA. 1994. The molecular basis of thyroid hormone action. N Engl J Med 331:847–853.

Brucker-Davis F. 1998. Effects of environmental synthetic chemicals on thyroid function. Thyroid 8:827–856.

Bujalska IJ, Kumar S, Stewart PM. 1997. Does central obesity reflect "Cushing's disease of the omentum"? Lancet 349:1210–1213.

Cantón RF, Scholten DE, Marsh G, de Jong PC, van den Berg M. 2008. Inhibition of human placental aromatase activity by hydroxylated polybrominated diphenyl ethers (OH-PBDEs). Toxicol Appl Pharmacol 227:68–75.

Cao Y, Winneke G, Wilhelm M, Wittsiepe J, Lemm F, Fürst P, et al. 2008. Environmental exposure to dioxins and polychlorinated biphenyls reduce levels of gonadal hormones in newborns: Results from the Duisburg cohort study. Int J Hyg Environ Health 211:30–39.

Clegg DJ, Brown LM, Woods SC, Benoit SC. 2006. Gonadal hormones determine sensitivity to central leptin and insulin. Diabetes 55:978–987.

Colles A, Koppen G, Hanot V, Nelen V, Dewolf MC, Noël E, et al. 2008. Fourth WHO-coordinated survey of human milk for persistent organic pollutants (POPs): Belgian results. Chemosphere 73:907–914.

Cooke PS, Naaz A. 2004. Role of estrogens in adipocyte development and function. Exp Biol Med (Maywood) 229:1127–1135.

Costopoulou D, Vassiliadou I, Papadopoulos A, Makropoulos V, Leondiadis L. 2006. Levels of dioxins, furans and PCBs in human serum and milk of people living in Greece. Chemosphere 65:1462–1469.

Covaci A, Koppen G, Van Cleuvenbergen R, Schepens P, Winneke G, van Larebeke N, et al. 2002. Persistent organochlorine pollutants in human serum of 50–65 years old women in the Flanders Environmental and Health Study (FLEHS). Part 2: Correlations among PCBs, PCDD/PCDFs and the use of predictive markers. Chemosphere 48:827–832.

Covaci A, Voorspoels S, Roosens L, Jacobs W, Blust R, Neels H. 2008. Polybrominated diphenyl ethers (PBDEs) and polychlorinated biphenyls (PCBs) in human liver and adipose tissue samples from Belgium. Chemosphere 73:170–175.

Decherf S, Seugnet I, Fini JB, Clerget-Froidevaux MS, Demeneix BA. 2010. Disruption of thyroid hormone-dependent hypothalamic set-points by environmental contaminants. Mol Cell Endocrinol 323:172–182.

Delport R, Bornman R, MacIntyre UE, Oosthuizen NM, Becker PJ, Aneck-Hahn NH, et al. 2011. Changes in retinol-binding protein concentrations and thyroid homeostasis with nonoccupational exposure to DDT. Environ Health Perspect 119: 647–651.

D'Eon TM, Souza SC, Aronovitz M, Obin MS, Fried SK, Greenberg AS. 2005. Estrogen regulation of adiposity and fuel partitioning. Evidence of genomic and non-genomic regulation of lipogenic and oxidative pathways. J Biol Chem 280: 35983–35991.

Deroo BJ, Korach KS. 2006. Estrogen receptors and human disease. J Clin Invest 116:561–570.

Ding L, Murphy MB, He Y, Xu Y, Yeung LW, Wang J, et al. 2007. Effects of brominated flame retardants and brominated dioxins on steroidogenesis in H295R human adrenocortical carcinoma cell line. Environ Toxicol Chem 26:764–772.

Dirinck E, Jorens PG, Covaci A, Geens T, Roosens L, Neels H, et al. 2011. Obesity and persistent organic pollutants: Possible obesogenic effect of organochlorine pesticides and polychlorinated biphenyls. Obesity (Silver Spring) 19:709–714.

Dong S, Terasaka S, Kiyama R. 2011. Bisphenol A induces a rapid activation of Erk1/2 through GPR30 in human breast cancer cells. Environ Pollut 159:212–218.

Elobeid MA, Padilla MA, Brock DW, Ruden DM, Allison DB. 2010. Endocrine disruptors and obesity: An examination of selected persistent organic pollutants in the NHANES 1999–2002 data. Int J Environ Res Public Health 7:2988–3005.

Epel E, Lapidus R, McEwen B, Brownell K. 2001. Stress may add bite to appetite in women: A laboratory study of stress-induced cortisol and eating behavior. Psychoneuroendocrinology 26:37–49.

Eriksson P, Fischer C, Fredriksson A. 2006. Polybrominated diphenyl ethers, a group of brominated flame retardants, can interact with polychlorinated biphenyls in enhancing developmental neurobehavioral defects. Toxicol Sci 94:302–309.

Everett CJ, Mainous AG, 3rd, Frithsen IL, Player MS, Matheson EM. 2008. Association of polychlorinated biphenyls with hypertension in the 1999–2002 National Health and Nutrition Examination Survey. Environ Res 108:94–97.

Ezechiáš M, Svobodová K, Cajthaml T. 2012. Hormonal activities of new brominated flame retardants. Chemosphere 87:820–824.

Fan W, Yanase T, Nomura M, Okabe T, Goto K, Sato T, et al. 2005. Androgen receptor null male mice develop late-onset obesity caused by decreased energy expenditure and lipolytic activity but show normal insulin sensitivity with high adiponectin secretion. Diabetes 54:1000–1008.

Fan W, Yanase T, Nishi Y, Chiba S, Okabe T, Nomura M, et al. 2008. Functional potentiation of leptin-signal transducer and activator of transcription 3 signaling by the androgen receptor. Endocrinology 149:6028–6036.

Feige JN, Gelman L, Rossi D, Zoete V, Métivier R, Tudor C, et al. 2007. The endocrine disruptor monoethyl-hexyl-phthalate is a selective peroxisome proliferator-activated receptor gamma modulator that promotes adipogenesis. J Biol Chem 282: 19152–19166.

Foster WG, Neal MS, Han M, Dominguez MM. 2008. Environmental contaminants and human infertility: Hypothesis or cause for concern? J Toxicol Environ Health B Crit Rev 11:162–176.

Freitas J, Cano P, Craig-Veit C, Goodson ML, Furlow JD, Murk AJ. 2011. Detection of thyroid hormone receptor disruptors by a novel stable *in vitro* reporter gene assay. Toxicol in Vitro 25:257–266.

Fried SK, Bunkin DA, Greenberg AS. 1998. Omental and subcutaneous adipose tissues of obese subjects release interleukin-6: Depot difference and regulation by glucocorticoid. J Clin Endocrinol Metab 83:847–850.

Gladen BC, Ragan NB, Rogan WJ. 2000. Pubertal growth and development and prenatal and lactational exposure to polychlorinated biphenyls and dichlorodiphenyl dichloroethene. J Pediatr 136:490–496.

Gladen BC, Klebanoff MA, Hediger ML, Katz SH, Barr DB, Davis MD, et al. 2004. Prenatal DDT exposure in relation to anthropometric and pubertal measures in adolescent males. Environ Health Perspect 112:1761–1767.

Glynn AW, Granath F, Aune M, Atuma S, Darnerud PO, Bjerselius R, et al. 2003. Organochlorines in Swedish women: Determinants of serum concentrations. Environ Health Perspect 111:349–355.

Goncharov A, Rej R, Negoita S, Schymura M, Santiago-Rivera A, Morse G, et al. 2009. Lower serum testosterone associated with elevated polychlorinated biphenyl concentrations in Native American men. Environ Health Perspect 117:1454–1460.

Grün F, Watanabe H, Zamanian Z, Maeda L, Arima K, Cubacha R, et al. 2006. Endocrine-disrupting organotin compounds are potent inducers of adipogenesis in vertebrates. Mol Endocrinol 20:2141–2155.

Gumy C, Chandsawangbhuwana C, Dzyakanchuk AA, Kratschmar DV, Baker ME, Odermatt A. 2008. Dibutyltin disrupts glucocorticoid receptor function and impairs glucocorticoid-induced suppression of cytokine production. PLoS ONE 3:e3545.

Hallgren S, Sinjari T, Håkansson H, Darnerud PO. 2001. Effects of polybrominated diphenyl ethers (PBDEs) and polychlorinated biphenyls (PCBs) on thyroid hormone and vitamin A levels in rats and mice. Arch Toxicol 75:200–208.

Hamers T, Kamstra JH, Sonneveld E, Murk AJ, Kester MH, Andersson PL, et al. 2006. *In vitro* profiling of the endocrine-disrupting potency of brominated flame retardants. Toxicol Sci 92:157–173.

Harju M, Hamers T, Kamstra JH, Sonneveld E, Boon JP, Tysklind M, et al. 2007. Quantitative structure-activity relationship modeling on *in vitro* endocrine effects and metabolic stability involving 26 selected brominated flame retardants. Environ Toxicol Chem 26:816–826.

Harris CA, Henttu P, Parker MG, Sumpter JP. 1997. The estrogenic activity of phthalate esters *in vitro*. Environ Health Perspect 105:802–811.

Hatch EE, Nelson JW, Qureshi MM, Weinberg J, Moore LL, Singer M, et al. 2008. Association of urinary phthalate metabolite concentrations with body mass index and waist circumference: A cross-sectional study of NHANES data, 1999–2002. Environ Health 7:27.

Hauner H, Entenmann G, Wabitsch M, Gaillard D, Ailhaud G, Negrel R, et al. 1989. Promoting effect of glucocorticoids on the differentiation of human adipocyte precursor cells cultured in a chemically defined medium. J Clin Invest 84:1663–1670.

Hauser R, Meeker JD, Duty S, Silva MJ, Calafat AM. 2006. Altered semen quality in relation to urinary concentrations of phthalate monoester and oxidative metabolites. Epidemiology 17:682–691.

Hauser R, Meeker JD, Singh NP, Silva MJ, Ryan L, Duty S, et al. 2007. DNA damage in human sperm is related to urinary levels of phthalate monoester and oxidative metabolites. Hum Reprod 22:688–695.

Heindel JJ, vom Saal FS. 2009. Role of nutrition and environmental endocrine disrupting chemicals during the perinatal period on the aetiology of obesity. Mol Cell Endocrinol 304:90–96.

Heine PA, Taylor JA, Iwamoto GA, Lubahn DB, Cooke PS. 2000. Increased adipose tissue in male and female estrogen receptor-alpha knockout mice. Proc Natl Acad Sci U S A 97:12729–12734.

Hertz-Picciotto I, Charles MJ, James RA, Keller JA, Willman E, Teplin S. 2005. In utero polychlorinated biphenyl exposures in relation to fetal and early childhood growth. Epidemiology 16:648–656.

Hirschberg AL. 2012. Sex hormones, appetite and eating behaviour in women. Maturitas 71:248–256.

Howell G, 3rd, Mangum L. 2011. Exposure to bioaccumulative organochlorine compounds alters adipogenesis, fatty acid uptake, and adipokine production in NIH3T3-L1 cells. Toxicol in Vitro 25:394–402.

Hue O, Marcotte J, Berrigan F, Simoneau M, Doré J, Marceau P, et al. 2007. Plasma concentration of organochlorine compounds is associated with age and not obesity. Chemosphere 67:1463–1467.

Hugo ER, Brandebourg TD, Woo JG, Loftus J, Alexander JW, Ben-Jonathan N. 2008. Bisphenol A at environmentally relevant doses inhibits adiponectin release from human adipose tissue explants and adipocytes. Environ Health Perspect 116: 1642–1647.

Inadera H, Shimomura A. 2005. Environmental chemical tributyltin augments adipocyte differentiation. Toxicol Lett 159:226–234.

Ishihara A, Nishiyama N, Sugiyama S, Yamauchi K. 2003. The effect of endocrine disrupting chemicals on thyroid hormone binding to Japanese quail transthyretin and thyroid hormone receptor. Gen Comp Endocrinol 134:36–43.

Jacobson JL, Jacobson SW, Humphrey HE. 1990. Effects of exposure to PCBs and related compounds on growth and activity in children. Neurotoxicol Teratol 12:319–326.

Janesick A, Blumberg B. 2011. Minireview: PPARγ as the target of obesogens. J Steroid Biochem Mol Biol 127:4–8.

Jones ME, Thorburn AW, Britt KL, Hewitt KN, Misso ML, Wreford NG, et al. 2001. Aromatase-deficient (ArKO) mice accumulate excess adipose tissue. J Steroid Biochem Mol Biol 79:3–9.

Kanayama T, Kobayashi N, Mamiya S, Nakanishi T, Nishikawa J. 2005. Organotin compounds promote adipocyte differentiation as agonists of the peroxisome proliferator-activated receptor gamma/retinoid X receptor pathway. Mol Pharmacol 67:766–774.

Kannan K, Falandysz J. 1997. Butyltin residues in sediment, fish, fish-eating birds, harbour porpoise and human tissues from the Polish coast of the Baltic Sea. Mar Pollut Bull 34:203–207.

Kannan K, Senthilkumar K, Giesy JP. 1999. Occurrence of butyltin compounds in human blood. Environ Sci Technol 33:1776–1779.

Karmaus W, Osuch JR, Eneli I, Mudd LM, Zhang J, Mikucki D, et al. 2009. Maternal levels of dichlorodiphenyl-dichloroethylene (DDE) may increase weight and body mass index in adult female offspring. Occup Environ Med 66:143–149.

Karmaus W, Osuch JR, Landgraf J, Taffe B, Mikucki D, Haan P. 2011. Prenatal and concurrent exposure to halogenated organic compounds and gene expression of CYP17A1, CYP19A1, and oestrogen receptor alpha and beta genes. Occup Environ Med 68:430–437.

Karpeta A, Rak-Mardyła A, Jerzak J, Gregoraszczuk EL. 2011. Congener-specific action of PBDEs on steroid secretion, CYP17, 17β-HSD and CYP19 activity and protein expression in porcine ovarian follicles. Toxicol Lett 206:258–263.

Kirchner S, Kieu T, Chow C, Casey S, Blumberg B. 2010. Prenatal exposure to the environmental obesogen tributyltin predisposes multipotent stem cells to become adipocytes. Mol Endocrinol 24:526–539.

Koch HM, Wittassek M, Brüning T, Angerer J, Heudorf U. 2011. Exposure to phthalates in 5–6 years old primary school starters in Germany—A human biomonitoring study and a cumulative risk assessment. Int J Hyg Environ Health 214:188–195.

Kraugerud M, Zimmer KE, Dahl E, Berg V, Olsaker I, Farstad W, et al. 2010. Three structurally different polychlorinated biphenyl congeners (Pcb 118, 153, and 126) affect hormone production and gene expression in the human H295R in vitro model. J Toxicol Environ Health A 73:1122–1132.

Kumari M, Chandola T, Brunner E, Kivimaki M. 2010. A nonlinear relationship of generalized and central obesity with diurnal cortisol secretion in the Whitehall II study. J Clin Endocrinol Metab 95:4415–4423.

Lamb MR, Taylor S, Liu X, Wolff MS, Borrell L, Matte TD, et al. 2006. Prenatal exposure to polychlorinated biphenyls and postnatal growth: A structural analysis. Environ Health Perspect 114:779–785.

Lanfranco F, Zitzmann M, Simoni M, Nieschlag E. 2004. Serum adiponectin levels in hypogonadal males: Influence of testosterone replacement therapy. Clin Endocrinol (Oxf) 60:500–507.

Lang IA, Galloway TS, Scarlett A, Henley WE, Depledge M, Wallace RB, et al. 2008. Association of urinary bisphenol A concentration with medical disorders and laboratory abnormalities in adults. JAMA 300:1303–1310.

Lans MC, Klasson-Wehler E, Willemsen M, Meussen E, Safe S, Brouwer A. 1993. Structure-dependent, competitive interaction of hydroxy-polychlorobiphenyls, -dibenzo-p-dioxins and -dibenzofurans with human transthyretin. Chem Biol Interact 88:7–21.

Lee HJ, Chattopadhyay S, Gong EY, Ahn RS, Lee K. 2003. Antiandrogenic effects of bisphenol A and nonylphenol on the function of androgen receptor. Toxicol Sci 75:40–46.

Lee DH, Lee IK, Song K, Steffes M, Toscano W, Baker BA, et al. 2006. A strong dose-response relation between serum concentrations of persistent organic pollutants and diabetes: Results from the National Health and Examination Survey 1999–2002. Diabetes Care 29:1638–1644.

Lee DH, Jacobs DR, Porta M. 2007. Association of serum concentrations of persistent organic pollutants with the prevalence of learning disability and attention deficit disorder. J Epidemiol Community Health 61:591–596.

Lee DH, Steffes MW, Sjödin A, Jones RS, Needham LL, Jacobs DR, Jr. 2011. Low dose organochlorine pesticides and polychlorinated biphenyls predict obesity, dyslipidemia, and insulin resistance among people free of diabetes. PLoS ONE 6: e15977.

Legler J. 2008. New insights into the endocrine disrupting effects of brominated flame retardants. Chemosphere 73:216–222.

Levin ER. 2009. Plasma membrane estrogen receptors. Trends Endocrinol Metab 20:477–482.

Li J, Li N, Ma M, Giesy JP, Wang Z. 2008. *In vitro* profiling of the endocrine disrupting potency of organochlorine pesticides. Toxicol Lett 183:65–71.

Lignell S, Aune M, Darnerud PO, Cnattingius S, Glynn A. 2009. Persistent organochlorine and organobromine compounds in mother's milk from Sweden 1996–2006: Compound-specific temporal trends. Environ Res 109:760–767.

Lim JS, Lee DH, Jacobs DR, Jr. 2008. Association of brominated flame retardants with diabetes and metabolic syndrome in the U.S. population, 2003–2004. Diabetes Care 31:1802–1807.

Lim JS, Son HK, Park SK, Jacobs DR, Jr., Lee DH. 2011. Inverse associations between long-term weight change and serum concentrations of persistent organic pollutants. Int J Obes 35:744–747.

Lo S, Alléra A, Albers P, Heimbrecht J, Jantzen E, Klingmüller D, et al. 2003. Dithioerythritol (DTE) prevents inhibitory effects of triphenyltin (TPT) on the key enzymes of the human sex steroid hormone metabolism. J Steroid Biochem Mol Biol 84:569–576.

Maloney EK, Waxman DJ. 1999. trans-Activation of PPARalpha and PPARgamma by structurally diverse environmental chemicals. Toxicol Appl Pharmacol 161: 209–218.

Masuzaki H, Paterson J, Shinyama H, Morton NM, Mullins JJ, Seckl JR, et al. 2001. A transgenic model of visceral obesity and the metabolic syndrome. Science 294: 2166–2170.

Mauvais-Jarvis F. 2011. Estrogen and androgen receptors: Regulators of fuel homeostasis and emerging targets for diabetes and obesity. Trends Endocrinol Metab 22: 24–33.

McKinney JD, Waller CL. 1994. Polychlorinated biphenyls as hormonally active structural analogues. Environ Health Perspect 102:290–297.

Meeker JD, Calafat AM, Hauser R. 2007. Di(2-ethylhexyl) phthalate metabolites may alter thyroid hormone levels in men. Environ Health Perspect 115:1029–1034.

Meerts IA, Letcher RJ, Hoving S, Marsh G, Bergman A, Lemmen JG, et al. 2001. In vitro estrogenicity of polybrominated diphenyl ethers, hydroxylated PDBEs, and polybrominated bisphenol A compounds. Environ Health Perspect 109(4):399–407.

Miller DB, Gray LE, Jr., Andrews JE, Luebke RW, Smialowicz RJ. 1993. Repeated exposure to the polychlorinated biphenyl (Aroclor 1254) elevates the basal serum levels of corticosterone but does not affect the stress-induced rise. Toxicology 81:217–222.

Mokdad AH, Ford ES, Bowman BA, Dietz WH, Vinicor F, Bales VS, et al. 2001. Prevalence of obesity, diabetes, and obesity-related health risk factors. JAMA 289:76–79.

Moriyama K, Tagami T, Akamizu T, Usui T, Saijo M, Kanamoto N, et al. 2002. Thyroid hormone action is disrupted by bisphenol A as an antagonist. J Clin Endocrinol Metab 87:5185–5190.

Morse DC, Groen D, Veerman M, van Amerongen CJ, Koëter HB, Smits van Prooije AE, et al. 1993. Interference of polychlorinated biphenyls in hepatic and brain thyroid hormone metabolism in fetal and neonatal rats. Toxicol Appl Pharmacol 122:27–33.

Mortensen AS, Arukwe A. 2007. Modulation of xenobiotic biotransformation system and hormonal responses in Atlantic salmon (*Salmo salar*) after exposure to tributyltin (TBT). Comp Biochem Physiol C Toxicol Pharmacol 145:431–441.

Mullerova D, Kopecky J, Matejkova D, Muller L, Rosmus J, Racek J, et al. 2008. Negative association between plasma levels of adiponectin and polychlorinated biphenyl 153 in obese women under non-energy-restrictive regime. Int J Obes 32:1875–1878.

Nakanishi T, Hiromori Y, Yokoyama H, Koyanagi M, Itoh N, Nishikawa J, et al. 2006. Organotin compounds enhance 17beta-hydroxysteroid dehydrogenase type I activity in human choriocarcinoma JAr cells: Potential promotion of 17beta-estradiol biosynthesis in human placenta. Biochem Pharmacol 71:1349–1357.

Nakayama S, Nishiyama M, Iwasaki Y, Shinahara M, Okada Y, Tsuda M, et al. 2011. Corticotropin-releasing hormone (CRH) transgenic mice display hyperphagia with increased Agouti-related protein mRNA in the hypothalamic arcuate nucleus. Endocr J 58:279–286.

Needham LL, Barr DB, Caudill SP, Pirkle JL, Turner WE, Osterloh J, et al. 2005. Concentrations of environmental chemicals associated with neurodevelopmental effects in U.S. population. Neurotoxicology 26:531–545.

Nichols BR, Hentz KL, Aylward L, Hays SM, Lamb JC. 2007. Age-specific reference ranges for polychlorinated biphenyls (PCB) based on the NHANES 2001–2002 survey. J Toxicol Environ Health A 70:1873–1877.

Nielsen JB, Strand J. 2002. Butyltin compounds in human liver. Environ Res 88: 129–133.

Ohshima M, Ohno S, Nakajin S. 2005. Inhibitory effects of some possible endocrine-disrupting chemicals on the isozymes of human 11beta-hydroxysteroid dehydrogenase and expression of their mRNA in gonads and adrenal glands. Environ Sci 12:219–230.

Ottosson M, Vikman-Adolfsson K, Enerbäck S, Olivecrona G, Björntorp P. 1994. The effects of cortisol on the regulation of lipoprotein lipase activity in human adipose tissue. J Clin Endocrinol Metab 79:820–825.

Palin SL, McTernan PG, Anderson LA, Sturdee DW, Barnett AH, Kumar S. 2003. 17Beta-estradiol and anti-estrogen ICI:compound 182,780 regulate expression of lipoprotein lipase and hormone-sensitive lipase in isolated subcutaneous abdominal adipocytes. Metabolism 52:383–388.

Parks LG, Ostby JS, Lambright CR, Abbott BD, Klinefelter GR, Barlow NJ, et al. 2000. The plasticizer diethylhexyl phthalate induces malformations by decreasing fetal testosterone synthesis during sexual differentiation in the male rat. Toxicol Sci 58:339–349.

Pasquali R, Vicennati V. 2000. The abdominal obesity phenotype and insulin resistance are associated with abnormalities of the hypothalamic-pituitary-adrenal axis in humans. Horm Metab Res 32:521–525.

Patandin S, Koopman-Esseboom C, de Ridder MA, Weisglas-Kuperus N, Sauer PJ. 1998. Effects of environmental exposure to polychlorinated biphenyls and dioxins on birth size and growth in Dutch children. Pediatr Res 44:538–545.

Patisaul HB, Adewale HB. 2009. Long-term effects of environmental endocrine disruptors on reproductive physiology and behavior. Front Behav Neurosci 3:10.

Pedersen SB, Kristensen K, Hermann PA, Katzenellenbogen JA, Richelsen B. 2004. Estrogen controls lipolysis by up-regulating alpha2A-adrenergic receptors directly in human adipose tissue through the estrogen receptor alpha. Implications for the female fat distribution. J Clin Endocrinol Metab 89:1869–1878.

Penza M, Jeremic M, Marrazzo E, Maggi A, Ciana P, Rando G, et al. 2011. The environmental chemical tributyltin chloride (TBT) shows both estrogenic and adipogenic activities in mice which might depend on the exposure dose. Toxicol Appl Pharmacol 255:65–75.

Plísková M, Vondrácek J, Canton RF, Nera J, Kocan A, Petrík J, et al. 2005. Impact of polychlorinated biphenyls contamination on estrogenic activity in human male serum. Environ Health Perspect 113:1277–1284.

Prasanth GK, Divya LM, Sadasivan C. 2010. Bisphenol-A can bind to human glucocorticoid receptor as an agonist: An in silico study. J Appl Toxicol 30:769–774.

Prentice AM. 2001. Overeating: The health risks. Obes Res 9(Suppl 4):234S–238S.

Ramos JG, Varayoud J, Kass L, Rodríguez H, Costabel L, Muñoz-De-Toro M, et al. 2003. Bisphenol A induces both transient and permanent histofunctional alterations of the hypothalamic-pituitary-gonadal axis in prenatally exposed male rats. Endocrinology 144:3206–3215.

Ravanan P, Harry GJ, Awada R, Hoareau L, Tallet F, Roche R, et al. 2011. Exposure to an organometal compound stimulates adipokine and cytokine expression in white adipose tissue. Cytokine 53:355–362.

Richardson VM, Staskal DF, Ross DG, Diliberto JJ, Devito MJ, Birnbaum LS. 2008. Possible mechanisms of thyroid hormone disruption in mice by BDE 47, a major polybrominated diphenyl ether congener. Toxicol Appl Pharmacol 226:244–250.

Rönn M, Lind L, van Bavel B, Salihovic S, Michaëlsson K, Lind PM. 2011. Circulating levels of persistent organic pollutants associate in divergent ways to fat mass measured by DXA in humans. Chemosphere 85:335–343.

Roosens L, D'Hollander W, Bervoets L, Reynders H, Van Campenhout K, Cornelis C, et al. 2010. Brominated flame retardants and perfluorinated chemicals, two groups of persistent contaminants in Belgian human blood and milk. Environ Pollut 158:2546–2552.

Ropstad E, Oskam IC, Lyche JL, Larsen HJ, Lie E, Haave M, et al. 2006. Endocrine disruption induced by organochlorines (OCs): Field studies and experimental models. J Toxicol Environ Health A 69:53–76.

Rosen ED, Sarraf P, Troy AE, Bradwin G, Moore K, Milstone DS, et al. 1999. PPAR gamma is required for the differentiation of adipose tissue in vivo and *in vitro*. Mol Cell 4:611–617.

Rylander L, Strömberg U, Dyremark E, Ostman C, Nilsson-Ehle P, Hagmar L. 1998. Polychlorinated biphenyls in blood plasma among Swedish female fish consumers in relation to low birth weight. Am J Epidemiol 147:493–502.

Sargis RM, Johnson DN, Choudhury RA, Brady MJ. 2010. Environmental endocrine disruptors promote adipogenesis in the 3T3-L1 cell line through glucocorticoid receptor activation. Obesity (Silver Spring) 18:1283–1288.

Schecter A, Johnson-Welch S, Tung KC, Harris TR, Päpke O, Rosen R. 2007. Polybrominated diphenyl ether (PBDE) levels in livers of U.S. human fetuses and newborns. J Toxicol Environ Health A 70:1–6.

Shen O, Du G, Sun H, Wu W, Jiang Y, Song L, et al. 2009. Comparison of *in vitro* hormone activities of selected phthalates using reporter gene assays. Toxicol Lett 191:9–14.

Shi ZX, Wu YN, Li JG, Zhao YF, Feng JF. 2009. Dietary exposure assessment of Chinese adults and nursing infants to tetrabromobisphenol-A and hexabromocyclododecanes: Occurrence measurements in food and human milk. Environ Sci Technol 43:4314–4319.

Shimada N, Yamauchi K. 2004. Characteristics of 3,5,3′-triiodothyronine (T3)-uptake system of tadpole red blood cells: Effect of endocrine-disrupting chemicals on cellular T3 response. J Endocrinol 183:627–637.

Silva JE. 2006. Thermogenic mechanisms and their hormonal regulation. Physiol Rev 86:435–464.

Singh R, Artaza JN, Taylor WE, Gonzalez-Cadavid NF, Bhasin S. 2003. Androgens stimulate myogenic differentiation and inhibit adipogenesis in C3H 10T1/2 pluripotent cells through an androgen receptor-mediated pathway. Endocrinology 144: 5081–5088.

Smink A, Ribas-Fito N, Garcia R, Torrent M, Mendez MA, Grimalt JO, et al. 2008. Exposure to hexachlorobenzene during pregnancy increases the risk of overweight in children aged 6 years. Acta Paediatr 97:1465–1469.

Soto AM, Sonnenschein C. 2010. Environmental causes of cancer: Endocrine disruptors as carcinogens. Nat Rev Endocrinol 6:363–370.

Stahlhut RW, van Wijngaarden E, Dye TD, Cook S, Swan SH. 2007. Concentrations of urinary phthalate metabolites are associated with increased waist circumference and insulin resistance in adult U.S. males. Environ Health Perspect 115:876–882.

Stapleton HM, Sjödin A, Jones RS, Niehüser S, Zhang Y, Patterson DG, Jr. 2008. Serum levels of polybrominated diphenyl ethers (PBDEs) in foam recyclers and carpet installers working in the United States. Environ Sci Technol 42(9): 3453–3458.

Stoker TE, Cooper RL, Lambright CS, Wilson VS, Furr J, Gray LE. 2005. In vivo and *in vitro* anti-androgenic effects of DE-71, a commercial polybrominated diphenyl ether (PBDE) mixture. Toxicol Appl Pharmacol 207:78–88.

Sugiyama S, Shimada N, Miyoshi H, Yamauchi K. 2005. Detection of thyroid system-disrupting chemicals using *in vitro* and in vivo screening assays in *Xenopus laevis*. Toxicol Sci 88:367–374.

Swan SH, Main KM, Liu F, Stewart SL, Kruse RL, Calafat AM, et al. 2005. Decrease in anogenital distance among male infants with prenatal phthalate exposure. Environ Health Perspect 113:1056–1061. Erratum in: Environ Health Perspect 2005 113:A583.

Takahashi S, Mukai H, Tanabe S, Sakayama K, Miyazaki T, Masuno H. 1999. Butyltin residues in livers of humans and wild terrestrial mammals and in plastic products. Environ Pollut 106:213–218.

Thomas P, Dong J. 2006. Binding and activation of the seven-transmembrane estrogen receptor GPR30 by environmental estrogens: A potential novel mechanism of endocrine disruption. J Steroid Biochem Mol Biol 102:175–179.

Tontonoz P, Spiegelman BM. 2008. Fat and beyond: The diverse biology of PPARgamma. Annu Rev Biochem 77:289–312.

Van Birgelen AP, Van der Kolk J, Fase KM, Bol I, Poiger H, Brouwer A, et al. 1994. Toxic potency of 3,3′,4,4′,5-pentachlorobiphenyl relative to and in combination with 2,3,7,8-tetrachlorodibenzo-p-dioxin in a subchronic feeding study in the rat. Toxicol Appl Pharmacol 127:209–221.

Vandenberg LN, Hauser R, Marcus M, Olea N, Welshons WV. 2007. Human exposure to bisphenol A (BPA). Reprod Toxicol 24:139–177.

Van Gaal LF, Mertens IL, Abrams PJ. 2003. Health risks of lipodystrophy and abdominal fat accumulation: Therapeutic possibilities with leptin and human growth hormone. Growth Horm IGF Res 13(Suppl A):S4–S9.

van der Ven LT, Verhoef A, van de Kuil T, Slob W, Leonards PE, Visser TJ, et al. 2006. A 28-day oral dose toxicity study enhanced to detect endocrine effects of hexabromocyclododecane in Wistar rats. Toxicol Sci 94:281–292.

Verhulst SL, Nelen V, Hond ED, Koppen G, Beunckens C, Vael C, et al. 2009. Intrauterine exposure to environmental pollutants and body mass index during the first 3 years of life. Environ Health Perspect 117:122–126.

Visser TJ, Kaptein E, van Toor H, van Raaij JA, van den Berg KJ, Joe CT, et al. 1993. Glucuronidation of thyroid hormone in rat liver: Effects of in vivo treatment with microsomal enzyme inducers and in vitro assay conditions. Endocrinology 133: 2177–2186.

Vos JG, Dybing E, Greim HA, Ladefoged O, Lambré C, Tarazona JV, et al. 2000. Health effects of endocrine-disrupting chemicals on wildlife, with special reference to the European situation. Crit Rev Toxicol 30:71–133.

Wadia PR, Vandenberg LN, Schaeberle CM, Rubin BS, Sonnenschein C, Soto AM. 2007. Perinatal bisphenol A exposure increases estrogen sensitivity of the mammary gland in diverse mouse strains. Environ Health Perspect 115:592–598.

Warner J, Osuch JR, Karmaus W, Landgraf JR, Taffe B, O'Keefe M, et al. 2012. Common classification schemes for PCB congeners and the gene expression of CYP17, CYP19, ESR1 and ESR2. Sci Total Environ 414:81–89.

Wauters M, Considine RV, Chagnon M, Mertens I, Rankinen T, Bouchard C, et al. 2002. Leptin levels, leptin receptor gene polymorphisms, and energy metabolism in women. Obes Res 10:394–400.

Wei J, Lin Y, Li Y, Ying C, Chen J, Song L, et al. 2011. Perinatal exposure to bisphenol A at reference dose predisposes offspring to metabolic syndrome in adult rats on a high-fat diet. Endocrinology 152:3049–3061.

Welshons WV, Nagel SC, vom Saal FS. 2006. Large effects from small exposures. III. Endocrine mechanisms mediating effects of bisphenol A at levels of human exposure. Endocrinology 147:S56–S69.

Wenzel A, Franz C, Breous E, Loos U. 2005. Modulation of iodide uptake by dialkyl phthalate plasticisers in FRTL-5 rat thyroid follicular cells. Mol Cell Endocrinol 244:63–71.

Wolff MS, Deych E, Ojo F, Berkowitz GS. 2005. Predictors of organochlorines in New York City pregnant women, 1998–2001. Environ Res 97:170–177.

Xu Y, Yu RM, Zhang X, Murphy MB, Giesy JP, Lam MH, et al. 2006. Effects of PCBs and MeSO2-PCBs on adrenocortical steroidogenesis in H295R human adrenocortical carcinoma cells. Chemosphere 63:772–784.

Ye L, Su ZJ, Ge RS. 2011. Inhibitors of testosterone biosynthetic and metabolic activation enzymes. Molecules 16:9983–10001.

Yen PM. 2001. Physiological and molecular basis of thyroid hormone action. Physiol Rev 81:1097–1142.

Zhang J, Zuo Z, He C, Cai J, Wang Y, Chen Y, et al. 2009. Effect of tributyltin on testicular development in *Sebastiscus marmoratus* and the mechanism involved. Environ Toxicol Chem 28:1528–1535.

Zhang Z, Hu Y, Zhao L, Li J, Bai H, Zhu D, et al. 2011. Estrogen agonist/antagonist properties of dibenzyl phthalate (DBzP) based on *in vitro* and in vivo assays. Toxicol Lett 207:7–11.

Zhao B, Chu Y, Huang Y, Hardy DO, Lin S, Ge RS. 2010. Structure-dependent inhibition of human and rat 11beta-hydroxysteroid dehydrogenase 2 activities by phthalates. Chem Biol Interact 183:79–84.

Zhou T, Ross DG, DeVito MJ, Crofton KM. 2001. Effects of short-term in vivo exposure to polybrominated diphenyl ethers on thyroid hormones and hepatic enzyme activities in weanling rats. Toxicol Sci 61:76–82.

Zimmer KE, Kraugerud M, Aleksandersen M, Gutleb AC, Ostby GC, Dahl E, et al. 2011. Fetal adrenal development: Comparing effects of combined exposures to PCB 118 and PCB 153 in a sheep model. Environ Toxicol 28:164–177.

Zitzmann M, Gromoll J, von Eckardstein A, Nieschlag E. 2003. The CAG repeat polymorphism in the androgen receptor gene modulates body fat mass and serum concentrations of leptin and insulin in men. Diabetologia 46:31–39.

Zoeller RT. 2007. Environmental chemicals impacting the thyroid: Targets and consequences. Thyroid 17:811–817.

Zuo Z, Chen S, Wu T, Zhang J, Su Y, Chen Y, et al. 2011. Tributyltin causes obesity and hepatic steatosis in male mice. Environ Toxicol 26:79–85.

CHAPTER 9

Effects and Predicted Consequences of Persistent and Bioactive Organic Pollutants on Thyroid Function

STEFANIE GIERA and R. THOMAS ZOELLER

ABSTRACT

Background: The thyroid gland and associated hormones regulate cellular metabolism and thus play an essential role in health and disease. A variety of external factors, including some organic chemicals, are known to alter thyroid function.

Objectives: The goal of this review is to summarize the regulation of thyroid hormones, to discuss the known effects of organic chemicals on thyroid function, and to identify to the degree possible the mechanisms whereby these effects occur.

Discussion: A variety of organic contaminants alter thyroid function, including polychlorinated biphenyls, polybrominated diphenyl ethers, bisphenol A, phthalates, perfluorinated chemicals, and triclosan.

Conclusion: While significant gaps in our knowledge remain, it is clear that at least some ubiquitous chemicals can interfere with thyroid hormone action in complex ways. Consideration of the effects on thyroid hormone action should be a part of the evaluation of any new chemical that is considered for wide distribution and potential exposure to humans.

Effects of Persistent and Bioactive Organic Pollutants on Human Health, First Edition.
Edited by David O. Carpenter.

INTRODUCTION

Persistent organic pollutants (POPs) are man-made chemicals, many of which are produced in large quantities and used in a variety of products ranging from industrial applications to consumer products (Crinnion 2010). Their stable chemical structure resists environmental degradation and often leads to bioaccumulation through the food chain. This is why POPs are detected ubiquitously in water, soil, animals, and human populations. They have been linked in humans to cancer (Fisher 1999; Schecter et al. 2005), diabetes (Everett et al. 2011; Ruzzin et al. 2010), and immune-related (Hertz-Picciotto et al. 2008) and endocrine-related diseases (Damstra 2002; Gilbert et al. 2011). A subgroup of POPs includes endocrine-disrupting chemicals (EDCs) because they interfere with hormone action. This disruption of endocrine systems can lead to dysregulation of metabolism and reproduction in adults. In fetuses and children, EDCs can interfere with the normal development.

It is potentially important that a number of effects of POPs on development are reminiscent of effects of thyroid disease during development (Zoeller 2001). In addition, studies focused on POPs as antithyroid agents have revealed a number of mechanisms by which POPs can interfere with thyroid hormone (TH) action. However, these studies present a complex and sometimes confusing picture of thyroid toxicology, and the goal of this chapter is to review this literature and to attempt to propose some clarifying perspectives.

REGULATION OF THE TH LEVELS

It is first important to provide a background about the thyroid system so that a discussion of the effects of POPs on various measures of thyroid function can be interpreted within a biological framework. The THs, thyroxine (T_4) and triiodothyronine (T_3), are produced in the thyroid gland. TH synthesis requires iodine, which is avidly sequestered in the thyroid gland by the action of the sodium/iodide symporter (NIS). The cascade of events from iodide uptake to T_4 and T_3 synthesis and release is controlled by thyrotropin (thyroid-stimulating hormone [TSH]) from the pituitary gland (Figure 9.1, node 1). After release from the thyroid, T_4 and T_3 circulate in the blood as dissolved (free) hormone or bound to one of three circulating proteins—thyroxine-binding globulin (TBG), transthyretin (TTR), or albumin. Most T_4 (75%) is bound to TBG; TTR and albumin bind the rest of T_4 (Figure 9.1, node 3). Although T_3 is secreted from the thyroid gland, about 80% of circulating T_3 is derived from the peripheral deiodination of T_4. THs gain access to tissues and cells through the action of specific membrane transporters. Within a tissue, these hormones bind to the TH receptor (TR) and cause gene activation or inhibition (Gilbert et al. 2011; Zoeller 2007).

Normally, the amount of T_4 found in serum is regulated by a balance between the stimulatory actions of the hypothalamus and the pituitary gland

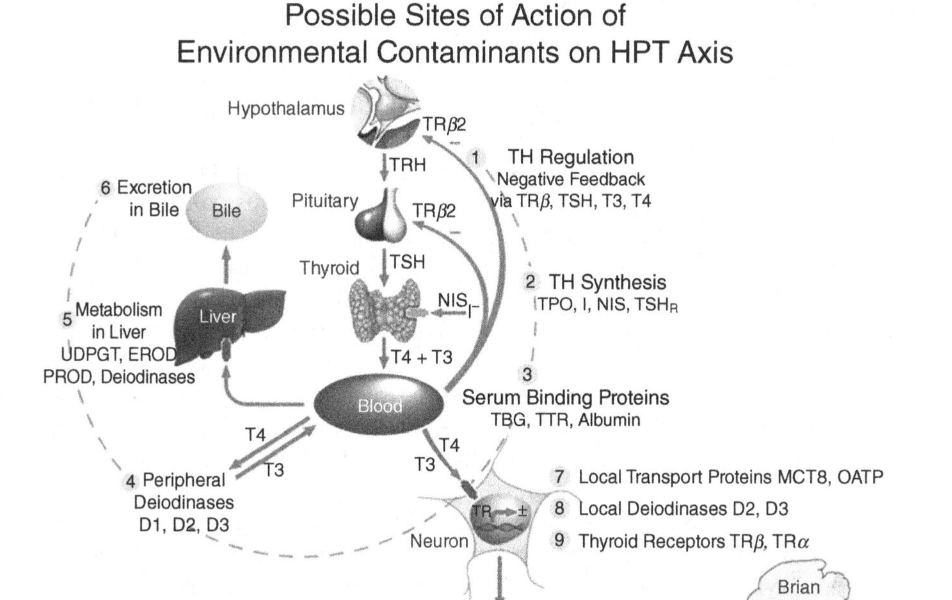

Figure 9.1. HPT axis and sites of TH action. At the center is the HPT axis with the negative feedback loop to control TH levels and target sites of TH action in different tissues, which can be influenced by POPs. *Source*: Figure replicated with permission from Gilbert et al. (2011).

and the inhibitory action of T_4 itself. This system of relationships among the various levels is known as the hypothalamic–pituitary–thyroid (HPT) axis. Thyrotropin-releasing hormone (TRH) is released from the paraventricular nucleus (PVN) and transported to the anterior pituitary, where it binds to the TRH receptor and stimulates thyroid-stimulating hormone (TSH) synthesis and release (Zoeller et al. 2007). TSH is then released into the blood from the anterior pituitary and stimulates thyroid function. T_4 is predominantly responsible for suppressing both TRH and TSH such that serum T_4 levels remain within a relatively narrow range. This conclusion is based on the observation that serum T_4 is more tightly correlated to serum TSH than is T_3 (Panicker et al. 2008). Early work in this field revealed that about 80% of T_3 in the brain and pituitary gland are derived from local deiodination (Larsen et al. 1981). This deiodination is accomplished by one of three types of deiodinase enzymes (Kopp 2008). Thus, the concept is that serum T_4 is most responsible for negative feedback, while serum T_3 is a reflection of T_4 metabolism (Figure 9.1, nodes 1 and 2) (Zoeller et al. 2007).

REGULATION OF TH ACTION

TH exerts its actions largely by regulating the activity of its receptors, which are ligand-controlled transcription factors (Tata 2013). It is generally believed that levels of T_4 in the serum are a good proxy measure of TH action in tissue. We have demonstrated that there is a very tight relationship between serum total T_4 and the number of oligodendrocytes in the corpus callosum of the developing brain (Sharlin et al. 2008). Because oligodendrocyte differentiation is under TH regulation (Dugas et al. 2012), the tight relationship between serum T_4 and oligodendrocyte number indicates that TH action at the receptor is proportional to T_4 levels in the blood. But there are several steps required for T_4 to move from the bloodstream to acting on receptors in cells, and each of these steps can be independently regulated such that there is a disconnect between changes in serum T_4 levels and changes in TH action (e.g., Fekete et al. 2005; Yasuo et al. 2006). Thus, if specific chemicals can selectively interfere with specific steps in the movement of TH from serum to cell, there may well be a disconnect between chemical exposure, serum TH levels, and TH action.

THs (T_4 and T_3) enter individual cells through specific transporters, known as monocarboxylate transporter 8 (MCT8) and monocarboxylate transporter 10 (MCT10). Another group of TH transporters belong to the family of organic anion transporter proteins (OATPs) (Figure 9.1, node 7). One member of the OATP family, OATP1C1, which is expressed in the endothelial cells in the brain, the choroid plexus, and cochlea, mediates the transport of T_4 into the brain (Friesema et al. 2005). Once inside cells, TH is metabolized by a group of enzymes called deiodinases. Deiodinases remove an iodide molecule from both T_4 and T_3 to either activate or inactivate their ability to act on the TR. Deiodinase type 1 (Dio1) and deiodinase type 2 (Dio2) remove an outer ring iodine, which converts T_4 into T_3. In contrast, deiodinase type 3 (Dio3) removes an inner ring iodine and converts T_4 and T_3 into the hormonally inactive forms rT_3 and T_2, respectively (Figure 9.1, nodes 4 and 8). The expression of deiodinases is tissue specific. Dio1 is most abundantly expressed in the liver and kidney. Dio2 is expressed in the heart, muscle, fatty tissue, and the central nervous system. Dio3 is expressed in the uterus and placenta, and during fetal and neonatal development in various tissues, like the pancreas, the cochlea, and the central nervous system (Gereben et al. 2008). Most T_4 is first converted to T_3 within target tissues by deiodinases. One estimate is that 80% of T_3 in the central nervous system is derived locally from T_4 (Larsen et al. 1981). This is why TH action in the brain is correlated with serum T_4 and not serum T_3. Deiodinase activity varies in tissues and as a response to changes in T_4 and T_3 levels (Bianco 2011).

Though both T_4 and T_3 can interact with the TRs, these forms have distinct biological roles. T_4 is more abundant and has a longer half-life in serum than T_3; however, it has a much weaker affinity for the TR and therefore a limited ability to exert changes in gene expression (Sandler et al. 2004). Two genes, THRA (NR1A1) and THRB (NR1A2), encode the three major ligand binding TRs, TRα1, TRβ1, and TRβ2. The gene *c-erbA-β* can be spliced into TRβ1 and

TRβ2 (Figure 9.1, node 9). These isoforms have a differential pattern of expression in the body. Note that the TRβ2 is almost exclusively expressed in the pituitary and hypothalamus, leading some to suggest that negative feedback to THs is mediated solely by this isoform (Hodin et al. 1989; Lechan et al. 1994). If that were the case, it would mean a chemical capable of mimicking TH by interacting with TRs would only exert the negative feedback effect if it could interact with TRβ2.

Both T_4 and T_3 can be removed from serum by glucuronidation catalyzed by UDP-glucuronosyltransferase (UGT) isoforms in the liver (Tong et al. 2007), but also in the intestine and the kidneys. Most of the thyroxine is metabolized in the liver by the UGT isoform UGT1A1 (Figure 9.1, node 5) (Yamanaka et al. 2007). Glucuronidated T_4 and T_3 are then excreted through the bile (Figure 9.1, node 6) (Sellin and Vassilopoulou-Sellin 2000).

ROLE OF TH IN BRAIN DEVELOPMENT

TH is essential for normal development of the brain and other organs (Zoeller et al. 2007). The clinical condition of cretinism, caused by iodine deficiency, was the first evidence provided that TH is essential for development (Brown et al. 1939; Chen and Hetzel 2010). In this condition, maternal hypothyroidism is severe and the outcome for the offspring is severe. Interestingly, maternal hypothyroidism during the first trimester produces a syndrome called neurological cretinism, whereas maternal and fetal hypothyroidism throughout gestation produces a syndrome called myxedematous cretinism (Boyages et al. 1988). The differences in symptoms between these two conditions are sufficiently large that it is clear that TH insufficiency produces different effects at different times during development (Zoeller and Rovet 2004).

If the symptoms of TH insufficiency are different depending on the timing of the insult (Zoeller and Rovet 2004), then TH must be acting in different ways and on different developing neural substrates as development proceeds. This issue has been approached in different ways and in animals and in humans. From a broad perspective, it is clear that TH plays a role in development of both the forebrain and the cerebellum. In the forebrain, TH is important for neural migration. Importantly, even mild and transient TH insufficiency in the dam will adversely affect neuronal migration in the fetus, resulting in an abnormal distribution of cells in the cerebral cortex of the adult offspring (Auso et al. 2004). This is important because defects in neuronal migration are believed to be one of the most significant causes of neurological disabilities (Guerrini and Parrini 2010; Valiente and Marin 2010; Verrotti et al. 2010). Interestingly, neuronal migration is regulated in part by an interaction between *Reelin* and *Notch* (Hashimoto-Torii et al. 2008) and TH may control the expression of both of these genes (Bansal et al. 2005; Pathak et al. 2011).

Within the forebrain itself, TH produces different effects on different neural substrates. For example, Royland et al. (2008) used a genomic analysis to identify TH-regulated genes in the cerebral cortex and hippocampus and found that the transcriptional profile is quite different in the two brain regions, although there was overlap as well. Moreover, as development proceeds, TH produces different effects in the same region. For example, TH controls neuronal migration in the cortex early and controls synaptogenesis later (Legrand et al. 1982; Nicholson and Altman 1972a,b; Rami and Rabie 1990). These findings demonstrate a basic feature of TH action in the developing brain—that it controls different processes at different times during development. However, the mechanisms controlling this are not well understood.

TH plays similar roles in the developing cerebellum (Koibuchi 2009). Early work showed that TH could influence granule cell proliferation (Nicholson and Altman 1972c), neuronal migration, and synaptogenesis (Nicholson and Altman 1972c; Vincent et al. 1982).

Hypothyroidism or hypothyroxinemia during development can be caused by the lack of iodine supply or thyroid disease. Another source of TH disruption is the exposure to environmental chemicals that can act on various points of TH regulation and therefore interfere with TH-dependent events during development. Through studying the effects of hypothyroidism on brain development, we know that disrupting TH action at different times during or after birth can have different outcomes (Zoeller and Rovet 2004).

In the first half of the pregnancy in both animals and humans, the maternal thyroid gland is the sole source of TH in the fetus, while in the second half of pregnancy, fetal thyroid function begins (Howdeshell 2002). Therefore, disruption of maternal thyroid function during the first half of pregnancy represents an important experimental model relevant to the issue of environmental contaminants. Neuronal migration and the formation of the CA1 region of the hippocampus have been shown to be affected by a moderate and transient reduction in maternal thyroid function before the onset of fetal thyroid function (Auso et al. 2004; Madeira et al. 1992). After the beginning of fetal TH production in the latter half of pregnancy, disrupting TH action leads to malformation of the hippocampal CA3 region and the dentate gyrus (Seress et al. 2001), and deficits in myelination and of glia cell differentiation (Figure 9.1, node 10) (Sharlin et al. 2008). During development, even small reductions in T_4, as seen in subclinical hypothyroidism, can lead to changes in gene expression and physiological and behavioral alterations (Figure 9.1, node 11) (Gilbert and Zoeller 2010). Changes of circulating TH levels due to exposure to environmental chemicals without raising TSH can therefore have subtle changes in TH-dependent developmental outcomes.

CONCLUSIONS

The review of TH action earlier demonstrates the complexity of the system. This complexity must be considered in studies of the impacts of environmental chemicals on TH action. Next we review the impacts of several important

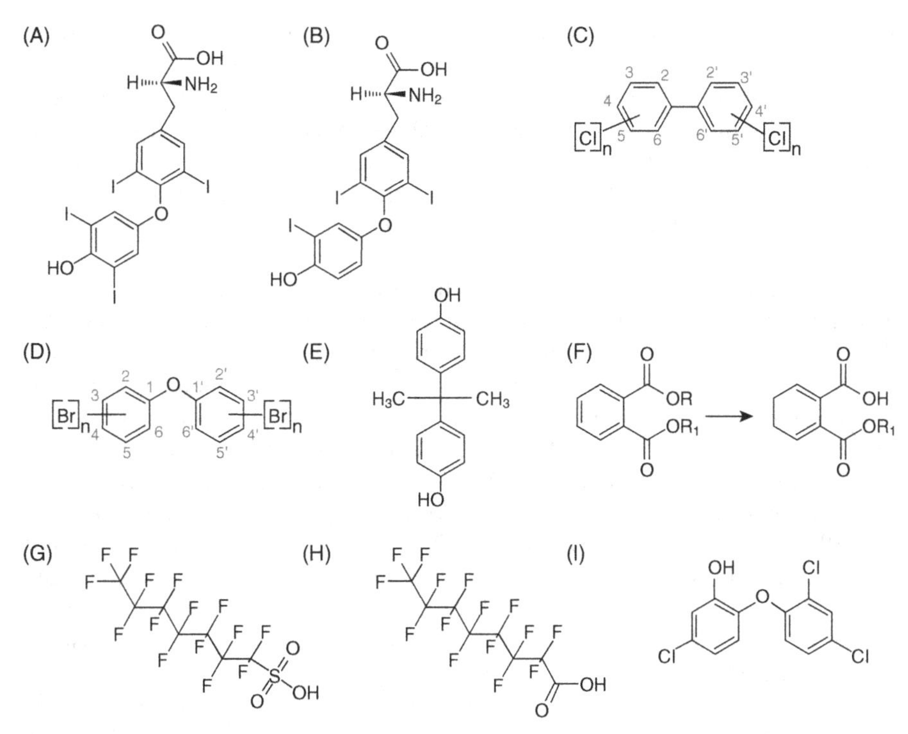

Figure 9.2. Chemical structures. (A) Thyroxine (T_4). (B) Triiodothyronine (T_3). (C) Polychlorinated biphenyl (PCB) basic structure; different congeners based on amount and position of chlorine atoms on the two phenyl rings. (D) Polybrominated diphenyl ether (PBDE) basic structure; different congeners based on the amount and position of bromine atoms on phenyl rings. (E) Bisphenol A. (F) Phthalate basic structure; moieties R and R1 as ethyhexyl is DEHP; as isononyl (DiNP); as ethyl (DEP); as methyl (DMP); as butyl (DBP). Metabolism removes either R or R1; for example, DEHP is metabolized to MEHP or DBP is metabolized to MBP. (G) Perfluorooctane sulfonic acid (PFOS). (H) Perfluorooctanic acid (PFOA). (I) Triclosan.

environmental chemicals on the thyroid system, paying particular attention to the complexity of this system (Figure 9.2).

Polychlorinated Biphenyls (PCBs)

PCBs are chemicals found ubiquitously in the environment, whose exposure cannot be avoided by the general population (CDC 2009). Over a billion pounds of these chemicals were produced worldwide, before the U.S. Congress banned their manufacture in the late 1970s. PCBs consist of 209 different individual congeners, depending on the chlorination number and position on the two linked phenyl rings. The biological effects of PCBs can be subdivided generally into three different structural classes. coplanar, non-*ortho*

PCBs are dioxin-like congeners that are able to bind to the aryl hydrocarbon receptor (AhR). Both mono- and di-*ortho* PCBs are noncoplanar and do not bind to AhR (Erickson 1986; Kodavanti and Tilson 1997; Tilson and Kodavanti 1997). Stability of PCBs, due to their chemical structure, makes them persistent environmental pollutants, and their lipophilic property causes them to bioaccumulate in the food chain. Their presence is ubiquitously found in humans and in animals (Erickson 2001) and therefore presents a health concern.

Epidemiological studies have shown that PCB exposure during gestation is associated with reduced cognitive function. A collaborative study showed decreased performance in a standardized test of cognitive function and verbal comprehension, predicted by the total level of PCBs in maternal or cord plasma (Patandin et al. 1999). Observations by Stewart et al. found total PCB levels in cord blood to be predictive of deficits in autonomic and habituation scores on the Brazelton Neonatal Assessment Scale (Stewart et al. 2000). A wide variety of exposure levels and neurophysiological assessments have consistently demonstrated a negative relationship between PCB exposure and cognitive function in perinatally exposed children (Schantz et al. 2003; Tee et al. 2003). These effects are similar to those observed in infants with hypothyroxinemic mothers, suggesting that they could be caused in part by disruption of the thyroid system.

PCBs are known to reduce circulating THs in humans and in animals (Zoeller 2001), but linking PCBs to TH disruption in humans is hindered by the heterogeneity of measured TH end points, populations, and levels of exposure. Osius et al. found a correlation between circulating levels of certain PCBs and a decrease in free T_3, but not free T_4, with a corresponding increase in TSH (Osius et al. 1999). Other studies have found a negative relationship between PCB exposure free T_4, again with a positive correlation to TSH, but not to free T_3 (Schell et al. 2008). Abdelouahab et al. reported an inverse relationship between PCBs and T_4 and a positive association with TSH in males, but only an inverse relationship between T_3 and PCBs for females (Abdelouahab et al. 2008). Other studies found total T_4 decrease correlated with PCB exposure in both males and females (Persky et al. 2001; Turyk et al. 2007). Sala et al. found no associations between PCBs and TSH, T_3, or T_4 (Sala et al. 2001).

Animal exposure to PCBs has revealed a potential paradoxical effect of these chemicals on the TH system. Rats dosed with PCB mixtures consistently show a decrease in T_4 and T_3 in a dose-dependent manner. Yet, PCBs can induce TR-dependent gene expression in various tissues (Bansal and Zoeller 2008; Giera et al. 2011; Zoeller et al. 2000). PCB-induced reduction of TH levels can also have a similar effect on organ development as hypothyroidism, such as hearing loss and advance in the timing of eye opening (Goldey and Crofton 1998; Goldey et al. 1995). However, in PCB-exposed animals, low TH levels do not cause a compensatory increase in TSH as it is observed in hypothyroid animals (Giera et al. 2011).

PCB mixtures appear to produce a mixed agonist activity and this may be due to metabolism of individual congeners. Once these individual congeners are metabolized into their hydroxylated metabolite, they can then bind to and affect TR function. This metabolism appears to require cytochrome P450 isoform 1A1 (CYP1A1), the expression of which can be induced by dioxin-like congeners (Gauger et al. 2007). Hydroxylated PCB 107 has been implicated in both *in vitro* and human studies as a potential metabolite responsible for affecting the TH system. Hydoxylated PCB 107 is a metabolite of PCB 105 and 118. In addition, this metabolite accumulates in the fetal liver and cerebellum, reduces fetal total T_4 and free T_4, increases fetal TSH, but does not change in D1, UGT, and CYP1A1 activity in the liver (Meerts et al. 2002). Hydoxylated PCB 107 levels in cord blood at birth were negatively associated with both cognitive and motor skill development in subsequent follow-up examinations 16 months later (Park et al. 2009). Hydroxyl groups in the para- and meta-positions on a phenyl ring are required to bind to TTR (Lans et al. 1993).

The mechanism of how PCBs and their metabolites interfere with the TR is not completely understood. PCBs may act as an antagonist on the TR (Iwasaki et al. 2002); this may be due to the ability of specific PCB congeners to cause the displacement of TR/retinoid X receptor (RXR) heterodimer complexes from the TRE (Miyazaki et al. 2004). This effect is DNA dependent. These observations do not explain potential TR agonistic properties of both PCBs and their metabolites (Gauger et al. 2007; You et al. 2006) not only *in vitro* but also *in vivo*. In rats, the exposure to PCBs in the form of a commercial mixture, Aroclor 1254, increases TH-target gene expression in both the cortex and hippocampus during fetal (Gauger et al. 2004) and neonatal development (Bansal and Zoeller 2008; Zoeller and Crofton 2000), respectively. In chicken embryos, the exposure to PCBs leads TH-responsive genes in the brain to display both hypothyroid and thyromimetic expression profiles (Roelens et al. 2005). White matter development is also dependent on TH and exposure to PCBs leads to a reduction of oligodendrocytes in white matter structures similar to the effect of hypothyroidism (Sharlin et al. 2006). In cell cultures of normal human neural progenitor cells, exposure to the individual PCB congener 118 and T_3 leads to the differentiation of progenitor cells into oligodendrocytes (Fritsche et al. 2005). Overall, these studies indicate that the impact of PCBs and their metabolites on TR action is multifaceted and this must be considered in the interpretation of studies in this field.

Overall, studies show a trend of decreased T_4 and increased TSH associated with human PCB exposure. This has led many to suggest that the action of PCBs is to lower the circulating level of T_4, resulting in the rise of TSH, and therefore possibly causing harmful effects that are associated with hypothyroidism. There are several indicated mechanisms by which this could occur. Some early reports indicate PCBs can act directly to damage the thyroid, reducing its ability to produce THs in response to TSH (Collins et al. 1977). However, other studies indicate that in cases of PCB exposure, the thyroid does not react in a way consistent with increased TSH, that is, with an increase in thyroid follicular cell growth (Hood et al. 1999b; Klaassen and Hood 2001).

Another mechanism for TH disruption by PCBs is through the induction of UGTs, which then metabolize both T_4 and T_3, increasing the turnover of TH and renal clearance. There are indeed many limitations to the designs of these studies in humans. Not all studies control for the same variables, like body mass index (BMI), smoking practices, demographics, gender, or exposure to other pollutants. An important consideration is that many of these studies observed only the total sum of PCB levels in serum while not taking into account that PCB congeners can have different biological actions, not only as individual congeners, for example (Fritsche et al. 2005), but also as mixtures (Gauger et al. 2007; Giera et al. 2011). The mixture and proportions of the different congeners vary between individuals in different environments (Beyer and Biziuk 2009; Pelletier et al. 2009; She et al. 2007; Turyk et al. 2006). With improvements in the sensitivity of PCB detection techniques, it should be possible in the future to have a more complete picture of the effect of PCB exposure (Salay and Garabrant 2009). These data indicate that the effects of PCB mixtures are heterogeneous in disrupting TH action in multiple ways. The TSH level may not be the most sensitive predictor of HPT axis interference. Additionally, the levels of circulating TH may not reflect the action of TH. Zoeller (2011) has suggested the need to identify new end points by which TH action can be assessed.

Polybrominated Diphenyl Ethers (PBDEs)

PBDEs are a group of chemicals that have been added to various consumer products (Frederiksen et al. 2008). They are still in widespread use today in household goods, often for the purpose of flame retardation, including many textiles and plastics. There are three different mixtures that were commercially available: decaBDE, pentaBDE, and octaBDE, which were based on the average degree of bromination. DecaBDE consists of the fully brominated congener 209 and represents about 80% of worldwide production (Wikoff et al. 2012). Over time, these chemicals diffuse from products, aerosolize, and disperse (Siddiqi et al. 2003). PBDEs have been identified in air samples, food sources, and in wildlife and humans, including breast milk (Schecter et al. 2003, 2010; She et al. 2007) and cord blood (Lin et al. 2010). PBDEs are persistent environmental toxins and are known to bioaccumulate. PBDEs share some structural similarities with PCBs and with T_4. Studies have indicated that PCB body burden is declining while PBDE levels were rising until the late 1990s (Noren and Meironyte 2000), but levels appear to stagnate since then in various parts of the world (Frederiksen et al. 2008). The widespread use of PBDEs makes it likely that exposure comes from a number of sources. Manufacturing sites, recycling plants, and landfills have all been identified as sources that contribute to environmental levels of PBDEs (Darnerud et al. 2001). In humans, however, the indoor environment and dietary sources are strong predictors of exposure (Wu et al. 2007).

There have been several assessments on the contribution of diet to exposure. One 1999 Swedish market basket study estimated a total intake of 51 ng/day with the highest exposures being linked to dairy, fish, and meat consumption (Darnerud et al. 2006). The same study also showed an infant intake of 110 ng/day from mother's milk. A more recent, questionnaire-based, U.S. study has also found a connection between dairy and meat intake and PBDE concentration in breast milk, with a 3.6-fold increase in PBDE concentration for 1 oz of dairy consumed and a 1.5-fold increase per serving (2–3 oz) of meat (Wu et al. 2007). Despite this, PBDE measurements in humans are much higher than suggested by diet.

House dust ingestion, and to a lesser extent indoor air inhalation, may also be important pathways of exposure. One study found a correlation between PBDEs in dust and in breast milk (Jones-Otazo et al. 2005). However, it remains unclear if house dust is a vector or simply an indirect indicator of PBDE exposure from another source. Johnson-Restrepo et al. estimated the daily dose from house dust to be 1.6–9.5 ng/kg-bw/day for toddlers (Johnson-Restrepo and Kannan 2009). These findings imply that PBDE exposure is directly related to the purchase and possession of products containing these chemicals. This route of exposure may also partially account for the much higher levels of PBDEs in toddlers, who are known to ingest dust at a much higher level than adults. Finally, these studies indicate that the route of exposure for PBDEs is different than that of PCBs, which could affect both how these chemicals should be studied and the effects seen from exposure.

In humans, the exposure to PBDEs is associated with a reduction of TSH levels in the serum of adult males (Hagmar et al. 2001) and pregnant women (Chevrier et al. 2010). Two other studies of pregnant women revealed a positive association with specific PBDE congeners and total and free T_4 and total T_3 (Stapleton et al. 2011) and with TSH (Zota et al. 2011). In a study of male sport fishers, their serum levels of PBDEs were also negatively associated with TSH and T_3 but had a positive association with T_4 (Turyk et al. 2008). Two studies in neonates showed no association between the serum levels of PBDE and TSH (Chevrier et al. 2011), free T_4, and cord TSH (Kim et al. 2012).

Many studies have assessed the ability of PBDEs to interfere with TH signaling in animals. In mice and rats, mixtures of PBDEs decrease serum T_4 (Ernest et al. 2012; Fowles et al. 1994; Hallgren et al. 2001; Stoker et al. 2004), reduce T_3, as well as increase TSH (Lee et al. 2010) and cause thyroid cell growth, with mixtures of lower-brominated PBDEs often having more potent effects (Hooper and McDonald 2000). In studies with fetal and postnatal exposure to PBDEs, the offspring displayed hypothyroxinemia in a dose-dependent way (Kodavanti et al. 2010; Zhou et al. 2002). A recent study examined the effect of coexposure to PCBs and PBDEs. The study by Miller et al. revealed that equimolar concentrations of both PCBs and PBDEs have an additive effect on reducing T_4 levels (Miller et al. 2012). Zebra fish larvae exposed to PBDE congener 209 or PBDE mixtures during development had increased T_4 and decreased T_3 levels. TR-dependent gene transcription of TSHβ, Dio1, Dio2, TRα, and TRβ was increased, but genes regulating TH transport and metabolism were reduced (Chen et al. 2012; Yu et al. 2010). Long-term exposure of zebra fish embryo to lower-brominated PBDEs resulted in increased levels of TH and reduced gene transcription of TSHβ, Dio1, and Dio2 (Yu et al. 2011a).

The exact mechanisms by which PBDEs are able to reduce T_4 are yet unknown. One possibility is that PBDEs induce UGTs, phase II enzymes in the liver, which increase glucuronidation and subsequent clearance of T_4, a mechanism also seen with PCB exposure. While PBDEs are associated with increases in UGT (Stoker et al. 2004), the extent to which T_4 metabolism occurs is dependent on the PBDE mixture (Zhou et al. 2001).

Another mechanism of PBDEs interfering with the thyroid system is at the receptor level. PBDEs, like PCBs, can dissociate TR from its DNA-binding site and suppress TR-mediated transcription and inhibit TH-mediated Purkinje cell dendrite arborization (Ibhazehiebo et al. 2011). Also, PBDEs and their metabolites may be able to mimic THs and displace the hormones from TTR, although Suvorov et al. failed to identify a competitive interaction between PBDEs and the TRβ receptor (Suvorov et al. 2011). Meerts et al. have shown that many PBDEs are metabolized by liver enzymes into hydroxyl-PBDEs that would compete for the TTR (Meerts et al. 2000). Because of the putative role of TTR in transporting T_4 across the placental barrier from mother to fetus, this may partially explain the elevated levels of PBDE metabolites seen in the fetus and raises additional concern for an already vulnerable population (Brouwer et al. 1998).

Overall, studies in humans, animals, and *in vitro* provide evidence that exposure to PBDEs and their hydoxylated metabolites interfere with TH action. They do this by reducing the serum levels of TH (Zota et al. 2011), most likely due to the increase in TH metabolism. PBDE may also interfere with the TR by acting as an agonist, although this needs to be more fully explored. The pattern by which PBDEs interfere with the TH system is reminiscent of that of PCBs. These two classes of chemicals act in an additive manner on circulating TH levels in animals (Miller et al. 2012). This could be an indication that PCBs and PBDEs may interact to affect TH action in humans.

Bisphenol A (BPA)

BPA is a widely used component of polycarbonate plastics, the interior coating of food cans, plastic bottles, and dental sealants (Vandenberg et al. 2007). Although BPA has been shown to have a rapid rate of metabolism as well as clearance and has a low potential for accumulation in the environment (Staples et al. 1998), levels in human blood are often detected at the nanogram per milliliter range, and it is expected that people in developed nations are exposed continuously. The predominant source of exposure is suspected to be dietary, as extensive studies have shown that BPA leeches into food and water from plastic food containers, can linings, water bottles, and many other everyday sources (Erler and Novak 2010; vom Saal et al. 2007). It is of special concern that BPA is detectable in placental tissue, amniotic fluid, and fetal serum (Vandenberg et al. 2007; Yamada et al. 2002) and may become concentrated in amniotic fluid due to low clearance (Ikezuki et al. 2002).

The Centers for Disease Control and Prevention assessed the exposure of BPA among the general U.S. population and found detectable levels of urinary BPA in 93% of participants. The geometric mean of BPA concentrations for participants over 20 was 2.5 μg/L, with higher levels in women, children, and low-income households (Calafat et al. 2008a), indicating these groups are at increased risk of adverse health outcomes.

BPA has been identified as a weak estrogen because of its low affinity with the estrogen receptor. In recent years, this term has come under debate, however, as it has been shown that concentrations of BPA in the picomolar to nanomolar range are enough to interact with membrane receptors (Welshons et al. 2006). However, manufacturers continue to produce products containing BPA, citing these results as inconclusive and remarking that industry-funded studies have not shown significant adverse effects at low doses (vom Saal and Hughes 2005).

In 2002, Moriyama et al. found the first indications that BPA acts as a TH antagonist (Moriyama et al. 2002). Using a competition assay, they first demonstrated that BPA is a weak ligand for the TR. BPA can inhibit the transcription of genes activated by the TR in cell culture (Freitas et al. 2011; Sun et al. 2009). *In vivo*, BPA inhibits T_3-inducible resorption of the tail segments in *Xenopus laevis* larvae, induces T_4-controlled metamorphic changes in tadpoles, and reduces the expression of the TH-mediated gene TRβ, supporting the conclusion that BPA acts as a TH antagonist (Iwamuro et al. 2003). A more recent study suggests that BPA can interfere with TH action on the TRβ receptor by a nongenomic mechanism (Sheng et al. 2012).

Zoeller et al. (2005) were further able to elucidate the effects of BPA on the thyroid system by examining rat pups dietarily exposed to BPA during pregnancy and lactation. They observed that BPA exposure caused elevated T_4 levels without affecting TSH levels, a pattern similar to that found in thyroid resistance syndrome. Seemingly paradoxically, they also found that the TH-responsive gene *RC3* was elevated in the dentate gyrus, in which TRα is expressed. This led them to suggest that BPA acts as an antagonist selectively on the TRβ isoform, causing disruption of the negative feedback mechanism, believed to be mediated by the TRβ (Yen et al. 2003), and increasing transcription of TH-activated genes where TRα is present as a response to elevated T_4 levels (Zoeller et al. 2005). If BPA creates a pattern of disruption that mimics thyroid resistance syndrome, it may affect the development of the cortex. It also may be responsible for the observation of symptoms of attention deficit/hyperactivity disorder (ADHD) in rats exposed to BPA (Ishido et al.

2004), a neurodevelopmental disorder associated with thyroid resistance syndrome (Siesser et al. 2005).

Most of the studies investigating the effect of BPA on TH action report that BPA acts as an antagonist on the TR, but there is also evidence that BPA can indirectly suppress TR-mediated transcription (Sheng et al. 2012). Most studies on the effects of BPA on human health focused on its estrogenic properties and its influence on breast cancer, obesity, and type 2 diabetes (vom Saal et al. 2007). The other controversy in BPA research has been the question about how to evaluate the EDC and its low-dose potency in risk assessment for public health (Myers et al. 2009a,b; vom Saal et al. 2010).

Phthalates

Phthalates are man-made chemicals used as plasticizers in materials such as plastic to improve material flexibility. The usage of various phthalates depends on the molecular weight. Phthalates with higher molecular weights, like di-(2-ethylhexyl) phthalate (DEHP), di-isononyl phthalate (DiNP), and di-isodecyl phthalate (DiDP) are found in a wide range of products such as construction material, clothing, furnishing, polyvinyl chloride (PVC) plastic, medical tubing, blood bags, food containers, electrical devices, and toys. Smaller phthalates, like diethyl phthalate (DEP), dimethyl phthalate (DMP), and dibutyl phthalate (DBP) are used in cosmetics, paints, insecticides, pharmaceuticals, and lacquers. Phthalates are not chemically bound to the plastic and can therefore leach into the surrounding environment, so the route of exposure can be through skin absorption, ingestion, injection, and inhalation (Latini 2005; Schettler 2006).

The average exposure of humans to DEHP in the United States is 3–30 μg/kg/day (ATSDR 2002). In Europe the exposure of DEHP through the environment is 2–67 μg/kg/day for adults and 20–312 μg/kg/day for children (TCS 2008). The metabolites of DEHP are mono-2-ethylhexyl phthalate (MEHP) and mono-2-ethyl-5-hydroxyhexyl phthalate (MEHHP). DBP is metabolized into mono-n-butyl phthalate (MBP). Some evidence suggests the action of phthalates on the TH system is meditated not by the parent compound but by their metabolites.

In a series of human studies, Meeker et al. reported an inverse association between urinary MEHP and free T_4 and total T_3 levels in adult men (Meeker et al. 2007) and an inverse relationship between urinary DEHP metabolite and total T_4, free T_4, total T_3 and TG, as well as a positive association with TSH in adults. In adolescents, however, they found a positive relationship between DEHP metabolites and total T_3 (Meeker and Ferguson 2011), which stands in contrast to a Danish study in children where DEHP and DBP metabolites were inversely related to total and free T_3 (Boas et al. 2010). Yet another study with pregnant women showed a negative association between urinary MBP and serum levels of T_4 and free T_4 (Huang et al. 2007). Both studies found different metabolites of DEHP and DBP having a possible effect on TH serum levels. One human study showed no changes in TH serum levels after dermal application of a mixture of different phthalates (Janjua et al. 2007).

Many animal studies have looked at the effects of phthalates on TH and TSH. Hinton and Mitchell et al. provided the first evidence that phthalates can disrupt TH in 1986. In their study, they treated rats with DEHP and observed increased activity in the thyroid gland as well as a reduction of serum T_4, but T_3 levels did not change (Hinton et al. 1986). Similar alterations in the thyroid and decreased T_4 levels were found in later studies (Howarth et al. 2001; Poon et al. 1997; Price et al. 1988). However, other studies with DEHP treatment at similar concentrations showed no change in the levels of THs and TSH (Bernal et al. 2002). In rats, low levels of DEHP can cause an increase in T_3 and T_4 serum levels but no change in TSH or hyperplasia in the thyroid gland. The effects of the DEHP treatment were only temporary and 7 days after treatment, the serum levels of T_3 and T_4 showed no significant change from the control group (Gayathri et al. 2004). Small phthalates, like DBP, also seem able to reduce serum T_3 and T_4 levels in a dose-dependent manner. Interestingly, a reduction of both T_3 and T_4 is observed only at high concentrations, while low concentrations had no effect on TH and TSH (Lee et al. 2008; O'Connor et al. 2002). Phthalates may influence circulating TH levels by inhibiting binding of T_4 to TTR, a major TH-binding protein in plasma (Ishihara et al. 2003).

While it could not be proven conclusively that phthalates change TH serum levels both in humans or in animals, there is strong evidence that they affect genes regulated by TR in different tissues. A variety of phthalates show TH-disrupting potential by affecting TH-dependent rat pituitary GH3 cell proliferation (Ghisari and Bonefeld-Jorgensen 2009). Phthalic acid monoester metabolites showed TR antagonistic activity in a luciferase reporter gene assay. The phthalic acid monoester MBP possesses a higher antagonist activity than either parent molecules, DBP and DEHP (Shen et al. 2009). Nevertheless, unmetabolized phthalates show T_3 antagonistic activity *in vitro* as well as inhibit the transcription of TRβ both *in vitro* and *in vivo* (Sugiyama et al. 2005).

More recent studies have shown that DBP and MBP reduce the expression of TRβ and RXRγ while increasing both TSHα and TSHβ expression in frog tadpoles. It was also shown that both phthalates enhance the interaction between SMRT (Silencing Mediator of Retinoic Acid Receptor and Thyroid Hormone Receptor) and TR in a dose-dependent way with MBP being slightly more potent than DBP (Shen et al. 2011). Also, *in vitro* studies found that DEHP and other phthalates change iodide uptake of thyroid follicular cells (Wenzel et al. 2005), and Breous et al. reported that selected phthalates can induce transcription of the sodium/iodide symporter, a transporter essential for iodide transport into the thyroid (Breous et al. 2005). However, another study with tadpole red blood cells showed that DBP and n-butylbenzyl phthalate (BBP), but not DEHP and DEP, inhibit T_3 uptake through amino acid transporters into red blood cells (Shimada and Yamauchi 2004).

In general, there is sufficient evidence to support concern that at least some phthalates may interfere with TH action in some way. However, a more uniform testing of a variety of phthalates and their metabolites might be helpful in elucidating their impact on the thyroid system. The different *in vivo* studies on the effects of phthalates on the HPT axis are not conclusive. Further studies are necessary to determine which phthalates have the ability to impact the thyroid system and by which mechanism. Additionally, looking at serum levels of T_4, T_3, and TSH might not be a reliable end point to measure the effects of phthalates on TH-mediated actions. There is strong evidence *in vitro* that phthalates might act as antagonists on the TR.

Perfluorinated Chemicals (PFCs)

PFCs are used in a wide variety of applications, from cooking utensils to firefighting foams and industrial lubricating agents (Jensen and Leffers 2008). The salts of perfluorooctane sulfonic acid (PFOS) and perfluorooctanic acid (PFOA) are water soluble (Prevedouros et al. 2006), but once dissociated, the long perfluorinated alkyl chains are not further degraded (Dinglasan et al. 2004; Lau et al. 2007). Human exposure is mostly through house dust, increasing exposure in children to 5–10 times relative to body weight than adults (Shoeib et al. 2005). The half-life of PFOA and PFOS in humans is 3.8–5.4 years, respectively (Olsen and Zobel 2007).

PFOS and PFOA are two of the most commonly found compounds in human epidemiological studies, ranging from microgram per liter in pregnant women (Inoue et al. 2004) to milligram per liter in manufacturer workers (Olsen et al. 2003). Most studies focused on thyroid effects find no association between T_4 or TSH serum levels and exposure to PFOS and/or PFOA (Bloom et al. 2010; Chan et al. 2011; Inoue et al. 2004; Olsen and Zobel 2007; Olsen et al. 2003), but T_3 is positively associated with PFOS and PFOA in two studies (Olsen and Zobel 2007; Raymer et al. 2011). However, one study by Dallaire et al. 2009 shows that PFOS is negatively associated with total T_3, TSH, and TGB while being positively associated with fT_4 (Dallaire et al. 2009).

Several studies in rats report hypothyroxinemia in both dams and their offspring after being exposed to low and high doses of PFOS or PFOA without leading to an increase in TSH (Lau et al. 2003; Luebker et al. 2005; Martin et al. 2007; Thibodeaux et al. 2003; Yu et al. 2009a,b; Yu et al. 2011b). The induction of TH metabolizing enzymes in the liver as well as the increase in ^{125}I turnover in PFOS-treated rats leads to the conclusion that the reduction of T_4 in serum is caused by increased T_4 metabolism (Chang et al. 2008; Martin et al. 2007; Yu et al. 2009a). However, this conclusion is somewhat weakened by the observation that some microsomal activators, especially phenobarbital, reduce serum T_4 and increase serum TSH (Hood et al. 1999a); thus, it is not clear how chemicals such as perfluorinated compounds can reduce serum T_4 without triggering an increase in serum TSH.

There are a few *in vitro* studies aimed to determine the mechanism of PFC-mediated disruption of the TH system. A binding study with TTR showed that PFOS and PFOA can dissociate T_4 from TTR, but the PFC compound perfluorohexansulfonate (PFHxS) is more potent (Weiss et al. 2009). PFHxS has a shorter carbon chain than PFOS and PFOA and, along with other short-chain PFCs, like perfluorohexanoic acid (PFHxA) and perfluorohexanoic acid (PFHpA), changes expression of TH target genes in primary avian neuronal cells that are related to TH action like *Dio2*, *Dio3*, *MBP*, and *RC3* (Vongphachan et al. 2011). The same group found in an *in ovo* study with PFHxS that it accumulated in the liver and cerebral cortex as well as causing a reduction in fT_4. In these tissues, expression of TH target genes was increased in a dose-dependent way (Cassone et al. 2012).

The few epidemiological studies in human populations do not show a consistent effect of PFCs on the TH system. Most of the studies were carried out in adults and do not take into account more susceptible subpopulations, like pregnant women and children, where disruption of the TH system leads to developmental impairments. PFOS and PFOA are being phased out by their manufacturers, but their ability to bioaccumulate will most likely cause further disruption of endocrine systems. While the mechanism is still unknown by which PFOS or PFOA interact with the TH system, PFHxS and other short-chain PFCs are beginning to replace them. Preliminary studies indicate these chemicals may have an impact on TH signaling that is very similar to that of PFOS and PFOA.

Triclosan

Triclosan is an antibiotic agent used in a variety of consumer products, like toothpaste and soaps, and in clothes and has been reviewed extensively in Dann and Hontela (2011). Several human studies found triclosan in both human plasma as well as human breast milk at similar concentrations (~ 0.8–300 ng/g) (Adolfsson-Erici et al. 2002; Allmyr et al. 2006, 2009; Calafat et al. 2008b; Crinnion 2010; Dayan 2007; Sandborgh-Englund et al. 2006). In men, triclosan has a half-life of about 4 days when exposed to a single dose (Sandborgh-Englund et al. 2006). Two human studies that investigated the effect of triclosan on TH serum levels found no correlation between TH and triclosan exposure. Both studies used triclosan-containing toothpaste as route of exposure, one for 2 weeks (Allmyr et al. 2009), the other over 4 years (Cullinan et al. 2012).

The first animal studies in bullfrogs demonstrated that triclosan influences the TH system. In tadpoles, environmentally relevant concentrations of triclosan reduced the expression of TRβ in the tail fin. Triclosan in combination with T_3 increased hind limb development, while tadpole bodyweight was reduced (Veldhoen et al. 2006). Other studies in *Xenopus* tadpoles also showed an adverse effect of triclosan on larvae growth, and TRβ expression in the tail fin is increased in a nonmonotonic dose–response during metamorphosis (Fort et al. 2010).

Several studies using weaned male and female rats reported a dose-dependent reduction of total T_4 and T_3 (Crofton et al. 2007; Paul et al. 2010a,b;

Rodriguez and Sanchez 2010; Stoker et al. 2010; Zorrilla et al. 2009), while TSH was unchanged in the two studies examining TSH levels (Paul et al. 2010b; Stoker et al. 2010). Two studies exposed pregnant rats to an array of triclosan doses, which reduced TH levels in the dams (Paul et al. 2010a; Stoker et al. 2010), but only one reported that perinatal exposure to triclosan reduces T_4 in the offspring (Paul et al. 2010a). Triclosan is thought to act by increasing levels of hepatic metabolizing enzymes, which in turn causes an increase in metabolism of TH causing the observed reduction in serum T_4 (Paul et al. 2010b).

In vitro studies have shown that triclosan inhibits Dio1 activity in a dose-dependent manner (Butt et al. 2011) as well as T_2 sulfotransferases, a metabolizing enzyme for T_3 (Schuur et al. 1998). Inhibiting TH metabolizing enzymes as well as reducing circulating levels of TH could lead to a very diverse set of actions of triclosan in different tissues depending on their metabolic properties. In the pituitary cell line GH3, methyl-triclosan, a triclosan metabolite, suppressed T_3-mediated transcription of growth hormone, Dio 1, and prolactin (Hinther et al. 2011).

The current information on triclosan suggests that it may have important but complex impacts on TH action. However, well-designed *in vivo* and *in vitro* work is required to determine the mechanism by which triclosan affects TH action, and the consequences of such effects. Also, more comprehensive human studies with various TH-sensitive end points should be conducted, especially in pregnant women, infants, and children who are more sensitive to developmental interruption of TH.

DISCUSSION AND CONCLUSION

There are significant gaps in our understanding of the action and outcomes of thyroid system disruption by environmental contaminants. In this chapter, the described POPs all have demonstrated the ability to influence circulating levels of either TH and/or TSH in humans. Not all human studies show an association between TSH or TH levels with the measured concentration of different contaminants. This may be due mostly to a small number of participants within the study. Also the biological markers for TH disruption vary between human studies, as well as which congeners of one group of pollutants and their metabolites are measured. These differences among studies make it difficult to establish that environmental chemicals can interfere with TH action in humans and to identify potential consequences. Human studies are lacking in measuring TH serum levels and TH-dependent outcomes in more sensitive populations like children and infants because, due to their reduced bodyweight, they might have higher concentrations of pollutants.

Another variable that has been overlooked in epidemiological studies is the supply of iodine. Sufficient iodine is necessary for normal TH function, not only in adults but also during development. Therefore, it is important for the fetus to be supplied with adequate amounts of both T_4 and T_3 during gestation. Having sufficient maternal iodine supply can prevent one of the most common reasons of insufficient TH supply during pregnancy (de Escobar et al. 2007). Gross deficits in iodine supply during pregnancy leads to a well-known form of mental and physical retardation called cretinism. In addition, mild or moderate iodine insufficiency can cause changes in differentiation of neurons and can interfere with cortex, hippocampus, and cerebellum development in animal studies (Auso et al. 2004; Dong et al. 2005, 2010; Martinez-Galan et al. 1997; Potter et al. 1982). In humans, mild iodine deficits lead to cognitive disabilities (Berbel et al. 2009; de Escobar et al. 2007). A deficiency in iodine is still one of the most common causes of hypothyroidism (Melse-Boonstra and Jaiswal 2010). Exposure to multiple compounds interfering with the TH system in iodine-deficient populations could increase the potential of adverse health outcome. These compounds may act by further inhibiting iodine uptake into the thyroid gland (e.g., perchlorate; Wolff 1998) by displacing T_4 and/or T_3 from serum-binding proteins (e.g., PCBs and PBDEs), by activating liver metabolism (e.g., POPs), and by interacting with TRs in ways that may be very complex. There is no information about how these chemicals, to all of which the human population is simultaneously exposed, interact to interfere with TH action and affect development or health.

In the case of PCBs and PBDEs, initial observations showed a reduction of serum TH levels that suggested the effect of PCBs was mimicking hypothyroidism, but looking at additional end points revealed that this was not uniformly the case. Experimental studies now indicate that mixtures of PCBs affect the action of THs through several mechanisms, creating a more complex pattern of disruption than was originally assumed.

To model human exposure, we use animal experiments, which usually employ higher concentrations of POPs than those found in the human population. Animal studies do not model the long-term low-dose exposure of the human population. The exposure to mixtures is not sufficiently taken into account as well as the metabolic interactions between compounds that individually are not necessarily impacting the TH system. Also an important aspect of predicting the potential harm for the human population in epidemiological studies is the exposure to a variety of individual compounds that can act in an additive or synergistic way.

In vitro work shows nonmonotonic dose–responses that do not show up well in validated regulatory toxicology assays and in risk assessment. Also *in vitro* studies, like binding studies with the TR isoforms, do not take into account the interactions of nuclear receptors with both activating and repressing cofactors and DNA-binding sites. The complex protein–protein and protein–DNA interactions are usually not simulated in *in vitro* studies. Even if EDCs do not bind to the receptor, it does not preclude the possibility that they can still interfere with proper receptor function.

These studies of POP actions on the TH system should be used to inform the testing of future emerging chemicals that are widely distributed into the environment and the human population. The assessment of disruption of the TH action based on changing TH and TSH serum levels is not sensitive enough to understand the complex mode of action of POPs.

REFERENCES

Abdelouahab N, Mergler D, Takser L, Vanier C, St-Jean M, Baldwin M, et al. 2008. Gender differences in the effects of organochlorines, mercury, and lead on thyroid hormone levels in lakeside communities of Quebec (Canada). Environ Res 107: 380–392.

Adolfsson-Erici M, Pettersson M, Parkkonen J, Sturve J. 2002. Triclosan, a commonly used bactericide found in human milk and in the aquatic environment in Sweden. Chemosphere 46:1485–1489.

Allmyr M, Adolfsson-Erici M, McLachlan MS, Sandborgh-Englund G. 2006. Triclosan in plasma and milk from Swedish nursing mothers and their exposure via personal care products. Sci Total Environ 372:87–93.

Allmyr M, Panagiotidis G, Sparve E, Diczfalusy U, Sandborgh-Englund G. 2009. Human exposure to triclosan via toothpaste does not change CYP3A4 activity or plasma concentrations of thyroid hormones. Basic Clin Pharmacol Toxicol 105:339–344.

ATSDR. 2002. Toxicological Profile for di(2-Ethylhexyl)phthalate (DEHP). Atlanta, GA: Agency for Toxic Substances and Disease Registry.

Auso E, Lavado-Autric R, Cuevas E, Escobar Del Rey F, Morreale De Escobar G, Berbel P. 2004. A moderate and transient deficiency of maternal thyroid function at the beginning of fetal neocorticogenesis alters neuronal migration. Endocrinology 145:4037–4047.

Bansal R, Zoeller RT. 2008. Polychlorinated biphenyls (Aroclor 1254) do not uniformly produce agonist actions on thyroid hormone responses in the developing rat brain. Endocrinology 149:4001–4008.

Bansal R, You SH, Herzig CT, Zoeller RT. 2005. Maternal thyroid hormone increases HES expression in the fetal rat brain: An effect mimicked by exposure to a mixture of polychlorinated biphenyls (PCBs). Brain Res Dev Brain Res 156:13–22.

Berbel P, Mestre JL, Santamaria A, Palazon I, Franco A, Graells M, et al. 2009. Delayed neurobehavioral development in children born to pregnant women with mild hypothyroxinemia during the first month of gestation: The importance of early iodine supplementation. Thyroid 19:511–519.

Bernal CA, Martinelli MI, Mocchiutti NO. 2002. Effect of the dietary exposure of rat to di(2-ethyl hexyl) phthalate on their metabolic efficiency. Food Addit Contam 19:1091–1096.

Beyer A, Biziuk M. 2009. Environmental fate and global distribution of polychlorinated biphenyls. Rev Environ Contam Toxicol 201:137–158.

Bianco AC. 2011. Minireview: Cracking the metabolic code for thyroid hormone signaling. Endocrinology 152:3306–3311.

Bloom MS, Kannan K, Spliethoff HM, Tao L, Aldous KM, Vena JE. 2010. Exploratory assessment of perfluorinated compounds and human thyroid function. Physiol Behav 99:240–245.

Boas M, Frederiksen H, Feldt-Rasmussen U, Skakkebaek NE, Hegedus L, Hilsted L, et al. 2010. Childhood exposure to phthalates: Associations with thyroid function, insulin-like growth factor I, and growth. Environ Health Perspect 118: 1458–1464.

Boyages SC, Halpern JP, Maberly GF, Eastman CJ, Morris J, Collins J, et al. 1988. A comparative study of neurological and myxedematous endemic cretinism in western China. J Clin Endocrinol Metab 67:1262–1271.

Breous E, Wenzel A, Loos U. 2005. The promoter of the human sodium/iodide symporter responds to certain phthalate plasticisers. Mol Cell Endocrinol 244:75–78.

Brouwer A, Morse DC, Lans MC, Schuur AG, Murk AJ, Klasson-Wehler E, et al. 1998. Interactions of persistent environmental organohalogens with the thyroid hormone system: Mechanisms and possible consequences for animal and human health. Toxicol Ind Health 14:59–84.

Brown AW, Bronstein IP, Kraines R. 1939. Hypothyroidism and cretinism in childhood. VI. Influence of thyroid therapy on mental growth. Am J Dis Child 57:517–523.

Butt CM, Wang D, Stapleton HM. 2011. Halogenated phenolic contaminants inhibit the in vitro activity of the thyroid-regulating deiodinases in human liver. Toxicol Sci 124:339–347.

Calafat AM, Ye X, Wong LY, Reidy JA, Needham LL. 2008a. Exposure of the U.S. population to bisphenol A and 4-tertiary-octylphenol: 2003–2004. Environ Health Perspect 116:39–44.

Calafat AM, Ye X, Wong LY, Reidy JA, Needham LL. 2008b. Urinary concentrations of triclosan in the U.S. population: 2003–2004. Environ Health Perspect 116:303–307.

Cassone CG, Vongphachan V, Chiu S, Williams KL, Letcher RJ, Pelletier E, et al. 2012. In ovo effects of perfluorohexane sulfonate and perfluorohexanoate on pipping success, development, mRNA expression and thyroid hormone levels in chicken embryos. Toxicol Sci 127:216–224.

CDC. 2009. Fourth National Report on Human Exposure to Environmental Chemicals. Atlanta, GA: Center for Disease Control and Prevention.

Chan E, Burstyn I, Cherry N, Bamforth F, Martin JW. 2011. Perfluorinated acids and hypothyroxinemia in pregnant women. Environ Res 111:559–564.

Chang SC, Thibodeaux JR, Eastvold ML, Ehresman DJ, Bjork JA, Froehlich JW, et al. 2008. Thyroid hormone status and pituitary function in adult rats given oral doses of perfluorooctanesulfonate (PFOS). Toxicology 243:330–339.

Chen ZP, Hetzel BS. 2010. Cretinism revisited. Best Pract Res Clin Endocrinol Metab 24:39–50.

Chen Q, Yu L, Yang L, Zhou B. 2012. Bioconcentration and metabolism of decabromodiphenyl ether (BDE-209) result in thyroid endocrine disruption in zebra fish larvae. Aquat Toxicol 110–111:141–148.

Chevrier J, Harley KG, Bradman A, Gharbi M, Sjodin A, Eskenazi B. 2010. Polybrominated diphenyl ether (PBDE) flame retardants and thyroid hormone during pregnancy. Environ Health Perspect 118:1444–1449.

Chevrier J, Harley KG, Bradman A, Sjodin A, Eskenazi B. 2011. Prenatal exposure to polybrominated diphenyl ether flame retardants and neonatal thyroid-stimulating hormone levels in the CHAMACOS study. Am J Epidemiol 174:1166–1174.

Collins WT, Capen CC, Kasza L, Carter C, Dailey RE. 1977. Effect of polychlorinated biphenyl (PCB) on the thyroid gland of rats. Ultrastructural and biochemical investigations. Am J Pathol 89:119–130.

Crinnion WJ. 2010. The CDC fourth national report on human exposure to environmental chemicals: What it tells us about our toxic burden and how it assist environmental medicine physicians. Altern Med Rev 15:101–109.

Crofton KM, Paul KB, DeVito MJ, Hedge JM. 2007. Short-term in vivo exposure to the water contaminant triclosan: Evidence for disruption of thyroxine. Environ Toxicol Pharmacol 24:194–197.

Cullinan MP, Palmer JE, Carle AD, West MJ, Seymour GJ. 2012. Long term use of triclosan toothpaste and thyroid function. Sci Total Environ 416:75–79.

Dallaire R, Dewailly E, Pereg D, Dery S, Ayotte P. 2009. Thyroid function and plasma concentrations of polyhalogenated compounds in Inuit adults. Environ Health Perspect 117:1380–1386.

Damstra T. 2002. Potential effects of certain persistent organic pollutants and endocrine disrupting chemicals on the health of children. J Toxicol Clin Toxicol 40: 457–465.

Dann AB, Hontela A. 2011. Triclosan: Environmental exposure, toxicity and mechanisms of action. J Appl Toxicol 31:285–311.

Darnerud PO, Eriksen GS, Johannesson T, Larsen PB, Viluksela M. 2001. Polybrominated diphenyl ethers: Occurrence, dietary exposure, and toxicology. Environ Health Perspect 109(Suppl 1):49–68.

Darnerud PO, Atuma S, Aune M, Bjerselius R, Glynn A, Grawé KP, et al. 2006. Dietary intake estimations of organohalogen contaminants (dioxins, PCB, PBDE and chlorinated pesticides, e.g., DDT) based on Swedish market basket data. Food Chem Toxicol 44:1597–1606.

Dayan AD. 2007. Risk assessment of triclosan [Irgasan] in human breast milk. Food Chem Toxicol 45:125–129.

Dinglasan MJ, Ye Y, Edwards EA, Mabury SA. 2004. Fluorotelomer alcohol biodegradation yields poly- and perfluorinated acids. Environ Sci Technol 38:2857–2864.

Dong H, Wade M, Williams A, Lee A, Douglas GR, Yauk C. 2005. Molecular insight into the effects of hypothyroidism on the developing cerebellum. Biochem Biophys Res Commun 330:1182–1193.

Dong J, Liu W, Wang Y, Xi Q, Chen J. 2010. Hypothyroidism following developmental iodine deficiency reduces hippocampal neurogranin, CaMK II and calmodulin and elevates calcineurin in lactational rats. Int J Dev Neurosci 28:589–596.

Dugas JC, Ibrahim A, Barres BA. 2012. The T3-induced gene KLF9 regulates oligodendrocyte differentiation and myelin regeneration. Mol Cell Neurosci 50:45–57.

Erickson MD. 1986. Analytical Chemistry of PCBs. Boston: Butterworth.

Erickson MD. 2001. PCB properties, uses, occurrence, and regulatory history. In: PCBs: Recent Advances in Environmental Toxicology and Health Effects (Robertson LW, Hansen LG, eds.). Lexington, KY: The University Press of Kentucky, pp. xii–xxx.

Erler C, Novak J. 2010. Bisphenol a exposure: Human risk and health policy. J Pediatr Nurs 25:400–407.

Ernest SR, Wade MG, Lalancette C, Ma YQ, Berger RG, Robaire B, et al. 2012. Effects of chronic exposure to an environmentally relevant mixture of brominated flame retardants on the reproductive and thyroid system in adult male rats. Toxicol Sci 127:496–507.

de Escobar GM, Obregon MJ, del Rey FE. 2007. Iodine deficiency and brain development in the first half of pregnancy. Public Health Nutr 10:1554–1570.

Everett CJ, Frithsen I, Player M. 2011. Relationship of polychlorinated biphenyls with type 2 diabetes and hypertension. J Environ Monit 13:241–251.

Fekete C, Sarkar S, Christoffolete MA, Emerson CH, Bianco AC, Lechan RM. 2005. Bacterial lipopolysaccharide (LPS)-induced type 2 iodothyronine deiodinase (D2) activation in the mediobasal hypothalamus (MBH) is independent of the LPS-induced fall in serum thyroid hormone levels. Brain Res 1056: 97–99.

Fisher BE. 1999. Most unwanted. Environ Health Perspect 107:A18–A23.

Fort DJ, Rogers RL, Gorsuch JW, Navarro LT, Peter R, Plautz JR. 2010. Triclosan and anuran metamorphosis: No effect on thyroid-mediated metamorphosis in *Xenopus laevis*. Toxicol Sci 113:392–400.

Fowles JR, Fairbrother A, Baecher-Steppan L, Kerkvliet NI. 1994. Immunologic and endocrine effects of the flame-retardant pentabromodiphenyl ether (DE-71) in C57BL/6J mice. Toxicology 86:49–61.

Frederiksen M, Vorkamp K, Thomsen M, Knudsen LE. 2008. Human internal and external exposure to PBDEs—A review of levels and sources. Int J Hyg Environ Health 212:109–134.

Freitas J, Cano P, Craig-Veit C, Goodson ML, Furlow JD, Murk AJ. 2011. Detection of thyroid hormone receptor disruptors by a novel stable in vitro reporter gene assay. Toxicol in Vitro 25:257–266.

Friesema EC, Jansen J, Milici C, Visser TJ. 2005. Thyroid hormone transporters. Vitam Horm 70:137–167.

Fritsche E, Cline JE, Nguyen NH, Scanlan TS, Abel J. 2005. Polychlorinated biphenyls disturb differentiation of normal human neural progenitor cells: Clue for involvement of thyroid hormone receptors. Environ Health Perspect 113:871–876.

Gauger KJ, Kato Y, Haraguchi K, Lehmler HJ, Robertson LW, Bansal R, et al. 2004. Polychlorinated biphenyls (PCBs) exert thyroid hormone-like effects in the fetal rat brain but do not bind to thyroid hormone receptors. Environ Health Perspect 112:516–523.

Gauger KJ, Giera S, Sharlin DS, Bansal R, Iannacone E, Zoeller RT. 2007. Polychlorinated biphenyls 105 and 118 form thyroid hormone receptor agonists after cytochrome P4501A1 activation in rat pituitary GH3 cells. Environ Health Perspect 115:1623–1630.

Gayathri NS, Dhanya CR, Indu AR, Kurup PA. 2004. Changes in some hormones by low doses of di (2-ethyl hexyl) phthalate (DEHP), a commonly used plasticizer in PVC blood storage bags & medical tubing. Indian J Med Res 119: 139–144.

Gereben B, Zavacki AM, Ribich S, Kim BW, Huang SA, Simonides WS, et al. 2008. Cellular and molecular basis of deiodinase-regulated thyroid hormone signaling. Endocr Rev 29:898–938.

Ghisari M, Bonefeld-Jorgensen EC. 2009. Effects of plasticizers and their mixtures on estrogen receptor and thyroid hormone functions. Toxicol Lett 189:67–77.

Giera S, Bansal R, Ortiz-Toro TM, Taub DG, Zoeller RT. 2011. Individual polychlorinated biphenyl (PCB) congeners produce tissue- and gene-specific effects on thyroid hormone signaling during development. Endocrinology 152: 2909–2919.

Gilbert ME, Zoeller RT. 2010. Thyroid hormone-impact on the developing brain: Possible mechanisms of neurotoxicity. In: Neurotoxicology, 3rd edition, Target Organ Toxicology Series (Harry GJ and Tilson HA, eds.). New York: Informa Healthcare USA, Inc, pp. 79–111.

Gilbert ME, Rovet J, Chen ZP, Koibuchi N. 2011. Developmental thyroid hormone disruption: Prevalence, environmental contaminants and neurodevelopmental consequences. Neurotoxicology 33:842–852.

Goldey ES, Crofton KM. 1998. Thyroxine replacement attenuates hypothyroxinemia, hearing loss, and motor deficits following developmental exposure to Aroclor 1254 in rats. Toxicol Sci 45:94–105.

Goldey ES, Kehn LS, Lau C, Rehnberg GL, Crofton KM. 1995. Developmental exposure to polychlorinated biphenyls (Aroclor 1254) reduces circulating thyroid hormone concentrations and causes hearing deficits in rats. Toxicol Appl Pharmacol 135:77–88.

Guerrini R, Parrini E. 2010. Neuronal migration disorders. Neurobiol Dis 38: 154–166.

Hagmar L, Bjork J, Sjodin A, Bergman A, Erfurth EM. 2001. Plasma levels of persistent organohalogens and hormone levels in adult male humans. Arch Environ Health 56:138–143.

Hallgren S, Sinjari T, Hakansson H, Darnerud PO. 2001. Effects of polybrominated diphenyl ethers (PBDEs) and polychlorinated biphenyls (PCBs) on thyroid hormone and vitamin A levels in rats and mice. Arch Toxicol 75:200–208.

Hashimoto-Torii K, Torii M, Sarkisian MR, Bartley CM, Shen J, Radtke F, et al. 2008. Interaction between Reelin and Notch signaling regulates neuronal migration in the cerebral cortex. Neuron 60:273–284.

Hertz-Picciotto I, Park HY, Dostal M, Kocan A, Trnovec T, Sram R. 2008. Prenatal exposures to persistent and non-persistent organic compounds and effects on immune system development. Basic Clin Pharmacol Toxicol 102:146–154.

Hinther A, Bromba CM, Wulff JE, Helbing CC. 2011. Effects of triclocarban, triclosan, and methyl triclosan on thyroid hormone action and stress in frog and mammalian culture systems. Environ Sci Technol 45:5395–5402.

Hinton RH, Mitchell FE, Mann A, Chescoe D, Price SC, Nunn A, et al. 1986. Effects of phthalic acid esters on the liver and thyroid. Environ Health Perspect 70: 195–210.

Hodin RA, Lazar MA, Wintman BI, Darling DS, Chin WW. 1989. Identification of a thyroid hormone receptor that is pituitary-specific. Science 244:76–79.

Hood A, Hashmi R, Klaassen CD. 1999a. Effects of microsomal enzyme inducers on thyroid-follicular cell proliferation, hyperplasia, and hypertrophy. Toxicol Appl Pharmacol 160:163–170.

Hood A, Liu YP, Gattone VH, 2nd, Klaassen CD. 1999b. Sensitivity of thyroid gland growth to thyroid stimulating hormone (TSH) in rats treated with antithyroid drugs. Toxicol Sci 49:263–271.

Hooper K, McDonald TA. 2000. The PBDEs: An emerging environmental challenge and another reason for breast-milk monitoring programs. Environ Health Perspect 108:387–392.

Howarth JA, Price SC, Dobrota M, Kentish PA, Hinton RH. 2001. Effects on male rats of di-(2-ethylhexyl) phthalate and di-n-hexylphthalate administered alone or in combination. Toxicol Lett 121:35–43.

Howdeshell KL. 2002. A model of the development of the brain as a construct of the thyroid system. Environ Health Perspect 110(Suppl 3):337–348.

Huang PC, Kuo PL, Guo YL, Liao PC, Lee CC. 2007. Associations between urinary phthalate monoesters and thyroid hormones in pregnant women. Hum Reprod 22:2715–2722.

Ibhazehiebo K, Iwasaki T, Kimura-Kuroda J, Miyazaki W, Shimokawa N, Koibuchi N. 2011. Disruption of thyroid hormone receptor-mediated transcription and thyroid hormone-induced Purkinje cell dendrite arborization by polybrominated diphenyl ethers. Environ Health Perspect 119:168–175.

Ikezuki Y, Tsutsumi O, Takai Y, Kamei Y, Taketani Y. 2002. Determination of bisphenol A concentrations in human biological fluids reveals significant early prenatal exposure. Hum Reprod 17:2839–2841.

Inoue K, Okada F, Ito R, Kato S, Sasaki S, Nakajima S, et al. 2004. Perfluorooctane sulfonate (PFOS) and related perfluorinated compounds in human maternal and cord blood samples: Assessment of PFOS exposure in a susceptible population during pregnancy. Environ Health Perspect 112:1204–1207.

Ishido M, Masuo Y, Kunimoto M, Oka S, Morita M. 2004. Bisphenol A causes hyperactivity in the rat concomitantly with impairment of tyrosine hydroxylase immunoreactivity. J Neurosci Res 76:423–433.

Ishihara A, Nishiyama N, Sugiyama S, Yamauchi K. 2003. The effect of endocrine disrupting chemicals on thyroid hormone binding to Japanese quail transthyretin and thyroid hormone receptor. Gen Comp Endocrinol 134:36–43.

Iwamuro S, Sakakibara M, Terao M, Ozawa A, Kurobe C, Shigeura T, et al. 2003. Teratogenic and anti-metamorphic effects of bisphenol A on embryonic and larval *Xenopus laevis*. Gen Comp Endocrinol 133:189–198.

Iwasaki T, Miyazaki W, Takeshita A, Kuroda Y, Koibuchi N. 2002. Polychlorinated biphenyls suppress thyroid hormone-induced transactivation. Biochem Biophys Res Commun 299:384–388.

Janjua NR, Mortensen GK, Andersson AM, Kongshoj B, Skakkebaek NE, Wulf HC. 2007. Systemic uptake of diethyl phthalate, dibutyl phthalate, and butyl paraben following whole-body topical application and reproductive and thyroid hormone levels in humans. Environ Sci Technol 41:5564–5570.

Jensen AA, Leffers H. 2008. Emerging endocrine disrupters: Perfluoroalkylated substances. Int J Androl 31:161–169.

Johnson-Restrepo B, Kannan K. 2009. An assessment of sources and pathways of human exposure to polybrominated diphenyl ethers in the United States. Chemosphere 76:542–548.

Jones-Otazo HA, Clarke JP, Diamond ML, Archbold JA, Ferguson G, Harner T, et al. 2005. Is house dust the missing exposure pathway for PBDEs? An analysis of the urban fate and human exposure to PBDEs. Environ Sci Technol 39:5121–5130.

Kim TH, Bang du Y, Lim HJ, Won AJ, Ahn MY, Patra N, et al. 2012. Comparisons of polybrominated diphenyl ethers levels in paired South Korean cord blood, maternal blood, and breast milk samples. Chemosphere 87:97–104.

Klaassen CD, Hood AM. 2001. Effects of microsomal enzyme inducers on thyroid follicular cell proliferation and thyroid hormone metabolism. Toxicol Pathol 29: 34–40.

Kodavanti PRS, Tilson HA. 1997. Structure-activity relationships of potentially neurotoxic PCB congeners in the rat. Neurotoxicology 18:425–442.

Kodavanti PR, Coburn CG, Moser VC, MacPhail RC, Fenton SE, Stoker TE, et al. 2010. Developmental exposure to a commercial PBDE mixture, DE-71: Neurobehavioral, hormonal, and reproductive effects. Toxicol Sci 116:297–312.

Koibuchi N. 2009. Animal models to study thyroid hormone action in cerebellum. Cerebellum 8:89–97.

Kopp PA. 2008. Reduce, recycle, reuse—Iodotyrosine deiodinase in thyroid iodide metabolism. N Engl J Med 358:1856–1859.

Lans MC, Klasson-Wehler E, Willemsen M, Meussen E, Safe S, Brouwer A. 1993. Structure-dependent, competitive interaction of hydroxy-polychlorobiphenyls, -dibenzo-p-dioxins and -dibenzofurans with human transthyretin. Chem Biol Interact 88:7–21.

Larsen PR, Silva JE, Kaplan MM. 1981. Relationships between circulating and intracellular thyroid hormones: Physiological and clinical implications. Endocr Rev 2: 87–102.

Latini G. 2005. Monitoring phthalate exposure in humans. Clin Chim Acta 361: 20–29.

Lau C, Thibodeaux JR, Hanson RG, Rogers JM, Grey BE, Stanton ME, et al. 2003. Exposure to perfluorooctane sulfonate during pregnancy in rat and mouse. II: Postnatal evaluation. Toxicol Sci 74:382–392.

Lau C, Anitole K, Hodes C, Lai D, Pfahles-Hutchens A, Seed J. 2007. Perfluoroalkyl acids: A review of monitoring and toxicological findings. Toxicol Sci 99:366–394.

Lechan RM, Qi Y, Jackson IMD, Mahdavi V. 1994. Identification of thyroid hormone receptor isoforms in thyrotropin-releasing hormone neurons of the hypothalamic paraventricular nucleus. Endocrinology 135:92–100.

Lee E, Kim HJ, Im JY, Kim J, Park H, Ryu JY, et al. 2008. Hypothyroidism protects di(n-butyl) phthalate-induced reproductive organs damage in Sprague-Dawley male rats. J Toxicol Sci 33:299–306.

Lee E, Kim TH, Choi JS, Nabanata P, Kim NY, Ahn MY, et al. 2010. Evaluation of liver and thyroid toxicity in Sprague-Dawley rats after exposure to polybrominated diphenyl ether BDE-209. J Toxicol Sci 35:535–545.

Legrand C, Clos J, Legrand J. 1982. Influence of altered thyroid and nutritional states on early histogenesis of the rat cerebellar cortex with special reference to synaptogenesis. Reprod Nutr Dev 22:201–208.

Lin SM, Chen FA, Huang YF, Hsing LL, Chen LL, Wu LS, et al. 2010. Negative associations between PBDE levels and thyroid hormones in cord blood. Int J Hyg Environ Health 214:115–120.

Luebker DJ, York RG, Hansen KJ, Moore JA, Butenhoff JL. 2005. Neonatal mortality from in utero exposure to perfluorooctanesulfonate (PFOS) in Sprague-Dawley rats: Dose-response, and biochemical and pharamacokinetic parameters. Toxicology 215:149–169.

Madeira MD, Sousa N, Lima-Andrade MT, Calheiros F, Cadete-Leite A, Paula-Barbosa MM. 1992. Selective vulnerability of the hippocampal pyramidal neurons to hypothyroidism in male and female rats. J Comp Neurol 322:501–518.

Martin MT, Brennan RJ, Hu W, Ayanoglu E, Lau C, Ren H. 2007. Toxicogenomic study of triazole fungicides and perfluoroalkyl acids in rat livers predicts toxicity and categorizes chemicals based on mechanisms of toxicity. Toxicol Sci 97:595–613.

Martinez-Galan JR, Pedraza P, Santacana M, Escobar del Ray F, Morreale de Escobar G, Ruiz-Marcos A. 1997. Early effects of iodine deficiency on radial glial cells of the hippocampus of the rat fetus. A model of neurological cretinism. J Clin Invest 99:2701–2709.

Meeker JD, Ferguson KK. 2011. Relationship between urinary phthalate and bisphenol A concentrations and serum thyroid measures in U.S. adults and adolescents from the National Health and Nutrition Examination Survey (NHANES) 2007–2008. Environ Health Perspect 119:1396–1402.

Meeker JD, Calafat AM, Hauser R. 2007. Di(2-ethylhexyl) phthalate metabolites may alter thyroid hormone levels in men. Environ Health Perspect 115:1029–1034.

Meerts IA, van Zanden JJ, Luijks EA, van Leeuwen-Bol I, Marsh G, Jakobsson E, et al. 2000. Potent competitive interactions of some brominated flame retardants and related compounds with human transthyretin in vitro. Toxicol Sci 56:95–104.

Meerts IA, Assink Y, Cenijn PH, Van Den Berg JH, Weijers BM, Bergman A, et al. 2002. Placental transfer of a hydroxylated polychlorinated biphenyl and effects on fetal and maternal thyroid hormone homeostasis in the rat. Toxicol Sci 68:361–371.

Melse-Boonstra A, Jaiswal N. 2010. Iodine deficiency in pregnancy, infancy and childhood and its consequences for brain development. Best Pract Res Clin Endocrinol Metab 24:29–38.

Miller VM, Sanchez-Morrissey S, Brosch KO, Seegal RF. 2012. Developmental co-exposure to polychlorinated biphenyls and polybrominated diphenyl ethers has additive effects on circulating thyroxine levels in rats. Toxicol Sci 127:76–83.

Miyazaki W, Iwasaki T, Takeshita A, Kuroda Y, Koibuchi N. 2004. Polychlorinated biphenyls suppress thyroid hormone receptor-mediated transcription through a novel mechanism. J Biol Chem 279:18195–18202.

Moriyama K, Tagami T, Akamizu T, Usui T, Saijo M, Kanamoto N, et al. 2002. Thyroid hormone action is disrupted by bisphenol A as an antagonist. J Clin Endocrinol Metab 87:5185–5190.

Myers JP, vom Saal FS, Akingbemi BT, Arizono K, Belcher S, Colborn T, et al. 2009a. Why public health agencies cannot depend on good laboratory practices as a criterion for selecting data: The case of bisphenol A. Environ Health Perspect 117: 309–315.

Myers JP, Zoeller RT, vom Saal FS. 2009b. A clash of old and new scientific concepts in toxicity, with important implications for public health. Environ Health Perspect 117:1652–1655.

Nicholson JL, Altman J. 1972a. The effects of early hypo- and hyperthyroidism on the development of rat cerebellar cortex. I. Cell proliferation and differentiation. Brain Res 44:13–23.

Nicholson JL, Altman J. 1972b. The effects of early hypo- and hyperthyroidism on the development of the rat cerebellar cortex. II. Synaptogenesis in the molecular layer. Brain Res 44:25–36.

Nicholson JL, Altman J. 1972c. Synaptogenesis in the rat cerebellum: Effects of early hypo- and hyperthyroidism. Science 176:530–532.

Noren K, Meironyte D. 2000. Certain organochlorine and organobromine contaminants in Swedish human milk in perspective of past 20–30 years. Chemosphere 40: 1111–1123.

O'Connor JC, Frame SR, Ladics GS. 2002. Evaluation of a 15-day screening assay using intact male rats for identifying antiandrogens. Toxicol Sci 69:92–108.

Olsen GW, Zobel LR. 2007. Assessment of lipid, hepatic, and thyroid parameters with serum perfluorooctanoate (PFOA) concentrations in fluorochemical production workers. Int Arch Occup Environ Health 81:231–246.

Olsen GW, Burris JM, Burlew MM, Mandel JH. 2003. Epidemiologic assessment of worker serum perfluorooctanesulfonate (PFOS) and perfluorooctanoate (PFOA) concentrations and medical surveillance examinations. J Occup Environ Med 45:260–270.

Osius N, Karmaus W, Kruse H, Witten J. 1999. Exposure to polychlorinated biphenyls and levels of thyroid hormones in children. Environ Health Perspect 107:843–849.

Panicker V, Wilson SG, Spector TD, Brown SJ, Falchi M, Richards JB, et al. 2008. Heritability of serum TSH, free T4 and free T3 concentrations: A study of a large UK twin cohort. Clin Endocrinol (Oxf) 68:652–659.

Park HY, Park JS, Sovcikova E, Kocan A, Linderholm L, Bergman A, et al. 2009. Exposure to hydroxylated polychlorinated biphenyls (OH-PCBs) in the prenatal period and subsequent neurodevelopment in eastern Slovakia. Environ Health Perspect 117:1600–1606.

Patandin S, Lanting CI, Mulder PG, Boersma ER, Sauer PJ, Weisglas-Kuperus N. 1999. Effects of environmental exposure to polychlorinated biphenyls and dioxins on cognitive abilities in Dutch children at 42 months of age. J Pediatr 134:33–41.

Pathak A, Sinha RA, Mohan V, Mitra K, Godbole MM. 2011. Maternal thyroid hormone before the onset of fetal thyroid function regulates Reelin and downstream signaling cascade affecting neocortical neuronal migration. Cereb Cortex 21:11–21.

Paul KB, Hedge JM, Devito MJ, Crofton KM. 2010a. Developmental triclosan exposure decreases maternal and neonatal thyroxine in rats. Environ Toxicol Chem 29: 2840–2844.

Paul KB, Hedge JM, DeVito MJ, Crofton KM. 2010b. Short-term exposure to triclosan decreases thyroxine in vivo via upregulation of hepatic catabolism in young Long-Evans rats. Toxicol Sci 113:367–379.

Pelletier G, Masson S, Wade MJ, Nakai J, Alwis R, Mohottalage S, et al. 2009. Contribution of methylmercury, polychlorinated biphenyls and organochlorine pesticides to the toxicity of a contaminant mixture based on Canadian Arctic population blood profiles. Toxicol Lett 184:176–185.

Persky V, Turyk M, Anderson HA, Hanrahan LP, Falk C, Steenport DN, et al. 2001. The effects of PCB exposure and fish consumption on endogenous hormones. Environ Health Perspect 109:1275–1283.

Poon R, Lecavalier P, Mueller R, Valli VE, Procter BG, Chu I. 1997. Subchronic oral toxicity of di-n-octyl phthalate and di(2-ethylhexyl) phthalate in the rat. Food Chem Toxicol 35:225–239.

Potter BJ, Mano MT, Belling GB, McIntosh GH, Hua C, Cragg BG, et al. 1982. Retarded fetal brain development resulting from severe dietary iodine deficiency in sheep. Neuropathol Appl Neurobiol 8:303–313.

Prevedouros K, Cousins IT, Buck RC, Korzeniowski SH. 2006. Sources, fate and transport of perfluorocarboxylates. Environ Sci Technol 40:32–44.

Price SC, Chescoe D, Grasso P, Wright M, Hinton RH. 1988. Alterations in the thyroids of rats treated for long periods with di-(2-ethylhexyl) phthalate or with hypolipidaemic agents. Toxicol Lett 40:37–46.

Rami A, Rabie A. 1990. Delayed synaptogenesis in the dentate gyrus of the thyroid-deficient developing rat. Dev Neurosci 12:398–405.

Raymer JH, Michael LC, Studabaker WB, Olsen GW, Sloan CS, Wilcosky T, et al. 2011. Concentrations of perfluorooctane sulfonate (PFOS) and perfluorooctanoate (PFOA) and their associations with human semen quality measurements. Reprod Toxicol 33:419–427.

Rodriguez PE, Sanchez MS. 2010. Maternal exposure to triclosan impairs thyroid homeostasis and female pubertal development in Wistar rat offspring. J Toxicol Environ Health A 73:1678–1688.

Roelens SA, Beck V, Aerts G, Clerens S, Vanden Bergh G, Arckens L, et al. 2005. Neurotoxicity of polychlorinated biphenyls (PCBs) by disturbance of thyroid hormone-regulated genes. Ann N Y Acad Sci 1040:454–456.

Royland JE, Parker JS, Gilbert ME. 2008. A genomic analysis of subclinical hypothyroidism in hippocampus and neocortex of the developing rat brain. J Neuroendocrinol 20:1319–1338.

Ruzzin J, Petersen R, Meugnier E, Madsen L, Lock EJ, Lillefosse H, et al. 2010. Persistent organic pollutant exposure leads to insulin resistance syndrome. Environ Health Perspect 118:465–471.

vom Saal FS, Hughes C. 2005. An extensive new literature concerning low-dose effects of bisphenol A shows the need for a new risk assessment. Environ Health Perspect 113:926–933.

vom Saal FS, Akingbemi BT, Belcher SM, Birnbaum LS, Crain DA, Eriksen M, et al. 2007. Chapel Hill bisphenol A expert panel consensus statement: Integration of mechanisms, effects in animals and potential to impact human health at current levels of exposure. Reprod Toxicol 24:131–138.

vom Saal FS, Akingbemi BT, Belcher SM, Crain DA, Crews D, Guidice LC, et al. 2010. Flawed experimental design reveals the need for guidelines requiring appropriate positive controls in endocrine disruption research. Toxicol Sci 115:612–613; author reply 614–620.

Sala M, Sunyer J, Herrero C, To-Figueras J, Grimalt J. 2001. Association between serum concentrations of hexachlorobenzene and polychlorobiphenyls with thyroid hormone and liver enzymes in a sample of the general population. Occup Environ Med 58:172–177.

Salay E, Garabrant D. 2009. Polychlorinated biphenyls and thyroid hormones in adults: A systematic review appraisal of epidemiological studies. Chemosphere 74: 1413–1419.

Sandborgh-Englund G, Adolfsson-Erici M, Odham G, Ekstrand J. 2006. Pharmacokinetics of triclosan following oral ingestion in humans. J Toxicol Environ Health A 69:1861–1873.

Sandler B, Webb P, Apriletti JW, Huber BR, Togashi M, Cunha Lima ST, et al. 2004. Thyroxine-thyroid hormone receptor interactions. J Biol Chem 279:55801–55808.

Schantz SL, Widholm JJ, Rice DC. 2003. Effects of PCB exposure on neuropsychological function in children. Environ Health Perspect 111:357–576.

Schecter A, Pavuk M, Papke O, Ryan JJ, Birnbaum L, Rosen R. 2003. Polybrominated diphenyl ethers (PBDEs) in U.S. mothers' milk. Environ Health Perspect 111: 1723–1729.

Schecter A, Papke O, Tung KC, Joseph J, Harris TR, Dahlgren J. 2005. Polybrominated diphenyl ether flame retardants in the U.S. population: Current levels, temporal trends, and comparison with dioxins, dibenzofurans, and polychlorinated biphenyls. J Occup Environ Med 47:199–211.

Schecter A, Colacino J, Sjodin A, Needham L, Birnbaum L. 2010. Partitioning of polybrominated diphenyl ethers (PBDEs) in serum and milk from the same mothers. Chemosphere 78:1279–1284.

Schell LM, Gallo MV, Denham M, Ravenscroft J, DeCaprio AP, Carpenter DO. 2008. Relationship of thyroid hormone levels to levels of polychlorinated biphenyls, lead, p,p′- DDE, and other toxicants in Akwesasne Mohawk youth. Environ Health Perspect 116:806–813.

Schettler T. 2006. Human exposure to phthalates via consumer products. Int J Androl 29:134–139; discussion 181–135.

Schuur AG, Legger FF, van Meeteren ME, Moonen MJ, van Leeuwen-Bol I, Bergman A, et al. 1998. In vitro inhibition of thyroid hormone sulfation by hydroxylated metabolites of halogenated aromatic hydrocarbons. Chem Res Toxicol 11: 1075–1081.

Sellin JH, Vassilopoulou-Sellin R. 2000. The gastrointestinal tract and liver in thyrotoxicosis. In: Werner and Ingbar's the Thyroid: A Fundamental and Clinical Text, 8th edition (Braverman LE, Utiger RD, eds.). Philadelphia: Lippincott Williams & Wilkins, pp. 622–626.

Seress L, Abraham H, Tornoczky T, Kosztolanyi G. 2001. Cell formation in the human hippocampal formation from mid-gestation to the late postnatal period. Neuroscience 105:831–843.

Sharlin DS, Bansal R, Zoeller RT. 2006. Polychlorinated biphenyls exert selective effects on cellular composition of white matter in a manner inconsistent with thyroid hormone insufficiency. Endocrinology 147:846–858.

Sharlin DS, Tighe D, Gilbert ME, Zoeller RT. 2008. The balance between oligodendrocyte and astrocyte production in major white matter tracts is linearly related to serum total thyroxine. Endocrinology 149:2527–2536.

She J, Holden A, Sharp M, Tanner M, Williams-Derry C, Hooper K. 2007. Polybrominated diphenyl ethers (PBDEs) and polychlorinated biphenyls (PCBs) in breast milk from the Pacific Northwest. Chemosphere 67:S307–S317.

Shen O, Du G, Sun H, Wu W, Jiang Y, Song L, et al. 2009. Comparison of in vitro hormone activities of selected phthalates using reporter gene assays. Toxicol Lett 191:9–14.

Shen O, Wu W, Du G, Liu R, Yu L, Sun H, et al. 2011. Thyroid disruption by di-n-butyl phthalate (DBP) and mono-n-butyl phthalate (MBP) in *Xenopus laevis*. PLoS ONE 6:e19159.

Sheng ZG, Tang Y, Liu YX, Yuan Y, Zhao BQ, Chao XJ, et al. 2012. Low concentrations of bisphenol a suppress thyroid hormone receptor transcription through a nongenomic mechanism. Toxicol Appl Pharmacol 259:133–142.

Shimada N, Yamauchi K. 2004. Characteristics of 3,5,3′-triiodothyronine (T3)-uptake system of tadpole red blood cells: Effect of endocrine-disrupting chemicals on cellular T3 response. J Endocrinol 183:627–637.

Shoeib M, Harner T, Wilford BH, Jones KC, Zhu J. 2005. Perfluorinated sulfonamides in indoor and outdoor air and indoor dust: Occurrence, partitioning, and human exposure. Environ Sci Technol 39:6599–6606.

Siddiqi MA, Laessig RH, Reed KD. 2003. Polybrominated diphenyl ethers (PBDEs): New pollutants-old diseases. Clin Med Res 1:281–290.

Siesser WB, Cheng SY, McDonald MP. 2005. Hyperactivity, impaired learning on a vigilance task, and a differential response to methylphenidate in the TRbetaPV knock-in mouse. Psychopharmacology (Berl) 181:653–663.

Staples CA, Dome PB, Klecka GM, Oblock ST, Harris LR. 1998. A review of the environmental fate, effects, and exposures of bisphenol A. Chemosphere 36:2149–2173.

Stapleton HM, Eagle S, Anthopolos R, Wolkin A, Miranda ML. 2011. Associations between polybrominated diphenyl ether (PBDE) flame retardants, phenolic metabolites, and thyroid hormones during pregnancy. Environ Health Perspect 119: 1454–1459.

Stewart P, Reihman J, Lonky E, Darvill T, Pagano J. 2000. Prenatal PCB exposure and neonatal behavioral assessment scale (NBAS) performance. Neurotoxicol Teratol 22:21–29.

Stoker TE, Laws SC, Crofton KM, Hedge JM, Ferrell JM, Cooper RL. 2004. Assessment of DE-71, a commercial polybrominated diphenyl ether (PBDE) mixture, in the EDSP male and female pubertal protocols. Toxicol Sci 78:144–155.

Stoker TE, Gibson EK, Zorrilla LM. 2010. Triclosan exposure modulates estrogen-dependent responses in the female Wistar rat. Toxicol Sci 117:45–53.

Sugiyama S, Shimada N, Miyoshi H, Yamauchi K. 2005. Detection of thyroid system-disrupting chemicals using in vitro and in vivo screening assays in *Xenopus laevis*. Toxicol Sci 88:367–374.

Sun H, Shen OX, Wang XR, Zhou L, Zhen SQ, Chen XD. 2009. Anti-thyroid hormone activity of bisphenol A, tetrabromobisphenol A and tetrachlorobisphenol A in an improved reporter gene assay. Toxicol in Vitro 23:950–954.

Suvorov A, Bissonnette C, Takser L, Langlois MF. 2011. Does 2,2′,4,4′-tetrabromodiphenyl ether interact directly with thyroid receptor? J Appl Toxicol 31:179–184.

Tata JR. 2013. The road to nuclear receptors of thyroid hormone. Biochim Biophys Acta 1830:3860–3866.

TCS TaCSECB. 2008. Bis (2-Ethylhexyl) Phthalate (DEHP). Ispra, VA, Italy: Institute for Health and Consumer Protection.

Tee PG, Sweeney AM, Symanski E, Gardiner JC, Gasior DM, Schantz SL. 2003. A longitudinal examination of factors related to changes in serum polychlorinated biphenyl levels. Environ Health Perspect 111:702–707.

Thibodeaux JR, Hanson RG, Rogers JM, Grey BE, Barbee BD, Richards JH, et al. 2003. Exposure to perfluorooctane sulfonate during pregnancy in rat and mouse. I: Maternal and prenatal evaluations. Toxicol Sci 74:369–381.

Tilson HA, Kodavanti PRS. 1997. Neurochemical effects of polychlorinated biphenyls: An overview and identification of research needs. Neurotoxicology 13:727–744.

Tong Z, Li H, Goljer I, McConnell O, Chandrasekaran A. 2007. In vitro glucuronidation of thyroxine and triiodothyronine by liver microsomes and recombinant human UDP-glucuronosyltransferases. Drug Metab Dispos 35:2203–2210.

Turyk M, Anderson HA, Hanrahan LP, Falk C, Steenport DN, Needham LL, et al. 2006. Relationship of serum levels of individual PCB, dioxin, and furan congeners and DDE with Great Lakes sport-caught fish consumption. Environ Res 100:173–183.

Turyk ME, Anderson HA, Persky VW. 2007. Relationships of thyroid hormones with polychlorinated biphenyls, dioxins, furans, and DDE in adults. Environ Health Perspect 115:1197–1203.

Turyk ME, Persky VW, Imm P, Knobeloch L, Chatterton R, Anderson HA. 2008. Hormone disruption by PBDEs in adult male sport fish consumers. Environ Health Perspect 116:1635–1641.

Valiente M, Marin O. 2010. Neuronal migration mechanisms in development and disease. Curr Opin Neurobiol 20:68–78.

Vandenberg LN, Hauser R, Marcus M, Olea N, Welshons WV. 2007. Human exposure to bisphenol A (BPA). Reprod Toxicol 24:139–177.

Veldhoen N, Skirrow RC, Osachoff H, Wigmore H, Clapson DJ, Gunderson MP, et al. 2006. The bactericidal agent triclosan modulates thyroid hormone-associated gene expression and disrupts postembryonic anuran development. Aquat Toxicol 80: 217–227.

Verrotti A, Spalice A, Ursitti F, Papetti L, Mariani R, Castronovo A, et al. 2010. New trends in neuronal migration disorders. Eur J Paediatr Neurol 14:1–12.

Vincent J, Legrand C, Rabie A, Legrand J. 1982. Effects of thyroid hormone on synaptogenesis in the molecular layer of the developing rat cerebellum. J Physiol (Paris) 78:729–738.

Vongphachan V, Cassone CG, Wu DM, Chiu SZ, Crump D, Kennedy SW. 2011. Effects of perfluoroalkyl compounds on mRNA expression levels of thyroid hormone-responsive genes in primary cultures of avian neuronal cells. Toxicol Sci 120: 392–402.

Weiss JM, Andersson PL, Lamoree MH, Leonards PE, van Leeuwen SP, Hamers T. 2009. Competitive binding of poly- and perfluorinated compounds to the thyroid hormone transport protein transthyretin. Toxicol Sci 109:206–216.

Welshons WV, Nagel SC, vom Saal FS. 2006. Large effects from small exposures. III. Endocrine mechanisms mediating effects of bisphenol A at levels of human exposure. Endocrinology 147:S56–S69.

Wenzel A, Franz C, Breous E, Loos U. 2005. Modulation of iodide uptake by dialkyl phthalate plasticisers in FRTL-5 rat thyroid follicular cells. Mol Cell Endocrinol 244:63–71.

Wikoff D, Fitzgerald L, Brinbaum L. 2012. Persistent organic pollutants: An overview. In: Dioxins and Health: Including Other Persistent Organic Pollutants and Endocrine Disruptors, 3rd edition (Schecter A, ed.). Hoboken, NJ: John Wiley & Sons, Inc, pp. 1–36.

Wolff J. 1998. Perchlorate and the thyroid gland. Pharmacol Rev 50:89–105.

Wu N, Herrmann T, Paepke O, Tickner J, Hale R, Harvey LE, et al. 2007. Human exposure to PBDEs: Associations of PBDE body burdens with food consumption and house dust concentrations. Environ Sci Technol 41:1584–1589.

Yamada H, Furuta I, Kato EH, Kataoka S, Usuki Y, Kobashi G, et al. 2002. Maternal serum and amniotic fluid bisphenol A concentrations in the early second trimester. Reprod Toxicol 16:735–739.

Yamanaka H, Nakajima M, Katoh M, Yokoi T. 2007. Glucuronidation of thyroxine in human liver, jejunum, and kidney microsomes. Drug Metab Dispos 35:1642–1648.

Yasuo S, Nakao N, Ohkura S, Iigo M, Hagiwara S, Goto A, et al. 2006. Long-day suppressed expression of type 2 deiodinase gene in the mediobasal hypothalamus of the Saanen goat, a short-day breeder: Implication for seasonal window of thyroid hormone action on reproductive neuroendocrine axis. Endocrinology 147:432–440.

Yen PM, Feng X, Flamant F, Chen Y, Walker RL, Weiss RE, et al. 2003. Effects of ligand and thyroid hormone receptor isoforms on hepatic gene expression profiles of thyroid hormone receptor knockout mice. EMBO Rep 4:581–587.

You SH, Gauger KJ, Bansal R, Zoeller RT. 2006. 4-Hydroxy-PCB106 acts as a direct thyroid hormone receptor agonist in rat GH3 cells. Mol Cell Endocrinol 257–258: 26–34.

Yu WG, Liu W, Jin YH. 2009a. Effects of perfluorooctane sulfonate on rat thyroid hormone biosynthesis and metabolism. Environ Toxicol Chem 28:990–996.

Yu WG, Liu W, Jin YH, Liu XH, Wang FQ, Liu L, et al. 2009b. Prenatal and postnatal impact of perfluorooctane sulfonate (PFOS) on rat development: A cross-foster study on chemical burden and thyroid hormone system. Environ Sci Technol 43:8416–8422.

Yu L, Deng J, Shi X, Liu C, Yu K, Zhou B. 2010. Exposure to DE-71 alters thyroid hormone levels and gene transcription in the hypothalamic-pituitary-thyroid axis of zebra fish larvae. Aquat Toxicol 97:226–233.

Yu L, Lam JC, Guo Y, Wu RS, Lam PK, Zhou B. 2011a. Parental transfer of polybrominated diphenyl ethers (PBDEs) and thyroid endocrine disruption in zebra fish. Environ Sci Technol 45:10652–10659.

Yu WG, Liu W, Liu L, Jin YH. 2011b. Perfluorooctane sulfonate increased hepatic expression of OAPT2 and MRP2 in rats. Arch Toxicol 85:613–621.

Zhou T, Ross DG, DeVito MJ, Crofton KM. 2001. Effects of short-term in vivo exposure to polybrominated diphenyl ethers on thyroid hormones and hepatic enzyme activities in weanling rats. Toxicol Sci 61:76–82.

Zhou T, Taylor MM, DeVito MJ, Crofton KM. 2002. Developmental exposure to brominated diphenyl ethers results in thyroid hormone disruption. Toxicol Sci 66: 105–116.

Zoeller RT. 2001. Polychlorinated biphenyls as disruptors of thyroid hormone action. In: PCBs: Recent Advances in the Environmental Toxicology and Health Effects of

PCBs (Fisher LJ, Hansen L, eds.). Lexington, KY: University of Kentucky Press, pp. 265–272.

Zoeller RT. 2007. Environmental chemicals impacting the thyroid: Targets and consequences. Thyroid 17:811–817.

Zoeller RT. 2011. Endocrine disruption of the thyroid and its consequences in development. In: Multi-system Endocrine Disruption (Bourguignon J-P, Jegou B, Kerdelhue B, Toppari J, and Christen Y, eds.). Berlin, Heidelberg: Springer, pp. 51–71.

Zoeller RT, Crofton KM. 2000. Thyroid hormone action in fetal brain development and potential for disruption by environmental chemicals. Neurotoxicology 21:935–945.

Zoeller RT, Rovet J. 2004. Timing of thyroid hormone action in the developing brain: Clinical observations and experimental findings. J Neuroendocrinol 16:809–818.

Zoeller RT, Dowling AL, Vas AA. 2000. Developmental exposure to polychlorinated biphenyls exerts thyroid hormone-like effects on the expression of RC3/neurogranin and myelin basic protein messenger ribonucleic acids in the developing rat brain. Endocrinology 141:181–189.

Zoeller RT, Bansal R, Parris C. 2005. Bisphenol-A, an environmental contaminant that acts as a thyroid hormone receptor antagonist in vitro, increases serum thyroxine, and alters RC3/neurogranin expression in the developing rat brain. Endocrinology 146:607–612.

Zoeller RT, Tan SW, Tyl RW. 2007. General background on the hypothalamic-pituitary-thyroid (HPT) axis. Crit Rev Toxicol 37:11–53.

Zorrilla LM, Gibson EK, Jeffay SC, Crofton KM, Setzer WR, Cooper RL, et al. 2009. The effects of triclosan on puberty and thyroid hormones in male Wistar rats. Toxicol Sci 107:56–64.

Zota AR, Park JS, Wang Y, Petreas M, Zoeller RT, Woodruff TJ. 2011. Polybrominated diphenyl ethers, hydroxylated polybrominated diphenyl ethers, and measures of thyroid function in second trimester pregnant women in california. Environ Sci Technol 45:7896–7905.

CHAPTER 10

An Overview of the Effects of Organic Compounds on Women's Reproductive Health and Birth Outcomes

SUSAN R. REUTMAN and JULIANA W. MEADOWS

ABSTRACT

Background: Female reproductive system perturbations may decrease a woman's likelihood of conceiving and carrying her baby to term, influence the future life course of a viable infant, and may also reflect and have an impact on her own gynecological and general health and well-being. This chapter surveys effects of persistent organic compounds (POCs) and other less persistent but often pervasive bioactive organic compounds (BOCs) on these health outcomes.

Objectives: To present an overview that highlights evidence gathered from studies and reviews, and to provide readers with background to interpret results of future studies.

Results: This chapter provides an overview of human BOC and POC research methods, issues, and findings pertaining to women's reproductive and offspring's developmental effects of exposure. Research gaps and additional information resources are also highlighted.

Discussion: Human evidence is mounting for adverse intrauterine exposure effects of (1) polychlorinated biphenyls (particularly estrogenic congeners) on lowered birth weight; (2) phthalates on testicular dysgenesis syndrome; (3) various solvents on fetal loss; (4) smoking on altered reproductive hormone

(*Continued*)

Effects of Persistent and Bioactive Organic Pollutants on Human Health, First Edition.
Edited by David O. Carpenter.

levels, lowered fertility, fecundity, birth weights, and menopausal age, and increased stillbirths, cleft lip malformations, and maternal (diastolic) hypertension; and (5) pesticides on childhood leukemia. There are strikingly few studies of other BOCs and POCs and reproductive outcomes, considering how many of these compounds are prevalent and that their potential effects and comorbidities may be severe with impacts across the life span.

Conclusions: Evidence of adverse female reproductive and developmental effects is mounting for some BOCs and POCs but is sparse for many others. Well-designed longitudinal human studies are needed.

INTRODUCTION

Perturbations in the female reproductive system may decrease a woman's likelihood of conceiving and carrying her baby to term, influence the future life course of a viable infant, and impact her own gynecological and general health and well-being.[1] The human research evidence base covering women's and offspring's potential reproductive and developmental harms from exposure to persistent organic (i.e., carbon containing) compounds (POCs) (see the abbreviation list in Table 10.1) and other less persistent but often pervasive bioactive organic compounds (BOCs) is broad but not consistently deep for all the reproductive health outcomes of interest. One objective of this chapter is to provide an overview with a focus on the human evidence of potential effects of BOCs and POCs on the well-being of women and offspring. We limit the emphasis to nonorganometallic POCs and BOCs for which persuasive, suggestive, or conflicting evidence of (primarily) human reproductive or developmental effects was found in research literature, that is, PubMed searches to identify and select key studies and recent review papers (cited herein). Because the evidence base is dynamic, a second objective is to present some general context and caveats for conducting and interpreting studies on the effects of BOC and POC exposures on women and offspring. Current research pertaining to many aspects of women's health is considered, including gynecological dysfunction (i.e., menstrual and reproductive endocrine abnormalities, early/premature menopause, endometriosis, uterine fibroids, and polycystic ovarian syndrome [PCOS]), reproductive function (fertility and fecundity), pregnancy (hypertension, miscarriage, and stillbirth), and neonatal outcomes (i.e., prematurity, fetal growth, sex ratio, immune function, thyroid function, and birth defects). Cancer-related outcomes are discussed only incidentally where fetal origins are suspected since cancer is addressed in other chapters. Because the neuroendocrine axis regulates the healthy as well as the dysfunctional female

[1]The findings and conclusions in this report are those of the authors and do not necessarily represent the views of the National Institute for Occupational Safety and Health.

TABLE 10.1. Abbreviation List

Exposure Abbreviations	
ATZ	Atrazine or 2-chloro-4-(ethylamino)-6-(isopropylamino)-s-triazine
BPA	Bisphenol A
BPB	Bisphenol B
BOC	Bioactive organic compound
BFR	Brominated flame retardant
Co-PCBs	Coplanar polychlorinated biphenyls
DDE	Dichlorodiphenyldichloroethylene
DDT	Dichlorodiphenyltrichloroethane
DBP	Dibutyl phthalate
DEHP	Bis(2-ethylhexyl) phthalate or di(2-ethylhexyl) phthalate
DES	Diethylstilbestrol
EDC	Endocrine-disrupting compound
HAC	Hormonally active compound
HCB	Hexachlorobenzene
α-HCH	Alpha-hexachlorocyclohexane
β-HCH	Beta-hexachlorocyclohexane
MBP	Monobutyl phthalate
MBzP	Monobenzyl phthalate
MEP	Monoethyl phthalate
MEHP	Mono(2-ethylhexyl) phthalate
OC	Organochlorine
OCP	Organochlorine pesticide
OH-PCB	Hydroxylated PCB metabolite
OP	Organophosphate
PBB	Polybrominated biphenyl
PBDE	Polybrominated diphenyl ether
PCB	Polychlorinated biphenyl
PCDD	Polychlorinated dibenzo-*p*-dioxin
PCDF	Polychlorinated dibenzofuran
PCE	Perchloroethylene, tetrachloroethylene
PFC	Perfluorochemical
PCP	Pentachlorophenol
PFOS	Perfluorooctane sulfonate
PFOA	Perfluorooctanoic acid
POC	Persistent organic compound
POP	Persistent organic pollutant
PVC	Polyvinyl chloride
TCS	Triclosan (2,4,4′-trichloro-2′-hydroxydiphenyl ether)
TCDD	2,3,7,8-tetrachlorodibenzo-*p*-dioxin
TCE	Trichloroethylene
THM	Trihalomethane

(*Continued*)

TABLE 10.1. (*Continued*)

Other Abbreviations	
ADJ	Adjusted for covariates and confounders
AGD	Anogenital distance
AhR	Aryl hydrocarbon receptor
BMI	Body mass index
CM	Congenital malformation
FR	Fecundability ratio
HR, HR_{ADJ}	Hazard ratio, adjusted hazard ratio
IUGR	Intrauterine growth retardation
MA	Menopausal age
(*N*=), (*n*=)	Full study sample size, subgroup sample size
NYSA	New York State Angler cohort
OR, OR_{ADJ}	Odds ratio, adjusted odds ratio
PCOS	Polycystic ovarian syndrome
PP	Precocious puberty
PR	Prevalence ratio
RR, RR_{ADJ}	Relative risk (or risk ratio), adjusted relative risk
SA	Spontaneous abortion
SGA	Small for gestational age
SSIG	Statistically significant
SSR	Secondary sex ratio
T4	Thyroxine
TD	Testicular dysgenesis
TTP	Time to pregnancy
Wt	Weight

reproductive system, special attention is given to publications describing hormonally active compounds (HACs) as potential endocrine-disrupting compounds (EDCs).

BACKGROUND

BOC and POCs

The number of organic chemicals in commercial use is substantial, estimated at over 22,000 in 2010 in the United States and Canada alone (Howard and Muir 2010). Potential human health risks associated with the vast majority of these compounds are currently not well characterized, and still less research has focused on women's reproductive health and offspring's developmental risks. Among the features of POCs are varying levels of resistance to degradation and propensity to accumulate in living organisms (i.e., bioaccumulate). In the absence of human data, information on exposure opportunity (e.g., production, use) and certain physical-chemical features of compounds offer clues regarding such attributes. For instance, new compounds of interest for environmental monitoring as potentially bioaccumulative and persistent have been identified from among those in commerce based on their use and inferred properties (Howard and Muir 2010). Generally, compounds were characterized as more biodegradable if they were straight-chain aliphatics, esters, acids, or had hydroxyl functional groups; bioaccumulative based on octanol–water (log K_{ow}) and octanol–air (log K_{oa}) coefficients (i.e., bioaccumulative if log $K_{ow} > 3$ or >2, if log $K_{oa} \geq 5$ and ≤ 12, or if $K_{ow} > 8$ but molecular weights were not very high); whereas more persistent compounds were highly branched, highly halogenated, or nitroaromatic, and many tended to be partially or totally ionic.

Findings from experimental animal, *in vitro*, and *in silico* models can reveal compounds and doses with potentially adverse biological activities (i.e., bioactivities) in humans. However, assumptions about differences due to species, gender, age, and reproductive status are typically required to extrapolate from animals to humans, introducing added uncertainty. Broadly, multiple bioactivities, alone or in concert, have been implicated at various sites and etiologic pathway levels for adverse female reproductive health outcomes in human and (mostly) animal tissues. Some examples include immune alterations, oxidative stress-induced inflammation, altered levels or function of enzymes and other proteins, and endocrine disruption. Genetic and epigenetic mechanisms are frequently implicated. The genomic DNA complement (i.e., throughout the genome) is the inherited template or "instructions" for the structure and function of the entire organism. Alterations in single genes (e.g., point mutations, insertions, deletions) and larger DNA units such as chromosomes (e.g., amplifications, deletions, inversions, translocations) may occur throughout life due to exogenous assaults from genotoxic exposures (e.g., adducts, cross-linkers) or endogenous ones such as inflammation and oxidative damage (which can also result from genotoxicant exposures). These genetic lesions subsequently may or may not be repaired, and the repairs themselves may contain errors or damage that may be promulgated, dormant, or lead to programmed cell death (i.e., apoptosis). Genetic damage during the first few weeks of pregnancy may result in fetal loss, while birth defects may ensue during organogenesis. Throughout the life span, net genetic differences among individuals across their genomes, whether inherited or acquired, are one source of variation in their susceptibility to inherited diseases and responses to environmental agents that exist between individuals and species. Epigenetic differences are another important source of variation. Epigenetic modifications are alterations in gene function (not nucleotide sequence) that occur via DNA methylation, histone modification, and aberrant microRNA expression (Zhang and Ho 2011). In the case of DNA methylation, epigenetic "reprogramming" of gene activation and silencing during developmental windows may produce immediate or latent overt effects. Also, sometimes transgenerational epigenetic transmission is believed to occur in response to environmental influences. The plasticity and reversibility of epigenetic effects are a topic of inquiry and interest from a prevention and treatment standpoint (Zhang and Ho 2011). Altered levels or functions of enzymes and other proteins may ensue downstream from genetic and epigenetic modifications. One such downstream effect with reproductive and developmental implications is endocrine disruption. An endocrine disruptor is defined as "a compound, either natural or synthetic, which, through environmental or inappropriate developmental exposures alters the hormonal and homeostatic systems that enable the organism to communicate with and respond to its environment" (Diamanti-Kandarakis et al. 2009).

Many BOCs and POCs that are the focus of observational reproductive health studies in women and offspring are those for which nonhuman studies have signaled that possible harm in humans may occur. A heavy reliance on nonhuman sentinels exists because the human *in vivo* evidence base for effects potentially caused by BOC and POC exposure is currently insufficient for many outcomes. One reason is that, unlike *in vitro* experiments and clinical trials where there may be a therapeutic benefit, exposure of humans to potentially harmful compounds such as BOCs and POCs for research purposes is unethical. Other explanations are pragmatic: inadequate numbers of exposed women or offspring for study recruitment and limitations of analyses using data collected for purposes not primarily related to studying female reproduction. Furthermore, each human is exposed to an individual mix of these compounds throughout the life cycle, unlike laboratory animals in which the environment is controlled and effects of a xenobiotic can be measured directly. The exception in humans in which causal links have been more clearly drawn is the case of accidental contamination or incidents when known or unknown population or workplace contamination has occurred in a local or widespread vacuum of human exposure or risk information. Such incidents may result in acute or chronic exposure. Generally, accidental and occupational exposures tend to be higher than those typically encountered by women otherwise, leaving vast data gaps regarding effects of low-dose and dose-rate exposures. Other human reproductive research barriers are often societal, including obstacles to participant recruitment and retention, risks of litigation to producers and users of potentially toxic compounds, and limited funding for reproductive research.

While restrictions and bans on persistent compounds (e.g., certain organochlorine pesticides [OCPs], hexachlorobenzenes [HCBs], polybrominated biphenyls [PBBs], polybrominated diphenyl ethers [PBDEs], polychlorinated biphenyls [PCBs], brominated flame retardants [BFRs], polychlorinated dibenzo-*p*-dioxins [PCDDs]/polychlorinated dibenzofurans [PCDFs], perfluorooctane sulfonate [PFOS]) (Howard and Muir 2010; Muir and Howard 2006) have become more pervasive in some countries, POCs have long half-lives in both humans (O'Grady-Milbrath et al. 2009) and biota in our food chain and so may linger in human bodies, even transgenerationally, long after the compounds have been retired commercially (Quinn et al. 2011). In addition, ongoing exposure to banned compounds may continue for some time from "reservoirs," that is, products with long life cycles that contain POCs (e.g., PBDEs, PFOS), and from environmental contamination (e.g., PBBs, OCPs) (Howard and Muir 2010). While some less persistent compounds remaining in commerce may be more rapidly eliminated from the body, they may still be bioavailable over long periods of time when exposure is chronic. Thus, possible public health implications ensue, especially when exposure is commonplace and the underlying attributable risk is high.

Women's and Offspring's Exposures

The impact of potentially toxic exposures is different for women than for men. Differences potentially exist in exposure histories (e.g., historically divergent occupations and smoking habits), toxicant biokinetics (i.e., absorption, transport, metabolism, storage, elimination), exposure effects (e.g., potential targets include both the woman's reproductive system and her offspring during pregnancy and lactation), and the behavioral, hormonal, and other biological factors that modulate toxicant kinetics and risks (Gochfeld 2007). Figure 10.1 illustrates, broadly, the nature and fate of BOC and POC exposures, together with selected female reproductive and developmental exposure effects that have been hypothesized in studies across the course of life. A number of persistent compounds have been detected in body tissues unique to women and their physiologically dependent offspring. This includes dichlorodiphenyldichloroethylene (DDE) and PCBs measured in ovarian follicular fluid (Meeker et al. 2009), PCDDs, PCDFs, and coplanar polychlorinated biphenyls (Co-PCBs) measured in the placenta (Chao et al. 2007; Suzuki et al. 2005), PCDDs, PCDFs, Co-PCBs, dichlorodiphenyltrichloroethane (DDT), DDE, and PBDEs measured in breast milk (Haraguchi et al. 2009; Suzuki et al. 2005), PBDEs, PCDDs, PCDFs, and PCBs, in umbilical cord (i.e., cord tissue) or cord blood (Kawashiro et al. 2008; Suzuki et al. 2005), alpha-hexachlorocyclohexane (α-HCH), and DDE in amniotic fluid (Foster et al. 2000), and PCBs in miscarried fetal tissues (Lanting et al. 1998). Other less persistent BOCs that are currently in use commercially (e.g., bisphenol A [BPA] and phthalates) have also been found in many of the same body compartments (Ikezuki et al. 2002; Maffini et al. 2006; Padmanabhan et al. 2008; Schönfelder et al. 2002; Silva et al. 2004a). Between 99% and 100% of pregnant women in a nationally representative sample of the U.S. population had detectable blood or urine levels of BOCs and POCs including PCBs, OCPs, PFCs, phenols, PBDEs, phthalates, and polycyclic aromatic hydrocarbons (PAHs), as well as other compounds (Woodruff et al. 2011). Among these, DDE was the highest of the lipophilic (i.e., chemical affinity for lipids) POCs in serum, whereas PFOS was the highest of the nonlipophilic POCs.

Many POCs are sequestered in adipose tissue because of their high nonpolarity/lipophilicity. Women have proportionately more adipose versus lean body mass than men and so are thought to generally have higher body burdens of lipophilic compounds (Gochfeld 2007). Stores of stable organic compounds that have accumulated over time in adipose tissue may later be released from storage when body fat is remobilized, such as occurs during weight loss (Lim et al. 2011), pregnancy, and lactation. Other POCs (e.g., PFOS

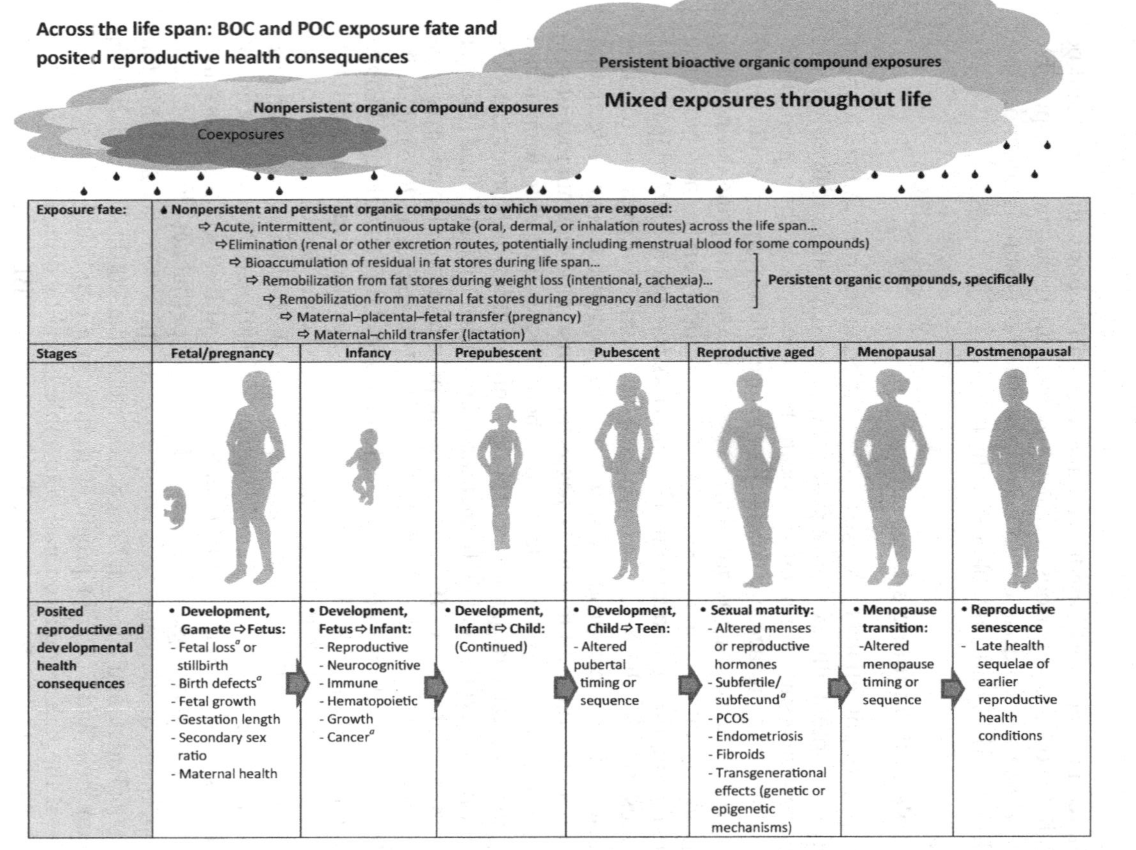

Figure 10.1. Across the life span: BOC and POC exposure fate and posited reproductive health consequences.

[a]May involve periconceptual (gametes) and postconceptual (embryo and fetus) exposure susceptibility

TABLE 10.2. Correlations between Maternal and Fetal BOCs and POCs in Selected Studies

Compounds	Correlations	References
Arochlor	$r_p = 0.78$	Anda et al. 2007
BPA	$r_s = -0.02$—$r = 0.63$	Chou et al. 2011; Kuroda et al. 2003
DDE	$r_p = 0.58$—$r_p = 0.94$	Bergonzi et al. 2009; Sala et al. 2001
DDT	$r_p = 0.76$—$r_s = 0.94$	Anda et al. 2007; Jaraczewska et al. 2006
HCB	$r_p = 0.49$—$r_p = 0.71$	Bergonzi et al. 2009; Covaci et al. 2002
HCHs	$r_p = 0.40$—$r_p = 0.71$	Anda et al. 2007; Sala et al. 2001
PBBs	$r_p = 0.81$	Jacobson et al. 1984
PBDEs	$r_p = 0.56$—$r_s = 0.91$	Kawashiro et al. 2008; Wan et al. 2010
PCBs	$r_p = 0.42$—$r_p = 0.99$	Jacobson et al. 1984; Needham et al. 2011
PCDD/Fs	$r_p = 0.44$—$r_p = 0.85$	Nakamura et al. 2008; Wang et al. 2004
PCP	$r_p = 0.73$—$r = 0.91$	Guvenius et al. 2003; Park et al. 2008b
PFCs	$r_p = 0.05$—$r_p = 0.91$	Fromme et al. 2010; Needham et al. 2011

r_p, Pearson correlation coefficient; r_s, Spearman correlation coefficient; r, r_p versus r_s unspecified.

and perfluorooctanoic acid [PFOA]) reportedly have an affinity for blood proteins and so it is plausible that their elimination may occur through additional routes specific to women during their reproductive years (i.e., elimination of albumin in menstrual blood and through the placenta to the fetus) (Harada et al. 2005; Thompson et al. 2010). Pregnant women may experience remobilization of compounds, altering bioavailability, redistribution, elimination, or transfer to the placenta and fetus, with potentially harmful fetal exposure. Maternal–fetal transfer of many POCs has been demonstrated in blood, serum, or plasma, usually near pregnancy term, by many investigators. Table 10.2 lists the correlations between maternal and fetal levels for several compounds measured in several studies. Some of these correlations are relatively high. Fetal (cord) levels of these compounds often are an appreciable fraction of levels measured in mothers' circulation and may even exceed levels in mothers. The efficient transfer and high levels of these compounds to the fetus is cause for concern since these tissues are developing and are potentially programmable at a time when fetal tissue targets may be more bioavailable due to immature metabolism, undeveloped barriers (e.g., blood–brain barrier), and undeveloped fetal storage compartments to blunt exposures (e.g., fat) (Jacobson et al. 1984). Cross-study comparisons of maternal–fetal transfer need to account for study differences in, for example, the type of biological sample analyzed (e.g., blood fraction), timing of sample collection (e.g., trimester), and statistical methods employed (e.g., non-normal distributions, imputation of low or undetectable measurements, and lipid adjustment vs. nonadjustment). Circulating lipophilic compounds are often sequestered by blood components including lipids. Accordingly, blood compound levels are often presented with lipid adjustments. Cord blood and placental tissue contain lower lipid levels, and so lipid adjustments are not as universally applied to those matrices.

Some women's health conditions that are associated with altered storage, metabolism, or elimination of compounds have hypothetically plausible effects on the net retention versus elimination of BOCs and POCs from the body. For instance, if appreciable elimination of a compound occurs through menstrual blood, then the rate of that compound's elimination would be dependent on reproductive health conditions that alter the volume of menstrual blood loss (e.g., PCOS, fibroids, menopause). Similarly, conditions sometimes associated with not lactating or reduced parity (e.g., infertility, fetal loss, PCOS, endometriosis) may increase retention of compounds. However, evidence linking parity with reduced body burden is equivocal (Quinn et al. 2011). Much remains to be learned about the variation of kinetics for compounds within and among women across various physiological and health states. Assumptions about kinetic shifts must be tempered by the observation that they are multifactorial and that the inherent variability is not well characterized by current predictive models (Ibarluzea et al. 2011). Investigators have described the potential for reverse causality (Weinberg and Wilcox 2008) such that the level of a compound may have been caused by a reproductive health outcome instead of the converse. This phenomenon must be considered in exposure–disease studies of persistent compounds when causal attribution may be blurred in full or part by underlying biokinetic differences between disease and comparison groups and when the longitudinal timeline of exposures and outcomes (i.e., temporality) is uncertain (e.g., cross-sectional studies). Reverse causation may also be an issue for pregnancy and birth outcome studies since exposure levels measured at the time of conception may differ from those measured later in gestation due to shifts in physiology (e.g., hemodynamic and body composition, fat mobilization) and behavior (e.g., diet, smoking cessation) (Chapin et al. 2004; Glynn et al. 2011; Hansen et al. 2010). Timing of exposure can affect pregnancy outcomes; timing of exposure assessment can affect interpretation of outcome causation.

Other factors, such as lactation history, body mass index (BMI), and age may also be independently related to both the exposure under study (Hsu et al. 2010; Knutsen et al. 2011) and the female reproductive disease of concern, and so may need to be considered as candidate statistical confounders; that is, study participant profiles for such factors may differ systematically between women with and without health outcomes of potential research concern (e.g., subfertility, PCOS, endometriosis, menopause) regardless of their BOC and POC exposure levels. For example, nulliparous women rendered infertile by their conditions will not have eliminated POCs from their bodies by maternal–fetal transfer or breast-feeding routes and so, hypothetically, may have higher POC burdens than parous women. Women with a history of long-term breast-feeding reportedly have lower circulating levels of organochlorine (OC) than those who have never breast-fed (Ibarluzea et al. 2011). Another example is women who are postmenopausal or have PCOS are more prone to obesity with a more abdominal (i.e., visceral and bioavailable) distribution that could, hypothetically, alter their POC uptake, storage, and elimination compared to unaffected women. Fat mass has been associated with altered circulating persistent organic pollutant (POP) levels; some POPs appear to vary inversely, while others vary directly with fat mass (Rönn et al. 2011). Time-related variables may also act as potential confounders to the extent they are independently associated with both exposure and the occurrence of the health outcome of interest. Examples of time-related variables include current age, time since exposure, exposure duration, exposure latency, age at first exposure, and birth cohort. For instance, an older birth cohort may have had both more time for latent reproductive effects to manifest and more opportunity for exposure to compounds that emerged at earlier or declined (or were banned) at later points in time. Age thresholds and disease latencies also must be considered in analyses when age-related outcomes, such as endometriosis and menopause, manifest only after a certain age threshold or may occur long after exposure.

Given the complexity of human exposure patterns, it is important to control for time-related confounders as well as possible. For example, consider the impact of confounding by time-related variables on conclusions of a large nonrandomized (observational) study of an intentional endocrine exposure, hormone replacement therapy (HRT) in postmenopausal women. The study found cardiovascular benefits from treatment that drove an uptick in prescriptions. A later randomized human clinical trial of this seemingly well-understood endocrine exposure found a contradictory increase in cardiovascular risk from HRT with broad public health implications (Prentice et al. 2005; Smith 2004). Subsequent reanalysis was conducted to reconcile the discrepancy in the cardiovascular disease findings of the observational study versus the randomized trial (Prentice et al. 2005; Willett et al. 2006). Reconciling the results entailed adjustment of observational results for the variables age at first HRT exposure (i.e., initiation) and HRT exposure duration plus attention to the role of disease latency. While this example pertains to a hormonal drug rather than a hormonally active pollutant exposure, it illustrates the importance of collecting and examining appropriate time-related variables as candidate confounders in noninterventional observational BOC and POC exposure studies as well.

Epidemiological Studies (Special Considerations)

Observational epidemiological studies examine distributions of diseases or physiological conditions in human populations together with factors (e.g., exposures) that influence these distributions (Lilienfeld 1976). They are generally nonexperimental in nature as participants are not randomly preassigned to exposed versus nonexposed groups as they would be in experiments or clinical trials. The body of epidemiological evidence regarding many potential BOC and POC health effects has evolved over time as duration of exposure increases, exposed cohorts mature, and study designs and methods increase in sophistication. The more common epidemiological study designs include cross-sectional, case-control, and cohort designs. Cross-sectional studies offer a "snapshot" of health effects and exposure(s) of interest concurrently, within one window of time. Cross-sectional studies are often considered exploratory. A purely cross-sectional design does not distinguish new cases of disease that occurred within a given time frame (i.e., incident cases) from all cases including new and preexisting cases (i.e., prevalent cases), nor does it establish whether an exposure occurred before or after the disease (i.e., temporal sequence).

In contrast, cohort studies follow defined groups over time (prospectively or retrospectively) to detect the underlying timeline when health effects arose relative to key exposures. Risks of exposed versus nonexposed cohort members are often reported as point estimates such as relative risks (i.e., risk ratios [RRs]) and, for incidence rate and time-to-event outcomes (e.g., survival analyses), as hazard ratios (HR) (Hernán 2010; Peacock and Peacock 2011). Cohorts in whom reproductive end points and exposures have been characterized across the life span are rare but valuable for several reasons. Exposure profiles and sequences may vary throughout life and many exposures occur as mixtures of individual compounds. Coexposures and other potentially important confounding variables (e.g., age, smoking status, BMI) may also vary across the life span. Cohort studies may potentially capture the temporal sequence of these events. Also, life span cohort studies may capture exposures that happen during critical neuroendocrine and reproductive developmental windows (e.g., fetal stages, puberty). This is important since several lines of evidence suggest such exposures may potentially trigger not only effects that are readily apparent at birth but also latent effects that will not appear until decades later.

Case-control studies compare exposures of those with health effects of interest (cases) with a sample of the population from which the cases arose (controls). Results of case-control studies are typically reported as odds ratios (ORs). The OR is a valid estimate of the relative risk when the sampling frame is properly designed. Nested case-control studies are those in which controls and sometimes cases are drawn from a larger defined cohort. This approach reduces the potential for bias resulting from the use of an inappropriate control group for comparison purposes.

The statistical significance of any individual study result is, among other things, a function of study power and, therefore, a reflection of sample size. Studies with insufficient sample size may not adequately capture human variation (e.g., exposure and biokinetic differences) and may also suggest clinically meaningful risks and yet may be underpowered to demonstrate statistically significant (SSIG) effects beyond chance. In such studies, large between group differences may be accompanied by higher p-values or elevated risk estimates may be bounded by wide confidence intervals (CIs). Conversely, studies with a generous sample size and power may detect SSIG effects that are small (e.g., RRs or ORs near one) and potentially so subtle as to not be considered clinically meaningful. Achieving a large sample size is resource intensive, and characterizing suspected interactions (e.g., by age or gender) can inflate sample size requirements. Historical occupational and clinical cohorts often contained few or no women subjects. Therefore, a more frequent challenge in terms of interpreting studies of women's health risks is insufficient statistical power, as opposed to too much statistical power.

Examining suspected interactions strengthens risk assessments. Additive, multiplicative, or other shades of interaction between different BOCs and POCs or between these compounds and other factors are conceivable considering the range of aggregate effects that might be expressed and since individual compounds may act as agonists, antagonists, or synergists. Further, such interactions could be dependent on dose and on the order of exposure. Findings of SSIG effects together with large risk estimate values, dose–response relationships, and convergent results across different human populations and other models lend evidential weight to epidemiological studies and their conclusions. Synthesis of results across epidemiology studies by "pooled" analysis of data from multiple cohorts or by "meta-analysis," that is, combined analysis of study results, is a tool to aid interpretation of the gestalt of data when sufficient numbers of quality studies permit.

Endocrine Disruption

Because the reproductive endocrine axis regulates so many aspects of women's health, from fetal development through reproductive senescence and beyond, endocrine disruption hypotheses form the basis for many human studies of reproductive exposures and effects. The view that certain compounds act as endocrine disruptors in humans has in the past been considered controversial by some. While suspected EDCs are hormonally active by definition, their capacity for various physiological disruptions at the doses typically encountered has been the focus of dispute. However, findings linking EDCs to reproductive effects in women and developmental effects in their offspring are mounting (see the following chapter sections). Consensus that EDCs harm various aspects of human health is growing in tandem with burgeoning evidence. It has been argued recently that EDCs do not fit the traditional toxicological "dose makes the poison" paradigm since low-dose EDC health effects "cannot be predicted by the effects observed at high doses" (Vandenberg et al. 2012). Dose–response curves for EDCs are often nonlinear and U or inverted U in shape rather than monotonic, reflecting the cumulative effects of actions on multiple targets (Gore 2010a,b). Multiple authors have, therefore, concluded that traditional methods used to determine the safety of chemicals must be modified in order to characterize EDC health risks. In efforts to consolidate what is suspected and known about EDC risks, in-depth reviews and analyses have been prepared by scientific bodies such as the Endocrine Society and the World Health Organization (WHO) (Diamanti-Kandarakis et al. 2009; World Health Organization (WHO), International Programme on Chemical Safety 2002). Recently, the U.S. Environmental Protection Agency (EPA) released a highly anticipated report on the reproductive and post-birth-related EDC effects of the dioxin 2,3,7,8-tetrachlorodibenzo-*p*-dioxin (TCDD) (U.S. EPA 2012).

One major contention with the paradigm that synthetic estrogens cause endocrine disruption in humans has been the view that the estrogenic activity of dietary phytochemicals (e.g., isoflavones) in soy and other plants greatly exceeded that of commonly encountered synthetic dietary estrogens (e.g., estrogenic pesticides) (Safe 1995, 2004). However, a recent *in vitro* study provided evidence contrary to this view. In that study, investigators reported that the total daily dietary "estrogenicity intake" (estradiol equivalents per kilogram based on the yeast estrogen screen) for a broad panel of foodstuffs was more than 1000-fold lower than previously thought (Behr et al. 2011; Safe 1995). Further, the estimated dietary estrogenic intake for the weak xenoestrogen, BPA (at the European Union's adult maximum tolerable intake limit), was approximately equal to that derived from a normal diet (maximum total daily intake). For infants (at weight 5 kg), the maximum tolerable daily estrogenicity intake from BPA leached from polycarbonate baby bottles was noted to be greater than the estrogenicity expected from the intake of milk-based formula (Behr et al. 2011). Estrogenicity derived from soy formula was 500-fold higher than from milk-based formula. These *in vitro* findings are made against the backdrop of the U.S. National Toxicology Program's (NTP) conclusion that limited health effects have been reported in soy-fed infants (McCarver et al. 2011) with a determination of "minimal concern" for adverse developmental effects in infants fed soy formula. This designation was based on evidence from human and nonhuman models. The human studies considered primarily growth-related parameters. However, the two studies of human infant exposure for which the NTP reported reproductive system outcomes showed associations between soy formula intake and breast bud development in the subset with thelarche onset before age 2 (Freni-Titulaer et al. 1986) and an increased rate of menstrual dysfunction in adulthood among women exposed to soy formula in infancy (Strom et al. 2001). Taken together, the impact of synthetic chemicals is not readily dismissed based on earlier assumptions about dietary estrogenicity intake, as dietary intake appears lower than previously thought. And while infants who consume soy formula have relatively high estrogenicity intakes, and "minimal concern" has been noted based primarily on growth parameter studies, the need for additional study of other developmental and female reproductive end points is evident given the breast bud and menstrual findings.

Potential EDCs are characterized based on their assumed mode(s) of action. Within the realms of the reproductive system, EDCs are classified as estrogenic (e.g., PCB, BPA, BFRs, diethylstilbestrol [DES], genistein), antiestrogenic (e.g., PCB), androgenic (e.g., trenbolone), antiandrogenic (e.g., DDE, dioxin, phthalates, BFR), gestagenic (e.g., control of estrous cycle by melengestrol), antithyroid (e.g., phthalates, PCB, dioxin, BFRs), antisteroidogenesis (PFOA, parabens, dibutylphthalate, di(2-ethylhexyl) phthalate [DEHP]), aromatase inhibiting (phytoestrogens) (Mouritsen et al. 2010), and aromatase inducing (e.g., atrazine [ATZ] in some but not all cell lines) (Simpkins et al. 2011). Various congeners of compounds may, however, exhibit pleiotropic properties; for example, PCB 138 exhibits both antiestrogenic and antiandrogenic effects (Bonefeld-Jørgensen et al. 2001). Congener groupings for epidemiological studies have been proposed for certain compounds (i.e., PCBs) based on their structural, biological, and pharmacokinetic properties (e.g., persistence, hormonal vs. dioxin-like properties, enzyme induction) (Wolff et al. 1997). Observable effects of endocrine disruptors on women and their offspring represent their diverse and multipathway aggregate effects on cells and tissues, about which much remains to be discovered. Endocrine actions of some compounds (e.g., phytoestrogens) may vary at different dose levels or may depend on the body's hormonal milieu. It has been suggested, for example, that TCDD's activity may vary across the life span since it displays antiestrogenicity when sufficient levels of estrogen are present, but estrogenicity in the absence of estrogen (Boverhof et al. 2006). In some cases, metabolites are more estrogenic or antiestrogenic than the parent compounds themselves (hydroxylated PCBs and hydroxylated PBDEs) (Hamers et al. 2011; Kojima et al. 2009).

Another complexity is that reproductive hormones and EDCs appear to both influence various nonendocrine biological activities and to also be influenced by them. Consider two examples of the interface of reproductive hormones with (1) immunity and, more generally, with (2) epigenetics. The immune system and sex steroids have an interconnected relationship. Estrogens, progesterone, and testosterone are involved in the distribution and function of innate and adaptive immune cells, whereas sex steroids and gender are implicated in a number of immune diseases (Muñoz-Cruz et al. 2011). Accordingly, a small body of evidence suggests human immune function may be impaired by EDC exposures (e.g., PCBs and PFCs developmentally, BPA and triclosan [TCS] among those age 6 through adults) (Clayton et al. 2011; Grandjean et al. 2012; Heilmann et al. 2006; Weisglas-Kuperus et al. 2000, 2004). Epigenetics is a proposed "missing link" between environment, genetics, and hormone function, perhaps accounting for much of the variability and plasticity of the endocrine system. It is thought that epigenetic processes both regulate hormone actions (i.e., synthesis and release, circulating and target tissue levels, and target organ responsiveness) and are also regulated by hormones that activate or repress transcription of different genes (Zhang and Ho 2011). Further, certain EDCs demonstrate epigenetic activity (i.e., DES, TCDD, DDT, PCBs, BPA, PFOA, and vinclozolin) as they alter methylation of various genes in humans and animals. Intriguingly, it has been shown that the potent pharmaceutical drug DES, endogenous estrogen (i.e., physiologic estrogen 17β-estradiol), and the certain exogenous estrogens (i.e., genistein, a phytoestrogen in soy) all alter expression of the same 179 genes in the immature mouse uterus; expression was not examined in the mature uterus in that study (Moggs et al. 2004). *In vivo*, however, other factors, such as biokinetics, and coexposures may come into play. Other EDCs may be relatively more bioavailable to tissues (e.g., uterine) than hormones as they are thought to be less apt than endogenous hormones to bind either blood proteins or the enzymes that metabolize endogenous hormones (Gore 2010a,b). It has been demonstrated that the activity of compounds considered to be estrogenic differs when considered alone versus as mixtures in the rodent uterus (Tinwell and Ashby 2004). In humans, it is thought that the activities of estrogens from dietary and synthetic sources may be additive (Behr et al. 2011); however, the net bioactivities of these and other mixtures encountered are often currently largely unknown. These uncertainties underscore the need for research on the combined actions and potential reproductive and developmental risks of mixtures commonly encountered by humans. The Endocrine Society has issued recommendations to address gaps in clinical, basic, and epidemiological research, and in clinical practice (Diamanti-Kandarakis et al. 2009).

SURVEY OF SELECTED BOC AND POC STUDIES

Gynecological Outcomes

Studies of the reproductive effects of tobacco and alcohol have been a research focus for decades as have, to a lesser extent, specific BOCs (e.g., solvents). The POCs have also garnered growing attention as they have been detected in humans. Some of these agents, including several EDCs described further, have been associated with altered menstrual function or female reproductive hormone levels. Trends in certain reproductive health indicators over time raise questions concerning whether environmental changes adversely affect gynecological health. An expert panel concluded there was sufficient evidence of a temporal trend of earlier breast development and menses onset between 1940 and 1994 among U.S. girls (Euling et al. 2008). Secular trends for most gynecological outcomes are generally less assessable due to the general lack and limitations of female reproductive health surveillance and exposure data.

Menstrual and Reproductive Endocrine Function *Pesticides (general):* Pesticides encompass a broad range of chemical compounds, including organic compounds such as OCs and others (see further). They are used worldwide primarily as insecticides, herbicides, fungicides, and growth regulators (Hanke and Jurewicz 2004). Over 400 pesticide chemicals have been approved for agricultural use, and these may be formulated into thousands of different consumer products. Among 3103 farm women in Iowa and North Carolina (United States, Agricultural Health Study cohort), those who used pesticides had longer menstrual cycles and increased odds of missed periods (OR = 1.5, 95% CI = 1.2, 1.9) compared with women (controls) who never used pesticides. The subset who used pesticides considered probable HACs had 60–100% increased odds of long cycles, missed periods, and intermenstrual bleeding compared with controls (Farr et al. 2004).

DDT is an OC that was first used to control malaria and typhus during the latter part of World War II. After the war, DDT was used extensively from

1950 to 1980 in agriculture as an insecticide until it was banned in the United States in 1972. In other countries, such as South Africa, Zambia, and Madagascar, DDT has been reintroduced in recent times with concurrent substantial reductions in malaria morbidity and, likely, mortality (Longnecker et al. 2005). DDE is the primary metabolite of DDT. DDT or DDE affect oocyte maturation, ovulation, and fertilization in several studies of humans or laboratory animals (Tiemann 2008). They have garnered the most attention because of their chemical stability and strong lipophilicity, slow degradation, and bioaccumulation in the food chain. DDE tends to build up in the body throughout life as it is rarely excreted from the body except being excreted in breast milk, thereby transferring the toxicant to the nursing infant.

High serum DDT concentrations in adult Chinese female textile workers were SSIG associated with earlier age at menarche (adjusted trend $p < 0.001$) and shorter menstrual cycles ($OR_{ADJ} = 2.78$, 95% CI = 1.07–7.14) (Ouyang et al. 2005). Windham et al. (2005) reported the mean cycle length of 48 adult, Southeast Asian immigrants with high DDE exposure was not SSIG different from the cycle length for women with low exposure (after adjustment). However, each doubling of DDE levels was linked to a 0.6-day shortening of the luteal phase (−1.1 to −0.2) and a consistent reduction in progesterone metabolite levels (Windham et al. 2005). Perry et al. (2006) found inverse relationships for serum levels of most forms of DDT with estrogen and progesterone metabolites in 287 reproductive-age women. Most SSIG inverse associations with the DDT metabolites were observed for progesterone metabolite during the luteal phase and for estrogen metabolite during the periovulatory phase, suggesting effects of DDT/DDE during the menstrual cycle phases that are crucial for ovulation and early pregnancy maintenance (Perry et al. 2006). In contrast, Chen et al. (2005) failed to detect any association between DDE or DDT levels and menstrual cycle length, duration of menses, or heaviness of menstrual flow in 60 young Chinese women.

TCDD, a PCDD, more commonly known as dioxin, is classified as a POP. It is one of the most potent congeners of the PCDDs and binds to the aryl hydrocarbon receptor (AhR), which mediates the effects of many EDCs and contributes to the loss of fertility in polluted environments. Dioxin exerts a multitude of adverse effects on organisms (teratogenesis, tumor promotion, immunosuppression, estrogenic action), mediated primarily by AhR (Baba et al. 2005). It occurs as a by-product in the manufacturing of some OCs or by incineration of chlorine-containing substances such as polyvinyl chloride (PVC). Dioxin contamination occurs mainly through industrial emission and/or accidents.

The Seveso Women's Health Study (SWHS) in Italy assessed subjects with the highest levels of TCDD exposure in this population as a result of a chemical explosion in 1976. A retrospective cohort study of exposed premenarcheal participants revealed no change in the risk of early onset of menarche with a 10-fold increase in TCDD (Warner et al. 2004) and a borderline SSIG association with an increased menstrual cycle length (Eskenazi et al. 2002b).

ATZ is notable as the most common herbicide used in the United States, but it is banned in the European Union due to widespread and persistent groundwater contamination. Women in Illinois where ATZ use was "extensive" were SSIG more apt to have irregular menstrual cycle lengths (OR_{ADJ} = 4.69, 95% CI = 1.58–13.95) and more than 6 weeks between bleedings (OR_{ADJ} = 6.16, 95% CI = 1.29–29.38) than women in Vermont where ATZ use was said to be used "sparingly" (Cragin et al. 2011).

PCBs were widely used as dielectric and coolant fluids. Due to their toxicities, PCBs have been identified as POPs. Production of PCBs was banned in the United States in 1979 and by the Stockholm Convention of Persistent Organic Pollutants in 2001. PCBs are lipophilic and have long half-lives (8–10 years). PCBs are stable and bioaccumulate in the food web. Human exposure to PCBs occurs via inhalation of contaminated air or through consumption of contaminated food. PCBs are EDCs and interfere with endocrine secretion and actions. The toxicity of PCBs vary among congeners. Co-PCBs tend to have dioxin-like properties, the most toxic of congeners. Certain congeners are known to be estrogenic agonists or antagonists of estradiol. Similar to dioxin, PCBs are thought to mediate by binding the AhR, thereby disrupting cell functions by altering gene transcription (Safe 1984; Safe et al. 1985). PCB exposure and their effects on menstrual cycle characteristics and function have received relatively little attention, and data to date are somewhat contradictory. PCBs affect the endocrine system by a variety of mechanisms. They are known to reduce secretion of the thyroid hormone thyroxine (T4) in animals (Brouwer et al. 1999). In humans, thyroid hormone underproduction (i.e., hypothyroidism) can cause prolonged duration of menses, while overproduction (i.e., hyperthyroidism) reduces menstrual flow and can result in oligomenorrhea (i.e., infrequent menstrual periods).

Total PCB levels measured in the third trimester of the pregnancies of 2314 U.S. women in the Collaborative Perinatal Project were positively associated with reported history of menstrual cycle length (adjusted difference five PCB exposure categories = 0.7 days, trend-test p = 0.02) (Cooper et al. 2005). In 1979, more than 2000 people were poisoned in Taiwan when they ingested PCB-contaminated cooking oil. Two decades later, women from this Yucheng ("oil disease") incident were interviewed and reported abnormally light menstrual bleeding, slightly reduced cycle length (p = 0.03–0.006), and borderline SSIG earlier age of menarche (Yang et al. 2005, 2011; Yu et al. 2000). PCB levels associated with earlier age of menarche were also observed in 138 Native American youth (Denham et al. 2005). Women from the New York State Angler (NYSA) cohort who consumed contaminated fish had moderate to high estimated PCB levels and SSIG cycle length reductions (−1.1 days, 95% CI = −1.87 to −0.35 overall) (Mendola et al. 1997). Women raised in Swedish fishing villages on the east coast off the Baltic sea exposed to contaminated fatty fish (presumably, *in utero*, during breastfeeding and/or through their diet) directly experienced menarche at an older age than women from less contaminated west coast villages (Axmon 2006). Delayed sexual maturation was also observed in girls living in Antwerp, Belgium, near waste incinerators that may have spewed pollutants including heavy metals and BOCs and POCs including dioxins, dioxin-like compounds, and aromatic hydrocarbons (Den Hond et al. 2002; Staessen et al. 2001).

PBBs are used as flame retardants and are added to plastics used in products such as home electrical appliances, textiles, and plastic foams. While PBB manufacturing ceased in 1976 due to slow degradation, the contaminant can still be found in the environment. Similar to PCBs, PBBs bioaccumulate in the food web and exposure is most likely through ingestion of contaminated food and drinks. Over 4000 individuals were exposed to PBB in Michigan in 1973 via accidental contamination of livestock feed. Mothers were exposed by ingestion of contaminated cow's milk and meat. Maternal exposures were passed onto the fetus and infant through the placenta and during breastfeeding (Blanck et al. 2002).

BPA is a synthetic compound that is used extensively as a monomer in polycarbonate plastics and in epoxy resins that are used to line food and beverage containers, and in dental sealants. BPA is one of the most omnipresent and high-profile EDCs (Talsness et al. 2009; Völkel et al. 2002), having received much attention from the scientific community and media. Exposure to BPA occurs primarily through the diet, but also by dermal contact and inhalation. BPA metabolites have been detected in more than 90% of people in representative populations in the United States and in Europe (Calafat et al. 2008b; Galloway et al. 2010). However, it is difficult to assess the full impact of human exposure since outcomes of interest manifest latently at later ages.

TCS is an antimicrobial and preservative that, for more than two decades, has been added to personal care products (e.g., hand soaps, dishwashing and laundry detergents, mouthwashes, toothpastes, wound disinfection solutions, deodorants) and other commonly encountered items (e.g., toys, facial tissues, textiles [e.g., underwear, socks], plastic kitchen utensils, and medical devices) (Calafat et al. 2008a; Clayton et al. 2011). It is detectable in human breast milk (Allmyr et al. 2006) and, in a population-based U.S. cohort (National Health and Nutrition Examination Survey [NHANES] 2003–2004), was detected in the urine of 75% of people sampled ($n = 2517$) (Calafat et al. 2008b). A recent review of a small number of *in vitro* TCS studies described findings consistent with possible endocrine disruption including weak AhR activity in estrogen and androgen receptors and dependent gene expression, and (human cell) findings of estrogenic and androgenic effects and lowered testosterone-induced transcriptional activity (Dann and Hontela 2011). In one human study, brushing twice daily with TCS toothpaste did not produce an adverse effect on thyroid hormones (Cullinan et al. 2012). No other published epidemiological studies of TCS exposure and reproductive related outcomes were noted.

Phthalates are mainly used as plasticizers. They are in consumer items from cosmetics and personal care products to children's toys to cosmetics, shower curtains, wallpaper, vinyl blinds, food packaging, plastic wraps (ATSDR 1995), and intravenous tubing. Phthalates were once used to make pacifiers, soft rattlers, and teethers, but U.S. manufacturers stopped using phthalates in these products in 1999. Phthalates have been identified as EDCs associated with developmental and reproductive toxicity. Although they do not bioaccumulate in the body like dioxins and other OC chemicals, phthalates are ubiquitous in the environment. They are rapidly metabolized by the body and have a half-life shorter than 24 hours (Koch et al. 2005; Silva et al. 2003). Dietary uptake is a major route of exposure, along with contact (e.g., cosmetics and personal care items) and inhalation (e.g., air spray, nail polish, volatized phthalate from PVC) (Jurewicz and Hanke 2011; Lyche et al. 2009). Phthalate metabolites monoethyl phthalate (MEP) and monobutyl phthalate (MBP) have been detected in virtually all urine samples from a nationally representative U.S. study population (Silva et al. 2004b) and an urban study population in China (Guo et al. 2011). There is increasing concern that phthalates may be harmful to human reproductive health and fertility. Thelarche, the premature development of breast tissue, has a very high occurrence in Puerto Rico. Approximately 68% of early thelarche patients in Puerto Rico had high levels of multiple estrogenic phthalates as opposed to only one control in an analytical chemistry study (i.e., gas chromatography/mass spectrometry) (Colón et al. 2000). In a Chinese study of 210 girls, serum phthalate levels were detected in a higher percentage of precocious puberty (PP) girls than in non-PP girls for both the phthalates dibutyl phthalate (DBP) (27% vs. 4%) and DEHP (22.7% vs. 3%) (Qiao et al. 2007). The PP girls had larger ovaries and uteri with SSIG positive correlations between both DBP and DEHP levels and the sizes of these organs. However, a multicenter cross-sectional study in the United States of 28 girls with PP and 28 age- and race-matched prepubertal females showed no association between any phthalate metabolites and puberty onset (Lomenick et al. 2010).

Trihalomethane (THM) is a disinfection by-product that is commonly found contaminating drinking water in the United States and in Europe. In a follow-up study of a cohort of tap water consumers in California, researchers reported a shortened menstrual cycle length and a decreased follicular phase with increasing THM (decrease in cycle and follicular phases by 0.18 days, 95% CI = −0.29 to −0.07 per 10 μg/L increase in total THM) (Windham et al. 2003).

Perfluorochemicals (PFCs) are man-made surfactants widely used as water, stain, and grease repellants. Some common uses for PFCs include nonstick cookware, stain-resistant carpets, and fabric. The two most common PFCs are PFOS and PFOA (or C8). PFCs have a relatively long half-life in humans (median of 4.6 years for PFOS and 3.4 years for PFOA) (Olsen et al. 2007). They can generally be found in the soil, sediments, and groundwater. They are not lipophilic and do not break down in the environment.

The C8 health project in West Virginia studied over 69,000 subjects exposed to PFC in drinking water. Average serum levels of PFCs (except PFOS) were higher among C8 participants than in the U.S. population (NHANES 2003–2004), with PFOA levels reported as being 500% higher among the C8 group (Calafat et al. 2007; Frisbee et al. 2009). In one study that investigated the effect of PFOA and PFOS on serum estradiol and menopause onset, PFOS was negatively (SSIG) associated with serum estradiol levels in the perimenopausal ($p < 0.0001$) and menopausal ($p < 0.007$) women, but an association was not found between PFOA and estradiol (Knox et al. 2011).

Solvent exposure is common. Approximately 30% of working women reported regular solvent exposure in a population-based study in Brittany, France (Garlantézec et al. 2009). Solvents are commonly encountered and studied as mixtures. Benzene is an aromatic hydrocarbon solvent that has been widely used in the United States and is formed from both natural processes and human activities. Natural sources of benzene include volcanoes and forest fires; it can also be found in the air from emission from burning coal and oil, car exhaust, and cigarette smoke. Benzene is also widely used in industries producing plastics, resins, synthetic fibers, lubricants, rubbers, detergents, and pesticides. A major source of benzene exposure is tobacco smoke, followed by emissions from gasoline/petrochemical industries (Zhang and Lioy 2002). Exposure to benzene vapors in the workplace has been associated with menstrual irregularity and spontaneous abortion (SA) (Cho et al. 2001; Thurston et al. 2000; Xu et al. 1998). Toluene, an aromatic solvent, is widely used in the manufacturing industries such as shoes, textiles, electronic components, and plastics, and in industrial applications of paints, thinners, adhesives, inks, and pharmaceutical products. Many women are exposed to toluene in occupational settings (long-term, low-level exposure) or by inhalant abuse (intermittent, high-level exposure) (Bukowski 2001; Hannigan and Bowen 2010). Assembly line workers in Singapore reported significant levels of residual toluene in blood after workplace exposure (Foo et al. 1988); however, no significant difference was observed in menstrual cycle frequency or flow of workers (Ng et al. 1992a). In one study, the reproductive hormone profiles of U.S. Air Force women ($n = 63$) were assessed in relation to their occupational exposure to jet fuel (Reutman et al. 2002). Those women whose internal doses (i.e., exhaled breath levels) of total aromatic hydrocarbons (primarily toluene, but also benzene, ethylbenzene, and *m*,*p*,*o*-xylenes) were above the median and had SSIG lower urinary preovulatory luteinizing hormone (LH) levels ($p = 0.007$), suggesting low-level exposure to compounds in fuels, and some solvents may act as neuroendocrine disruptors.

The risk of deleterious effects of toluene exposure in an occupational setting can be greatly reduced in well-regulated workplaces. In contrast, inhalation abuse of toluene and other organic solvents is of greater concern due to repeated high-level exposures (Hannigan and Bowen 2010). Accurate assessment of inhalant abuse is difficult due to the varying contents of toluene in different consumer products. Despite the many case studies on occupational and toluene abuse, the reproductive toxicology of this organic solvent remains underinvestigated.

Tobacco (cigarette) smoke contains a mixture of over 7000 chemicals and additives (U.S. DHHS 2010), many of which are considered organic compounds, but also includes inorganics such as heavy metals. The PAHs, nicotine, and carbon monoxide found in tobacco smoke have been implicated in adverse female reproductive and developmental outcomes. Low levels of several BOCs and POCs described earlier (i.e., benzene, toluene, dioxin, and dioxin-like compounds) (Wilson et al. 2008) have also been detected among the many other compounds in tobacco smoke. Exposure to tobacco smoke occurs through both active and passive smoking, including transplacental fetal exposure during pregnancy (Rogers 2009). The 2010 U.S. Surgeon General Report found "consistent evidence that increases in follicle-stimulating hormone levels and decreases in estrogen and progesterone are associated with cigarette smoking in women, at least in part due to effects of nicotine on the endocrine system." The Report also concluded there was consistent evidence that tobacco smoke diminished oviduct function, which might impair fertilization (U.S. DHHS 2010). Additional reading on smoking-related reproductive and developmental risks is available in other U.S. Surgeon General reports (U.S. Department of Health and Human Services (U.S. DHHS), Office on Smoking and Health (U.S.) 2001; U.S. Department of Health and Human Services, Centers for Disease Control and Prevention, National Center for Chronic Disease Prevention and Health Promotion, Office on Smoking and Health 2004; U.S. DHHS 2006).

Early/Premature Menopause Natural menopause is the conclusion of a developmental transition that marks the end of ovulation and menstruation due to what is widely regarded as the permanent loss of ovarian follicular activity. The WHO defines natural menopause as having occurred after 12 consecutive months of no menstrual periods, given there is no obvious pathological or physiological cause (e.g., pregnancy, lactation) (WHO 1996). In studies of women in industrialized countries, "early" menopause is typically considered cessation of menses before age 46 years, while "premature" menopause is often defined as cessation before age 40 years (NAMS 2012). "Induced" menopause generally refers to both ovarian ablation by agents such as radiation or chemotherapy and removal of both ovaries (i.e., bilateral oophorectomy) with or without hysterectomy (WHO 1996). Women with induced menopause may have different risk profiles than those undergoing natural menopause. This is an important consideration for study design and analysis.

A multitude of factors appear to hasten or delay the timing of natural menopause. Factors consistently associated with earlier menopause include active smoking, hormonal contraceptive use, low parity, and low socioeconomic status (Gold 2011). Diet, body mass, race/ethnicity, physical activity, genetics (Gold 2011), and residence (e.g., country, rural/urban) may also affect menopausal age (MA). Women who are atypically young at menopause may have heightened risks of cardiovascular disease and osteoporosis with attending reductions in overall survival and life expectancy. Concerns about infertility, together with psychological and social complications, have also been reported among those with premature ovarian failure (Nippita and Baber

2007; Singer et al. 2011). Most studies of reproductive toxicant effects compare exposed and reference groups to determine more subtle effects on MA rather than the less frequent outcomes of "premature" or "early" menopause per se. It is not clear if these subtle effects on MA engender the same health risks. Still, the reproductive window will have narrowed perceptibly for some and may impact the subset of women planning pregnancies in the latter reproductive years. Later age at menopause is associated with different risks, including increased ovarian, endometrial, and breast cancers (Gold 2011).

OCs The few studies of the effects on MA of other POP compounds measured among U.S. women during adulthood provide inconclusive evidence. TCDD exposure quintiles among the Seveso accident study cohort reportedly followed a monotonically increasing trend with reduced MA up to a dose threshold of 100 ppt ([$n = 616$]: SSIG trend for quintiles 1–4 [$p = 0.04$]) (Eskenazi et al. 2005). Among a cross-sectional sample of naturally menopausal women in the Hispanic NHANES study ($n = 219$), four of seven OCPs measured (serum DDT, DDE, β-HCH, and *trans*-nonachlor) were associated with mean MA reductions of 5.7, 3.4, and 5.2 years, respectively, between women within the highest versus lowest (nondetects) dose categories (Akkina et al. 2004). Similarly, among a subset of women from the North Carolina breast cancer case-control study ($n = 1407$), the median MA of women with the highest DDE levels was 1 year earlier than controls. Accordingly, women with higher serum DDE levels had non-SSIG ($HR_{ADJ} = 1.4$, 95% CI = 0.9–2.1) higher rate of menopause, similar to the rate seen among smokers (Cooper et al. 2002). In contrast, the median MA of farm women ($n = 5013$) who reportedly used and mixed a broad group of pesticides, including ovotoxins and HACs, was approximately 3 months older at menopause than controls overall. The subset that used HAC pesticides was 5 months older at MA. Both findings were SSIG ($HR_{ADJ} = 0.87$, 95% CI = 0.78–0.97 and $HR_{ADJ} = 0.77$, 95% CI = 0.65–0.92, respectively) (Farr et al. 2006). Two analyses examining the timing of menopause relative to levels of PCBs and PBBs (Taiwan Yucheng Study; $n = 118$) and another of PCBs alone (Michigan PCB study; $n = 874$) failed to find an association between exposure and timing of menopause (Blanck et al. 2004; Yu et al. 2000).

PFCs Among a cross section of women exposed to PFO-contaminated water, higher PFOS and PFOA levels were SSIG associated with the odds of menopause among both perimenopausal (for PFOS, PFOA: $OR_{ADJ} = 1.4$, 95% CI = 1.1–1.8) and menopausal subgroups (PFOS: $OR_{ADJ} = 2.1$, 95% CI = 1.6–2.8; PFOA: $OR_{ADJ} = 1.7$, 95% CI = 1.3–2.3) (Knox et al. 2011).

Smoking The association between smoking and early menopause is considered fairly well established (Parente et al. 2008). Women who smoke have MAs approximately 1–2 years earlier than nonsmokers (Gold 2011).

Endometriosis Endometriosis occurs when tissue that is histologically similar to the inner uterine lining of reproductive-age women becomes ectopic, growing outside the uterus. Adverse endometriosis symptoms may include pelvic pain (particularly during menses or intercourse), infertility, and other possible symptoms and complications. Endometriosis, which occurs in 10–15% of women and requires estrogen to progress, typically presents during the adult reproductive years and subsides with reproductive senescence (Crain et al. 2008). Family history of endometriosis appears to be a risk factor. Altered responses to the extremely common phenomenon of retrograde menstruation are generally believed to be involved in the initiation of this condition, but the reason why progression only occurs in some women and the complete etiologic cascade are not well understood. Tissue samples from women undergoing surgery showed changes in mRNA expression for AhR and aryl hydrocarbon receptor nuclear translocator (ARNT) in diseased uterine tissues (endometriosis and fibroids). AhR and ARNT modify transcription of genes involved in cell differentiation and proliferation (Khorram et al. 2002). Developmental exposures to EDCs such as TCDD under experimental conditions appear to lead to a progesterone-resistant phenotype and hyperactive immune sensitivity to inflammatory stimuli, symptoms observed in women with endometriosis (Herington et al. 2011). Currently posited mechanistic roles of TCDD and other BOCs and POCs in the developmental and adult etiopathology of endometriosis are reviewed in detail elsewhere (Bruner-Tran and Osteen 2010; Crain et al. 2008; Guo 2004; Heilier et al. 2008; Herington et al. 2011; Mendola et al. 2008). Many published epidemiological studies of endometriosis have focused on EDCs.

OCs High residential exposure to TCDD, the most potent HAC, occurred in Seveso, Italy, when a chemical plant exploded in 1976. Risk of endometriosis was doubled 20 years after exposure among the 316 women with the highest (>100 ppt) versus lowest (<20 ppt) serum TCDD levels, although the risk and dose–response trend did not reach SSIG (RR_{ADJ} = 2.1, 90% CI = 0.5–8.0) in that historical cohort (Eskenazi et al. 2002a). Female Seveso residents exposed *in utero* and in the youngest age groups were too young for inclusion at the time of that analysis. The authors also noted the results may have presented "an underestimate of the true risk" since the sample was small and misclassification of cases as noncases may have occurred, as surgical or ultrasound confirmation was required for designation as a case. Several authors have reviewed the Seveso study together with results of a combined total of 26 other studies which only captured adult exposures to potential EDCs including mainly dioxin (TCDD), PCDD/DFs, and PCBs (Bruner-Tran and Osteen 2010; Crain et al. 2008; Heilier et al. 2008; Mendola et al. 2008). There was general agreement across the reviews that the lack of developmental data in exposed and longitudinally tracked human populations beyond Seveso limits investigation of the fetal origin EDC exposure endometriosis hypothesis. Also, there was consensus that an association between adulthood OC exposure and endometriosis is currently speculative since findings have been equivocal. However, a possible link was not ruled out. One review stressed that the deep nodular endometriosis phenotype appeared to be related to higher levels of marker PCBs and dioxin-like compounds (Heilier et al. 2008).

Bisphenols and Phthalates Few published studies of relationships between other EDCs (i.e., BPA, bisphenol B [BPB], DEHP, MBP, mono(2-ethylhexyl) phthalate [MEHP]) and endometriosis were identified. Findings from two analyses of fertile Italian women evaluated laproscopically suggested a direct association between endometriosis and serum BPA and BPB (Cobellis et al. 2009 [$n = 69$]) and an SSIG ($p = 0.0047$) direct SSIG association with DEHP (Cobellis et al. 2003 [$N = 79$]). In a U.S. (NHANES) study based on self-reported endometriosis, a positive association with MBP and a negative association with MEHP were found to be SSIG, but only when endometriosis or fibroid outcomes were combined ($OR_{ADJ} = 1.71$, 95% CI = 1.07–2.75 and $OR_{ADJ} = 0.59$, 95% CI = 0.37–0.95, respectively) (Weuve et al. 2010 [$n = 238$]). Neither MEP, MEPH, nor monobenzyl phthalate (MBzP) was SSIG in that study when endometriosis was considered alone.

Dietary Estrogens In their review, Heilier et al. (2008) hypothesized that phytoestrogens may be linked to endometriosis since one study found them to be associated with endometrial hyperplasia (Unfer et al. 2004) and another demonstrated their ability to stimulate aromatase activity in human endometrial stroma *in vitro* (Edmunds et al. 2005).

Several of the above-mentioned reviewers have speculated methodological dissimilarities such as different comparison groups, exposure profiles, congener groupings, uncontrolled confounders, case versus noncase definitional criteria (i.e., phenotypic and diagnostic), and the potential for reverse causation may hamper individual or aggregate interpretation of EDC endometriosis study results.

Fibroids Uterine fibroids (i.e., leiomyomata) are benign hormonally reactive uterine tumors that originate in the uterus (myometrium) among as many as half of menopausal women (Haney 2008). Fibroid growths may be asymptomatic or can decrease fertility and may cause pelvic pain, including excessive pain and bleeding during menstruation. Infrequently, complications such as blood loss can be life threatening. Risk factors for fibroids include obesity and elevated BMI, but the risk varies inversely with time interval since last delivery. Women in the United States undergo roughly 200,000 hysterectomies and 20,000 myomectomies per year to relieve uterine fibroid symptoms and complications (Haney 2008), and these surgeries also impart health risks.

Dietary Estrogens In one review (Crain et al. 2008), dietary phytoestrogens were reported to have been consistently protective against fibroids across several studies that examined the effects of their consumption by reproductively mature women. More recently, in the National Institute of Environmental Health Sciences (NIEHS) Sister Study breast cancer cohort, fibroids were positively but non-SSIG associated with having been fed soy formula in infancy ($RR_{ADJ} = 1.26$, 95% CI = 0.83–1.89 [$n = 2583$]) (D'Aloisio et al. 2011).

OCs Exposure to TCDD has been linked to a lowered risk of fibroids in 410 women in the Seveso cohort 20 years after the explosion (Eskenazi et al. 2007). The investigators proposed that this may reflect antiestrogenic effects in the myometrium. With the paucity of human studies, there are calls for more research on the relationship between BOCs and POCs and fibroids (Crain et al. 2008; Mendola et al. 2008).

PCOs Women with PCOS have clinical or biochemical androgen excess and ovarian dysfunction, which manifest as polycystic ovaries and/or infrequent, irregular, or absent ovulation, although there is much heterogeneity in the phenotype, and the definitional criteria are evolving (Azziz et al. 2009). This common condition affects approximately 4–8% of reproductive-age women, although higher prevalences approaching 20% have been reported using broader PCOS criteria (Teede et al. 2010). PCOS is often accompanied by metabolic syndrome symptoms such as abdominal obesity, insulin resistance, and lipid alterations, with elevated risks of diabetes, nonalcoholic fatty liver disease, coronary heart disease, stroke, and endometrial cancer (Crain et al. 2008; de Groot et al. 2011; Hossain et al. 2011). Reproductive risks include premature pubarche, anovulatory infertility, gestational diabetes, pregnancy-related hypertension, and miscarriage (Crain et al. 2008). The pathogenesis is unclear, but genetic predisposition, diet, and other environmental exposures of susceptible individuals during developmental windows are suspected (Crain et al. 2008; Diamanti-Kandarakis et al. 2012).

Bisphenols Adult BPA levels have been positively and SSIG associated with PCOS and androgen levels in two clinic-based cross-sectional epidemiology studies ($p < 0.001$ for PCOS and $p < 0.05$ for hyperandrogenism, $n = 171$, Kandarakis et al. 2011; $p < 0.05$ for PCOS and $p < 0.001$ for androgens, $n = 45$, Takeuchi et al. 2004). A positive SSIG relationship between BPA and insulin resistance in the PCOS group was also noted in one study ($p < 0.05$; Diamanti-Kandarakis et al. 2012). Review authors remarked that the potential for reverse causation must be considered when interpreting these study results as the elevated testosterone levels that attend PCOS may decrease BPA clearance (Crain et al. 2008). That few human studies have been published on the effects of BPA and other compounds on PCOS is striking, given the relatively high prevalence and comorbidities associated with this poorly understood condition. Refinements of and consensus on the most relevant phenotypic criteria of PCOS for research will strengthen future research quality.

Infertility/Subfecundity Infertility has been defined by the WHO as failure to conceive after at least 1 year of unprotected intercourse. Infertility rates differ by region (Vayena et al. 2002). Of couples who are at risk for pregnancy (e.g., stop contraception), about 10–15% will take more than 12 months to conceive (Baird and Strassmann 2000). Note, however, that a higher percentage of 30- to 40-year-old women will have experienced at least one infertility episode. Female infertility has been linked to tubal and pelvic factors (40%), ovulatory dysfunction (40%), and cervical and uterine cancers (5%), while 15% of infertility is unexplained (Levens and Decherney 2011). Fetal losses that occur very early in pregnancy are often unrecognized and thus contribute to perceived infertility. Studies of time to pregnancy (TTP) estimate fecundity, or the biological capacity of women, men, or couples to reproduce. A prolonged TTP suggests impaired fecundity (i.e., a reduced probability of conceiving each cycle or fecundability ratio [FR]).

OCs Two decades after the 1979 Yucheng PCB-contaminated cooking oil incident, the exposed women reported longer TTP ($p = 0.019$) together with higher odds of infertility ($OR_{ADJ} = 2.34$, 95% CI = 1.23–4.59) (Yang et al. 2008). The NYSA women who consumed fish contaminated with moderate to high levels of PCB later reported shorter menstrual cycles compared with nonexposed women (–1.03 days, 95% CI = –1.88 to –0.19) (Mendola et al. 1997). In addition, a subset of this study population, those who consumed contaminated fish the longest ($n = 732$), reported a small but non-SSIG reduction in fecundity (McGuinness et al. 2001).

Pesticides (General) Epidemiological studies of infertility and TTP among agricultural workers have been examined in several reviews, but few of the studies (described further) focused on women's exposures (Hanke and Jurewicz 2004; Sanford et al. 2007). One such study of female agricultural workers and controls ($N = 497$) reported SSIG increased odds of infertility among women in who worked in agriculture before conceiving, in industries associated with agriculture, or who resided on a farm (OR = 11.3, 95% CI = 2.6–48.8; OR = 7.0, 95% CI = 2.3–20.8; and OR = 1.8, 95% CI = 1.2–2.7, respectively) (Fuortes et al. 1997). Another was of 2012 pregnancies among Ontario (Canada) farm couples in whom the TTP was increased for 6 of 13 pesticide exposures during the intervals in which women participated in pesticide activities (as did most of the men) (conditional FR range = 0.51–0.80) (Curtis et al. 1999). A TTP study of 492 female greenhouse workers found SSIG increased TTP only when considering the subset who did not use gloves ($FR_{ADJ} = 0.67$, 95% CI = 0.46–0.98) (Abell et al. 2000). Several later TTP studies of female greenhouse workers did not find SSIG longer TTP (Bretveld et al. 2006; Lauria et al. 2006), although the odds were slightly elevated in one of these analyses (OR 1.9, 95% CI = 0.8–4.4) (Bretveld et al. 2008).

PFCs In a study of the Danish National Birth Cohort of 895 women who became pregnant, higher maternal serum concentrations of PFOS and PFOA were associated with SSIG longer TTP ($p < 0.001$), hence reduced fecundity (Fei et al. 2009). Infertility ORs were directly linked to PFOS and PFOA levels in that study ($p = 0.025$ and $p < 0.006$, respectively). In contrast, no SSIG association was found between TTP and exposure to PFCs (at similar levels measured in the Danish National Birth Cohort) among a national study of Danish couples planning a first pregnancy who were recruited from trade union rosters (Vestergaard et al. 2012). A case-control analysis of 910 women from the Norwegian Mother and Child Cohort Study found discrepant relationships between fecundity and serum levels of PFOS and PFOA among those who had been pregnant (prolonged TTP) and those who had never been pregnant (shortened TTP) after age and BMI adjustments (PFOS: $OR_{ADJ} = 2.1$, 95% CI = 1.2–3.8 vs. 0.7, 0.4–1.3, respectively; PFOA: $OR_{ADJ} = 2.1$, 95% CI = 1.0–4.0 vs. 0.5, 0.2–1.2, respectively) (Whitworth et al. 2012). As body burdens of compounds fluctuate with pregnancies and lactation, the authors commented that the analysis of TTP among never-pregnant (i.e., nulliparous) women may be the most informative.

Solvents In a study evaluating the relationship between toluene exposure and TTP, male and female employees from 14 different printing companies in Germany were recruited. All women performed jobs that had low-level exposure, while men worked in areas with low, medium, and high exposures to toluene. Results showed a strong association between toluene exposure and reduced fecundity; the FR was 0.48 (95% CI = 0.24–0.97) for these female employees compared to women not in the printing industry (Plenge-Bonig and Karmaus 1999). No association between toluene exposure and male infertility was observed, which suggests that women may be more susceptible to the adverse reproductive effects of toluene. Similar studies on female workers in European shoe manufacturing and laboratory workers exposed to solvents also show reduced fecundity (fertility density $ratio_{ADJ} = 0.55$, 95% CI = 0.40–0.70 for "low" exposure among shoe workers; $FR_{ADJ} = 0.79$, 95% CI = 0.68–0.93 for organic solvent exposure among laboratory workers, respectively) (Sallmén et al. 2008; Wennborg et al. 2001). Also, an SSIG increase in the odds of subfertility was found among the wives of pesticide applicators ($n = 2112$) in the Agricultural Health Study Cohort who were themselves exposed to solvents at least once per month (OR_{ADJ}: 1.42, 95% CI = 1.15–1.75) (Sallmén et al. 2006).

Smoking A consistent overall association (OR = 1.60, 95% CI = 1.34–1.91) between smoking and increased infertility risk was reported in a meta-analysis of 12 studies and was supported by a concurrent review of nine *in vitro* fertilization studies in which smoking women were also less fecund (Augood et al. 1998).

Pregnancy and Fetal Development Outcomes

Fetal Loss Fetal loss is a broad term for unintended pregnancy loss, which encompasses very early, often undetected pregnancies as well as clinically recognized miscarriages and stillbirths. Fetal losses are frequent, with most occurring during the early weeks of fertilization and implantation, often before the pregnancy is known. In a given menstrual cycle, it has been estimated that only 36% of attempting women will manifest chemically detectable implantations (i.e., detectable human chorionic gonadotropin levels). The deficit is attributable to fertilization failure and early (i.e., pre- and early postimplantation) pregnancy losses (Baird and Strassmann 2000). It was estimated that another 10% of these early detectable pregnancies will have been lost before 6 weeks after the last menstrual period. Estimates from Western countries indicate 10–20% of clinically recognized pregnancies end in SA (i.e., miscarriage) and 2% end in stillbirth (Baird and Strassmann 2000; Chard 1991). Many maternal and several paternal factors have been linked to fetal loss (e.g., medical and reproductive conditions and history, alcohol, smoking, and drug use, nutritional status, age, job and iatrogenic exposures, genetics, sociodemographic factors) (Reutman and LeMasters 2007). Miscarriages are defined here as losses occurring before 20 weeks of gestation, after which losses are termed stillbirths. However, some studies have used other stillbirth definitions (e.g., ≥28 weeks) (Wigle et al. 2008). The number of stillbirths approached 26,000 in the mid-2000s in the United States (MacDorman and Kirmeyer 2009). Risk factors linked to stillbirths across low-, middle-, and high-income countries include medical (e.g., previous stillbirth, high/low [primi-]parity, multiples, congenital anomalies, infections, AB blood type, intrauterine growth restriction, diabetes, placental pathologies, and eclampsia/preeclampsia), socioeconomic (i.e., no partner or education, lacking prenatal care), lifestyle (smoking or illicit drugs), elevated BMI, race/ethnicity, and advanced maternal age (Faiz et al. 2012; McClure et al. 2011; The Stillbirth Collaborative Research Network Writing Group 2011).

Pesticides (General) Studies of parental exposures to pesticides in general and SA or stillbirth have been reviewed elsewhere (Hanke and Jurewicz 2004; Sanborn et al. 2007), and two reviews are of particular interest in that they were focused on maternal occupational (Nurminen 1995) and residential (Shirangi et al. 2011) exposures. Based on positive findings in five of six SA and two of three stillbirth analyses, the Nurminen (1995) review concluded there was evidence to support an association between maternal agricultural occupation and pesticide exposure and SA and stillbirth. Several subsequent studies have been identified, including findings of no SSIG association between stillbirth and pesticides among Indonesian women exposed as spray operators or rice farmers (Murphy et al. 2000). Also, the odds of SA among 973 pregnancies of female greenhouse workers were found to be SSIG elevated among those who reentered within 24 hours of pesticide application (OR_{ADJ} = 3.2, 95%

CI = 1.3–7.7) (Settimi et al. 2008). In a systematic review of birth outcomes among women living near agricultural pesticide applications, two SA studies showed no SSIG associations and four of five stillbirth studies reported suggestive increases RR in one or more subgroups (RR = 1.2–2.0), but none reached SSIG (Shirangi et al. 2011). An SSIG exposure–response gradient was reported in one of the stillbirth studies, however (White et al. 1988).

OCs DDT and its metabolite, DDE, are reproductive toxicants associated with fetal loss (Hruska et al. 2000). Among 388 newly married Chinese women in whom preconception serum DDT levels were measured, the subsequent odds of pregnancy loss were higher for the highest DDT tertile compared with those for the lowest (OR_{adj} = 2.12, 95% CI = 1.26–3.57); a 10-ng/g increase in total DDT was accompanied by relative odds of early pregnancy loss of 1.17 (95% CI = 1.05–1.29) in that study (Venners et al. 2005). One recent multicountry study found risks of fetal loss associated with both DDE and PCB 153 exposures, but the investigators cautioned that firm conclusions could not be drawn since there was a lack of dose–response and inconsistencies between countries (OR_{ADJ} = 2.4, 95% CI = 1.1–5.5, n = 678; Toft et al. 2010). During the two decades after the Yucheng PCB mass ingestion poisoning incident, the percentage of reported stillbirths among exposed women was double that of the control group, a finding that approached, but did not reach, SSIG (p = 0.068; 4.2% vs. 1.7%, n = 668; Yu et al. 2000). Other large PCB studies of female cohorts in Boston (827 assisted reproduction cycles; Meeker et al. 2011), Michigan (n = 1344 pregnancies; Small et al. 2007), and Australia (n = 200 pregnancies; Khanjani and Sim 2007) did not detect an association between SA and PCB exposure. Neither did a German clinic-based study of 89 women referred for repeated SA find associations between SA and blood levels of PCP, PCB, DDE, HCHs, and HCB (Gerhard et al. 1998).

THMs One U.S. study reported that women consuming large volumes of tap water (≥5 glasses/day) containing high levels of total THMs (>75 μg/L) had a SSIG higher risk of SA (OR_{ADJ} = 1.8, 95% CI = 1.1–3.0) (Waller et al. 1998).

Solvents Occupational exposure of women to solvents during pregnancy is SSIG associated with increased risk of SA, based on a meta-analysis of 19 formaldehyde studies (overall RR = 1.76, 95% CI = 1.20–2.59) (Duong et al. 2011). Subgroups with higher versus lower maternal occupational toluene exposure reportedly had SSIG increased SAs across several individual studies reviewed (Lindbohm 1995; Ng et al. 1992b [OR_{ADJ} = 4.80, 95% CI = 1.01–22.86, N = 173 pregnancies]). A meta-analysis of five studies of various solvents revealed an increase in the odds of SA across those studies that bordered but did not reach SSIG (overall OR = 1.25, 95% CI = 0.99–1.58) (McMartin et al. 1998). The earliest study of SA after exposure to toluene and other solvents was part of a cross-sectional survey on pregnancy history and occupational exposure by female laboratory workers in Sweden. The study showed a non-SSIG increase in the risk of SA among exposed workers (Axelsson et al. 1984). Subsequent studies reported that female workers and wives of workers exposed to toluene had consistently higher rates of SA compared with control groups (Lindbohm et al. 1990; Ng et al. 1992b; Taskinen et al. 1989, 1994). Adjusted ORs for maternal toluene exposure in these studies were noteworthy and SSIG ranged from 2.79 (95% CI = 1.32–5.88) to 4.80 (1.01–22.86) (Ng et al. 1992b). Xylene (OR_{ADJ} = 3.1, 95% CI = 1.3–7.5) and formalin (OR_{ADJ} = 3.5, 95% CI = 1.1–11.2) were each SSIG associated with SA in one of these larger studies (N = 535) (Taskinen et al. 1994).

In Beijing, a higher percentage of female petrochemical factory production workers than nonchemical plant workers reported SA (8.8% vs. 2.2%; overall OR_{ADJ} = 2.7, 95% CI = 1.8–3.9) (Xu et al. 1998). Benzene and gasoline exposure were individually associated with SSIG elevated odds of SA in that study. Similarly, women working at petrochemical plants in Sweden had SSIG heightened odds of miscarriage (OR = 6.6, 95% CI = 2.3–19.2) (Axelsson and Molin 1988). In a subanalysis of chemical workers from that study, SA was SSIG ($p < 0.05$) elevated specifically among the laboratory workers (Axelsson and Rylander 1989).

Smoking Smoking is an established risk factor for stillbirths (Rogers 2009). The smoking population's attributable risk has been estimated at 4–7% in high-income countries to as high as 20% in disadvantaged subgroups within those countries in one meta-analysis of 96 population-based studies (Flenady et al. 2011). A 2010 U.S. Surgeon General's Report concluded there was consistent evidence that linked "maternal smoking to interference in the physiological transformation of spiral arteries and thickening of the villous membrane in forming the placenta" and suggested "smoking leads to immunosuppressive effects, including dysregulation of the inflammatory response"(U.S. DHHS 2010). Risk of fetal loss was suggested to be heightened by both of these mechanisms.

Maternal Hypertension Pregnancy-induced hypertension is linked to stillbirths and preterm births (PTBs), and sometimes portents the life-threatening (maternal and fetal) condition, preeclampsia.

PFCs A court-appointed scientific panel assessed harm caused by PFOA contamination of six Ohio and West Virginia water districts by manufacturers. They determined that PFOA was "more likely than not" to have caused pregnancy-induced hypertension in that settlement class (Holtcamp 2012). The panel noted data on prematurity, stillbirth, and preeclampsia were insufficient to draw firm conclusions, but that a moderate potential "signal" of a link between PFOA and preeclampsia was suggested.

Smoking A 2010 Surgeon General's Report on smoking relayed that maternal smoking during pregnancy transiently raises maternal heart rates and blood pressure (mainly diastolic) (U.S. DHHS 2010).

Live Birth Outcomes

Prematurity Premature (aka, preterm) births are those occurring before 37 weeks (259 days) of gestation. Infants born prematurely are at near- and long-term risk for adverse outcomes. Those with the shortest gestations are most at risk for infant death and other adverse complications such as cerebral palsy, respiratory problems, intellectual disabilities, vision and hearing loss, and feeding and digestive problems (U.S. Department of Health and Human Services, Centers for Disease Control and Prevention, National Center for Chronic Disease Prevention and Health Promotion, Division of Reproductive Health and National Center for Birth Defects and Developmental Disabilities 2011). Known risk factors include multiple births, previous PTB, black race, uterus or cervical problems, chronic maternal hypertension, clotting disorders or diabetes, certain infections in pregnancy, and lifestyle factors (smoking, alcohol, and illicit drug use).

OCs Studies of OC exposures (e.g., PCBs, DDE, DDT, HCB) have demonstrated mixed results, but several of the largest studies we reviewed reported SSIG or near-SSIG direct relationships. Odds of prematurity increased in tandem with increasing DDE exposure quintiles, that is, $OR_{ADJ} = 1, 1.5, 1.6, 2.5, 3.1$ (SSIG trend, $p < 0.0001$), among eligible U.S. Collaborative Perinatal Project births ($n = 2380$) (Longnecker et al. 2001). Risk of prematurity among 70 neonates born to parents residing near an electrochemical plant in Spain was also directly associated with umbilical cord levels of DDE (2.40 vs. 0.80 ng/mL in prematures and nonprematures, respectively; $p < 0.05$) and PCBs (0.70 vs. 0.14 ng/mL, $p < 0.05$) and approached SSIG for HCB (1.94 vs. 1.10 ng/mL, $p < 0.10$) (Ribas-Fitó et al. 2002). In another study (Wojtyniak et al. 2010), maternal serum DDE levels among women in Kharkiv, Poland ($N = 661$), and Greenland Inuit women ($N = 572$), as well as PCB 153 levels among Inuits only, were SSIG associated ($p < 0.05$) with decreased gestation length but not prematurity. Gestation length was inversely associated with both PCB levels in cord blood ($p < 0.05$) and self-reported maternal consumption of PCB-contaminated fish ($p < 0.01$) for women who delivered in hospitals near Lake Michigan, based on 313 newborns (Fein et al. 1984). Conversely, several studies failed to detect SSIG associations between OCs and shortened gestation (Berkowitz et al. 1996 [$n = 40$]; Farhang et al. 2005 [$n = 420$ male newborns]; Khanjani and Sim 2006 [$n = 200$]; Longnecker et al. 2005 [$N = 1043$]; Torres-Arreola et al. 2003 [$N = 233$]), yet two of these showed elevated ORs or borderline SSIG (Khanjani and Sim 2006; Torres-Arreola et al. 2003). The statistical power of some of these studies was limiting. Future studies with ample power are indicated to clarify the potential role of BOC and POC exposures on the length of gestation.

Solvents In their 2001 review, Scheeres and Chudley (2002) cited clear evidence that toluene abuse during pregnancy increases the risk of prematurity, but low-dose human studies are needed.

Smoking One review noted that dose–response relationships between maternal smoking and PTB have been reported across several studies (Rogers 2009). A recent 2010 Surgeon General's Report implicated smoking-induced placental abnormalities and immunosuppressive dysregulation of the inflammatory response (U.S. DHHS 2010) in the relationship between smoking and increased risk of prematurity (Rogers 2009). The report also concluded there was consistent evidence linking genetic variations in metabolizing enzymes (e.g., GSTT1) and prenatal tobacco smoke exposure with increased risk of adverse pregnancy outcomes such as reduced gestation and lowered birth weight.

Fetal Growth Fetal growth is influenced both by genetics and the intrauterine environment. Prenatally, fundal height and ultrasound imaging are applied to estimate gestational age and to clinically monitor fetal growth. Birth size metrics (i.e., birth weight, length, ponderal index as birth weight/length$^3 \times 100$, head circumference) are most often used to characterize fetal growth in epidemiology studies, with birth weight being the most common end point. Head circumference can be altered by molding during vaginal births, as opposed to C-section deliveries (Apelberg et al. 2007). Placental weight is sometimes also considered. Often, gestational age estimates at birth are taken into account in the analysis or interpretation of birth weight studies to isolate the effects of growth (i.e., small for gestational age [SGA] births) versus prematurity. It is important to note that birth weight studies may capture subtle growth shifts; however, the extremes of fetal growth curves are where the greatest risks reside. Broadly, many factors known or suspected to increase the risk of fetal loss also increase the risk of intrauterine growth restriction and SGA births (Reutman and LeMasters 2007). Risks of cardiovascular disease, osteoporosis, and type 2 diabetes are heightened among those who are small at birth and had poor growth during infancy, especially preceded by excessive childhood weight gain (Godfrey et al. 2011).

Pesticides (General) Studies of parental exposures to pesticides in general and fetal growth have been reviewed elsewhere (Hanke and Jurewicz 2004; Nurminen 1995; Sanborn et al. 2007; Shirangi et al. 2011). Some of these studies specifically addressed maternal pesticide exposures, most of which were occupational (Dabrowski et al. 2003; Heidam 1984; Lima et al. 1999; McDonald et al. 1988; Murphy et al. 2000; Schwartz et al. 1986; Xiang et al. 2000; Zhang et al. 1992) or residential in proximity to agricultural pesticide applications

(Grether et al. 1987; Levario-Carrillo et al. 2004; Thomas et al. 1992; Willis et al. 1993). Pesticide exposure was associated with SSIG findings, that is, lowered birth weight by 117 g ($p = 0.05$) and low birth weight (RR = 3.6) and ($p < 0.05$, beet crops) in only three of these studies. However, non-SSIG increases in intrauterine growth retardation (RR = 2.3–2.9) and low birth weight (RR = 2.3) were reported in three more of these studies. In more recent studies, early pregnancy maternal exposure to pesticides was not SSIG associated with birth weight in a study of births to 2246 Iowa and North Carolina (United States) farm women (with the exception of carbaryl) (Sathyanarayana et al. 2010). Fetal weight and head circumference were non-SSIG reduced, and placental weight were reported among 4680 infants born to mothers in the Netherlands, and placental weight reduction reached SSIG ($p < 0.05$) (Snijder et al. 2012). Inconsistencies in findings between studies likely reflect differences in exposures (e.g., specific pesticides used, levels, exposure timing in relationship to births, coexposures), study designs (e.g., individual vs. aggregate exposure data), and regional differences, to name just a few.

OCs A recent meta-analysis of European studies in which PCBs and DDE were quantified in maternal or cord samples concluded that low-level PCB exposures (or other correlated exposures) are associated with lowered birth weight, whereas DDE exposures were not (Govarts et al. 2012). Birth weight decreased with increasing PCB 153 after confounder adjustment in 12 of the 15 studies examined. This conclusion is consistent with many additional individual PCB birth weight studies (Fein et al. 1984; Halldorsson et al. 2008; Hertz-Picciotto et al. 2005a; Murphy et al. 2010; Patandin et al. 1998; Rylander et al. 1995, 1998; Sonneborn et al. 2008; Wojtyniak et al. 2010) but is inconsistent with several others (Givens et al. 2007; Gladen et al. 2003; Grandjean et al. 2001; Longnecker et al. 2005; Vartiainen et al. 1998). Smaller head circumference was also reported with increasing PCB levels among offspring of fishermen's wives who consumed OC (including PCB)-contaminated fish from the Baltic Sea (Fein et al. 1984; Hertz-Picciotto et al. 2005a; Rylander et al. 1995). Several of these PCB studies also examined DDE, as did a number of other DDE studies described in a recent paper (Lopez-Espinosa et al. 2011). DDE exposure was linked inversely to birth weight and SGA births in a number of these studies, while one study reported an exposure-related birth weight increase. However, consistent with the above-mentioned meta-analysis, most of the studies failed to reveal any DDE effect. Several studies of DDT, HCB, and HCH exposure also failed to find a birth weight effect, with some exceptions (Lopez-Espinosa et al. 2011).

Organophosphates (OPs) have been widely used as neurotoxic insecticides in agriculture. They were banned for residential use for over a decade (Williams et al. 2008). Prior to that, they were used extensively for residential use, during which time low-level exposure was ubiquitous. One recent review included studies of four U.S. epidemiological cohorts (i.e., the CCCEH, CHAMACOS, Mt. Sinai CCEHDPR, and a New Jersey cohort) in which associations between fetal growth and prenatal exposure to the OP chlorpyrifos were examined (Mink et al. 2012). Reviewers evaluated these studies in terms of the potential implications of their findings for risk assessment but concluded that no strong associations with consistent exposure–response patterns were shown. Another review of three of the same cohorts similarly concluded that the study results were not consistent (Koureas et al. 2012).

ATZ Potential fetal growth effects of the herbicide ATZ, a highly prevalent drinking water contaminant, were examined across several studies. Borderline (non-SSIG) associations between maternal pregnancy levels of an ATZ metabolite and outcomes of fetal growth restriction and small head circumferences were reported among 579 live births included in a nested case-cohort study in Britanny, France (i.e., PELAGIE Cohort) (OR_{ADJ} = 1.5, 95% CI = 1.0–2.2 and OR_{ADJ} = 1.7, 95% CI = 1.0–2.7, respectively) (Chevrier et al. 2011). Results of two ecological studies of agricultural community drinking water supplies suggest the third trimester of pregnancy might be the most vulnerable exposure window for putted effects of ATZ on SGA. Specifically, SGA was SSIG associated with mean ATZ water concentrations (i.e., above 0.644 μg/L) across all trimesters (PR_{ADJ} = 1.14, 95% CI = 1.03–1.24) among 24,154 Indiana (United States) births. In the same study during the third trimester, exposure to even lower ATZ concentrations was associated with an SSIG increase in SGA prevalence compared to controls (PR_{ADJ} = 1.17, 95% CI = 1.03–1.24) (Ochoa-Acuña et al. 2009). Similarly, based on a subanalysis of 238 SGA births in Finistère, France, when high ATZ water contamination (occurred May–September) coincided with women's entire third pregnancy trimester, the risk of SGA deliveries was elevated (OR_{ADJ} = 1.54, 95% CI = 1.11–2.13) (Villanueva et al. 2005). In that study, however, no relationship was found between ATZ levels measured in municipal water supplies and SGA or low birth weight. Another study of births to mothers in Iowa (United States) reported rural communities with water contaminated with ATZ and other herbicides had SSIG elevated risk of intrauterine growth retardation outcomes compared to control communities (RR 1.8, 95% CI = 1.3–2.7) (Munger et al. 1997). Potential limitations cited by authors of ecological ATZ studies included design constraints, such as aggregate ATZ water measurements and limited confounder data.

PFCs PFOA, but not PFOS, in maternal plasma was SSIG associated with lowered birth weight in a nationally representative analysis of 1400 births in Denmark (β_{ADJ} = –10.63 g, 95% CI = –20.79 to –0.47 g) (Fei et al. 2007). Increases in exposures to PFOA and PFOS were each associated with SSIG decreases in newborns' ponderal index (in $g/cm^3 \times 100$ units: –0.070, 95% CI = –0.138 to –0.001; and –0.074, 95% CI = –0.123 to –0.025, respectively, after adjustment). Reductions (non-SSIG) in simple birth with increased PFOA and PFOS exposures were also reported in a smaller (293 cord serums) cross-sectional hospital-based study in Maryland (United States) (per 2.7-fold unit cord level increase corresponded to –104 g, 95% CI = –213 to 5 g; and

–69 g, 95% CI = –149 to 10 g, respectively, after adjustment) (Apelberg et al. 2007). Birth weight also was SSIG decreased with PFOS in a hospital-based prospective cohort study of 428 Japanese newborns (10-fold unit PFOS increase associated with –148 g, 95% CI = –297.0 to –0.5 g), which appeared driven by birth weight reductions among females (Washino et al. 2009). Another PFOS analysis of 421 live births to women in an occupational mortality cohort found no association with birth weight (Grice et al. 2007). Head circumference was also SSIG reduced with PFOA and PFOS in adjusted analyses in the U.S. birth weight study (cord blood unit increases of 2.7-fold corresponded to –0.32 cm, 95% CI = –0.56 to –0.07 cm; and –0.41 cm, 95% CI = –0.76 to –0.07 cm, respectively, after adjustment) (Apelberg et al. 2007). A review paper with additional perspectives on the topic of perfluorinated acids, including PFOS and PFOA, has been published by industry experts (Olsen et al. 2009).

Smoking Evidence that active and passive maternal smoking represent risk factors for low-birth-weight births has been described as unequivocal (Rogers 2009). It was suggested in the 2010 Surgeon General's Report that smoking-induced placental problems may contribute to low birth weight (U.S. DHHS 2010).

A broad review published in 2008 described evidence (including that cited from the Collaborative on Heath and the Environment) for adverse effects of BOCs and POCs on either fetal growth or gestational age as follows: "limited" evidence for perfluorinated acids, carbon tetrachloride, dioxins/TCDD, perchlorethylene and trichloroethylene in water, phenoxyacetic herbicides, and phthalates; "moderate" evidence for air pollution, herbicides, nicotine, OC and OP pesticides, pentachlorophenol, PCBs, solvents, and water disinfection by-products; and "strong" evidence for carbon monoxide, cocaine, ethanol, and tobacco smoke (Windham and Fenster 2008).

Sex Ratios Secondary sex ratio (SSR) alterations (i.e., altered boy to girl ratios at birth) are thought to be caused by factors that modify the illusive "primary" sex ratio at conception or selectively affect the rate of lethal conditions postconception (e.g., nonviable anomalies) for one gender versus the other. It has been proposed that alterations in SSRs are mediated by "periconception and intrauterine hormone levels, insemination timing relative to ovulation/oocyte maturity, sexual behavior (e.g., family size and age of mate), adaptive responses to environmental stressors, and actions of toxicants" (Pergament et al. 2002). Different formulas are used to derive SSRs, so conversion may be required for comparisons across groups. The contemporary mass rise of selective abortion of females in Asia and elsewhere, contributing to an estimated 160 million "missing girls" (Douthat 2011), may complicate comparisons of SSRs across different time periods or with societies where that practice is widespread (Wilcox and Baird 2011).

DES Altered SSRs have been reported in populations where SSR distortions from selective abortions were improbable. Such was the case when an SSIG excess of baby boys (F2) was born to a subset of 1775 women (F1) exposed *in utero* to DES when their mothers (F0) took that drug to prevent miscarriage. Pregnant women were prescribed DES, a synthetic estrogen, between 1940 and 1971 as it was thought to prevent premature labor and miscarriage. Tragically, DES is now known as a transplacental endocrine disruptor (ED), teratogen and carcinogen. The subset of women whose mothers took their first DES dose earliest in gestation (<13 weeks) at the highest cumulative doses (≥5 g) had SSIG higher odds of a male birth versus unexposed controls ($OR_{ADJ} = 1.24$, 95% CI = 1.04–1.48), but no SSIG SSR effects with DES were described overall (Wise et al. 2007).

OCs and Polybrominated Compounds Neither would prenatal sex selection account for the SSIG ($p < 0.001$) excess in females among 74 births to those accidentally exposed to TCDD in Seveso, Italy, who gave birth within 9 months to 7 years after that event. No SSIG alteration in SSR was reported among births occurring in later years (Mocarelli et al. 1996). No association between TCDD levels in breast milk and SSR was apparent in a follow-up analysis of 15 births to a small Kazakhstan cohort exposed to high doses of TCDD (Hooper et al. 1998). In contrast with the female newborn preponderance in Seveso, five of six Kazakhstan newborns with the highest TCDD levels were males. Authors of the Kazakhstan study speculated culturally influenced heightened participation by mothers of sons may have biased their results.

No SSIG SSR effects of accidental PCBs or PCDF ingestion were seen within the F1 generation (children), including 137 children of highly exposed women who accidentally consumed PCBs and PCDF in the Yucheng, Taiwan, tragedy. Nor was there an SSR effect overall among the F1 generation of victims who accidentally ingested PCBs and PCDF in Yusho, Japan (Rogan et al. 1999). However, when the Yusho cohort analysis was restricted to births to parents under age 20 years at the time of the incident, there was a trend toward fewer male births to exposed women in both the F1 and F2 (grandchildren) generations ($p = 0.06$ and $p = 0.02$, respectively) (Tsukimori et al. 2012). Since the most marked sex ratio effect was a reduction in the number of girls born to early-exposed Yusho mothers, the authors posited that there may have been greater female susceptibility and transmission through the female line.

Results of other analyses of PCBs in F1 generations have also found a decline in male births (i.e., more females) in some, but not all, studies. This may not be surprising, given the variety of populations and, thus, PCB congeners likely encountered. Among 99 births to women in the New York Angler Cohort, there was a non-SSIG increase in the number of boys born to women with medium ($OR_{ADJ} = 1.29$) and high ($OR_{ADJ} = 1.48$) serum levels of estrogenic (mostly lower chlorination) PCB congeners. On the other hand, in this study, there was a non-SSIG increase in the number of girls ($OR_{ADJ} = 0.70$) born to women with high levels of antiestrogenic (somewhat higher chlorination) PCB congener (Taylor et al. 2007).

Another analysis of 399 newborns in the San Francisco Child Health and Development cohort also found SSIG more girls ($p < 0.02$) born when exposed *in utero* to (total) PCBs and a non-SSIG increase in the number of girls born when exposed *in utero* to the mostly higher chlorinated individual PCB congeners (Hertz-Picciotto et al. 2008). The authors cautioned the associations could also be due to PCB metabolites or contaminants. Two additional analyses found SSIG larger proportions of girls. Among 208 births to U.S. Great Lake women, more girls were born to mothers within the top quintile of maternal serum PCB levels than were born to mothers within the lowest quintiles (male birth $OR_{ADJ} = 0.18$, 95% CI = 0.06–0.59) (Weisskopf et al. 2003). Similar results were seen among 5054 births to PCB-exposed Baltic Sea mothers compared to those from Sweden's southwest coast ($p = 0.05$) (Rylander et al. 1995). Conversely, an analysis of 865 Michigan long-term PBB cohort births found a non-SSIG trend for more boys born to parents with elevated PBB and PCB levels ($OR_{ADJ} = 1.43$ and 1.53, respectively) (Terrell et al. 2009). No SSIG relationship between cumulative PCBs and SSR was found among 2595 births to women who worked in capacitor plants (Rocheleau et al. 2011).

Birth Defects Birth defects (i.e., congenital disorders or anomalies) are defined by the WHO as "structural or functional abnormalities, including metabolic disorders, which are present from birth." Incidence and mortality estimates for birth defects are uncertain, particularly in regions with less complete surveillance (WHO 2008). In Europe during 2003–2007, "major" congenital anomalies including chromosomal disorders were detected in approximately 2.4% of registered conceptuses (i.e., of all live births, fetal deaths after 20 weeks' gestation, and induced abortions after prenatal detection of fetal anomalies) (Dolk et al. 2010). Similarly, state and local U.S. data on 45 birth defects suggest they occur in approximately 3% of births (CDC 2006). Congenital heart defects, neural tube defects, and Down syndrome are the most common serious "major" anomalies (WHO 2008). "Major" birth defects cause approximately 7% of neonatal deaths worldwide and up to 25% of neonatal deaths in the European Region, where competing causes of neonatal death rates are lower (WHO 2008). "Minor" congenital anomalies (e.g., nevi, angiomas, preauricular tags, cryptorchidism [i.e., undescended testicle{s}]) are collectively very common (these were excluded from the above European and North American anomaly estimates). Later sequelae among survivors with major anomalies may be negligible or range from cosmetic concerns or functional disabilities to heightened disease susceptibility or reduced life expectancy. The specific causes of most birth defects are unknown (CDC 2008), but general causes include genetic disorders (at gene or chromosomal levels), multifactorial inheritance, micronutrient deficiencies, and periconceptual or prenatal exposure to environmental teratogens (WHO 2008). Gene–gene and gene–environment interactions may modify their phenotypic expression. A gene–environment interaction occurs when the combined effect of a particular gene variant and an environmental exposure exceeds what would be expected on the basis of the effects of the gene and the exposure measured in the absence of the other. Epigenetics also plays a role in several genetic congenital syndromes. The rarity of many birth defects and large sample sizes required to study such interactions are among the research challenges (Kartiko and Finnell 2009).

Pesticides (General) Studies of parental exposures to pesticides and birth defects have been reviewed elsewhere (Hanke and Jurewicz 2004; Nurminen 1995; Sanborn et al. 2007; Shirangi et al. 2011; Thulstrup and Bonde 2006). A subset of these studies has focused more specifically on women exposed to pesticides, primarily occupationally (García et al. 1999; Heeren et al. 2003; Hemminki et al. 1980; Lacasaña et al. 2006; McDonald et al. 1988; Medina-Carrilo et al. 2002; Nurminen et al. 1995; Restrepo et al. 1990; Shaw et al. 1999, Zhang et al. 1992) or through domestic application (e.g., garden, field, professional home application) (Loffredo et al. 2001; Shaw et al. 1999). The overall adjusted odds of for birth defects reported in these studies ranged from a slight but non-SSIG increase among offspring of Finnish women exposed through agricultural work versus referents ($N = 2612$) to SSIG among South African women exposed in their domestic gardens and fields ($OR_{ADJ} = 1.4$ [95% CI = 0.9–2.07] and OR = 7.18 [95% CI = 3.99–13.25, matched analysis], respectively) (Heeren et al. 2003; Nurminen et al. 1995). At least marginal or greater SSIG was found in one or more of the above-mentioned studies for nervous system defects (i.e., neural tube [$OR_{ADJ} = 1.6$], anencephaly [$OR_{ADJ} = 4.57$], central nervous system [observed/expected = 7.5]), musculoskeletal defects (OR = 5.0), hemangiomas (RR = 6.66), oral-facial clefts ($OR_{ADJ} = 1.9$), transposition of the great arteries (OR = 2.10), and limb defects (SSIG PR = 2.6 and non-SSIG $OR_{ADJ} = 1.6$). (Engel et al. 2000; García et al. 1999; Hemminki et al. 1980; Lacasaña et al. 2006; Loffredo et al. 2001; Nurminen et al. 1995; Restrepo et al. 1990; Shaw et al. 1999; Zhang et al. 1992). These specific malformations did not reach SSIG in many studies, some of which were likely underpowered statistically. Further, a meta-analysis of 19 studies that applied covariate adjustments found modestly elevated odds of oral-facial clefts among births to female pesticide workers (OR = 1.37, 95% CI = 1.04–1.81) (Romitti et al. 2007). Also, a recent systematic review of 20 studies in which exposure was estimated based on proximity of residence to agricultural pesticide applications suggested an association with congenital malformations (CMs), but the evidence was considered inconclusive, weakened by insufficient exposure and confounder data (Shirangi et al. 2011).

Solvents Several studies have linked individual solvents and "solvents" as a broad class to congenital anomalies. A 1998 meta-analysis of five studies of maternal occupational exposure to various solvents found an SSIG increase in the prevalence of major malformations (OR = 1.64, 95% CI = 1.16–2.30) (McMartin et al. 1998). A subsequent review of studies of pregnant women's exposures to organic solvents provided "no convincing evidence" of a link to birth defects (Thulstrup and Bonde 2006).

Two population-based, prospective occupational studies of solvent-exposed working women were also identified. One of those studies reported a non-SSIG but noteworthy increase in the odds of circulatory and genital malformations among 11,273 offspring of (mostly) painters compared to controls (OR_{ADJ}; 95% CI = 2.03; 0.85–4.84; and 2.24; 0.95–5.31, respectively) (Vaktskjold et al. 2011). In the other study, maternal solvent exposure showed positive SSIG associations with oral clefts (OR_{ADJ} = 11.22), urinary malformations (OR_{ADJ} = 2.18–4.11), and male genital malformations (OR_{ADJ} = 1.98–3.47) (Garlantézec et al. 2009). Also, for the highest versus no-solvent exposure category, an SSIG association was reported with major malformations overall (OR_{ADJ} = 2.48–3.48) with an SSIG dose–response trend (p = 0.002–0.005) among offspring of women in solvent-exposed compared to nonexposed jobs (n = 3005). Toluene exposure of women was also associated with increased CMs for their pregnancies in two case-control studies (McDonald et al. 1987; Taskinen et al. 1989). The authors of both papers reported that pregnancies for women with a history of aromatic solvent exposure were at increased risk for CM births, primarily renal-urinary and gastrointestinal defects. Despite the many case studies on occupational and toluene abuse, the reproductive toxicology and the teratogenic effects of this organic solvent remain understudied.

OC Solvent Elevated ORs, albeit with wide, non-SSIG CIs, were found between prenatal exposure to tetrachloroethylene-contaminated drinking water near the time of conception and increased neural tube defects (OR 3.5, 95% CI = 0.8 – 14.0), oral clefts (OR 3.2, 95% CI = 0.7 – 15.0), and, among only those in the upper exposure quartile, "all anomalies" (OR 1.5, 95% CI = 0.9 – 2.5) in a Cape Cod, Massachusetts, retrospective cohort study subset of 1658 births (Aschengrau et al. 2009). Authors of a 2006 review of human and nonhuman studies of trichloroethylene-contaminated drinking water and congenital heart defects concluded the literature did not support the notion that TCE is a teratogen at environmentally relevant doses (Watson et al. 2006).

THMs Cross-sectional analysis of three studies of the relationships between THM exposure and birth defects showed consistently elevated odds of ventricular septal defects among prenatally exposed groups (summary OR = 1.59, 95% CI = 1.21 – 2.07) (Hwang et al. 2008). THMs are formed as by-products when water is disinfected with chlorine.

Smoking Maternal smoking is linked to CMs in the 2010 Surgeon General's Report, which found consistent evidence of an association between periconceptual smoking and cleft lip with or without cleft palate (U.S. DHHS 2010). The report suggests that genetic polymorphisms (e.g., transforming growth factor-alpha variants) may modify maternal smoking-mediated oral clefting risks.

Reproductive System Development The hypothalamic–pituitary–gonadal (HPG) axis feedback loops regulate (among other things) gonadal development and the mature reproductive system. Fetal period development and programming of the HPG axis are thought to render it particularly vulnerable to dysregulation by EDCs, which may result in permanent homeostatic endocrine alterations (Fisher 2004). Some such fetal effects may be readily apparent at birth (e.g., altered gonad morphology), while others may manifest latently after the newborn period through the reproductive years and beyond (e.g., effects on adult female gynecological or male andrological outcomes).

Among newborn boys, signs suggestive of reduced sexual dimorphism may include the fairly common major anomalies hypospadias (i.e., urethra opening on the undersurface of the penis) and cryptorchidism, or relatedy, other more subtle signs such as shortened anogenital distance (AGD). Testicular dysgenesis (TD) syndrome-related (male) newborn outcomes such as hypospadias, cryptorchidism, or later, poor semen quality and testicular germ cell cancer, as well as related non-TD signs such as reduced AGD, have been associated with prenatal EDC exposures. While individual study quality and findings are mixed, the pattern emerging from human and nonhuman studies raises the index of suspicion that certain intrauterine EDC exposures are risk factors for TD-related anomalies. Decades during which the incidence of hypospadias has increased have been documented for many countries (e.g., Norway, Denmark, England, Hungary, Norway, Finland, China) during the last half-century (Li et al. 2012; Toppari et al. 2010). Evidence of such trends is inconclusive, however (Thorup et al. 2010). The large incidence differences between these countries could reflect differences in exposure trends or in registry-based case ascertainment and reporting over time and between counties (Toppari et al. 2010).

Pesticides (General) A recent meta-analysis of nine human studies found maternal exposure to pesticides to be SSIG associated with hypospadias risk (Rocheleau et al. 2009).

OCs Developmental exposure to the OC TCDD may also produce latent male reproductive effects. The 21 breast-fed sons of TCDD-exposed Seveso cohort mothers, assessed at an average age of 22 years, had reduced sperm concentrations ($p = 0.002$), total counts ($p = 0.02$), progressive motility ($p = 0.03$), and total motile counts ($p = 0.01$), and reduced serum inhibin B levels ($p < 0.02$) and elevated serum follicle-stimulating hormone levels ($p < 0.04$) compared to 36 breast-fed controls and (for hormone outcomes) compared to 18 formula-fed controls (Mocarelli et al. 2011)

Phthalates Several review authors have concluded that exposure to phthalates (particularly antiestrogenic phthalates) during male development is implicated in some TD disorders (Chacko and Barthold 2009; Fisher 2004; Kalfa et al. 2011; Main et al. 2009, 2010; Swan et al. 2005; Toppari et al. 2010; Vidaeff and Sever 2005; Weselak et al. 2007; Wohlfahrt-Veje et al. 2009; Yiee and Baskin 2010).

BPA The effects of BPA on reproduction and development in studies of male and female animals were recently reviewed. The authors cited effects on "sex differentiation of exploratory and affective behavior" at lower BPA doses and "gender-differentiated morphology" as key cross-study findings (Golub et al. 2010). The NTP also reported the body animal literature is "sufficiently consistent" to suggest that exposure to "low" doses of BPA during perinatal or pubertal development causes neural and behavior alterations in rats and mice, "especially related to the development of normal sex-based differences between males and females ('sexual dimorphism')" (National Institutes of Health (NIH), National Toxicology Program (NTP) 2008). Observations of BPA wildlife exposure effects and controlled studies of animal models have been extensive, but their relevance to human health continues to be controversial (Maffini et al. 2006).

Among newborn girls who may have more subtle signs of altered sexual dimorphism at birth and in whom reproductive organs are not readily accessible for examination, studies of the effects of prenatal EDC exposure are sparser. Animal studies provide some insights because of the shorter time required for observation of maturational and transgenerational effects, and so more discussion of those studies is included here. For example, developmental exposures (i.e., during gestation through early life) of animals to certain EDCs have been shown to produce lifelong molecular reprogramming of the hypothalamus and premature reproductive aging, which suggests human MA may also be affected by certain prenatal EDC exposures (Gore et al. 2011). Therefore, both human and animal evidence is discussed further.

DES Several millions of daughters were exposed to the drug DES *in utero*, with many experiencing adverse reproductive health consequences. Reports of reproductive abnormalities and cancer in DES daughters have evoked the concept that other HAC exposures during pregnancy, particularly at high doses, might be capable of producing adverse reproductive organ and function effects in adults. In addition to developmental genital tract defects associated with prenatal DES exposure, many latent adverse reproductive effects have also been documented. Cumulative risks of functional effects among DES versus non-DES-exposed women reportedly include infertility (HR = 2.37, 95% CI = 2.05 – 2.75), SA (HR = 1.64, 95% CI = 1.42 – 1.88), loss of second trimester pregnancy (HR = 3.77, 95% CI = 2.56 – 5.54), stillbirth (HR = 2.45, 95% CI = 1.33 – 4.54), ectopic pregnancy (HR = 3.72, 95% CI = 2.58 – 5.38), preterm delivery (HR = 4.68, 95% CI = 3.74 – 5.86), and preeclampsia (HR = 1.42, 95% CI = 1.07 – 1.89) (Hoover et al. 2011). DES-exposed daughters also had an 80% higher risk of laparoscopically confirmed endometriosis than nonexposed women in the prospective Nurse's Health Study II cohort (RR_{ADJ} = 1.8, 95% CI = 1.2 – 2.8, *n* = 767) (Missmer et al. 2004). Contrasting effects of DES on uterine fibroids were observed in two studies of women exposed prenatally to DES: One study of the NIEHS Uterine Fibroid Study cohort found an SSIG positive association with prevalence and size of uterine fibroids using ultrasound criteria (n = 1323; Baird and Newbold 2005); the other study of the National Cooperative Diethylstilbestrol and Dieckmann cohorts found no association between DES exposure and diagnosis of fibroid using histology (n = 179; Wise et al. 2005). Histological detection, it was noted, requires surgery. Therefore, misclassification would have occurred if fibroids were undetected among controls in the latter study. Further suggestion of influence by early-life exposure on uterine fibroids comes from a recent analysis of black women in the NIEHS Sister Study breast cancer cohort in which intrauterine DES exposure was SSIG linked to early onset of fibroids (RR_{ADJ} = 2.02, 95% CI = 1.28 – 3.18 [n = 2394]). Thus, a possible relationship between women's exposure to DES and fibroids is supported by some, but not consistently all, studies. A combined secondary analysis of four large DES daughter cohorts and the Sister Study cohort of breast cancer found that DES-exposed women were about 50% (SSIG) more likely to go through menopause at any given age compared with nonexposed women (DES cohorts [HR_{ADJ} = 1.49, 95% CI = 1.28 – 1.74; combined n = 6017]; Sister Study [HR_{ADJ} = 1.45, 95% CI = 1.27 – 1.65; n = 19,103]) (Hatch et al. 2006; Steiner et al. 2010).

While cancers are covered in other chapters of this text, it seems important to note here that women exposed to DES *in utero* were shown to have an SSIG latent increase in their risk of breast cancer after age 40 (HR = 1.82, 95% CI = 1.04 – 3.18) and specific reproductive organ cancers. These included clear cell adenocarcinoma of the vagina and cervix in a prospective follow-up study conducted in the Netherlands (standardized incidence ratio = 24.23, 95% CI = 8.89 – 52.47) (Verloop et al. 2010). Animal studies suggest that developmental exposures to other estrogenic EDCs (e.g., genistein, nonylphenol, equivocally BPA) tested at equivalent estrogenic doses produce similar increases in uterine cancer and that DES cancer risk may be transmitted transgenerationally (Newbold et al. 2006). While the mechanisms remain to be clarified, genetic alterations and epigenetic induction of cell proliferation and apoptosis are suggested by animal studies (Newbold et al. 2006; Sato et al. 2004).

OCs In one human study, daughters from the California Child Health and Development study cohort had prolonged exposure to PCBs *in utero* (Cohn et al. 2011). Levels of PCB congeners 187, 156, and 99 in the mother's serum were SSIG associated with longer TTP in their daughters, while serum levels of PCB congeners 105, 138, and 183 were SSIG associated with shorter TTP (borderline SSIG for congener 183) (Cohn et al. 2011). The probability of pregnancy fell by 38% and infertility was higher in the group whose mothers had a higher proportion of PCB congeners that are associated with longer TTP; this suggests that exposure to some congeners of PCB *in utero* will affect human reproduction, either by increasing or decreasing TTP. Animal evidence also suggests that developmental exposure to PCBs, including PCBs having the highest body burden in humans, "profoundly impairs sexual differentiation of the female hypothalamus," causing masculinization/defeminization of the female neuroendocrine system (Dickerson et al. 2011). Pubertal onset was advanced and estrous cyclicity was irregular in PCB endocrine-disrupted females. Developmental exposure to PCB also disrupts reproduction in animals. Rats exposed to PCBs during late pregnancy exhibited compromised reproductive physiology in the exposed female fetuses (F1 generation) and their female offspring (F2 generation) (Steinberg et al. 2008). The exposure skewed liter sex ratios (F1 and F2 considered together) to be females and F1 had significantly altered circulating LH concentrations; more profound effects were observed on F2, with reduced LH and progesterone levels, and correspondingly smaller uterine and ovarian weights in estrus. In another study, F1 rats exposed to PCBs had aberrant estrous cycles and changes in hypothalamic gene and protein expression (Dickerson et al. 2011). Further, in mice, developmental (i.e., intrauterine and lactational) TCDD exposure of either parent results in reduced uterine sensitivity to progesterone and increased risk of PTB (<37 weeks' gestation) in F1 mice exposed *in utero* (Bruner-Tran and Osteen 2011; Ding et al. 2011).

PBBs Daughters, exposed to PBB prenatally and during breastfeeding, of PBB-exposed Michigan mothers had SSIG earlier pubic hair staging, while their *in utero* exposures were associated with earlier menarche (Blanck et al. 2000). Recent analyses of data from the Michigan cohort reported that *in utero* PBB exposure was not associated with a change in TTP in women (Small et al. 2011). A recent study reported that *in utero* exposure to PBB of women within the Michigan cohort was associated with an increased risk of SA as adults among their 135 viable and nonviable pregnancies that were not electively aborted (p = 0.05 for trend between low-, mid-, and high-exposure ranges) (Small et al. 2011). The respective ORs of 2.75 (95% CI = 0.64 – 11.79) and 4.08 (95% CI = 0.94 – 17.70) for those with moderate and high-level exposure compared to those with low exposure were non-SSIG but noteworthy. A high SA rate (i.e., 35%) was observed among the small subset of female infants in the highest intrauterine exposure group who were also breast-fed (n = 23) (Small et al. 2011).

Bisphenol Prenatal BPA exposure in various mouse studies has been reported to induce changes long after exposure, such as altered ovarian, uterine, and vaginal morphology, increased estrogen and progesterone positive receptors in the endometrium, altered mammary gland morphogenesis, early sexual maturity, and altered patterns of estrous cyclicity in adulthood (Markey et al. 2001, 2003, 2005; Muñoz-de-Toro et al. 2005; Rubin et al. 2001). Rats exposed to low-dose BPA prenatally developed some similar changes (e.g., altered estrous cyclicity ovarian changes) as well as reduced serum LH levels after ovarectomy in adulthood, findings suggested to support potential hypothalamus–pituitary–ovarian axis disruption (Markey et al. 2003; Muñoz-de-Toro et al. 2005; Rubin et al. 2001). For example, neonatal exposure to BPA induced PCOS-like alterations in the ovaries and hypothalamic–pituitary–endocrine axis in an animal study (Fernández et al. 2010).

Solvents Animal studies provide a controlled atmosphere where subjects are not exposed to "polysolvents." In one study, pregnant Wistar rats were exposed to inhaled 300–1200 ppm toluene for 6 hours each day during gestational days 9–21. The F1 rats exposed prenatally displayed no differences in mating, fertility, or pregnancy compared to control animals (Thiel and Chahoud 1997).

Smoking Evidence linking developmental (intrauterine) exposure of daughters to tobacco and lowered fecundity was recently reviewed and reported to be inconclusive. On the other hand, in by far the largest cohort included in that review, a "small to modest" association was reported between Norwegian women's exposure to tobacco smoke *in utero* and reduced fecundity as adults (fecundability $OR_{ADJ} = 0.96$, 95% CI = 0.93 – 0.98, $n = 48{,}319$) (Ye et al. 2010). Daughters of women who smoked during pregnancy may also experience earlier MA, suggesting fetal reprogramming, based on SSIG findings within a DES daughters cohort (nonsmoking daughters HR = 1.38, 95% CI = 1.10 – 1.74; overall $HR_{ADJ} = 1.21$, 95% CI = 1.02 – 1.43, $n = 1719$) (Strohsnitter et al. 2008).

Neurodevelopment Converging lines of evidence have been proposed to suggest EDCs primarily target the hypothalamic–pituitary axis of the neuroendocrine system with downstream effects across multiple systems and, it is suspected, transmission of transgenerational epigenetic effects (Gore 2010a,b). Prenatal maternal consumption of PCB-contaminated food has been linked to lowered IQ and behavioral outcomes (Stillerman et al. 2008). Prenatal thyroid hormone derangements are a proposed mechanism since normal maternal thyroid levels are important for optimal fetal growth and development, especially neurodevelopment. Even subtle disruption of thyroid gland function during pregnancy may produce adverse fetal outcomes. Concern has been raised regarding potential interactions of EDCs with other susceptibility factors on thyroid homeostasis during pregnancy, particularly in vulnerable subgroups such as those living in areas where iodine deficiency is endemic (Hartoft-Nielsen et al. 2011).

OCs The chemical structure of TCDD metabolites is thought to be sufficiently similar to that of thyroxine so as to interfere with the hormone's functions (Hartoft-Nielsen et al. 2011). Studies of the effects of PCBs on thyroid hormone effects in pregnant women, newborns, infants, and adolescents were recently reviewed (Hartoft-Nielsen et al. 2011; Langer 2010). Elevated cord PCB levels were associated with changes in elements of thyroid hormone secretion and action across several studies (i.e., thyroid-stimulating hormone, thyroid hormone-binding globulin, free thyroxine, total tri-idothyronine, and total thyroxine), while other studies reported no SSIG thyroid effects. Authors of these reviews acknowledged that most current evidence supports a link between human prenatal exposure to certain OCs and adverse developmental and health outcomes among offspring, but with caveats that this generalization applies to higher maternal exposures (Langer 2010) and that conclusions are based on relatively few available studies (Hartoft-Nielsen et al. 2011).

OPs Two U.S. studies of prenatal OP pesticide exposure found evidence of neurobehavioral effects including abnormal neonatal reflexes (Young et al. 2005), and at age 3 years, SSIG more delays in psychomotor and developmental plus maternally reported attention problems (Rauh et al. 2006).

Solvents *In utero* exposure to toluene, which has structural similarities to BPA, has been linked to developmental delays and neurobehavioral problems in children (Badham and Winn 2010a).

Smoking A 2010 Surgeon General's Report on smoking found "consistent evidence that carbon monoxide leads to birth weight deficits and may play a role in neurological deficits (cognitive and neurobehavioral end points) in the offspring of smokers" (U.S. DHHS 2010).

Immune and Hematopoietic Development Immune development begins before birth with hematopoiesis, stem cell migration and progenitor cell expansion, and thymus and bone marrow colonization (Dietert et al. 2000). Mounting evidence suggests certain BOC and POC exposures may alter immune system development, manifesting later as cellular and humoral (i.e., antibody mediated) immune perturbations in neonates and children. The concept of prenatal origin (i.e., initiation) of hematopoietic cancer in childhood is consistent with the observation that specific chromosomal rearrangements associated with childhood leukemia are already present in the blood of newborn infants (Van Maele-Fabry et al. 2011).

OCs Prenatal PCB levels have been associated with smaller estimated thymic volume ($p = 0.047$) at delivery, which the investigators suggested may reflect immunologic development impairment (Park et al. 2008a). In another study, prenatal PCB levels were SSIG correlated with the number of lymphocytes ($p = 0.05$), T-cell markers CD3CD8 + ($p = 0.04$), CD4 + CD45RO + ($p = 0.02$) and CD3 + HLA-DR + ($p = 0.005$), T-cell receptor $\alpha\beta$ + ($p = 0.08$), lower antibody levels among preschoolers for measles ($p = 0.03$), and mumps ($p = 0.04$) (Weisglas-Kuperus et al. 2000). Maternal serum levels of PCBs and PFCs during pregnancy and/or lactation have been associated with reduced humoral immune reactions to routine vaccinations in their children (Grandjean et al. 2012; Heilmann et al. 2006; Weisglas-Kuperus et al. 2000). Another study found little evidence of such an association with pre- or postnatal PCB exposure (Jusko et al. 2010).

Pesticides Prenatal residential exposure to pesticides was associated with childhood leukemia in two separate meta-analyses of 13 and 8 partially overlapping studies (OR = 2.05, 95% CI = 1.80 – 2.32; and meta-RR = 2.19, 95% CI = 1.45 – 2.09, respectively) (Turner et al. 2010; Van Maele-Fabry et al. 2011). The strongest risk signal was for insecticide exposure (prenatal and childhood), which was SSIG in both meta-analyses, herbicides showing a positive SSIG relationship with leukemia in only one of two meta-analyses (Van Maele-Fabry et al. 2011). Similarly, in a meta-analysis of 40 studies of various pesticide exposures (i.e., residential, household, professional), the odds leukemia and lymphoma were SSIG elevated among children exposed prenatally whose mothers used pesticides in the home or garden (OR = 1.48, 95% 1.26 – 1.75 and OR = 1.53, 95% CI = 1.22 – 1.91, respectively) (Vinson et al. 2011). Two other meta-analyses that included studies where mothers were exposed to pesticides occupationally also found associations with childhood leukemia. One of these meta-analyses of 16 studies found SSIG increased odds of childhood leukemia in association with maternal prenatal exposure to pesticides overall (OR = 2.09, 95% CI = 1.51 – 2.88) and, specifically, for exposure to insecticides and to herbicides (Wigle et al. 2009). In the other meta-analysis in which strata of women exposed before (three studies) and during (eight studies) pregnancy were examined, SSIG elevated RRs of childhood leukemia were reported in each exposure strata (RR = 2.24, 95% CI = 1.34 – 3.72 and RR = 2.00, 95% CI = 1.11 – 3.62, respectively), but only the prepregnancy exposure strata studies showed consistency (Van Maele-Fabry et al. 2010).

Bisphenols One review of animal studies found consistent effects by BPA on "immune hyperresponsiveness at lower doses" (Golub et al. 2010).

Solvents/PAHs Benzene is a carcinogen, genotoxicant, and hematotoxicant (Khan 2007). Long-term or chronic exposure to benzene has been linked to various blood disorders, including aplastic anemia and cancer (leukemia). Benzene also causes several types of malignancies in animals. Prenatal and childhood exposure to benzene has been associated with the development of disorders such as aplastic anemia and leukemia (Badham and Winn 2010a; Shu et al. 1988; Steffen et al. 2004). Researchers found that *in utero* exposure to benzene causes a significant increase in reactive oxygen species (ROS) and significantly altered erythroid differentiation, potentially leading to the development of blood disorders (Badham and Winn 2010b). Prenatal exposure to pollutants (PAH and fine particles) were associated with reduced human cord blood immune cells (T-lymphocyte fractions CD3+, CD4+, CD8+) and increased B-lymphocyte CD19+ cells in a Czech study of 1397 deliveries (adjusted percentage decreases of –3.3% [95% CI = –5.6 to –1.0], –3.1%

[−4.9% to −1.3%], −1.0% [−1.8% to −0.2%], and 1.7% [0.4% to 3.0%], respectively) (Hertz-Picciotto et al. 2005b).

DISCUSSION AND IMPLICATIONS

The bulk of studies of women's potential reproductive health risks from exposures to the more widespread bioactive and persistent compounds has focused on assessing pregnancy outcomes. Clearly, many such compounds are detectable in pregnant and nonpregnant women, cross the placental barrier, and are excreted in breast milk. Evidence from long-term epidemiological studies of women exposed to the estrogenic drug DES *in utero* also lends support to the concept that latent adverse effects on gynecological, reproductive, pregnancy, and neonatal outcomes could ensue from early exposures to other EDCs. Beyond DES, converging lines of evidence from human and animal studies conducted during the critical fetal development window support emerging conclusions regarding latent adverse health effects for several common pre- and perinatal exposures. Across reviews, documents, and studies surveyed in this chapter, the strength of human evidence is mounting for adverse intrauterine exposure effects of (1) PCBs (particularly estrogenic congeners) on lowered birth weight; (2) phthalates on TD syndrome; (3) various solvents on fetal loss; (4) smoking on altered reproductive hormone levels, lowered fertility, fecundity, birth weight, and MA, stillbirths, cleft lip malformations, and among smokers, increased maternal (diastolic) hypertension; and (5) pesticides on childhood leukemia risk.

Beyond these more established risks, growing evidence was described regarding other biologically active organic compounds and POCs suspected of adversely affecting women's gynecological health or reproductive success or of harming their offspring. In general, major strengths across these studies include the use of biomarkers to characterize exposure and outcomes. A number of positive significant and suggestive findings from epidemiological studies of women and their offspring have been of compounds considered endocrine disruptors, particularly the studies of developmental exposure-related outcomes. To date, however, the potential effects of many of the individual compounds across the full spectrum of women's and offspring's outcomes are understudied, including potential effects of those found in foods, homes, schools, and workplaces. These, to mention a few, include BPA, pesticides such as ATZ, phthalates, flame retardants, and perfluorinated acids. Published studies of gynecological outcomes are particularly rare. This lack of human studies for many outcomes is striking considering the prevalence of some exposures and outcomes, their possible severity and comorbidities, and the potential for adverse impacts across the life span. Infertility, for example, affects many couples. Accordingly, the Centers for Disease Control and Prevention (CDC) has drafted the specific needs for research and policies favoring improved understanding of environmental compound effects on fertility (i.e., final Federal Register Notice at this writing) (U.S. Department of Health and Human Services, Centers for Disease Control and Prevention, National Center for Chronic Disease Prevention and Health Promotion, Division of Reproductive Health 2012).

It seems reasonable that, given awareness and resources, some individuals may be able to reduce their personal BOC and POC exposures by taking preventative or protective precautions. For some women, avoidance of known external sources of these compounds—especially during vulnerable developmental windows—may be achievable. The evidence base for prevention at the individual level is incipient, but there are some findings that modifiable lifestyle factors influence exposure risk. For instance, a systematic review of dietary habits showed they influence POP levels in humans (Gasull et al. 2011); results of a dietary intervention study demonstrated that consumption of an organic diet provides a dramatic and immediate reduction in children's organophosphorus pesticide levels (Lu et al. 2006). Organic produce is, however, relatively expensive. Women with fewer resources, awareness, or options to reduce their exposures from their diets, occupations, or residences may not take such precautions, even during their pregnancies.

Further, many women are not aware they are pregnant during the critical window of fetal organogenesis. Women with fewer social and economic resources may be particularly vulnerable for other potentially tractable reasons such as poor nutrition, a lack of prenatal care and education, and poorer general health. Other women may be more susceptible genetically or disproportionately affected due to confluence of adverse outcomes in individuals and families, just as multiple adverse health effects may manifest in one unfortunate individual from exposure to smoking or DES exposure.

There is a clear risk identification and educational role for women's advocates, such as health-care providers, who are positioned to offer preventive or intervention outreach and support. This illustrates the need to expand efforts to identify if, when, and where bioactive and persistent exposure engender the greatest exposure and health risks, and how to best mitigate risks from both population and individual health standpoints. It also demonstrates the importance of training the trainers (doctors, nurses, others) to routinely assess work and environmental exposures and to intervene to help women avoid potential harm. Broad application of succinct clinical queries for environmental exposures (e.g., the "environmental and occupational health history profile") (Chalupka and Chalupka 2010) coupled with tools that provide clinicians with guidance based on consolidated and scientifically vetted evidence (e.g., the "Navigation Guide") (Sutton et al. 2010) are also indicated. Development of a deeper toolkit of evidence-based preventive interventions is also critical to effective preventive education and intervention. Many agencies and organizations maintain web-based information resources that may provide valuable assistance to women and advocates seeking to determine individualized risk profiles and preventive strategies. Table 10.3 lists a number of these resources with current links.

TABLE 10.3. Current Web-Based Resources for Additional Information on Women's Reproductive Health

Source	Comments	Website
Collaborative on Health and the Environment	Environmental health-related information resource on fertility, children's health, learning and developmental disabilities, and so on	http://www.healthandenvironment.org/index.php
Environmental Working Group (EWG)	Information on environment and health (including reproductive and developmental) from documents, studies and EWG laboratory tests—chemical index site is available	http://www.ewg.org
La Leche League	Searchable index with science and opinion on breast-feeding benefits given breast milk contaminants	http://www.llli.org/resources.html
Lact-Med	Peer-reviewed database of drugs to which breast-feeding mothers could be exposed	http://toxnet.nlm.nih.gov/cgibin/sis/htmlgen?LACT
March of Dimes	Birth defect prevention information	http://www.marchofdimes.com
Motherisk	Program created by the Hospital for Sick Children in Toronto to provide evidence-based content and guidance concerning potential risks to the developing fetus or infant, from exposure to chemicals, drugs, and other environmental agents	http://www.motherisk.org/women/index.jsp
National Library of Medicine (NLM), Toxicology Data Network (Toxnet)	Contain references and links to toxicology reports, documents, and literature within various databases: Developmental and Reproductive Toxicology Database (DART), Toxic Literature Online (Toxline), Hazardous Substances Data Bank (HSDB), Genetox, IRIS, Household Products Database, and so on	http://toxnet.nlm.nih.gov/cgi-bin/sis/htmlgen?TOXLINE
Natural Defense Council Resources Council	"Environmental and Health" section contains information on lowering exposures and touches on reproductive and developmental hazards	http://www.nrdc.org/
Organization of Teratology Information Specialists (OTIS)	Information includes news, research, and resources including fact sheets for patients pertaining to teratogens, including environmental agents	http://www.otispregnancy.org/
Our Stolen Future website	Authors of book by same name provide updates on science and ongoing policy debates on endocrine disruptors and suggest risk minimization strategies	http://www.ourstolenfuture.org/

Pediatric Environmental Health Specialty Units	Training and resources geared primarily to pediatric health professionals but with some relevance to prenatal prevention	http://aoec.org/PEHSU/index.html
Physicians for Social Responsibility	Environment and health site contains information, resource guides, and clinician toolkits relevant to reproductive health	http://www.psr.org/environment-and-health/
Silent Spring Institute website	Primary focus is on breast cancer. Provides information on the research and publications of the institute plus tools such as individual and community action kits	http://www.silentspring.org/
TEDX Endocrine Disruption Exchange	Contains information on low-dose functional and developmental effects of exposure to endocrine disruptors	http://www.endocrinedisruption.com
(U.S. CDC) Agency for Toxic Substances and Disease Registry	Reproductive site contains links to ASTDR reports describing reproductive effects of various environmental compounds	http://www.atsdr.cdc.gov/substances/toxorganlisting.asp?sysid=21
(U.S. CDC) National Institute for Occupational Safety and Health (NIOSH)	Occupational reproductive site with a links including NIOSHTIC-2, a searchable database of NIOSH supported health publications, documents, and reports	http://www.cdc.gov/niosh/topics/repro/
(U.S. CDC) National Center on Birth Defects and Developmental Disabilities	Links to information on alcohol exposure and birth defects	http://www.cdc.gov/ncbddd/index.html
(U.S. Environmental Protection Agency [EPA]) Endocrine Disruptor Screening Program	Describes the U.S. EPA's approach toward and progress on screening chemicals for endocrine effects	http://www.epa.gov/endo/
(U.S. FDA) Food and Drug Administration's Endocrine Disruptor Knowledge Base	Resources for predicting estrogenic and androgenic activity; for researchers/regulators to "foster the development of computational predictive toxicology models and reduce dependency on slow and expensive animal experiments"	http://www.fda.gov/ScienceResearch/BioinformaticsTools/EndocrineDisruptorKnowledgebase/default.htm
(U.S. NIH) National Institute of Environmental Health Sciences, National Toxicology Program	Information on animal "reproductive assessment by continuous breeding" and links to reproductive toxicology result abstracts	http://ntp.niehs.nih.gov/?objectid=54ADC224-F1F6-975E-7EAD62A55DD96034
University of California, San Francisco	Program on Reproductive Health and the Environment offers bilingual (English and Spanish) pdf patient brochure	http://prhe.ucsf.edu/prhe/toxicmatters.html

Broader societal awareness and actions are also required to lower the level and ubiquity of these exposures in the population. Many women and developing offspring are unwittingly exposed to low levels of the more common organic compounds. Even those who are aware of their potential exposures may assume chemicals in commerce have been tested for safety with the same diligence as drugs and foodstuffs, but this has rarely been the case, especially for female reproductive-related outcomes including adverse effects in women and development-related sequelae in their infants and children. Failures of effective scrutiny of and protections from common chemical exposures for the past several decades under the U.S. Toxic Substance Control Act system have been well documented with calls for reform or modernization from the American Medical Association, the American Public Health Association, the American Nurses Association, and the American Chemistry Council (Council on Environmental Health, American Academy of Pediatrics 2011). U.S. Farm Bill remediation to protect reproductive health has also been advocated (Sutton et al. 2011). Recent regulatory framework reforms have been sought in the United States with the EPA's Reform of Chemicals Management Legislation (2009) and implemented in Europe under "REACH," the European Community Regulation on chemicals and their safe use (2007).

The American Academy of Pediatrics has published further specific recommendations for government advocacy that would further strengthen protections by, for example, recognizing the special needs of women and children, particularly during critical developmental windows (Council on Environmental Health, American Academy of Pediatrics 2011). While remedies are underway in wealthier nations, public and workplace protections have likely been even less effective in less developed countries where the bulk of manufacturing and waste stream management problems now reside. Inequities in distribution of consumer benefits and exposure risks may exist, therefore, to the extent that fewer consumer benefits of products made from these compounds have accrued to those who experience greater exposures and health effects, with environmental justice implications.

Because many women's reproductive health problems are intimate and perhaps latent—silent for long intervals after their initiation—exposure–disease-related clusters are difficult to identify and characterize and may be less apt to garner attention than nonreproductive health problems. This highlights the importance of surveillance and diligent cross-discipline collaborations to identify present and emerging risks to women's reproductive health. Development and synthesis of the human and experimental evidence base will enhance the quality of future risk assessments of various organic compound exposures, doses, and vulnerable life stages. In an era of diminished reproductive health research funding, much of the women's reproductive health research agenda appears driven by the availability of existing data for secondary analyses. Funding is required to permit a more proactive approach to human studies so that exposures and potential effects can be documented temporally from the periconception period and beyond. Long-term follow-up, while costly, is essential to track links between early exposures and downstream outcomes that may manifest decades later. One such ambitious effort in the United States is the National Children's Study, in which BOCs/POCs have been measured longitudinally.

Ultimately, one of the major goals of research on BOCs and POCs is to provide valid inferences about health risks so that women, parents, advocates, and policy makers can make informed and astute exposure-related decisions. The BOC and POC human evidence base is currently too small to support robust risk assessments for many women's reproductive and developmental outcomes for which there is at least some preliminary suggestion of risk. This is especially true within the lower-dose ranges that are most relevant to human populations and for interpretation of the nonmonotonic dose–response relationships that are often observed. Thus, the benefits of the products and processes that create opportunities for human BOC and POC exposures are more defined than are the potential health risk costs. Some compounds that may inflict potential harm also confer great benefit (e.g., pesticides control malaria and increase food production while lowering food costs). DDT is one specific example. There is a pressing need to expand the small epidemiological literature base available to predict DDT-related risks of adverse pregnancy outcomes (e.g., pregnancy loss, prematurity) for comparison to established risks of morbidity and mortality from post-DDT ban malaria, malaria being a disease that kills one million annually worldwide (Longnecker 2005).

Clearly, there is a need to reduce data gaps and other sources of uncertainty to permit accurate accounting for exposure-related health costs associated with outcomes such as fetal loss, infertility, prematurity, birth defects, PCOS, endometriosis, and uterine fibroids, and for the sequelae that can result. Studies with ample power to characterize risks among subgroups who may be vulnerable due to endogenous or nonendogenous factors (e.g., genetic makeup, other health conditions) are also needed. The Endocine Society has recommended future epidemiological studies in large, highly exposed, prospectively followed cohorts. Important elements of future epidemiology studies include the use of validated exposure biomarkers, assessment of dose–response relationships, and assessment of effects of mixtures (Diamanti-Kandarakis et al. 2009). In the face of current uncertainty about such health risks, there have been calls for "precautionary principle" centered action from many preeminent organizations including The Endocrine Society, supported by the American Medical Association.

ACKNOWLEDGMENTS

We thank Kathleen Connick and John A. Batey for accessing critical publications, documents, and other library services, and Katherine Marlow for her assistance in managing citations.

REFERENCES

Abell A, Juul S, Bonde JPE. 2000. Time to pregnancy among female greenhouse workers. Scand J Work Environ Health 26:131–136.

Akkina JE, Reif JS, Keefe TJ, Bachand AM. 2004. Age at natural menopause and exposure to organochlorine pesticides in Hispanic women. J Toxicol Environ Health A 67:1407–1422.

Allmyr M, Adolfsson-Erici M, McLachlan MS, Sandborgh-Englund G. 2006. Triclosan in plasma and milk from Swedish nursing mothers and their exposure via personal care products. Sci Total Environ 372:87–93.

Anda EE, Nieboer E, Dudarev AA, Sandanger TM, Odland JØ. 2007. Intra- and intercompartmental associations between levels of organochlorines in maternal plasma, cord plasma and breast milk, and lead and cadmium in whole blood, for indigenous peoples of Chukotka, Russia. J Environ Monit 9:884–893.

Apelberg BJ, Witter FR, Herbstman JB, Calafat AM, Halden RU, Needham LL, et al. 2007. Cord serum concentrations of perfluoroocatane sulfonate (PFOS) and perfluorooctanoate (PFOA) in relation to weight and size at birth. Environ Health Perspect 115:1670–1676.

Aschengrau A, Weinberg JM, Janulewicz PA, Gallagher LG, Winter MR, Vieira VM, et al. 2009. Prenatal exposure to tetrachloroethylene-contaminated drinking water and the risk of congenital anomalies: A retrospective cohort study. Environ Health 8:44.

ATSDR (Agency for Toxic Substances and Disease Registry). 1995. Toxicology Profile for Diethyl Phthalate. Atlanta, GA: Division of Toxicology.

Augood C, Duckitt K, Templeton AA. 1998. Smoking and female infertility: A systematic review and meta-analysis. Hum Reprod 13:1532–1539.

Axelsson G, Molin I. 1988. Outcome of pregnancy among women living near petrochemical industries in Sweden. Int J Epidemiol 17:363–369.

Axelsson G, Rylander R. 1989. Outcome of pregnancy in women engaged in laboratory work at a petrochemical plant. Am J Ind Med 16:539–545.

Axelsson G, Lütz C, Rylander R. 1984. Exposure to solvents and outcome of pregnancy in university laboratory employees. Br J Ind Med 41:305–312.

Axmon A. 2006. Menarche in women with high exposure to persistent organochlorine pollutants in utero and during childhood. Environ Res 102:77–82.

Azziz R, Carmina E, Dewailly D, Diamanti-Kandarakis E, Escobar-Morreale HF, Futterweit W, et al. 2009. The androgen excess and PCOS Society criteria for the polycystic ovary syndrome: The complete task force report. Fertil Steril 91: 456–488.

Baba T, Mimura J, Nakamura N, Harada N, Yamamoto M, Morohashi K, et al. 2005. Intrinsic function of the aryl hydrocarbon (dioxin) receptor as a key factor in female reproduction. Mol Cell Biol 25:10040–10051.

Badham HJ, Winn LM. 2010a. *In utero* exposure to benzene disrupts fetal hematopoietic progenitor cell growth via reactive oxygen species. Toxicol Sci 113:207–215.

Badham HJ, Winn LM. 2010b. *In utero* and *in vitro* effects of benzene and its metabolites on erythroid differentiation and the role of reactive oxygen species. Toxicol Appl Pharmacol 244:273–279.

Baird DD, Newbold R. 2005. Prenatal diethylstilbestrol (DES) exposure is associated with uterine leiomyoma development. Reprod Toxicol 20:81–84.

Baird DD, Strassmann BI. 2000. Women's fecundability and factors affecting it. In: Women & Health (Goldman MB, Hatch MC, eds.). New York: Academic Press, p. 129.

Behr M, Oehlmann J, Wagner M. 2011. Estrogens in the daily diet: *In vitro* analysis indicates that estrogenic activity is omnipresent in foodstuff and infant formula. Food Chem Toxicol 49:2681–2688.

Bergonzi R, Specchia C, Dinolfo M, Tomasi C, De Palma G, Frusca T, et al. 2009. Distribution of persistent organochlorine pollutants in maternal and foetal tissues: Data from an Italian polluted urban area. Chemosphere 76:747–754.

Berkowitz GS, Lapinski RH, Wolff MS. 1996. The role of DDE and polychlorinated biphenyl levels in preterm birth. Arch Environ Contam Toxicol 30:139–141.

Blanck HM, Marcus M, Tolbert PE, Rubin C, Henderson AK, Hertzberg VS, et al. 2000. Age at menarche and tanner stage in girls exposed *in utero* and postnatally to polybrominated biphenyl. Epidemiology 11:641–647.

Blanck HM, Marcus M, Rubin C, Tolbert PE, Hertzberg VS, Henderson AK, et al. 2002. Growth in girls exposed *in utero* and postnatally to polybrominated biphenyls and polychlorinated biphenyls. Epidemiology 13:205–210.

Blanck HM, Marcus M, Tolbert PE, Schuch C, Rubin C, Henderson AK, et al. 2004. Time to menopause in relation to PBBs, PCBs, and smoking. Maturitas 49: 97–106.

Bonefeld-Jørgensen EC, Andersen HR, Rasmussen TH, Vinggaard AM. 2001. Effect of highly bioaccumulated polychlorinated biphenyl congeners on estrogen and androgen receptor activity. Toxicology 158:141–153.

Boverhof DR, Kwekel JC, Humes DG, Burgoon LD, Zacharewski TR. 2006. Dioxin induces estrogen-like, estrogen receptor-dependent gene expression response in the murine uterus. Mol Pharmacol 69:1599–1606.

Bretveld RW, Thomas CMG, Scheepers PTJ, Zielhuis GA, Roeleveld N. 2006. Pesticide exposure: The hormonal function of the female reproductive system disrupted? Reprod Biol Endocrinol 4:30.

Bretveld RW, Hooiveld M, Zielhuis GA, Pellegrino A, van Rooij IALM, Roeleveld N. 2008. Reproductive disorders among male and female greenhouse workers. Reprod Toxicol 25:107–114.

Brouwer A, Longnecker MP, Birnbaum LS, Cogliano J, Kostyniak P, Moore J, et al. 1999. Characterization of potential endocrine-related health effects at low-dose levels of exposure to PCBs. Environ Health Perspect 107(Suppl 4):639–649.

Bruner-Tran KL, Osteen KG. 2010. Dioxin-like PCBs and endometriosis. Syst Biol Reprod Med 56:132–146.

Bruner-Tran KL, Osteen KG. 2011. Developmental exposure to TCDD reduces fertility and negatively affects pregnancy outcomes across multiple generations. Reprod Toxicol 31:344–350.

Bukowski JA. 2001. Review of the epidemiological evidence relating toluene to reproductive outcomes. Regul Toxicol Pharmacol 33:147–156.

Calafat AM, Wong L-Y, Kuklenyik Z, Reidy JA, Needham LL. 2007. Polyfluoroalkyl chemicals in the U.S. population: Data from the National Health and Nutrition

Examination Survey (NHANES) 2003–2004 and comparisons with NHANES 1999–2000. Environ Health Perspect 115:1596–1602.

Calafat AM, Ye X, Wong L-Y, Reidy JA, Needham LL. 2008a. Exposure of the U.S. population to bisphenol A and 4-*tertiary*-octylphenol: 2003–2004. Environ Health Perspect 116:39–44.

Calafat AM, Ye X, Wong L-Y, Reidy JA, Needham LL. 2008b. Urinary concentrations of triclosan in the US population: 2003–2004. Environ Health Perspect 116: 303–307.

CDC (Centers for Disease Control and Prevention). 2006. Improved national prevalence estimates for 18 selected major birth defects—United States, 1999–2001. MMWR Morb Mortal Wkly Rep 54(51/52):1301–1305.

CDC (Centers for Disease Control and Prevention). 2008. Update on overall prevalence of major birth defects—Atlanta, Georgia, 1978–2005. MMWR Morb Mortal Wkly Rep 57:1–5.

Chacko JK, Barthold JS. 2009. Genetic and environmental contributors to cryptorchidism. Pediatr Endocrinol Rev 6:476–480.

Chalupka S, Chalupka AN. 2010. The impact of environmental and occupational exposures on reproductive health. J Obstet Gynecol Neonatal Nurs 39:84–100.

Chao H-R, Wang S-L, Lin L-Y, Lee W-J, Päpke O. 2007. Placental transfer of polychlorinated dibenzo-*p*-dioxins, dibenzofurans, and biphenyls in Taiwanese mothers in relation to menstrual cycle characteristics. Food Chem Toxicol 45:259–265.

Chapin RE, Robbins WA, Schieve LA, Sweeney AM, Tabacova SA, Tomashek KM. 2004. Off to a good start: The influence of pre- and periconceptional exposures, parental fertility, and nutrition on children's health. Environ Health Perspect 112: 69–78.

Chard T. 1991. Frequency of implantation and early pregnancy loss in natural cycles. Baillieres Clin Obstet Gynaecol 5:179–189.

Chen A, Zhang J, Zhou L, Gao E-S, Chen L, Rogan WJ, et al. 2005. DDT serum concentration and menstruation among young Chinese women. Environ Res 99: 397–402.

Chevrier C, Limon G, Monfort C, Rouget F, Garlantézec R, Petit C, et al. 2011. Urinary biomarkers of prenatal atrazine exposure and adverse birth outcomes in the PELAGIE birth cohort. Environ Health Perspect 119:1034–1041.

Cho SI, Damokosh AI, Ryan L, Chen D, Hu Y, Smith T, et al. 2001. Effect of exposure to organic solvents on menstrual cycle length. J Occup Environ Med 43: 567–575.

Chou W-C, Chen J-L, Lin C-F, Chen Y-C, Shih F-C, Chuang C-Y. 2011. Biomonitoring of bisphenol A concentrations in maternal and umbilical cord blood in regard to birth outcomes and adipokine expression: A birth cohort study in Taiwan. Environ Health 10:94.

Clayton EM, Todd M, Dowd JB, Aiello AE. 2011. The impact of bisphenol A and triclosan on immune parameters in the US population, NHANES 2003–2006. Environ Health Perspect 119:390–396.

Cobellis L, Latini G, DeFelice C, Razzi S, Paris I, Ruggieri F, et al. 2003. High plasma concentrations of di-(2-ethylhexyl)-phthalate in women with endometriosis. Hum Reprod 18:1512–1515.

Cobellis L, Colacurci N, Trabucco E, Carpentiero C, Grumetto L. 2009. Measurement of bisphenol A and bisphenol B levels in human blood sera from healthy and endometriotic women. Biomed Chromatogr 23:1186–1190.

Cohn BA, Cirillo PM, Sholtz RI, Ferrara A, Park J-S, Schwingl PJ. 2011. Polychlorinated biphenyl (PCB) exposure in mothers and time to pregnancy in daughters. Reprod Toxicol 31:290–296.

Colón I, Caro D, Bourdony CJ, Rosario O. 2000. Identification of phthalate esters in the serum of young Puerto Rican girls with premature breast development. Environ Health Perspect 108:895–900.

Cooper GS, Savitz DA, Millikan R, Kit TC. 2002. Organochlorine exposure and age a natural menopause. Epidemiology 13:729–733.

Cooper GS, Klebanoff MA, Promislow J, Brock JW, Longnecker MP. 2005. Polychlorinated biphenyls and menstrual cycle characteristics. Epidemiology 16: 191–200.

Council on Environmental Health, American Academy of Pediatrics. 2011. Policy statement—Chemical-management policy: Prioritizing children's health. Pediatrics 127:983–990.

Covaci A, Jorens P, Jacquemyn Y, Schepens P. 2002. Distribution of PCBs and organochlorine pesticides in umbilical cord and maternal serum. Sci Total Environ 298: 45–53.

Cragin LA, Kesner JS, Bachand AM, Barr DB, Meadows JW, Krieg EF, et al. 2011. Menstrual cycle characteristics and reproductive hormone levels in women exposed to atrazine in drinking water. Environ Res 111:1293–1301.

Crain DA, Janssen SJ, Edwards TM, Heindel J, Ho S-M, Hunt P, et al. 2008. Female reproductive disorders: The roles of endocrine-disrupting compounds and developmental timing. Fertil Steril 90:911–940.

Cullinan MP, Palmer JE, Carle AD, West MJ, Seymour GJ. 2012. Long term use of triclosan toothpaste and thyroid function. Sci Total Environ 416:75–79.

Curtis KM, Savitz DA, Weinberg CR, Arbuckle TE. 1999. The effect of pesticide exposure on time to pregnancy. Epidemiology 10:112–117.

Dabrowski S, Hanke W, Polańska K, Makowiec-Dabrowska T, Sobala W. 2003. Pesticide exposure and birthweight: An epidemiological study in Central Poland. Int J Occup Med Environ Health 16:31–39.

Dann AB, Hontela A. 2011. Triclosan: Environmental exposure, toxicity and mechanisms of action. J Appl Toxicol 31:285–311.

D'Aloisio AA, Baird DD, DeRoo LA, Sandler DP. 2011. Early-life exposures and early onset uterine leiomyomata in black women in the Sister Study. Environ Health Perspect 118:375–381.

Denham M, Schell LM, Deane G, Gallo MV, Ravenscroft J, DeCaprio AP. 2005. Relationship of lead, mercury, mirex, dichlorodiphenyldichloroethylene, hexachlorobenzene, and polychlorinated biphenyls to timing of menarche among Akwesasne Mohawk girls. Pediatrics 115:e127–e134.

Den Hond E, Roels HA, Hoppenbrouwers K, Nawrot T, Thijs L, Vandermeulen C, et al. 2002. Sexual maturation in relation to polychlorinated aromatic hydrocarbons: Sharpe and Skakkebaek's hypothesis revisited. Environ Health Perspect 110: 771–776.

Diamanti-Kandarakis E, Bourguignon J-P, Giudice LC, Hauser R, Prins GS, Soto AM, et al. 2009. Endocrine-disrupting chemicals: An endocrine society scientific statement. Endocr Rev 30:293–342.

Diamanti-Kandarakis E, Christakou C, Marinakis E. 2012. Phenotypes and environmental factors: Their influence in PCOS. Curr Pharm Des 18:270–282.

Dickerson SM, Cunningham SL, Patisaul HB, Woller MJ, Gore AC. 2011. Endocrine disruption of brain sexual differentiation by developmental PCB exposure. Endocrinology 152:581–594.

Dietert RR, Etzel RA, Chen D, Halonen M, Holladay SD, Jarabek AM, et al. 2000. Workshop to identify critical windows of exposure for children's health: Immune and respiratory systems work group summary. Environ Health Perspect 108(Suppl 3):483–490.

Ding T, McConaha M, Boyd KL, Osteen KG, Bruner-Tran KL. 2011. Developmental dioxin exposure of either parent is associated with an increased risk of preterm birth in adult mice. Reprod Toxicol 31:351–358.

Dolk H, Loane M, Garne E. 2010. The prevalence of congenital anomalies in Europe. In: Rare Diseases Epidemiology, Advances in Experimental Medicine and Biology (Posada de la Paz M, Groft SC, eds.). Dordrecht: Springer Science+Business Media B.V., pp. 349–364.

Douthat R 2011. 160 Million and counting. New York Times, June 26: A21.

Duong A, Steinmaus C, McHale CM, Vaughan CP, Zhang L. 2011. Reproductive and developmental toxicity of formaldehyde: A systematic review. Mutat Res 728: 118–138.

Edmunds KM, Holloway AC, Crankshaw DJ, Agarwal SK, Foster WG. 2005. The effects of dietary phytoestrogens on aromatase activity in human endometrial stromal cells. Reprod Nutr Dev 45:709–720.

Engel LS, O'Meara ES, Schwartz SM. 2000. Maternal occupation in agriculture and risk of limb defects in Washington State, 1980–1993. Scand J Work Environ Health 26: 193–198.

Eskenazi B, Mocarelli P, Warner M, Samuels S, Vercellini P, Olive D, et al. 2002a. Serum dioxin concentrations and endometriosis: A cohort study in Seveso, Italy. Environ Health Perspect 110:629–634.

Eskenazi B, Warner M, Mocarelli P, Samuels S, Needham LL, Patterson DG, Jr., et al. 2002b. Serum dioxin concentrations and menstrual cycle characteristics. Am J Epidemiol 156:383–392.

Eskenazi B, Warner M, Marks AR, Samuels S, Gerthoux PM, Vercellini P, et al. 2005. Serum dioxin concentrations and age at menopause. Environ Health Perspect 113: 858–862.

Eskenazi B, Warner M, Samuels S, Young J, Gerthoux PM, Needham L, et al. 2007. Serum dioxin concentrations and risk of uterine Leiomyoma in the Seveso Women's Health Study. Am J Epidemiol 166:79–87.

Euling SY, Herman-Giddens ME, Lee PA, Selevan SG, Anders J, Srensen TA, et al. 2008. Examination of US puberty-timing data from 1940 to 1994 for secular trends: Panel findings. Pediatrics 121:S172–S191.

Faiz AS, Demissie K, Rich DQ, Kruse L, Rhoads GG. 2012. Trends and risk factors of stillbirth in New Jersey 1997–2005. J Matern Fetal Neonatal Med 25:699–705.

Farhang L, Weintraub JM, Petreas M, Eszenazi B, Bhatia R. 2005. Association of DDT and DDE with birth weight and length of gestation in the Child Health and Development Studies, 1959–1967. Am J Epidemiol 162:717–725.

Farr SL, Cooper GS, Cai J, Savitz DA, Sandler DP. 2004. Pesticide use and menstrual cycle characteristics among premenopausal women in the Agricultural Health Study. Am J Epidemiol 160:1194–1204.

Farr SL, Cai J, Savitz DA, Sandler DP, Hoppin JA, Cooper GS. 2006. Pesticide exposure and timing of menopause. Am J Epidemiol 163:731–742.

Fei C, McLaughlin JK, Tarone RE, Olsen J. 2007. Perfluorinated chemicals and fetal growth: A study within the Danish National Birth Cohort. Environ Health Perspect 115:1677–1682.

Fei C, McLaughlin JK, Lipworth L, Olsen J. 2009. Maternal levels of perfluorinated chemicals and subfecundity. Hum Reprod 24:1200–1205.

Fein GG, Jacobson JL, Jacobson SW, Schwartz PM, Dowler JK. 1984. Prenatal exposure to ploychlorinated biphenyls: Effects on birth size and gestational age. J Pediatr 105:315–320.

Fernández M, Bourguignon N, Lux-Lantos V, Libertun C. 2010. Neonatal exposure to bisphenol A and reproductive and endocrine alterations resembling the polycystic ovarian syndrome in adult rats. Environ Health Perspect 118:1217–1222.

Fisher JS. 2004. Environmental anti-androgens and male reproductive health: Focus on phthalates and testicular dysgenesis syndrome. Reproduction 127:305–315.

Flenady V, Koopmans L, Middleton P, Frøen JF, Smith GC, Gibbons K, et al. 2011. Major risk factors for stillbirth in high-income countries: A systematic review and meta-analysis. Lancet 377:1331–1340.

Foo SC, Phoon WO, Khoo NY. 1988. Toluene in blood after exposure to toluene. Am Ind Hyg Assoc J 49:255–258.

Foster W, Chan S, Platt L, Hughes C. 2000. Detection of endocrine disrupting chemicals in samples of second trimester human amniotic fluid. J Clin Endocrinol Metab 85:2954–2957.

Freni-Titulaer LW, Cordero JF, Haddock L, Lebrón G, Martínez R, Mills JL. 1986. Premature thelarche in Puerto Rico. A search for environmental factors. Am J Dis Child 140:1263–1267.

Frisbee SJ, Brooks PA, Jr., Maher A, Flensborg P, Arnold S, Fletcher T. 2009. The C8 health project: Design, methods, and participants. Environ Health Perspect 117: 1873–1882.

Fromme H, Mosch C, Morovitz M, Alba-Alejandre I, Boehmer S, Kiranoglu M, et al. 2010. Pre- and postnatal exposure to perfluorinated compounds (PFCs). Environ Sci Technol 44:7123–7129.

Fuortes L, Clark MK, Kirchner HL, Smith EM. 1997. Association between female infertility and agricultural work history. Am J Ind Med 31:445–451.

Galloway T, Cipelli R, Guralnik J, Ferrucci L, Bandinelli S, Corsi AM, et al. 2010. Daily bisphenol A excretion and associations with sex hormone concentrations: Results from the InCHIANTI adult population study. Environ Health Perspect 118: 1603–1608.

García AM, Fletcher T, Benavides FG, Orts E. 1999. Parental agricultural work and selected congenital malformations. Am J Epidemiol 149:64–74.

Garlantézec R, Monfort C, Rouget F, Cordier S. 2009. Maternal occupational exposure to solvents and congenital malformations: A prospective study in the general population. Occup Environ Med 66:456–463.

Gasull M, de Basea MB, Puigdomènech E, Pumarega J, Porta M. 2011. Empirical analyses of the influence of diet on human concentrations of persistent organic pollutants: A systematic review of all studies conducted in Spain. Environ Int 37: 1226–1235.

Gerhard I, Daniel V, Link S, Mong B, Runnebaum B. 1998. Chlorinated hydrocarbons in women with repeated miscarriages. Environ Health Perspect 106:675–681.

Givens ML, Small CM, Terrell ML, Cameron LL, Blanck HM, Tolbert PE, et al. 2007. Maternal exposure to polybrominated and polychlorinated biphenyls: Infant birth weight and gestational age. Chemosphere 69:1295–1304.

Gladen BC, Shkiryak-Nyzhnyk ZA, Chyslovska N, Zadorozhnaja TD, Little RE. 2003. Persistent organochlorine compounds and birth weight. Ann Epidemiol 13: 151–157.

Glynn A, Larsdotter M, Aune M, Darnerud PO, Bjerselius R, Bergman Å. 2011. Changes in serum concentrations of polychlorinated biphenyls (PCBs), hydroxylated PCB metabolites and pentachlorophenol during pregnancy. Chemosphere 83:144–151.

Gochfeld M. 2007. Framework for gender differences in human and animal toxicology. Environ Res 104:4–21.

Godfrey KM, Inskip HM, Hanson MA. 2011. The long-term effects of prenatal development on growth and metabolism. Semin Reprod Med 29:257–265.

Gold EB. 2011. The timing of the age at which natural menopause occurs. Obstet Gynecol Clin North Am 38:425–440.

Golub MS, Wu KL, Kaufman FL, Li L-H, Moran-Messen F, Zeise L, et al. 2010. Bisphenol A: Developmental toxicity from early prenatal exposure. Birth Defects Res B Dev Reprod Toxicol 89:441–466.

Gore AC. 2010a. Neuroendocrine targets of endocrine disruptors. Hormones 9: 16–27.

Gore AC. 2010b. Introduction to endocrine-disrupting chemicals. In: Endocrine Disrupting Chemicals (Gore AC, ed.). Totowa, NJ: Humana, pp. 3–7.

Gore AC, Walker DM, Zama AM, Armenti AE, Uzumcu M. 2011. Early life exposure to endocrine-disrupting chemicals causes lifelong molecular reprogramming of the hypothalamus and premature reproductive aging. Mol Endocrinol 25: 2157–2168.

Govarts E, Nieuwenhuijsen M, Schoeters G, Ballester F, Bloemen K, de Boer M, et al. 2012. Birth weight and prenatal exposure to polychlorinated biphenyls (PCBs) and dichlorodiphenyldichloroethylene (DDE): A meta-analysis within 12 European birth cohorts. Environ Health Perspect 120:162–170.

Grandjean P, Bjerve KS, Weihe P, Steuerwald U. 2001. Birthweight in a fishing community: Significance of essential fatty acids and marine food contaminants. Int J Epidemiol 30:1272–1278.

Grandjean P, Andersen EW, Budtz-Jørgensen E, Nielsen F, Mølbak K, Weihe P, et al. 2012. Serum vaccine antibody concentrations in children exposed to perfluorinated compounds. JAMA 307:391–397.

Grether JK, Harris JA, Neutra R, Kizer KW. 1987. Exposure to aerial malathion application and the occurrence of congenital anomalies and low birthweight. Am J Public Health 77:1009–1010.

Grice MM, Alexander BH, Hoffbeck R, Kampa DM. 2007. Self-reported medical conditions in perfluorooctanesulfonyl fluoride manufacturing workers. J Occup Environ Med 49:722–729.

de Groot PCM, Dekkers OM, Romijn JA, Dieben SWM, Helmerhorst FM. 2011. PCOS, coronary heart disease, stroke and the influence of obesity: A systematic review and meta-analysis. Hum Reprod Update 17:495–500.

Guo S-W. 2004. The link between exposure to dioxin and endometriosis: A critical reappraisal of primate data. Gynecol Obstet Invest 57:157–173.

Guo Y, Wu Q, Kannan K. 2011. Phthalate metabolites in urine form China, and implications for human exposures. Environ Int 37:893–898.

Guvenius DM, Aronsson A, Ekman-Ordeberg G, Bergman Å, Norén K. 2003. Human prenatal and postnatal exposure to polybrominated diphenyl ethers, polychlorinated biphenyls, polychlorobiphenylols and pentachlorophenol. Environ Health Perspect 111:1235–1241.

Halldorsson TI, Thorsdottir I, Meltzer HM, Nielsen F, Olsen SF. 2008. Linking exposure to polychlorinated biphenyls with fatty fish consumption and reduced fetal growth among Danish pregnant women: A cause for concern? Am J Epidemiol 168: 958–965.

Hamers T, Kamstra JH, Cenijn PH, Pencikova K, Palkova L, Simeckova P, et al. 2011. *In vitro* toxicity profiling of ultrapure non-dioxin-like polychlorinated biphenyl congeners and their relative toxic contribution to PCB mixtures in humans. Toxicol Sci 121:88–100.

Haney AF. 2008. Leiomyomata. In: Danforth's Obstetrics and Gynecology, 10th edition (Gibs RS, Karlan BY, Haney AF, Nygaard IE, eds.). Philadelphia, PA: Lippincott Williams & Wilkins, a Wolters Kluwer business, p. 916.

Hanke W, Jurewicz J. 2004. The risk of adverse reproductive and developmental disorders due to occupational pesticide exposure: An overview of current epidemiological evidence. Int J Occup Med Environ Health 17:223–243.

Hannigan JH, Bowen SE. 2010. Reproductive toxicology and teratology of abused toluene. Syst Biol Reprod Med 56:184–200.

Hansen S, Nieboer E, Odland JØ, Wilsgaard T, Veyhe AS, Sandanger TM. 2010. Levels of organochlorines and lipids across pregnancy, delivery and postpartum periods in women from Northern Norway. J Environ Monit 12:2128–2137.

Harada K, Inoue K, Morikawa A, Yoshinaga T, Saito N, Koizumi A. 2005. Renal clearance of perfluorooctane sulfonate and perfluorooctanoate in humans and their species-specific excretion. Environ Res 99:253–261.

Haraguchi K, Koizumi A, Inoue K, Harada KH, Hitomi T, Minata M, et al. 2009. Levels and regional trends of persistent organochlorines and polybrominated diphenyl ethers in Asian breast milk demonstrate POPs signatures unique to individual countries. Environ Int 35:1072–1079.

Hartoft-Nielsen M-L, Boas M, Bliddal S, Rasmussen ÅK, Main K, Feldt-Rasmussen U. 2011. Do thyroid disrupting chemicals influence foetal development during pregnancy? J Thyroid Res. DOI: 10.4061/2011/342189 [Epub 2011 Sep 11].

Hatch EE, Troisi R, Wise LA, Hyer M, Palmer JR, Titus-Ernstoff L, et al. 2006. Age at natural menopause in women exposed to diethylstilbestrol in utero. Am J Epidemiol 164:682–688.

Heeren GA, Tyler J, Mandeya A. 2003. Agricultural chemical exposures and birth defects in the Eastern Cape Province, South Africa: A case-control study. Environ Health 2:11.

Heidam LZ. 1984. Spontaneous abortions among dental assistants, factory workers, painters, and gardening workers: A follow-up study. J Epidemiol Community Health 38:149–155.

Heilier J-F, Donnez J, Lison D. 2008. Organochlorines and endometriosis: A mini-review. Chemosphere 71:203–210.

Heilmann C, Grandjean P, Weihe P, Nielsen F, Budtz-Jørgensen E. 2006. Reduced antibody responses to vaccinations in children exposed to polychlorinated biphenyls. PLoS Med 3:e311.

Hemminki K, Mutanen P, Luoma K, Saloniemi I. 1980. Congenital malformations by the parental occupation in Finland. Int Arch Occup Environ Health 46:93–98.

Herington JL, Bruner-Tran KL, Lucas JA, Osteen KG. 2011. Immune interactions in endometriosis. Expert Rev Clin Immunol 7:611–626.

Hernán MA. 2010. The hazards of hazard ratios. Epidemiology 21:13–15.

Hertz-Picciotto I, Charles MJ, James RA, Keller JA, Willman E, Teplin S. 2005a. *In utero* polychlorinated biphenyl exposures in relation to fetal and early childhood growth. Epidemiology 16:648–656.

Hertz-Picciotto I, Herr CEW, Yap P-S, Dostál M, Shumway RH, Ashwood P, et al. 2005b. Air pollution and lymphocyte phenotype proportions in cord blood. Environ Health Perspect 113:1391–1398.

Hertz-Picciotto I, Jusko TA, Willman EJ, Baker RJ, Keller JA, Teplin SW, et al. 2008. A cohort study of *in utero* polychlorinated biphenyl (PCB) exposures in relation to secondary sex ratio. Environ Health 7:37.

Holtcamp W. 2012. Pregnancy-induced hypertension probably linked to PFOA contamination. Environ Health Perspect 120:A59.

Hooper K, Petreas MX, Chuvakova T, Kazbekova G, Druz N, Seminova G, et al. 1998. Analysis of breast milk to assess exposure to chlorinated contaminants in Kazakstan: High levels of 2,3,7,8-tetrachlorodibenzo-*p*-dioxin (TCDD) in agricultural villages of Southern Kazakstan. Environ Health Perspect 106:797–806.

Hoover RN, Hyer M, Pfeiffer RM, Adam E, Bond B, Cheville AL, et al. 2011. Adverse health outcomes in women exposed in utero to diethylstilbestrol. N Engl J Med 365:1304–1314.

Hossain N, Stepanova M, Afendy A, Nader F, Younossi Y, Rafiq N, et al. 2011. Non-alcoholic steatohepatitis (NASH) in patients with polycystic ovarian syndrome (PCOS). Scand J Gastroenterol 46:479–484.

Howard PH, Muir DCG. 2010. Identifying new persistent and bioaccumulative organics among chemicals in commerce. Environ Sci Technol 44:2277–2285.

Hruska KS, Furth PA, Seifer DB, Sharara FI, Flaws JA. 2000. Environmental factors in infertility. Clin Obstet Gynecol 43:821–829.

Hsu J-F, Chang Y-C, Liao P-C. 2010. Age-dependent congener profiles of polychlorinated dibenzo-*p*-dioxins and dibenzofurans in the general population of Taiwan. Chemosphere 81:469–477.

Hwang B-F, Jaakkola JJK, Guo H-R. 2008. Water disinfection by-products and the risk of specific birth defects: A population-based cross-sectional study in Taiwan. Environ Health 7:23.

Ibarluzea J, Alvarez-Pedrerol M, Guxens M, Marina LS, Basterrechea M, Lertxundi A, et al. 2011. Sociodemographic, reproductive and dietary predictors of organochlorine compounds levels in pregnant women in Spain. Chemosphere 82:114–120.

Ikezuki Y, Tsutsumi O, Takai Y, Kamei Y, Taketani Y. 2002. Determination of bisphenol A concentrations in human biological fluids reveals significant early prenatal exposure. Hum Reprod 17:2839–2841.

Jacobson JL, Fein GG, Jacobson SW, Schwartz PM, Dowler JK. 1984. The transfer of polychlorinated biphenyls (PCBs) and polybrominated biphenyls (PBBs) across the human placenta and into maternal milk. Am J Public Health 74:378–379.

Jaraczewska K, Lulek J, Covaci A, Voorspoels S, Kaluba-Skotarczak A, Drews K, et al. 2006. Distribution of polychlorinated biphenyls, organochlorine pesticides and polybrominated diphenyl ethers in human umbilical cord serum, maternal serum and milk from Wielkopolska region, Poland. Sci Total Environ 372:20–31.

Jurewicz J, Hanke W. 2011. Exposure to phthalates: Reproductive outcome and children health. A review of epidemiological studies. Int J Occup Med Environ Health 24: 115–141.

Jusko TA, De Roos AJ, Schwartz SM, Lawrence BP, Palkovicova L, Nemessanyi T, et al. 2010. A cohort study of developmental polychlorinated biphenyl (PCB) exposure in relation to post-vaccination antibody response at 6-months of age. Environ Res 110:388–395.

Kalfa N, Philibert P, Baskin LS, Sultan C. 2011. Hypospadias: Interactions between environment and genetics. Mol Cell Endocrinol 335:89–95.

Kandaraki E, Chatzigeorgiou A, Livadas S, Palioura E, Economou F, Koutsilieris M, et al. 2011. Endocrine disruptors and polycystic ovary syndrome (PCOS): Elevated serum levels of bisphenol A in women with PCOS. J Clin Endocrinol Metab 96:E480–E484.

Kartiko ZH, Finnell RH. 2009. Importance of gene-environment interactions in the etiology of selected birth defects. Clin Genet 75:409–423.

Kawashiro Y, Fukata H, Omori-Inoue M, Kubonoya K, Jotaki T, Takigami H, et al. 2008. Perinatal exposure to brominated flame retardants and polychlorinated biphenyls in Japan. Endocr J 55:1071–1084.

Khan HA. 2007. Benzene's toxicity: A consolidated short review of human and animal studies. Hum Exp Toxicol 26:677–685.

Khanjani N, Sim MR. 2006. Reproductive outcomes of maternal contamination with cyclodiene insecticides, hexachlorobenzene and β-benzene hexachloride. Sci Total Environ 368:557–564.

Khanjani N, Sim MR. 2007. Maternal contamination with PCBs and reproductive outcomes in an Australian population. J Expo Sci Environ Epidemiol 17:191–195.

Khorram O, Garthwaite M, Golos T. 2002. Uterine and ovarian aryl hydrocarbon receptor (AHR) and aryl hydrocarbon receptor nuclear translocator (ARNT) mRNA expression in benign and lamignant gynaecological conditions. Mol Hum Reprod 8:75–80.

Knox SS, Jackson T, Javins B, Frisbee SJ, Shankar A, Ducatman AM. 2011. Implications of early menopause in women exposed to perfluorocarbons. J Clin Endocrinol Metab 96:1747–1753.

Knutsen HK, Kvalem HE, Haugen M, Meltzer HM, Brantsæter AL, Alexander J, et al. 2011. Sex, BMI and age in addition to dietary intakes influence blood concentrations and congener profiles of dioxins and PCBs. Mol Nutr Food Res 55:772–782.

Koch HM, Bolt HM, Preuss R, Angerer J. 2005. New metabolites of di(2-ethylhexyl) phthalate, (DEHP) in human urine and serum after single oral doses of deuterium-labeled DEHP. Arch Toxicol 79:367–376.

Kojima H, Takeuchi S, Uramaru N, Sugihara K, Yoshida T, Kitamura S. 2009. Nuclear hormone receptor activity of polybrominated diphenyl ethers and their hydroxylated metabolites in transactivation assays using Chinese hamster ovary cells. Environ Health Perspect 117:1210–1218.

Koureas M, Tsakalof A, Tsatsakis A, Hadjichristodoulou C. 2012. Systematic review of biomonitoring studies to determine the association between exposure to organophosphorus and pyrethroid insecticides and human health outcomes. Toxicol Lett 210:155–168.

Kuroda N, Kinoshita Y, Sun Y, Wada M, Kishikawa N, Nakashima K, et al. 2003. Measurement of bisphenol A levels in human blood serum and ascitic fluid by HPLC using a fluorescent labeling reagent. J Pharm Biomed Anal 30:1743–1749.

Lacasaña M, Vázquez-Grameix H, Borja-Aburto VH, Blanco-Muñoz J, Romieu I, Aguilar-Garduño C, et al. 2006. Maternal and paternal occupational exposure to agricultural work and the risk of anencephaly. Occup Environ Med 63:649–656.

Langer P. 2010. The impacts of organochlorines and other persistent pollutants on the thyroid and metabolic health. Front Neuroendocrinol 31:497–518.

Lanting CI, Huisman M, Muskiet FAJ, Van der Paauw CG, Essed CE, Boersma ER. 1998. Polychlorinated biphenyls in adipose tissue, liver, and brain from nine stillborns of varying gestational ages. Pediatr Res 44:222–225.

Lauria L, Settimi L, Spinelli A, Figá-Talamanca I. 2006. Exposure to pesticides and time to pregnancy among female greenhouse workers. Reprod Toxicol 22:425–430.

Levario-Carrillo M, Amato D, Ostrosky-Wegman P, González-Horta C, Corona Y, Sanin LH. 2004. Relation between pesticide exposure and intrauterine growth retardation. Chemosphere 55:1421–1427.

Levens ED, Decherney AH 2011. Infertility. In APC Medicine, 2012. (Nabel EG, Federman DD, eds.). Available at http://www.acpmedicine.com/acpmedicine/institutional/checkUser.action [accessed February 3, 2012].

Li Y, Mao M, Dai L, Li K, Li X, Zhou G, et al. 2012. Time trends and geographic variations in the prevalence of hypospadias in China. Birth Defects Res A Clin Mol Teratol 94:36–41.

Lilienfeld AM. 1976. The epidemiologic approach to disease: General purposes, content, and reasoning. In: Foundations of Epidemiology. New York: Oxford University Press, p. 3.

Lim JS, Son H-K, Park S-K, Jacobs DR, Jr., Lee D-H. 2011. Inverse associations between long-term weight change and serum concentrations of persistent organic pollutants. Int J Obes 35:744–747.

Lima M, Ismail S, Ashworth A, Morris SS. 1999. Influence of heavy agricultural work during pregnancy on birthweight in northeast Brazil. Int J Epidemiol 28:469–474.

Lindbohm ML. 1995. Effects of parental exposure to solvents on pregnancy outcome. J Occup Environ Med 37:908–914.

Lindbohm M-L, Taskinen H, Sallmén M, Hemminki K. 1990. Spontaneous abortions among women exposed to organic solvents. Am J Ind Med 17:449–463.

Loffredo CA, Silbergeld EK, Ferencz C, Zhang J. 2001. Association of transposition of the great arteries in infants with maternal exposures to herbicides and rodenticides. Am J Epidemiol 153:529–536.

Lomenick JP, Calafat AM, Melguizo Castro MS, Mier R, Stenger P, Foster MB, et al. 2010. Phthalate exposure and precocious puberty in females. J Pediatr 156:221–225.

Longnecker MP. 2005. Invited commentary: Why DDT matters now. Am J Epidemiol 162:726–728.

Longnecker MP, Klebanoff MA, Zhou H, Brock JW. 2001. Association between maternal serum concentration of the DDT metabolite DDE and preterm and small-for-gestational-age babies at birth. Lancet 358:110–114.

Longnecker MP, Klebanoff MA, Brock JW, Guo X. 2005. Maternal levels of polychlorinated biphenyls in relation to preterm and small-for-gestational-age birth. Epidemiology 16:641–647.

Lopez-Espinosa M-J, Murcia M, Iñiguez C, Vizcaino E, Llop S, Vioque J, et al. 2011. Prenatal exposure to organochlorine compounds and birth size. Pediatrics 128: e127–e134.

Lu C, Toepel K, Irish R, Fenske RA, Barr D, Bravo R. 2006. Organic diets significantly lower children's dietary exposure to organophosphorus pesticides. Environ Health Perspect 114:260–263.

Lyche JL, Gutleb AC, Bergman A, Eriksen GS, Murk ATJ, Ropstad E, et al. 2009. Reproductive and developmental toxicity of phthalates. J Toxicol Environ Health 12:225–249.

MacDorman MF, Kirmeyer S 2009. The Challenge of Fetal Mortality: NCHS Data Brief, No 16. Hyattsville, MD: National Center for Health Statistics.

Maffini MV, Rubin BS, Sonnenschein C, Soto AM. 2006. Endocrine disruptors and reproductive health: The case of bisphenol-A. Mol Cell Endocrinol 254–255:179–186.

Main KM, Skakkebæk NE, Toppari J. 2009. Cryptorchidism as part of the testicular dysgenesis syndrome: The environmental connection. In: Endocrine Involvement in Developmental Syndromes, Endocr Dev, Vol. 14 (Cappa M, Loche S, Bottazzo GF, eds.). Basel: Karger, pp. 167–173.

Main KM, Skakkebæk NE, Virtanen HE, Toppari J. 2010. Genital anomalies in boys and the environment. Best Pract Res Clin Endocrinol Metab 24:279–289.

Markey CM, Luque EH, Munoz de Toro MM, Sonnenschein C, Soto AM. 2001. *In utero* exposure to bisphenol A alters the development and tissue organization of the mouse mammary gland. Biol Reprod 65:1215–1223.

Markey CM, Coombs MA, Sonnenschein C, Soto AM. 2003. Mammalian development in a changing environment: Exposure to endocrine disruptors reveals the developmental plasticity of steroid-hormone target organs. Evol Dev 5:67–75.

Markey CM, Wadia PR, Rubin BS, Sonnenschein C, Soto AM. 2005. Long-term effects of fetal exposure to low doses of the xenoestrogen bisphenol-A in the female mouse genital tract. Biol Reprod 72:1344–1351.

McCarver G, Bhatia J, Chambers C, Clarke R, Etzel R, Foster W, et al. 2011. NTP-CERHR Expert panel report on the developmental toxicity of soy infant formula. Birth Defects Res B Dev Reprod Toxicol 92:421–468.

McClure EM, Pasha O, Goudar SS, Chomba E, Garces A, Tshefu A, et al. 2011. Epidemiology of stillbirth in low-middle income countries: A global network study. Acta Obstet Gynecol Scand 90:1379–1385.

McDonald JC, Lavoie J, Côté R, McDonald AD. 1987. Chemical exposures at work in early pregnancy and congenital defect: A case-referent study. Br J Ind Med 44: 527–533.

McDonald AD, McDonald JC, Armstrong B, Cherry NM, Côté R, Lavoie J, et al. 1988. Congenital defects and work in pregnancy. Br J Ind Med 45:581–588.

McGuinness BM, Buck GM, Mendola P, Sever LE, Vena JE. 2001. Infecundity and consumption of polychlorinated biphenyl-contaminated fish. Arch Environ Health 56:250–253.

McMartin KI, Chu M, Kopecky E, Einarson TR, Koren G. 1998. Pregnancy outcome following maternal organic solvent exposure: A meta-analysis of epidemiologic studies. Am J Ind Med 34:288–292.

Medina-Carrilo L, Rivas-Solis F, Fernández-Argüelles R. 2002. Risk for congenital malformations in pregnant women exposed to pesticides in the state of Nayarit, Mexico. Ginecol Obstet Mex 70:538–544.

Meeker JD, Missmer SA, Altshul L, Vitonis AF, Ryan L, Cramer DW, et al. 2009. Serum and follicular fluid organochlorine concentrations among women undergoing assisted reproduction technologies. Environ Health 8:32.

Meeker JD, Maity A, Missmer SA, Williams PL, Mahalingaiah S, Ehrlich S, et al. 2011. Serum concentrations of polychlorinated biphenyls in relation to *in vitro* fertilization outcomes. Environ Health Perspect 119:1010–1016.

Mendola P, Buck GM, Sever LE, Zielezny M, Vena JE. 1997. Consumption of PCB-contaminated freshwater fish and shortened menstrual cycle length. Am J Epidemiol 146:955–960.

Mendola P, Messer LC, Rappazzo K. 2008. Science linking environmental contaminant exposures with fertility and reproductive health impacts in the adult female. Fertil Steril 89(Suppl 1):e81–e94.

Mink PJ, Kimmel CA, Li AA. 2012. Potential effects of chlorpyrifos on fetal growth outcomes: Implications for risk assessment. J Toxicol Environ Health B Crit Rev 15:281–316.

Missmer SA, Hankinson SE, Spiegelman D, Barbieri RL, Michels KB, Hunter DJ. 2004. *In utero* exposures and the incidence of endometriosis. Fertil Steril 82: 1501–1508.

Mocarelli P, Brambilla P, Gerthoux PM, Patterson JRDG, Needham LL. 1996. Change in sex ratio with exposure to dioxin. Lancet 348:409.

Mocarelli P, Gerthoux PM, Needham LL, Patterson JRDG, Limonta G, Falbo R, et al. 2011. Perinatal exposure to low doses of dioxin can permanently impair human semen quality. Environ Health Perspect 119:713–718.

Moggs JG, Ashby J, Tinwell H, Lim FL, Moore DJ, Kimber I, et al. 2004. The need to decide if all estrogens are intrinsically similar. Environ Health Perspect 112: 1137–1142.

Mouritsen A, Aksglaede L, Sørensen K, Mogensen SS, Leffers H, Main KM, et al. 2010. Hypothesis: Exposure to endocrine-disrupting chemicals may interfere with timing of puberty. Int J Androl 33:346–359.

Muir DCG, Howard PH. 2006. Are there other persistent organic pollutants? A challenge for environmental chemists. Environ Sci Technol 40:7157–7166.

Munger R, Isacson P, Hu S, Burns T, Hanson J, Lynch CF, et al. 1997. Intrauterine growth retardation in Iowa communities with herbicide-contaminated drinking water supplies. Environ Health Perspect 105:308–314.

Muñoz-Cruz S, Togno-Pierce C, Morales-Montor J. 2011. Non-reproductive effects of sex steroids: Their immunoregulatory role. Curr Top Med Chem 11:1714–1727.

Muñoz-de-Toro M, Markey CM, Wadia PR, Luque EH, Rubin BS, Sonnenschein C, et al. 2005. Perinatal exposure to bisphenol-A alters peripubertal mammary gland development in mice. Endocrinology 146:4138–4147.

Murphy HH, Sanusi A, Dilts DR, Djajadisastra M, Hirschhorn N, Yuliantiningshi S. 2000. Health effects of pesticide use among Indonesian Women Farmers. II: Reproductive health outcomes. J Agromedicine 6:27–43.

Murphy LE, Gollenberg AL, Buck Louis GM, Kostyniak PJ, Sundaram R. 2010. Maternal serum preconception polychlorinated biphenyl concentrations and infant birth weight. Environ Health Perspect 118:297–302.

Nakamura T, Nakai K, Matsumura T, Suzuki S, Saito Y, Satoh H. 2008. Determination of dioxins and polychlorinated biphenyls in breast milk, maternal blood and cord blood from residents of Tohoku, Japan. Sci Total Environ 394:39–51.

NAMS (North American Menopause Society). 2012. Premature Menopause: Fertility Freedom or Freak Out? Available at http://www.regardinghealth.com/nam/RHO/2009/07/Article.aspx?bmkEMC=52356 [accessed May 24, 2013].

National Institutes of Health (NIH), National Toxicology Program (NTP). 2008. NTP-CERHR monograph on potential reproductive and developmental effects of bisphenol A. NTP CERHR MON. 22:v, vii–ix, 1–64 passim. NIH Publication 08-5994.

Needham LL, Grandgean P, Heinzow B, Jørgensen PJ, Nielsen F, Patterson DG, Jr., et al. 2011. Partition of environmental chemicals between maternal and fetal blood and tissues. Environ Sci Technol 45:1121–1126.

Newbold RR, Padilla-Banks E, Jefferson WN. 2006. Adverse effects of the model environmental estrogen diethylstilbestrol are transmitted to subsequent generations. Endocrinology 147:S11–S17.

Ng TP, Foo SC, Yoong T. 1992a. Menstrual function in workers exposed to toluene. Br J Ind Med 49:799–803.

Ng TP, Foo SC, Yoong T. 1992b. Risk of spontaneous abortion in workers exposed to toluene. Br J Ind Med 49:804–808.

Nippita TA, Baber RJ. 2007. Premature ovarian failure: A review. Climacteric 10: 11–22.

Nurminen T. 1995. Maternal pesticide exposure and pregnancy outcome. J Occup Environ Med 37:935–940.

Nurminen T, Rantala K, Kurppa K, Holmberg PC. 1995. Agricultural work during pregnancy and selected structural malformations in Finland. Epidemiology 6: 23–30.

Ochoa-Acuña H, Frankenberger J, Hahn L, Carbajo C. 2009. Drinking-water herbicide exposure in Indiana and prevalence of small-for-gestational-age and preterm delivery. Environ Health Perspect 117:1619–1624.

O'Grady-Milbrath M, Wenger Y, Chang C-W, Emond C, Garabrant D, Gillespie BW, et al. 2009. Apparent half-lives of dioxins, furans, and polychlorinated biphenyls as a function of age, body fat, smoking status, and breast-feeding. Environ Health Perspect 117:417–425.

Olsen GW, Burris JM, Ehresman DJ, Froehlich JW, Seacat AM, Butenhoff JL, et al. 2007. Half-life of serum elimination of perfluorooctanesulfonate, perfluorohexane-sulfonate, and perfluorooctanoate in retired fluorochemical production workers. Environ Health Perspect 115:1298–1305.

Olsen GW, Butenhoff JL, Zobel LR. 2009. Perfluoroalkyl chemicals and human fetal development: An epidemiologic review with clinical and toxicological perspectives. Reprod Toxicol 27:212–230.

Ouyang F, Perry MJ, Venners SA, Chen C, Wang B, Yang F, et al. 2005. Serum DDT, age at menarche, and abnormal menstrual cycle length. Occup Environ Med 62:878–884.

Padmanabhan V, Siefert K, Ransom S, Johnson T, Pinkerton J, Anderson L, et al. 2008. Maternal bisphenol-A levels at delivery: A looming problem? J Perinatol 28: 258–263.

Parente RC, Faerstein E, Celeste RK, Werneck GL. 2008. The relationship between smoking and age at the menopause: A systematic review. Maturitas 61:287–298.

Park H-Y, Hertz-Picciotto I, Petrik J, Palkovicova L, Kocan A, Trnovec T. 2008a. Prenatal PCB exposure and thymus size at birth in Neonates in Eastern Slovakia. Environ Health Perspect 116:104–109.

Park J-S, Bergman Å, Linderholm L, Athanasiadou M, Kocan A, Petrik J, et al. 2008b. Placental transfer of polychlorinated biphenyls, their hydroxylated metabolites and pentachlorophenol in pregnant women from eastern Slovakia. Chemosphere 70:1676–1684.

Patandin S, Koopman-Esseboom C, de Ridder MAJ, Weisglas-Kuperus N, Sauer PJ. 1998. Effects of environmental exposure to polychlorinated biphenyls and dioxins on birth size and growth in Dutch children. Pediatr Res 44:538–545.

Peacock J, Peacock P. 2011. Oxford Handbook of Medical Statistics, 1st edition. New York: Oxford University Press, Inc.

Pergament E, Todydemir PB, Fiddler M. 2002. Sex ratio: A biological perspective of "Sex and the City." Reprod Biomed Online 5:43–46.

Perry MJ, Ouyang F, Korrick SA, Venners SA, Chen C, Xu X, et al. 2006. A prospective study of serum DDT and progesterone and estrogen levels across the menstrual cycle in nulliparous women of reproductive age. Am J Epidemiol 164:1056–1064.

Plenge-Bonig A, Karmaus W. 1999. Exposure to toluene in the printing industry is associated with subfecundity in women but not in men. Occup Environ Med 56: 443–448.

Prentice RL, Langer R, Stefanick ML, Howard BV, Pettinger M, Anderson G, et al. 2005. Combined postmenopausal hormone therapy and cardiovascular disease: Toward resolving the discrepancy between observational studies and the Women's Health Initiative Clinical Trial. Am J Epidemiol 162:404–414.

Qiao L, Zheng L, Cai D. 2007. [Study on the di-n-butyl phthalate and di-2-ethylhexyl phthalate level of girl serum related with precocious puberty in Shanghai]. Wei Sheng Yan Jiu 36:93–95.

Quinn CL, Wania F, Czub G, Breivik K. 2011. Investigating intergenerational differences in human PCB exposure due to variable emissions and reproductive behaviors. Environ Health Perspect 119:641–646.

Rauh VA, Garfinkel R, Perera FP, Andrews HF, Hoepner L, Barr DB. 2006. Impact of prenatal chlorpyrifos exposure on neurodevelopment in the first 3 years of life among inner-city children. Pediatrics 118:1845–1859.

Restrepo M, Muñoz N, Day N, Parra JE, Hernandez C, Blettner M, et al. 1990. Birth defects among children born to a population occupationally exposed to pesticides in Colombia. Scand J Work Environ Health 16:239–246.

Reutman SR, LeMasters GK. 2007. Evaluation of occupational exposures and effects on male and female reproduction. In: Environmental & Occupational Medicine, 4th edition (Rom WN, Markowitz SB, eds.). Philadelphia, PA: Wolters Kluwer-Lippincott Williams & Wilkins, pp. 146–150.

Reutman SR, LeMasters GK, Knecht EA, Shukla R, Lockey JE, Burroughs GE, et al. 2002. Evidence of reproductive endocrine effects in women with occupational fuel and solvent exposures. Environ Health Perspect 110:805–811.

Ribas-Fitó N, Sala M, Cardo E, Mazón C, De Muga ME, Verdú A, et al. 2002. Association of hexachlorobenzene and other organochlorine compounds with anthropometric measures at birth. Pediatr Res 52:163–167.

Rocheleau CM, Romitti PA, Dennis LK. 2009. Pesticides and hypospadias: A meta-analysis. J Pediatr Urol 5:17–24.

Rocheleau CM, Bertke SJ, Deddens JA, Ruder AM, Lawson CC, Waters MA, et al. 2011. Maternal exposure to ploychlorinated biphenyls and the secondary sex ratio: An occupational cohort study. Environ Health 10:20.

Rogan WJ, Gladen BC, Guo Y-LL, Hsu C-C. 1999. Sex ratio after exposure to dioxin-like chemicals in Taiwan. Lancet 353:206–207.

Rogers JM. 2009. Tobacco and pregnancy. Reprod Toxicol 28:152–160.

Romitti PA, Herring AM, Dennis LK, Wong-Gibbons DL. 2007. Meta-analysis: Pesticides and orofacial clefts. Cleft Palate Craniofac J 44:358–365.

Rönn M, Lind L, Van Bavel B, Salihovic S, Michaëlsson K, Lind PM. 2011. Circulating levels of persistent organic pollutants associate in divergent ways to fat mass measured by DXA in humans. Chemosphere 85:335–343.

Rubin BS, Murray MK, Damassa DA, King JC, Soto AM. 2001. Perinatal exposure to low doses of bisphenol A affects body weight, patterns of estrous cyclicity, and plasma LH levels. Environ Health Perspect 109:675–680.

Rylander L, Strömberg U, Hagmar L. 1995. Decreased birth weight in infants born to women with a high dietary intake of fish contaminated with persistent organochlorine compounds. Scand J Work Environ Health 21:368–375.

Rylander L, Strömberg U, Dyremark E, Östman C, Nilsson-Ehle P, Hagmar L. 1998. Polychlorinated biphenyls in blood plasma among Swedish female fish consumers in relation to low birth weight. Am J Epidemiol 147:493–502.

Safe S. 1984. Polychlorinated biphenyls (PCBs) and polybrominated biphenyls (PBBs): Biochemistry, toxicology, and mechanism of action. Crit Rev Toxicol 13: 319–395.

Safe SH. 1995. Environmental and dietary estrogens and human health: Is there a problem? Environ Health Perspect 103:346–351.

Safe S. 2004. Endocrine disruptors and human health: Is there a problem. Toxicology 205:3–10.

Safe S, Bandiera S, Sawyer T, Robertson L, Safe L, Parkinson A, et al. 1985. PCBs: Structure-function relationships and mechanism of action. Environ Health Perspec 60:47–56.

Sala M, Ribas-Fitó N, Cardo E, de Muga ME, Marco E, Mazón C, et al. 2001. Levels of hexachlorobenzene and other organochlorine compounds in cord blood: Exposure across placenta. Chemosphere 43:895–901.

Sallmén M, Baird DD, Hoppin JA, Blair A, Sandler DP. 2006. Fertility and exposure to solvents among families in the Agricultural Health Study. Occup Environ Med 63:469–475.

Sallmén M, Neto M, Mayan ON. 2008. Reduced fertility among shoe manufacturing workers. Occup Environ Med 65:518–524.

Sanborn M, Kerr KJ, Sanin LH, Cole DC, Bassil KL, Vakil C. 2007. Non-cancer health effects of pesticides: Systematic review and implications for family doctors. Can Fam Physician 53:1712–1720.

Sanford M, Kerr KJ, Sanin LH, Cole DC, Bassil KL, Vakil C. 2007. Non-cancer health effects of pesticides: Systematic review and implications for family doctors. Can Fam Physician 53:1712–1720.

Sathyanarayana S, Basso O, Karr CJ, Lozano P, Alavanja M, Sandler DP, et al. 2010. Maternal pesticide use and birth weight in the agricultural health study. J Agromedicine 15:127–136.

Sato T, Fukazawa Y, Ohta Y, Iguchi T. 2004. Involvement of growth factors in induction of persistent proliferation of vaginal epithelium of mice exposed neonatally to diethylstilbestrol. Reprod Toxicol 19:43–51.

Scheeres JJ, Chudley AE. 2002. Solvent abuse in pregnancy: A perinatal perspective. J Obstet Gynaecol Can 24:22–26.

Schönfelder G, Wittfoht W, Hopp H, Talsness CE, Paul M, Chahoud I. 2002. Parent bisphenol A accumulation in the human maternal-fetal-placental unit. Environ Health Perspect 110:A703–A707.

Schwartz DA, Newsum LA, Heifetz RM. 1986. Parental occupation and birth outcome in an agricultural community. Scand J Work Environ Health 12:51–54.

Settimi L, Spinelli A, Lauria L, Miceli G, Pupp N, Angotzi G, et al. 2008. Spontaneous abortion and maternal work in greenhouses. Am J Ind Med 51:290–295.

Shaw GM, Wasserman CR, O'Malley CD, Nelson V, Jackson RJ. 1999. Maternal pesticide exposure from multiple sources and selected congenital anomalies. Epidemiology 10:60–66.

Shirangi A, Nieuwenhuijsen M, Vienneau D, Holman CD. 2011. Living near agricultural pesticide applications and the risk of adverse reproductive outcomes: A review of the literature. Paediatr Perinat Epidemiol 25:172–191.

Shu XO, Gao YT, Brinton LA, Linet MS, Tu JT, Zheng W, et al. 1988. A population-based case-control study of childhood leukemia in Shanghai. Cancer 62:635–644.

Silva MJ, Barr DB, Reidy JA, Kato K, Malek NA, Hodge CC, et al. 2003. Glucoronidation patterns of common urinary and serum monoester phthalate metabolites. Arch Toxicol 77:561–567.

Silva MJ, Reidy JA, Herbert AR, Preau JL, Jr., Needham LL, Calafat AM. 2004a. Detection of phthalate metabolites in human amniotic fluid. Bull Environ Contam Toxicol 72:1226–1231.

Silva MJ, Barr DB, Reidy JA, Malek NA, Hodge CC, Caudill SP, et al. 2004b. Urinary levels of seven urinary phthalate metabolites in the U.S. population from the National Health and Nutrition Examination Survey (NHANES) 1999–2000. Environ Health Perspect 112:331–338.

Simpkins JW, Swenberg JA, Weiss N, Brusick D, Eldridge JC, Stevens JT, et al. 2011. Atrazine and breast cancer: A framework assessment of the toxicological and epidemiological evidence. Toxicol Sci 123:441–459.

Singer D, Mann E, Hunter MS, Pitkin J, Panay N. 2011. The silent grief: Psychosocial aspects of premature ovarian failure. Climacteric 14:428–437.

Small CM, Cheslack-Postava K, Terrell M, Blanck HM, Tolbert P, Rubin C, et al. 2007. Risk of spontaneous abortion among women exposed to polybrominated biphenyls. Environ Res 105:247–255.

Small CM, Murray D, Terrell ML, Marcus M. 2011. Reproductive outcomes among women exposed to a brominated flame retardant *in utero*. Arch Environ Occup Health 66:201–208.

Smith GD. 2004. Classics in epidemiology: Should they get it right? Int J Epidemiol 33:441–442.

Snijder CA, Roeleveld N, Te Velde E, Steegers EA, Raat H, Hofman A, et al. 2012. Occupational exposure to chemicals and fetal growth: The Generation R Study. Hum Reprod 27:910–920.

Sonneborn D, Park H-Y, Petrik J, Kocan A, Palkovicova L, Trnovec T, et al. 2008. Prenatal polychlorinated biphenyl exposures in eastern Slovakia modify effects of social factors on birthweight. Paediatr Perinat Epidemiol 22:202–213.

Staessen JA, Nawrot T, Den Hond E, Thijs L, Fagard R, Hoppenbrouwers K, et al. 2001. Renal function, cytogenetic measurements, and sexual development in adolescents in relation to environmental pollutants: A feasibility study of biomarkers. Lancet 357:1660–1669.

Steffen C, Auclerc MF, Auvrignon A, Baruchel A, Kebaili K, Lambilliotte A, et al. 2004. Acute childhood leukaemia and environmental exposure to potential sources of benzene and other hydrocarbons; a case-control study. Occup Environ Med 61: 773–778.

Steinberg RM, Walker DM, Juenger TE, Woller MJ, Gore AC. 2008. Effects of perinatal polychlorinated biphenyls on adult female rat reproduction: Development, reproductive physiology, and second generational effects. Biol Reprod 78:1091–1101.

Steiner AZ, D'Aloisio AA, DeRoo LA, Sandler DP, Baird DD. 2010. Association of intrauterine and early-life exposures with age at menopause in the Sister Study. Am J Epidemiol 172:140–148.

The Stillbirth Collaborative Research Network Writing Group. 2011. Association between stillbirth and risk factors known at pregnancy confirmation. JAMA 306:2469–2479.

Stillerman KP, Mattison DR, Giudice LC, Woodruff TJ. 2008. Environmental exposures and adverse pregnancy outcomes: A review of the science. Reprod Sci 15:631–650.

Strohsnitter WC, Hatch EE, Hyer M, Troisi R, Kaufman RH, Robboy SJ, et al. 2008. The association between in utero cigarette smoke exposure and age at menopause. Am J Epidemiol 167:727–733.

Strom BL, Schinnar R, Zeigler EE, Barnhart KT, Sammel MD, Macones GA. 2001. Exposure to soy-based formula in infancy and endocrinological and reproductive outcomes in young adulthood. JAMA 286:807–814.

Sutton P, Guidice LC, Woodruff TJ. 2010. Reproductive environmental health. Curr Opin Obstet Gynecol 22:517–524.

Sutton P, Wallinga D, Perron J, Gottlieb M, Sayre L, Woodruff T. 2011. Reproductive health and the industrialized food system: A point of intervention for health policy. Health Aff 30:888–897.

Suzuki G, Nakano M, Nakano S. 2005. Distribution of PCDDs/PCDFs and Co-PCBs in human maternal blood, cord blood, placenta, milk and adipose tissue: Dioxins showing high toxic equivalency factor accumulate in the placenta. Biosci Biotechnol Biochem 69:1836–1847.

Swan SH, Main KM, Liu F, Stewart SL, Kruse RL, Calafat AM, et al. 2005. Decrease in anogenital distance among male infants with prenatal phthalate exposure. Environ Health Perspect 113:1056–1061.

Takeuchi T, Tsutsumi O, Ikezuki Y, Takai Y, Taketani Y. 2004. Positive relationship between androgen and the endocrine disruptor, bisphenol A, in normal women and women with ovarian dysfunction. Endocr J 51:165–169.

Talsness CE, Andrade AJ, Kuriyama SN, Taylor JA, vom Saal FS. 2009. Components of plastic: Experimental studies in animals and relevance for human health. Philos Trans R Soc Lond B Biol Sci 364:2079–2096.

Taskinen H, Anttila A, Lindbohm M-L, Sallmén M, Hemminki K. 1989. Spontaneous abortions and congenital malformations among the wives of men occupationally exposed to organic solvents. Scand J Work Environ Health 15:345–352.

Taskinen H, Kyyrönen P, Hemminki K, Hoikkala M, Lajunen K, Lindbohm ML. 1994. Laboratory work and pregnancy outcome. J Occup Med 36:311–319.

Taylor KC, Jackson LW, Lynch CD, Kostyniak PJ, Buck Louis GM. 2007. Preconception maternal polychlorinated biphenyl concentrations and the secondary sex ratio. Environ Res 103:99–105.

Teede H, Deeks A, Moran L. 2010. Polycystic ovary syndrome: A complex condition with psychological, reproductive and metabolic manifestations that impacts on health across the life span. BMC Med 8:41.

Terrell ML, Berzen AK, Small CM, Cameron LL, Wirth JH, Marcus M. 2009. A cohort study of the association between secondary sex ratio and parental exposure to polybrominated biphenyl (PBB) and ploychlorinated biphenyl (PCB). Environ Health 8:35.

Thiel R, Chahoud I. 1997. Postnatal development and behaviour of Wistar rats after prenatal toluene exposure. Arch Toxicol 71:258–265.

Thomas DC, Petitti DB, Goldhaber M, Swan SH, Rappaport EB, Hertz-Picciotto I. 1992. Reproductive outcomes in relation to malathion spraying in the San Francisco Bay Area, 1981–1982. Epidemiology 3:32–39.

Thompson J, Lorber M, Toms L-ML, Kato K, Calafat AM, Mueller JF. 2010. Use of simple pharmacokinetic modeling to characterize exposure of Australians to perfluorooctanoic acid and perfluorooctane sulfonic acid. Environ Int 36: 390–397.

Thorup J, McLachland R, Cortes D, Nation TR, Balic A, Southwell BR, et al. 2010. What is new in cryptorchidism and hypospadias—A critical review on the testicular dysgenesis hypothesis. J Pediatr Surg 45:2074–2086.

Thulstrup AM, Bonde JP. 2006. Maternal occupational exposure and risk of specific birth defects. Occup Med 56:532–543.

Thurston SW, Ryan L, Christiani D, Snow R, Carlson J, You L, et al. 2000. Petrochemical exposure and menstrual disturbances. Am J Ind Med 38:555–564.

Tiemann U. 2008. *In vivo* and *in vitro* effects of the organochlorine pesticides DDT, TCPM, methoxychlor, and lindane on the female reproductive tract of mammals: A review. Reprod Toxicol 25:316–326.

Tinwell H, Ashby J. 2004. Sensitivity of the immature rat uterotrophic assay to mixtures of estrogens. Environ Health Perspect 112:575–582.

Toft G, Thulstrup AM, Jönsson BA, Pedersen HS, Ludwicki JK, Zvezday V, et al. 2010. Fetal loss and maternal serum levels of 2,2′,4,4′,5,5′-hexachlorbiphenyl (CB-153) and 1,1-dichloro-2,2-bis(p-chlorophenyl)ethylene (p,p′-DDE) exposure: A cohort study in Greenland and two European populations. Environ Health 9:22.

Toppari J, Virtanen HE, Main KM, Skakkebæk NE. 2010. Cryptorchidism and hypospadias as a sign of testicular dysgenesis syndrome (TDS): Environmental connection. Birth Defects Res A Clin Mol Teratol 88:910–919.

Torres-Arreola L, Berkowitz G, Torres-Sánchez L, López-Cervantes M, Cebrián ME, Uribe M, et al. 2003. Preterm birth in relation to maternal organochlorine serum levels. Ann Epidemiol 13:158–162.

Tsukimori K, Yasukawa F, Uchi H, Furue M, Morokuma S. 2012. Sex ratio in two generations of the Yusho cohort. Epidemiology 23:349–350.

Turner MC, Wigle DT, Krewski D. 2010. Residential pesticides and childhood leukemia: A systematic review and meta-analysis. Environ Health Perspect 118:33–41.

Unfer V, Casini ML, Costabile L, Mignosa M, Gerli S, Di Renzo GC. 2004. Endometrial effects of long-term treatment with phytoestrogens: A randomized, double-blind, placebo-controlled study. Fertil Steril 82:145–148.

U.S. Department of Health and Human Services, Centers for Disease Control and Prevention, National Center for Chronic Disease Prevention and Health Promotion, Division of Reproductive Health. 2012. A National Public Health Action Plan for the Prevention, Detection and Management of Infertility. Atlanta, GA. [Published Draft]. Available at http://www.cdc.gov/reproductivehealth/Infertility/PublicHealth.htm [accessed June 21, 2012].

U.S. Department of Health and Human Services, Centers for Disease Control and Prevention, National Center for Chronic Disease Prevention and Health Promotion, Division of Reproductive Health and National Center for Birth Defects and Developmental Disabilities. 2011. Premature Birth. Atlanta, GA. Available at http://www.cdc.gov/Features/PrematureBirth/ [accessed August 15, 2012].

U.S. Department of Health and Human Services, Centers for Disease Control and Prevention, National Center for Chronic Disease Prevention and Health Promotion, Office on Smoking and Health. 2004. The Health Consequences of Smoking: A

Report of the Surgeon General. Atlanta, GA. Available at http://www.cdc.gov/tobacco/data_statistics/sgr/2004/errata/index.htm [accessed April 25, 2012].

U.S. DHHS (U.S. Department of Health and Human Services). 2006. The Health Consequences of Involuntary Exposure to Tobacco Smoke: A Report of the Surgeon General. Atlanta, GA: U.S. Department of Health and Human Services, Centers for Disease Control and Prevention, Coordinating Center for Health Promotion, National Center for Chronic Disease Prevention and Health Promotion, Office on Smoking and Health.

U.S. DHHS (U.S. Department of Health and Human Services). 2010. How Tobacco Smoke Causes Disease. The Biology and Behavioral Basis for Smoking-Attributable Disease: A Report of the Surgeon General. Atlanta, GA: U.S. Department of Health and Human Services, Centers for Disease Control and Prevention, National Center for Chronic Disease Prevention and Health Promotion, Office on Smoking and Health. Available at http://www.cdc.gov/tobacco/data_statistics/sgr/2010/index.htm [accessed June 21, 2012].

U.S. Department of Health and Human Services (U.S. DHHS), Office on Smoking and Health (U.S.). 2001. Women and Smoking: A Report of the Surgeon General. Atlanta, GA: Centers for Disease Control and Prevention (US). Available at http://www.ncbi.nlm.nih.gov/books/NBK44303/ [accessed March 28, 2012].

U.S. EPA (U.S. Environmental Protection Agency). 2012. EPA's Reanalysis of Key Issues Related to Dioxin Toxicity and Response to NAS Comments, Vol. 1. Available at http://www.epa.gov/iris/supdocs/dioxinv1sup.pdf [accessed April 25, 2012].

Vaktskjold A, Talykova LV, Nieboer E. 2011. Congenital anomalies in newborns to women employed in jobs with frequent exposure to organic solvents—A register-based prospective study. BMC Pregnancy Childbirth 11:83.

Vandenberg LN, Colborn T, Hayes TB, Heindel JJ, Jacobs DR, Jr., Lee D-H, et al. 2012. Hormones and endocrine-disrupting chemicals: Low-dose effects and nonmonotonic dose responses. Endocr Rev 33:378–455.

Van Maele-Fabry GV, Lantin A-C, Hoet P, Lison D. 2010. Childhood leukemia and parental occupational exposure to pesticides: A systematic review and meta-analysis. Cancer Causes Control 21:787–809.

Van Maele-Fabry GV, Lantin A-C, Hoet P, Lison D. 2011. Residential exposure to pesticides and childhood leukemia: A systematic review and meta-analysis. Environ Int 37:280–291.

Vartiainen T, Jaakkola JJK, Saarikoski S, Tuomisto J. 1998. Birth weight and sex of children and the correlation to the body burden of PCDDs/PCDFs and PCBs of the mother. Environ Health Perspect 106:61–66.

Vayena E, Rowe PJ, Peterson HB. 2002. Assisted reproductive technology in developing countries: Why should we care? Fertil Steril 78:13–15.

Venners SA, Korrick S, Xu X, Chen C, Guang W, Huang A, et al. 2005. Preconception serum DDT and pregnancy loss: A prospective study using a biomarker of pregnancy. Am J Epidemiol 162:709–716.

Verloop J, van Leeuwen FE, Helmerhorst TJM, van Boven HH, Rookas MA. 2010. Cancer risk in DES daughters. Cancer Causes Control 21:999–1007.

Vestergaard S, Nielsen F, Andersson AM, Hjøllund NH, Grandjean P, Andersen HR. 2012. Association between perfluorinated compounds and time to pregnancy in a prospective cohort of Danish couples attempting to conceive. Hum Reprod 27:873–880.

Vidaeff AC, Sever LE. 2005. *In utero* exposure to environmental estrogens and male reproductive health: A systemic review of biological and epidemiologic evidence. Reprod Toxicol 20:5–20.

Villanueva CM, Durand G, Coutté M-B, Chevrier C, Cordier S. 2005. Atrazine in municipal drinking water and risk of low birth weight, preterm delivery, and small-for-gestational-age status. Occup Environ Med 62:400–405.

Vinson F, Merhi M, Baldi I, Raynal H, Gamet-Payrastre L. 2011. Exposure to pesticides and risk of childhood cancer: A meta-analysis of recent epidemiological studies. Occup Environ Med 68:694–702.

Völkel W, Colnot T, Csanády GA, Filser JG, Dekant W. 2002. Metabolism and kinetics of bisphenol A in humans at low doses following oral administration. Chem Res Toxicol 15:1281–1287.

Waller K, Swan SH, DeLorenze G, Hopkins B. 1998. Trihalomethanes in drinking water and spontaneous abortion. Epidemiology 9:134–140.

Wan Y, Choi K, Kim S, Ji K, Chang H, Wiseman S, et al. 2010. Hydroxylated polybrominated diphenyl ethers and bisphenol A in pregnant women and their matching fetuses: Placental transfer and potential risks. Environ Sci Technol 44:5233–5239.

Wang S-L, Lin C-Y, Guo YL, Lin L-Y, Chou W-L, Chang LW. 2004. Infant exposure to polychlorinated dibenzo-*p*-dioxins, dibenzofurans and biphenyls (PCDD/Fs, PCBs)—Correlation between prenatal and postnatal exposure. Chemosphere 54: 1459–1473.

Warner M, Samuels S, Mocarelli P, Gerthoux PM, Needham L, Patterson DG, Jr., et al. 2004. Serum dioxin concentrations and age at menarche. Environ Health Perspect 112:1289–1292.

Washino N, Saijo Y, Sasaki S, Kato S, Ban S, Konishi K, et al. 2009. Correlations between prenatal exposure to perfluorinated chemicals and reduced fetal growth. Environ Health Perspect 117:660–667.

Watson RE, Jacobson CF, Williams AL, Howard WB, DeSesso JM. 2006. Trichloroethylene-contaminated drinking water and congenital heart defects: A critical analysis of the literature. Reprod Toxicol 21:117–147.

Weinberg CR, Wilcox AJ. 2008. Methodologic issues in reproductive epidemiology. In: Modern Epidemiology, 3rd edition (Rothman KJ, Greenland S, Lash TL, eds.). Philadelphia, PA: Wolters Kluwer-Lippincott Williams & Wilkins, p. 622.

Weisglas-Kuperus N, Patandin S, Berbers GAM, Sas TCJ, Mulder PGH, Sauer PJJ, et al. 2000. Immunologic effects of background exposure to polychlorinated biphenyls and dioxin in Dutch preschool children. Environ Health Perspect 108: 1203–1207.

Weisglas-Kuperus N, Vreugdenhil HJI, Muder PGH. 2004. Immunological effects of environmental exposure to polychlorinated biphenyls and dioxins in Dutch school children. Toxicol Lett 149:281–285.

Weisskopf MG, Anderson HA, Hanrahan LP, Great Lakes Consortium. 2003. Decreased sex ratio following maternal exposure to polychlorinated biphenyls from contaminated Great Lakes sport-caught fish: A retrospective cohort study. Environ Health 2:2.

Wennborg H, Bodin L, Vainio H, Axelsson G. 2001. Solvent use and time to pregnancy among female personnel in biomedical laboratories in Sweden. Occup Environ Med 58:225–231.

Weselak M, Arbucle TE, Foster W. 2007. Pesticide exposures and developmental outcomes: The epidemiological evidence. J Toxicol Environ Health B Crit Rev 10:41–80.

Weuve J, Hauser R, Calafat AM, Missmer SA, Wise LA. 2010. Association of exposure to phthalates with endometriosis and uterine leiomyomata: Findings from NHANES, 1999–2004. Environ Health Perspect 118:825–832.

White FM, Cohen FG, Sherman G, McCurdy R. 1988. Chemicals, birth defects and stillbirths in New Brunswick: Associations with agricultural activity. CMAJ 138: 117–124.

Whitworth KW, Haug LS, Baird DD, Becher G, Hoppin JA, Skjaerven R, et al. 2012. Perfluorinated compounds and subfecundity in pregnant women. Epidemiology 23:257–263.

Wigle DT, Arbuckle TE, Turner MC, Bérubé A, Yang Q, Liu S, et al. 2008. Epidemiologic evidence of relationships between reproductive and child health outcomes and environmental chemical contaminants. J Toxicol Environ Health B Crit Rev 11:373–517.

Wigle DT, Turner MC, Krewski D. 2009. A systematic review and meta-analysis of childhood leukemia and parental occupational pesticide exposure. Environ Health Perspect 117:1505–1513.

Wilcox AJ, Baird DD. 2011. Invited commentary: Natural versus unnatural sex ratios-a quandary of modern times. Am J Epidemiol 174:1332–1334.

Willett WC, Manson JE, Grodstein F, Stampfer MJ, Colditz GA. 2006. Re: "combined postmenopausal hormone therapy and cardiovascular disease: Toward resolving the discrepancy between observational studies and the Women's Health Initiative Clinical Trial." Am J Epidemiol 163:1067–1069.

Williams MK, Rundle A, Holmes D, Reyes M, Hoepner LA, Barr DB, et al. 2008. Changes in pest infestation levels, self-reported pesticide use, and permethrin exposure during pregnancy after the 2000–2001 US Environmental Protection Agency restriction of organophosphates. Environ Health Perspect 116: 1681–1688.

Willis WO, de Peyster A, Molgaard CA, Walker C, MacKendrick T. 1993. Pregnancy outcome among women exposed to pesticides through work or residence in an agricultural area. J Occup Med 35:943–949.

Wilson CL, Bodnar JA, Brown BG, Morgan WT, Potts RJ, Borgerding MF. 2008. Assessment of dioxin and dioxin-like compounds in mainstream smoke from selected US cigarette brands and reference cigarettes. Food Chem Toxicol 46:1721–1733.

Windham G, Fenster L. 2008. Environmental contaminants and pregnancy outcomes. Fertil Steril 89(Suppl 1):e111–e116.

Windham GC, Waller K, Anderson M, Fenster L, Mendola P, Swan S. 2003. Chlorination by-products in drinking water and menstrual cycle function. Environ Health Perspect 111:935–941; discussion A409.

Windham GC, Lee D, Mitchell P, Anderson M, Petreas M, Lasley B. 2005. Exposure to organochlorine compounds and effects on ovarian function. Epidemiology 16: 182–190.

Wise LA, Palmer JR, Rowlings K, Kaufman RH, Herbst AL, Noller KL, et al. 2005. Risk of benign gynecologic tumors in relation to prenatal diethylstilbestrol exposure. Obstet Gynecol 105:167–173.

Wise LA, Palmer JR, Hatch EE, Triosi R, Titus-Ernstoff L, Herbst AL, et al. 2007. Secondary sex relation among women exposed to diethylstilbestrol *in utero*. Environ Health Perspect 115:1314–1319.

Wohlfahrt-Veje C, Main KM, Skakkebæk NE. 2009. Testicular dysgenesis syndrome: Foetal origin of adult reproductive problems. Clin Endocrinol (Oxf) 71:459–465.

Wojtyniak BJ, Rabczenko D, Jönsson BAG, Zvezday V, Pedersen HS, Rylander L, et al. 2010. Association of maternal serum concentrations of 2,2′,4,4′,5,5′-hexachlorobiphenyl (CB-153) and 1,1-dichloro-2,2-bis (p-chlorophenyl)-ethylene (p,p′-DDE) levels with birth weight, gestational age and preterm births in Inuit and European populations. Environ Health 9:56.

Wolff MS, Camann D, Gammon M, Stellman SD. 1997. Proposed PCB congener groupings for epidemiological studies. Environ Health Perspect 105:13.

Woodruff TJ, Zota AR, Schwartz JM. 2011. Environmental chemicals in pregnant women in the United States: NHANES 2003–2004. Environ Health Perspect 119:878–885.

WHO (World Health Organization), Scientific Group on Research on the Menopause in the 1990s. 1996. Research on the Menopause: Report of a WHO Scientific Group. WHO technical report series, 866, Geneva, Switzerland.

WHO (World Health Organization). 2008. The Global Burden of Disease: 2004 Update. Geneva. Available at http://www.who.int/healthinfo/global_burden_disease/2004_report_update/en/index.html [accessed April 25, 2012].

WHO (World Health Organization). 2010. Birth Defects, Report by the Secretariat, 63rd World Health Assembly, April 1, 2010. Available at http://apps.who.int/gb/ebwha/pdf_files/WHA63/A63_10-en.pdf [accessed July 26, 2012].

World Health Organization (WHO), International Programme on Chemical Safety. (Damstra T, Barlow S, Bergman A, Kavlock R, Van Der Kraak G, eds.). 2002. Global Assessment of the State-of-the-Science of Endocrine Disruptors. Available at http://www.who.int/ipcs/publications/new_issues/endocrine_disruptors/en/ [accessed April 25, 2012].

Xiang H, Nuckols JR, Stallones L. 2000. A geographic information assessment of birth weight and crop production patterns around mother's residence. Environ Res 82: 160–167.

Xu X, Cho S-II, Sammel M, You L, Cui S, Huang Y, et al. 1998. Association of petrochemical exposure with spontaneous abortion. Occup Environ Med 55:31–36.

Yang C-Y, Yu M-L, Guo H-R, Lai T-J, Hsu C-C, Lambert G, et al. 2005. The endocrine and reproductive function of the female Yucheng adolescents prenatally exposed to PCBs/PCDFs. Chemosphere 61:355–360.

Yang C-Y, Wang Y-J, Chen P-C, Tsai S-J, Guo YL. 2008. Exposure to a mixture of polychlorinated biphenyls and polychlorinated dibenzofurans resulted in a prolonged time to pregnancy in women. Environ Health Perspect 116:599–604.

Yang C-Y, Huang T-S, Lin K-C, Kuo P, Tsai P-C, Guo YL. 2011. Menstrual effects among women exposed to polychlorinated biphenyls and dibenzofurans. Environ Res 111: 288–294.

Ye X, Skjaerven R, Basso O, Baird D, Eggesbo M, Cupul Uicab LA, et al. 2010. *In utero* exposure to tobacco smoke and subsequent reduced fertility in females. Hum Reprod 25:2901–2906.

Yiee JH, Baskin LS. 2010. Environmental factors in genitourinary development. J Urol 184:34–41.

Young JG, Eskenazi B, Gladstone EA, Bradman A, Pedersen L, Johnson C. 2005. Association between in utero organophosphate pesticide exposure and abnormal reflexes in neonates. Neurotoxicology 26:199–209.

Yu M-L, Guo YL, Hsu C-C, Rogan WJ. 2000. Menstruation and reproduction in women with polychlorinated biphenyl (PCB) poisoning: Long-term follow-up interviews of the women from the Taiwan Yucheng cohort. Int J Epidemiol 29:672–677.

Zhang X, Ho S-M. 2011. Epigenetics meets endocrinology. J Mol Endocrinol 46:R11–R32.

Zhang JJ, Lioy PJ. 2002. Human exposure assessment in air pollution systems. Scientificworldjournal 2:497–513.

Zhang J, Cai WW, Lee DJ. 1992. Occupational hazards and pregnancy outcomes. Am J Ind Med 21:397–408.

CHAPTER 11

Effects of Organic Chemicals on the Male Reproductive System

LARS RYLANDER and ANNA RIGNELL-HYDBOM

ABSTRACT

Background: It has been questioned whether male reproductive health has been impaired over time. There is very strong evidence that the incidence of testicular cancer has increased over time in Western countries, whereas a time-related impairment regarding sperm quality and congenital malformations in the male reproductive organs is much more controversial. In parallel with the time trend discussion, it has been hypothesized that environmental pollutants with hormone-disrupting properties might affect male reproductive health.

Objectives: This chapter aims to give an overview of the male reproductive system, to introduce the testicular dysgenesis syndrome hypothesis, and to present results regarding four groups of environmental pollutants (namely, chlorinated, brominated, and fluorinated compounds and phthalates) and their associations with male reproductive systems.

Discussion: The chapter focuses on the literature dealing with human studies regarding the association between the four groups of environmental pollutants and male reproductive systems. The level of ambition is to describe some studies that support and refute the hypothesized association. The number of studies within the four groups of environmental pollutants varied a lot and there is no clear picture regarding the associations.

Conclusion: Although a number of animal, experimental, and epidemiological studies show associations, it is still too early to draw firm conclusions on a causal association between environmental pollutants and male reproductive health. There are a number of data gaps and a number of research questions to be evaluated.

Effects of Persistent and Bioactive Organic Pollutants on Human Health, First Edition.
Edited by David O. Carpenter.

HAS MALE REPRODUCTIVE HEALTH BEEN IMPAIRED OVER TIME?

A possible decline in human sperm quality was first suggested in 1974 (Nelson and Bunge 1974). The question was highlighted in an article by Carlsen and coworkers in 1992 (Carlsen et al. 1992). Based on 61 studies from all parts of the world, performed between 1938 and 1991, the authors concluded from a meta-analysis that there had been a rapid time-related decline in sperm quality. The results have been criticized, and in other studies, no time-related changes were found (Axelsson et al. 2011; Bujan et al. 1996; Fisch et al. 1996; Olsen et al. 1995). On the other hand, the hypothesis of a time-related decline has gained support in studies from France, Scotland, Belgium, and Denmark (Auger et al. 1995; Bonde et al. 1998; Irvine et al. 1996; Nordkap et al. 2011; Swan et al. 2000; Van Waeleghem et al. 1996). It has also been suggested that the incidence of male reproductive abnormalities has increased over time (Baskin et al. 2001; Chilvers et al. 1984; Lund et al. 2009; Pierik et al. 2002). However, as for sperm quality, there are studies suggesting no time-related increase (Fisch et al. 1996, 2009). A less controversial finding is the increased time-related incidence of testicular cancer in several Western countries (Chia et al. 2010; Huyghe et al. 2003).

Despite the contradictory results for some of these male reproductive outcomes, the question of the possible cause was raised. It was proposed that these interrelated disorders could be linked to the widespread use of chemicals with hormone-disrupting properties, also known as endocrine disruptors, described further in more detail (Sharpe and Skakkebaek 1993).

OVERVIEW OF THE MALE REPRODUCTIVE SYSTEM

Male Sex Hormones

The entire male reproductive system is dependent on hormones, which are chemicals that stimulate or regulate the activity of cells or organs. Male sex hormones are also known as androgens. The main androgen is testosterone, and in some tissues, testosterone is chemically changed into dihydrotestosterone in order to have an effect. Ninety-five percent of the testosterone is produced and secreted by the Leydig cells in the testes. Testosterone is essential for the development and function of the testes, the maturation of secondary sexual characteristics, and the masculinization of bones and muscles, libido and stimulation of spermatogenesis. Gonadotropin-releasing hormone (GnRH) secretion from the hypothalamus acts upon its receptor in the anterior pituitary gland to regulate the production and release of the gonadotropins, follicle-stimulating hormone (FSH) and luteinizing hormone (LH). FSH affects the Sertoli cells in the testis and is necessary for sperm production, and LH acts on the Leydig cells and stimulates the synthesis and secretion of testosterone, which is necessary to continue the process of spermatogenesis.

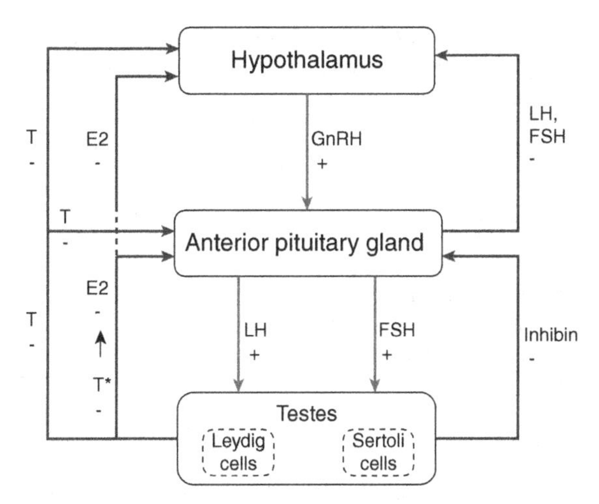

Figure 11.1. Hormonal control of male reproduction. E2, estradiol; FSH, follicle-stimulating hormone; GnRH, gonadotropin-releasing hormone; LH, luteinizing hormone; T, testosterone; T*, testosterone converted to estradiol.

Regulation of FSH and LH secretion is mediated through a negative feedback system (illustrated in Figure 11.1). A hormone, inhibin B, from the Sertoli cells inhibits FSH and testosterone, and estradiol inhibits LH as well as FSH secretion (Sharpe 2001).

The Hormone-Dependent System

The endocrine system (i.e., the hormone-dependent system) of the body plays an essential role in the regulation of metabolic processes. Nutritional, behavioral, and reproductive processes are intricately regulated by endocrine systems, as are growth, gut, cardiovascular, and kidney functions. For most endocrine systems, the primary goal is to maintain some form of balance (homeostasis), avoiding huge fluctuations in hormone levels (Robaire and Viger 1995; Viger et al. 2005). Disorders in any of the endocrine systems, involving either increasing or decreasing hormone secretion, may result in disease. An increasing body of evidence indicates that certain chemicals in the environment, known as endocrine disruptors, cause developmental and reproductive abnormalities in humans and in animals (Colborn et al. 1993; Main et al. 2007; Sharpe and Skakkebaek 1993). In wildlife populations, associations have been reported between reproductive and developmental effects and endocrine-disrupting chemicals. In the aquatic environment, effects have been observed in mammals, birds, reptiles, fish, and mollusks from Europe, North America, and other areas. The observed abnormalities vary from subtle changes to permanent alterations, including disturbed sex differentiation with feminized or masculinized sex organs and changed sexual behavior (Guillette 2000; Sonne et al. 2008).

TABLE 11.1. WHO's Definition (2010) of Normal Sperm Quality Presented as Lower Reference Limits (Fifth Percentile with 95% Confidence Interval [CI]) for Different Semen Characteristics

	Lower Reference Limit	
Parameter	5th Percentile	(95% CI)
Semen volume (mL)	1.5	(1.4–1.7)
Total sperm count (10^6 per ejaculate)	39	(33–46)
Sperm concentration (10^6 per mL)	15	(12–16)
Total motility (% showing good forward movement)	40	(38–42)
Sperm morphology (normal forms, %)	4	(3.0–4.0)

Semen Quality

Semen quality is generally considered to be a proxy measure of male fertility, and changes in semen quality can, for instance, occur after exposure to toxic agents. A semen analysis evaluates certain characteristics of a male's semen and the sperm contained in the semen. Examples of parameters measured in a semen analysis are sperm concentration, sperm count, motility, morphology, and volume. Sperm concentration measures the concentration of sperms per milliliter semen in a man's ejaculate, and sperm count measures the total number of sperms per ejaculate. The total sperm count is achieved by multiplying semen volume by sperm concentration. Sperm concentrations over 15 million sperms per milliliter is considered normal, according to the WHO in 2010 (Table 11.1) (Cooper et al. 2010). The average sperm count today is between 20 and 40 million per milliliter in the Western world. Sperm motility describes the ability of the sperm to move properly toward an egg. The WHO defines normal motility as 60% of observed sperm, or at least 8 million per milliliter, showing good forward movement. When interpreting the results from semen analyses, it is necessary to be aware that there are various methodological factors that may influence the results, giving rise to intermethod variation.

Congenital Malformations

Hypospadias is a congenital malformation where the urethra opens in an abnormal place, on the underside of the penis rather than at the tip (Figure 11.2). It is one of the most common genital deformities affecting newborn boys. About 1 in 300 infants in the United States is born with the condition. However, data on the prevalence of hypospadias are inconsistent. The etiology of hypospadias remains largely unknown. One possible explanation could be that exposure to environmental agents with an estrogenic or an antiandrogenic effect causes changes in concentrations of sex hormones regulating fetal genital development during weeks 8–14 of pregnancy. In animal studies, several endocrine disruptors with antiandrogenic or estrogenic properties caused hypospadias in experimental animals (Toppari 2008).

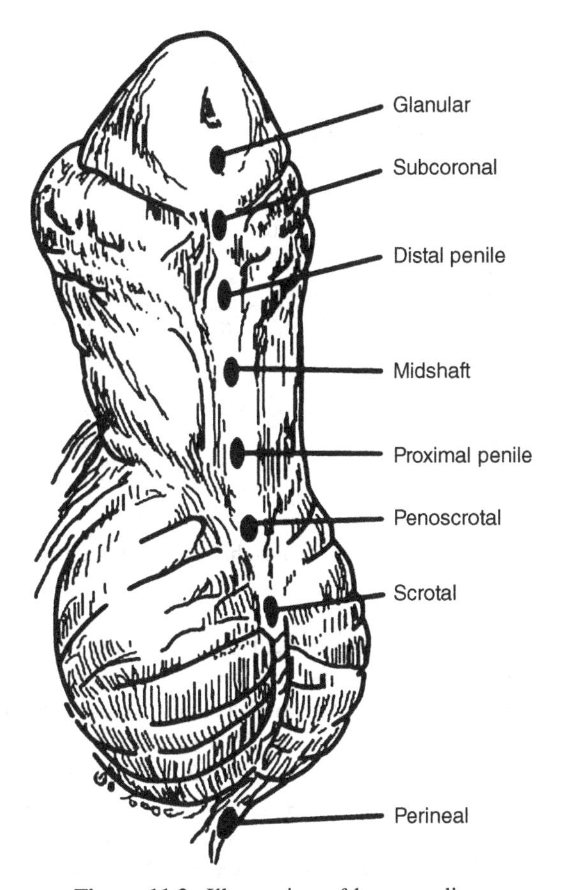

Figure 11.2. Illustration of hypospadias.

Cryptorchidism or undescended testis is the failure of one or both testes to descend into the scrotum (Figure 11.3). The process of testis descent is dependent on a combination of growth processes and hormonal influences (Bay et al. 2007; Hutson et al. 2010; Kubota et al. 2002). It used to be said that this condition affected around 2–4 boys in every 100, but now it seems that in some countries, rates are much higher. For example, around 9 boys per 100 are affected in Denmark and around 6 boys per 100 in the United Kingdom. Cryptorchidism is often considered a mild malformation, but it represents the best-characterized risk factor for infertility and testicular cancer in adulthood (Dieckmann and Pichlmeier 2004; Virtanen et al. 2007).

Testicular Cancer

Although testicular cancer is the most common malignancy in men of reproductive age, it only accounts for about 1–2% of all tumors in men (Maneckshа

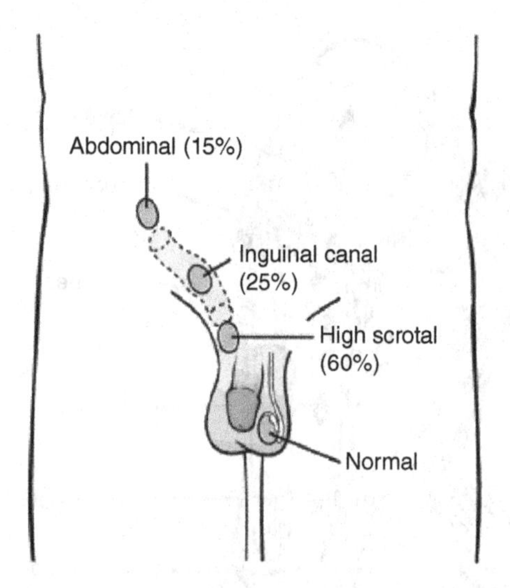

Figure 11.3. Illustration of cryptorchidism (undescended testis).

and Fitzpatrik 2009). Numerous publications have shown an increase in the incidence of testicular cancer in the last 40 years. However, in the past few decades, survival rates for many cancers have improved significantly. Of men treated for testicular cancer, more than 95% can now be cured. But there is great variation in the incidence of testicular cancer between populations and geographical regions. The incidence is highest in developed nations with primarily Caucasian populations and lowest in Asia and Africa. Men from Denmark and Norway have the highest age-standardized incidence rates at around 9–10 per 100,000 man-years, whereas the incidence in Estonia and Spain, for instance, is only about 2–3. The reasons for such a large difference between populations within Europe are unknown. Interestingly, migrant studies from Sweden and Denmark indicate that testicular cancer incidences move toward those of the adopted country after one generation, but that incidences in first-generation migrants are unchanged in relation to their countries of origin. This might suggest that lifestyle and environmental exposures in the "new" country are of importance.

TESTICULAR DYSGENESIS SYNDROME (TDS)

It has been proposed that environmental antiandrogens, in addition to the previously suggested estrogenic compounds, could act as endocrine-disrupting chemicals with potential adverse effects on male reproductive health, including cryptorchidism, hypospadias, low semen quality, and testicular cancer. This hypothesis was based on animal exposure studies and on the observation that sons of women treated during pregnancy with diethylstilbestrol (DES) had substantially increased incidences of cryptorchidism, hypospadias, poor semen quality, and possibly testicular cancer (Toppari et al. 1996). The hypothesis was initially suggested by Skakkebaek and colleagues and was termed "TDS" (Skakkebaek et al. 2001). While the TDS hypothesis is not generally accepted, especially concerning the inclusion of hypospadias, certain aspects of it are supported by both laboratory and epidemiological studies.

ENVIRONMENTAL POLLUTANTS AND THE MALE REPRODUCTIVE SYSTEM

The objective of this chapter is to present results regarding four groups of environmental pollutants (namely, chlorinated, brominated, and fluorinated compounds and phthalates) and their associations with male reproductive systems. For all four groups of environmental pollutants, associations with male reproductive functions have been shown in animal studies, but we will focus here on the literature dealing with human studies. The level of ambition is not to present a complete review of the literature but rather to illustrate that, for all these four groups of environmental pollutants, there are studies that both support and refute the TDS hypothesis. A complete review demands, for instance, a much more detailed critical assessment of the study designs used in each study. The number of studies with respect to the different groups of environmental pollutants and with respect to the different male reproductive outcomes varies widely. There are, for instance, several studies that have evaluated the association between chlorinated compounds and semen quality, whereas for other compounds, there are very few studies published.

Persistent Organochlorine Compounds (POPs)

POPs including polychlorinated biphenyls (PCBs), dioxins (including polychlorinated dibenzofurans, [PCDFs]), and the insecticide DDT and its major metabolite, 2,2-bis(p-chlorophenyl)-1,1-dichloroethylene (DDE), have common properties such as long-term persistence, widespread diffusion in the environment, and bioaccumulation in fatty tissues of living organisms. Even though most POPs were banned in most countries several years ago, POP residues continue to be found in both animals and human tissues. PCBs have a potential estrogenic, antiestrogenic, or antiandrogenic effect depending on their structure; DDT exhibits estrogenic effects, whereas its major metabolite, p,p′-DDE, exhibits antiandrogenic effects (Bonefeld-Jørgensen et al. 2001; Kelce et al. 1998; Wetterauer et al. 2012).

Two accidental episodes with very high exposures to PCBs and PCDFs occurred in 1968 in Japan (Yusho) and in 1979 in Taiwan (Yucheng). More than 4000 people were exposed to contaminated rice oil and showed symptoms of intoxication. In 1998, a small number of Yucheng boys who were exposed during their fetal period were compared to matched boys regarding semen quality. The Yucheng boys had a higher frequency of abnormal sperm morphology and decreased sperm motility (Guo et al. 2000). In addition, among boys exposed after birth during the disaster, similar changes in semen quality were found, except that sperm motility was unaffected (Hsu et al. 2003).

Studies among populations with much lower POP exposure levels from different parts of the world have also shown negative associations between POP exposure and sperm quality. However, the pattern is far from consistent (Cook et al. 2011). An example of a large-scale project dealing with the hypothesized association between POP exposure and fertility is the "INUENDO project" (http://www.inuendo.dk), which was performed over the period 2002–2006 and was financed by the European Union. This project included about 800 men from four populations, and its objective was to unravel the impact of environmental exposure to xenobiotic compounds with hormone-like actions on human fertility (Bonde et al 2008; Toft et al. 2005). In the initial phase, a crucial part of the study was to evaluate the impact of POP exposure on semen quality. The selected populations were recruited from four countries and were aimed at representing a wide range of POP exposure. One group comprised Greenlandic Inuits, previously known to have the highest body burdens of POPs in the world. The second group comprised men from Kharkiv, Ukraine, an area probably more exposed to pesticides. The third group from Warsaw, Poland, was included to represent a group with low levels of POP. These three groups consisted of spouses of pregnant women, and their mean age was about 30 years. The fourth group comprised fishermen from the Swedish east coast as well as the west coast, of whom the former had been consuming highly POP-polluted fish from the Baltic Sea (off the east coast) (Rignell-Hydbom et al. 2004). The Swedish fishermen were older than the men from the three other groups (mean age was about 45 years). While the men from Ukraine and Poland were tested at the hospital, the Swedish fishermen were tested at their homes, which necessitated a mobile laboratory (Figure 11.4). p,p′-DDE and PCB 153 (one out of 209 theoretical PCB variants) were used as biomarkers for POP exposure. Overall, the most pronounced effect within the INUENDO project was a negative effect on sperm motility. A statistically significant decrease in progressive sperm motility with increasing serum concentrations of PCB 153 was observed in all four populations. It is, however, important to stress that the explained variance in motility due to PCB 153 seemed to be relatively low. A relatively strong association between serum concentrations of PCB 153 and sperm chromatin integrity (measuring the fraction of sperms with DNA damage) was observed among the Caucasian populations but was absent among the Inuits (Spanò et al. 2005). The POP concentrations did not indicate any associations with semen volume, sperm concentration, total sperm count, or morphology. The INUENDO project has now been followed by the CLEAR project (http://www.inuendo.dk/clear). One aim of the CLEAR project is to investigate associations between other endocrine disruptors and semen quality within the three INUENDO populations, which comprised spouses of pregnant women.

Figure 11.4. The mobile laboratory unit used in the Swedish part of the INUENDO project.

Regarding the association between POP exposure and male genital malformations, that is, hypospadias and cryptorchidism, there are few human studies. The majority of these studies have measured DDT, p,p′-DDE, and/or PCB as the POP exposure, and none of these exposures were associated with male genital malformations (Cook et al. 2011). In general, the studies were hampered by a small number of cases. However, a small study from Italy showed a statistically significant association between hexachlorobenzene (HCB) and hypospadias (Giordano et al. 2010). This finding was not replicated in a study from the United States (Carmichael et al. 2010).

Servicemen's Testicular Tumor Environmental and Endocrine Determinants (STEED) is the largest study dealing with the hypothesized association between POP exposure and testicular cancer (McGlynn et al. 2008, 2009). The study included some 750 cases and showed that increased p,p′-DDE levels were significantly associated with an increased risk of developing testicular cancer. The STEED study also showed that some other POPs (*trans*-nonachlor, *cis*-nonachlor, and total chlordanes) were positively associated with testicular cancer, whereas the majority of the PCBs were negatively associated (i.e., "protective") with the risk for developing testicular cancer. The patterns for other, much smaller, studies do not provide a clear picture (Hardell et al. 2006; Purdue et al. 2009).

Brominated Compounds

Polybrominated diphenyl ethers (PBDEs) are an important class of flame retardants, widely used in a variety of consumer products. In the past several years, PBDEs have become widespread environmental pollutants and have been detected in water, soil, air, animals, and human tissues. Exposure occurs in particular through the diet and the indoor environment. PBDEs have endocrine-disrupting effects as they have been shown to interact as antagonists or agonists at androgen, progesterone, and estrogen receptors (Stoker et al. 2005).

Very little is known about the association between PBDE exposure and effects on male reproductive function in humans. In a recently published small study comprising men recruited from a fertility clinic, semen mobility was negatively related to PBDE variants (BDE-47, BDE-100, and ΣBDE) (Abdelouahab et al. 2011). No relations were observed for other semen parameters. In a very small pilot study comprising only 10 Japanese males, a significant inverse association was observed between BDE-153 and sperm concentration (Akutsu et al. 2008). Some PBDEs may alter the testosterone/estrogen balance in the fetus, and results from a Danish study found that the concentration of BDE congeners 47, 153, 99, 100, 28, 66, and 154 was higher in breast milk from mothers who gave birth to boys with cryptorchidism than in controls (Main et al. 2007). Exposure to endocrine-disrupting chemicals during the fetal period has been postulated to be a risk factor for testis cancer. In a Swedish study, PBDE serum levels from 58 cases with testicular cancer and age-matched controls were analyzed. Differences were found between cases and controls (Hardell et al. 2006).

Fluorinated Compounds

Perfluorinated compounds (PFCs) have been produced in the range of several hundred metric tons per year (Lau et al. 2004). They are used in a large variety of applications, especially in surface coatings, making products water and oil resistant. Several of the compounds are approved for use in food packaging materials, food containers, and cooking pans and may thus be transferred to food (Renner 2001). PFCs are highly biopersistent, with a half-life in humans of about 5 years (Olsen et al. 2007). The compounds mostly investigated are perfluorooctane sulfonate (PFOS) and perfluorooctanoic acid (PFOA). The main exposure route for the general population is via contaminated food and water intake (Vestergren et al. 2008).

Overall, there are very few human studies regarding the hypothesized association between PFCs and the male reproductive system. These studies have mainly focused on semen quality. One study among young Danish men (median age 19 years) showed an association between combined PFOS and PFOA exposure and the proportion of morphologically normal sperm cells, and effects, although not statistically significant, on several other semen parameters and levels of reproductive hormones (Joensen et al. 2009). The authors concluded that these compounds may contribute to the otherwise unexplained low semen quality often seen in young men. However, another study evaluated semen quality among infertility patients in relation to PFOS and PFOA in serum and semen and found no association between PFOS or PFOA levels and sperm concentration, volume, or motility (Raymer et al. 2012). A third study, the CLEAR project, evaluated the possible association between PFCs and biomarkers of male reproductive health in a larger population of almost 600 partners of pregnant women from Greenland, Poland, and Ukraine, representing individuals with considerable variation in these exposures (Lindh et al. 2012), allowing for evaluation of potential effects across a range of exposure levels that can be expected in nonoccupationally exposed men (Toft et al. 2012). The proportion of morphologically normal cells was 35% lower for the third tertile of PFOS exposure than for the first. On the other hand, at the third PFOA exposure tertile, the percentage of motile spermatozoa was 19% higher than in the first. Across countries, sperm concentration, total sperm count, and semen volume were not consistently associated with the PFCs analyzed. As stated by the authors, due to the multiple statistical tests performed, it cannot be excluded that these associations represent chance findings.

In a study from the United States, the mortality pattern was investigated within a cohort of employees of an ammonium perfluorooctanoate (APFO) manufacturing facility (Lundin et al. 2009). APFO rapidly dissociates to PFOA in blood. The study did not observe an association between exposure and testicular cancer.

Although there are some findings regarding the association between PFCs and male reproductive functions, it is far too early to conclude that these findings are causal.

Phthalates

Phthalates are among the most widely used man-made chemicals released to the environment over the last several decades. They are used as plasticizers to increase the flexibility of polyvinyl chloride (PVC) products such as toys, vinyl flooring, and electricity cables or medical devices, and as solvents or fixing agents in perfumes, body lotions, and other cosmetics. Despite the short half-lives of these compounds, they have been detected in urine in more than 95% of humans who have been investigated (Wittassek et al. 2011). The main exposure routes for humans are through diet, medical devices, and consumer products.

Regarding semen quality, there are three studies from the United States where urine phthalate metabolites have been negatively associated with different sperm quality parameters (Duty et al. 2003; Hauser et al. 2006; Wirth et al. 2008). The participants in these studies were recruited among either men from subfertile couples or men from infertility clients. In addition, a study from India including men who attended an infertility clinic found negative correlations between semen phthalate levels and sperm quality parameters (Pant et al. 2008). A recent pilot study from Japan, including male partners of subfertile couples who had infertility consultations at a gynecology clinic, observed an association between phthalate concentrations in the urine and sperm concentration (Toshima et al. 2011). On the other hand, among men from subfertile couples in Germany, no associations were found between phthalate and sperm parameters (Herr et al. 2009). In addition, a study among men from the general population in Sweden found weak associations between one phthalate metabolite (monoethyl phthalate [MEP]) and sperm parameters, but no associations with the other phthalate metabolites were analyzed (Jönsson et al. 2005).

A case-control study from the United Kingdom included 471 boys with hypospadias and randomly selected controls (Ormond et al. 2009). Telephone interviews were conducted with the mothers. Occupational exposure to phthalates (based on a job exposure matrix) was associated with a threefold increased risk of having a boy with hypospadias. A weakness with this study was that there were no biological measurements of phthalate exposure. It is not, however, possible to find any studies where biological measurements of exposure (in urine or blood) have been associated with hypospadias.

In summary, a number of associations have been observed between different phthalates and male reproductive functions, especially among men from subfertile couples and among infertile men, but there is still too much uncertainty to draw firm conclusions. More studies among men from the general population are needed.

CONCLUSIONS SO FAR AND WHAT'S NEXT

The number of studies investigating the hypothesized association between exposure to endocrine disruptors and effects on the male reproductive system is steadily growing. Although a number of animal studies, experimental studies, and epidemiological studies show associations, it is still too early to draw firm conclusions on a causal association. There are still a number of data gaps and a number of research questions to be evaluated, for instance,

- Which exposure window is the most critical? Might it differ for different outcomes? Most often, the studies focus on current exposure, but it might, for instance, be the fetal exposure that is critical. Studies should focus on different exposure windows.
- What is the significance of mixtures of endocrine disruptors and other environmental contaminants? Up to now, the vast majority of studies have focused on single exposures, but there is accumulating evidence showing the additivity of the effects of these chemicals (Kortenkamp and Faust 2010).
- How representative is a single exposure measurement? Repeated measurements of exposures and outcomes will most probably increase our understanding.

- Do we need new exposure measures? One way might be to discover new high-throughput surrogate measures based on biological activity.
- How much does sensitivity vary between individuals? It is, of course, important to take genetics into account.

REFERENCES

Abdelouahab N, Ainmelk Y, Takser L. 2011. Polybrominated diphenyl ethers and sperm quality. Reprod Toxicol 31:546–550.

Akutsu K, Takatori S, Nozawa S, Yoshiike M, Nakazawa H, Hayakawa K, et al. 2008. Polybrominated diphenyl ethers in human serum and sperm quality. Bull Environ Contam Toxicol 80:345–350.

Auger J, Kunstmann JM, Czyglik F, Jouannet P. 1995. Decline in semen quality among fertile men in Paris during the past 20 years. N Engl J Med 332:281–285.

Axelsson J, Rylander L, Rignell-Hydbom A, Giwercman A. 2011. No secular trend over the last decade in sperm counts among Swedish men from the general population. Hum Reprod 26:1012–1016.

Baskin LS, Colborn T, Aimes K. 2001. Hypospadias and endocrine disruption: Is there a connection? Environ Health Perspect 109:1175–1183.

Bay K, Virtanen HE, Hartung S, Ivell R, Main KM, Skakkebaek NE, et al. 2007. Insulin-like factor 3 levels in cord blood and serum from children: Effects of age, postnatal hypothalamic-pituitary-gonadal axis activation, and cryptorchidism. J Clin Endocrinol Metab 92:4020–4027.

Bonde JP, Kold Jensen T, Brixen Larsen S, Abell A, Scheike T, Hjollund NH, et al. 1998. Year of birth and sperm count in 10 Danish occupational studies. Scand J Work Environ Health 24:407–413.

Bonde JP, Toft G, Rylander L, Rignell-Hydbom A, Giwercman A, Spano M, et al. 2008. Fertility and markers of male reproductive function in Inuit and European populations spanning large contrasts in blood levels of persistent organochlorines. Environ Health Perspect 116:269–277.

Bonefeld-Jørgensen EC, Andersen HR, Rasmussen TH, Vinggaard AM. 2001. Effect of highly bioaccumulated polychlorinated biphenyl congeners on estrogen and androgen receptor activity. Toxicology 14:141–153.

Bujan L, Mansat A, Pontonnier F, Mieusset R. 1996. Time series analysis of sperm concentration in fertile men in Toulouse, France between 1977 and 1992. BMJ 312:471–472.

Carlsen E, Giwercman A, Keiding N, Skakkebaek NE. 1992. Evidence for decreasing quality of semen during past 50 years. BMJ 305:609–613.

Carmichael SL, Herring AH, Sjödin A, Jones R, Needham L, Ma C, et al. 2010. Hypospadias and halogenated organic pollutant levels in maternal mid-pregnancy serum samples. Chemosphere 80:641–646.

Chia VM, Quraishi SM, Devesa SS, Purdue MP, Cook MB, McGlynn KA. 2010. International trends in the incidence of testicular cancer, 1973–2002. Cancer Epidemiol Biomarkers Prev 19:1151–1159.

Chilvers C, Pike MC, Forman D, Fogelman K, Wadsworth ME. 1984. Apparent doubling of frequency of undescended testis in England and Wales in 1962–81. Lancet 2:330–332.

Colborn T, vom Saal FS, Soto AM. 1993. Developmental effects of endocrine disrupting chemicals in wildlife and humans. Environ Health Perspect 101:378–384.

Cook MB, Trabert B, McGlynn KA. 2011. Organochlorine compounds and testicular dysgenesis syndrome: Human data. Int J Androl 2011 34:e68–84.

Cooper TG, Noonan E, von Eckardstein S, Auger J, Baker HW, Behre HM, et al. 2010. World Health Organization reference values for human semen characteristics. Hum Reprod Update 16:231–245.

Dieckmann KP, Pichlmeier U. 2004. Clinical epidemiology of testicular germ cell tumors. World J Urol 22:2–14.

Duty SM, Silva MJ, Barr DB, Brock JW, Ryan L, Chen Z, et al. 2003. Phthalate exposure and human semen parameters. Epidemiology 14:269–277.

Fisch H, Goluboff ET, Olson JH, Feldshuh J, Broder SJ, Barad DH. 1996. Semen analyses in 1283 men from the United States over a 25-year period: No decline in quality. Fertil Steril 65:1009–1014.

Fisch H, Lambert SM, Hensle TW, Hyun G. 2009. Hypospadias rates in New York State are not increasing. J Urol 181:2291–2294.

Giordano F, Abballe A, De Felip E, di Domenico A, Ferro F, Grammatico P, et al. 2010. Maternal exposures to endocrine disrupting chemicals and hypospadias in the offspring. Birth Defects Res A Clin Mol Teratol 88:241–250.

Guillette LJ. 2000. Contaminant-induced endocrine disruption in wildlife. Growth Horm IGF Res 10(Suppl B):S45–S50.

Guo YL, Hsu PC, Hsu CC, Lambert GH. 2000. Semen quality after prenatal exposure to polychlorinated biphenyls and dibenzofurans. Lancet 356:1240–1241.

Hardell L, van Bavel B, Lindström G, Eriksson M, Carlberg M. 2006. In utero exposure to persistent organic pollutants in relation to testicular cancer risk. Int J Androl 29:228–234.

Hauser R, Meeker JD, Duty S, Silva MJ, Calafat AM. 2006. Altered semen quality in relation to urinary concentrations phthalate monoester and oxidative metabolites. Epidemiology 17:682–691.

Herr C, zur Nieden A, Koch HM, Schuppe HC, Fieber C, Angerer J, et al. 2009. Urinary di(2-ethylhexyl)phthalate (DEHP)—Metabolites and male human markers of reproductive function. Int J Hyg Environ Health 212:648–653.

Hsu PC, Huang W, Yao WJ, Wu MH, Guo YL, Lambert GH. 2003. Sperm changes in men exposed to polychlorinated biphenyls and dibenzofurans. JAMA 289: 2943–2944.

Hutson JM, Balic A, Nation T, Southwell B. 2010. Cryptorchidism. Semin Pediatr Surg 19:215–224.

Huyghe E, Matsuda T, Thonneau P. 2003. Increasing incidence of testicular cancer worldwide: A review. J Urol 170:5–11.

Irvine S, Cawood E, Richardson D, MacDonald E, Aitken J. 1996. Evidence of deteriorating semen quality in the United Kingdom: Birth cohort study in 577 men in Scotland over 11 years. BMJ 312:467–471.

Joensen UN, Bossi R, Leffers H, Jensen AA, Skakkebaek NE, Jørgensen N. 2009. Do perfluoroalkyl compounds impair human semen quality? Environ Health Perspect 117:923–927.

Jönsson BA, Richthoff J, Rylander L, Giwercman A, Hagmar L. 2005. Urinary phthalate metabolites and biomarkers of reproductive function in young men. Epidemiology 16:483–487.

Kelce WR, Gray LE, Wilson EM. 1998. Antiandrogens as an environmental endocrine disruptor. Reprod Fertil Dev 10:105–111.

Kortenkamp A, Faust M. 2010. Combined exposures to anti-androgenic chemicals: Steps towards cumulative risk assessment. Int J Androl 33:463–474.

Kubota Y, Temelcos C, Bathgate RA, Smith KJ, Scott D, Zhao C, et al. 2002. The role of insulin 3, testosterone, Mullerian inhibiting substance and relaxin in rat gubernacular growth. Mol Hum Reprod 8:900–905.

Lau C, Butenhoff JL, Rogers JM. 2004. The developmental toxicity of perfluoroalkyl acids and their derivatives. Toxicol Appl Pharmacol 198:231–241.

Lindh CH, Rylander L, Toft G, Axmon A, Rignell-Hydbom A, Giwercman A, et al. 2012. Blood serum concentrations of perfluorinated compounds in men from Greenlandic Inuit and European populations. Chemosphere 88:1269–1275.

Lund L, Engebjerg MC, Pedersen L, Ehrenstein V, Norgaard M, Sorensen HT. 2009. Prevalence of hypospadias in Danish boys: A longitudinal study, 1977–2005. Eur Urol 55:1022–1026.

Lundin JI, Alexander BH, Olsen GW, Church TR. 2009. Ammonium perfluooctanoate production and occupational mortality. Epidemiology 20:921–928.

Main K, Kiviranta H, Virtanen HE, Sundqvist E, Tuomisto JT, Tuomisto J, et al. 2007. Flame retardants in placenta and breast milk and cryptorchism in new born boys. Environ Health Perspect 115:1519–1526.

Manecksha RP, Fitzpatrik JM. 2009. Epidemiology of testicular cancer. BJU Int 104:1329–1333.

McGlynn KA, Quraishi SM, Graubard BI, Weber JP, Rubertone MV, Erickson RL. 2008. Persistent organochlorine pesticides and risk of testicular germ cell tumors. J Natl Cancer Inst 100:663–671.

McGlynn KA, Quraishi SM, Graubard BI, Weber JP, Rubertone MV, Erickson RL. 2009. Polychlorinated biphenyls and risk of testicular germ cell tumors. Cancer Res 69:1901–1909.

Nelson CM, Bunge RG. 1974. Semen analysis: Evidence for changing parameters of male fertility potential. Fertil Steril 25:503–507.

Nordkap L, Joensen UN, Blomberg Jensen M, Jörgensen N. 2012. Regional differences and temporal trends in male reproductive health disorders: Semen quality may be a better sensitive marker of environmental exposures. Mol Cell Endocrinol 355:221–230.

Olsen GW, Bodner KM, Ramlow JM, Ross CE, Lipshultz LI. 1995. Have sperm counts been reduced 50 percent in 50 years? A statistical model revisited. Fertil Steril 63:887–893.

Olsen GW, Burris JM, Ehresman DJ, Froehlich JW, Seacat AM, Butenhoff JL, et al. 2007. Half-life of serum elimination of perfluorooctanesulfonate, perfluorohexanesulfonate, and perfluorooctanoate in retired fluorochemical production workers. Environ Health Perspect 115:1298–1305.

Ormond G, Nieuwenhuijsen MJ, Nelson P, Toledano MB, Iszatt N, Geneletti S, et al. 2009. Endocrine disruptors in the workplace, hair spray, folate supplementation, and risk of hypospadias: Case-control study. Environ Health Perspect 117:303–307.

Pant N, Shukla M, Kumar Patel D, Shukla Y, Mathur N, Kumar Gupta Y, et al. 2008. Correlation of phthalate exposure with semen quality. Toxicol Appl Pharmacol 15:112–116.

Pierik FH, Burdorf A, Nijman JM, de Muinck Keizer-Schrama SM, Weber RF. 2002. A high hypospadias rate in the Netherlands. Hum Reprod 17:1112–1115.

Purdue MP, Engel LS, Langseth H, Needham LL, Andersen A, Barr DB, et al. 2009. Prediagnostic serum concentrations of organochlorine compounds and risk of testicular germ cell tumors. Environ Health Perspect 117:1514–1519.

Raymer JH, Michael LC, Studabaker WB, Olsen GW, Sloan CS, Wilcosky T, et al. 2012. Concentrations of perfluorooctane sulfonate (PFOS) and perfluorooctanoate (PFOA) and their associations with human semen quality measurements. Reprod Toxicol 33:419–427.

Renner R. 2001. Growing concern over perfluorinated chemicals. Environ Sci Technol 35:154A–160A.

Rignell-Hydbom A, Rylander L, Giwercman A, Jönsson B, Nilsson-Ehle P, Hagmar L. 2004. Effect of dietary exposure to CB-153 and p,p′-DDE on reproductive function in Swedish fishermen. Hum Reprod 19:2066–2075.

Robaire B, Viger RS. 1995. Regulation of epididymal epithelial cell functions. Biol Reprod 52:226–236.

Sharpe RM. 2001. Hormones and testis development and the possible adverse effects of environmental chemicals. Toxicol Lett 120:221–232.

Sharpe RM, Skakkebaek NE. 1993. Are oestrogens involved in falling sperm counts and disorders of the male reproductive tract? Lancet 341:1392–1395.

Skakkebaek NE, Rajpert-De Meyts E, Main KM. 2001. Testicular dysgenesis syndrome: An increasingly common developmental disorder with environmental aspects. Hum Reprod 16:972–978.

Sonne C, Dietz R, Born EW. 2008. Is there a link between hypospadias and organochlorine exposure in East Greenland sledge dogs (*Canis familiaris*)? Ecotoxicol Environ Saf 69:391–395.

Spanò M, Toft G, Hagmar L, Eleuteri P, Rescia M, Rignell-Hydbom A, et al.; INUENDO. 2005. Exposure to PCB and p,p-DDE in European and Inuit population: Impact on human sperm chromatin integrity. Hum Reprod 20:3488–3499.

Stoker TE, Cooper RL, Lambright CS, Wilson VS, Furr J, Gray LE. 2005. *In vivo* and *in vitro* anti-androgenic effects of DE-71, a commercial polybrominated diphenyl ether (PBDE) mixture. Toxicol Appl Pharmacol 207:78–88.

Swan SH, Elkin EP, Fenster L. 2000. The question of declining sperm density revisited: An analysis of 101 studies published 1934–1996. Environ Health Perspect 108: 961–966.

Toft G, Axmon A, Giwercman A, Thulstrup AM, Rignell-Hydbom A, Pedersen HS, et al. 2005. Fertility in four regions spanning large contrasts in serum levels of widespread persistent organochlorines: A cross-sectional study. Environ Health 4:26.

Toft G, Jönsson BAG, Lindh C, Giwercman A, Spano M, Heederik D, et al. 2012. Exposure to perfluorinated compounds and human semen quality in Arctic and European populations. Hum Reprod 27:2532–2540.

Toppari J, Larsen JC, Christiansen P, Giwercman A, Grandjean P, Guillette LJ Jr., et al. 1996. Male reproductive health and environmental xenoestrogens. Environ Health Perspect 104 Suppl 4:741–803.

Toppari J. 2008. Environmental endocrine disrupters. Sex Dev 2:260–267.

Toshima H, Suzuki Y, Imai K, Yoshinaga J, Shiraishi H, Mizumoto Y, et al. 2011. Endocrine disrupting chemicals in urine of Japanese male partners of subfertile couples: A pilot study on exposure and semen quality. Int J Hyg Environ Health 215:502–506.

Van Waeleghem K, De Clercq N, Vermeulen L, Schoonjans F, Comhaire F. 1996. Deterioration of sperm quality in young healthy Belgian men. Hum Reprod 11:325–329.

Vestergren R, Cousins IT, Trudel D, Wormuth M, Scheringer M. 2008. Estimating the contribution of precursor compounds in consumer exposure to PFOS and PFOA. Chemosphere 73:1617–1624.

Viger RS, Silverside DW, Tremblay JJ. 2005. New insights into the regulation of mammalian sex determination and male sex differentiation. New insights into the regulation of mammalian sex determination and male sex differentiation. Vitam Horm 70:387–413.

Virtanen HE, Bjerknes R, Cortes D, Jørgensen N, Rajpert-De Meyts E, Thorsson AV, et al. 2007. Cryptorchidism: Classification, prevalence and long-term consequences. Acta Paediatr 96:611–616.

Wetterauer B, Ricking M, Otte JC, Hallare AV, Rastall A, Erdinger L, et al. 2012. Toxicity, dioxin-like activities, and endocrine effects of DDT metabolites—DDA, DDMU, DDMS, and DDCN. Environ Sci Pollut Res Int 19:403–415.

Wirth JJ, Rossano MG, Potter R, Puscheck E, Daly DC, Paneth N, et al. 2008. A pilot study associating urinary concentrations of phthalate metabolites and semen quality. Syst Biol Reprod Med 54:143–154.

Wittassek M, Koch HM, Angerer J, Bruning T. 2011. Assessing exposure to phthalates—The human biomonitoring approach. Mol Nutr Food Res 55:7–31.

CHAPTER 12

Effects of Endocrine-Disrupting Substances on Bone and Joint

CHI-HSIEN CHEN and YUELIANG LEON GUO

ABSTRACT

Background: Bones and joints are living tissues regulated by a series of hormones, immune factors, and vitamins. Thus, even though they are composed primarily of minerals, they are vulnerable to influences by organic chemicals.

Objectives: The goal of this chapter is to summarize what is known about bone and joint development and function, and how normal physiology and function are altered by several classes of organic chemicals by both direct and indirect actions through disruption of endocrine and immune system function.

Discussion: Early life exposure to some organic chemicals alters bone and joint development, but exposure later in life also changes bone and joint function. The best-documented effects are from persistent organic pollutants, such as polychlorinated biphenyls, dioxins, chlorinated pesticides, and perfluorinated compounds, but less persistent organics like bisphenol A and phthalates also alter bone and joint function. These actions may increase risk of osteoporosis, osteoarthritis, rheumatoid arthritis, and susceptibility to fracture.

Conclusions: Bones and joints are tissues that are vulnerable to diseases caused at least in part by exposure to organic chemicals. While much more research is needed in this area, results to date clearly show that organic chemicals influence human diseases of bones and joints.

Effects of Persistent and Bioactive Organic Pollutants on Human Health, First Edition.
Edited by David O. Carpenter.

INTRODUCTION

Bone provides an essential framework and functional rigidity to the body. A joint is located between two or more contacted bones, allowing movement and providing mechanical support for related skeletal structures. Articular cartilage, synovial membrane, and fibrous capsule are the three main components of joints. Bone and articular cartilage are highly specialized tissues with osteocytes and chondrocytes embedded in complex matrices containing minerals and macromolecules, including collagen, proteoglycan, and glycosaminoglycans. During a lifetime, skeletal tissues are active on resorption and remodeling to maintain mineral homeostasis and repair microdamaged bones without change of the gross shape of bones. Unlike remodeling, bone modeling, which determines the skeletal architecture, is a process occurring principally during development and growth, including the differentiation of articular cartilages. Both modeling and remodeling are complex processes under the control of hormonal and nutritional factors and may be modified by exposure to environmental agents.

There is a significant literature in teratology that indicates that prenatal development of bones and joints is to chemical toxicity. Some examples include the association of corticosteroid and hypervitaminosis A with cleft palate formation, thalidomide exposure with long-bone deformity, and high-dose salicylates with chondrodystrophy and skeletal abnormality. However, compared to the information on toxic effects on the kidney, liver, and other internal organs, such information on bone and joints is rather scarce. Despite the knowledge that bone tissue serves as a good reservoir for many xenobiotics, including heavy metals, some pesticides, agrochemicals, and plant dyes, the understanding of the skeletal and cartilage toxicity caused by postnatal exposure to these chemicals is still limited.

There are three principal mechanisms of toxic damage to bones and joints. The first is developmental defects, which result from exposure to toxic materials during prenatal development or throughout postnatal maturation. The second is through the impairment in the nutritional requirements of growing or mature bone or cartilage, including alterations in metabolism of calcium, phosphate, zinc, copper, and manganese. The third is by alterations in hormonal profiles influencing mineral exchange and the cellular environment. Although the net effect of environmental toxicants on the complex remodeling processes and homeostasis of bones and joints may not be immediately fatal, serious disability or structural damage after long-term cumulative exposure may result.

THE DEVELOPMENT OF BONES AND JOINTS

The main origin of cells in bones and joints is the mesoderm of the early embryo. The differentiation and development of skeletal tissues are still imperfectly understood, as are the teratological mechanism of environmental toxicants for bone modeling. The development of bone during the fetal stage occurs by two processes, intramembranous ossification and endochondral ossification. Intramembranous ossification mainly occurs during formation of the flat bones of the skull, mandible, maxilla, and clavicles. During this process, bone is formed from connective tissue such as mesenchyme tissue rather than from cartilage. Most of the rest of the bone in the body, including long bones, are developed by endochondral ossification. Endochondral bone is formed in cartilage, which is started with a primary ossification center in the middle diaphysis and is followed by two secondary ossification centers in the epiphysis. Finally, the diaphysis and the epiphysis are separated by a metaphysial plate of growing cartilage. The osseous formation in the growth plate increases the length of long bones and ceases after the plate becomes completely ossified. In the meantime, a sleeve of osseous tissue is formed around the middle diaphysis by intramembranous ossification in the subperiosteal area, which increases the cortical diameter of long bones.

Most joints are formed early in the process of limb bud differentiation of an embryo. After mesenchymal chondrification of future bones, synovial mesenchyme begins to form in the peripheral portion of the interzone between the two cartilaginous bones. Later, cavitation occurs centrally in the interzone and is completed early in the fetal period. In late adolescence and adulthood, the cartilaginous tissue is maintained in the articular surfaces of long bones and joint regions but becomes thinner with advancing age.

PHYSIOLOGY OF BONES AND JOINTS

The major role of bones and joints is to provide mechanical support of the body. In addition, they are also metabolically active tissues. Bones reserve a tremendous amount of calcium phosphate and other minerals and maintain mineral homeostasis. Blood calcium balance is modulated by three major organs (kidney, intestine, and bone) and three major hormones (vitamin D, parathyroid hormone [PTH], and calcitonin). Other elements rich in bone, such as zinc and magnesium, can be released into the circulation in response to hormonal or nutritional conditions. Mineral homeostasis is closely related to the remodeling process of the bone, which is performed by three types of cells: osteoblasts, osteocytes, and osteoclasts. Osteoblasts are bone-forming cells derived from mesenchymal stem cells and produce a protein mixture known as osteoid, which is mineralized to become bone. Osteocytes originate from osteoblasts and serve several functions, including maintenance of bone matrix and calcium homeostasis. Osteoclasts are derived from pluripotential hematopoietic cells and are responsible for bone resorption. Osteoblasts and osteocytes are target cells for PTH, and they communicate with PTH-insensitive osteoclasts by paracrine signaling. Low serum calcium concentration stimulates the secretion of PTH, which activates bone turnover. On the other side, calcitonin directly inhibits osteoclasts, which slows down bone resorption, and acts to lower serum calcium. The formation and destruction of bones occur simultaneously. The negative skeletal balance in most postmenopausal women occurs because bone resorption exceeds bone formation. This imbalance may result from an increase in osteoclast number or activity, a decrease in osteoblast number or activity, or a combination of the two.

There are three types of joints. First, the fibrous joints connect bones by fibrous connective tissue, such as the suture of the skull and teeth secured into alveoli. Second, the cartilaginous joints connect bones by fibrocartilage or hyaline cartilage, such as intervertebral joints and sacroiliac joints. Third, the synovial joints connect long bones, which contain synovial membranes and synovial fluid. Synovial membranes are a thin layer of tissue lining the joint surface. Synovial fluid secreted by synovial membrane serves many important functions, including reducing friction by lubricating the joint, absorbing shocks, regulating synovial cell growth, and supplying oxygen and nutrients to and removing carbon dioxide and metabolic wastes from the chondrocytes within the articular cartilage.

ENDOCRINE EFFECTS ON BONES AND JOINTS

Bone is a sensitive target tissue to endocrine regulation. The modeling and remodeling of bone are continuously in response to hormones. Bone growth is primarily influenced by the growth hormone (GH)/insulin-like growth factor 1 (IGF-1) axis, which is a mixed endocrine–paracrine–autocrine system. Other hormones, including PTH, thyroxine, glucocorticosteroids, and sex steroids, play a role as well. GH is secreted by the pituitary gland, under the regulation of stimulatory (growth hormone-releasing hormone [GHRH]) and inhibitory (somatostatin) hypothalamic hormones. Circulatory GH directly acts on many tissues and leads to the transcription of many target genes, including IGF-1. The liver is the main origin of circulatory IGF-1. IGF-1 enhances the differentiated function of the osteoblast and prevents osteoblast apoptosis, referring to its anabolic effect on bone mass. In addition, IGF-1 synthesis is mediated by other hormones, including PTH. PTH is an important anabolic hormone of bone and affects osteoblast directly and indirectly through stimulating IGF-1 secretion by bone cells.

Thyroxine acts both on bone modeling and remodeling. Thyroxine binds to thyroid hormone receptors (TRs), which can be identified on chondrocytes and osteoblasts. Thyroxine regulates chondrocyte proliferation, promotes terminal differentiation, and induces mineralization and angiogenesis. Thyroid hormone stimulates the production of type II and X collagen and alkaline phosphatase (ALP), a marker of bone mineralization. Hypothyroidism leads to growth arrest of the growth plate, epiphyseal dysgenesis, delayed bone age, and short stature. Thyroid hormone directly stimulates osteoblasts. Osteoclast activity is increased by thyroid hormone but only in the presence of osteoblasts, indicating that the stimulation is mediated by paracrine signaling. In hyperthyroidism, the rate of bone resorption is normal, but the rate of bone formation is reduced, leading to the loss of bone mass. In addition, hyperthyroid patients demonstrate increased dietary calcium intake but have reduced absorption and increased fecal and dermal calcium loss, leading to a negative calcium balance.

Of sex hormones, estrogen and testosterone play the major role in bone growth. Both estrogen and testosterone are present in women and men and decrease in level with age. The effects of sex hormones on growth are complex. Elevated levels of estrogen and testosterone in infancy and childhood increase bone growth but also stimulate skeletal maturation, so that the growth plate is closed earlier and body height is reduced. Estrogen and testosterone also play important roles in maintaining bone mass throughout a person's lifetime. An increased level of estrogen induces bone formation, while a decreased level of estrogen decreases bone formation. In women, approximately 10% of bone mass is lost during menopause. Estrogen use during menopause prevents bone loss. Estrogen receptors (ERs) are present on osteoblasts, osteoclasts, and chrondroblasts, indicating that bone and cartilage are target tissues for sex hormones. In addition to direct effects, estrogen can play an indirect role through being a regulatory link between the immune system and bone homeostasis. Several cytokines, including interleukin 1 (IL-1), interleukin 6 (IL-6), and interleukin 7 (IL-7), have been found responsible for bone loss after menopause. Estrogens provide a proapoptotic effect on osteoclasts. Ovariectomy enhances bone resorption by osteoclasts and increases the apoptosis of osteoblasts and osteocytes.

In men, testosterone deficiency can cause osteoporosis. Androgen receptors have been identified on osteoblasts, osteocytes, and chondrocytes, indicating a direct effect of androgen on bone and cartilage. In mouse osteoblast cells, testosterone increases osteoprotegerin expression, which is a soluble receptor for the receptor activator of nuclear factor kappa B ligand (RANKL) and is able to inhibit both differentiation and function of osteoclasts. Thus, androgen deficiency in males leads to an increase in osteoclastic bone resorption and a progressive decrease in bone mineral density (BMD). Testosterone stimulates proliferation and differentiation of osteoblasts. In animal studies, orchiectomy increases apoptosis of osteoblasts and osteocytes.

In addition to articular cartilage, the synovium plays a major role in maintaining the normal function of joints, which might be influenced by endocrine systems. Synovial macrophages, monocytes, lymphocytes, and fibroblasts possess androgen and ERs, indicating that sex hormones may directly modulate certain pathophysiologic processes in joints. Estrogen and testosterone have been found to exert anti-inflammatory effects on rat arthritic synovial fibroblasts at physiological concentrations. *In vitro*, estrogen stimulates rheumatoid synovial fibroblasts to produce osteoprotegerin, which in turn inhibits osteoclasts and plays substantial a role in periarticular bone erosion of rheumatoid arthritis (RA). Furthermore, the GH/IGF-1 axis has been associated with several rheumatic diseases, including osteoarthritis (OA) and RA. In OA and RA, synovial fluid GH exceeds serum GH levels. Clinical trials of RA therapy using somatostatin, an inhibitory hormone of GH, found improved clinical symptoms and decreased thickness of the synovial membrane of RA. This suggests that the GH/IGF-1 axis may be involved in the functional balance of synovial joints.

NUTRITION EFFECTS ON BONES AND JOINTS

With prolonged life expectancy, osteoporosis and its related fractures have become a global health problem. Nutrition is an important modifiable factor in the development and maintenance of bone mass. Approximately 80–90% of bone mineral content is composed of calcium and phosphorus. Therefore, adequate intake of these two elements is critical for bone health. Other dietary components, such as protein, magnesium, zinc, copper, iron, fluoride, and vitamins D, A, C, and K are required for normal bone metabolism. They may affect bone by various mechanisms, including alterations in bone structure, rate of bone metabolism, balance of endocrine and/or paracrine system, and homeostasis of minerals. Furthermore, these essential nutrients of bone may interact with and be modulated by environmental exposures, including endocrine-disrupting chemicals (EDCs).

PATHOPHYSIOLOGY AND TOXICOLOGY OF EDCs ON BONES AND JOINTS

EDCs are substances in our environment, food, and consumer products that, when entering the human and animal body, can interfere with hormone biosynthesis, metabolism, or action resulting in a deviation from normal homeostatic control. The mechanisms of EDCs involve divergent pathways and actions including (but not limited to) effects on estrogenic, androgenic, thyroid, peroxisome proliferator-activated receptor γ, and retinoid systems, which then may modulate normal development and function of bones and joints. EDCs may act via the nuclear receptors of hormones mentioned earlier, on membranous steroid receptors (such as membranous ER), and on nonsteroid receptors (such as the aryl hydrocarbon receptor [AhR]). A wide range of industrial chemicals have been identified as EDCs, including polychlorinated biphenyls (PCBs), polybrominated biphenyls (PBBs), dioxins, plastics (bisphenol A [BPA]), plasticizers (phthalates), pesticides (methoxychlor, chlorpyrifos, dichlorodiphenyltrichloroethane [DDT]), fungicides (vinclozolin), and pharmaceutical agents (diethylstilbestrol [DES]). Natural chemicals found in human and animal food (e.g., phytoestrogens, including genistein and coumestrol) can also act as endocrine disruptors. A number of EDCs are estrogenic, for example, several lowly chlorinated PCBs, and antiestrogenic, for example, 2,3,7,8-tetrachlorodibenzo-*p*-dioxin (TCDD) and certain coplanar PCB congeners, for example, 3,3′,4,4′,5-pentachlorobiphenyl (PCB 126) (Astroff and Safe 1990; Biegel and Safe 1990; Krishnan and Safe 1993).

Unfortunately, not much is known of the interactions between EDCs and PTH.

In addition to the direct endocrine-disrupting capabilities on bone, EDCs interfere with the homeostasis of vitamins, which are essential for bone. For example, circulating vitamin D is depressed in animals or humans exposed to PCBs (Alvarez-Lloret et al. 2009; Lilienthal et al. 2000), DDT (Yang et al. 2012), and other persistent organic pollutants (POPs) (Routti et al. 2008). Studies on wildlife have shown an association between the exposure to environmental organohalogen contaminants and lower plasma and hepatic retinol (vitamin A) concentrations (Jenssen et al. 2003; Kirkegaard et al. 2010; Nyman et al. 2002, 2003; Skaare et al. 2001; de Swart et al. 1996). Similar findings have been confirmed in experimental studies on rats (Brouwer and van den Berg 1986; Chen et al. 1992; Martin et al. 2006; Morse and Brouwer 1995; van der Plas et al. 2001; Van Birgelen et al. 1994). Organohalogen contaminants, especially PCBs, interfere with transport, storage, metabolism, and excretion of retinoids (Novák et al. 2008). Vitamin C and E are known to lower oxidative stress caused by PCBs and dioxin (Ashida et al. 1996; Banudevi et al. 2006; Krishnamoorthy et al. 2007; Martino et al. 2009; Murugesan et al. 2005; Muthuvel et al. 2006; Venkataraman et al. 2007; Wang et al. 2009; Yin et al. 2012), but paradoxically, the former augments oxidative injury in rats caused by BPA, nonylphenol, and octylphenol (Aydoğan et al. 2008, 2010; Korkmaz et al. 2011).

The mechanisms of the toxic effects on joints include immune-mediated and non-immune-mediated processes. EDCs have been associated with arthritis in human studies (Guo et al. 1999; Lee et al. 2007; Okumura 1984). Some experimental studies have tried to uncover the underlining mechanisms. For example, TCDD and PCBs have been found to induce apoptosis of articular chondrocytes in culture (Yang and Lee 2010). Destruction of cartilage matrix may accelerate the progression of arthritis, in particular, OA (Kim and Blanco 2007). AhR, a receptor with high affinity for dioxin-like EDCs, is highly expressed in synoviocytes of RA through the upregulation by tumor necrosis factor-α (TNF-α). TCDD stimulates the secretion of IL-1α, IL-6, and IL-8 in synoviocytes via AhR, which can amplify local inflammation (Kobayashi et al. 2008). Thus, EDCs may affect joint health either directly or indirectly through immune-mediated inflammation.

Fetuses and infants, being exposed via placenta and breast milk, are generally more susceptible to the effects of toxic compounds since exposure occurs during development and phases of rapid growth. Effects of lipophilic and persistent EDCs on the fetus are probably the most important of all environmental chemicals because they are excreted relatively slowly in the human body and are among those organic chemicals that most readily penetrate the placenta and enter fetal blood.

The following sections will discuss the bone and joint effects of each group of individual EDCs. Phytoestrogens will not be covered in this chapter because they are mostly natural products, and exposures to these chemicals are frequently voluntary.

PCBs

PCBs are halogenated aromatic hydrocarbons that have been produced for a multitude of industrial purposes and are now recognized as persistent environmental pollutants (Carpenter 1998). They have been widely used as industrial fluids, flame retardants, diluents, and fluids for capacitors and transformers. Although their production and application have been banned in most industrialized countries (Carpenter 1998), many PCB-containing industrial products and equipment are still in use and thus continue to pose a threat to the environment and human health. They tend to bioaccumulate both in the environment and in living organisms due to their lipophilicity and chemical stability (Safe 1994). It is estimated that approximately 33% of the PCBs ever produced are still environmentally available (Tanabe 1988). Contamination with these persistent toxic substances occurs primarily by ingestion of contaminated food and the food chain.

There are 209 possible PCB congeners, differing in the position of chlorine atoms and the degree of chlorination. These differences affect their physicochemical properties and biological activities. Some of the effects might be explained by the fact that PCBs exhibit estrogenic or antiestrogenic properties. Some PCB congeners resemble estradiol-17-β(E2) in terms of chemical structure and thus can bind to ERs and mimic the effects of the endogenous ligand (Massaad et al. 2002). Lower-chlorinated PCB congeners tend to have estrogenic properties. PCB congeners with more planar structure and thus resembling TCDD may display antiestrogenic actions by binding and activating the cytoplasmic AhR (Safe and Wormke 2003). Formation of ligand-bound AhR and AhR nuclear translocator (ARNT) protein triggers transcription of cytochrome P450 1A1 and 1B1 on the DNA, which in turn metabolize intracellular estrogen. Highly chlorinated PCB mixtures such as Aroclor 1254 may exert antiestrogenic effects. The divergent effects observed in animals exposed to PCB congeners seem to be related to the affinity to ER and AhR. For example, PCB 153 has high affinity to ER but low affinity to AhR, while PCB 126 has high affinity to AhR but variable estrogenic properties depending on the estrogen status of individuals, being estrogenic in estrogen-deprived tissues. Perinatal exposure to PCB 153, but not to PCB 126, was associated with the increase in trabecular BMD and the decrease in total cross-sectional area at diaphyseal bone in goat (Lundberg et al. 2006). On the other hand, PCB 126 showed weak estrogen agonistic activity in estrogen-deprived tissue like the uteri of ovariectomized rats but antiestrogenic effects in intact rats (Lind et al. 1999, 2004). Another mechanism may involve thyroid functioning. PCBs have been known to reduce thyroid hormone (Brouwer 1991; Byrne et al. 1987;

Roth-Härer et al. 2001), which plays an important role in bone physiology, especially during development and maturation of skeletal tissues (Bassett et al. 2007). PCBs may alter thyroxin levels in many ways, including the interference with plasma-binding proteins by PCB metabolites (Rickenbacher et al. 1986; Van den Berg et al. 1991) and the disturbance of thyroid hormone metabolism (Van Birgelen et al. 1994).

There may be a gender-specific difference in response to PCB exposure. In a study of sheep, prenatal treatment of the dioxin-like PCB 118 depressed bone mineral content in male fetuses, while the non-dioxin-like PCB 153 resulted in a decrease in the trabecular cross-sectional area at the metaphysis in females, and an increase in BMD and cortical thickness in both genders (Gutleb et al. 2010) . More studies are needed to elucidate the interactions between gender, endocrines, and exposure of PCBs, as well as their combined effects on bone health.

In human epidemiologic studies, one of the most important PCB effects on bone was osteoporosis and related health outcomes. However, investigations found inconsistent results. There are several epidemiological studies on the bone health of Swedish fishermen. The Baltic sea was highly contaminated by persistent organic compounds (POCs), including dioxin-like chemicals (mainly PCBs), DDT, and other organochlorines (Asplund et al. 1994). In Sweden, east coast fishermen have higher plasma levels of dioxin-like POC (290 pg/g lipid in average) than west coast fishermen (139 pg/g lipid in average) (Svensson et al. 1995). Alveblom et al. (2003) reported an increased incidence ratio of osteoporotic vertebral fractures in women exposed to high levels of organochlorines, including PCBs and other dioxin-like chemicals, through dietary intake of contaminated fish. However, Wallin et al. (2004, 2005) did not find significant detrimental effects in Swedish east coast fishermen on risk of fracture, BMD, and bone biomarkers. Using BMD as an outcome indicator, some human studies have reported an association between PCB exposure and low BMD (Alveblom et al. 2003; Glynn et al. 2000; Hodgson et al. 2008), while others found no association between BMD and plasma concentration of POPs, including PCBs (Côté et al. 2006; Rignell-Hydbom et al. 2009). Even in the Japanese Yusho cohort, a population that was highly exposed to PCBs and polychlorinated dibenzofurans, BMD was not found associated with serum levels of the contaminants (Yoshimura et al. 2009).

Regarding the health effects on joints, increased incidence of arthritis and herniated intervertebral disks have been documented in people highly exposed to PCBs (Guo et al. 1999; Okumura 1984). In the general population, serum levels of PCBs and other organochlorine pesticides were associated with increases in self-reported arthritis, especially the rheumatoid form (Lee et al. 2007). Among Taiwanese women highly exposed to PCBs and their pyrolytic by-products, the mortality rate due to systemic lupus erythematosus was highly elevated (Tsai et al. 2007). The mechanisms of the above-mentioned effects have been unclear but are likely related to autoimmunity.

The reported effects of prenatal exposure to PCBs are different from those in adults. Funatsu et al. (1972) reported that among the four babies born to mothers highly exposed to PCBs due to ingestion of contaminated rice oil, three had spotted calcification on the parieto-occipital skull and large or wide frontal and occipital fontanels and sagittal sutures. This suggests that the prenatal effects of PCBs on bone development might be more prominent than those in grown-ups. In human epidemiologic studies on adults or fetuses, the individuals were exposed to mixtures of PCB congeners. Therefore, it is difficult to determine which specific agents caused the observed effects. Only in controlled exposure scenarios, such as in laboratory animal studies, can effects of any specific agent be elucidated.

The results from cell and animal studies do support the observed effects of PCBs on bones and joints in humans. Changes in bone properties have been documented in many wildlife animals living in areas highly contaminated with POPs, including otters (Roos et al. 2010), deer mice (Johnson et al. 2009), voles (Murtomaa et al. 2007), polar bears (Sonne et al. 2004), and seals (Routti et al. 2008). Prenatal exposure to PCBs leads to changes in bone morphometry and mechanical strength in experimental rats (Elabbas et al. 2011), goats (Lundberg et al. 2006), and sheep (Gutleb et al. 2010).

Animal studies regarding PCB effects on joint health are relatively limited. Lee and Yang (2012) found that PCB 126 could induce chondrocyte apoptosis, which was blocked by inhibitors of reactive oxygen species and inducible nitric oxide synthase. AhR, a target receptor for dioxin-like PCBs, is upregulated in synoviocytes of RA. Through the activation of AhR, expression of some proinflammatory cytokines, such as IL-1β, IL-6, and IL-8, is simulated, which might exacerbate the inflammation in joints (Kobayashi et al. 2008). These results provided some explanation of the immune and non-immune-mediated mechanisms of PCB toxicity on joints.

Table 12.1 summarizes the findings of PCB exposure to bones and joints from laboratory studies.

Dioxin-Like Chemicals

Dioxins are a class of chemical contaminants that are formed during combustion processes such as waste incineration, forest fires, and backyard trash burning, as well as during some industrial processes such as paper pulp bleaching and herbicide manufacturing. The most toxic chemical in the class is TCDD, which exerts most of its toxic effects through activation of the AhR. Currently, almost every living creature has been exposed to dioxins or dioxin-like compounds. People are exposed to dioxins primarily by eating food, in particular, animal products contaminated by these chemicals.

TCDD is known to cause bone toxicity, particularly during animal development. Exposure to TCDD during gestation and lactation results in decreased bone length and area, as well as reduced BMD and bone strength in offspring

TABLE 12.1. Summary of PCB-Related Bone and Joint Effects by Experimental Studies

Author	Year	Cell/Animal	Exposure	Outcome Assessment	Result
In vitro					
Lee	2012	Rabbit articular chondrocytes	Cells were treated with DMSO (0.1%), 0.01, 0.1, 1 mM PCB 126 for 24 hours	Productions of reactive oxygen species (ROS) and nitric oxide (NO), binding activity of nuclear factor-kappa β (NF-κB), apoptotic cell death	PCB 126: increases in production of ROS, NO, NF-κB binding activity and indicators for apoptosis of chondrocytes
In vivo					
Hoffman	1996	Posthatching kestrel	Treated orally for 10 days with 5 μL/g bw of corn oil (controls) or PCB 126 at concentrations of 0.050, 0.25, or 1 mg/kg bw	Developmental change	PCB 126: decrease in bone growth at the dosage of 0.25 mg/kg bw
Lind	1999	Female intact and ovariectomized rat	Intraperitoneal injection of corn oil (control) and PCB 126 for 3 months (total dose, 0.384 mg/kg bw)	Bone density and morphology	PCB 126: decrease in bone length and increase in BMD of tibia in ovariectomized rat; increase in osteoid surface, cortical thickness, and organic content, but no change in BMD or trabecular bone volume of tibia in intact rat
Lind	2000	Female intact and ovariectomized rat	Intraperitoneal injection of corn oil (control) and PCB 126 for 3 months (total dose, 0.384 mg/kg bw)	BMD, geometry, and bone composition analysis, serum retinoid	PCB 126: shorter bone length, lower water content, and a decreased torsional stiffness in both ovariectomized and normal rats; lower collagen concentration but higher pyridinoline concentration of cortical bone; lower level of retinoid in liver but higher levels in serum and kidney

Lind	2000	Female rat	Three groups exposed to intraperitoneal injection of PCB 126 (total dose 0.32 mg/kg bw), either alone or in combination with vitamin C added to the drinking water (1 and 10 g/L); one group received increased level of vitamin A (600,000 IU/kg) in diet and one group with no treatment (control)	BMD, geometry and serum osteocalcin	PCB 126: increase in trabecular density and cortical thickness but reduction in the trabecular area, maximum torque and stiffness of the humerus, and serum osteocalcin levels Addition of vitamin C: only inhibits the reduction of serum osteocalcin
Render	2000	Male mink	Oral exposure to 0.024 ppm PCB 126, or none (control), for 1~2 months; euthanized when there are clinical signs of toxicity development	Gross examination of upper and lower jaws, histology examination of maxilla and mandible	PCB 126: proliferation of gingival and loose teeth, mandibular and maxillary nodular proliferation of the gingiva, loss of alveolar bone in maxilla and mandible due to proliferation of squamous cells that formed infiltrating cords
Lind	2004	Female intact and ovariectomized rat	Intraperitoneal injection of vehicle, PCB 126 or PCB 126 (total dose 0.384 mg/kg bw) plus estradiol (0.023 mg/kg bw days weekly) for 3 months	BMD and geometry	In ovariectomized rats: PCB 126 + E2: increase in trabecular bone volume, comparing to control and PCB 126 only PCB 126 only: no significant change In intact rats: PCB 126 + E2: decrease in trabecular bone volume, comparing to control and PCB 126 only PCB 126 only: no significant change

(Continued)

TABLE 12.1. (*Continued*)

Author	Year	Cell/Animal	Exposure	Outcome Assessment	Result
Yilmaz	2006	Female intact and ovariectomized rat	Subcutaneous injection of DMSO (control), Aroclor 1221 (10 mg/kg), Aroclor 1254 (10 mg/kg), estradiol (0.03 mg/kg) every other day for 6 weeks; technical PCB mixtures, Aroclor 1221 and 1254 with chlorine percentages of approximately 21% and 42%, respectively	Serum parathyroid hormone (PTH), calcitonin, osteocalcin, alkaline phosphatase (ALP), calcium, and inorganic phosphate; urinary deoxypyridinoline (DPD) Histology of lumbar vertebrae by light microscopy	Aroclor 1221: reduction in urinary DPD and reverse adverse effects of ovariectomy on vertebrae histology in ovariectomized rats; increase in serum ALP and calcium in intact rats Aroclor 1254: increase in urinary DPD, increase in serum PTH and expansion of necrotic area on vertebrae histology in ovariectomized rats; increase in serum calcium and phosphate in intact rats
Lundberg	2006	Goat	Prenatal exposure (corn oil, PCB 126 0.049 mg/kg bw/day, and PCB 153 0.098 mg/kg bw/day) from gestational day 60 until delivery; postnasal exposure through mother's milk for 6 weeks	BMD, geometry, and biomechanics	PCB 153: decrease in the total cross-sectional area, decrease in the marrow cavity, decrease in the moment of resistance at the diaphysis, increase in the trabecular BMD at the metaphysis, and no effect on biomechanical testing of the bones PCB 126: no observable changes in bone tissue

			1254 (2 mg/kg bw), Aroclor 1254 + vitamin C (100 mg/kg bw) and Aroclor 1254 + vitamin E (50 mg/kg bw)	markers (ALP, collagen), bone resorption marker (TRAP), antioxidant enzymes (SOD, GPX, and GST), and lipid peroxidation in the femur	activity and collagen, increases in TRAP activity, decreases in SOD and GPX activity, but no change in GST activity Aroclor 1254 + vitamin C or E: prevents the adverse effects of Aroclor 1254 in the femur
Alvarez-Lloret	2009	Rat	Intraperitoneal injection of corn oil (control) and PCB 126 for 3 months (total dose, 0.384 mg/kg bw)	BMD, geometry, and bone composition analysis, serum levels of thyroid hormone and vitamin D	PCB 126: lower degree of mineralization, increase in trabecular density, decrease in the size and crystallinity of apatite crystals of vertebral bone; lower vitamin D and thyroxin (free and total T4) in serum
Gutleb	2010	Pregnant ewes and their fetuses	Oral exposure (corn oil 0.1 mL/kg bw/day), PCB 118 (0.049 mg/kg bw/day and PCB 153 (0.098 mg/kg bw/day), starting at conception and ending a week before expected delivery	BMD, geometry and biomechanics	PCB 118: shorter bone length in both sexes; 28% lower trabecular bone mineral content at metaphysis in male fetuses, but not in female. PCB 153: shorter bone length in both sex; 24% reduction of marrow cavity at diaphysis in male fetuses; 19% smaller trabecular cross-sectional area at metaphysis, 15% higher body mineral density and 20% thicker cortical bone at diaphysis, 13% and 24% smaller endosteal circumference and marrow cavity in female fetuses.

(Continued)

TABLE 12.1. (*Continued*)

Author	Year	Cell/Animal	Exposure	Outcome Assessment	Result
Elabbas	2011	Pregnant rats and their fetuses	Oral exposure (corn oil, Northern contaminant mixture at dose levels of 0.05, 0.5, or 5 mg/kg bw) from gestational day 1 to postnatal day 23 (NCM contains 14 PCB congeners, 12 organochlorine pesticides, and methyl mercury)	BMD, geometry, and biomechanics	NCM at 5 mg/kg wt: short and thin femur with reduced mechanical strength in offspring at PND 35; treatment-related bone differences were not detected at PND 350
Elabbas	2011.	Rat	Oral exposure (corn oil, Aroclor 1254 15 mg/kg bw/day) from gestational day 1 to postnatal day 23	BMD, geometry, and biomechanics	Aroclor 1254: Shorter and thinner femur and tibia, weaker femoral head at PND 35. The effects of A1254 on bone were not detected at PND 77 and 350.

Aroclor 1254 is a technical PCB mixture containing 54% chlorine by weight.
bw, body weight; PND, postnatal day; TRAP, tartrate-resistant acid phosphatase; SOD, superoxide dismutase; GPX, glutathione peroxidase; GST, glutathione S-transferase.

(Finnilä et al. 2010) . TCDD downregulates the expression of osteopontin in osteoblast-like rat cells. Studies on bone cell cultures have indicated that both osteoblasts and osteoclasts strongly express the AhR. However, only the former has been documented to be directly affected by TCDD (Ilvesaro et al. 2005; Korkalainen et al. 2009; Ryan et al. 2007; Singh et al. 2000). TCDD primarily inhibits the differentiation, rather than the proliferation, of osteoblasts (Gierthy et al. 1994; Korkalainen et al. 2009). *In vitro*, 3-methylcholanthrene, another AhR activator, retards proliferation and differentiation of osteoblasts (Naruse et al. 2002), while the AhR antagonist, resveratrol, reverses dioxin-induced inhibition of osteodifferentiation (Singh et al. 2000).

The mechanisms of bone toxicities caused by dioxin-like chemicals are not totally understood. However, the AhR may play significant roles (Ryan et al. 2007; Singh et al. 2000). In two strains of rats with a different sensitivity of AhR, TCDD impaired bone growth and mechanical strength in a dose–response manner, especially in those with high AhR sensitivity (Herlin et al. 2010; Jämsä et al. 2001). In addition, estrogen and ERs likely play roles as well. *In utero* and lactational exposure to TCDD leads to estrogenic bone effects (decreased bone length and decreased cross-sectional area) as well as antiestrogenic bone effects (decreased cortical BMD) in rats with the dioxin-sensitive AhR allele but not in rats with the dioxin-resistant AhR allele (Miettinen et al. 2005). Another possible mechanism of TCDD-induced bone mineralization impairment is upregulation of the active form of vitamin D (Nishimura et al. 2009).

The effects of prenatal and perinatal dioxin exposure on bone growth and mechanical strength are reversible (Miettinen et al. 2005). Such reversibility is seen in rats and occurs before 1 year in age (Miettinen et al. 2005). The bone effects caused by dioxins are gender dependent in rodents and in nonhuman primates (Enan et al. 1996). For example, in monkeys, prenatal exposure to low-dose TCDD caused increases in BMD and cortical cross-sectional area in female offspring, but an increase in fragility of bone in the male offspring (Hermsen et al. 2008).

Concerning effects on joints, TCDD (Yang and Lee 2010) and dioxin-like compounds (Lee and Yang 2012) have been documented to enhance apoptosis of chondrocytes *in vitro*. Chondrocyte apoptosis is responsible for cartilage damage and the development of OA. TCDD enhances chondrocyte apoptosis in a dose-dependent manner, which is blocked by inhibitors of reactive oxygen species and nitric oxide (Yang and Lee 2010). In addition, TCDD enhances the inflammatory process of RA via the AhR (Kobayashi et al. 2008). Thus, dioxins affect joint health potentially through both immune and non-immune-mediated mechanisms. Although current studies support the conclusion that dioxin plays an important role in progression of joint diseases, more studies are needed to clarify its effect on the development of arthritis in the human population.

Table 12.2 summarizes the findings of dioxin exposure to bones and joints from laboratory studies.

TABLE 12.2. Summary of Dioxin-Related Bone and Joint Effects by Experimental Studies

Author	Year	Cell/Animal	Exposure	Outcome Assessment	Result
In vitro					
Gierthy	1994	Rat calvaria osteoblast	Untreated control cultured cells, cells treated with 0.1% DMSO alone or 10 nM TCDD in 0.1% DMSO	Alkaline phosphatase activity, mineral deposition, quantification of bone nodule formation	TCDD: inhibition of formation of multicelluar bone nodules, suppression of alkaline phosphatase and osteoclacin activity, no effect on the proliferation of osteoblasts
Singh	2000	Chick periosteal explants and rat bone marrow cell	Cells were treated with 1 nM TCDD, 1 μM resveratrol, TCDD + resveratrol, or vehicle control	Bone biomarkers (alkaline phospatase, type I collagen, osteopontin, bone sialoprotein), histologic evaluation	TCDD: inhibition of bone formation in chicken periosteal explant culture; inhibition of mineralization in rat bone marrow cell culture; all the negative effects of TCDD in the study were reversed by resveratrol
Ilvesaro	2005	Rat osteoclast	Treated with control and 10 nM TCDD	Pit formation assay for activity of osteoclasts	TCDD: no change in the activity of the osteoclast, which strongly expressed AHR
Korkalainen	2009	Bone marrow stem cells from rats with wild and null Ahr	Control and TCDD at doses of 100 nM, 1 nM, 10 pM, and 100 fM	Gene expression of markers of osteoblastic differentiation byRT-PCR, cell proliferation and calcium quantification assay, pit formation assay	TCDD: dose–response suppression of markers for osteoblastic differentiation (RUNX2, alkaline phosphatase and osteoclacin); suppression of calcium deposition in osteoblast; dose–response decreases in osteoclastogenesis and resorption pit area; the effects of TCDD were not found in cells from AhR knockout mice

Yang	2012	Rabbit articular chondrocytes	Cells were treated with DMSO (0.1%), 1, 10, or 100 nM TCDD for 24 hours	Productions of reactive oxygen species (ROS) and nitric oxide (NO), activity of apoptotic cell death	TCDD: increases in production of ROS, NO, and indicators for apoptosis of chondrocytes
In vivo					
Jamsa	2001	TCDD-sensitive Long–Evans (L-E) rats and TCDD-insensitive Han/Wistar (H/W) rats (different TCDD sensitivity resulted from the structure of AhR)	Subcutaneous injection of corn oil (control) and TCDD for 20 weeks; the total dose of TCDD were 0.17, 1.7, 17, and 170 μg/kg (H/W only)	BMD, geometry, and biomechanics, serum alkaline phosphatase (ALP) activity	TCDD: reduction of total bone volume in L-E rats at a dose of 17 μg/kg, but no change in H/W rats; ALP activities were dose-dependently increased in L-E rats at 1.7 and 17 μg/kg and in H/W rats at 17 μg/kg and 170 μg/kg; decrease in cross-sectional area in L-E rats only; decrease breaking force in both rats at highest dose; decreases in tibial length, circumferences at mid-diaphysis, and ash weight of tibia in L-E rats at 1.7 and 17 μg/kg dose, but in H/W rats at 170 μg/kg. No change in BMD between all groups
Miettinen	2005	Rat A with mutated AHR (AhR^{hw}), rat B with another mutated AHR allele (B^{hw}), and rat C with intact AhR structure	Pregnant dams were given a single oral dose of 0.03, 0.1, 0.3, or 1 μg/kg TCDD, or corn oil on gestational day 15.	BMD, geometry, and biomechanics	TCDD: bone changes only in rat C; decreases in bone length, cross-sectional area of cortex, periosteal circumference, endosteal circumference, BMD, and bone strength; bone change due to *in utero* and lactational exposure recovered 1 year after birth

(*Continued*)

TABLE 12.2. (*Continued*)

Author	Year	Cell/Animal	Exposure	Outcome Assessment	Result
Hermsen	2008	Pregnant monkeys and their offspring	Subcutaneous injections of TCDD with a total dose of 40.5–42.0 or 405–420 ng/kg bw starting at gestational day 20 and followed by injections every 30 days until 90 days after delivery	BMD, geometry, and biomechanics, bone biomarkers, immunohistochemistry	TCDD: increases in trabecular bone mineral content and cortical cross-sectional area in female offspring; increases in bone fragility and displacement at failure in male offspring; no changes in bone biomarkers ALP, CTX-1, and 25-OH vitamin D; no changes in tissue expression of vWF, ICAM-1, osteocalcin and AhR
Lind	2009	2-month-old male rats	Single intraperitoneal injection of 50 μg TCDD/kg bw in corn oil or vehicle only	BMD, geometry, biomechanics, and bone composition analysis, serum bone biomarkers (5 days after the TCDD exposure)	Short-term TCDD exposure: decrease in trabecular bone cross-sectional area and increase in BMD; more mature bones, presented as higher proportion of crystalline phosphate and lower acid phosphate content; decrease in bone formation marker (PINP) and increase in bone resorption marker (CTX)
Herlin	2010	TCDD-sensitive L-E rats and TCDD-insensitive H/W rats (different TCDD sensitivity resulted from structure of AhR)	Subcutaneous injections of TCDD once per week for 20 weeks, at doses corresponding to calculated daily doses of 0, 1, 10, 100 and 1000 ng TCDD/kg bw (H/W only)	BMD, geometry, and biomechanics	TCDD: dose–response decreases in cortical area, length, cross-sectional area, trabecular area, periosteal circumference, endosteal circumference, and mechanical breaking force of long bones, more significant changes in L-E rats, while bone mineral change was not so obvious

bw, body weight; PINP, type I collagen N-terminal propeptide; CTX, collagen type 1 cross-linked C-telopeptide.

Hexachlorobenzene (HCB)

HCB was widely used as a pesticide to protect crops against fungus, as well as making fireworks, ammunition, and synthetic rubber. Currently, there are no commercial uses for HCB. However, HCB can be formed as a by-product during the manufacture of solvents, pesticides, and other chlorine-containing compounds. Small amounts of HCB can also be produced during combustion processes such as burning of city wastes. HCB is very stable in the environment, with a half-life of 0.6–6.3 years in air, 3.0–6.0 years in soil, 2.7–5.7 years in surface water, and 5.3–11.4 years in groundwater (Ariano and Panzani 2012).

In population studies, HCB has been associated with osteoporosis and painless arthritis (swelling of the joints distinct from RA) in patients exposed to the chemical from consumption of bread prepared from contaminated grain (Cripps et al. 1984; Peters et al. 1982, 1987). After 2–3 years exposure, the orthopedic effect of HCB can persist for 20 or 30 years. The mechanisms for this finding remain unknown. The increase of arthritis as observed in the exposed human population has not been observed in animal studies. In laboratory animals, exposure to HCB induces dose-related increases in femur density (osteosclerosis) and cortical area in male rats. The suspected mechanism of the skeletal effect involves reduction of bone resorption (Andrews et al. 1989, 1990). The apparent discrepancies in findings between laboratory-exposed animals and accidentally exposed humans warrant further investigation.

Bisphenol A (BPA)

BPA is widely used as the base compound in the manufacture of polycarbonate plastic and the resin lining of food and beverage cans, and as an additive in many other plastics such as polyvinyl chloride and polyethylene terephthalate (Alonso-Magdalena et al. 2012). It has been reported to be estrogenic by many researchers. Up to 2012, a Medline search by the authors did not find any published articles on bone and joint effects of BPA exposure in humans. In goldfish, BPA significantly suppressed both tartrate-resistant acid phosphatase and ALP activities, which were used as markers of osteoclasts and osteoblasts, respectively (Suzuki and Hattori 2003). In another fish model, the fathead minnow, BPA exposure from the egg stage to 25 days' posthatch did not significantly impair skeletal development or induce vertebral malformations. In a mouse osteoblast-like cell line, MC3T3-E1, BPA increased ALP activity and the cellular content of calcium and phosphate, which are indicators of mineralization (Kanno et al. 2004). Thus, BPA can potentially affect bone homeostasis without causing an observable clinical effect on skeletal development. Whether adverse effects will be seen after long-term exposure to BPA is unclear and warrants further investigation.

Phthalates

Phthalate esters are widely used industrial chemicals, serving mainly as plasticizers, which are added to plastics to increase their flexibility, transparency, durability, and longevity. Although used primarily in polyvinylchloride (PVC) resins, phthalate esters have been added at varying degrees in other resins such as polyvinyl acetates, cellulosics, and polyurethanes. Because there is no covalent bond between phthalates and plastics in which they are mixed, phthalates are easily released into the environment, as plastics age and break down.

Many animal studies have shown that some phthalate esters have teratogenicity and increase skeletal malformations, including cleft palate (Ema et al. 1993, 1997), deformity of the cervical vertebrae (Ema et al. 1993, 1994, 1996, 1997), deformity of thoracic vertebrae [94–97] (Ema et al. 1993, 1994, 1996, 1997; Tyl et al. 1988), retarded ossification of bones (Saillenfait et al. 2006, 2008, 2009), deformities of long bones (Parkhie et al. 1982), deformity of the ribs (Ema et al. 1996, 1997; Saillenfait et al. 2006, 2009; Tyl et al. 1988), and fusion of sternebrae (Ema et al. 1993, 1994). The mechanism of phthalate-induced skeletal malformation is not totally clear. An *in vitro* study showed that phthalate esters could induce apoptosis of mouse osteoblasts, partially through p53 activation (Sabbieti et al. 2009). However, no studies have documented clear bone and joint effects due to phthalate exposure in humans. There have essentially no publications on the effects of phthalates on joints in animal studies either.

Perfluorocarbons (PFCs)

PFCs are a group of human-made chemicals composed of only carbon and fluorine, emitted as by-products of industrial processes and used in manufacturing. The applications of PFCs have been restricted due to their persistence in the environment and being powerful greenhouse gases. Among PFCs, perfluorooctane sulfonate (PFOS) and perfluorooctanoic acid (PFOA) are the most evaluated compounds for health effects. The main routes of exposure to PFOS and PFOA are via inhalation of contaminated air or by ingestion of contaminated water or food. PFCs have very long half-lives in the environment and in the human body (Olsen et al. 2007).

Currently, only a few investigations have studied bone and joint effects of PFCs in humans. Innes et al. (2011) found the association between OA and serum levels of PFOA and PFOS in a large population. Although the mechanism of action was unknown, they found that PFOA was positively associated with OA risk, while PFOS was negatively associated with OA risk (Innes et al. 2011). As endocrine disruptors, PFOA and PFOS could disrupt the effects of thyroid hormone, sex steroid, and GH/IGF-1 axis, which are related to bone development and homeostasis. However, in a limited number of animal studies, the detrimental effects of PFCs on skeletal development have not been clearly demonstrated (Fuentes et al. 2006; Harris and Birnbaum 1989; Spachmo and Arukwe 2012).

DDT (1,1,1-Trichloro-2,2-Bis(P-Chlorophenyl)Ethane)

DDT is a pesticide that was once widely used to control insects on agricultural crops and insects that carry diseases like malaria and typhus. 1,1-Dichloro-2,2-bis(p-chlorophenyl)ethylene (DDE) and 1,1-dichloro-2,2-bis(p-chloro-phenyl) ethane (DDD) are metabolites of DDT and are frequently measureable in the environmental media. Due to their persistence in the environment and detrimental effects on wildlife and humans, DDT and its metabolites are currently used in only a few countries to control malaria. However, because of the long half-lives of these chemicals in soil and other media, contaminated particles could be transported long distances. They also bioaccumulate due to high lipophilicity and become concentrated in the top of the food chain. Therefore, DDT is still an issue of global concern.

DDT and its metabolites act on estrogen and androgen receptors, leading to estrogenic and antiandrogenic effects (Bitman and Cecil 1970; Kelce et al. 1995). Human studies have yielded mixed results regarding bone effects of DDT and its residues. In women, DDE was unrelated to either BMD of the lumbar spine or the rate of bone loss in one study (Bohannon et al. 2000) but was weakly associated with reduced BMD in another (Beard et al. 2000). In men, a weak association was found between serum concentrations of DDE and BMD (Glynn et al. 2000). As for prenatal exposure, DDT compounds were found to be unrelated to anthropometrics in fetuses in one study (Gladen et al. 2004) and were found to cause detrimental effects on anthropometric development in another (Al-Saleh et al. 2012).

Information from animal studies examining the relationship between DDT exposure and bone changes has been limited. In animal studies from the National Cancer Institute of the United States, exposure to high-dose DDT compounds (30~45 mg DDT/kg/day, 49~59 mg DDE/kg/day, or 142~231 mg DDD/kg/day) did not cause significant adverse skeletal effects. Lundberg et al. (2007) found reduced BMD in frogs after exposure to 1 mg/kg/day of DDE, but such an effect was not seen in groups with higher or lower dosages. Thus, existing information does not support strong toxic effects of DDT/DDE on bones and joints.

CONCLUSION

In conclusion, from the limited literature up to 2012, EDCs are known to affect bones and joints. However, information concerning this effect is relatively lacking on several known EDCs. In the meantime, the mechanisms behind bone and joint effects of EDCs through several endocrine systems, such as GH, PTH, thyroid hormone, and sex hormones, are still not totally understood. Studies are needed in these areas to strengthen human knowledge to protect bone and joint health against EDCs.

REFERENCES

Alonso-Magdalena P, Ropero AB, Soriano S, García-Arévalo M, Ripoll C, Fuentes E, et al. 2012. Bisphenol-A acts as a potent estrogen via non-classical estrogen triggered pathways. Mol Cell Endocrinol 355:201–207.

Al-Saleh I, Al-Doush I, Alsabbaheen A, Mohamed Gel D, Rabbah A. 2012. Levels of DDT and its metabolites in placenta, maternal and cord blood and their potential influence on neonatal anthropometric measures. Sci Total Environ 416: 62–74.

Alvarez-Lloret P, Lind PM, Nyberg I, Orberg J, Rodríguez-Navarro AB. 2009. Effects of 3,3′,4,4′,5-pentachlorobiphenyl (PCB126) on vertebral bone mineralization and on thyroxin and vitamin D levels in Sprague-Dawley rats. Toxicol Lett 187:63–68.

Alveblom AK, Rylander L, Johnell O, Hagmar L. 2003. Incidence of hospitalized osteoporotic fractures in cohorts with high dietary intake of persistent organochlorine compounds. Int Arch Occup Environ Health 76:246–248.

Andrews JE, Courtney KD, Stead AG, Donaldson WE. 1989. Hexachlorobenzene-induced hyperparathyroidism and osteosclerosis in rats. Fundam Appl Toxicol 12:242–251.

Andrews JE, Jackson LD, Stead AG, Donaldson WE. 1990. Morphometric analysis of osteosclerotic bone resulting from hexachlorobenzene exposure. J Toxicol Environ Health 31:193–201.

Ariano R, Panzani RC. 2012. Late onset asthma in the elderly and its relationship with atopy. Eur Ann Allergy Clin Immunol 44:35–41.

Ashida H, Enan E, Matsumura F. 1996. Protective action of dehydroascorbic acid on the Ah receptor-dependent and receptor-independent induction of lipid peroxidation in adipose tissue of male guinea pig caused by TCDD administration. J Biochem Toxicol 11:269–278.

Asplund L, Svensson BG, Nilsson A, Eriksson U, Jansson B, Jensen S, et al. 1994. Polychlorinated biphenyls, 1,1,1-trichloro-2,2-bis(p-chlorophenyl)ethane (p,p′-DDT) and 1,1-dichloro-2,2-bis(p-chlorophenyl)-ethylene (p,p′-DDE) in human plasma related to fish consumption. Arch Environ Health 49:477–486.

Astroff B, Safe S. 1990. 2,3,7,8-Tetrachlorodibenzo-p-dioxin as an antiestrogen: Effect on rat uterine peroxidase activity. Biochem Pharmacol 39:485–488.

Aydoğan M, Korkmaz A, Barlas N, Kolankaya D. 2008. The effect of vitamin C on bisphenol A, nonylphenol and octylphenol induced brain damages of male rats. Toxicology 249:35–39.

Aydoğan M, Korkmaz A, Barlas N, Kolankaya D. 2010. Pro-oxidant effect of vitamin C coadministration with bisphenol A, nonylphenol, and octylphenol on the reproductive tract of male rats. Drug Chem Toxicol 33:193–203.

Banudevi S, Krishnamoorthy G, Venkataraman P, Vignesh C, Aruldhas MM, Arunakaran J. 2006. Role of alpha-tocopherol on antioxidant status in liver, lung and kidney of PCB exposed male albino rats. Food Chem Toxicol 44:2040–2046.

Bassett JH, Nordström K, Boyde A, Howell PG, Kelly S, Vennström B, et al. 2007. Thyroid status during skeletal development determines adult bone structure and mineralization. Mol Endocrinol 21:1893–1904.

Beard J, Marshall S, Jong K, Newton R, Triplett-McBride T, Humphries B, et al. 2000. 1,1,1-Trichloro-2,2-bis (p-chlorophenyl)-ethane (DDT) and reduced bone mineral density. Arch Environ Health 55:177–180.

Biegel L, Safe S. 1990. Effects of 2,3,7,8-tetrachlorodibenzo-p-dioxin (TCDD) on cell growth and the secretion of the estrogen-induced 34-, 52- and 160-kDa proteins in human breast cancer cells. J Steroid Biochem Mol Biol 37:725–732.

Bitman J, Cecil HC. 1970. Estrogenic activity of DDT analogs and polychlorinated biphenyls. J Agric Food Chem 18:1108–1112.

Bohannon AD, Cooper GS, Wolff MS, Meier DE. 2000. Exposure to 1,1-dichloro-2,2-bis(p-chlorophenyl)ethylene (DDT) in relation to bone mineral density and rate of bone loss in menopausal women. Arch Environ Health 55:386–391.

Brouwer A. 1991. Role of biotransformation in PCB-induced alterations in vitamin A and thyroid hormone metabolism in laboratory and wildlife species. Biochem Soc Trans 19:731–737.

Brouwer A, van den Berg KJ. 1986. Binding of a metabolite of 3,4,3′,4′-tetrachlorobiphenyl to transthyretin reduces serum vitamin A transport by inhibiting the formation of the protein complex carrying both retinol and thyroxin. Toxicol Appl Pharmacol 85:301–312.

Byrne JJ, Carbone JP, Hanson EA. 1987. Hypothyroidism and abnormalities in the kinetics of thyroid hormone metabolism in rats treated chronically with polychlorinated biphenyl and polybrominated biphenyl. Endocrinology 121:520–527.

Carpenter DO. 1998. Polychlorinated biphenyls and human health. Int J Occup Med Environ Health 11:291–303.

Chen LC, Berberian I, Koch B, Mercier M, Azais-Braesco V, Glauert HP, et al. 1992. Polychlorinated and polybrominated biphenyl congeners and retinoid levels in rat tissues: Structure-activity relationships. Toxicol Appl Pharmacol 114: 47–55.

Côté S, Ayotte P, Dodin S, Blanchet C, Mulvad G, Petersen HS, et al. 2006. Plasma organochlorine concentrations and bone ultrasound measurements: A cross-sectional study in peri-and postmenopausal Inuit women from Greenland. Environ Health 5:33.

Cripps DJ, Peters HA, Gocmen A, Dogramici I. 1984. Porphyria turcica due to hexachlorobenzene: A 20 to 30 year follow-up study on 204 patients. Br J Dermatol 111:413–422.

Elabbas LE, Herlin M, Finnilä MA, Rendel F, Stern N, Trossvik C, et al. 2011. In utero and lactational exposure to Aroclor 1254 affects bone geometry, mineral density and biomechanical properties of rat offspring. Toxicol Lett 207:82–88.

Ema M, Itami T, Kawasaki H. 1993. Teratogenic phase specificity of butyl benzyl phthalate in rats. Toxicology 79:11–19.

Ema M, Amano H, Ogawa Y. 1994. Characterization of the developmental toxicity of di-n-butyl phthalate in rats. Toxicology 86:163–174.

Ema M, Harazono A, Miyawaki E, Ogawa Y. 1996. Developmental toxicity of mono-n-benzyl phthalate, one of the major metabolites of the plasticizer n-butyl benzyl phthalate, in rats. Toxicol Lett 86:19–25.

Ema M, Harazono A, Miyawaki E, Ogawa Y. 1997. Developmental effects of di-n-butyl phthalate after a single administration in rats. J Appl Toxicol 17:223–229.

Enan E, Overstreet JW, Matsumura F, VandeVoort CA, Lasley BL. 1996. Gender differences in the mechanism of dioxin toxicity in rodents and in nonhuman primates. Reprod Toxicol 10:401–411.

Finnilä MA, Zioupos P, Herlin M, Miettinen HM, Simanainen U, Håkansson H, et al. 2010. Effects of 2,3,7,8-tetrachlorodibenzo-p-dioxin exposure on bone material properties. J Biomech 43:1097–1103.

Fuentes S, Colomina MT, Rodriguez J, Vicens P, Domingo JL. 2006. Interactions in developmental toxicology: Concurrent exposure to perfluorooctane sulfonate (PFOS) and stress in pregnant mice. Toxicol Lett 164:81–89.

Funatsu I, Yamashita F, Ito Y, Tsugawa S, Funatsu T. 1972. Polychlorbiphenyls (PCB) induced fetopathy. I. Clinical observation. Kurume Med J 19:43–51.

Gierthy JF, Silkworth JB, Tassinari M, Stein GS, Lian JB. 1994. 2,3,7,8-Tetrachlorodibenzo-p-dioxin inhibits differentiation of normal diploid rat osteoblasts in vitro. J Cell Biochem 54:231–238.

Gladen BC, Klebanoff MA, Hediger ML, Katz SH, Barr DB, Davis MD, et al. 2004. Prenatal DDT exposure in relation to anthropometric and pubertal measures in adolescent males. Environ Health Perspect 112:1761–1767.

Glynn AW, Michaëlsson K, Lind PM, Wolk A, Aune M, Atuma S, et al. 2000. Organochlorines and bone mineral density in Swedish men from the general population. Osteoporos Int 11:1036–1042.

Guo YL, Yu ML, Hsu CC, Rogan WJ. 1999. Chloracne, goiter, arthritis, and anemia after polychlorinated biphenyl poisoning: 14-year follow-up of the Taiwan Yucheng cohort. Environ Health Perspect 107:715–719.

Gutleb AC, Arvidsson D, Orberg J, Larsson S, Skaare JU, Aleksandersen M, et al. 2010. Effects on bone tissue in ewes (*Ovis aries*) and their foetuses exposed to PCB 118 and PCB 153. Toxicol Lett 192:126–133.

Harris MW, Birnbaum LS. 1989. Developmental toxicity of perfluorodecanoic acid in C57BL/6N mice. Fundam Appl Toxicol 12:442–448.

Herlin M, Kalantari F, Stern N, Sand S, Larsson S, Viluksela M, et al. 2010. Quantitative characterization of changes in bone geometry, mineral density and biomechanical properties in two rat strains with different Ah-receptor structures after long-term exposure to 2,3,7,8-tetrachlorodibenzo-p-dioxin. Toxicology 273: 1–11.

Hermsen SA, Larsson S, Arima A, Muneoka A, Ihara T, Sumida H, et al. 2008. In utero and lactational exposure to 2,3,7,8-tetrachlorodibenzo-p-dioxin (TCDD) affects bone tissue in rhesus monkeys. Toxicology 253:147–152.

Hodgson S, Thomas L, Fattore E, Lind PM, Alfven T, Hellström L, et al. 2008. Bone mineral density changes in relation to environmental PCB exposure. Environ Health Perspect 116:1162–1166.

Ilvesaro J, Pohjanvirta R, Tuomisto J, Viluksela M, Tuukkanen J. 2005. Bone resorption by aryl hydrocarbon receptor-expressing osteoclasts is not disturbed by TCDD in short-term cultures. Life Sci 77:1351–1366.

Innes KE, Ducatman AM, Luster MI, Shankar A. 2011. Association of osteoarthritis with serum levels of the environmental contaminants perfluorooctanoate and perfluorooctane sulfonate in a large Appalachian population. Am J Epidemiol 174: 440–450.

Jämsä T, Viluksela M, Tuomisto JT, Tuomisto J, Tuukkanen J. 2001. Effects of 2,3,7,8-tetrachlorodibenzo-p-dioxin on bone in two rat strains with different aryl hydrocarbon receptor structures. J Bone Miner Res 16:1812–1820.

Jenssen BM, Haugen O, Sørmo EG, Skaare JU. 2003. Negative relationship between PCBs and plasma retinol in low-contaminated free-ranging gray seal pups (*Halichoerus grypus*). Environ Res 93:79–87.

Johnson KE, Knopper LD, Schneider DC, Ollson CA, Reimer KJ. 2009. Effects of local point source polychlorinated biphenyl (PCB) contamination on bone mineral density in deer mice (*Peromyscus maniculatus*). Sci Total Environ 407:5050–5055.

Kanno S, Hirano S, Kayama F. 2004. Effects of phytoestrogens and environmental estrogens on osteoblastic differentiation in MC3T3-E1 cells. Toxicology 196:137–145.

Kelce WR, Stone CR, Laws SC, Gray LE, Kemppainen JA, Wilson EM. 1995. Persistent DDT metabolite p,p′-DDE is a potent androgen receptor antagonist. Nature 375:581–585.

Kim HA, Blanco FJ. 2007. Cell death and apoptosis in osteoarthritic cartilage. Curr Drug Targets 8:333–345.

Kirkegaard M, Sonne C, Jakobsen J, Jenssen BM, Letcher RJ, Dietz R. 2010. Organohalogens in a whale-blubber-supplemented diet affects hepatic retinol and renal tocopherol concentrations in greenland sled dogs (*Canis familiaris*). J Toxicol Environ Health A 73:773–786.

Kobayashi S, Okamoto H, Iwamoto T, Toyama Y, Tomatsu T, Yamanaka H, et al. 2008. A role for the aryl hydrocarbon receptor and the dioxin TCDD in rheumatoid arthritis. Rheumatology (Oxford) 47:1317–1322.

Korkalainen M, Kallio E, Olkku A, Nelo K, Ilvesaro J, Tuukkanen J, et al. 2009. Dioxins interfere with differentiation of osteoblasts and osteoclasts. Bone 44:1134–1142.

Korkmaz A, Aydoğan M, Kolankaya D, Barlas N. 2011. Vitamin C coadministration augments bisphenol A, nonylphenol, and octylphenol induced oxidative damage on kidney of rats. Environ Toxicol 26:325–337.

Krishnamoorthy G, Venkataraman P, Arunkumar A, Vignesh RC, Aruldhas MM, Arunakaran J. 2007. Ameliorative effect of vitamins (alpha-tocopherol and ascorbic acid) on PCB (Aroclor 1254) induced oxidative stress in rat epididymal sperm. Reprod Toxicol 23:239–245.

Krishnan V, Safe S. 1993. Polychlorinated biphenyls (PCBs), dibenzo-p-dioxins (PCDDs), and dibenzofurans (PCDFs) as antiestrogens in MCF-7 human breast cancer cells: Quantitative structure-activity relationships. Toxicol Appl Pharmacol 120:55–61.

Lee HG, Yang JH. 2012. PCB126 induces apoptosis of chondrocytes via ROS-dependent pathways. Osteoarthritis Cartilage 20:1179–1185.

Lee DH, Steffes M, Jacobs DR. 2007. Positive associations of serum concentration of polychlorinated biphenyls or organochlorine pesticides with self-reported arthritis, especially rheumatoid type, in women. Environ Health Perspect 115:883–888.

Lilienthal H, Fastabend A, Hany J, Kaya H, Roth-Härer A, Dunemann L, et al. 2000. Reduced levels of 1,25-dihydroxyvitamin D(3) in rat dams and offspring after exposure to a reconstituted PCB mixture. Toxicol Sci 57:292–301.

Lind PM, Eriksen EF, Sahlin L, Edlund M, Orberg J. 1999. Effects of the antiestrogenic environmental pollutant 3,3′,4,4′, 5-pentachlorobiphenyl (PCB #126) in rat bone and uterus: Diverging effects in ovariectomized and intact animals. Toxicol Appl Pharmacol 154:236–244.

Lind PM, Eriksen EF, Lind L, Orberg J, Sahlin L. 2004. Estrogen supplementation modulates effects of the endocrine disrupting pollutant PCB126 in rat bone and uterus: Diverging effects in ovariectomized and intact animals. Toxicology 199: 129–136.

Lundberg R, Lyche JL, Ropstad E, Aleksandersen M, Rönn M, Skaare JU, et al. 2006. Perinatal exposure to PCB 153, but not PCB 126, alters bone tissue composition in female goat offspring. Toxicology 228:33–40.

Lundberg R, Jenssen BM, Leiva-Presa A, Rönn M, Hernhag C, Wejheden C, et al. 2007. Effects of short-term exposure to the DDT metabolite p,p′-DDE on bone tissue in male common frog (*Rana temporaria*). J Toxicol Environ Health A 70:614–619.

Martin PA, Mayne GJ, Bursian S, Palace V, Kannan K. 2006. Changes in thyroid and vitamin A status in mink fed polyhalogenated-aromatic-hydrocarbon-contaminated carp from the Saginaw River, Michigan, USA. Environ Res 101:53–67.

Martino L, Novelli M, Masini M, Chimenti D, Piaggi S, Masiello P, et al. 2009. Dehydroascorbate protection against dioxin-induced toxicity in the beta-cell line INS-1E. Toxicol Lett 189:27–34.

Massaad C, Entezami F, Massade L, Benahmed M, Olivennes F, Barouki R, et al. 2002. How can chemical compounds alter human fertility? Eur J Obstet Gynecol Reprod Biol 100:127–137.

Miettinen HM, Pulkkinen P, Jämsä T, Koistinen J, Simanainen U, Tuomisto J, et al. 2005. Effects of in utero and lactational TCDD exposure on bone development in differentially sensitive rat lines. Toxicol Sci 85:1003–1012.

Morse DC, Brouwer A. 1995. Fetal, neonatal, and long-term alterations in hepatic retinoid levels following maternal polychlorinated biphenyl exposure in rats. Toxicol Appl Pharmacol 131:175–182.

Murtomaa M, Tervaniemi OM, Parviainen J, Ruokojärvi P, Tuukkanen J, Viluksela M. 2007. Dioxin exposure in contaminated sawmill area: The use of molar teeth and bone of bank vole (*Clethrionomys glareolus*) and field vole (*Microtus agrestis*) as biomarkers. Chemosphere 68:951–957.

Murugesan P, Muthusamy T, Balasubramanian K, Arunakaran J. 2005. Studies on the protective role of vitamin C and E against polychlorinated biphenyl (Aroclor 1254)-induced oxidative damage in Leydig cells. Free Radic Res 39:1259–1272.

Muthuvel R, Venkataraman P, Krishnamoorthy G, Gunadharini DN, Kanagaraj P, Jone Stanley A, et al. 2006. Antioxidant effect of ascorbic acid on PCB (Aroclor 1254) induced oxidative stress in hypothalamus of albino rats. Clin Chim Acta 365:297–303.

Naruse M, Ishihara Y, Miyagawa-Tomita S, Koyama A, Hagiwara H. 2002. 3-Methylcholanthrene, which binds to the arylhydrocarbon receptor, inhibits proliferation and differentiation of osteoblasts in vitro and ossification in vivo. Endocrinology 143:3575–3581.

Nishimura N, Nishimura H, Ito T, Miyata C, Izumi K, Fujimaki H, et al. 2009. Dioxin-induced up-regulation of the active form of vitamin D is the main cause for its inhibitory action on osteoblast activities, leading to developmental bone toxicity. Toxicol Appl Pharmacol 236:301–309.

Novák J, Benísek M, Hilscherová K. 2008. Disruption of retinoid transport, metabolism and signaling by environmental pollutants. Environ Int 34:898–913.

Nyman M, Koistinen J, Fant ML, Vartiainen T, Helle E. 2002. Current levels of DDT, PCB and trace elements in the Baltic ringed seals (*Phoca hispida baltica*) and grey seals (*Halichoerus grypus*). Environ Pollut 119:399–412.

Nyman M, Bergknut M, Fant ML, Raunio H, Jestoi M, Bengs C, et al. 2003. Contaminant exposure and effects in Baltic ringed and grey seals as assessed by biomarkers. Mar Environ Res 55:73–99.

Okumura M. 1984. Past and current medical states of yusho patients. Prog Clin Biol Res 137:13–18.

Olsen GW, Burris JM, Ehresman DJ, Froehlich JW, Seacat AM, Butenhoff JL, et al. 2007. Half-life of serum elimination of perfluorooctanesulfonate, perfluorohexanesulfonate, and perfluorooctanoate in retired fluorochemical production workers. Environ Health Perspect 115:1298–1305.

Parkhie MR, Webb M, Norcross MA. 1982. Dimethoxyethyl phthalate: Embryopathy, teratogenicity, fetal metabolism and the role of zinc in the rat. Environ Health Perspect 45:89–97.

Peters HA, Gocmen A, Cripps DJ, Bryan GT, Dogramaci I. 1982. Epidemiology of hexachlorobenzene-induced porphyria in Turkey: Clinical and laboratory follow-up after 25 years. Arch Neurol 39:744–749.

Peters H, Cripps D, Göcmen A, Bryan G, Ertürk E, Morris C. 1987. Turkish epidemic hexachlorobenzene porphyria. A 30-year study. Ann N Y Acad Sci 514:183–190.

van der Plas SA, Lutkeschipholt I, Spenkelink B, Brouwer A. 2001. Effects of subchronic exposure to complex mixtures of dioxin-like and non-dioxin-like polyhalogenated aromatic compounds on thyroid hormone and vitamin A levels in female Sprague-Dawley rats. Toxicol Sci 59:92–100.

Rickenbacher U, McKinney JD, Oatley SJ, Blake CC. 1986. Structurally specific binding of halogenated biphenyls to thyroxine transport protein. J Med Chem 29: 641–648.

Rignell-Hydbom A, Skerfving S, Lundh T, Lindh CH, Elmståhl S, Bjellerup P, et al. 2009. Exposure to cadmium and persistent organochlorine pollutants and its association with bone mineral density and markers of bone metabolism on postmenopausal women. Environ Res 109:991–996.

Roos A, Rigét F, Orberg J. 2010. Bone mineral density in Swedish otters (*Lutra lutra*) in relation to PCB and DDE concentrations. Ecotoxicol Environ Saf 73: 1063–1070.

Roth-Härer A, Lilienthal H, Bubser M, Kronthaler U, Mundy WR, Ward TR, et al. 2001. Neurotransmitter concentrations and binding at dopamine receptors in rats after maternal exposure to 3,4,3′,4′-tetrachlorobiphenyl: The role of reduced thyroid hormone concentrations. Environ Toxicol Pharmacol 9:103–115.

Routti H, Nyman M, Jenssen BM, Bäckman C, Koistinen J, Gabrielsen GW. 2008. Bone-related effects of contaminants in seals may be associated with vitamin D and thyroid hormones. Environ Toxicol Chem 27:873–880.

Ryan EP, Holz JD, Mulcahey M, Sheu TJ, Gasiewicz TA, Puzas JE. 2007. Environmental toxicants may modulate osteoblast differentiation by a mechanism involving the aryl hydrocarbon receptor. J Bone Miner Res 22:1571–1580.

Sabbieti MG, Agas D, Santoni G, Materazzi S, Menghi G, Marchetti L. 2009. Involvement of p53 in phthalate effects on mouse and rat osteoblasts. J Cell Biochem 107:316–327.

Safe SH. 1994. Polychlorinated biphenyls (PCBs): Environmental impact, biochemical and toxic responses, and implications for risk assessment. Crit Rev Toxicol 24:87–149.

Safe S, Wormke M. 2003. Inhibitory aryl hydrocarbon receptor-estrogen receptor alpha cross-talk and mechanisms of action. Chem Res Toxicol 16:807–816.

Saillenfait AM, Sabaté JP, Gallissot F. 2006. Developmental toxic effects of diisobutyl phthalate, the methyl-branched analogue of di-n-butyl phthalate, administered by gavage to rats. Toxicol Lett 165:39–46.

Saillenfait AM, Gallissot F, Sabaté JP. 2008. Evaluation of the developmental toxicity of diallyl phthalate administered orally to rats. Food Chem Toxicol 46: 2150–2156.

Saillenfait AM, Gallissot F, Sabaté JP. 2009. Differential developmental toxicities of di-n-hexyl phthalate and dicyclohexyl phthalate administered orally to rats. J Appl Toxicol 29:510–521.

Singh SU, Casper RF, Fritz PC, Sukhu B, Ganss B, Girard B, Jr., et al. 2000. Inhibition of dioxin effects on bone formation in vitro by a newly described aryl hydrocarbon receptor antagonist, resveratrol. J Endocrinol 167:183–195.

Skaare JU, Bernhoft A, Wiig O, Norum KR, Haug E, Eide DM, et al. 2001. Relationships between plasma levels of organochlorines, retinol and thyroid hormones from polar bears (*Ursus maritimus*) at Svalbard. J Toxicol Environ Health A 62:227–241.

Sonne C, Dietz R, Born EW, Riget FF, Kirkegaard M, Hyldstrup L, et al. 2004. Is bone mineral composition disrupted by organochlorines in east Greenland polar bears (*Ursus maritimus*). Environ Health Perspect 112:1711–1716.

Spachmo B, Arukwe A. 2012. Endocrine and developmental effects in Atlantic salmon (*Salmo salar*) exposed to perfluorooctane sulfonic or perfluorooctane carboxylic acids. Aquat Toxicol 108:112–124.

Suzuki N, Hattori A. 2003. Bisphenol A suppresses osteoclastic and osteoblastic activities in the cultured scales of goldfish. Life Sci 73:2237–2247.

Svensson BG, Nilsson A, Jonsson E, Schütz A, Akesson B, Hagmar L. 1995. Fish consumption and exposure to persistent organochlorine compounds, mercury, selenium and methylamines among Swedish fishermen. Scand J Work Environ Health 21: 96–105.

de Swart RL, Ross PS, Vos JG, Osterhaus AD. 1996. Impaired immunity in harbour seals (*Phoca vitulina*) exposed to bioaccumulated environmental contaminants: Review of a long-term feeding study. Environ Health Perspect 4:823–828.

Tanabe S. 1988. PCB problems in the future: Foresight from current knowledge. Environ Pollut 50:5–28.

Tsai PC, Ko YC, Huang W, Liu HS, Guo YL. 2007. Increased liver and lupus mortalities in 24-year follow-up of the Taiwanese people highly exposed to polychlorinated biphenyls and dibenzofurans. Sci Total Environ 374:216–222.

Tyl RW, Price CJ, Marr MC, Kimmel CA. 1988. Developmental toxicity evaluation of dietary di(2-ethylhexyl)phthalate in Fischer 344 rats and CD-1 mice. Fundam Appl Toxicol 10:395–412.

Van Birgelen AP, Van der Kolk J, Fase KM, Bol I, Poiger H, Brouwer A, et al. 1994. Toxic potency of 3,3′,4,4′,5-pentachlorobiphenyl relative to and in combination with 2,3,7,8-tetrachlorodibenzo-p-dioxin in a subchronic feeding study in the rat. Toxicol Appl Pharmacol 127:209–221.

Van den Berg KJ, van Raaij JA, Bragt PC, Notten WR. 1991. Interactions of halogenated industrial chemicals with transthyretin and effects on thyroid hormone levels in vivo. Arch Toxicol 65:15–19.

Venkataraman P, Muthuvel R, Krishnamoorthy G, Arunkumar A, Sridhar M, Srinivasan N, et al. 2007. PCB (Aroclor 1254) enhances oxidative damage in rat brain regions: Protective role of ascorbic acid. Neurotoxicology 28:490–498.

Wallin E, Rylander L, Hagmar L. 2004. Exposure to persistent organochlorine compounds through fish consumption and the incidence of osteoporotic fractures. Scand J Work Environ Health 30:30–35.

Wallin E, Rylander L, Jönssson BA, Lundh T, Isaksson A, Hagmar L. 2005. Exposure to CB-153 and p,p′-DDE and bone mineral density and bone metabolism markers in middle-aged and elderly men and women. Osteoporos Int 16:2085–2094.

Wang XH, Zhou XQ, Xu JP, Wang Y, Lu J. 2009. The effects of vitamin E on NK cell activity and lymphocyte proliferation in treated mice by 2,3,7,8-tetrachlorodibenzo-p-dioxin. Immunopharmacol Immunotoxicol 31:432–438.

Yang JH, Lee HG. 2010. 2,3,7,8-Tetrachlorodibenzo-p-dioxin induces apoptosis of articular chondrocytes in culture. Chemosphere 79:278–284.

Yang JH, Lee YM, Bae SG, Jacobs DR, Jr., Lee DH. 2012. Associations between organochlorine pesticides and vitamin D deficiency in the U.S. population. PLoS ONE 7:e30093.

Yin HP, Xu JP, Zhou XQ, Wang Y. 2012. Effects of vitamin E on reproductive hormones and testis structure in chronic dioxin-treated mice. Toxicol Ind Health 28:152–161.

Yoshimura T, Nakano J, Masuda T, Tokuda M, Sakakibara A, Kataoka H, et al. 2009. [Bone mineral density, PCB, PCQ and PCDF in Yusho]. Fukuoka Igaku Zasshi 100:136–140.

CHAPTER 13

Organic Chemicals and the Immune System

DAVID O. CARPENTER

ABSTRACT

Background: The immune system is critical for fighting infection and cancer, and abnormalities of immune system function, whether it is suppressed or hyperactive, increase the risk of many different human diseases.

Objectives: This chapter reviews what is known concerning perturbations of immune system function secondary to exposure to organic chemicals.

Results: Many different organic chemicals alter immune system function. The most frequent effects are immune suppression of innate and cellular immunity and increased risk of infections and cancer. Some organics reduce humoral immunity and lower antibody responses to immunization, which also increases the risk of infections. Other organics promote allergy, asthma and perhaps autoimmune disease. The expression of many genes that are related to immune system function is altered by organic chemicals.

Conclusions: The influence of organic chemicals on immune system function is considerable and often goes unrecognized because it is reflected only in an increased frequency and severity of infections, allergy, or asthma and increased risk of cancer.

INTRODUCTION

The immune system is complex, and there are many different possible target sites for alteration of immune system function. Therefore, a brief but superficial introduction to the organization of the immune system is essential before

Effects of Persistent and Bioactive Organic Pollutants on Human Health, First Edition.
Edited by David O. Carpenter.

a discussion of perturbation of immune function by organic substances. An excellent reference for this background material is the chapter by Goronzy and Weyand (2008).

There are two major components of the immune system. The cells of the innate immune system are always there and respond immediately to infection and injury, but do not have "memory." The most abundant cells in the immune system are polymorphonuclear neutrophils. These are phagocytic cells that eliminate pathogens primarily by generation of reactive oxygen species. Macrophages are responsible for surveillance and in the lung call in the neutrophils directly via stimulating epithelial cells to secrete chemoattractants. Other cells in the innate immune system include eosinophils, basophils, monocytes, mast cells, dendritic cells, and natural killer (NK) cells, a specialized lymphocyte that identifies and removes virus-infected cells. These cells are activated by a variety of serum proteins and generate relatively nonspecific immune responses. However, each of these cell types has different specific roles and functions, responds to different signals, and releases and responds to different humoral agents and cytokines.

In contrast, the adaptive immune system consists of lymphocytes, both the T cells that differentiate in the thymus and the B cells generated and differentiated in bone marrow. These cells are "adaptive" in that they respond and react to specific signals in their environment following exposure. Each cell differentiates to respond to a single specific antigen and then becomes a clone of multiple cells responding to that specific antigen. In the process of differentiation, many different subtypes develop. These are usually identified based on cell surface markers (CDs). There are large numbers of immune system cells distinguished by their surface proteins. T lymphocytes, the agents of cellular immunity, differentiate into many broad classes with different surface protein markers. For example, $CD4^+$ T-helper cells produce many cytokines and express surface markers important for activation of phagocytic cells. $CD4^+$ T cells are further classified as Th1 or Th2 lymphocytes, based on the types of cytokines and cytotoxic substances they produce. Th1 lymphocytes produce primarily interferon γ (INF-γ) and tumor necrosis factor-α (TNF-α), whereas Th2 lymphocytes produce primarily interleukins IL-4, IL-5, and IL-13, and have proinflammatory actions. Helper T cells release cytokines to stimulate B cells to differentiate. $CD8^+$ cytotoxic T cells have a major function to lyse antigen-bearing target cells. They bind to specific surface receptors and kill with the release of digestive enzymes. There is a class of T cells called suppressor cells that are important in immune tolerance, and also regulatory T cell (T_{reg}) having different surface markets.

B cells secrete antibodies and are thus the agents of humoral immunity. B cells also may express a variety of surface receptors. The activation of B cells requires assistance from macrophages, which present antigen to the B cells, and usually requires the action of helper T cells that release cytokines that stimulate the B cells, which then differentiate into plasma cells that produce the antibodies. There are five classes of antibodies, all immunoglobins, named

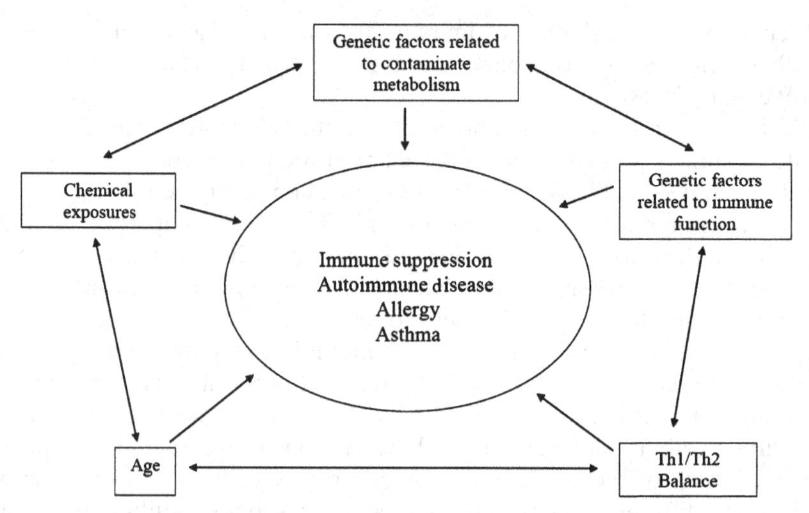

Figure 13.1. Factors that alter immune system function, leading to either suppression of the immune system and increased risk of infection or overactivity of the immune system, leading to autoimmune disease, allergy, or asthma.

IgA, IgD, IgE, IgG, and IgM. However, within these five classes, there are millions of different antibodies, which is the key to specificity of response to particular antigens. IgE is especially important in allergic diseases, including asthma.

Another important concept is that the immune system can be altered in many different ways. These changes may take the form of immunosuppression or allergy and autoimmunity, whereby the body responds inappropriately to "self-antigens." Environmental exposure can trigger each of these different responses, as discussed further. If the immune system is suppressed, there is increased infection and greater susceptibility to diseases such as cancer. If there is a predominance of Th2 lymphocyes, there is likely to be a greater incidence of allergics and asthma. If the body develops antibodies to its own tissues, one has diseases such as type 1 diabetes, lupus, rheumatoid arthritis, and others. Figure 13.1 shows in diagrammatic form these complicated interactions between exposure and other critical variables, including especially age and genetics.

EFFECTS OF DEVELOPMENTAL STAGE ON RESPONSE TO ORGANIC POLLUTANTS

Another consideration is the age of exposure. It is clear that early life determines much of the responsiveness of the immune system for life (Dietert 2008; Holladay and Smialowicz 2000). There are both protective and risk factors early in life. Growing up on a farm with livestock is protective against allergy and hay fever (Smit et al. 2007), and the "hygiene hypothesis" suggests that developing in too sterile an environment early in life skews the immune system to a Th2 profile rather than the normal Th1 profile, resulting in increased risk of asthma and allergy (Douwes et al. 2009; Renz et al. 2006). In addition, exposures to toxicants in the fetal and perinatal stages often have been found to have effects different from those in adults, and often will result in permanent, lifelong perturbation of the immune system function (Dietert and Piepenbrink 2006). The developing immune system is often more sensitive to chemicals than the more mature system. Luebke et al. (2006) reported that sensitivity to drugs and environmental chemicals is an order of magnitude greater in early life as compared with adults. In addition, the specific effects may not be the same as those seen in adults, and the severity may be greater. There are cases where a sublethal exposure early in life produces a change that is not recognized until later in life when there is a subsequent stress on immune function.

Perturbation of immune system function, whether early in life or in adulthood, can result in altered susceptibility to many different diseases beyond just infection. Allergy and autoimmunity (which can affect different organs) are totally dependent upon immune function. Sensitization to airborne environmental allergens in early childhood may lead to atopic asthma years later (Holt et al. 1999). The immune system is the first line of defense against cancer (Beyer and Schultze 2006). This is because many infections, either directly or secondarily as a consequence of inflammation, cause cancer. Lippman and Hong (2008) report that 1.6 million people worldwide develop cancer as a consequence of infection, especially in developing countries. Both the innate and adaptive immune systems influence the development of atherosclerosis (Hansson 2005). Vasculitis (Stone 2008) and several types of kidney diseases, such as glomerulonephritis (Appel 2008), are immune system disorders.

ORGANIC CHEMICAL EXPOSURES AND IMMUNOSUPPRESSION

A large number of organic chemicals have been found to result in immunosuppression. These include persistent organic pollutants (POPs) such as dioxins, furans, polychlorinated biphenyls (PCBs) and chlorinated pesticides, several volatile organics (benzene, toluene, xylol), and some polyaromatic hydrocarbons such as benzo[a]pyrene (Vos and van Loveren 1995).

In some regards, describing the effects of many of these chemicals as "immunosuppression" is not exactly correct since what is seen is more a perturbation of immune system function, often with some components of cellular immunity suppressed and others increased (suppression of Th1 but stimulation of Th2, which has the net effect of reducing resistance to infection but promoting allergy, asthma, and autoimmune disease) (Duramad et al. 2007), and with humoral immunity either unaffected or increased. For example, Phillips (2000) demonstrated that while Th1 cellular immunity, measured by cytokine profiles, was reduced in children exposed to chlordane and heptachlor, there were elevations in proinflammatory cytokines. Daniel et al. (2002) showed that farmers with high exposure showed suppression of Th1 cytokines, but elevation in Th2 cytokines in relation to serum levels of dichlorodiphenyltrichloroethane (DDT) and its metabolites, 2,3,4,5,6-pentachlorophenol, PCBs, hexachlorobenzene (HCB), and γ-hexachlorocyclohexane (HCH). Styrene-exposed workers showed suppression of cellular immunity with little or no change in humoral immunity (Tulinska et al. 2000). Diesel exhaust also induced a transient T-cell depression but increased IgE production, stimulating pulmonary inflammatory responses. There is an abundant literature demonstrating elevations in risk of infectious disease in children in relation to perinatal exposure to several organic chemicals. Byers et al. (1988) found elevated incidence of infections in children exposed *in utero* to water supplies containing elevated concentrations of trichloroethylene (TCE), perchloroethylene, and 1,2-transdichloroethylene. In Yucheng children, Chao et al. (1997) reported those with higher levels of 2,3,4,7,8-pentachloro- and 1,2,3,4,7,8-hexachlorodibenzofurans had a higher incidence of middle ear disease as documented by otolaryngologists. Weisglas-Kuperus et al. (2000, 2004) found that Dutch children with perinatal exposure to PCBs and dioxins showed a higher prevalence of recurrent middle ear infections and chicken pox but a reduced risk of allergic diseases. Dewailly et al. (2000) found that Inuit infants exposed to organochlorines had more frequent otitis media. Compared to the lowest tertile of serum concentration, infants in the highest tertile of 2,2-bis(*p*-chlorophenyl)-1,1-dichloroethylene (DDE), the major metabolite of DDT, showed an odds ratio (OR) = 1.87 [95% confidence interval (CI) = 1.07–3.26]. Later study of the same population (Dallaire et al. 2004) found that prenatal exposure to PCBs and DDE, monitored by cord blood concentrations, was associated with elevated risks of otitis media, upper and lower respiratory tract infections, and gastrointestinal infections. Miyashita et al. (2011) reported a significantly increased risk of otitis media (OR = 5.3, 95% CI = 1.5–1.9) in relation to maternal serum levels of 2,3,4,7,8-pentachlorodibenzofuran.

Strong evidence for immunosuppression after exposure to POPs comes from studies of PCBs or mixtures of PCBs and other POPs. Park et al. (2008) reported that Slovakian children exposed to PCBs prenatally had significantly reduced thymic volume on the third or fourth postnatal day. A more recent and larger study of the same population (Jusko et al. 2012) confirmed the relationship between maternal PCB levels and thymic volume in the child early after birth, but also found some evidence that thymus volume at 16 months was correlated with current serum PCB levels.

It is a mistake, however, to consider that all POPs have identical effects on immune system function. Karmaus et al. (2005) reported that while PCBs were significantly associated with increased IgM levels, HCB was inversely related to IgM. DDE, but not PCBs, was associated with significant increases in IgE, IgG, and IgA levels. In Flemish adolescents, Van Den Heuvel et al. (2002) showed that dioxin-like activity, monitored by the chemical activated luciferase gene expression (CALUX) assay, was negatively associated with asthma, while asthma was positively associated with PCB concentrations.

Studies in Sweden of individuals with high consumption of POP-contaminated Baltic Sea fish (Svensson et al. 1994) reported that fish consumption was associated with lower proportions and numbers of NK cells. Hagmar et al. (1995) investigated immune parameters in a Latvian population who ate fatty fish from the Baltic Sea. High rates of fish consumption were positively correlated with B-cell levels and $CD4^+/CD8^+$ ratios but were negatively correlated with levels of cytotoxic $CD8^+$ cells. Daniel et al. (2001) reported on immune system effects in 146 patients occupationally exposed to PCBs and other chemicals and found weak changes in both cellular and humoral immunity. The most striking result was a suppression of interferon γ in relation to HCB concentration and of IL-4 and PCB 138 levels.

Innate immunity may also be altered. Siegel et al. (2004) reviewed the effects of diesel exhaust in humans and animal models and found that innate alveolar macrophage function was reduced, and demonstrated that this was dependent upon the organic compounds, not the carbon core of the particulates. Levin et al. (2005) studied the effects of PCBs on phagocytosis in human leukocytes and found that the non-dioxin-like PCBs suppressed phagocytosis. This is consistent with other evidence that different classes of PCBs have quite different actions. Non-dioxin-like, *ortho*-substituted PCBs (Voie et al. 2000) and polybrominated biphenyls (PBBs) (Kristoffersen et al. 2002) were found to trigger respiratory bursts with elevations of intracellular calcium in human neutrophils. A somewhat similar and cytotoxic response to lower-chlorinated *ortho*-substituted PCBs was seen in thymocytes (Tan et al. 2003) and was demonstrated to result from a change in cell membrane fluidity (Tan et al. 2004). Dioxin-like PCBs did not show these effects (Tan et al. 2004; Yilmaz et al. 2006). Lower-chlorinated *ortho*-substituted, but not coplanar, PCBs have also been demonstrated to inhibit lipopolysaccharide (LPS)-induced splenocyte proliferation (Smithwick et al. 2004). However, since humans are never exposed to only single PCB congeners or groups of congeners, it has not been possible to determine the degree to which these actions have physiological impact. However, these effects of non-dioxin-like PCB congeners are important to recognize because if only dioxin-like toxic equivalency is monitored, these immunosuppressive actions will not be detected.

IMMUNOSUPPRESSION MEDIATED VIA ACTIVATION OF THE ARYL HYDROCARBON RECEPTOR (AhR)

In rodents, the best-studied action of organic substances on the immune system is that of dioxin, especially the most potent congener, 3,4,7,8-tetrachloro-*p*-dibenzodioxin (TCDD). TCDD triggers apoptotic cell death of thymocytes, and this results in thymic atrophy (McConkey et al. 1988). TCDD causes strong suppression of cellular- and humoral-mediated immunity (Vos et al.

1973). In fact, suppressed humoral and cellular immunity is one of the most sensitive responses to TCDD in rodents and is associated with an increased susceptibility to infection (Holsapple et al. 1991). It is likely that the reduced thymic volume seen in children exposed to PCBs (Jusko et al. 2012; Park et al. 2008) reflects this action of dioxin-like PCBs. The immunosuppressive actions of TCDD are completely absent in AhR knockout mice (Vorderstrasse et al. 2001), although earlier work by Kerkvliet et al. (1990) had suggested that there were AhR-dependent and AhR-independent mechanisms of immunosuppression by TCDD. However, there are differences in response between mice and rats (Smialowicz et al. 1994). Neubert et al. (1991) reported that TCDD concentrations as low as 1×10^{-13} M reduce mitogen-stimulated proliferation and differentiation of human lymphocytes, with T-helper cells being the most sensitive target.

Kerkvliet (2009) reviewed the variety of effects of TCDD. TCDD exposure leads to a dose-dependent increase in the number of neutrophils but inconsistent effects on numbers of macrophages and NK cells. One important effect is on dendritic cells, where TCDD alters function in ways that may result in altered T-cell activation and immune suppression. TCDD potently suppresses all types of adaptive immune responses and reduces risk of autoimmune diseases. However, she notes that other organic chemicals that bind the AhR may not have exactly the same effects.

Hennig et al. (2002) examined the effects of dioxin-like PCBs on the function of vascular endothelial cells and reported that PCBs 77, 126, and 169 caused AhR-dependent oxidative stress and an inflammatory response. The non-dioxin-like PCB 153 did not cause these changes but, unlike the dioxin-like congeners, did cause increased endothelial production of IL-6.

In spite of the potency of TCDD on isolated immune system cells, changes in human populations have not shown effects as dramatic. No correlation between dioxin and furan concentrations and occurrence of infections or immune system parameters were found comparing workers in a German pesticide factory to controls (Jung et al. 1998). No consistent immune abnormalities were seen in Ranch Hand Vietnam Air Force personnel exposed to dioxin through the use of Agent Orange (Michalek et al. 1999). No changes in either cellular or humoral immunity were found in Yucheng children at ages 8–16 (Yu et al. 1998), and these authors suggest that the earlier reports of increased infections may indicate a transient, but not maintained, immunosuppression. A slight but statistically significant decrease in levels of IgG in persons exposed to TCDD was seen by Neubert et al. (2000), but no changes in other immunoglobulins or cytokines were detected. Results from Seveso, where local residents were exposed to dioxin from an industrial accident, found that increasing TCDD levels were associated with decreased levels of IgG, but with no effect on other immune system parameters (Baccarelli et al. 2002). Dutch children studied at age 8 years showed a decrease in allergy and increased $CD4^+$ T-helper and CD45RA+ cells in relation to perinatal dioxins levels (ten Tusscher et al. 2003), but not the striking immunosuppression seen in rodents.

PESTICIDES OTHER THAN THE ORGANOCHLORINES AND THE IMMUNE SYSTEM

There has been considerable interest in the question of the degree to which exposure to pesticides other than organochlorines alter immune system function, particularly as exposure to these pesticides is so ubiquitous in our environment and food. Corsini et al. (2008) reviewed the human literature on this subject and reported a great variety of different effects of different pesticides, but with no clear pattern. Most studies do report changes, but often the effects are small and have not been replicated in other studies. Li (2007) and Li et al. (2002) reviewed the literature on organophosphate pesticides, drawing particularly on studies done of Japanese victims of the sarin subway attack in 1995. He concludes that organophosphates inhibit NK cells, lymphokine-activated killer cells, and cytotoxic T lymphocytes, and induce apoptosis. Galloway and Handy (2003) reviewed the wildlife and laboratory rodent literature on the immune system effects of organophosphates and concluded that while there are many reports of altered nonspecific immune system changes, decreased host resistance, hypersensitivity, and autoimmunity, the overall patterns suggest decreased immune function following acute exposure but enhanced immune activity at low concentrations that do not cause acetylcholinesterase inhibition. One major question is the degree to which stress in a major factor.

HUMORAL IMMUNITY AND ORGANIC CHEMICALS

Humoral immunity is also altered by exposure to organics, and the effects of POPs have been studied most because their persistence makes exposure assessment relatively easier. Yucheng patients in Taiwan, exposed to PCBs and furans through consumption of contaminated rice oil, showed decreased concentrations of IgA and IgM but unaltered IgG (Lü and Wu 1985), and similar changes in immunoglobins were also found in Japanese Yusho patients (Nakanishi et al. 1985). Cord blood obtained from women who practiced subsistence fishing along the St. Lawrence River in Canada contained PCB concentrations threefold higher than that of a control population (Belles-Isles et al. 2002), and showed reduced IgM and increased IgG relative to controls. Weisglas-Kuperus et al. (2000) reported that prenatal PCB exposure was associated with lower antibody levels to mumps and measles vaccine. Heilmann et al. (2010) reported that antibody response to vaccination for diphtheria and tetanus was inversely associated with serum PCB concentrations. The response to diphtheria toxoid decreased by 24.4% (95% CI = 1.63–41.9) for each doubling of PCB concentration at the time of examination. The tetanus toxoid response decreased by 16.5% (95% CI = 1.51–29.3) for each doubling of prenatal exposure. Grandjean et al. (2012) found a similar reduction in antibody concentrations against tetanus and diphtheria in children in the Faroe Islands in relation to exposure to perfluorinated compounds. Karmaus et al. (2001) reported that children exposed to DDE and either PCBs or HCB showed a significantly elevated risk of otitis media (OR = 3.71, 95% CI = 1.10–12.56) and a proportionately elevated level of IgE, while Sunyer et al. (2005) found that children with elevated DDE were at greater risk of asthma. Williamson et al. (2006) monitored serum immunoglobulins in people living near hazardous waste sites as compared to controls, and found a consistent elevation in IgA, but not other immunoglobins. Even polar bears show reduced antibody responses to influenza and reovirus immunization in relation to serum PCB levels (Lie et al. 2004).

The evidence that antibody responses to immunization are reduced in children with elevated concentrations of at least some organic chemicals suggests that this is a very serious problem with major public health consequences. Immunization has been in great part responsible for the dramatic decrease in morbidity and mortality of viral diseases that were once major killers, and failure to develop antibodies to vaccines will leave the individual vulnerable to dangerous diseases.

ORGANICS THAT PROMOTE ALLERGY

Chemicals can not only suppress the immune system but some also increase risks of allergic diseases and asthma (see Chapter 14, Rumchev). Choi et al. (2010) found that levels of propylene glycol and glycol ethers in bedroom air were associated with an increased risk of rhinitis (OR = 2.8, 95% CI = 1.6–4.7) and eczema (OR = 1.6, 95% CI = 1.1–2.3), as well as asthma (OR = 1.5, 95% CI = 1.0–2.3). Organic acid anhydrides, used in the production of alkyd, epoxy, and polyester resins, trigger respiratory hypersensitivity reactions of three different types, depending on whether they are triggered by IgG, IgE, or cell-mediated immune changes (Zhang et al. 2002).

ORGANICS AND AUTOIMMUNE DISEASE

Autoimmune diseases are dependent on the adaptive immune system and are a failure of the ability to distinguish self and nonself. In spite of considerable speculation, there is only weak evidence for a role of organic chemicals and most forms of autoimmune disease.

The strongest suggestion that organics can trigger autoimmune disease comes from the use of medications. Hess (2002) identified more than 70 medications that have been associated with development of autoimmune diseases and also identified vinyl chloride and organic solvents as risk factors. Pollard et al. (2010) reported that two medications, procainamide and hydralazine, trigger lupus in some patients and that the symptoms go away after the patients were taken off the drugs. In addition, contaminated cooking oil consumption in Spain in the early 1980s resulted in syndrome of pseudoscleroderma, apparently caused by rapeseed oil.

Tsai et al. (2007) reported a study that followed the Yucheng cohort of PCB- and furan-exposed persons in Taiwan, reporting on mortality in a 24-year follow-up. They found a highly significant elevation in deaths from systemic lupus erythematosus (OR = 14.7, 95% CI 4–7–34.4) in exposed women as compared to controls. Lee et al. (2007), using data from the National Health and Nutrition Examination Survey in the United States, reported that there was a significant association between arthritis in women and both dioxin-like and non-dioxin-like PCB concentrations. For rheumatoid arthritis, by quartiles, the ORs were 1.0, 7.6, 6.1, and 8.5 for dioxin-like PCBs, and 1.0, 2.2, 4.4, and 5.4 for non-dioxin-like PCBs. These results with PCBs, including dioxin-like PCBs, stand in contrast to clear evidence that TCDD suppresses allergic and autoimmune diseases such as experimental allergic encephalitis, a model for multiple sclerosis and type 1 diabetes (Kerkvliet 2009).

Cooper et al. (2009) reviewed the information on the relation between exposure to TCE and autoimmune diseases in animals and in humans. Human occupational studies indicate elevations in scleroderma and a generalized hypersensitivity skin disorder. Subjects exposed to TCE and related compounds in well water showed increased frequency of antinuclear autoantibodies as compared to less exposed persons (Kilburn and Warshaw 1992). Lupus-prone mice developed accelerated lupus after exposure to TCE at low concentrations (Sobel et al. 2005).

EFFECTS OF ENDOCRINE-DISRUPTIVE CHEMICALS ON IMMUNE SYSTEM FUNCTION

Diethylstilbestrol (DES) is a potent estrogenic agent that was given to many pregnant women up until 1971 to prevent spontaneous abortion, after which it was discontinued because of a correlation with the development of vaginal cancer in young girls. Both male and female children born to mothers who took DES have been found to show a higher incidence of autoimmune diseases and asthma (Baird et al. 1996; Noller et al. 1988). Animal studies have shown that prenatal exposure to DES has long-lasting effects on thymocytes and may alter the T cells that are derived from the thymocytes (Fenaux et al. 2004).

EFFECTS OF ORGANIC CHEMICALS ON IMMUNE SYSTEM FUNCTION AND CANCER

The immune system is the first order of defense against cancer as well as infection. Therefore, it is not surprising that some of the carcinogenic effects of organic chemicals may be a reflection of altered immune system function.

These relations have been particularly well studied for non-Hodgkin lymphoma and probably apply also to other immune cell cancers. Grulich et al. (2007) reviewed the evidence for risk of non-Hodgkin lymphoma in relation to immune function and showed clearly that immune system deficiency is a strong risk factor. Atopic conditions (eczema, hay fever, and asthma) have proven to be protective in most studies, but autoimmune conditions (rheumatoid arthritis, systemic lupus erythematosis, Sjogren's syndrome, and psoriasis) lead to increased risk.

Colt et al. (2009) examined the relationship between non-Hodgkin lymphoma and serum levels of PCB 180, serum toxic equivalency quotient, and levels of α-chlordane in relation to immune gene variations. They found that risk in relation to exposure was increased only in individuals with the same genotypes for IFN-γ for all three exposure groups and for PCB 180 also to the same genotypes for IL-16, IL-8, and IL-10. Kramer et al. (2012) in a very recent review conclude that the weight of evidence supports a causal role of PCBs in lymphomagenesis mediated via immune system dysfunction.

MECHANISMS WHEREBY ORGANIC CHEMICALS ALTER IMMUNE SYSTEM FUNCTION

Studies of human immune system alteration after exposure to organic chemicals in most cases are descriptive, and therefore, in order to determine mechanisms, one usually must study cellular or animal systems. It is clear that both the innate and adaptive immune systems are altered by organic chemicals. Ganey et al. (1993) reported that non-dioxin-like PCBs stimulate the generation of superoxide anion in rat neutrophils. Levin et al. (2005) demonstrated significant depression of phagocytosis in human neutrophils mediated by non-dioxin-like PCB congeners, but with no effect of dioxin-like PCBs. In a review, Girard (2003) noted that chlordane, toxaphene, and dieldrin all increase O_2^- production in isolated neutrophils. In an experimental study of harbor seals fed with contaminated Baltic Sea herring, neutrophils were increased, but all measures of adaptive immunity were decreased (Van Loveren et al. 2000). In contrast to the stimulation of the action of neutrophils, NK cells were decreased in individuals who consume high amounts of POP-contaminated Baltic Sea fish (Svensson et al. 1994). These studies indicate how complex the actions of organic chemicals are on immune system function, with different chemicals, and even different congeners of the same class of chemical, having different actions or no action at all on cells of the innate immune system.

The fundamental mechanisms behind the alteration of both innate and adaptive immune system function by organic chemicals are likely via gene induction. Bezdecny et al. (2005) showed that the mechanism behind the stimulation of superoxide anion by non-dioxin-like PCBs (Ganey et al. 1993) was through the induction of cyclooxygenase-2. Gene induction caused by dioxin-like substances that activate the AhR has been studied by several laboratories (Johnson et al. 2004; Maier et al. 2007; Vezina et al. 2004). A very large number of genes are either up- or downregulated. Furthermore, the gene induction is organ specific, such that the pattern of gene expression in liver is very different from those in glioma cells (Maier et al. 2007). While a number of the affected genes in these studies relate to immune system function, altered genes regulating almost every cellular function are found. Non-dioxin-like PCBs with phenobarbital-like action activate and are metabolized by cytochrome P450-1B1. These alter a similarly large number of immune system and nonimmune system genes (Johnson et al. 2004; Vezina et al. 2004), but the specific genes and direction of induction are not the same as those induced by the AhR.

While many studies of gene induction by organic compounds do not focus specifically on immune system genes, Ezendam et al. (2004) have reported on the toxicogenomics of subchronic exposure of rats to two concentrations of HCB in spleen, liver, blood, thymus, and mesenteric lymph nodes. Table 13.1

TABLE 13.1. Representative Splenic Genes that Change Significantly ($p < 0.001$) after 4 Weeks' Feeding of Brown Norway Rats with a Diet Supplemented with HCB at Concentrations of 150 or 450 mg/kg as Compared to Controls (Reproduced with Permission from Ezendam et al. 2004[a])

Accession Number	Gene Name	HCB Low Dose	HCB High Dose
Granulocytes and macrophages			
AA957003	S100 calcium-binding protein A8	2.8	34
U50353	Defensin 3a	2.5	32
AA946503	Lipocalin 2	1.7	24
L18948	S100 calcium-binding protein A9	3.2	19
L06040	12-lipoxygenase	1.9	5.7
M32062	Fc receptor, IgG, low affinity III	1.4	2.0
AA894004	ESTs, highly similar to Capg mouse macrophage capping protein	1.2	1.4
X73579	Fc receptor, IgE, low affinity II	−1.1	−2.3
Mast cells			
U67913	Mast cell protease 10	12	42
U67888	Mast cell protease 3	3.4	20
U67907	Mast cell protease 4 precursor	1.5	8.7
M21622	High-affinity IgE receptor	3.2	7.0
U67914	Mast cell carboxypeptidase A precursor	1.8	6.8
U67908	Mast cell protease 5 precursor	1.2	6.0
M38759	Histidine decarboxylase	3.7	4.3
Pattern recognition molecules			
AF087943	CD14 antigen	1.1	1.7
Complement			
AF036548	Response gene to complement	−1.3	20

(Continued)

TABLE 13.1. (*Continued*)

Accession Number	Gene Name	HCB Low Dose	HCB High Dose
AA818025	CD59 antigen precursor	1.1	1.7
Cell adhesion molecules			
X05834	Fibronectin 1	1.8	3.5
AJ009698	Embigin	1.4	3.3
Chemokines			
U90448	CXC chemokine LIX	1.0	1.9
U17035	Chemokine (CXC motif) ligand 10	1.0	−2.3
Cytokines and cytokine-associated genes			
M63122	Tumor necrosis factor receptor	1.3	1.3
AF075382	Suppression of cytokine signaling	1.3	−1.3
M98820	Interleukin 1 beta	1.5	−1.2
M55050	Interleukin 2 receptor beta chain	1.2	−1.4
L00981	Lymphotoxin, tumor necrosis factor-α	−1.1	−1.4
M34253	Interferon regulatory factor 1	−1.1	−1.6
U14647	Interleukin 1 beta converting enzyme	1.1	−1.6
U69272	Interleukin 15	−1.1	−1.7
U48596	MAPK kinase kinase 1	1.0	−1.8
U03491	Transforming growth factor, beta 3	−2.9	−3.0
Genes associated with T and B cells and MHCII expression			
U39609	Antinerve growth factor 30 antibody light chain	1.3	3.8
L22654	Antiacetylcholine receptor antibody rearranged immunoglobulin gamma-2a chain	3.2	1.6
L07398	Immunoglobulin rearranged gamma-chain V region	1.0	2.4
M18526	Immunoglobulin germline kappa chain	1.2	1.6
X13016	MRC OX-45 surface antigen	1.1	−1.3
U11681	Rapamycin and FKBP12 target-1 protein	−1.0	−1.3
D13555	T-cell receptor CD3, subunit zeta	−1.1	−1.4
U31599	MHC class II-like beta chain RT1.Mb	−1.0	−1.4
L14004	Polymeric immunoglobulin receptor	1.0	−1.4
D10728	Lymphocyte antigen CD5	−1.2	−1.6
M85193	RT6.2	−1.3	−1.6
U24652	Linker of T-cell receptor pathways	−1.0	−1.7
X14319	T-cell receptor active beta-chain, V region	−1.2	−2.1

[a]Table contains GenBank accession numbers (http://www.ncbi.nih.gov/entrez/query.fcgi?db=nucleotide) of the cDNA fragments present on Affymetrix RG U34A gene chips, gene name, and average fold change in expression of both low-dose and high-dose versus control. Fold changes are calculated with data from five to six rats per group. A one-way analysis of variance (ANOVA) was used to determine significance; only probe sets that changed significantly with $p < 0.001$ are shown.

EST, expressed sequence tag.

(from Ezendam et al. 2004) lists some of the splenic genes significantly changed after exposure of rats to 150 or 450 mg/kg of HCB. It is clear that this one chemical alters critical genes related to different components of the immune system, and does so in different ways. Some genes are upregulated, while others are downregulated. Some are downregulated at 150 mg/kg but upregulated by 450 mg/kg (e.g., the response gene to complement). Other tables show induction of immune system genes in mesenteric lymph nodes, thymus, blood, liver, and kidney, some of which are the same as those in the spleen, but many of which are different.

These authors conclude that HCB induces a systemic inflammatory response with oxidative stress. Figure 13.2, also from Ezendam et al. (2004), shows the authors' diagrammatic representation of the various immune system actions of HCB that lead to the ultimate inflammatory response. While these pathways are hypothesized, and not all specifically documented, the diagram clearly indicates how complex the interactions can be.

The pathway whereby HCB causes induction of such a wide variety of immune system genes is not clear. At least in bird hepatocytes, highly purified HCB does induce cytochrome P4501A (Mundy et al. 2012), although the World Health Organization has not assigned a toxic equivalency factor (TEF) value to HCB because of concerns that commercial mixtures were contaminated with dioxin-like compounds. But the gene induction pattern from HCB

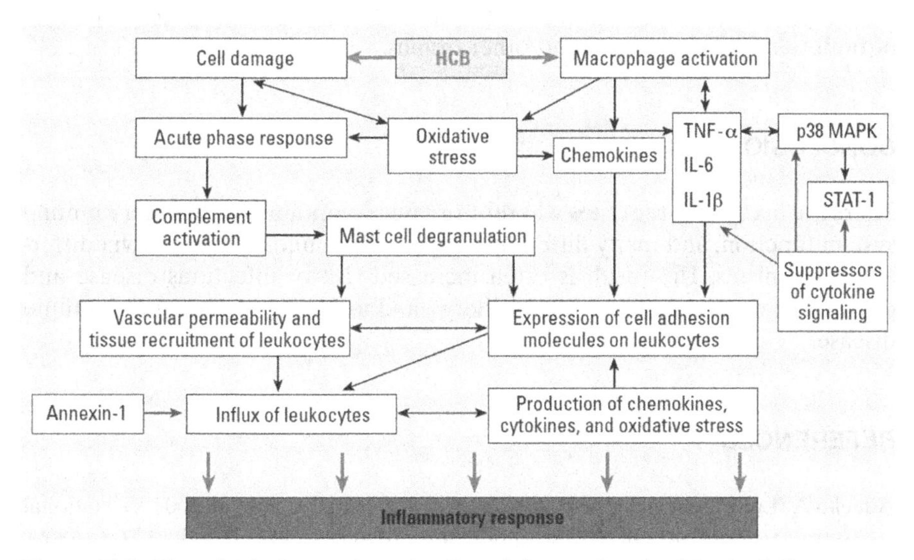

Figure 13.2. Hypothetical overview of cells and factors involved in the inflammatory response initiated by hexachlorobenzene (HCB). Based on the assumption that HCB activates macrophages, this would lead to a cascade of reactions, activating immune cells and pro- and anti-inflammatory mediators, eventually leading to inflammation. Reproduced with permission from Ezendam et al. 2004.

is different from that of TCDD. Adeeko et al. (2003) administered a mixture of POPs to pregnant rats and tested gene induction in pup and maternal liver. They reported that HCB had both AhR- and phenobarbital-type induction. There may well be other pathways also involved.

A number of other organic compounds that are not activators of the AhR also induce genes. Kim et al. (2011) have reported altered levels of 23 genes in workers exposed to benzene, toluene, ethylbenzene, and xylene. Although the genes induced were not specifically related to immune system function, some were associated with oxidative stress. Sarma et al. (2010) investigated gene induction in a human promyelocytic leukemia cell line (HL-60) and found dose-dependent changes in expression in several immune response and apoptosis-related genes. Organophosphate pesticides cause gene induction in *Caenorhabditis elegans* (Lewis et al. 2009).

Organic chemicals may also induce epigenetic changes in gene expression. Gestational exposure of rats to bisphenol A resulted in altered expression of several endocrine receptors for several generations (Wolstenholme et al. 2012). In a study of Greenlandic Inuits, Rusiecki et al. (2008) reported a significant reduction of global DNA methylation in individuals with elevated DDT, DDE, β-hexachlorocyclohexane, oxychlordane, mirex, and several PCB congeners. Lee et al. (2009) proposed that DNA hypomethylation may be an extremely important component of human responses to exposure to multiple chemicals, even if at low concentrations. Because many organic chemicals, especially POPs, are found in essentially all animal fats in food, human nutrition may be a sufficient source of exposure so as to trigger epigenetic changes of both the immune system and other organs.

CONCLUSION

There are many different sites whereby organic compounds act to alter immune system function, and many different organic compounds that do so via different mechanisms. The result is often increased risk of infectious disease and cancer, but others increase risk of allergy and asthma and even autoimmune disease.

REFERENCES

Adeeko A, Li D, Doucet J, Cooke GM, Trasler JM, Robaire B, et al. 2003. Gestational exposure to persistent organic pollutants: Maternal liver residues, pregnancy outcome, and effects on hepatic gene expression profiles in the dam and fetus. Toxicol Sci 72:242–252.

Appel GB. 2008. Glomerular disorders and nephrotic syndromes. In: Cecil Medicinie, 23rd edition (Goldman L, Ausiello D, eds.). Philadelphia, PA: Saunders Elsevier, pp. 866–876.

Baccarelli A, Mocarelli P, Patterson DG, Jr., Bonzini M, Pesatori AC, Caporaso N, et al. 2002. Immunologic effects of dioxin: New results from Seveso and comparison with other studies. Environ Health Perspect 110:1169–1173.

Baird DD, Wilcox AJ, Herbst AL. 1996. Self-reported allergy, infection and autoimmune diseases among men and women exposed in utero to diethylstilbestrol. J Clin Epidemiol 49:263–266.

Belles-Isles M, Ayotte P, Dewailly E, Weber JP, Roy R. 2002. Cord blood lymphocyte functions in newborns from a remote maritime population exposed to organochlorines and methylmercury. J Toxicol Environ Health A 65:165–182.

Beyer M, Schultze JL. 2006. Regulatory T cells in cancer. Blood 108:804–811.

Bezdecny SA, Roth RA, Ganey PE. 2005. Effects of 2,2′,4,4′-tetrachlorobiphenyl on granulocytic HL-60 cell function and expression of cyclooxygenase-2. Toxicol Sci 84:328–334.

Byers VS, Levin AS, Ozonoff DM, Baldwin RW. 1988. Association between clinical symptoms and lymphocyte abnormalities in a population with chronic domestic exposure to industrial solvent-contaminated domestic water supply and a high incidence of leukaemia. Cancer Immunol Immunother 27:77–81.

Chao WY, Hsu CC, Guo YL. 1997. Middle-ear disease in children exposed prenatally to polychlorinated biphenyls and polychlorinated dibenzofurans. Arch Environ Health 52:257–262.

Choi H, Schmidbauer N, Sundell J, Hasselgren M, Spengler J, Bornehag CG. 2010. Common household chemicals and the allergy risks in pre-school age children. PLoS ONE 5:e13423.

Colt JS, Rothman N, Severson RK, Hartge P, Cerhan JR, Chatterjee N, et al. 2009. Organochlorine exposure, immune gene variation, and risk of non-Hodgkin lymphoma. Blood 113:1899–1905.

Cooper GS, Makris SL, Nietert PJ, Jinot J. 2009. Evidence of autoimmune-related effects of trichloroethylene exposure from studies in mice and humans. Environ Health Perspect 117:696–702.

Corsini E, Liusivuoir J, Vergieva T, Van Loveren H, Colosia C. 2008. Effects of pesticide exposure on the human immune system. Hum Exp Toxicol 27:671–680.

Dallaire F, Dewailly E, Muckle G, Vézina C, Jacobson SW, Jacobson JL, et al. 2004. Acute infections and environmental exposure to organochlorines in Inuit infants from Nunavik. Environ Health Perspect 112:1359–1365.

Daniel V, Huber W, Bauer K, Suesal C, Conradt C, Opelz G. 2001. Associations of blood levels of PCB, HCHS, and HCB with numbers of lymphocyte subpopulations, in vitro lymphocyte response, plasma cytokine levels, and immunoglobulin autoantibodies. Environ Health Perspect 109:173–178.

Daniel V, Huber W, Bauer K, Suesal C, Conradt C, Opelz G. 2002. Associations of dichlorodiphenyltrichloroethane (DDT) 4.4 and dichlorodiphenyldichloroethylene (DDE) 4.4 blood levels with plasma IL-4. Arch Environ Health 57:541–547.

Dewailly E, Ayotte P, Bruneau S, Gingras S, Belles-Isles M, Roy R. 2000. Susceptibility to infections and immune status in Inuit infants exposed to organochlorines. Environ Health Perspect 108:205–211.

Dietert RR. 2008. Developmental immunotoxicology (DIT): Windows of vulnerability, immune dysfunction and safety assessment. J Immunotoxicol 5:401–412.

Dietert RR, Piepenbrink MS. 2006. Perinatal immunotoxicity: Why adult exposure assessment fails to predict risk. Environ Health Perspect 114:477–483.

Douwes J, Brooks C, Pearce N. 2009. Protective effects of farming on allergies and asthma: Have we learnt anything since 1873? Expert Rev Clin Immunol 5:213–219.

Duramad P, Tager IB, Holland NT. 2007. Cytokines and other immunological biomarkers in children's environmental health studies. Toxicol Lett 172:48–59.

Ezendam J, Staedtler F, Pennings J, Vandebriel RJ, Pieters R, Harleman JH, et al. 2004. Toxicogenomics of subchronic hexachlorobenzene exposure in Brown Norway rats. Environ Health Perspect 112:782–791.

Fenaux JB, Gogal RM, Jr., Ahmed SA. 2004. Diethylstilbestrol exposure during fetal development affects thymus: Studies in fourteen-month-old mice. J Reprod Immunol 64:75–90.

Galloway T, Handy R. 2003. Immunotoxicity of organophosphorous pesticides. Ecotoxicology 12:345–363.

Ganey PE, Sirois JE, Denison M, Robinson JP, Roth RA. 1993. Neutrophil function after exposure to polychlorinated biphenyls in vitro. Environ Health Perspect 101:430–434.

Girard D. 2003. Activation of human polymorphonuclear neutrophils by environmental contaminants. Rev Environ Health 18:75–89.

Goronzy JJ, Weyand CM. 2008. The innate and adaptive immune systems. In: Cecil Medicine, 23rd edition (Goldman L, Ausiello D, eds.). Philadelphia, PA: Saunders Elsevier, pp. 249–259.

Grandjean P, Andersen EW, Budtz-Jørgensen E, Nielsen F, Mølbak K, Weihe P, et al. 2012. Serum vaccine antibody concentrations in children exposed to perfluorinated compounds. JAMA 307:391–397.

Grulich AE, Vajdic CM, Cozen W. 2007. Altered immunity as a risk factor for non-Hodgkin lymphoma. Cancer Epidemiol Biomarkers Prev 16:405–408.

Hagmar L, Hallberg T, Leja M, Nilsson A, Schütz A. 1995. High consumption of fatty fish from the Baltic Sea is associated with changes in human lymphocyte subset levels. Toxicol Lett 77:335–342.

Hansson GK. 2005. Inflammation, atherosclerosis, and coronary artery disease. N Engl J Med 352:1685–1695.

Heilmann C, Budtz-Jørgensen E, Nielsen F, Heinzow B, Weihe P, Grandjean P. 2010. Serum concentrations of antibodies against vaccine toxoids in children exposed perinatally to immunotoxicants. Environ Health Perspect 118:1434–1438.

Hennig B, Meerarani P, Slim R, Toborek M, Daugherty A, Silverstone AE, et al. 2002. Proinflammatory properties of coplanar PCBs: In vitro and in vivo evidence. Toxicol Appl Pharmacol 181:174–183.

Hess EV. 2002. Environmental chemicals and autoimmune disease: Cause and effect. Toxicology 181–182:65–70.

Holladay SD, Smialowicz RJ. 2000. Development of the murine and human immune system: Differential effects of immunotoxicants depend on time of exposure. Environ Health Perspect 108:463–473.

Holsapple MP, Snyder NK, Wood SC, Morris DL. 1991. A review of 2,3,7,8-tetrachlrorodibenzo-p-dioxin-induced changes in immunocompetence: 1991 update. Toxicology 69:219–255.

Holt PG, Macaubas C, Stumbles PA, Sly PD. 1999. The role of allergy in the development of asthma. Nature 402:B12–B17.

Johnson CD, Balagurunathan Y, Tadesse MG, Falahatpisheh MH, Brun M, Walker MK, et al. 2004. Unraveling gene-gene interactions regulated by ligands of the aryl hydrocarbon receptor. Environ Health Perspect 112:403–412.

Jung D, Berg PA, Edler L, Ehrenthal W, Fenner D, Flesch-Janys D, et al. 1998. Immunologic findings in workers formerly exposed to 2,3,7,8-tetrachlorodibenzo-p-dioxin and its congeners. Environ Health Perspect 106(Suppl 2):689–695.

Jusko TA, Sonneborn D, Palkovicova L, Kocan A, Drobna B, Trnovec T, et al. 2012. Pre- and postnatal polychlorinated biphenyl concentrations and longitudinal measures of thymus volume in infants. Environ Health Perspect 120:595–600.

Karmaus W, Kuehr J, Kruse H. 2001. Infections and atopic disorders in childhood and organochlorine exposure. Arch Environ Health 56:485–492.

Karmaus W, Brooks KR, Nebe T, Witten J, Obi-Osius N, Kruse H. 2005. Immune function biomarkers in children exposed to lead and organochlorine compounds: A cross-sectional study. Environ Health 4:5.

Kerkvliet NI. 2009. AHR-mediated immunomodulation: The role of altered gene transcription. Biochem Pharmacol 77:746–760.

Kerkvliet NI, Steppan LB, Brauner JA, Deyo JA, Henderson MC, Tomar RS, et al. 1990. Influence of the Ah locus on the humoral immunotoxicity of 2,3,7,8-tetrachlorodibenzo-p-dioxin: Evidence for Ah-receptor-dependent and Ah-receptor-independent mechanisms of immunosuppression. Toxicol Appl Pharmacol 105:26–36.

Kilburn KH, Warshaw RH. 1992. Prevalence of symptoms of systemic lupus erythematosus (SLE) and of fluorescent antinuclear antibodies associated with chronic exposure to trichloroethylene and other chemicals in well water. Environ Res 57:1–9.

Kim JH, Moon JY, Park EY, Lee KH, Hong YC. 2011. Changes in oxidative stress biomarker and gene expression levels in workers exposed to volatile organic compounds. Ind Health 49:8–14.

Kramer S, Hikel SM, Adams K, Hinds D, Moon K. 2012. Current status of the epidemiologic evidence linking polychlorinated biphenyls and non-hodgkin lymphoma, and the role of immune dysregulation. Environ Health Perspect 120: 1067–1075.

Kristoffersen A, Voie ØA, Fonnum F. 2002. Ortho-substituted polybrominated biphenyls activate respiratory burst in granulocytes from humans. Toxicol Lett 129: 161–166.

Lee DH, Steffes M, Jacobs DR. 2007. Positive associations of serum concentration of polychlorinated biphenyls or organochlorine pesticides with self-reported arthritis, especially rheumatoid type, in women. Environ Health Perspect 115: 883–888.

Lee DH, Jacobs DR, Jr., Porta M. 2009. Hypothesis: A unifying mechanism for nutrition and chemicals as lifelong modulators of DNA hypomethylation. Environ Health Perspect 117:1799–1802.

Levin M, Morsey B, Mori C, Nambiar PR, De Guise S. 2005. Non-coplanar PCB-mediated modulation of human leukocyte phagocytosis: A new mechanism for immunotoxicity. J Toxicol Environ Health A 68:1977–1993.

Lewis JA, Szilagyi M, Gehman E, Dennis WE, Jackson DA. 2009. Distinct patterns of gene and protein expression elicited by organophosphorus pesticides in *Caenorhabditis elegans*. BMC Genomics 10:202.

Li Q. 2007. New mechanism of organophosphorus pesticide-induced immunotoxicity. J Nippon Med Sch 74:92–105.

Li Q, Nagahara N, Takahashi H, Takeda K, Okumura K, Minami M. 2002. Organophosphorus pesticides markedly inhibit the activities of natural killer, cytotoxic T lymphocyte and lymphokine-activated killer: A proposed inhibiting mechanism via granzyme inhibition. Toxicology 172:181–190.

Lie E, Larsen HJ, Larsen S, Johansen GM, Derocher AE, Lunn NJ, et al. 2004. Does high organochlorine (OC) exposure impair the resistance to infection in polar bears (*Ursus maritimus*)? Part I: Effect of OCs on the humoral immunity. J Toxicol Environ Health A 67:555–582.

Lippman SM, Hong WK. 2008. Cancer prevention. In: Cecil Medicine, 23rd edition (Goldman L, Ausiello D, eds.). Philadelphia, PA: Saunders Elsevier, pp. 367–1370.

Lü YC, Wu YC. 1985. Clinical findings and immunological abnormalities in Yu-Cheng patients. Environ Health Perspect 59:17–29.

Luebke RW, Chen DH, Dietert R, Yang Y, King M, Luster MI, et al. 2006. The comparative immunotoxicity of five selected compounds following developmental or adult exposure. J Toxicol Environ Health B Crit Rev 9:1–26.

Maier MS, Legare ME, Hanneman WH. 2007. The aryl hydrocarbon receptor agonist 3,3′,4,4′,5-pentachlorobiphenyl induces distinct patterns of gene expression between hepatoma and glioma cells: Chromatin remodeling as a mechanism for selective effects. Neurotoxicology 28:594–612.

McConkey DJ, Hartzell P, Duddy SK, Håkansson H, Orrenius S. 1988. 2,3,7,8-Tetrachlorodibenzo-p-dioxin kills immature thymocytes by Ca^{2+}-mediated endonuclease activation. Science 242:256–259.

Michalek JE, Ketchum NS, Check IJ. 1999. Serum dioxin and immunologic response in veterans of Operation Ranch Hand. Am J Epidemiol 149:1038–1046.

Miyashita C, Sasaki S, Saijo Y, Washino N, Okada E, Kobayashi S, et al. 2011. Effects of prenatal exposure to dioxin-like compounds on allergies and infections during infancy. Environ Res 111:551–558.

Mundy LJ, Crump D, Jones SP, Konstantinov A, Utley F, Potter D, et al. 2012. Induction of cytochrome P4501A by highly purified hexachlorobenzene in primary cultures of ring-necked pheasant and Japanese quail embryo hepatocytes. Comp Biochem Physiol C Toxicol Pharmacol 155:498–505.

Nakanishi Y, Shigematsu N, Kurita Y, Matsuba K, Kanegae H, Ishimaru S, et al. 1985. Respiratory involvement and immune status in yusho patients. Environ Health Perspect 59:31–36.

Neubert R, Jacob-Müller U, Helge H, Stahlmann R, Neubert D. 1991. Polyhalogenated dibenzo-p-dioxins and dibenzofurans and the immune system. 2. In vitro effects of 2,3,7,8-tetrachlorodibenzo-p-dioxin (TCDD) on lymphocytes of venous blood from man and a non-human primate (*Callithrix jacchus*). Arch Toxicol 65:213–219.

Neubert R, Maskow L, Triebig G, Broding HC, Jacob-Müller U, Helge H, et al. 2000. Chlorinated dibenzo-p-dioxins and dibenzofurans and the human immune system:

3. Plasma immunoglobulins and cytokines of workers with quantified moderately-increased body burdens. Life Sci 66:2123–2142.

Noller KL, Blair PB, O'Brian HA, Staskawicz MO. 1988. Increased occurrence of autoimmune disease among women exposed in utero to diethylstilberstrol. Fertil Steril 49:1080–1087.

Park HY, Hertz-Picciotto I, Petrik J, Palkovicova L, Kocan A, Trnovec T. 2008. Prenatal PCB exposure and thymus size at birth in neonates in Eastern Slovakia. Environ Health Perspect 116:104–109.

Phillips TM. 2000. Assessing environmental exposure in children: Immunotoxicology screening. J Expo Anal Environ Epidemiol 10:769–775.

Pollard KM, Hultman P, Kono DH. 2010. Toxicology of autoimmune diseases. Chem Res Toxicol 23:455–466.

Renz H, Blümer N, Virna S, Sel S, Garn H. 2006. The immunological basis of the hygiene hypothesis. Chem Immunol Allergy 91:30–48.

Rusiecki JA, Baccarelli A, Bollati V, Tarantini L, Moore LE, Bonefeld-Jorgensen EC. 2008. Global DNA hypomethylation is associated with high serum-persistent organic pollutants in Greenlandic Inuit. Environ Health Perspect 116:1547–1552.

Sarma SN, Kim YJ, Ryu JC. 2010. Gene expression profiles of human promyelocytic leukemia cell lines exposed to volatile organic compounds. Toxicology 271: 122–130.

Siegel PD, Saxena RK, Saxena QB, Ma JK, Ma JY, Yin XJ, et al. 2004. Effect of diesel exhaust particulate (DEP) on immune responses: Contributions of particulate versus organic soluble components. J Toxicol Environ Health A 67:221–231.

Smialowicz RJ, Riddle MM, Williams WC, Diliberto JJ. 1994. Effects of 2,3,7,8-tetrachlorodibenzo-p-dioxin (TCDD) on humoral immunity and lymphocyte subpopulations: Differences between mice and rats. Toxicol Appl Pharmacol 124:248–256.

Smit LA, Zuurbier M, Doekes G, Wouters IM, Heederik D, Douwes J. 2007. Hay fever and asthma symptoms in conventional and organic farmers in The Netherlands. Occup Environ Med 64:101–107.

Smithwick LA, Quensen JF, 3rd, Smith A, Kurtz DT, London L, Morris PJ. 2004. The inhibition of LPS-induced splenocyte proliferation by ortho-substituted and microbially dechlorinated polychlorinated biphenyls is associated with a decreased expression of cyclin D2. Toxicology 204:61–74.

Sobel ES, Gianini J, Butfiloski EJ, Croker BP, Schiffenbauer J, Roberts SM. 2005. Acceleration of autoimmunity by organochlorine pesticides in (NZB x NZW)F1 mice. Environ Health Perspect 113:323–328.

Stone JH. 2008. The systemic vasculitides. In: Cecil Medicine, 23rd edition (Goldman L, Ausiello D, eds.). Philadelphia, PA: Saunders Elsevier, pp. 2049–2059.

Sunyer J, Torrent M, Muñoz-Ortiz L, Ribas-Fitó N, Carrizo D, Grimalt J, et al. 2005. Prenatal dichlorodiphenyldichloroethylene (DDE) and asthma in children. Environ Health Perspect 113:1787–1790.

Svensson BG, Hallberg T, Nilsson A, Schütz A, Hagmar L. 1994. Parameters of immunological competence in subjects with high consumption of fish contaminated with persistent organochlorine compounds. Int Arch Occup Environ Health 65: 351–358.

Tan Y, Li D, Song R, Lawrence D, Carpenter DO. 2003. Ortho-substituted PCBs kill thymocytes. Toxicol Sci 76:328–337.

Tan Y, Chen C-H, Lawrence D, Carpenter DO. 2004. Ortho-substituted PCBs kill cells by altering membrane structure. Toxicol Sci 80:54–59.

Tsai PC, Ko YC, Huang W, Liu HS, Guo YL. 2007. Increased liver and lupus mortalities in 24-year follow-up of the Taiwanese people highly exposed to polychlorinated biphenyls and dibenzofurans. Sci Total Environ 374:216–222.

Tulinska J, Dusinska M, Jahnova E, Liskova A, Kuricova M, Vodicka P, et al. 2000. Changes in cellular immunity among workers occupationally exposed to styrene in a plastics lamination plant. Am J Ind Med 38:576–583.

ten Tusscher GW, Steerenberg PA, van Loveren H, Vos JG, von dem Borne AE, Westra M, et al. 2003. Persistent hematologic and immunologic disturbances in 8-year-old Dutch children associated with perinatal dioxin exposure. Environ Health Perspect 111:1519–1523.

Van Den Heuvel RL, Koppen G, Staessen JA, Hond ED, Verheyen G, Nawrot TS, et al. 2002. Immunologic biomarkers in relation to exposure markers of PCBs and dioxins in Flemish adolescents (Belgium). Environ Health Perspect 110:595–600.

Van Loveren H, Ross PS, Osterhaus AD, Vos JG. 2000. Contaminant-induced immunosuppression and mass mortalities among harbor seals. Toxicol Lett 112–113: 319–324.

Vezina CM, Walker NJ, Olson JR. 2004. Subchronic exposure to TCDD, PeCDF, PCB126, and PCB153: Effect on hepatic gene expression. Environ Health Perspect 112:1636–1644.

Voie OA, Tysklind M, Andersson PL, Fonnum F. 2000. Activation of respiratory burst in human granulocytes by polychlorinated biphenyls: A structure-activity study. Toxicol Appl Pharmacol 167:118–124.

Vorderstrasse BA, Steppan LB, Silvertone AE, Kerkvliet NI. 2001. Aryl hydrocarbon receptor-deficient mice generate normal immune response to model antigens and are resistant to TCDD-induced immune suppression. Toxicol Appl Pharmacol 171: 157–164.

Vos JG, van Loveren H. 1995. Markers for immunotoxic effects in rodents and man. Toxicol Lett 82–83:385–394.

Vos JG, Moore JA, Zinkl JG. 1973. Effect of 2,3,7,8-tetrachlorodibenzo-p-dioxin on the immune system of laboratory animals. Environ Health Perspect 5:149–164.

Weisglas-Kuperus N, Patandin S, Berbers GA, Sas TC, Mulder PG, Sauer PJ, et al. 2000. Immunologic effects of background exposure to polychlorinated biphenyls and dioxins in Dutch preschool children. Environ Health Perspect 108:1203–1207.

Weisglas-Kuperus N, Vreugdenhil HJ, Mulder PG. 2004. Immunological effects of environmental exposure to polychlorinated biphenyls and dioxins in Dutch school children. Toxicol Lett 149:281–285.

Williamson DM, White MC, Poole C, Kleinbaum D, Vogt R, North K. 2006. Evaluation of serum immunoglobulins among individuals living near six Superfund sites. Environ Health Perspect 114:1065–1071.

Wolstenholme JT, Edwards M, Shetty SR, Gatewood JD, Taylor JA, Rissman EF, et al. 2012. Gestational exposure to bisphenol A produces transgenerational changes in behaviors and gene expression. Endocrinology 153:3828–3838.

Yilmaz B, Sandal S, Chen CH, Carpenter DO. 2006. Effects of PCB 52 and PCB 77 on cell viability, [Ca(2+)](i) levels and membrane fluidity in mouse thymocytes. Toxicology 217:184–193.

Yu ML, Hsin JW, Hsu CC, Chan WC, Guo YL. 1998. The immunologic evaluation of the Yucheng children. Chemosphere 37:1855–1865.

Zhang XD, Siegel PD, Lewis DM. 2002. Immunotoxicology of organic acid anhydrides (OAAs). Int Immunopharmacol 2:239–248.

CHAPTER 14

Exposures to Organic Pollutants and Respiratory Illnesses in Adults and Children

KRASSI RUMCHEV, JEFF SPICKETT, and JANELLE GRAHAM

ABSTRACT

Background: During the last decades, concern has increased about adverse health effects resulting from exposure to indoor organic pollutants. As a class of contaminants, organic air pollutants indoors are extremely diverse and the range of its sources is extensive, including building materials and household and personal products. There have been numerous reports of increased prevalence of respiratory symptoms, including asthma among adults and children, due to elevated indoor concentrations of organic pollutants.

Objectives: In this chapter we conducted a review of the current literature with the aim of providing additional insights into the association between organic pollutants and respiratory health among adults and children.

Discussion: It is concluded that indoor air pollution, in particular, organic contaminants, may pose a significant risk for our respiratory health, and the potential health impacts of building design may require greater recognition and further research.

Conclusion: Exposure to organic chemical pollutants is likely to continue with newer compounds being introduced. Therefore, an approach to risk management must respond in a timely and effective manner.

Effects of Persistent and Bioactive Organic Pollutants on Human Health, First Edition.
Edited by David O. Carpenter.

INTRODUCTION

It is a fundamental human right that all people should have free access to air of acceptable quality. However, in our everyday lives, people become exposed to various air pollutants emitted from both natural and man-made sources. The impact of air pollution is extensive in humans, and the neurological, dermal, and pulmonary deposition and absorption of inhaled chemicals can have direct consequences to our health. The relative contribution of emission sources and exposure to air pollution may vary according to regional and lifestyle factors, but for some pollutants such as organic compounds, indoor air pollution is likely to be of greater health significance than outdoor air pollution.

Given the amount of time people spend indoors, investigations on indoor air environment have been an increasingly important area of research over the last 30–40 years. It has been estimated that people in developed nations spend up to 90% of their time indoors; mainly at home, at the workplace, day care centers, retirement homes, entertainment and shopping centers, and at schools (Brasche and Bischof 2005; Klepeis et al. 2001). Problems associated with indoor air quality (IAQ) are recognized as important risk factors for human health, especially among certain population groups that may be particularly vulnerable including young children, the elderly, and those with predisposing medical conditions (WHO 2010). Children require special protection because they are more vulnerable to the effects of environmental hazards. They receive greater exposure per unit of body weight than adults, and they are more susceptible to these effects because of the immature and developing systems.

While outdoor air pollution has been largely studied due to popular and high interest issues such as health impacts from climate change and growing urbanization, IAQ is increasingly becoming a focal point from the scientific community. In a recent report, WHO (2010) stated that research assessing health effects associated with indoor air pollution has lagged behind that of outdoor air pollution.

One of the first studies related to IAQ was undertaken in 1965, when Biersteker and colleagues measured nitrogen dioxide in homes and realized that exposure levels to this pollutant were higher in homes with gas-fired combustion devices (Biersteker et al. 1965). However, it was in the late 1970s and early 1980s when studies identified the presence of organic compounds at high concentrations indoors (Berglund et al. 1982; Molhave 1979). A few years later, links between volatile organic compounds (VOCs) and sensory health effects were reported by Mølhave et al. (1986). These health effects were similar to those included in the World Health Organization (WHO) definition of sick building syndrome (SBS) and VOCs emerged as a possible cause of SBS (WHO 1989). Hence, the interest in studying indoor organic compounds and potential health effects has continued to be of interest to this day.

One of the main health effects associated with exposure to organic compounds includes asthma and respiratory symptoms whose prevalence has been increasing since the mid 1970s. Phase III of the International Study of Asthma and Allergies in Childhood (ISAAC) study (Pearce et al. 2007) indicated that, although in the last few years the international differences in asthma symptom prevalence have reduced, the global burden of asthma is continuing to rise. The cause of the increased prevalence has not been established yet. Genetic factors appear to be an important risk factor for asthma development, but it cannot be responsible for the observed increased prevalence. Therefore, interactions between genetic and environmental factors are most likely to explain the burden of asthma worldwide (Weisel 2002). Wheezing, which is one of the most common respiratory causes of medical consultation and hospital admission during the first years of life, has also been reported to be associated with several genetic and environmental factors (Belanger et al. 2003; Emenius et al. 2004). The environmental factors include perinatal tobacco smoke exposure, housing conditions, mold stains on household walls, and hazardous air pollutants (HAPs). Exposure to some HAPs has been associated with cancer, but others have noncancer end points including effects on the respiratory system (Caldwell et al. 1998; Morello-Frosch et al. 2000). According to Caldwell and colleagues, the HAPs with noncancer, acute end points include oxygenated VOCs. This is consistent with the findings of a study of Kjaergaard et al. (1991) conducted more than a decade ago which showed that indoor exposures to mixtures of VOCs, including many aromatic and chlorinated organic compounds, could irritate the mucus membrane in the respiratory tract in both healthy and sensitive individuals.

Organic compounds are defined as chemical compounds containing at least one carbon atom and hydrogen atom in their molecular structure and they occur in a greater number of forms when compared with inorganic pollutants. In one of the first detailed studies of exposure to organic pollutants, Mølhave (1982) identified between 23 and 32 organic compounds indoors, which included a large variety of hydrocarbons, cycloalkanes, terpenes, alcohols, glycols, aldehydes, ketones, halocarbons, acids, and esters.

In this chapter, we discuss the nature of organic pollutants, which are ubiquitous in indoor environments, and also the evidence for their association with respiratory symptoms and asthma among adults and children.

VOLATILE ORGANIC COMPOUNDS

VOCs are some of the most common air pollutants found indoors (Brown et al. 1994; Maroni et al. 1995). Over the past 40 years, the home environment has changed with the introduction of soft furniture, fitted carpets, air conditioning, and central hearing. Since the energy crisis in the 1970s, people have

TABLE 14.1. WHO Classification System for Organic Indoor Pollutants

Group	Boiling Point (°C)
VVOC—very volatile organic compound	<0 to 50–100
VOC—volatile organic compound	50–100 to 240–260
SVOC—semivolatile organic compound	240–260 to 380–400
POM—organic compounds associated with particulate matter	>380

Adapted from Maroni et al. 1995

attempted to increase energy efficiency, which resulted in airtight buildings with lower rates of natural ventilation. All these transformations have led to significant increased concentrations of air contaminants indoors including those of VOCs (Hyndman et al. 1994; Klepeis et al. 2001).

Physicochemical Properties

VOC contaminants are classified into subgroups in several ways. The WHO Classification System (1989) is based on boiling points and is presented in Table 14.1.

The relatively low boiling point allows evaporation or sublimation of VOCs from solid phase into the air and this characteristic is sometimes referred to as "off-gassing." VOCs exist mainly in the gas phase in temperature and humidity ranges encountered indoors.

In 2001, Wolkoff and Nielsen (2001) made suggestions for alternatives to the VOC classification and considered other properties than boiling point being more relevant with regard to health effect evaluation. Thus, a range of other organic compounds was encompassed that included a boiling point range from 0 to 400°C and molecular weights <500–1000 Da.

Occurrence and Sources

The most significant sources of exposure to most VOCs are identified indoors. Research has identified levels of about a dozen common organic pollutants to be two to five times higher inside buildings than outside, regardless of whether the buildings are located in rural or highly industrial areas (Brown 2002; Ilgen et al. 2001; Lai 2003). Fourteen different studies indicated higher indoor levels for 48 of 49 VOCs with the indoor : outdoor ratios ranging from 2 to 80 (Brown et al. 1994). In a study of the implications for exposure research and environmental policy in Europe (EXPOLIS) conducted in six European cities, exposures to 30 VOCs from personal, indoor, and outdoor sources were measured and in almost all cases, the concentrations of the VOCs were higher indoors than outdoors, with the most significant exposure measured inside of homes (Saarela et al. 2003).

Almost any material in a building can release organic compounds into the air and samples of VOCs may contain up to hundreds of different organic compounds.

Maroni et al. (1995) classified the sources of VOCs into major groups, which included

- consumer and commercial buildings
- paints and associated supplies
- pesticides
- cosmetic care products
- automotive products
- furnishing and clothing
- building materials
- heating, ventilation, and air-conditioning systems
- personal sources
- outdoor sources.

Brown et al. (1994), reported that the mean concentrations of *individual* VOCs in established buildings (greater than 3 months old) are generally below 50 μg/m^3, with most falling below 5 μg/m^3, while concentrations of VOCs in new buildings (less than 3 months) are considerably higher than in established buildings. The concentrations of individual compounds measured by Brown et al. (1994) in a new building was up to 100 μg/m^3, while the total volatile organic compound (TVOC) concentrations were significantly higher at approximately 4000 μg/m^3. Newer building constructions use modern building materials, which may pose another hazard by introducing new compounds not found in older or established buildings. Several studies were conducted involving established buildings with reported occupant complaints. Although the nature and incidence of such complaints were not investigated, concentrations of some individual VOCs were greater when compared with buildings with no complaints (Bayer and Black 1988; Berglund et al. 1990; Hodgson et al. 1989; Norbäck et al. 1990; Zweers et al. 1990). Further, in a more recent study conducted in America, cooking activities, air freshener use, and mothball storage were identified as predictors for emission rates of VOCs in urban homes of Chicago (Van Winkle and Scheff 2001).

The concentration and rate of VOC emission found in indoor air are dependent on various physical factors including temperature, relative humidity, and ventilation rate. According to the American Society of Heating, Refrigeration, and Air Conditioning Engineers (ASHRAE), temperatures above the thermal comfort range of 23–26°C might increase the emission rate of VOCs (Bayer and Black 1988). The VOC emission rate from building products is also influenced by the total amount of constituents in the material, age, and surface area of materials, and also chemical reactions in the source (Spengler et al. 2001). Wet materials such as paints and adhesives emit most of their VOCs in the first few hours or days following application, while dry building products such as floor and ceiling products and pressed wood products are likely to emit VOCs at low levels for longer periods, usually from weeks to years (Tichenor and Sparks 1996). Porous materials such as soft furnishings can absorb VOCs emitted from other sources, subsequently acting as secondary sources of VOCs and potentially prolonging the exposure (Hodgson and Levin 2003). Newer constructed homes and buildings are often built for energy efficiency, which decreases the air exchange rate and retains the buildup of pollutants leading to a higher indoor : outdoor ratio of VOC concentrations.

Health Effects

The chemical diversity of VOCs is reflected in the diversity of health effects that an individual organic compound may cause to people. These effects range from sensory, irritation, and allergic effects to respiratory illnesses and carcinogenic effects (Boeglin et al. 2006; Maroni et al. 1995; Mølhave 2003; Rumchev et al. 2004). Table 14.2 summarizes the studies conducted among adults and children in relation to indoor exposures to VOCs and respiratory symptoms. One of the first studies on VOCs and associated adverse health effects was conducted in Canada in the late 1980s by Hosein et al. (1989); however, due to lack of evidence, no conclusions were made. Some years later, Ware et al. (1993) reported an association between exposure to VOCs and respiratory symptoms in children. Another study involving children and conducted in Leipzig, Germany, showed that higher concentrations of styrene and benzene were associated with increased risk of pulmonary and airway infections, respectively (Diez et al. 2000). A more recent study by Delfino et al. (2003) found a significant association between exposure of toluene, xylene, and benzene and increased prevalence of respiratory symptoms in children. This result was consistent with the findings of Kim et al. (2005). In a case-control study conducted by Rumchev et al. (2004), the results indicated that children with asthma were exposed to significantly higher indoor concentrations of benzene, toluene, dichlorobenzene, ethylbenzene, and TVOCs. In addition, this study demonstrated that exposure to selected organic compounds was a significant risk factor for the development of childhood asthma.

Respiratory illnesses in adults have also been associated with exposures to different VOCs and have been reported in numerous studies (Arif and Shah 2007; Elliott et al. 2006; Pappas et al. 2000). A study in Norway (Norbäck et al. 1995) established an association between nocturnal breathlessness and decreased lung function in adults and exposure to VOCs. Later in 1997,

TABLE 14.2. Studies Investigating the Association between VOC Exposures and Respiratory Health in Adults and Children

Study	Health Outcome	Year and Country of Study
Hosem et al. *Int J Epidemiol*	Respiratory symptoms in children	1989, Canada
Ware et al. *Am J Epidemiol*	Chronic respiratory symptoms, irritant effects in children	1993, United States
Norbäck et al. *Occup Environ Medicine*	Nocturnal breathlessness and decreased lung function in adults	1995, Norway
Weislander et al. *Int Arch Occup Environ Health*	Inflammation reactions and asthma in adults	1997, Norway
Diez et al. *Int J Hyg Environ Health*	Pulmonary and airway infection in children	2000, Germany
Pappas et al. *Int J Occup Environ Health*	Lower and upper respiratory symptoms in adults	2000, United States
Delfino et al. *J Espo Anal Environ Epidemiol*	Respiratory symptoms and asthma in children	2003, United States
Rumchev et.al. *Thorax*	Respiratory symptoms and asthma in children	2004, Australia
Kim et al. *Yonsei Med J*	Asthma in children	2005, Korea
Elliot et al. *Environ Health Perspective*	Reduced pulmonary function in adults	2006, United States
Arif and Shah *Int Arch Occup Environ Health*	Adverse respiratory effects, including asthma in adults	2007, United States

Wieslander et al. (1997) found that VOC exposures caused inflammatory reactions in the airways. In more recent studies in America conducted by Elliott et al. (2006) and Arif and Shah (2007), adverse respiratory effects, including asthma, were related to exposures to VOCs.

FORMALDEHYDE

Formaldehyde is a reactive VOC that may induce airway irritation even at low concentrations. Although it is a volatile compound, formaldehyde is considered and studied separately due to differences in measurement techniques and methods for analysis that applied to other VOCs.

Physicochemical Properties

Formaldehyde is a colorless and pungent-smelling gas and is one of the most common aldehydes found in the indoor environment. The International Agency for Research on Cancer has designated formaldehyde under Group 1, as a known human carcinogen (IARC 2004). Formaldehyde is soluble in water and decomposes into methanol and carbon monoxide at temperatures above 150°C.

Occurrence and Sources

Similar to VOCs, the main sources of formaldehyde are found in the indoor environment. Some of the most common sources include cigarette smoke, urea-formaldehyde resins, and wooden products such as particle board, new carpet, or furniture (Maroni et al. 1995). Combustion sources, cosmetics, paints, and building materials made with adhesives such as plastic surfaces are also considered as significant sources of formaldehyde. Indoor climate can affect the concentrations of formaldehyde indoors as higher ventilation rates can decrease the exposure levels of this pollutant, but when the temperature and relative humidity rise, more formaldehyde is released from the source (Godish 2001). Therefore, indoor formaldehyde levels can change with seasons and also throughout the day. Concentrations of formaldehyde may be high on hot days and low on cool and dry days.

Health Effects

Similar to VOCs, exposure to formaldehyde can affect people differently as some people are more sensitive than others. It may cause eye, nose, and throat irritation, and in some people, in higher concentrations (above 0.1 ppm), may cause breathing difficulties, or may trigger respiratory attacks among people with asthma (Norbäck et al. 1995; Rumchev et al. 2004; Wieslander et al. 1997). One of the first studies demonstrating association between asthma-like symptoms and formaldehyde was published in by Helwig (1977). Later in 1990, Samet reported a relationship between induced airway irritation and formaldehyde. In 1995, Norbäck and colleagues showed that nocturnal breathlessness, bronchial hyperresponsiveness, and asthma-like symptoms among adults were associated with indoor exposures to formaldehyde. Similar findings were published 2 years later in a study by Wieslander et al. (1997). Similar to VOCs, formaldehyde has also been associated with respiratory health in children. A significantly higher prevalence of asthma and chronic bronchitis was recorded among those children exposed to increased levels of formaldehyde compared to those exposed to lower formaldehyde concentrations (Krzyzanowski et al. 1990). Garrett et al. (1999) found a significant relationship between atopy in children and higher formaldehyde exposures. Some years later, in a case-control study conducted in Australia, Rumchev et al. (2002) established formaldehyde as a risk factor for the development of asthma in young children, aged between 6 months and 3 years A year later, Venn et al. (2003) showed that formaldehyde exposures increased prevalence of respiratory symptoms among children with persistent wheeze.

Although the data are still limited, there is enough evidence to demonstrate that concentrations of VOCs and formaldehyde must be retained low in the indoor environment in order to protect adults and children from adverse respiratory effects.

PESTICIDES

Pesticides are one of the broad areas of concern in relation to IAQ due to their physical, chemical, volatile properties and their common usage in and around buildings.

Physicochemical Properties

There are hundreds of pesticide compounds in use in domestic environments that fall into the following groups of chemical compounds including organophosphates, pyrethroids, and carbamates. The organochlorine group of pesticides has been banned in most developed countries as they are very persistent in the environment. Many pesticides are considered as semivolatile organic compounds (SVOCs).

Exposure and Occurrence

Studies have indicated that for pesticides typically used indoors, the levels are higher than levels found outside (Maroni et al. 1995). There are approximately 20,000 different household pesticides products which include 300 active ingredients and 1700 so-called inert ingredients. Pesticide products used indoors include a large variety of substances and applications including insecticides to control cockroaches, flies, ants, spiders, and moths. It has also been shown that there are seasonal variations in levels for many pesticides, which is a complex process and is influenced by variables such as temperature, humidity, pesticide usage patterns, the physiochemical properties of the compound, and the physical activities in the home (Godish and Rouch 1986).

Health Effects

Young children are more susceptible to the adverse effects of exposure to many environmental contaminants as a result of their physiological characteristics.

Research has shown that pesticide residues can persist inside houses both in the air and on surfaces after application. For chlorpyrifos, a broad spectrum of organophosphorus (OP) insecticides formulated to control ants, spiders, ticks, lice, fleas, and cockroaches, has been shown that residues can persist for 2 weeks after a single application. This could lead to young children being exposed to levels well in excess of the U.S. Environmental Protection Agency (USEPA)-recommended exposure levels (Fenske et al. 1990; Gurunathan et al. 1998). Pesticide exposures have been associated with respiratory symptoms even if they do not live in an agricultural environment (Senthiselvan et al. 1992; Sprince et al. 2000).

As many of the pesticides used are semivolatile, it is likely that much of the exposures are as a result of disturbed settled particles.

There has been a shift in home pesticide use from SVOCs, such as chlorpyrifos to pyrethroids, which are essentially nonvolatile and suggest that inhalation exposure may be less relevant compared to direct and indirect exposures (Stout and Mason 2003).

POLYNUCLEAR AROMATIC HYDROCARBONS (PAHs)

PAHs are a large group of organic compounds containing two or more aromatic (benzene) rings joined together in a variety of different arrangements. They are lipophilic and are not very soluble in water. Many of the PAH compounds are known to be carcinogenic (NRC 1983). The main sources of PAH indoors are combustion activities such as smoking, heating, cooking, and tobacco smoke. In addition, there can be intrusion from outdoor sources most commonly from motor vehicles (Hildemann et al. 1991). PAHs are distributed between the vapor phase and particulate phase in the air with the higher-molecular-weight compounds being more associated with particles (Lu et al. 2008). Indoor air concentrations of PAH compounds have been determined in homes in the United States (Barro et al. 1991) and the results indicated that in homes where smoking occurred, the PAH concentrations were dominated by tobacco smoke. A more recent study in the United States has indicated that the use of mothballs was associated with the presence of a low-molecular-weight PAH, mainly naphthalene (Van Winkle and Scheff 2001). It was also reported that cooking food, mainly frying, was also associated with PAH exposure.

FRAGRANCED CONSUMER PRODUCTS

Scented consumer goods can often emit hundreds of VOCs, which can be classified as toxic or hazardous by federal laws (Steinemann et al. 2010). These products include items with a perfume or scent that might be found in common everyday products such as air fresheners, candles, deodorizers, laundry detergents, fabric softeners, dishwashing detergents, hand sanitizers, personal care products, baby shampoo, and cleaning supplies. Those goods described as "green" or "organic" may also emit as many VOCs as standard goods. Although relatively little is known about the composition of many of these products due to disclosure laws, there is much deliberation about the safety of these chemicals in indoor environments due to the widespread and sheer scale of use of these products.

PRACTICAL SOLUTIONS FOR REDUCING ORGANIC POLLUTANTS INDOORS

Over the past 40 organic compounds emerged as a potential health risk, which mainly is a reflection of people's lifestyle in developed nations. Despite the existing evidence in relation to exposure of organic compounds and the associated potential negative health effects, community awareness and knowledge of what exactly these compounds are, how people become exposed to them, and ways to protect themselves are relatively low. The most common public perception of air pollution is associated with emissions related to industries and motor vehicles. Images of indoor environments where people spend most of their time, such as homes, offices, schools, and shopping centers, rarely come to mind when thinking of poor air quality.

The use of chemicals containing organic compounds is likely to continue into the future and, therefore, our approach to exposure management must respond in a timely and effective manner. This requires a combined effort in sectors such as government, construction, industry, and professionals involved in health and planning, as well as individual consumer choices. By increasing public awareness of the potential harmful health effects of organic chemicals, and in particular VOCs found in indoor environments, a momentum of change can be encouraged for the demand and adaption of measures to ensure better IAQ.

Basic measures can be taken to prevent and reduce potential harmful exposures found in the home, officc, and public buildings. Although it is impossible to prevent all organic indoor pollutants, the prevention of as many sources as possible is paramount in reducing exposure. Activities involved in the construction of new buildings and renovations have implications for using newer products with low to minimal VOC emissions. Appropriate design and planning can provide the opportunity to increase aspects such as ventilation, taking into account outdoor sources that may compromise IAQ. Consumer choices for sourcing and purchasing personal care products and cleaning agents in the home and in business will also contribute to overall air quality.

Humans have the capacity to respond, adapt, and recover in variable degrees from complex chemically polluted environments, but not under all circumstances (Lebowitz 1996). The potential for adaption in humans to exposures in a low, mixed environment is still poorly understood, varying widely with the susceptibility of the individual. As continued research improves understanding of the links between VOCs and health effects, it is paramount that this is translated to better legislation, education, and communication strategies to the wider public. Benefits realized are not only to the immediate and long-term health impacts, but also in regard to financial and improved amenities in the community.

REFERENCES

Arif AA, Shah SM. 2007. Association between personal exposure to volatile organic compounds and asthma among US adult population. Int Arch Occup Environ Health 80:711–719.

Barro R, Regueiro J, Llompart M, Garcia-Jares C. 1991. Sampling polycyclic aromatic hydrocarbons and related semi-volatile organic compounds in indoor air. Indoor Air 4:513–521.

Bayer CW, Black MS 1988. Indoor air quality evaluations of three office buildings, In: Proceedings of ASHRAE Conference IAQ 88, "Engineering Solutions to Indoor Air Problems," Atlanta, ASHRAE, pp. 294–316.

Belanger K, Beckett W, Triche E, Bracken MB, Holford T, Ren P, et al. 2003. Symptoms of wheeze and persistent cough in the first year of life: Associations with indoor allergens, air contaminants, and maternal history of asthma. Am J Epidemiol 158:195–202.

Berglund B, Johansson I, Lindvall T. 1982. A longitudinal study of air contaminants in a newly built preschool. Environ Int 8:111–115.

Berglund B, Johansson I, Lindvall T, Lundin L 1990. A longitudinal study of airborne chemical compounds in a sick library building. In: Proceedings of Indoor Air '90, Ottawa, Canada Mortgage and Housing Corporation, Vol. 2 (Walkinshaw DS, ed.) p. 677482.

Biersteker K, DeGraaf H, Nass CA. 1965. Indoor air pollution in Rotterdam homes. Air Water Pollut 9:343–350.

Boeglin ML, Wessels D, Henshel D. 2006. An investigation of the relationship between air emissions of volatile organic compounds and the incidence of cancer in Indiana counties. Environ Res 100:242–254.

Brasche S, Bischof W. 2005. Daily time spent indoors in German homes—Baseline data for the assessment of indoor exposure of German occupants. Int J Hyg Environ Health 208:247–253.

Brown SK. 2002. Volatile organic pollutants in new and established buildings in Melbourne, Australia. Indoor Air 12:55–63.

Brown SK, Sim MR, Abramson MJ, Gray CN. 1994. Concentrations of VOCs in indoor air—A review. Indoor Air 4:123–134.

Caldwell JC, Woodruff TJ, Morello-Frosch R, Axelrad DA. 1998. Application of health information to hazardous air pollutants modeled in EPA's cumulative exposure project. Toxicol Ind Health 14:429–454.

Delfino RJ, Gong H, Linn WS, Hu Y, Pellizzari ED. 2003. Respiratory symptoms and peak expiratory flow in children with asthma in relation to volatile organic compounds in exhaled breath and ambient air. J Expo Anal Environ Epidemiol 13:348–363.

Diez U, Kroessner T, Rehwagen M, Richter M, Wetzig H, Schulz R, et al. 2000. Effects of indoor painting and smoking on airway symptoms in atopy risk children in the first year of life results of the LARS-study. Leipzig Allergy High-Risk Children Study. Int J Hyg Environ Health 203:23–28.

Elliott L, Longnecker MP, Kissling GE, London SJ. 2006. Volatile organic compounds and pulmonary function in the Third National Health and Nutrition Examination Survey, 1988–1994. Environ Health Perspect 114:1210–1214.

Emenius G, Svartengren M, Korsgaard J, Nordvall L, Pershagen G, Wickman M. 2004. Indoor exposures and recurrent wheezing in infants: A study in the BAMSE cohort. Acta Paediatr 93:899–905.

Fenske RA, Black KG, Elkner KP, Lee CL, Menthner MM, Soto R. 1990. Potential exposure and health risk of infants following indoor pesticide application. Am J Public Health 80:589–693.

Garrett M, Hooper M, Hooper B, Rayment PR, Abramson MJ. 1999. Increased risk of allergy in children due to formaldehyde exposure in homes. Allergy 54: 330–337.

Godish T. 2001. Indoor Environmental Quality. New York: CRC Press.

Godish T, Rouch J. 1986. Mitigation of residential formaldehyde by indoor climate control. Am Ind Hyg Assoc J 47:792–797.

Gurunathan S, Robson M, Freeman N, Buckley B, Roy A, Meyer R, et al. 1998. Accumulation of chlorpyrifos on residential surfaces and toys accessible to children. Environ Health Perspect 106:9–16.

Helwig H. 1977. How safe is formaldehyde. Dtsch Med Wochenschr 102:1612–1613.

Hildemann LM, Markowski GR, Cass GR. 1991. Chemical composition of emissions from urban sources of fine organic aerosols. Environ Sci Technol 25:744–759.

Hodgson A, Levin H 2003. Volatile Organic Compounds in Indoor Air: A Review of Concentrations Measured in North America Since 1990. Lawrence Berkerly National Laboratory.

Hodgson AT, Daisey JJ, Grot RA 1989. Sources and strengths of volatile organic compounds in a new office building. In: Proceedings of 82nd Annual Meeting of the Air Waste Management Association, Pittsburg, 89: pp. 80–87.

Hosein HR, Corey P, Robertson J. 1989. The effect of domestic factors on respiratory symptoms and FEV1. Int J Epidemiol 18:390–396.

Hyndman SJ, Brown DL, Ewan PW, Higenbottam TW, Maunder JW, Williams DRR. 1994. Humidity regulation in the management of asthma patients sensitized to house dust mites. Q J Med 87:367.

IARC. 2004. IARC Monographs on the Evaluation of Carcinogenic Risks to Humans.

Ilgen E, Karfich N, Levsen K, Angerer J, Schneider P, Heinrich J, et al. 2001. Aromatic hydrocarbons in the atmospheric environment: Part I: Indoor versus outdoor sources, the influence of traffic. Atmos Environ 35:1235–1252.

Kim JH, Kim JK, Son BK, Oh JE, Lim DH, Lee KH, et al. 2005. Effects of air pollutants on childhood asthma. Yonsei Med J 46:239–244.

Kjaergaard SJ, Molhave L, Pedersen OF. 1991. Human reactions to a mixture of indoor air volatile organic compounds. Atmos Environ 25a:1417–1426.

Klepeis N, Nelson WC, Ott WR, Robinson JP, Tsang AM, Switzer P. 2001. The national human activity pattern survey (NHAPS): A resource for assessing exposure to environmental pollutants. J Expo Anal Environ Epidemiol 11:231–252.

Krzyzanowski M, Quackenboss J, Lebowitz M. 1990. Chronic respiratory effects of indoor formaldehyde exposure. Environ Res 52:117–125.

Lai HK 2003. Modelling of the Determinants of Indoor $PM_{2.5}$ and 38 VOCs of the Seven Participated Cities in the EXPOLIS Study. [MPhil upgrade transfer report submitted in partial fulfilment of the requirement for PhD]. London: Imperial College.

Lebowitz M. 1996. Epidemiological studies of the respiratory effects of air pollution. Eur Respir J 9:1029–1054.

Lu H, Zhu L, Chen S. 2008. Pollution level, phase distribution and health risk of polycyclic aromatic hydrocarbons in indoor air at public places of Hangzhou, China. Environ Pollut 152:569–575.

Maroni M, Seifert B, Lindvall T. 1995. Indoor Air Quality. A Comprehensive Reference Book. Amsterdam, The Netherlands: Elsevier.

Molhave LM. 1979. The atmospheric environment in modern Danish dwellings: Measurements in 39 flats. In: Indoor Climate (Fanger PO, Valbjo O, eds.). Copenhagen: Danish Building Research Institute, pp. 171–186.

Mølhave L. 1982. Indoor air pollution due to organic gases and vapours of solvents in building materials. Environ Int 8:117–127.

Mølhave L. 2003. Organic compounds as indicators of air pollution. Indoor Air 13(Suppl 6):12–19.

Mølhave L, Bach B, Pederson OF. 1986. Human reactions to low concentrations of volatile organic compounds. Environ Int 12:167–175.

Morello-Frosch RA, Woodruff RJ, Axelrad DA, Caldwell JC. 2000. Air toxics and health risks in California: The public health implications of outdoor concentrations. Risk Anal 20:273–291.

NRC (National Research Council). 1983. Polyacyclic Aromatic Hydrocarbons: Evaluation of Sources and Effects, Commission on Life Sciences. Washington, DC: National Academy Press.

Norbäck D, Torgén M, Edling C. 1990. Volatile organic compounds, respirable dust, and personal factors related to the prevalence and incidence of sick building syndrome in primary schools. Br J Ind Med 47:733–741.

Norbäck D, Bjornsson E, Janson C, Widstrom J, Boman G. 1995. Asthmatic symptoms and volatile organic compounds, formaldehyde, and carbon dioxide in dwellings. Occup Environ Med 52:388–395.

Pappas G, Herbert R, Henderson W, Koenig J, Stover B, Barnhart S. 2000. The respiratory effects of volatile organic compounds. Int J Occup Environ Health 6: 1–8.

Pearce N, Aït-Khaled N, Beasley R, Mallol J, Keil U, Mitchell E, et al. 2007. Worldwide trends in the prevalence of asthma symptoms: Phase III of the International Study of Asthma and Allergies in Childhood (ISAAC). Thorax 62:758–766.

Rumchev K, Spickett J, Bulsara M, Phillips M, Stick S. 2002. Domestic exposure to formaldehyde significantly increases the risk of asthma in young children. Eur Respir J 20:403–408.

Rumchev K, Spickett J, Bulsara M, Phillips M, Stick S. 2004. Association of domestic exposure to volatile organic compounds with asthma in young children. Thorax 59:746–751.

Saarela K, Tirkkonen T, Laine-Ylijoki J, Jurvelin J, Nieuwenhuijsen MJ, Jantunen M. 2003. Exposure of population and microenvironmental distributions of volatile organic compound concentrations in the EXPOLIS study. Atmos Environ 37: 5563–5575.

Senthiselvan A, McDuffie HH, Dosman JA. 1992. Association of asthma with the use of pesticides. Am Rev Respir Dis 1466:844–847.

Spengler JJ, McCarthy JF, Samet JM. 2001. Indoor Air Quality Handbook. New York: McGraw-Hill.

Sprince NL, Lewis MQ, Whitten PS, Reynols SJ, Zwerling C. 2000. Respiratory symptoms: Associations with pesticides, silos and animal confinement in the IOWA Farm Family Health and Hazard Surveillance Project. Am J Ind Med 38: 455–462.

Steinemann A, MacGregor I, Gordon M, Gallagher L, Davis A, Ribiero D, Wallace L. 2010. Fragranced consumer products: Chemicals emitted ingredients unlisted. Environ Impact Assess Rev 31(3):328–333.

Stout DM, Mason MA. 2003. The distribution of chlorpyrifos following a crack and crevice type application in the USEPA Indoor Air Quality Research House. Atmos Environ 37:5539–5549.

Tichenor B, Sparks LE. 1996. Managing exposure to indoor air pollutants in residential and office environments. Indoor Air 6:259–270.

Van Winkle MR, Scheff PA. 2001. Volatile organic compounds, polycyclic aromatic hydrocarbons and elements in the air of ten urban homes. Indoor Air 11: 49–64.

Venn AJ, Cooper M, Antoniak M, Laughlin C, Britton J, Lewis SA. 2003. Effects of volatile organic compounds, damp, and other environmental exposures in the home on wheezing illness in children. Thorax 58:955–960.

Ware JH, Spengler JD, Neas LM, Samet JM, Wagner GR, Coultas D, et al. 1993. Respiratory and irritant health effects of ambient volatile organic compounds. The Kanawha County Health Study. Am J Epidemiol 15(137):1287–1301.

Weisel CP. 2002. Assessing exposure to air toxins relative to asthma. Environ Health Perspect 110(Suppl 4):527–537.

WHO (World Health Organization). 1989. Indoor air quality: Organic pollutants. EURO Reports and Studies 111, WHO Regional Office for Europe, Copenhagen, Denmark.

WHO (World Health Organization). 2010. WHO guidelines for indoor air quality: Selected pollutants. Available: http://www.euro.who.int/__data/assets/pdf_file/0009/128169/e94535.pdf [accessed May 24, 2013].

Wieslander G, Norbäck D, Björnsson E, Janson C, Boman G. 1997. Asthma and the indoor environment: The significance of emission of formaldehyde and volatile organic compounds from newly painted surfaces. Int Arch Occup Environ Health 69:115–124.

Wolkoff PI, Nielsen GD. 2001. Organic compounds in indoor air—Their relevance for perceived indoor air quality? Atmos Environ 35:4407–4417.

Zweers T, Skov P, Valbjom 0, Mdhave L. 1990. The effect of ventilation and air pollution on perceived air quality in five town halls. Energ Buildings 14:175–181.

CHAPTER 15

Cognitive Function

LIZBETH LÓPEZ-CARRILLO and MARIANO E. CEBRIÁN

ABSTRACT

Background: The impact of environmental toxicants on children's neurodevelopment is recognized as highly significant since their brain is uniquely sensitive to exposure at levels far below those known to harm adults. Alterations in neurodevelopment have led to the recognition that an early insult can initiate a cascade of changes that may not be detected structurally or functionally until later in life. Moreover, the social consequences of cognition damage include increased likelihood of school failure and diminished economic productivity.

Objective: Our primary objective is to summarize the relationships between children's cognitive function and exposure to selected organic compounds aiming to increase our current understanding of the consequences of low-level exposures.

Results: Suggestive but inconclusive evidence shows a relationship between exposure to polychlorinated biphenyls (PCBs) as well as the organochlorine insecticide dichlorodiphenyltrichloroethane (DDT) with children's cognitive impairment. Insufficient information is available in regard to polychlorinated dibenzofurans (PCDFs), polychlorinated dibenzo-*p*-dioxins (PCDDs), polybrominated diphenyl ethers (PBDEs), hexachlorobenzenes (HCBs), organophosphate pesticides (OPs), polycyclic aromatic hydrocarbons (PAHs), benzene, bisphenol A (BPA), and phthalates.

Conclusions: The effects of early exposures to environmental chemicals and children's neurodevelopment is a challenging field of research that has been constrained by methodological issues in exposure assessment and in the types of cognitive functions evaluated, along with the incomplete characterization of key additional cofactors as diet and genetic susceptibility. Differences in exposure levels and nonmeasured concurrent exposure to other neurotoxic contaminants could account for inconsistencies across studies. Further markers of exposure need to be developed to assess the potential impact of recently synthetized compounds on human pregnancy.

Effects of Persistent and Bioactive Organic Pollutants on Human Health, First Edition.
Edited by David O. Carpenter.

INTRODUCTION

Cognitive function is an intellectual process by which a person becomes aware, perceives, or comprehends ideas; this process involves all aspects of perception, thinking, reasoning, and remembering (Mosby's Medical Dictionary 2008). Although the use of the term varies in different disciplines, it includes processes such as memory, association, concept formation, pattern recognition, language, attention, perception, mental imagery, action, problem solving, and making decisions. Attention and executive functions (EFs) are unique aspects of human cognition because they modulate all other cognitive functions; therefore, their impairment might indirectly affect other components of cognitive functioning (Boucher et al. 2009).

Cognition Development

Human brain development is an extraordinary complex process that includes cell proliferation, migration, differentiation, synaptogenesis, apoptosis, and myelination. Both the extent and temporal windows of these neurogenic processes vary within different regions of the brain, beginning early in gestation and continuing into the postnatal period, thus broadening the temporal window of vulnerability and the number of developmental processes that may be affected by exposure to xenobiotics (Rice and Barone 2000). For example, synapses, the neurobiological substrates of almost all cell-to-cell communication, grow rapidly over the first 2 years and mature through adolescence. Neurotransmitters have a special place in this regard since they are pleiotropic and play fundamentally different roles during development than during adulthood (Levitt 2003). Thus, characterization of these events is necessary to understand normal nervous system development and to elucidate how neurotoxic agents perturb this process and result in altered cell number and neural function after exposure. Diverse neurotoxicants can affect the same developmental processes depending on dose and duration, but in many cases, the effects may be mediated through different action mechanisms. In addition, the specific effects will depend not only on the timing and duration of the exposure but also on the processes affected by specific neurotoxicants. The identification of alterations in neurodevelopment induced by xenobiotics has led to the recognition that an insult can initiate a cascade of changes that may not be detected structurally or functionally until later in life. Thus, these effects may be manifested at a time distant from the critical developmental window when the exposure occurred.

There are a number of standardized clinical tests to assess the integrity of nervous system function in the developing human from infancy through the school years (Rice and Barone 2000). The Neonatal Behavioral Assessment Scale (NBAS) is available for assessing intactness of the nervous system in young infants, measuring such functions as neuromuscular and motor reflexes, reaction to sensory stimuli, habituation to repeated stimuli, and autonomic function (Lester 1984). The assessment of neurological and cognitive development may be performed in children by the use of age-adjusted scales such as the Bayley Scales of Infant Development (BSID) in very young children, which includes mental developmental index (MDI) and psychomotor developmental index (PDI), the McCarthy Scales of Children's Ability, which consists of 18 items derived from six different areas: general cognitive, verbal, perceptual performance and quantitative, memory and motor skills, as well as the Kaufman Assessment Battery for Children in older children. These instruments include tests for psychomotor, memory, verbal, perceptual, and other components, depending on the instrument and age range being assessed. In older children, the Wechsler Intelligence Scales for Children-Revised is typically used to assess cognitive function; this instrument is divided into a number of subscales to partition out various aspects of cognitive performance. The preferential-looking test (Fagan test) has been extremely useful in the assessment of the cognitive abilities of infants, taking advantage of the fact that an infant looks longer at a novel stimulus (e.g., a picture of a face) than at a familiar one, allowing to assess short-term memory. This test is better correlated with later performance on intelligence quotient (IQ) tests than sensorimotor tests. Preterm infants do less well on tests of preferential looking at 6 and 12 months of age than full-term infants (Fagan 1990). The psychometric intelligence (IQ) score is considered a measure of aptitude and a predictor of school achievement. This clinical and educational diagnostic tool is one of the most widely used end points for functional consequences to both the individual and the society (Stewart et al. 2008).

The impact of environmental toxicants on children's neurodevelopment is recognized as highly significant since the brain of infants and children is uniquely sensitive to environmental neurotoxicants at levels far below those known to harm adults (Miodovnik 2011). Infancy is also a highly vulnerable period of exposure to persistent environmental pollutants since children are proportionally more exposed, per unit of body weight, than adults (World Health Organization 2010). Moreover, the social consequences of cognition damage include increased likelihood of school failure and diminished economic productivity (Julvez and Grandjean 2009).

Although environmental levels of polychlorinated biphenyls (PCBs) and certain organochlorine pesticides, such as hexachlorobenzene (HCB), dichlorodiphenyltrichloroethane (DDT), and its primary metabolite, dichlorodiphenyl dichloroethene (DDE), are in general terms declining, the levels of other contaminants, such as phthalates, flame retardants, polycyclic aromatic hydrocarbons (PAHs), and other pesticides are on the rise. Therefore, early-life exposures to these prevalent neurotoxic contaminants still continue.

The primary objective of this chapter is to summarize the relationships between children's cognitive function and exposure to selected organic compounds with the purpose of increasing our current understanding of the potential neurodevelopmental consequences of low-level exposures to these contaminants.

PCBs, Polychlorinated Dibenzofurans (PCDFs), and Polychlorinated Dibenzo-*p*-dioxins (PCDDs)

PCBs are a family of chlorinated hydrocarbons possessing electrical insulating properties, nonflammability, chemical stability at normal temperature, and having a high boiling point. First manufactured in 1929 and banned in 1970 due to their persistence in the environment, they are still present in many old products, such as transformers, capacitors, and other electrical equipment (Miodovnik 2011). In relation with neurotoxicity in humans, impaired cognitive development, behavior damage, growth retardation, and other outcomes were observed in children whose mothers consumed PCBs and their heat-degradation products, mainly PCDFs, from the ingestion of contaminated rice oil in Taiwan during 1978–1979 (Guo et al. 2004). Therefore, concern arose in relation to potential neurodevelopment effects at prenatal low-exposure levels.

To date, 11 cohort studies looking at PCB exposure and cognitive function have been performed. In the United States, the following cohort studies have been performed: Michigan (Schartz et al. 1983), North Carolina (Rogan et al. 1986), Oswego (Darvill et al. 2000), the Collaborative Perinatal Project (CPP) (Daniels et al. 2003), and the Pregnancy, Infection, and Nutrition Babies Study. Cohorts in other countries include Nunavik in Canada (Muckle et al. 2001), Düsseldorf and Duisburg in Germany (Winneke et al. 1998), Faroe Islands in Denmark (Grandjean et al. 2001), Groningen/Rotterdam in the Netherlands (Patandin et al. 1999), and Hokkaido in Japan (Nakajima et al. 2006). In these studies, mothers were recruited before pregnancy, at birth, or prenatally. Sample populations varied from 135 in Japan to 1207 women in the CPP study. Exposures were assessed by measuring PCBs in different biological matrices: maternal blood during pregnancy, cord blood, or breast milk. The median values of PCB 153 in maternal serum, considered a sound indicator of exposure (Longnecker et al. 2003), varied from 23 in Japan to 450 (ng/g lipid basis) in the Faroe Islands (Boucher et al. 2009; Goodman et al. 2010).

Although most studies did not find significant effects of prenatal PCB exposure on the global mental development of infants (6–30 months of age), only the cohort study performed in Germany was able to detect a reduction in both mental and psychomotor indexes (Walkowiak et al. 2001). However, effects on global IQ tests were consistently observed during childhood (3–11 years of age) in several cohorts (Oswego, the Netherlands, Michigan, and Germany). An explanation for such findings might be that infant mental development measures have different psychometric properties than childhood IQ tests in terms of reliability and validity, especially for predictive validity of later cognitive functioning.

A recent study has compared PCB exposure and developmental effects in the Duisburg and Düsseldorf cohorts. Prenatal exposure in Duisburg was about two- to threefold lower than in Düsseldorf cohort. The mental and motor development at 18 and 30 months of age were negatively associated with PCB exposure in Düsseldorf's study, whereas no association was observed in the Duisburg study. These results suggested that exposure to PCBs at lower exposure levels does not impair neurodevelopment of infants in Duisberg (Wilhelm et al. 2008). These findings tentatively support the view of a recent meta-analysis indicating that results of persistent organic pollutant (POP)-related effects on neurodevelopment in infants are subtle and may resolve with time as POP exposure continues to decrease (Longnecker 2006).

A recent review (Boucher et al. 2009) concluded that the most consistent effects of prenatal PCB exposure on cognitive function observed across studies were impaired EFs, which are crucial for learning and retrieval strategies and are solicited to adapt to new testing situations. The four aspects of executive functioning considered were response inhibition, planning, set shifting, and the executive component of working memory. In several studies, adverse effects on information processing speed and visual recognition memory were also reported (Michigan, the Netherlands, and Faroe Islands cohorts). Vocabulary and verbal comprehension, both highly related to language skills, have been consistently associated with prenatal PCB exposure in all cohorts in which they were part of the study design. However, there is relatively little evidence of effects on visual-spatial abilities, episodic memory, and sustained attention. Prenatal PCB exposure was also associated with increased impulsive responding and commission errors due to impaired response inhibition and not impaired sustained attention (Stewart et al. 2005). The impairments associated with prenatal exposure were more evident in non-breast-fed and/or more socioeconomically disadvantaged children in two cohort studies (the Netherlands and Michigan), suggesting that optimal child stimulation can help compensate for subtle brain insults related to these exposures, decreasing the likelihood of observing significant effects in more advantaged subgroups of the population (Jacobson and Jacobson 2004). Thus, this suggests that parental characteristics and home environment might modulate the effects of these pollutants on cognitive development (Vreugdenhil et al. 2002).

The comparison and interpretation of human data on low-level and early-life exposure to PCBs have proved to be challenging because of methodological differences in the assessment of both exposure and cognitive function. The latter is particularly important since it is likely that prenatal PCB exposure affects a specific profile of cognitive functions in children rather than all of its aspects.

PCDFs are a family of chemicals containing one to eight chlorine atoms attached to the carbon atoms of dibenzofuran, the parent compound. There are at least 135 different types of PCDFs with varying harmful health and environmental effects. PCDDs are a family of 75 chemically related compounds commonly known as chlorinated dioxins and 2,3,7,8-tetrachlorodibenzo-*p*-dioxin (TCDD) is among the most toxic chemical compounds ever produced. PCDDs and PCDFs are produced unintentionally during many processes such as combustion in municipal and industrial incinerators, and throughout the use of certain chlorinated chemical products (e.g., phenols) and improper waste disposal. They may also be produced during high-temperature industrial processes, such as copper smelting, electrical arc furnaces in steel mills, production of metallic magnesium and refined nickel, chlorine bleaching of pulp and paper, and accidental fires or malfunction of PCB-filled transformers and capacitors, among others (ATSDR 1994, 1998). Little information is available regarding the potential association between exposure to PCDDs and polychlorinated dibenzofurans (PCDFs) and neurodevelopment in infants. Nakajima et al. (2006) examined the influence of PCBs, PCDFs, and dioxins on the mental and motor development of infants in Sapporo, Japan. Exposure was assessed by measuring these compounds in maternal blood after the second trimester of their last pregnancy. Bayley Scales of Infant Development–Revised (BSID-II) was used to assess mental and psychomotor development at 6 months of age. Their results showed that the PCDD isomer 1,2,3,4,6,7,8-heptachlorodibenzo-*p*-dioxin (HpCDD), total PCDDs, and total PCDDs/PCDFs were significantly and negatively associated with the MDI. On the other hand, PCDD isomers 1,2,3,7,8,9-hexachlorodibenzo-p-dioxin (HxCDD) and HpCDD, 2,3,7,8-tetrachlorodibenzofuran (TCDF), 1,2,3,7,8-pentaCDF, and PCDF isomer 1,2,3,6,7,8-hexaCDF were significantly and negatively associated with PDI. The total levels of PCBs and dioxins were not associated with PDI, and the total toxic equivalent values were not associated with MDI or PDI. The authors concluded that background-level exposure to several dioxin isomers during the prenatal period probably affects more the motor than the mental development of infants.

POLYBROMINATED DIPHENYL ETHERS (PBDEs)

PBDEs are chemical additives widely used as flame retardants applied to a wide array of textiles, building materials, and electronic equipment, such as computers and televisions. They are released into the environment and some of their congeners are highly persistent and have been commonly detected in human tissues (Sjödin et al. 2008). PBDEs have a chemical structure similar to thyroid hormones, most notably thyroxine (T4). Developmental neurotoxicity has been observed after prenatal exposure to PBDEs in experimental animal models (Darnerud 2008). An exploratory study performed in the Netherlands investigated the influence of prenatal exposure to organohalogen compounds, including brominated flame retardants (BDE-47, BDE-99, BDE-100, BDE-153, BDE-154, and hexabromocyclododecane) on motor, cognitive, and behavioral outcomes in 62 children aged 5–6 years. Exposure was assessed by measuring these chemicals in their mother's serum in the 35th week of pregnancy. Total, verbal, and performance intelligence were assessed using a revised short form of the Wechsler Preschool and Primary Scale of Intelligence-Revised (WPPSI-R). Brominated flame retardants correlated with worse fine manipulative abilities, worse attention, better coordination, better visual perception, and better behavior. However, it was difficult in this study to determine how many of these effects can reliably be assigned to specific contaminants (Roze et al. 2009).

A longitudinal study in New York reported inverse associations between elevated cord blood concentrations of PBDEs (congeners 47, 99, or 100) and adverse neurodevelopmental test scores at 12–48 and 72 months. Associations were significant for 12-month PDI with BDE-47; 24-month MDI with the three congeners; 36-month MDI only with BDE-100; 48-month full-scale and verbal IQ with the three congeners; 48- and 72-month performance IQ were associated only with BDE-100 (Herbstman et al. 2010). Chevrier et al. (2010) examined PBDE levels and thyroid hormones in serum collected from pregnant women during their second trimester. They observed an inverse association between PBDEs and thyroid-stimulating hormone (TSH), but not with T4. The odds of subclinical hyperthyroidism, low TSH, and normal T4 were significantly higher in women showing the highest PBDE levels, as compared to those with the lowest levels. In a recent study performed on a pregnancy cohort, primarily of African American women in North Carolina, PBDE congeners 47, 99, and 100 were significantly and positively associated with free and total thyroxine (T4) and suggested an inverse association between 4′OH-BDE-49 and total triiodothyronine levels (TT3) (Stapleton et al. 2011). Further research is needed to determine the mechanisms by which PBDEs affect thyroid hormone regulation during pregnancy and how this, in turn, affects fetus development.

Organochlorine Pesticides

These synthetic highly lipophilic chemicals persist in the environment and accumulate in the food chain and high-lipid content human tissues. DDT was widely used to control insects on agricultural crops and insects carrying diseases like malaria and typhus, but it is now used in only a few countries to control malaria (ATSDR 2002). DDE is the main metabolite of DDT, and both compounds cross the placenta and also pass into breast milk. Studies on experimental animals indicate that DDT is a neurodevelopmental toxicant, but no information on DDE is available. Animal and *in vitro* studies have shown that DDE exposure significantly alters the homeostasis of thyroid hormones (Liu et al. 2011).

In humans, perinatal exposure to DDE and children's neurodevelopment has been evaluated in prospective birth cohorts that have been developed in different countries: United States, Canada, Denmark (Faroe Islands), Mexico, the Netherlands, and Spain (Korrick and Sagiv 2008). Most of those studies enrolled women in their first gestational weeks, except for the study performed in Mexico where couples were recruited before getting married. In most studies, DDE exposure was estimated by measurements in maternal serum during pregnancy, but in some studies, levels in placenta, cord serum, and/or breast milk were determined. The magnitude of DDE exposure across studies is difficult to compare because levels were not uniformly reported in wet or lipid basis. However, in the study performed among farmworker families living in California (CHAMACOS birth cohort), DDE levels were higher than those observed in the Mexican cohort in Morelos, which allow further comparisons regarding the DDE dose and its relationship with neurodevelopment outcomes (Eskenazi et al. 2008; Torres-Sánchez et al. 2007). Several tests have been used in the aforementioned studies aiming to identify subtle changes in neurodevelopment according to children's age. Most studies did not show neonatal neurodevelopment damage due to perinatal DDE exposure using BSID, a finding that has been confirmed even applying three different neonatal tests at 1 month of age (Bahena-Medina et al. 2011). In general, some, but not all, birth cohorts have shown that prenatal DDE exposure impairs motor development in older infants and cognitive development in children (Eskenazi et al. 2008), as it was the case in the Mexican cohort (Table 15.1). Although most studies have focused on the effects of DDE, Eskenazi et al. (2006) have provided evidence suggesting that DDT may be more strongly associated with neurodevelopmental delays than DDE in young children, but further confirmation is needed in view of the implications for countries that are reconsidering or continuing the use of DDT for malarial control. Among the limitations of the studies described here is that very few have considered important covariables that determine children's neurodevelopment, such as stimulation of children at home, usually measured with the Home Observation for Measurement of the Environment scale.

Pan et al. (2009) examined the associations between lactational exposure to PCBs, DDT, and DDE and infant development at 12 months, using data from the Pregnancy, Infection, and Nutrition Babies Study (2004–2006) in North Carolina. No consistent associations were observed between lactational exposure to PCBs, DDT, and DDE through the first 12 months and the measures of infant development. However, DDE was associated only among boys with scoring below average on the gross motor scale of the Mullen Scales of Early Learning.

Thus, there is suggestive but not conclusive evidence that DDE/DDT alters neurodevelopment. Further longitudinal studies are needed to confirm that the reported damage is permanent.

HCB was widely used as a fungicide and it is not currently manufactured as a commercial product. Most of the general population is not likely to be exposed to large amounts of HCB, but some exposure is likely to occur, as many studies have detected small amounts in food and in human tissue samples. Animal and human studies have demonstrated that HCB crosses the placenta, accumulates in fetal tissues, and is also transferred in breast milk (ATSDR 2002). Studies on experimental animals have shown that HCB impairs neurological development and reduces growth and neonatal viability. The poisoning epidemic in Anatolia, Turkey, demonstrated that HCB is a neurodevelopmental toxin. However, there is still uncertainty regarding the long-term neurodevelopmental risks in humans associated with early-life and low-level exposure to HCB (Korrick and Sagiv 2008).

TABLE 15.1. Mexico Perinatal Cohort Study

Pregnancy	Prenatal DDE Exposure (ng/mL) First Trimester	Second Trimester	Third Trimester	
Neurodevelopment Test	Odds Ratio	Odds Ratio	Odds Ratio	Reference
(At 1 month of age)				Bahena-Medina et al. (2011)
Abnormal reflexes	0.98	0.86	0.92	
Graham–Rosenblith (at 1 month of age)				
Soft neurological signs	1.06	1.08	0.90	
Bayley	β	β	β	
At 1 month of age				
Psychomotor developmental index (PDI)	-0.02	0.46	0.03	
Mental developmental index(MDI)	0.14	–0.03	–0.19	
During first year of life				
PDI	–0.52[a]	0.23	0.16	Torres-Sánchez et al. (2007)
MDI	–0.06	–0.12	0.07	
12–30 months of age				
PDI	0.08	0.39	0.37	Torres-Sánchez et al. (2009)
MDI	0.05	0.02	0.56	
Pea body (36 months of age)				
Gross motor	0.60	0.95	0.99	
Fine motor	1.17	–0.09	0.88	
McCarthy (42–60 months of age)				
General cognitive index	0.73	–0.03	–2.01[a]	Torres-Sánchez (2013)
Perceptual performance	0.51	0.45	-1.02	
Quantitative	–0.74	–0.63	–2.06[a]	
Verbal	0.34	–0.21	–1.14[a]	
Memory	–0.29	–0.51	–1.26[a]	
Motor	1.06	0.70	–0.10	

[a] $p < 0.05$.

Organophosphate Pesticides (OPs)

The spraying of OPs is widely used to combat plagues not only in agricultural fields but also in residential settings. Exposure to OPs (such as chlorpyrifos, methyl parathion, malathion, diazinon, and dimethoate) or their residues may also occur via consumption of conventionally grown fruits and vegetables.

Numerous animal studies have demonstrated that prenatal or early exposure affects neurodevelopment, probably because OPs irreversibly inactivate acetylcholinesterase, an effect reported to disrupt cell replication and differentiation, synaptogenesis, and axonogenesis (Slotkin 1999). Children are also considered a high-risk group since they have lower activity and levels of enzymes that detoxify activated forms of certain OP pesticides (Holland et al. 2006).

Several studies have investigated the relationship between prenatal or child OP exposure with children's neurodevelopment. Rauh et al. (2006) studied the effects of chlorpyrifos prenatal exposure and children whose mothers had higher chlorpyrifos blood levels were more likely to experience PDI and MDI (BSID-II) delays, signs of attention deficit/hyperactivity disorder (ADHD), and pervasive developmental disorder (PDD) at 3 years of age. Eskenazi et al. (2007) investigated the relationship between maternal prenatal or child OP urinary metabolite levels (markers of OP exposure) and children's neurodevelopment. Their results showed that dialkyl phosphate (DAP) metabolite levels during pregnancy, particularly from dimethyl phosphate pesticides, were negatively associated at 24 months with MDI on the Bayley Scales and an increase in risk of maternally reported PDD. In addition, abnormal primitive reflexes, considered critical markers of neurological integrity, were positively associated with prenatal OP exposure, measured as DAP maternal urinary levels in New York City (the Mount Sinai Children's Environmental Health Cohort study) without adverse associations with PCBs or DDE blood levels (Engel et al. 2007). Recently, Engel et al. (2011) studied in their cohort the relationship between biomarkers of OP exposure, PON1 polymorphisms, and cognitive development at ages 12 and 24 months and 6–9 years. PON1 is a key enzyme in OP metabolism and is considered a biomarker of susceptibility to the toxic effects of OP. Prenatal total DAP metabolite levels were associated with a decrement in mental development beginning at 12 months and continuing through early childhood among blacks and Hispanics. These associations appeared to be enhanced among children of mothers who carried the PON1 Q192R QR/RR genotype. The authors concluded that prenatal OP exposure is negatively associated with cognitive development, particularly with perceptual reasoning, and that PON1 appears to be an important susceptibility factor for these deleterious effects. In a recent report from a birth cohort study (CHAMACOS) performed in an agricultural community in California, maternal urinary DAPs during pregnancy were associated with poorer scores for working memory, processing speed, verbal comprehension, perceptual reasoning (WISC-IV), and full-scale IQ in children at 7 years of age. Children in the highest quintile of maternal DAP concentrations had an average deficit of 7.0 IQ points compared with those in the lowest quintile. However, children's urinary DAP concentrations were not consistently associated with cognitive scores. This study suggests that prenatal, but not postnatal, urinary DAP concentrations were associated with poorer intellectual development in 7-year-olds (Bouchard et al. 2011).

PAHs AND BENZENE

PAHs, such as benzo[*a*]pyrene, are ubiquitous air pollutants released to ambient and indoor air from combustion sources such as coal-burning power plants, diesel- and gasoline-powered vehicles, home heating, and cooking and are also present in tobacco smoke and charred foods (ATSDR 1995). PAHs were shown to be neurodevelopmental toxicants in experimental studies, rising concerns about their effects on the human developing brain (Wormley et al. 2004).

In a prospective cohort study of nonsmoking African American and Dominican mothers and children in New York City, Perera et al. (2006) reported on the effects of prenatal exposure to airborne PAHs on mental and psychomotor development of children through 36 months of age. Exposure was estimated through personal air monitoring for the mothers during pregnancy. Vapors and particles of ≤2.5 μm in diameter were collected and analyzed for benz[a]anthracene, chrysene, benzo[b]fluoranthene, benzo-[k]fluoranthene, benzo[a]pyrene, indeno[1,2,3-cd]pyrene, disbenz[a,h]anthracene, and benzo[g,h,i]perylene. A composite variable was calculated from the concentrations of these intercorrelated PAHs. The BSID-II was used to assess cognitive and psychomotor development at 12, 24, and 36 months of age. Behavioral problems were measured by maternal report on the 99-item Child Behavior Checklist (CBCL) for children 1.5–5 years of age. Prenatal exposure to PAHs was not associated with psychomotor development index or behavioral problems. However, high prenatal exposure to PAHs was associated with lower mental development index at age 3, and the odds of cognitive developmental delay were also significantly greater, in addition to a significant age × PAH effect. The results suggest that environmental PAHs at levels encountered in New York City air may adversely affect children's cognitive development at 3 years of age.

Perera et al. (2009) continued to study the relationship between prenatal exposure to airborne PAHs and child intelligence by using the WPPSI-R in 249 children of 5 years of age in their cohort. Full-scale IQ and verbal IQ scores in children in the high-exposure group (above the median of 2.26 ng/m^3) were inversely associated with prenatal PAH exposure levels, providing evidence that environmental PAHs at levels encountered in New York City air adversely affect children's IQ.

Edwards et al. (2010) paralleled the studies of the New York City cohort summarized earlier in Krakow, Poland. To assess exposure to airborne PAHs, women were monitored over a 48-hour period during the second or third trimester of pregnancy. The eight airborne PAHs mentioned were determined and a composite variable was calculated. Nonverbal reasoning ability was assessed in 214 children 5 years of age using the Raven Coloured Progressive Matrices (RCPM) test. Higher prenatal exposure to airborne PAHs (above the median of 17.96 ng/m^3) was associated with decreased RCPM scores. These results were in agreement with those obtained in the New York City parallel cohort described earlier, suggesting that prenatal exposure to airborne PAHs adversely affects children's cognitive development by 5 years of age, with potential implications for school performance, since RCPM scores measured during the preschool period have been shown to correlate with academic achievement later in life.

Recently, Guxens et al. (2012) assessed in several regions of Spain the influence of antioxidant/detoxification factors on potential associations between prenatal exposure to residential air pollution and impaired infant mental development. NO_2 and benzene were measured with passive samplers and were considered markers of toxic air pollutants rather than potential causative agents. Land use regression models were developed for each pollutant to predict average outdoor air pollution levels for the entire pregnancy at each residential address. Maternal diet was obtained at the first trimester through a validated food frequency questionnaire. Infant mental development was assessed in 1889 children at 14 months of age using BSID-II. Mean exposure during pregnancy was 29.0 $\mu g/m^3$ for NO_2 and 1.5 $\mu g/m^3$ for benzene. Exposure was not significantly associated with mental development. However, strong inverse associations were observed among infants whose mothers reported low intakes of fruits/vegetables during pregnancy. The authors concluded that prenatal exposure to residential air pollutants may adversely affect infant mental development, but potential effects may be limited to infants whose mothers reported low antioxidant intakes.

PHTHALATES

Several phthalates, such as diethyl phthalate (DEP) and dibutyl phthalate (DBP), are widely used in cosmetic and personal care products for infants, children, and adults. Other phthalates, including diisobutyl phthalate (DiBP), butylbenzyl phthalate (BBzP), di(2-ethylhexyl) phthalate (DEHP), and di-n-octyl phthalate (DOP), are used as plasticizers in the manufacture of flexible vinyl plastic in consumer products, flooring and wall coverings, food contact applications, medical devices, and pharmaceutical products (Wormuth et al. 2006). Although there is strong evidence of the adverse effects of phthalates on animals, less information is available in humans and particularly on their effects on children's health (National Research Council, Committee on the Health Risks of Phthalates 2008). Concern about the possibility that environmental exposure may contribute to poor neurodevelopmental outcomes in humans has emerged from studies in rodents reporting that phthalates have toxic effects on the thyroid gland, causing decreased levels of thyroxine (T4) in plasma (Hinton et al. 1986). In addition, negative associations between urinary MBP levels and (T4) and free T4 serum levels were shown in pregnant women (Huang et al. 2007). The relationship between hypothyroidism and subsequent effects on neurodevelopment is well known (Haddow et al. 1999). These findings raised the possibility that phthalates may alter brain development similarly to other endocrine disruptors known to affect cognitive function. However, few studies have reported associations between phthalate exposure and neurological effects in humans.

In a prospective multiethnic cohort that enrolled primiparous women with singleton pregnancies (Mount Sinai Children's Environmental Health Cohort), the associations between maternal urinary phthalate metabolite concentrations in the third trimester and neonatal behavior were studied. Exposure was assessed by measuring the metabolites of seven phthalate esters in maternal urine collected between 25 and 40 gestation weeks. Metabolites were grouped into two categories defined as high molecular weight phthalates (HMWP) and low molecular weight phthalates (LMWP) according to the molecular weight of the monoesters. Most metabolites were detected in over 95% of the samples and concentrations were within the range reported for the 1999–2000 North American Health and Nutrition Examination Survey (NHANES) for U.S. adults. Behavior was assessed within 5 days of birth using the Brazelton Neonatal Behavioral Assessment Scale (BNBAS). Overall, there were few associations between individual phthalate metabolites or their molar sums and most BNBAS domains. There were significant sex–phthalate metabolite interactions for the orientation and motor domains and the overall quality of alertness score. There were strong inverse associations between HMWP metabolite concentrations and the orientation and quality of alertness scores among girls, whereas a slight positive association between LMWP concentrations and motor performance was observed among boys. These findings were consistent with the hypothesis that phthalates are hormonally active and may exert sex-specific effects (Engel et al. 2009).

Engel et al. (2010) continued to study the consequences of prenatal phthalate exposure on neurobehavioral development during childhood in their cohort by examining the association of prenatal phthalate exposure with behavior and executive functioning at 4–9 years of age. Exposure was estimated by measuring 10 phthalate metabolites in maternal urine. Mothers completed the parent report forms of the Behavior Rating Inventory of Executive Function (BRIEF) and the Behavior Assessment System for Children-Parent Rating Scales (BASC-PRS). HMWPs were not associated with most of the BASC or BRIEF domains, except that increased HMWP levels were associated with poorer scores on the adaptability scale of the BASC. In addition, higher LMWP concentrations were associated with poorer scores on the aggression, attention problems, conduct problems, depression, and externalizing problem scales, and for the overall Behavioral Symptoms Index (BSI) on the BASC. Similarly, poorer executive functioning was indicated by elevated scores on the emotional control scale and on the Global Executive Composite (GEC) index of the BRIEF. All of these effects were consistent with a dose–response gradient. These findings were consistent with domains typically affected in childhood: oppositional defiant disorder, conduct disorder, and ADHD.

Cho et al. (2010) investigated the relationships between the urinary concentrations of phthalate metabolites and children's neurocognitive functioning. The study enrolled 667 children in five South Korean cities. Exposure was estimated by measuring the urinary levels of two DEHP metabolites (mono-2-ethylhexyl phthalate [MEHP] and mono(2-ethyl-5-oxohexyl)phthalate [MEOHP]), and a DBP metabolite (mono-n-butyl phthalate [MBP]) in children. Intellectual functioning was assessed by means of the abbreviated Korean validated form of the Wechsler Intelligence Scale for Children (KEDI-WISC), which consists of vocabulary, arithmetic, picture arrangement, and block design tests. The geometric means of MEHP, MEOHP, and MBP were 21.3, 18.0, and 48.9 μg/L, respectively. The full-scale IQ and verbal IQ scores were negatively associated with DEHP metabolites but not with DBP metabolites. A significant negative relationship between DEHP and DBP metabolites and children's vocabulary subscores was also found. A negative association between increasing concentrations of MEHP and the sum of DEHP metabolites and WISC vocabulary score in boys, but not in girls, was also reported. The latter differential pattern by sex was in agreement with previous studies showing other sex-related differential effects (Hatch et al. 2008; Wolff et al. 2008).

BISPHENOL A (BPA)

BPA is a high-production synthetic monomer, weakly estrogenic, and used in the manufacture of polycarbonate plastics and epoxy resins. These materials can be found in reusable water bottles, food container linings, medical tubing, toys, compact disks, flooring, and paints, and there are indications of the dermal transferability of BPA from recycled and thermal paper. Polymers made from BPA can be hydrolyzed in high-temperature, acidic, and basic conditions, leading to BPA leaching into water and food containers and baby feeding bottles (Biedermann et al. 2010). BPA has been found in urine and serum samples from pregnant women (Geens et al. 2011; Rubin 2011). Concern about the possibility that environmental exposure may contribute to poor neurodevelopmental outcomes in humans has emerged from studies in rodents, reporting that prenatal BPA exposure was associated with cognitive deficits and that this compound may affect neonatal development through disruption of thyroid hormone pathways (Tian et al. 2010; Wolstenholme et al. 2011).

Braun et al. (2009) examined the association between prenatal BPA exposure and behavior in mothers and their children (2-year-olds) participating in the Health Outcomes and Measures of the Environment Study, an ongoing prospective birth cohort in Cincinnati, Ohio. Exposure was assessed by measuring prenatal urinary BPA concentrations (around 16 and 26 weeks of gestation and at birth) in 249 women. Levels were similar to those from women of childbearing age from a nationally representative U.S. sample. Child behavior was assessed at 2 years of age using the Behavioral Assessment System for Children (BASC-2) Parent Rating Scale for preschoolers. Composite scores for externalizing behaviors consist of hyperactivity and aggression scales, and internalizing composite scores consist of depression, anxiety, and somatization scales. The BSI is a composite score that reflects the overall level of problem behaviors and consists of scales for aggression, hyperactivity, depression, and attention. No associations between BPA concentrations and behavior scores among all children were observed. However, an association with externalizing behaviors in 2-year-old girls driven by BPA concentrations measured at 16 weeks of gestation was reported. However, a recent study in the same cohort examined the association of prenatal exposure to BPA and select common phthalates with infant neurobehavior measured at 5 weeks, suggesting that prenatal exposure to BPA was not associated with notable neurobehavioral outcomes during early infancy (Yolton et al. 2011). These findings were not consistent with previous studies of behavioral and cognitive associations with gestational exposure to BPA (Braun et al. 2009). These differences could be due to the low levels of exposure or to the nature of the outcome measured. However, significant associations with DBP and DEHP exposure only occurred with exposure measured at 26 weeks and suggested both favorable and detrimental outcomes. Increased levels of DBP metabolites were significantly associated with decreased arousal and decreased handling required, and trends toward improved regulation and movement quality. However, increased DEHP metabolites were associated with increased nonoptimal reflexes in boys only.

Exposure to all the compounds described in this chapter has been shown to interfere with thyroid function. Brain development *in utero* is dependent upon normal levels of thyroid hormones, and their absence reduces neuronal growth and differentiation in the cerebral cortex, hippocampus, and cerebellum. Approximately 25% of adult brain function is achieved by age 1, 90% by age 6, and the rest in adolescence (Mori and Todaka 2008). Epidemiological evidence has indicated that even marginally low thyroxine levels in pregnant women adversely affect cognitive functions in the offspring (Haddow et al. 1999; Pop et al. 2003). Therefore, the interference with the hypothalamic–pituitary–thyroid axis, by various mechanisms, may constitute a common pathway by which these compounds interfere with normal neurodevelopment (Porterfield 2000).

FINAL COMMENTS

The effects of early exposures to environmental chemicals and children's neurodevelopment are a challenging field of research that has been constrained by methodological issues in exposure assessment and in the types of cognitive functions evaluated, along with the incomplete characterization of dietary and genetic factors.

Differences in exposure levels and nonmeasured concurrent exposure to other neurotoxic contaminants could account for inconsistencies across studies. The insufficient information on exposure to recently synthetized compounds during human pregnancy challenges us to determine the best marker of exposure. The presence of diverse mixtures of organic compounds with different congeners/metabolites across studies may result in dissimilar outcomes. The potential interactions between contaminants complicate the observation of independent relationships with cognitive outcomes even after documenting and statistically controlling for other contaminants. For example, the quantitative interactions among PCBs, MeHg, and organochlorine pesticide exposure during childhood are still the subject of study.

Most studies on the effects of organic contaminants had relied on broad measures of global cognitive functioning (i.e., McCarthy Scales, Kaufman Assessment Battery for Children, and WISC) and were able to predict important outcomes such as school performance, but much less information is available on the effects on specific processes (e.g., attention and response inhibition). There is a great need to move beyond global cognitive measures and to identify the processes affected. For example, failure to assess specific aspects of EFs may dilute the magnitude of the effects by amalgamating affected and unaffected domains in the calculation of global IQ, potentially leading to studies reporting the absence of significant relationships between prenatal organic contaminant exposure and cognitive development. The age at assessment may also influence the outcomes of birth cohort studies because maturation of certain brain areas, particularly the prefrontal cortex, lasts until late adolescence; therefore, it is possible that some effects may appear during development, while others may disappear.

Dietary and genetic factors modulate the neurotoxicity of the contaminant(s) of interest. For example, several studies have shown that impairments associated with prenatal exposure to organic compounds were more significant in non-breast-fed or in socioeconomically disadvantaged children, suggesting the need for controlling for the appropriate vulnerability factors. Regarding genetic factors, polymorphisms in the *ACHE/PON1* locus have been studied as factors involved in the neurophysiological response to OP exposure. However, much less information is available on the interactions among other genes involved in neurodevelopment and exposure to different categories of organic contaminants.

Therefore, there is a need for establishing consensus standards for the conduct, analysis, and reporting of epidemiological studies evaluating the effects of potential neurotoxic exposures.

REFERENCES

ATSDR (Agency for Toxic Substances and Disease Registry). 1994. Toxicological Profile for Chlorodibenzofurans (CDFs). Atlanta, GA: U.S. Department of Health and Human Services, Public Health Service.

ATSDR (Agency for Toxic Substances and Disease Registry). 1995. Toxicological Profile for Polycyclic Aromatic Hydrocarbons (PAHs). Atlanta, GA: Agency for Toxic Substances and Disease Registry.

ATSDR (Agency for Toxic Substances and Disease Registry). 1998. Toxicological Profile for Chlorinated Dibenzo-p-dioxins (CDDs). Atlanta, GA: U.S. Department of Health and Human Services, Public Health Service.

ATSDR (Agency for Toxic Substances and Disease Registry). 2002. Toxicological Profile for DDT, DDE, DDD. Atlanta, GA: U.S. Department of Health and Human Services, Public Health Service.

Bahena-Medina LA, Torres-Sánchez L, Schnaas L, Cebrián ME, Chávez CH, Osorio-Valencia E, et al. 2011. Neonatal neurodevelopment and prenatal exposure to dichlorodiphenyldichloroethylene (DDE): A cohort study in Mexico. J Expo Sci Environ Epidemiol 21:609–614.

Biedermann S, Tschudin P, Grob K. 2010. Transfer of bisphenol A from thermal printer paper to the skin. Anal Bioanal Chem 398:571–576.

Bouchard MF, Chevrier J, Harley KG, Kogut K, Vedar M, Calderon N, et al. 2011. Prenatal exposure to organophosphate pesticides and IQ in 7-year-old children. Environ Health Perspect 119:1189–1195.

Boucher O, Muckle G, Bastien C. 2009. Prenatal exposure to polychlorinated biphenyls: A neuropsychologic analysis. Environ Health Perspect 117:7–16.

Braun JM, Yolton K, Dietrich KN, Hornung R, Ye X, Calafat AM, et al. 2009. Prenatal bisphenol A exposure and early childhood behavior. Environ Health Perspect 117:1945–1952.

Chevrier J, Harley KG, Bradman A, Gharbi M, Sjödin A, Eskenazi B. 2010. Polybrominated diphenyl ether (PBDE) flame retardants and thyroid hormone during pregnancy. Environ Health Perspect 118:1444–1449.

Cho SC, Bhang SY, Hong YC, Shin MS, Kim BN, Kim JW, et al. 2010. Relationship between environmental phthalate exposure and the intelligence of school-age children. Environ Health Perspect 118:1027–1032.

Daniels JL, Longnecker MP, Klebanoff MA, Gray KA, Brock JW, Zhou H, et al. 2003. Prenatal exposure to low-level polychlorinated biphenyls in relation to mental and motor development at 8 months. Am J Epidemiol 157:485–492.

Darnerud PO. 2008. Brominated flame retardants as possible endocrine disruptors. Int J Androl 31:152–160.

Darvill T, Lonky E, Reihman J, Stewart P, Pagano J. 2000. Prenatal exposure to PCBs and infant performance on the fagan test of infant intelligence. Neurotoxicology 21:1029–1038.

Edwards SC, Jedrychowski W, Butscher M, Camann D, Kieltyka A, Mroz E, et al. 2010. Prenatal exposure to airborne polycyclic aromatic hydrocarbons and children's intelligence at 5 years of age in a prospective cohort study in Poland. Environ Health Perspect 118:1326–1331.

Engel SM, Berkowitz GS, Barr DB, Teitelbaum SL, Siskind J, Meisel SJ, et al. 2007. Prenatal organophosphate metabolite and organochlorine levels and performance on the Brazelton Neonatal Behavioral Assessment Scale in a multiethnic pregnancy cohort. Am J Epidemiol 165:1397–1404.

Engel SM, Zhu C, Berkowitz GS, Calafat AM, Silva MJ, Miodovnik A, et al. 2009. Prenatal phthalate exposure and performance on the Neonatal Behavioral Assessment Scale in a multiethnic birth cohort. Neurotoxicology 30:522–528.

Engel SM, Miodovnik A, Canfield RL, Zhu C, Silva MJ, Calafat AM, et al. 2010. Prenatal phthalate exposure is associated with childhood behavior and executive functioning. Environ Health Perspect 118:565–571.

Engel SM, Wetmur J, Chen J, Zhu C, Barr DB, Canfield RL, et al. 2011. Prenatal exposure to organophosphates, paraoxonase 1, and cognitive development in childhood. Environ Health Perspect 119:1182–1188.

Eskenazi B, Marks AR, Bradman A, Fenster L, Johnson C, Barr DB, et al. 2006. *In utero* exposure to dichlorodiphenyltrichloroethane (DDT) and dichlorodiphenyldichloroethylene (DDE) and neurodevelopment among young Mexican American children. Pediatrics 118:233–241.

Eskenazi B, Marks AR, Bradman A, Harley K, Barr DB, Johnson C, et al. 2007. Organophosphate pesticide exposure and neurodevelopment in young Mexican-American children. Environ Health Perspect 115:792–798.

Eskenazi B, Rosas LG, Marks AR, Bradman A, Harley K, Holland N, et al. 2008. Pesticide toxicity and the developing brain. Basic Clin Pharmacol Toxicol 102:228–236.

Fagan JF, III. 1990. The paired-comparison paradigm and infant intelligence. Ann N Y Acad Sci 608:337–364.

Geens T, Goeyens L, Covaci A. 2011. Are potential sources for human exposure to bisphenol-A overlooked? Int J Hyg Environ Health 214:339–347.

Goodman M, Squibb K, Youngstrom E, Anthony LG, Kenworthy L, Lipkin PH, et al. 2010. Using systematic reviews and meta-analyses to support regulatory decision making for Neurotoxicants: Lessons learned from a case study of PCBs. Environ Health Perspect 118:727–734.

Grandjean P, Weihe P, Burse VW, Needham LL, Storr-Hansen E, Heinzow B, et al. 2001. Neurobehavioral deficits associated with PCB in 7-year-old children prenatally exposed to seafood neurotoxicants. Neurotoxicol Teratol 23:305–317.

Guo YL, Lambert GH, Hsu CC, Hsu MM. 2004. Yucheng: Health effects of prenatal exposure to polychlorinated biphenyls and dibenzofurans. Int Arch Occup Environ Health 77:153–158.

Guxens M, Aguilera I, Ballester F, Estarlich M, Fernández-Somoano A, Lertxundi A, et al. 2012. Prenatal exposure to residential air pollution and infant mental development: Modulation by antioxidants and detoxification factors. Environ Health Perspect 1201:144–149.

Haddow JE, Palomaki GE, Allan WC, Williams JR, Knight GJ, Gagnon J, et al. 1999. Maternal thyroid deficiency during pregnancy and subsequent neuropsychological development of the child. N Engl J Med 341:549–555.

Hatch EE, Nelson JW, Qureshi MM, Weinberg J, Moore LL, Singer M, et al. 2008. Association of urinary phthalate metabolite concentrations with body mass index and waist circumference: A cross-sectional study of NHANES data, 1999–2002. Environ Health 7:27.

Herbstman JB, Sjödin A, Kurzon M, Lederman SA, Jones RS, Rauh V, et al. 2010. Prenatal exposure to PBDEs and neurodevelopment. Environ Health Perspect 118:712–719.

Hinton RH, Mitchell FE, Mann A, Chescoe D, Price SC, Nunn A, et al. 1986. Effects of phthalic acid esters on the liver and thyroid. Environ Health Perspect 70:195–210.

Holland N, Furlong C, Bastaki M, Richter R, Bradman A, Huen K, et al. 2006. Paraoxonase polymorphisms, haplotypes, and enzyme activity in Latino mothers and newborns. Environ Health Perspect 114:985–991.

Huang PC, Kuo PL, Guo YL, Liao PC, Lee CC. 2007. Associations between urinary phthalate monoesters and thyroid hormones in pregnant women. Hum Reprod 22:2715–2722.

Jacobson JL, Jacobson SW. 2004. Breast-feeding and gender as moderators of teratogenic effects on cognitive development. Neurotoxicol Teratol 24:349–358.

Julvez J, Grandjean P. 2009. Neurodevelopmental toxicity riks due to occupational exposure to industrial chemicals during pregnancy. Ind Health 47:459–468.

Korrick SA, Sagiv SK. 2008. Polychlorinated biphenyls, organochlorine pesticides and neurodevelopment. Curr Opin Pediatr 20:198–204.

Lester B. 1984. Data analysis and prediction. In: Neonatal Behavioral Assessment (Brazelton TB, ed.). Philadelphia, PA: Lippincott, pp. 85–96.

Levitt P. 2003. Structural and functional maturation of the developing primate brain. J Pediatr 143(4 Suppl):S35–S45.

Liu C, Li H, Wang Y, Yang K. 2011. p,p′-DDE disturbs the homeostasis of thyroid hormones via thyroid hormone receptors, transthyretin, and hepatic enzymes. Horm Metab Res 43:391–396.

Longnecker M 2006. POPs and neurodevelopment in humans: What is the evidence? In: 26th International Symposium on Halogenated Persistent Organic Pollutants, Plenary Lecture Abstracts and Session Summaries—Dioxin (Thomsen C, Becher G, eds.), Oslo, August 21–25, 2006, pp. 13–17.

Longnecker MP, Wolff MS, Gladen BC, Brock JW, Grandjean P, Jacobson JL, et al. 2003. Comparison of polychlorinated biphenyl levels across studies of human neurodevelopment. Environ Health Perspect 111:65–70.

Miodovnik A. 2011. Environmental neurotoxicants and developing brain. Mt Sinai J Med 78:58–77.

Mori C, Todaka E. 2008. Environmental Contaminants and Children's Health. Sustainable Health Science for Future Generations. Tokyo, Japan: Maruzen Planet Co., Ltd.

Mosby's Medical Dictionary. 8th edition. 2008. St. Louis, MO: Elsevier.

Muckle G, Ayotte P, Dewailly EE, Jacobson SW, Jacobson JL. 2001. Prenatal exposure of the northern Québec Inuit infants to environmental contaminants. Environ Health Perspect 109:1291–1299.

Nakajima S, Saijo Y, Kato S, Sasaki S, Uno A, Kanagami N, et al. 2006. Effects of prenatal exposure to polychlorinated biphenyls and dioxins on mental and motor development in Japanese children at 6 months of age. Environ Health Perspect 114:773–778.

National Research Council, Committee on the Health Risks of Phthalates. 2008. Phthalates and Cumulative Risk Assessment: The Task Ahead. Washington, DC: National Academies Press.

Pan IJ, Daniels JL, Goldman BD, Herring AH, Siega-Riz AM, Rogan WJ. 2009. Lactation exposure to polychlorinated biphenyls, dichlorodiphenyltrichloroethane, and dichlorodiphenyldichloroethylene and infant neurodevelopment: An analysis of the pregnancy, infection, and nutrition babies study. Environ Health Perspect 117: 488–494.

Patandin S, Lanting CI, Mulder PGH, Boersma ER, Sauer PJJ, Weisglas-Kuperus N. 1999. Effects of environmental exposure to polychlorinated biphenyls and dioxins on cognitive abilities in Dutch children at 42 months of age. J Pediatr 134:33–41.

Perera FP, Rauh V, Whyatt RM, Tsai WY, Tang D, Diaz D, et al. 2006. Effect of prenatal exposure to airborne polycyclic aromatic hydrocarbons on neurodevelopment in the first 3 years of life among inner-city children. Environ Health Perspect 114:1287–1292.

Perera FP, Li Z, Whyatt R, Hoepner L, Wang S, Camann D, et al. 2009. Prenatal polycyclic aromatic hydrocarbon exposure and child intelligence at age 5. Pediatrics 124:e195–e202.

Pop VJ, Brouwers EP, Vader HL, Vulsma T, van Baar AL, de Vijlder JJ. 2003. Maternal hypothyroxinaemia during early pregnancy and subsequent child development: A 3-year follow-up study. Clin Endocrinol (Oxf) 59:282–288.

Porterfield SP. 2000. Thyroidal dysfunction and environmental chemicals-potential impact on brain development. Environ Health Perspect 108(Suppl 3):433–438.

Rauh VA, Garfinkel R, Perera FP, Andrews HF, Hoepner L, Barr DB, et al. 2006. Impact of prenatal chlorpyrifos exposure on neurodevelopment in the first 3 years of life among inner-city children. Pediatrics 118:e1845–e1859.

Rice D, Barone S, Jr. 2000. Critical periods of vulnerability for the developing nervous system: Evidence from humans and animal models. Environ Health Perspect 108(Suppl 3):511–533.

Rogan WJ, Gladen BC, McKinney JD, Carreras N, Hardy P, Thullen J, et al. 1986. Neonatal effects of transplacental exposure to PBCs and DDE. J Pediatr 109: 335–341.

Roze E, Meijer L, Bakker A, Koenraad NJA, Braeckel V, Sauer PJJ, et al. 2009. Prenatal exposure to organohalogens, including brominated flame retardants, influences motor, cognitive, and behavioral performance at school age. Environ Health Perspect 117:1953–1958.

Rubin BS. 2011. Bisphenol A: An endocrine disruptor with widespread exposure and multiple effects. J Steroid Biochem Mol Biol 127:27–34.

Schartz P, Jacobson WS, Fein G, Jacobson JL, Price HA. 1983. Lake Michigan fish consumption as a source of polychlorinated biphenyls in human cord serum, maternal serum, and milk. Am J Public Health 73:293–296.

Sjödin A, Wong LY, Jones RS, Park A, Zhang Y, Hodge C, et al. 2008. Serum concentrations of polybrominated diphenyl ethers (PBDEs) and polybrominated biphenyl (PBB) in the United States population: 2003–2004. Environ Sci Technol 42: 1377–1384.

Slotkin TA. 1999. Development cholinotoxicants: Nicotine and chlorpyrifos. Environ Health Perspect 107(Suppl 1):71–80.

Stapleton HM, Eagle S, Anthopolos R, Wolkin A, Miranda ML. 2011. Associations between polybrominated diphenyl ther (PBDE) flame retardants, phenolic metabolites, and thyroid hormones during pregnancy. Environ Health Perspect 119: 1454–1459.

Stewart P, Reihman J, Gump B, Lonky E, Darvill T, Pagano J. 2005. Response inhibition at 8 and 9 1/2 years of age in children prenatally exposed to PCBs. Neurotoxicol Teratol 27:771–780.

Stewart PW, Lonky E, Reihman J, Pagano J, Gump BB, Darvill T. 2008. The Relationship between prenatal PCB exposure and intelligence (IQ) in 9-year-old children. Environ Health Perspect 116:1416–1422.

Tian Y, Baek J, Lee S, Jang C. 2010. Prenatal and postnatal exposure to bisphenol A induces anxiolytic behaviors and cognitive deficits in mice. Synapse 64:432–439.

Torres-Sánchez L, Rothenberg SJ, Schnaas L, Cebrián ME, Osorio E, Hernández MC, et al. 2007. *In Utero* p-p′-DDE exposure and infant neurodevelopment: A perinatal cohort in Mexico. Environ Health Perspect 115:435–443.

Torres-Sánchez L, Schnaas L, Cebrián ME, Hernández Mdel C, Valencia EO, García Hernández RM, et al. 2009. Prenatal dichlorodiphenyldichloroethylene (DDE) exposure and neurodevelopment: A follow-up from 12 to 30 months of age. Neurotoxicology 30:1162–1165.

Torres-Sánchez L, Schnaas L, Cebrián ME, Valencia-Osorio E, Hernández M del C, García Hernández RM, et al. 2013. Prenatal p,p′-DDE exposure and neurodevelopment among children 3.5 to 5 years of age. Environ Health Perspect 121:263–268.

Vreugdenhil HJI, Lanting CI, Mulder PGH, Boersma ER, Weisglas-Kuperus N. 2002. Effects of prenatal PCB and dioxin background exposure on cognitive and motor abilities in Dutch children at school age. J Pediatr 140:48–86.

Walkowiak J, Wiener JA, Fastabend A, Heinzow B, Krämer U, Schmidt E, et al. 2001. Environmental exposure to polychlorinated biphenyls and quality of the home environment: Effects on psychodevelopment in early childhood. Lancet 358:1602–1607.

Wilhelm M, Ranft U, Krämer U, Wittsiepe J, Lemm F, Fürst P, et al. 2008. Lack of neurodevelopmental adversity by prenatal exposure of infants to current lowered PCB levels: Comparison of two German birth cohort studies comparison of two German birth cohort studies. J Toxicol Environ Health A 71:700–702.

Winneke G, Bucholski A, Heinzow B, Krämer U, Schmidt E, Walkowiak J, et al. 1998. Developmental neurotoxicity of polychlorinated biphenyls (PCBs): Cognitive and psychomotor functions in 7-month old children. Toxicol Lett 102-103:423–428.

Wolff MS, Engel SM, Berkowitz GS, Ye X, Silva MJ, Zhu C, et al. 2008. Prenatal phenol and phthalate exposures and birth outcomes. Environ Health Perspect 116: 1092–1097.

Wolstenholme JT, Rissman EF, Connelly JJ. 2011. The role of bisphenol A in shaping the brain, epigenome and behavior. Horm Behav 59:296–305.

World Health Organization. 2010. Persistent Organic Pollutants: Impact on Child Health. Geneva, Switzerland: WHO Press.

Wormley DD, Ramesh A, Hood DB. 2004. Environmental contaminant-mixture effects on CNS development, plasticity, and behavior. Toxicol Appl Pharmacol 197:49–65.

Wormuth M, Scheringer M, Vollenweider M, Hungerbuhler K. 2006. What are the sources of exposure to eight frequently used phthalic acid esters in Europeans? Risk Anal 26:803–824.

Yolton K, Xu Y, Strauss D, Altaye M, Calafat AM, Khoury J. 2011. Prenatal exposure to bisphenol A and phthalates and infant neurobehavior. Neurotoxicol Teratol 33: 558–566.

CHAPTER 16

Intellectual Developmental Disability Syndromes and Organic Chemicals

DAVID O. CARPENTER

ABSTRACT

Background: Organic chemicals cause and/or are suspected to cause a number of different syndromes that are characterized by anatomical and physiological changes in the brain and cognitive and behavioral abnormalities.

Objectives: This review will discuss peer-reviewed publications on what is known of the etiology of three syndromes: fetal alcohol syndrome disorder (FASD), attention deficit/hyperactivity disorder (ADHD), and autism spectrum disorder (ASD).

Discussion: FASD is caused only by early prenatal exposure to alcohol. It varies in severity and is accompanied by significant abnormalities in brain and body structure and altered cognitive function and neurobehavior. While the effects of exposure to several organic compounds, such as polychlorinated biphenyls (PCBs), give rise to symptoms of ADHD, it is still unclear whether exposure to environmental contaminants is the only cause of the syndrome. In contrast, while there are many reasons to suspect that ASDs are caused by exposure to environmental contaminants, there is little hard evidence that this is the case. Both ADHD and autism are also accompanied by alterations in brain structure as well as function.

Conclusions: Prenatal or early-life exposure to a number of different organic chemicals causes abnormalities in brain development that result in anatomic changes, cognitive decrements, and behavioral changes. However, the mechanistic basis for these syndromes is still incompletely understood.

Effects of Persistent and Bioactive Organic Pollutants on Human Health, First Edition.
Edited by David O. Carpenter.

INTRODUCTION

The neurobehavioral diseases of early life have captured the attention of both the scientific community and the public in the past several decades. In the past, children who demonstrated what was considered to be "abnormal" behavior were usually considered to have some form of mental retardation. More recently, the term "mental retardation" has been replaced by "intellectual disability." However, there are a number of developmental neurological disorders, not all of which are accompanied by "retardation." A recent World Health Organization working group (Salvador-Carulla et al. 2011) has proposed the term "intellectual developmental disorders" and has defined this as "a group of developmental conditions characterized by significant impairment of cognitive function, which are associated with limitations of learning, adaptive behavior and skills." They note that very different patterns of cognitive impairment are found within this broad grouping, but there are general difficulties with verbal communication, perceptual reasoning, working memory, and processing speed. Individuals within this broad category often have difficulty managing their behavior, emotions, and interpersonal relations, and maintaining motivation in learning. Some of these syndromes are accompanied by clear defects in brain structure.

Many of the most severe cases of intellectual developmental disability are a consequence of inherited disorders or ischemia during development, including individuals with very severe retardation and diseases such as cerebral palsy. Little is known of possible environmental causes of severe cognitive dysfunction, and these diseases will not be discussed further here beyond a brief discussion of organic chemicals known to cause both brain birth defects and cognitive impairment. However, there are several distinct syndromes of behavioral and cognitive differences from the norm that have distinct features. While reduced cognitive function is commonly a characteristic in these syndromes, many children display remarkable abilities in certain domains while being very different from the "normal" in many others. Although recent awareness of these symptom complexes certainly has a lot to do with the rapid increase is diagnoses, there is still strong evidence that incidence rates are rapidly increasing independent of simply more diagnosis.

The term "learning disability" is commonly used as a distinct diagnosis but will not be discussed in detail here for several reasons. There is a separate chapter in this book on the adverse effects of a variety of organic chemicals on cognitive function (Chapter 15). Few recent studies have focused on learning disability for specific study, although Lee et al. (2007) used the NHANES dataset to examine the association between levels of seven persistent organic pollutants (POPs) and learning disabilities. They reported significantly elevated odds ratios (ORs) for one dioxin and one furan, and nonsignificant elevations with several of the other compounds.

Cognitive ability is determined by a variety of influences. There is a strong genetic component to intelligence, but intelligence is also a function of education, family nurturing, socioeconomic considerations, and exposure to environmental chemicals. Many people do not achieve the level of success in life that they would be capable of had they had greater opportunities for intellectual development. However, this cannot be considered to be an "intellectual developmental disability" but is rather a function of the level of stimulation in family life, education, and motivation. What we are considering here is exogenous influences on cognitive function that reduce cognitive function quite independently of lifestyle, socioeconomic status, and education level.

This review will focus primarily on three such syndromes: fetal alcohol spectrum disorder (FASD), attention deficit/hyperactivity disorder (ADHD), and autism spectrum disorder (ASD). The evidence for a role for organic chemicals varies among these syndromes. FASD is clearly directly caused by prenatal exposure to alcohol, but there are variations in genetic susceptibility. ADHD has been linked relatively strongly to exposure to certain organic chemicals but is also linked to metals such as lead. It is likely that other exposures and influences remain to be discovered, and it is by no means certain that exposure to environmental chemicals is the only cause of ADHD. Autism has not been definitively linked to any specific environmental exposure, although there are strong reasons to suspect that, like FASD, some exposure during development is the cause of the disease.

While each of these syndromes is distinct, they share a number of common features. There is a genetic component to each, but only a small fraction of the incidence in the general population can be explained solely by genetics. There is a frequent comorbidity in any individual. Furthermore, there are a variety of forms of the disease within each category. This is particularly true for ASD. While there is evidence for a number of different environmental contaminants influencing the expression of ADHD, this chapter will focus on organic compounds and will only briefly mention metals, which also play a major role. There are also structural changes in the brain associated with these syndromes, although they vary in severity.

FETAL ALCOHOL SPECTRUM DISORDER (FASD)

Alcohol consumption during pregnancy increases the risk of adverse birth outcomes, birth defects and low birth weight, and a series of neurobehavioral outcomes. Fetal alcohol syndrome (FAS), as described first by Lemoine et al. (1968) and named by Jones and Smith (1973), consists of infants of chronic alcoholic mothers that are born with serious dysmorphogenesis of the brain and with distinctive facial features of short palpebral fissures, maxillary hypoplasia, joint and palmar crease anomalies, and cardiac defects, together with growth retardation, developmental delay, or mental retardation (Riley et al. 2011). However, as more studies were made of infants born to mothers who consumed alcohol during pregnancy, it became apparent that many infants who lacked the clear physical effects still exhibited abnormal cognition and behavior. Thus, the term FASD has been applied to the full range of alcohol-induced physical and mental deformities.

Not all alcohol-exposed infants show abnormal cognition or behavior, which suggests variations in genetic susceptibility. The effects on weight and growth appear to be transient since they were not seen after 18 months of age, whereas the effects on cognition and behavior appear to be permanent and irreversible (Streissguth 2007). In the United States, the prevalence of FAS is estimated to be 0.5–2.0 cases per 1000 births, while FASD is thought to be at least about three times more common and is often not diagnosed (CDC 2009). Sampson et al. (1997) estimated the total rate of FAS and FASD to be near to 1 in every 100 live births. Together, these syndromes are the leading cause of preventable mental retardation in the Western world.

In full blown FAS, there are major abnormalities in the cerebral cortex and corpus callosum. The volume of the skull and overall brain size are reduced to a degree that is obvious. The basal ganglia volumes are disproportionately reduced, while the hippocampus is relatively less affected (Archibald et al. 2001). Application of high-resolution imaging of children with FASD as compared to sex and age-matched controls reported a 7.6% reduction in intracranial vault, a 8.6% reduction in total white matter, a 7.8% reduction in total cortical gray matter, and 13.1% reduction in total deep gray matter. The caudate (16% reduced) and globus pallidus (18% reduced) were most affected (Nardelli et al. 2011). Given the extent of alteration of the brain structure, it is not surprising that such individuals show significant decrements in neurological function. FASD children also have an elevated prevalence of epilepsy (Bell et al. 2010).

Children with FASD, even when they do not show obvious physical abnormalities, do more poorly on IQ tests, and the decrements in IQ are larger in those with more pronounced physical defects (Figure 16.1) (from Streissguth et al. 2004). The decrements in IQ are also associated with neurobehavioral defects. Their behaviors reflect immaturity, argumentativeness, inattention, and general disobedience (Nash et al. 2011). In a study of 415 patients with FAS or FASD, Streissguth et al. (2004) reported that 61% had disruptive school experiences; 60% had trouble with the law; 50% were jailed or hospitalized for alcohol or drug use; 49% showed repeated inappropriate sexual behavior; and 35% had alcohol or drug problems.

There is a strong impact of genetic variation in susceptibility to FAS and FASD (Ramsay 2010). Some strains of rodents are very sensitive to alcohol, while others are not. Equal prenatal exposure in humans does not result in equal risk of development of the syndrome. Siblings of FAS children have a very highly elevated risk for showing the same FAS or FASD. Ramsay (2010) suggests that there are epigenetic changes following alcohol consumption and that these changes result in transgenerational risks of the disease. These epigenetic changes can occur in sperm DNA as well as general global epigenetic remodeling during gametogenesis, preimplantation, and gastrulation, primarily due to alcohol-induced hypomethylation of the DNA. Ramsay concludes

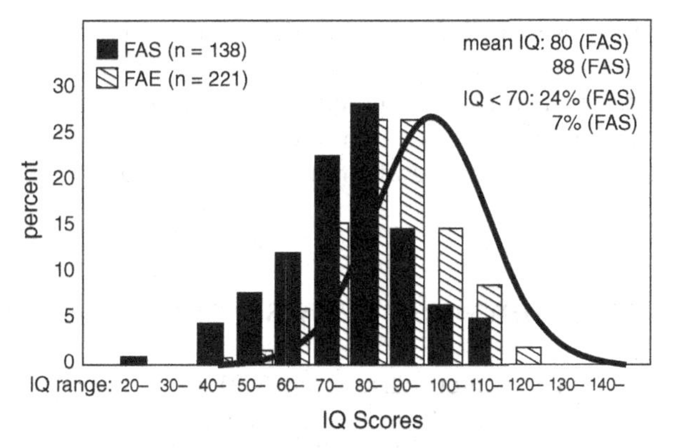

Figure 16.1. IQ scores of patients with fetal alcohol syndrome (FAS) and fetal alcohol effects (FAE) superimiposed on the normative curve for IQ. An IQ score of 100 is the mean of the population for IQ. The sample size for IQ is below 415 because IQ testing was not funded for this study but was retrieved from research records as available. Reproduced with permission from Streissguth et al. 2004.

that studies "are consistent with the transmission of environmentally acquired changes in epigenetic status, which persist to variable extents in subsequent generations and depend on both the stochastic deposition of epigenetic marks during development and a changing environment." The observation that these epigenetic changes occur so early after fertilization illustrates another important factor. Since many women are unaware of their pregnancy at this stage, even if they are informed about the risk of FAS, they often are not taking precautions regarding alcohol consumption at this time.

Given that different brain regions are not affected equally, it is perhaps not surprising that FASD individuals demonstrate greater decrements in some functions than others. There are significant decrements in attention, communication, adaptive functioning, and executive functioning, but FASD individuals are less impaired on academic achievement (other than math) and intellect (Rasmussen et al. 2010). One particular concern is the high rates of conduct disorder and oppositional defiant disorder codiagnosis. This is probably one major reason that FASD individuals are overrepresented in the justice system.

Clearly, while the effects of prenatal alcohol on growth rate may be transient, the effects on nervous system function persist for life. This is reflected in the elevated rates of detention, alcohol and drug dependence, and general misconduct seen in adolescents and adults with FASD (Streissguth et al. 2004). There is a well-known relationship between IQ and income and job stability, and an inverse relationship between IQ and risk of antisocial behavior resulting in arrest and incarceration (Carpenter and Nevin 2010). In spite of clear evidence that early diagnosis and treatment will reduce the disabilities resulting from prenatal alcohol exposure (Riley et al. 2003), there is frequently a reluctance to diagnose the disease for several reasons. In the first place, if there are no clear physical characteristics, it is not easy to diagnose. However, a diagnosis of FAS or FASD is an indictment of the mother, something many clinicians are unwilling to do. As a consequence, children who would benefit from interventions if diagnosed are not identified and do not benefit from therapy.

ATTENTION DEFICIT/HYPERACTIVITY SYNDROME (ADHD)

Sometimes called "hyperkinetic disorder," ADHD is the most common psychiatric disease of childhood. Worldwide estimates indicate that over 5% of children suffer from ADHD (Polanczyk et al. 2007). The symptoms include inattentiveness, lack of concentration, overactivity, and impulsiveness (Rappley 2005). While the symptoms are most prominent during childhood, they often persistent into adult years (Aguiar et al. 2010). Children with ADHD are at greater risk of developing other social projects related to adjustment during education and work, antisocial behavior, and substance abuse (Harpin 2005). Children with ADHD often but not always will show other psychiatric disorders, especially oppositional defiant disorder, but also autism, dyslexia, conduct disorder, anxiety, depression, and/or bipolar disorder (Stergiakouli and Thapar 2010).

ADHD is more common in boys than in girls, and there tend to be somewhat different profiles of behavior by gender (Sagvolden et al. 2005). In boys, hyperactivity and impulsiveness are usually the major symptoms, while in girls, it is more commonly inattentiveness. The latter syndrome is sometimes called attention deficit disorder (ADD). Other children will show all of these symptoms, with varying degrees of learning and conduct disorders, inattentiveness, and hyperactivity (Castellanos and Tannock 2002). While symptoms often regress somewhat with age, about half of ADHD children will still show symptoms as adults (Aguiar et al. 2010) and Simon et al. (2009) have reported that the prevalence in adults is 2.5%. Crocker et al. (2009) compared adaptive behavior of children with prenatal alcohol exposure to those with ADHD and found that while there was improvement over time in children with ADHD, that was not the case for children prenatally exposed to alcohol.

There is a strong genetic component to ADHD, but this is most obvious at young ages (Ilott et al. 2010). Monozygotic twins have greater than 50% concordance for ADHD, whereas dizygotic twins show about 33% concordance (Bradley and Golden 2001). There is clear evidence that ADHD and anxiety disorders cosegregate within families (Biederman et al. 1991). However, the search for specific genes involved in ADHD has, to date, not been very successful. Since so much of the evidence for neuropsychiatric diseases has involved dopamine receptors, and because stimulant medication is effective in treating ADHD, many studies have focused on genes related to dopaminergic transmission and some have found modest associations. There have been a

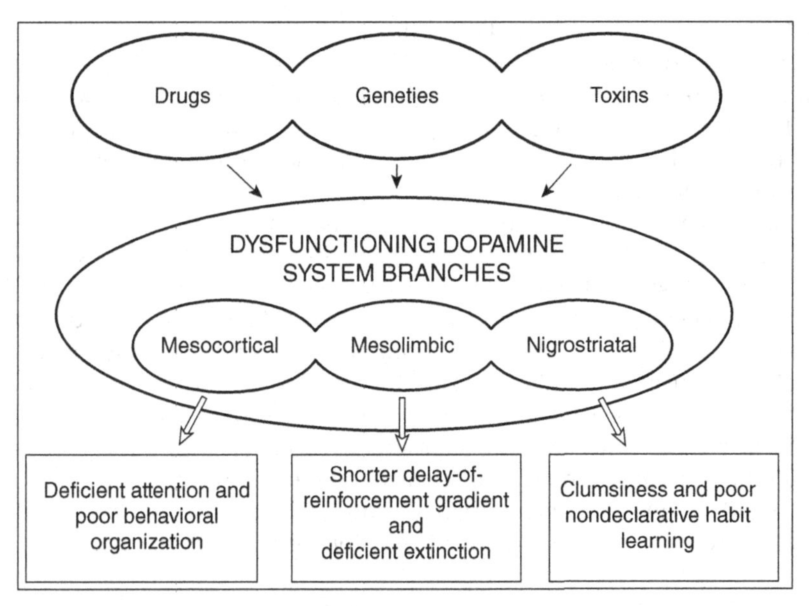

Figure 16.2. Dysfunction of dopaminergic systems resulting from drug abuse, genetic transmission, or environmental pollutants may cause ADHD symptoms by interacting with frontostriatal circuits of the brain. Reproduced with permission from Sagvolden et al. 2005.

number of studies on the dopamine D4 and D5 receptors, the dopamine transporter gene *VNTR*, and *catechol-O-methyltransferase (COMT) val 158/108 met* variant. The COMT variant has been particularly associated with conduct disorders. Genome-wide association studies have identified some genes of interest but, to date, have had limited success (Stergiakouli and Thapar 2010). Some have proposed that dysfunction of brain dopaminergic systems is a central component of ADHD (see Figure 16.2, from Sagvolden et al. 2005). This model is consistent with many of the behavioral aspects of ADHD and the influences of toxins and drugs that are discussed further, but it does not directly account for the cognitive deficits.

However, the clear evidence that even monozygotic twins may not both show ADHD is strong evidence for an environmental influence and for gene–environment interactions. Most of the evidence indicates that, as with other developmental psychiatric diseases, epigenetic processes are important in explaining susceptibility (Mill and Petronis 2008). Because the rate of DNA synthesis is very high during development, alterations in DNA methylation during this period of time are much more likely to result in phenotypic changes than if the methylation occurs later in life.

There are a number of established risk factors for ADHD in addition to exposure to environmental chemicals. These include small size at birth, poor maternal diet, and maternal stress. Low parental education is an established risk factor (St Sauver et al. 2004). But these risk factors may also be surrogates for exposure to chemicals. Koger et al. (2005) list seven organic compounds that elevate the risk of learning disabilities, including ethanol, styrene, xylene, DDT, and metabolites, organophosphate pesticides, dioxins, and polychlorinated biphenyls (PCBs). They also note that trichloroethylene increases the risk of hyperactivity.

There is significant comorbidity between FASD and ADHD. In one study, 63% of children with FASD had a comorbid diagnosis of ADHD (Rasmussen et al. 2010). However, Coles et al. (1997) compared a group of FASD children to one of ADHD children and concluded that, while both showed reduced intellectual abilities, the ADHD children performed more poorly on conventional tests sensitive to attention and conduct disorder, while FASD children did better in terms of academic achievement. ADHD children were best identified using behavior checklists and measures of ability to sustain attention, while FASD children showed deficits in visual and spatial skills, encoding of information, and flexibility in problem solving. They concluded that the two diagnoses were distinct patterns of deficits and therefore are distinct disorders.

ADHD, NEURODEVELOPMENTAL ABNORMALITIES, AND ORGANIC CONTAMINANTS

There is strong evidence that exposure to PCBs, especially exposure before birth or early in life, results in an increased risk of ADHD (Boucher et al. 2009; Eubig et al. 2010; Lai et al. 2002). PCB exposure not only results in reduced cognitive function but also is accompanied by reduced attention, reduced processing time, reduced visual recognition, reduced somatic and short-term memory, vocabulary, verbal comprehension, reading comprehension, and visuomotor integration in many, but not all, studies, as detailed in the above-mentioned reviews.

Lee et al. (2007) were among the first to use the National Health and Nutrition Examination Survey (NHANES) data to analyze the association between levels of POPs and ADHD. NHANES is a near-random sampling of Americans where extensive information on disease history and blood samples for measurement of various chemical pollutants was obtained. Lee et al. examined the relationship between diagnosis of ADHD with reported concentrations of seven POPs (one PCB, two dioxins, one furan, and three chorinated pesticides). They reported a significant association between risk of ADHD and a heptachlorodioxin, and nonsignificant elevations with other POPs.

Sagiv et al. (2010) studied 607 children ages 7–11 on the Connor's Rating Scale for Teachers in relation to their concentrations of four PCB congeners and 2,2-bis(*p*-chlorophenyl)-1,1-dichloroethylene (p,p′ DDE). They found a significant elevation in ADHD-like symptoms in relation to the sum of the four PCB congeners (OR = 1.76, 95% CI = 1.06–2.92) and similar elevations for DDE. In a later report Sagiv et al. (2012) followed children with cord blood measures of PCBs and studied the children at approximately 8 years of age. Boys, but not girls, showed significant decrements of performance on the Continuous Performance Test and components of the Wechsler Intelligence Scale for Children. Similar but smaller associations were found for p,p′-DDE. It is interesting that the effects of prenatal exposure to PCBs are stronger in boys than in girls, which may be a partial explanation for the gender difference in incidence of ADHD.

Other studies have reported that PCB exposure is correlated with specific components of the full ADHD syndrome. Stewart et al. (2005) studied 212 children from whom prenatal PCB levels were obtained from cord blood. They found a strong relationship between prenatal PCB exposure and impulsive responding. However, they concluded that this was secondary to response inhibition, not inattention. This effect was associated with a reduction in the size of the splenium of the corpus callosum, the brain structure known to mediate this effect. Boucher et al. (2012), in a study of Inuit children, found that exposure to PCB 153 was associated with an alteration in error monitoring, an aspect of behavioral regulation required to adequately adapt to the changing demands of the environment. Sly et al. (2012) found that Australian adolescents with elevated serum PCB levels showed significantly more externalizing behavior than did adolescents with lower PCB concentrations. They did not see a similar relationship with the chlorinated pesticides.

The polybrominated diphenyl ethers (PBDEs), widely used as flame retardants, are also POPs with many structural and toxicological similarities to PCBs. While the adverse effects of PBDEs have not been as widely studied as have PCBs, Gascon et al. (2012) found that postnatal exposure to PBDEs was significantly associated with increased risk of symptoms of ADHD on the attention scale but not on the hyperactivity scale. This effect was not mediated by alternation of the thyroid hormones.

Colborn (2006) has reviewed the evidence of neurobehavioral effects of early-life exposure to other types of pesticides, including chlorpyrifos, various herbicides, insecticides, and fungicides. Exposure to chlorpyrifos has been clearly demonstrated to reduce cognitive function (Rauh et al. 2011) and to be accompanied by structural changes in the brain of exposed children (Rauh et al. 2012), but relations to ADHD have not been specifically noted. Almost all of the chemicals that have been studied in animal model systems have shown effects on neurodevelopment. There have been fewer studies in humans for several reasons. Unlike the organochlorine pesticides, others are less persistent in the human body, and therefore it is less easy to monitor past exposure. In addition, there are so many different chemicals used as herbicides, pesticides, and fungicides that all of us are exposed to one degree or another to many different chemicals. Guillette et al. (1998) did a study of Yaqui preschool children in Mexico who lived in valleys with high exposure to a variety of pesticides as compared to those living on the foothills and had much less exposure. They found that the children with greater exposure showed reduced motor function, memory, and ability to draw pictures.

Studies have reported on neurodevelopment in relation to human exposure to organophosphate pesticides through measurement of urinary metabolites. Eskenazi et al. (2007) measured six nonspecific dialkyl phosphate (DAP) metabolites in maternal and child urine. They found that both pre- and postnatal DAPs were associated with increased risk of pervasive developmental disorder, but there were no changes in attention. When these children were studied at 7 years of age, those with elevated maternal urinary DAPs had reduced IQ. Bouchard et al. (2010), using NHANES data, reported that children ages 8–15 who had elevated levels of dimethyl alkyl phosphate (DMAP), a major metabolite of organophosphate pesticides, had a significantly elevated risk of showing ADHD (OR = 1.55, 95% CI = 1.14–2.10) after adjustment for age, gender, race, socioeconomic status, fasting, and urinary creatinine. Those children whose DMAP level was higher than the mean had an even more elevated OR of 1.93 (95% CI = 1.23–3.02). In a later study, Bouchard et al. (2011) investigated the relationship between maternal urinary excretion of DAP and reduced IQ in children. In the highest quintile of exposure, there was a 7.0 point IQ deficit in the children. Interestingly, there was no evidence for a relationship between postnatal excretion of DAP by the child, suggesting that the decrement resulted from prenatal exposure during brain development. This suggests that exposure later in life is not nearly as effective in reducing cognitive function as it is during development.

In the early 1990s, there was illegal spraying of methyl parathion for pest control in Ohio and Mississippi. Ruckart et al. (2004) reported neurobehavioral effects in children 6 years of age or younger at the time of the spraying, as compared to local unexposed children, using the Pediatric Environmental Neurobehavioral Test Battery. Exposed children did more poorly on tasks requiring short-term memory and attention. Parents of exposed children reported more behavioral and motor skill problems. While they did not consistently show cognitive deficits, they were reported by parents to misbehave, show problems with anger, act impulsively and have problems relating to other children.

Xu et al. (2011) used the NHANES data to investigate ADHD in children in relation to urinary levels of chlorophenols. They reported an elevated OR (1.77 [1.18–2.66]) in children with urinary concentrations of 2,4,6-trichlorophenol greater than or equal to 3.58 μg/g as compared to children with levels less than the level of detection. They did not find any relationship with urinary levels of 2,3,5-trichlorophenol.

Kim et al. (2009) studied 261 Korean children ages 8–11 years and looked at the association between teacher-rated ADHD scores and metabolites of several phthalates. They found significant relations between indicators of hyperactivity and inattention and urinary concentrations of metabolites of di-2-ethylhexyl phthalate but no association with metabolites of dibutylphthalates. There were also significant associations with omission and commission errors on standard tests. Another chemical used as a plasticizer, bisphenol A (BPA), also alters behavior after prenatal exposure, although the effects and the gender susceptibility are not the same as with ADHD. Braun et al. (2009, 2011) reported that prenatal but not postnatal exposure to BPA resulted in hyperactivity, aggression, more anxious and depressed behavior, and poorer emotional control and inhibition in girls, but not in boys, when studied at 2 and 3 years of age. Quite different results were reported, however, by Perera et al. (2012a) in a study of 198 children followed from pregnancy to 5 years of age, with multiple spot urine samples analyzed for BPA. Boys but not girls showed significantly higher scores on the Child Behavior Checklist for Emotionally Reactive, Aggressive Behavior and Internalizing Problems subsections in relation to prenatal exposure.

Organomercury compounds are well documented to cause reductions in IQ in children exposed prenatally or early in life as a consequence of maternal ingestion of contaminated seafood (Axelrad et al. 2007). Methylmercury exposure of children in the Faroe Islands, exposed through consumption of contaminated pilot whales, showed most of the symptoms of ADHD (Debes et al. 2006). (Figure 16.3 from Grandjean et al. 1997). Clearly, exposure results in a variety of effects on intelligence, motor function, attention, and visuospatial abilities. Methylmercury and POPs are often found together in high concentrations in fish and other seafoods, and even the presence of the supposedly

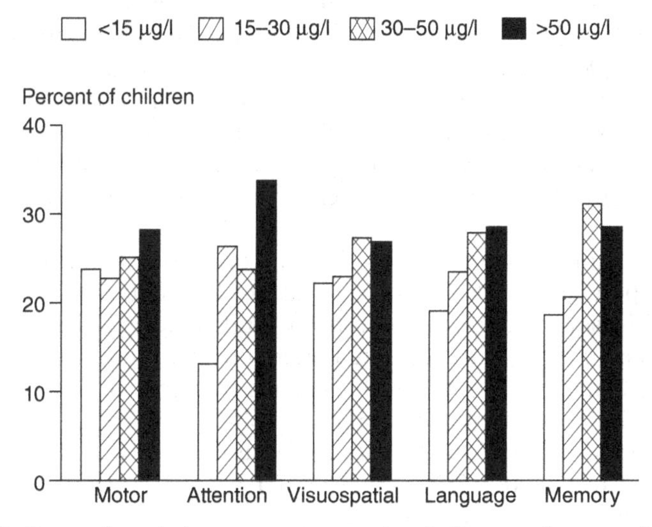

Figure 16.3. Prenatal methylmercury exposure levels (in quartile groups) of Faroese children with scores in the lowest quartile after adjustment for confounders. For each of five major cognitive functions, one neuropsychological test with a high psychometric validity was selected. Motor: NES2 Finger Tapping with preferred hand (p-value for trend = 0.23). Attention: reaction time on the NES2 Continued Performance Test ($p = 0.0003$). Visuospatial: Bender Visual Motor Gestalt Test error score ($p = 0.16$). Language: Boston Naming Test score after cues ($p = 0.02$). Memory: California Verbal Learning Test (children) long-delay recall ($p = 0.004$). Reproduced with permission from Grandjean et al. 1997.

healthy omega-3 fatty acids in seafood may not be sufficient to counteract the adverse neurological effects of the contaminants (Bushkin-Bedient and Carpenter 2010).

A number of other organic compounds have been reported to elevate risk of at least some components of ADHD. Koger et al. (2005) list seven organic compounds that elevate risk of learning disabilities, including ethanol, styrene, xylene, DDT and metabolites, organophosphate pesticides, dioxins, and PCBs. They also note that trichloroethylene increases risk of hyperactivity. Hoffman et al. (2010) reported a significant relationship between elevated levels of perfluoroalkyl chemicals and ADHD in 13- to 15-year-old children, using NHANES data. Stein and Savitz (2011) reported an adjusted OR of 1.59 (95% CI = 1.21–2.08) between levels of perfluorohexane sulfonate and ADHD. Perera et al. (2012b) investigated the relation between prenatal polycyclic aromatic hydrocarbon exposure and child behavior at 6–7 years of age. They found that high prenatal exposure, monitored both by personal air monitoring and maternal and cord blood adducts, was positively associated with symptoms of being anxious/depressed and attention problems on the child behavior checklist.

One major concern has been the possibility that food additives, especially artificial coloring, promote hyperactivity, and possibly ADHD. There are few careful studies that adequately control for dietary intake, however. Schab and Trinh (2004) reviewed data available up to that date and concluded that it was likely that food additives promote a small increase in hyperactivity.

Till et al. (2001) tested the cognitive function and behavior of 33 children whose mothers were occupationally exposed to organic solvents during pregnancy as compared to 28 matched children who were not prenatally exposed. While overall cognitive function did not differ, there were cognitive deficits in specific areas. In addition, the exposed children showed mild to severe problem behaviors.

Tobacco smoke contains many different organic compounds as well as metals and particulates. There is strong evidence that maternal smoking during pregnancy is associated with neurobehavioral abnormalities in the child, including increased activity, decreased attention, and reduced cognitive function (Weitzman et al. 2002). The decrements in cognitive function occur at remarkably low levels of urinary cotinine (Yolton et al. 2005). Froehlich et al. (2009) reported that prenatal tobacco exposure posed a significant risk for development of ADHD (OR = 2.4, 95% CI = 1.5–3.7). They also found, as have other investigators, that prenatal exposure to lead significantly increased risk, and that children exposed prenatally to both tobacco and lead had a much greater elevation in risk than seen with either alone. In a study of German children, Rückinger et al. (2010) found that children exposed to tobacco smoke both pre- and postnatally had twice the risk of being classified as "abnormal" on the Strength and Difficulties Questionnaire at age 10 years. Cho et al. (2010), in a study of Korean children, found that urinary cotinine levels were inversely associated with several neurobehavioral outcomes, including response inhibition, and both selective and sustained attention; Kabir et al. (2011) reported

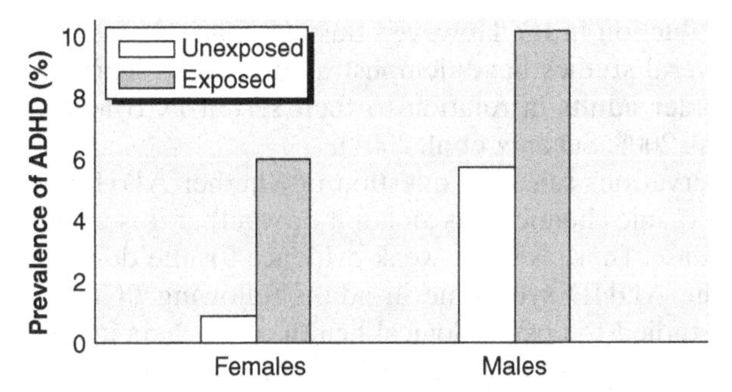

Figure 16.4. Prevalence of ADHD among U.S. children by prenatal environmental tobacco smoke (ETS) exposure and sex. While the prevalence of ADHD in unexposed children is greater for males, the increase in risk for ADHD among ETS-exposed children was greater in females. Females who were prenatally exposed to tobacco were at 4.6-fold higher risk for ADHD compared with unexposed females (OR = 4.6, 95% CI = 1.7–12.4), whereas exposed males were at twofold higher risk for ADHD compared with unexposed males (OR = 2.1, 95% CI = 0.9–4.7) ($p = 0.141$ for sex by prenatal EHS exposure interaction). Model adjusted for race/ethnicity, sex, age, blood lead level, ferritin level, presence of a smoker in the home, preschool attendance, and insurance status. Reproduced with permission from Braun et al. 2006.

that children exposed to secondhand smoke at home had a 50% increased risk of showing adverse neurobehavioral outcomes. Figure 16.4 (from Braun et al. 2006) shows the prevalence of ADHD in U.S. children in relation to prenatal exposure to tobacco smoke and by gender. While both males and females are at increased risk if exposed to smoke, the gender difference in incidence remains clear for both exposed and unexposed children.

Children born to mothers who use addictive drugs during pregnancy show transient changes in behavior, but most of these disappear over a period of time (Robins and Mills 1993). There is some evidence that prenatal exposure to drugs of abuse (cocaine, marijuana) results in effects on attention and impulsivity in children (Leech et al. 1999), and more frequent delinquency (Minnes et al. 2010), but these effects are difficult to separate from other social factors.

There are a number of established risk factors for ADHD in addition to exposure to environmental chemicals. These include small size at birth, poor maternal diet, and maternal stress. But these risk factors may also be surrogates for exposure to chemicals.

DO ORGANIC CONTAMINANTS CAUSE ADHD-LIKE SYMPTOMS IN ADULT?

There is clear evidence (Jacobson and Jacobson 1996; Newman et al. 2006; Stewart et al. 2008; Chapter 15) that prenatal and early-life exposure to PCBs results in a reduction in IQ. However, this effect is not restricted to early-life exposure. Several studies have demonstrated reduced learning and memory function in older adults in relation to their serum PCB levels (Haase et al. 2009; Lin et al. 2008; Schantz et al. 2001).

These observations raise the question of whether ADHD can result from exposure to organic chemicals as an adult or whether it is exclusively a developmental disease. There is some weak evidence for the development of components of the ADHD syndrome in adults following PCB exposure. Peper et al. (2005) studied the psychological health of teachers in a school that was highly contaminated with PCBs and where the route of exposure was inhalation. They report subtle changes in attention and emotional complaints, which were associated with elevations in serum levels of PCBs 28 and 101. These are relatively low-chlorinated congeners that are more volatile than those with a greater number of chlorines. There is some evidence that the same is true following adult exposure to methylmercury (Silbernagel et al. 2011; Yokoo et al. 2003), although the primary deficits are in memory and motor function. However, the weight of evidence indicates that ADHD is a developmental disease and that the most vulnerable time is the prenatal and early postnatal periods.

DO CHEMICALS THAT RESULT IN REDUCED IQ ALSO ALWAYS CAUSE OTHER ADHD SYMPTOMS?

One important question that cannot be definitively answered at present is whether exposure to those organic compounds that result in a reduction in IQ also causes the other symptoms of ADHD, and conversely, whether all organic chemicals that cause hyperactivity and shortened attention span cause deficits in IQ.

The two chemicals that have been most studied in relation to cognitive deficits and ADHD are lead and PCBs. It is striking how similar these two very different chemicals are in their actions. Both cause a shift downward in the IQ distribution curve of about 5–7 IQ points (Chen et al. 1992; Needleman and Gatsonis 1990). Both reduce long-term potentiation (LTP), an electrophysiological measure of cognitive potential in animals (Carpenter et al. 2002). Both trigger the hyperactivity, reduced attention span, and impulsiveness characteristic of ADHD (Eubig et al. 2010). Methylmercury, an organometallic compound very different from either lead or PCBs, does very similar things to attention span, memory, motor function, and visual-spatial ability (Grandjean et al. 1997).

We now know that exposure to a great number of chemicals, including both metals and organics, results in reduced IQ. This includes lead (Canfield et al. 2003), arsenic (Wasserman et al. 2007), manganese (Menezes-Filho et al. 2011), fluoride (Wang et al. 2007), PCBs (Jacobson and Jacobson 1996), dioxins (Barrett et al. 2001), chlorinated pesticides like DDT (Ribas-Fitó et al. 2006), organophosphate pesticides (Bouchard et al. 2011), PBDEs (Gascon et al. 2011), environmental tobacco smoke (Yolton et al. 2005), airborne polycyclic aromatic hydrocarbons (Edwards et al. 2010), and even *in utero* exposure to phenobarbital (Reinisch et al. 1995). Do all of these exposures increase risk of ADHD?

While there is at present insufficient evidence to give a clear answer to this question, it appears very likely that many of these chemicals are acting on common neuronal pathways that couple intelligence with behavior. The reduction in IQ that has been reported for each of these substances is similar (~5–7 IQ points) and subtle such that it is only clear when comparing populations of exposed and less exposed persons. What these common pathways are and how cognitive function is coupled to hyperactivity, attentiveness, and impulsivity are far from being understood. Much more research is needed to elucidate fully the role of this large number of environmental contaminants, organic and inorganic, in relation to risk of ADHD. However, it is not unreasonable to presume, until proven otherwise, that most chemicals that reduce IQ also increase risk of the full ADHD syndrome.

ASDs

Autism is not a single disease but a spectrum of related developmental abnormalities, usually referred to as ASD. However, in common usage, all are often referred to as "autism." There are five distinct subgroups of ASDs, autistic disorder (AD); Asperger's syndrome (AS), where there are social and behavioral deficits without major cognitive or language delays; Rett's disorder; childhood disintegrative disorder (CDD), where there is normal initial development followed by rapid decline; and pervasive developmental disorder, not otherwise specified (PDD-NOS) (Korvatska et al. 2002). While AD, AS, and CDD are similar but vary in degree and timing of symptoms, Rett's disorder is now believed to be a distinct disease in that it affects almost entirely females and is an X-linked disease, but like CDD is characterized by an initial normal development followed by rapid loss. ASDs are characterized by severely impaired social interactions and communication, frequently with repetitive movements and stereotyped behavior. Autism is often comorbid with mental retardation and epilepsy (Lord et al. 2000). Autism becomes apparent within the first 3 years of life, and in most cases, behavior is abnormal from birth. In a small percentage of children, there is an apparent normal early development followed by a rapid deterioration.

The cause(s) of autism is much less well understood than ADHD, although there are many common features. There is a strong genetic component to autism, with monozygotic twins showing a high rate of concordance (but less than 100%) (Turner et al. 2000) and dizygotic twins showing concordance rates between 0% and 31% (Ronald and Hoekstra 2011). However, genetic factors explain only about 7–8% of cases of autism (Landrigan 2010). Rates of diagnosis of autism have risen dramatically, now affecting about 1 in every 110 children (Weintraub 2011). While increased awareness and knowledge of the syndrome is certainly one factor, most experts agree that there is a real increase in incidence (DeSoto 2009; Hertz-Picciotto and Delwiche 2009). While a number of genes have been found to be correlated with autism through the study of families with high rates (Morrow et al. 2008), most cases of autism are not explained by single gene mutations.

Autism is associated with physical changes in brain structure and immune system parameters that suggest either an infectious process or changes secondary to some environmental exposures (see review by Cohly and Panja 2005). Children with autism tend to have small brains at birth followed by unusually rapid brain growth and asymmetrical development of white matter. The corpus callosum and cerebellum are particularly affected. Up to 70% of children with autism show some degree of cognitive impairment, and about 40% show severe mental retardation (Fombonne 2003). About a quarter will have seizures (Canitano 2007). Since prenatal viral infections have been found to be associated with a number of developmental abnormalities, some have suggested that autism is a consequence of maternal exposure to a virus during pregnancy, particularly measles virus. However, viral infection is unlikely to explain the majority of cases of autism.

One of the major controversies concerning autism is whether there is a relation between immunization, especially with the measles, mumps, and rubella (MMR) vaccine, and development of autism. Much of the concern about safety of vaccines has focused on the presence of thimerosal, an organomercurial compound (ethyl mercury) widely used in the past as a preservative in many vaccines. There is strong evidence that prenatal and early-life exposure to methylmercury from maternal consumption of fish or marine mammals is associated with a reduction in IQ in children (Grandjean et al. 1997) but no evidence that such consumption has increased the risk of ASD.

The issue of thimerosal as a risk factor for autism remains controversial. Geier and Geier (2004) examined adverse immunization reports from the Centers for Disease Control, comparing reports of autism and mental retardation in children receiving MMR vaccines containing or not containing thimerosal. They reported an RR of 6.0 for autism, 6.1 for mental retardation, and 2.2 for speech defects in children receiving the vaccine containing thimerosal, while for autism but not for mental retardation, the incidence was greater for boys than for girls. In a later study, the same authors (Geier and Geier 2006) reported that nonchelated autistic children have elevated levels of urinary coproporphyrin and that the severity of the disease was related to the porphyrinuria. In spite of these reports and an earlier report implicating vaccines in general that was subsequently withdrawn, the weight of evidence indicates that there is no relationship between receiving the MMR or any other vaccine and autism (DeStefano and Thompson 2004; Fombonne and Chakrabarti 2001; Madsen et al. 2002; Price et al. 2010). The controversy has generated significant concern about the safety of vaccines in some segments of the general public, whether or not that is at all justified. Public health professionals are worried that the controversy over the safety of vaccines will result in parents not allowing their children to be immunized for serious and vaccine-preventable diseases (Demicheli et al. 2012).

Hertz-Picciotto et al. (2010) and Ip et al. (2004) measured blood mercury in children with and without autism and found no differences. However, a subsequent study by the latter group (Stamova et al. 2011) looked at genes whose expression was altered by mercury in autistic children and controls. While mercury concentrations were not different in the cases and controls, there were 189 genes in the autistic children whose expression was correlated with mercury levels but where no correlation was present in the controls. These findings would be consistent with, but do not prove, the conclusion that there are genetic differences in autistic children that make them particularly vulnerable to exposure to mercury compounds. In both of these studies, total mercury was measured, and levels were highly correlated with fish consumption, which contains methylmercury. However, others have suggested that inorganic mercury (Palmer et al. 2009) and/or lead (Lakshmi Priya and Geetha 2011) are elevated in autistic children, although other studies (Abdullah et al. 2011) find no differences in concentrations of any toxic metal. Thus, at present, the possible role of either organic mercury or inorganic metals of any kind remains uncertain, although the weight of evidence does not indicate that exposure to any form of mercury is the etiologic agent.

Autism is more frequently diagnosed in children conceived during the winter months (Zerbo et al. 2011). Frequency of diagnosis of autism is not geographically equally distributed. This has led many to believe that there must be environmental exposures that are critical to the expression of the disease. A number of studies have focused on exposure to organic chemicals other than organomercurial compounds. DeSoto (2009) found that in the state of Minnesota, there was a statistically significant greater incidence of autism among children living within 10 or 20 miles of an Environmental Protection Agency (EPA)-designated National Priority List (Superfund) site. Roberts et al. (2007) found that rates of ASD increased in relation to the amount of organochlorine pesticides applied and the distance from the field sites. Windham et al. (2006) reported a correlation between rates of autism and air contaminants in the San Francisco Bay area. Volk et al. (2011) reported that rates of ASD were significantly elevated among individuals living <309 m from a California freeway as compared with those who did not (OR = 1.86, 95% CI = 1.04–3.45), and that the OR was even greater if the residence was in the third trimester. Together, the above studies implicate chemicals found in the home during the winter months when houses are more closed and chemicals associated with hazardous waste sites and particulate air pollutants. PBDEs are one of the few organic compounds for which there is clear evidence that it is not involved in autism, as Hertz-Picciotto et al. (2011) found no difference in levels in autistic and control children. Kalkbrenner et al. (2012) found no association between maternal smoking and ASD.

There is some reason to consider the possibility that autism is an autoimmune disease, or perhaps caused by maternal antibodies to the fetus (Cohly and Panja 2005). Comi et al. (1999) found that 46% of autistic children have two or more family members with autoimmune diseases and suggested that common environmental factors might be involved in both diseases. Goines et al. (2011) reported finding autoantibodies to cerebellar proteins some in children with autism and found that the presence of these autoantibodies correlated with lower adaptive and cognitive function as compared with autistic children without these autoantibodies. Furthermore, they suggest the possibility that antibodies to two different proteins (45 and 62 kDa) may have associations with different aspects of the disease. Braunschweig et al. (2011) reported the presence of maternal antibodies to fetal brain proteins in autistic children and concluded that there was a strong association between these antifetal brain antibodies and the diagnosis of autism.

It is worth noting that mercury compounds are well known to promote hyperimmunty and allergy, and could possibly be one factor that bridges the environmental exposure and autoimmunity features of autism. Other chemical exposures could also act via similar pathways, although at present, there is little specific information to support this hypothesis. This, in addition to the demonstration that autistic children show different patterns of mercury-induced genes, may provide one pathway whereby environmental exposures influence a disease that is primarily due to immune system dysfunction.

CONCLUSIONS

Much remains to be learned about the cause of intellectual developmental disorders. The questions include even the basic issue of whether or not environmental exposures cause these diseases. This is most apparent with regard to autism as, at present, there is no convincing evidence for any specific chemical exposure. This is in spite of fact that rates are increasing dramatically, that incidence of the disease is not the same everywhere, and that many signs are highly suggestive of some environmental exposure. For ADHD, in contrast, it seems that rates of the disease are correlated with exposure to almost every organic compound that has been studied, as well as some metals that are beyond the scope of this review. The major unanswered question regarding ADHD is the degree to which any other factor than exposure to environmental contaminants contributes to the disease. For FASD, we have only one organic chemical responsible. For all three diseases, there are clear indications of gene–environment interactions, but genetics alone explains very little of the incidence.

There also remain major unknowns with regard to mechanisms, both in the basic neurobiology of the diseases and about the mechanisms whereby organic contaminants exert their adverse effects. The central nervous system pathways that integrate cognitive function to behaviors such as attention span, distractibility, and ability to deal with frustration are not well understood. We have only limited understanding of the mechanisms of interaction of organic chemicals on nervous system function. In addition, our understanding of which chemicals are associated with intellectual developmental disorders is incomplete and is limited to a great degree by our ability to do an adequate exposure assessment. We know much more about the toxic actions of chemicals that are persistent simply because we can monitor levels in blood or other body fluids or tissues and can obtain some indication of long-term exposure. But just because a chemical does not stay in the body for years does not mean that it does not have toxic effects on neurodevelopment.

Another concern is the degree to which these exposures early in life are irreversible. For FASD, there is strong evidence for little or no improvement over time, which for ADHD and autism the degree of disability often does decrease with age and appropriate levels of intervention. However, it is tragic when exposure to organic chemicals during development leads to lifelong reduction of cognitive function and behavioral changes that reduce the ability to live independent and productive lives.

REFERENCES

Abdullah MM, Ly AR, Goldberg WA, Clarke-Stewart KA, Dudgeon JV, Mull CG, et al. 2011. Heavy metal in children's tooth enamel: Related to autism and disruptive behaviors? J Autism Dev Disord 42:929–936.

Aguiar A, Eubig PA, Schantz SL. 2010. Attention deficit/hyperactivity disorder: A focused overview for children's environmental health researchers. Environ Health Perspect 118:1646–1653.

Archibald SL, Fennema-Notestine C, Gamst A, Riley EP, Mattson SN, Jernigan TL. 2001. Brain dysmorphology in individuals with severe prenatal alcohol exposure. Dev Med Child Neurol 43:148–154.

Axelrad DA, Bellinger DC, Ryan LM, Woodruff TJ. 2007. Dose-response relationship of prenatal mercury exposure and IQ: An integrative analysis of epidemiologic data. Environ Health Perspect 115:609–615.

Barrett DH, Morris RD, Akhtar FZ, Michalek JE. 2001. Serum dioxin and cognitive functioning among veterans of Operation Ranch Hand. Neurotoxicology 22: 491–502.

Bell SH, Stade B, Reynolds JN, Rasmussen C, Andrew G, Hwang PA, et al. 2010. The remarkably high prevalence of epilepsy and seizure history in fetal alcohol spectrum disorders. Alcohol Clin Exp Res 34:1084–1089.

Biederman J, Faraone SV, Keenan K, Steingard R, Tsuang MT. 1991. Familial association between attention deficit disorder and anxiety disorders. Am J Psychiatry 148:251–256.

Bouchard MF, Bellinger DC, Wright RO, Weisskopf MG. 2010. Attention-deficit/hyperactivity disorder and urinary metabolites of organophosphate pesticides. Pediatrics 125:e1270–e1277.

Bouchard MF, Chevrier J, Harley KG, Kogut K, Vedar M, Calderon N, et al. 2011. Prenatal exposure to organophosphate pesticides and IQ in 7-year-old children. Environ Health Perspect 119:1189–1195.

Boucher O, Muckle G, Bastien CH. 2009. Prenatal exposure to polychlorinated biphenyls: A neuropsychologic analysis. Environ Health Perspect 117:7–16.

Boucher O, Burden MJ, Muckle G, Saint-Amour D, Ayotte P, Dewailly E, et al. 2012. Response inhibition and error monitoring during a visual go/no-go task in Inuit children exposed to lead, polychlorinated biphenyls, and methylmercury. Environ Health Perspect 120:608–615.

Bradley JD, Golden CJ. 2001. Biological contributions to the presentation and understanding of attention-deficit/hyperactivity disorder: A review. Clin Psychol Rev 21: 907–929.

Braun JM, Kahn RS, Froehlich T, Auinger P, Lanphear BP. 2006. Exposures to environmental toxicants and attention deficit hyperactivity disorder in U.S. children. Environ Health Perspect 114:1904–1909.

Braun JM, Kalkbrenner AE, Calafat AM, Yolton K, Ye X, Dietrich KN, et al. 2011. Impact of early-life bisphenol A exposure on behavior and executive function in children. Pediatrics 128:873–882.

Braun JM, Yolton K, Dietrich KN, Hornung R, Ye X, Calafat AM, et al. 2009. Prenatal bisphenol A exposure and early childhood behavior. Environ Health Perspect 117:1945–1952.

Braunschweig D, Duncanson P, Boyce R, Hansen R, Ashwood P, Pessah IN, et al. 2011. Behavioral correlates of maternal antibody status among children with autism. J Autism Dev Disord 42:1435–1445.

Bushkin-Bedient S, Carpenter DO. 2010. Benefits versus risks associated with consumption of fish and other seafood. Rev Environ Health 25:161–191.

Canfield RL, Henderson CR, Jr., Cory-Slechta DA, Cox C, Jusko TA, Lanphear BP. 2003. Intellectual impairment in children with blood lead concentrations below 10 microg per deciliter. N Engl J Med 348:1517–1526.

Canitano R. 2007. Epilepsy in autism spectrum disorders. Eur Child Adolesc Psychiatry 16:61–66.

Carpenter DO, Nevin R. 2010. Environmental causes of violence. Physiol Behav 99: 260–268.

Carpenter DO, Hussain RJ, Berger DF, Lombardo JP, Park HY. 2002. Electrophysiologic and behavioral effects of perinatal and acute exposure of rats to lead and polychlorinated biphenyls. Environ Health Perspect 110(Suppl 3):377–386.

Castellanos FX, Tannock R. 2002. Neuroscience of attention-deficit/hyperactivity disorder: The search for endophenotypes. Nat Rev Neurosci 3:617–628.

CDC (Centers for Disease Control and Prevention). 2009. Alcohol use among pregnant and nonpregnant women of childbearing age—United States, 1991–2005. MMWR Morb Mortal Wkly Rep 58:529–532.

Chen YC, Guo YL, Hsu CC, Rogan WJ. 1992. Cognitive development of Yu-Cheng ("oil disease") children prenatally exposed to heat-degraded PCBs. JAMA 268: 3213–3218.

Cho S-C, Kim B-N, Hong Y-C, Shia M-S, Yoo H-J, Kim J-W, et al. 2010. Effect of environmental exposure to lead and tobacco smoke on inattentive and hyperactive symptoms and neurocognitive performance in children. J Child Psychol Psychiatry 51:1050–1057.

Cohly HH, Panja A. 2005. Immunological findings in autism. Int Rev Neurobiol 71:317–341.

Colborn T. 2006. A case for revisiting the safety of pesticides: A closer look at neurodevelopment. Environ Health Perspect 114:10–17.

Coles CD, Platzman KA, Raskind-Hood CL, Brown RT, Falek A, Smith IE. 1997. A comparison of children affected by prenatal alcohol exposure and attention deficit, hyperactivity disorder. Alcohol Clin Exp Res 21:150–161.

Comi AM, Zimmerman AW, Frye VH, Law PA, Peeden JN. 1999. Familial clustering of autoimmune disorders and evaluation of medical risk factors in autism. J Child Neurol 14:388–394.

Crocker N, Vaurio L, Riley EP, Mattson SN. 2009. Comparison of adaptive behavior in children with heavy prenatal alcohol exposure or attention-deficit/hyperactivity disorder. Alcohol Clin Exp Res 33:2015–2023.

Debes F, Budtz-Jørgensen E, Weihe P, White RF, Grandjean P. 2006. Impact of prenatal methylmercury exposure on neurobehavioral function at age 14 years. Neurotoxicol Teratol 28:363–375.

Demicheli V, Rivetti A, Debalini MG, Di Pietrantonj C. 2012. Vaccines for measles, mumps and rubella in children. Cochrane Database Syst Rev (2):CD004407.

DeSoto MC. 2009. Ockham's razor and autism: The case for developmental neurotoxins contributing to a disease of neurodevelopment. Neurotoxicology 30:331–337.

DeStefano F, Thompson WW. 2004. MMR vaccine and autism: An update of the scientific evidence. Expert Rev Vaccines 3:19–22.

Edwards SC, Jedrychowski W, Butscher M, Camann D, Kieltyka A, Mroz E, et al. 2010. Prenatal exposure to airborne polycyclic aromatic hydrocarbons and children's intelligence at 5 years of age in a prospective cohort study in Poland. Environ Health Perspect 118:1326–1331.

Eskenazi B, Marks AR, Bradman A, Harley K, Barr DB, Johnson C, et al. 2007. Organophosphate pesticide exposure and neurodevelopment in young Mexican-American children. Environ Health Perspect 115:792–798.

Eubig PA, Aguiar A, Schantz SL. 2010. Lead and PCBs as risk factors for attention deficit/hyperactivity disorder. Environ Health Perspect 118:1654–1667.

Fombonne E. 2003. Epidemiological surveys of autism and other pervasive developmental disorders: An update. J Autism Dev Disord 33:365–382.

Fombonne E, Chakrabarti S. 2001. No evidence for a new variant of measles-mumps-rubella-induced autism. Pediatrics 108:E58.

Froehlich TE, Lanphear BP, Auinger P, Hornung R, Epstein JN, Braun J, et al. 2009. Association of tobacco and lead exposures with attention-deficit/hyperactivity disorder. Pediatrics 124:e1054–e1063.

Gascon M, Vrijheid M, Martínez D, Forns J, Grimalt JO, Torrent M, et al. 2011. Effects of pre and postnatal exposure to low levels of polybromodiphenyl ethers on neurodevelopment and thyroid hormone levels at 4 years of age. Environ Int 37:605–611.

Gascon M, Fort M, Martínez D, Carsin AE, Forns J, Grimalt JO, et al. 2012. Polybrominated diphenyl ethers (PBDEs) in breast milk and neuropsychological development in infants. Environ Health Perspect 120:1760–1765.

Geier D, Geier MR. 2004. Neurodevelopmental disorders following thimerosal-containing childhood immunizations: A follow-up analysis. Int J Toxicol 23: 369–376.

Geier DA, Geier MR. 2006. A prospective assessment of porphyrins in autistic disorders: A potential marker for heavy metal exposure. Neurotox Res 10:57–64.

Goines P, Haapanen L, Boyce R, Duncanson P, Braunschweig D, Delwiche L, et al. 2011. Autoantibodies to cerebellum in children with autism associate with behavior. Brain Behav Immun 25:514–523.

Grandjean P, Weihe P, White RF, Debes F, Araki S, Yokoyama K, et al. 1997. Cognitive deficit in 7-year-old children with prenatal exposure to methylmercury. Neurotoxicol Teratol 19:417–428.

Guillette EA, Meza MM, Aquilar MG, Soto AD, Garcia IE. 1998. An anthropological approach to the evaluation of preschool children exposed to pesticides in Mexico. Environ Health Perspect 106:347–353.

Haase RF, McCaffrey RJ, Santiago-Rivera AL, Morse GS, Tarbell A. 2009. Evidence of an age-related threshold effect of polychlorinated biphenyls (PCBs) on neuropsychological functioning in a Native American population. Environ Res 109: 73–85.

Harpin VA. 2005. The effect of ADHD on the life of an individual, their family, and community from preschool to adult life. Arch Dis Child 90(Suppl 1):i2–i7.

Hertz-Picciotto I, Delwiche L. 2009. The rise in autism and the role of age at diagnosis. Epidemiology 20:84–90.

Hertz-Picciotto I, Green PG, Delwiche L, Hansen R, Walker C, Pessah IN. 2010. Blood mercury concentrations in CHARGE Study children with and without autism. Environ Health Perspect 118:161–166.

Hertz-Picciotto I, Bergman A, Fängström B, Rose M, Krakowiak P, Pessah I, et al. 2011. Polybrominated diphenyl ethers in relation to autism and developmental delay: A case-control study. Environ Health 10:1.

Hoffman K, Webster TF, Weisskopf MG, Weinberg J, Vieira VM. 2010. Exposure to polyfluoroalkyl chemicals and attention deficit/hyperactivity disorder in U.S. children 12–15 years of age. Environ Health Perspect 118:1762–1767.

Ilott NE, Saudino KJ, Asherson P. 2010. Genetic influences on attention deficit hyperactivity disorder symptoms from age 2 to 3: A quantitative and molecular genetic investigation. BMC Psychiatry 10:102.

Ip P, Wong V, Ho M, Lee J, Wong W. 2004. Mercury exposure in children with autistic spectrum disorder: Case-control study. J Child Neurol 19:431–434.

Jacobson JL, Jacobson SW. 1996. Intellectual impairment in children exposed to polychlorinated biphenyls *in utero*. N Engl J Med 335:783–789.

Jones KL, Smith DW. 1973. Recognition of the fetal alcohol syndrome in early infancy. Lancet 302:999–1001.

Kabir Z, Connolly GN, Alpert HR. 2011. Secondhand smoke exposure and neurobehavioral disorders among children in the United States. Pediatrics 128:263–270.

Kalkbrenner AE, Braun JM, Durkin MS, Maenner MJ, Cunniff C, Lee LC, et al. 2012. Maternal smoking during pregnancy and the prevalence of autism spectrum disorders, using data from the autism and developmental disabilities monitoring network. Environ Health Perspect 120:1042–1048.

Kim BN, Cho SC, Kim Y, Shin MS, Yoo HJ, Kim JW, et al. 2009. Phthalates exposure and attention-deficit/hyperactivity disorder in school-age children. Biol Psychiatry 66:958–963.

Koger SM, Schettler T, Weiss B. 2005. Environmental toxicants and developmental disabilities: A challenge for psychologists. Am Psychol 60:243–255.

Korvatska E, Van de Water J, Anders TF, Gershwin ME. 2002. Genetic and immunologic considerations in autism. Neurobiol Dis 9:107–125.

Lai TJ, Liu X, Guo YL, Guo NW, Yu ML, Hsu CC, et al. 2002. A cohort study of behavioral problems and intelligence in children with high prenatal polychlorinated biphenyl exposure. Arch Gen Psychiatry 59:1061–1066.

Lakshmi Priya MD, Geetha A. 2011. Levels of trace elements (copper, zinc, magnesium and selenium) and toxic elements (lead and mercury) in the hair and nail of children with autism. Biol Trace Elem Res 142:148–158.

Landrigan PJ. 2010. What causes autism? Exploring the environmental contribution. Curr Opin Pediatr 22:219–225.

Lee DH, Jacobs DR, Porta M. 2007. Association of serum concentrations of persistent organic pollutants with the prevalence of learning disability and attention deficit disorder. J Epidemiol Community Health 61:591–596.

Leech SL, Richardson GA, Goldschmidt L, Day NL. 1999. Prenatal substance exposure: Effects on attention and impulsivity of 6-year-olds. Neurotoxicol Teratol 21: 109–118.

Lemoine P, Harousseau H, Borteyru JP, Menuet JC. 1968. Les enfant des parents alcooliques: Anomolies obervees, a propos de 127 cas. Ouest Med 21: 476–482.

Lin KC, Guo NW, Tsai PC, Yang CY, Guo YL. 2008. Neurocognitive changes among elderly exposed to PCBs/PCDFs in Taiwan. Environ Health Perspect 116: 184–189.

Lord C, Cook EH, Leventhal BL, Amaral DG. 2000. Autism spectrum disorders. Neuron 28:355–363.

Madsen KM, Hviid A, Vestergaard M, Schendel D, Wohlfahrt J, Thorsen P, et al. 2002. A population-based study of measles, mumps, and rubella vaccination and autism. N Engl J Med 347:1477–1482.

Menezes-Filho JA, Novaes Cde O, Moreira JC, Sarcinelli PN, Mergler D. 2011. Elevated manganese and cognitive performance in school-aged children and their mothers. Environ Res 111:156–163.

Mill J, Petronis A. 2008. Pre- and peri-natal environmental risks for attention-deficit hyperactivity disorder (ADHD): The potential role of epigenetic processes in mediating susceptibility. J Child Psychol Psychiatry 49:1020–1030.

Minnes S, Singer LT, Kirchner HL, Short E, Lewis B, Satayaathum S, et al. 2010. The effects of prenatal cocaine exposure on problem behavior in children 5–10 years. Neurotoxicol Teratol 32:443–451.

Morrow EM, Yoo SY, Flavell SW, Kim TK, Lin Y, Hill RS, et al. 2008. Identifying autism loci and genes by tracing recent shared ancestry. Science 321:218–223.

Nardelli A, Lebel C, Rasmussen C, Andrew G, Beaulieu C. 2011. Extensive deep gray matter volume reductions in children and adolescents with fetal alcohol spectrum disorders. Alcohol Clin Exp Res 35:1404–1417.

Nash K, Koren G, Rovet J. 2011. A differential approach for examining the behavioural phenotype of fetal alcohol spectrum disorders. J Popul Ther Clin Pharmacol 18: e440–e453.

Needleman HL, Gatsonis CA. 1990. Low-level lead exposure and the IQ of children. A meta-analysis of modern studies. JAMA 263:673–678.

Newman J, Aucompaugh AG, Schell LM, Denham M, DeCaprio AP, Gallo MV, et al. 2006. PCBs and cognitive functioning of Mohawk adolescents. Neurotoxicol Teratol 28:439–445.

Palmer RF, Blanchard S, Wood R. 2009. Proximity to point sources of environmental mercury release as a predictor of autism prevalence. Health Place 15:18–24.

Peper M, Klett M, Morgenstern R. 2005. Neuropsychological effects of chronic low-dose exposure to polychlorinated biphenyls (PCBs): A cross-sectional study. Environ Health 4:22.

Perera F, Vishnevetsky J, Herbstman JB, Calafat AM, Xiong W, Rauh V, et al. 2012a. Prenatal bisphenol A exposure and child behavior in an inner-city cohort. Environ Health Perspect 120:1190–1194.

Perera FP, Tang D, Wang S, Vishnevetsky J, Zhang B, Diaz D, et al. 2012b. Prenatal polycyclic aromatic hydrocarbon (PAH) exposure and child behavior at age 6–7 years. Environ Health Perspect 120:921–926.

Polanczyk G, de Lima MS, Horta BL, Biederman J, Rohde LA. 2007. The worldwide prevalence of ADHD: A systematic review and metaregression analysis. Am J Psychiatry 164:954–948.

Price CS, Thompson WW, Goodson B, Weintraub ES, Croen LA, Hinrichsen VL, et al. 2010. Prenatal and infant exposure to thimerosal from vaccines and immunoglobulins and risk of autism. Pediatrics 126:656–664.

Ramsay M. 2010. Genetic and epigenetic insights into fetal alcohol spectrum disorders. Genome Med 2:27.

Rappley MD. 2005. Attention deficit-hyperactivity disorder. N Engl J Med 352: 165–173.

Rasmussen C, Benz J, Pei J, Andrew G, Schuller G, Abele-Webster L, et al. 2010. The impact of an ADHD co-morbidity on the diagnosis of FASD. Can J Clin Pharmacol 17:e165–e176.

Rauh V, Arunajadai S, Horton M, Perera F, Hoepner L, Barr DB, et al. 2011. Severn-year neurodevelopmental scores and prenatal exposure to chlorpyrifos, a common agricultural pesticide. Environ Health Perspect 119:1196–1201.

Rauh VA, Perera FP, Horton MK, Whyatt RM, Bansal R, Hao X, et al. 2012. Brain anomalies in children exposed prenatally to a common organophosphate pesticide. Proc Natl Acad Sci U S A 109:7871–7876.

Reinisch JM, Sanders SA, Mortensen EL, Rubin DB. 1995. *In utero* exposure to phenobarbital and intelligence deficits in adult men. JAMA 274:1518–1525.

Ribas-Fitó N, Torrent M, Carrizo D, Muñoz-Ortiz L, Júlvez J, Grimalt JO, et al. 2006. *In utero* exposure to background concentrations of DDT and cognitive functioning among preschoolers. Am J Epidemiol 164:955–962.

Riley EP, Mattson SN, Li TK, Jacobson SW, Coles CD, Kodituwakku PW, et al. 2003. Neurobehavioral consequences of prenatal alcohol exposure: An international perspective. Alcohol Clin Exp Res 27:362–373.

Riley EP, Infante MA, Warren KR. 2011. Fetal alcohol spectrum disorders: An overview. Neuropsychol Rev 21:73–80.

Roberts EM, English PB, Grether JK, Windham GC, Somberg L, Wolff C. 2007. Maternal residence near agricultural pesticide applications and autism spectrum disorders among children in the California Central Valley. Environ Health Perspect 115: 1482–1489.

Robins LN, Mills JL. 1993. Effects of *in utero* exposure to street drugs. Am J Public Health 83:S1–S32.

Ronald A, Hoekstra RA. 2011. Autism spectrum disorders and autistic traits: A decade of new twin studies. Am J Med Genet B Neuropsychiatr Genet 156:255–274.

Ruckart PZ, Kakolewski K, Bove FJ, Kaye WE. 2004. Long-term neurobehavioral health effects of methyl parathion exposure in children in Mississippi and Ohio. Environ Health Perspect 112:46–51.

Rückinger S, Rzehak P, Chen CM, Sausenthaler S, Koletzko S, Bauer CP, et al. 2010. Prenatal and postnatal tobacco exposure and behavioral problems in 10-year-old children: results from the GINI-plus prospective birth cohort study. Environ Health Perspect 118:150–154.

Sagiv SK, Thurston SW, Bellinger DC, Tolbert PE, Altshul LM, Korrick SA. 2010. Prenatal organochlorine exposure and behaviors associated with attention deficit hyperactivity disorder in school-aged children. Am J Epidemiol 171:593–601.

Sagiv SK, Thurston SW, Bellinger DC, Altshul LM, Korrick SA. 2012. Neuropsychological measures of attention and Impulse control among 8-year-old children exposed prenatally to organochlorines. Environ Health Perspect 120:904–909.

Sagvolden T, Johansen EB, Aase H, Russell VA. 2005. A dynamic developmental theory of attention-deficit/hyperactivity disorder (ADHD) predominantly hyperactive/impulsive and combined subtypes. Behav Brain Sci 28:397–419.

Salvador-Carulla L, Reed GM, Vaez-Azizi LM, Cooper SA, Martinez-Leal R, Bertelli M, et al. 2011. Intellectual developmental disorders: Towards a new name, definition and framework for "mental retardation/intellectual disability" in ICD-11. World Psychiatry 10:175–180.

Sampson PD, Streissguth AP, Bookstein FL, Little RE, Clarren SK, Dehaene P, et al. 1997. Incidence of fetal alcohol syndrome and prevalence of alcohol-related neurodevelopmental disorder. Teratology 56:317–326.

Schab DW, Trinh NH. 2004. Do artificial food colors promote hyperactivity in children with hyperactive syndromes? A meta-analysis of double-blind placebo-controlled trials. J Dev Behav Pediatr 25:423–434.

Schantz SL, Gasior DM, Polverejan E, McCaffrey RJ, Sweeney AM, Humphrey HE, et al. 2001. Impairments of memory and learning in older adults exposed to polychlorinated biphenyls via consumption of Great Lakes fish. Environ Health Perspect 109:605–611.

Silbernagel SM, Carpenter DO, Gilbert SG, Gochfeld M, Groth E, 3rd, Hightower JM, et al. 2011. Recognizing and preventing overexposure to methylmercury from fish and seafood consumption: Information for physicians. J Toxicol. DOI: 10.1155/2011/983072 [Online July 13, 2011].

Simon V, Czobor P, Bálint S, Mészáros A, Bitter I. 2009. Prevalence and correlates of adult attention-deficit hyperactivity disorder: Meta-analysis. Br J Psychiatry 194: 204–211.

Sly PD, Chen L, Cangell CL, Scott J, Zubrick S, Whitehouse A, et al. 2012. Polychlorinated biphenyls but not chlorinated pesticides are associated with externalizing behavior in adolescents. Organohal Comp. Available at http://www.dioxin20xx.org/pdfs/2012/1329.pdf. Accessed May 24, 2013.

Stamova B, Green PG, Tian Y, Hertz-Picciotto I, Pessah IN, Hansen R, et al. 2011. Correlations between gene expression and mercury levels in blood of boys with and without autism. Neurotox Res 19:31–48.

Stein CR, Savitz DA. 2011. Serum perfluorinated compound concentration and attention deficit/hyperactivity disorder in children 5–18 years of age. Environ Health Perspect 119:1466–1471.

Stergiakouli E, Thapar A. 2010. Fitting the pieces together: Current research on the genetic basis of attention deficit/hyperactivity disorder (ADHD). Neuropsychiatr Dis Treat 6:651–660.

Stewart P, Reihman J, Gump B, Lonky E, Darvill T, Pagano J. 2005. Response inhibition at 8 and 9 1/2 years of age in children prenatally exposed to PCBs. Neurotoxicol Teratol 27:771–780.

Stewart PW, Lonky E, Reihman J, Pagano J, Gump BB, Darvill T. 2008. The relationship between prenatal PCB exposure and intelligence (IQ) in 9-year-old children. Environ Health Perspect 116:1416–1422.

Streissguth A. 2007. Offspring effects of prenatal alcohol exposure from birth to 25 years: The Seattle prospective longitudinal study. J Clin Psychol Med Settings 14:81–101.

Streissguth AP, Bookstein FL, Barr HM, Sampson PD, O'Malley K, Young JK. 2004. Risk factors for adverse life outcomes in fetal alcohol syndrome and fetal alcohol effects. J Dev Behav Pediatr 25:228–238.

St Sauver JL, Barbaresi WJ, Katusic SK, Colligan RC, Weaver AL, Jacobsen SJ. 2004. Early life risk factors for attention-deficit/hyperactivity disorder: A population-based cohort study. Mayo Clin Proc 79:1124–1131.

Till C, Koren G, Rovet JF. 2001. Prenatal exposure to organic solvents and child neurobehavioral performance. Neurotoxicol Teratol 23:235–245.

Turner M, Barnby G, Bailey A. 2000. Genetic clues to the biological basis of autism. Mol Med Today 6:238–244.

Volk HE, Hertz-Picciotto I, Delwiche L, Lurmann F, McConnell R. 2011. Residential proximity to freeways and autism in the CHARGE study. Environ Health Perspect 119:873–877.

Wang SX, Wang ZH, Cheng XT, Li J, Sang ZP, Zhang XD, et al. 2007. Arsenic and fluoride exposure in drinking water: Children's IQ and growth in Shanyin county, Shanxi province, China. Environ Health Perspect 115:643–647.

Wasserman GA, Liu X, Parvez F, Ahsan H, Factor-Litvak P, Kline J, et al. 2007. Water arsenic exposure and intellectual function in 6-year-old children in Araihazar, Bangladesh. Environ Health Perspect 115:285–289.

Weintraub K. 2011. The prevalence puzzle: Autism counts. Nature 479:22–24.

Weitzman M, Byrd RS, Aligne CA, Moss M. 2002. The effects of tobacco exposure on children's behavioral and cognitive functioning: Implications for clinical and public health policy and future research. Neurotoxicol Teratol 24:397–406.

Windham GC, Zhang L, Gunier R, Croen LA, Grether JK. 2006. Autism spectrum disorders in relation to distribution of hazardous air pollutants in the San Francisco Bay area. Environ Health Perspect 114:1438–1444.

Xu X, Nembhard WN, Kan H, Kearney G, Zhang ZJ, Talbott EO. 2011. Urinary trichlorophenol levels and increased risk of attention deficit hyperactivity disorder among US school-aged children. Occup Environ Med 68:557–561.

Yokoo EM, Valente JG, Grattan L, Schmidt SL, Platt I, Silbergeld EK. 2003. Low level methylmercury exposure affects neuropsychological function in adults. Environ Health 2:8.

Yolton K, Dietrich K, Auinger P, Lanphear BP, Hornung R. 2005. Exposure to environmental tobacco smoke and cognitive abilities among U.S. children and adolescents. Environ Health Perspect 113:98–103.

Zerbo O, Iosif AM, Delwiche L, Walker C, Hertz-Picciotto I. 2011. Month of conception and risk of autism. Epidemiology 22:469–475.

CHAPTER 17

Mechanisms of the Neurotoxic Actions of Organic Chemicals

DAVID O. CARPENTER

ABSTRACT

Background: Many different organic chemicals cause reduced IQ, neurobehavioral abnormalities, and peripheral and central nervous system diseases.

Objective: The goal of this chapter is to present what is known about fundamental mechanisms whereby organic chemicals alter the physiology and biochemistry of neurons that might explain the human effects, based on studies from cellular and animal research.

Discussion: Organic chemicals have actions at many different levels. Some alter activity of voltage-dependent or transmitter-activated ion channels, while others alter cellular calcium concentration and more biochemical processes. Much, however, remains unknown. A further complication is that while most cellular and animal studies are done with a single chemical, humans are always exposed to chemical mixtures, which allows for additive, less-than-additive, and/or synergistic effects. These interactions have not been well studied.

Conclusions: While we know some of the sites of action of organic chemicals on the nervous system, there is much that is not yet understood. In part, this is because the detailed mechanisms behind cognition, behaviour, and general nervous system function are imperfectly understood. But clearly, organic chemicals alter many aspects of higher nervous system function in adverse ways.

Effects of Persistent and Bioactive Organic Pollutants on Human Health, First Edition.
Edited by David O. Carpenter.

INTRODUCTION

Organic chemicals can interact with nervous tissue at multiple levels. They can alter brain development and migration of neurons, as is the case with prenatal exposure to methyl mercury (Clarkson 1983) and alcohol (see Chapter 16). Other organics interfere with ion channel function, alter neurotransmitter receptors, alter rates of synthesis or metabolism of neurotransmitters, and/or affect neuronal biochemical mechanisms. Still other organics effect specific processes known to be involved in cognitive function, memory, and behavior. Others trigger nerve cell death and may contribute to the development of neurodegenerative diseases. Because of the great number of different organic chemicals and the variety of possible sites of neuronal interactions, this chapter will not attempt to be encyclopedic but will briefly review the variety of possible sites of action, then focus on a more limited list of subjects that have been relatively well studied and for which there is strong evidence of human disease resulting from exposure. We will also focus on the effects of a limited number of chemicals for which a considerable amount of information is available, primarily the persistent organic pollutants (POPs).

ORGANIC SUBSTANCES THAT ALTER NEURONAL VOLTAGE-ACTIVATED ION CHANNELS

Nerve cells communicate to each other primarily through electrical signals. Neurons and muscle cells differ from many other cell types in that there is a significant potential difference across the plasma membrane which is a result of the establishment of ion concentration gradients via the action of active transport processes and the presence of a high membrane resistance. This high resting resistance is due to a relative impermeability of the resting membrane to cations. As a consequence, the internal medium contains much less sodium and calcium and much more potassium than is found in the extracellular fluid. Communication over long distances is via action potentials, dependent upon the sequential opening of voltage-dependent sodium, calcium, and potassium channels. The rising phase of the action potential is due to the opening of sodium channels, allowing external sodium to enter the cell and causing a depolarization. This is followed by a slower opening of calcium channels, then an opening of potassium channels to cause potassium to flow down its concentration gradient and repolarize the cell. These ion channels are called "voltage dependent" because the trigger for initiative of the whole process is a change in the voltage across the membrane. Muscle contraction is due to a similar process whereby membrane depolarization triggers the generation of an action potential that rapidly propagates and triggers contraction.

There are a number of specific organic toxins that act at the level of ion channels. As aficionados of Ian Fleming and *From Russia With Love* may remember, the knife hidden in the shoe of the nasty lady agent was covered with tetrodotoxin, the puffer fish poison, which is one of several animal toxins that bind to and block voltage-active sodium channels, causing paralysis and death. There are also a variety of specific antagonists of calcium and potassium channels, many of which are useful research tools, and some of which have clinical use, especially those that block calcium channels. However, except when eating puffer fish, handling certain frogs in tropical rain forests, or getting too close to certain cone snails while SCUBA diving (Olivera et al. 1985), there is relatively little exposure to humans from these animal toxins acting at the level of ion channels.

Voltage-activated sodium channels in insects are the primary site of action of the organochlorine pesticides dichlorodiphenyltrichloroethane (DDT), dieldrin, and aldrin, and pyrethroid pesticides (Hille 1984; Soderlund 2012). DDT and the pyrethroids act on sodium channels to keep them in the open configuration, leading to hyperexcitation, which leads to insect death (Narahashi 2002). The pyrethroids have been best studied, and they alter the rapid kinetic transitions between the open and closed states of sodium channels (Cao et al. 2011). Interestingly, pyrethroids also alter the properties of calcium and chloride channels, although the degree to which these actions contribute to the neurotoxic effects is uncertain. Mammalian sodium channels are much less sensitive to all of these pesticides than are those in insects, which is the basis for the utility of these substances. There is, however, a subclass of human sodium channels that are pyrethroid sensitive (Soderlund 2012). In rodents, two types of behavioral response are seen with different pyrethroids: One involves increased aggression, hypersensitivity, tremor, and convulsions, while the other leads to choreoathetosis with excessive salivation.

While some of the symptoms of acute human poisoning with these pesticides may be due to action at sodium channels, it is very important to remember that no chemical necessarily acts at only one site. DDT may be very effective in altering activity at insect sodium channels, but these are not the targets of greatest concern for humans. DDT has actions at many other sites that are described in different chapters in this book.

Other organic substances alter ion channel activity either directly or indirectly. Alcohol is known to increase the response of applied acetylcholine in isolated rat neurons (Moriguchi et al. 2007), probably secondary to dissolving in the membrane and changing fluidity.

ORGANIC SUBSTANCES THAT ALTER TRANSMITTER RECEPTORS AND ASSOCIATED IONOTROPIC CHANNELS

Nerve cell-to-nerve cell communication occurs through release of neurotransmitter substances, which in turn activate membrane receptors. These receptors are coupled to specific ion channels that selectively open permeability of the membrane to one or more ions, which then flow down their concentration gradient. The ionotropic ion channels differ from the voltage-dependent ion channels in that it is the binding of a neurotransmitter substance to a specific receptor that causes the ion channel to open, not a change in voltage. There are a great variety of substances that have specific receptors on neurons. Acetylcholine is the excitatory transmitter at the neuromuscular junction. Release of acetylcholine from the synapse of the nerve onto the muscle fiber opens an ion channel that is selectively permeable to both sodium and potassium. This results in a depolarization, which then triggers the voltage-dependent channels that underlie the action potential, described earlier. The active protein of neurotoxic snake venom, such as that of cobras, binds to and blocks acetylcholine receptors, causing respiratory paralysis and death. Organics such as curare, used on arrowheads in the Amazon to paralyze small animals, block the acetylcholine-activated nicotinic receptor.

In the mammalian brain, the major fast excitatory neurotransmitter is glutamic acid. Like the nicotinic acetylcholine receptor channels on skeletal muscle, the glutamate-activated channels on mammalian neurons are permeable to both sodium and potassium, resulting in a fast membrane depolarization. In addition, there are inhibitory receptors coupled to ion channels selective for chloride that, when activated, cause membrane hyperpolarization. In mammals, these channels are opened by specific receptors for either glycine or gamma-aminobutyric acid (GABA). In insects, a major fast inhibitor neurotransmitter is glutamic acid, unlike the situation in mammals. Some pesticides act at the levels of these ionotropic channels. Fipronil, a phenylpyrazole insecticide, blocks both GABA and glutamate chloride channels in insects with a potency 315 times greater than mammalian inhibitory ion channels. As for the voltage-activated channels, there are a variety of different specific organic toxins or drugs that block specific channels. Most are of great use for the study of the basic biology of neurons or for the treatment of specific diseases. Strychnine is an antagonist at glycine-activated chloride channels and when given to an intact animal results in convulsions and hyperexcitability, leading to death. It is widely used as a rodenticide and sometimes causes convulsions in children who are accidentally exposed.

While transmitter ion channels are in general not major targets for organic chemicals of environmental concern to humans, some do elicit changes at this level. Several chlorinated pesticides produce hyperexcitability and convulsions in mammals. Katz et al. (1997) reported that chlorpyrifos and parathion bind to and desensitize nicotinic acetylcholine receptors. Vale et al. (2003) found that γ-hexachlorocyclohexane (lindane) inhibited both GABA and glycine responses, while α-endosulfan and dieldrin were more effective at GABA responses. Cyclodiene insecticides are potent inhibitors of chloride movement through $GABA_A$ ion channels (the ionotropic GABA channel) in rat brain (Gant et al. 1987). Polychlorinated biphenyls (PCBs) have been found to reduce chloride influx through $GABA_A$ channels by 50% at low micromolar concentrations (Inglefield and Shafer 2000). Ivermectin, an antiparasitic agent, acts by reducing desensitization of glutamate-sensitive chloride channels (Priel and Silberberg 2004), which are common in parasitic worms but not in mammals.

There are a number of demonstrations of altered synaptic activity in the presence of various POPs. Hong et al. (1998) reported that tetrachlorodibenzo-*p*-dioxin (TCDD) suppressed excitatory synaptic responses in some parts of the hippocampus but had no effect in other areas. Pessah and colleagues have performed a number of studies of the actions of several individual PCB congeners, with a focus of PCB 95 (2,2′3,5′,6-pentachlorobiphenyl). Kim et al. (2009) reported that both excitatory and inhibitory synaptic transmission were altered by perfusion of PCB 95 and PCB 170 (2,2′,3,3′,4,4′,5 heptachlorobiphenyl) onto slices of rodent hippocampus, the area of the brain most associated with cognitive function. They found that PCB 95 increased the size of the population response in these slices that reflected activation of glutamate ionotropic receptors. This action, they demonstrate, is a consequence of noncoplanar PCBs that stimulate calcium release from ryanodine-sensitive calcium stores in the endoplasmic reticulum (Wong et al. 2001).

ORGANICS AND METABOTROPIC TRANSMITTER AND NEUROHORMONE RECEPTORS

There are, however, other receptors, usually called metabotropic receptors, that elicit changes in intracellular biochemistry. Most catecholamine, hormone, and peptide receptors are of this type, but there are also metabotropic muscarinic acetylcholine, glutamate, and GABA receptors ($GABA_B$). These receptors are coupled to G proteins and act via triggering intracellular changes mediated by a variety of "second messengers," such as cyclic nucleotides, which regulate many intracellular kinases, or phospholipase C, diacylglycerol, and inositol-triphosphate, which have major roles in regulating intracellular calcium concentration (Berridge 1993), and may secondarily alter gene expression. These are the sites at which many different organic pollutants act, as described further. Again, there are a number of animal toxins that target this type of transmitter/peptide/hormone receptor (Servent and Fruchart-Gaillard 2009).

Activation of metabotropic receptors often activate different cellular kinases and/or transcription factors. For example, nonplanar PCBs with *ortho*-substituted chlorines cause translocation of protein kinase C (Kodavanti and Ward 1998), which in turn alters many aspects of neuronal cellular functioning (Hussain and Carpenter 2003) including motor behavior, learning, and memory (Birnbaum et al. 2004). In a later study, Kodavanti et al. (2003) found that metabolic products of these *ortho*-substituted congeners were even more potent than the parent PCBs. This is an important principle in that while metabolites of organic chemicals may not be as persistent as the parent compounds, this does not mean that they do not have biological effects. Aroclor 1254 has been demonstrated to increase phosphorylation of cyclic AMP-responsive element binding protein (CREB) (Inglefield et al. 2002), which also is important in the developing nervous system.

Shafer et al. (1996) studied cultured cerebellar granule cells and found that PCB 4 (2,2′-diclorobiphenyl) disrupted calcium homeostatis by actions on inositol phosphate signaling mediated by muscarinic actylcholine receptors. Later studies from the same laboratory (Kodavanti et al. 1998) showed at least some of these actions involved alterations in levels of protein kinase C. The organophosphate pesticide chlorpyrifos alters the function of many different metabotropic receptors by action on adenylyl cyclase activity (Aldridge et al. 2003; Auman et al. 2000), which is the second messenger for many of these receptors.

ORGANICS THAT ACT AT NEUROTRANSMITTER SYNTHESIS AND/OR UPTAKE

Ortho-substituted PCBs, but not dioxin-like congeners, inhibit the uptake of dopamine, serotonin, GABA, and glutamate (Mariussen and Fonnum 2001). Since dopamine and serotonin play critical roles in psychiatric disease, this effect may have significant consequences. Parkinson's disease (Chapter 18) results from loss of dopaminergic neurons in the substantia nigra, and there is some evidence for a role of dieldrin and heptachlor, both chlorinated pesticides, in the etiology of Parkinson's disease. Purkerson-Parker et al. (2001) reported that heptachlor and dieldrin bind to the dopamine transporter in rats. Kitazawa et al. (2001) studied PC12 cells (a dopaminergic cell line) and found that dieldrin caused generation of reactive oxygen species (ROS) and a rapid release of dopamine. ROS generation is a widely documented pathway to neuronal cell death. Kanthasamy et al. (2005) have reviewed the evidence for dieldrin neurotoxocity and concluded that dieldrin is selectively toxic to dopaminergic neurons, increasing production of ROS, inhibiting the dopamine transporter and vesicular monoamine transporter in presynaptic terminals, and impairing mitochondrial function.

Seegal et al. (1991a,b) reported significant reduction in dopamine concentration in the brains of rat and monkeys, respectively, when exposed to Aroclor mixtures. However, the reductions were more pronounced in some brain regions than others. Later, Seegal et al. (1997) studied the effects of two PCB congeners, PCB 47 (2,2′4,4′-tetrachlorobiphenyl) and PCB 77 (3,3′4, 4′-tetrachlorobiphenyl). PCB 47 is a di-*ortho* congener, with PCBs at the 2 and 2′ positions, whereas PCB 77 is a dioxin-like, coplanar congener. They found that PCB 77 caused increases in dopamine concentration, while PCB 47 caused decreases in dopamine concentration in the frontal cortex. Bemis and Seegal (2004) and Mariussen et al. (1999) demonstrated that the effects of *ortho*-substituted PCBs were mediated by the inhibition of the vesicular dopamine transporter. These results are important because they show clearly that not every PCB congener has identical actions and that the non-dioxin-like congeners have clear toxic effects on the brain.

ORGANICS THAT ACT BY ALTERING NEUROTRANSMITTER METABOLISM

Environmental organic substances may also alter the rates of metabolism of neurotransmitters and neurohormones. Organophosphate and carbamate pesticides exert their actions on pests by inhibiting acetylcholinesterase, the enzyme that breaks down acetylcholine (Pope 1999). As a result, acetylcholine accumulates in the synaptic cleft and overexcites neurons, which can be fatal to pests. Acute poisoning of humans show hyperexcitability symptoms similar to those of pests (Kamel and Hoppin 2004). However, as shown in Figure 17.1, organophosphate pesticides have a large number of secondary targets (Casida and Quistad 2004). It is often these secondary actions that are important to human health from exposure to pesticides. There is an intermediate syndrome that occurs 2–4 days after the acute poisoning and consists of muscle weakness and myopathy (Abdollahi and Karami-Mohajeri 2012). It is likely related to the excessive stimulation during the acute cholinergic phase, and the generation of ROS. Following severe acute poisoning, there is often the development of a peripheral neuropathy 10–20 days later with greater damage to motor than sensory nerves (Jokanović and Kosanović 2010; Keifer and Firestone 2007). Only some organophosphate pesticides cause this neuropathy, and it appears to be a consequence of inhibition of a membrane protein called "neuropathy target esterase."

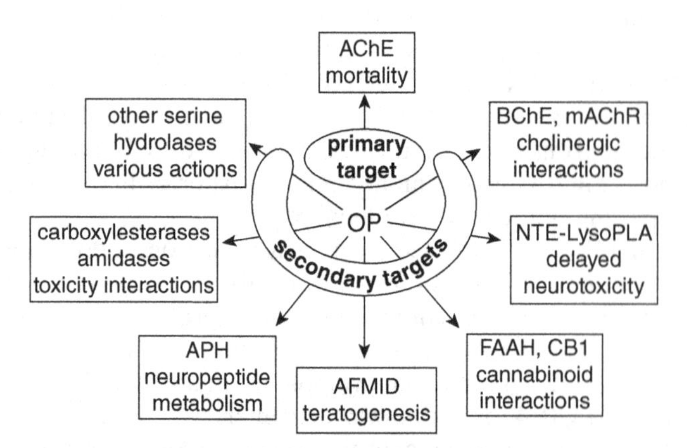

Figure 17.1. Primary and secondary targets of organophosphate serine hydrolase inhibitor pesticides (OP) (from Casida and Quistad 2004). While the primary effect is inhibition of acetylcholinesterase (AChE), these pesticides also alter activity of other serine hydrolases, carboxylesterases, amidases, acylpeptide hydrolase (APH), arylformamidase (AFMID), fatty acid amide hydrolase (FAAH) and its interactions with the cannabinoid CB1 receptor, neuropathy target esterase (NTE), lysophospholipase (LysoPLA), and butyrylcholinesterase (BChE) and its interaction with the muscarinic acetylcholine receptor (mAChR).

While the above-mentioned symptoms follow acute and high exposures, there is less understanding of the effects of low-level, chronic exposure to pesticides that inhibit acetylcholinesterase, some of which will be discussed further (Ray and Richards 2001).

Peripheral neuropathies are also triggered by other organic molecules. Acrylamide and 2,5-hexadione, the metabolite of n-hexane (DeCaprio et al. 2009), are best studied, but similar distal axonopathies are seen with other common solvents, agricultural and industrial chemicals (LoPachin 2000). The mechanisms of action involved the formation of protein adducts which disrupt protein structure and/or function (LoPachin and DeCaprio 2005). With acrylamide toxicity, there is a loss of the ionic concentration gradients across the axonal membrane, probably secondary to inhibition of Na^+/K^+-ATPase, the enzyme responsible for Na^+ and K^+ transport. Failure of sodium transport causes secondary dysfunction of calcium regulation, which ultimately results in axonal degeneration (LoPachin 2000).

CYTOTOXIC ACTIONS OF ORGANIC CHEMICALS

Parkinson's disease is caused by a loss of dopaminergic neurons in the substantia nigra (see Chapter 18). Mice treated with the herbicide paraquat were found to have loss of substantia nigra neurons and developed a neurobehavioral syndrome similar to Parkinson's (Brooks et al. 1999). Sanchez-Ramos et al. (1998) found that dieldrin was a potent cytotoxic agent against dopaminergic neurons in culture.

Several laboratories have reported that *in vitro* exposure of isolated neurons to PCBs disrupts intracellular calcium homeostasis and triggers a rapid cell death (Mariussen et al. 2002). Kodavanti et al. (1993) studied cerebellar granule cells dissociated from 6- to 8-day-old rat pups and maintained in tissue culture for 7 days. They reported that di-*ortho* PCB 4 (200 μM) was cytotoxic, as reflected by leakage of lactic dehydrogenase (LDH), after a 2-hour exposure, whereas the non-*ortho* PCB 126 (3,3′,4,4′,5-pentaclororbiphenyl) did not increase LDH leakage at the same concentration for up to 4 hours' exposure. Both congeners caused an increase in intracellular Ca^{2+} concentration, but the magnitude was greater with PCB 4. They found that PCB 4 inhibited Ca^{2+}-ATPase and concluded that some PCB congeners disrupt Ca^{2+} homeostasis in neurons.

Tan et al. (2004b) studied the cytotoxicity of several PCB congeners on acutely isolated cerebellar granule cells using flow cytometry. They measured cell viability using propidium iodide, a DNA-binding dye normally excluded from passing the membrane of intact neurons. Figure 17.2 (from Tan et al. 2004b) shows changes in cell viability over time of cells exposed to 10 μM concentrations of several different PCB congeners. The most potent was PCB 52 (2,2′5,5′-tetrachlorobiphenyl), followed by PCB 47 (2,2′,4,4′-tetrachlorobiphenyl), PCB 8 (2,4-dichlorobiphenyl), and PCB 28 (2,4,4′-trichlorobiphenyl). The non-*ortho* congeners (PCB 77, PCB 80 [3,3′5,5′-tetrachlorobiphenyl] and

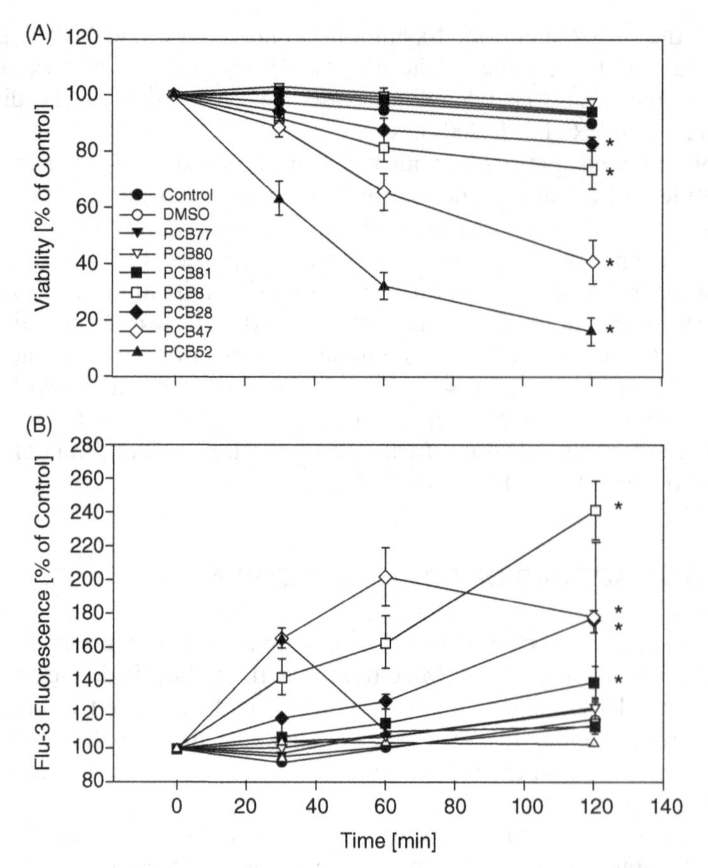

Figure 17.2. Effects of seven PCB congeners (10 μM) on membrane integrity (which reflects cell viability) (A) and intracellular calcium concentration (B) in acutely isolated rat cerebellar granule cells studied by flow cytometry. Membrane integrity was assessed by propidium iodide uptake, while intracellular calcium concentration was monitored using fluo-3 fluorescence. The apparent decrease in fluo-3 fluorescence with higher concentrations of PCB 52 and 47 reflects leakage of the dye due to loss of membrane integrity. * indicates $p < 0.05$ compared to control (0 minute). All results are mean values ± SEM. DMSO, dimethyl sulfoxide. Reproduced with permission from Tan et al. 2004b.

PCB 81 [3,4,4′,5-tetrachlorobipenyl]), had little, if any, effect. Figure 17.2B shows that the loss of cell viability was accompanied by an elevation of intracellular Ca^{2+} at all but the most cytotoxic concentrations when the fluo-3 dye escaped from the neurons because of loss of membrane integrity. These results show clearly that it is not the numbers of chlorines around the biphenyl ring that is important but whether there are chlorines in the *ortho* positions. However, Tan et al. (2004b) demonstrated that while calcium homeostasis was clearly altered by these *ortho*-substituted PCB congeners, it was not the elevation in intracellular calcium concentration that caused the cell death.

Immature neurons like cerebellar granule cells tolerate high Ca^{2+} concentrations to a greater degree than mature neurons. While incubation in Ca^{2+}-free solution prevented the elevation in intracellular calcium concentration, it did not prevent the cytotoxicity when *ortho*-substituted congeners were applied. In a later study, Tan et al. (2004a) demonstrated that the cause of the cytotoxicity was that the *ortho*-substituted PCB congeners dissolved in the membrane and increased membrane fluidity, altering plasma, mitochondrial, and endoplasmic reticulum membranes and disrupting cellular function.

The physiological significance of these *in vitro* studies is uncertain, but these results are especially important in documenting the toxicity of lower-chlorinated, *ortho*-substituted PCB congeners on the nervous system. Too often there is the assumption that all toxicity of PCB mixtures is mediated by the dioxin-like congeners, and this is clearly not the case.

Neuronal cell death is of at least two types, necrotic or apoptotic. The neuronal cell death described earlier is too rapid to be apoptotic, but rather appears to be a consequence of altered calcium homeostasis, generation of ROS, and increased membrane fluidity. However, dioxins and PCBs are also known to induce apoptotic cell death in neurons. Subchronic exposure of rats to TCDD or PCB 126, a dioxin-like congener, resulted in markers of oxidative stress in brain tissue (Hassoun et al. 2000). Sánchez-Alonso et al. (2003) reported that PCB 77 (dioxin-like), but not PCB 153 (2,2′,4,4′,5,5′-hexachlorobiphenyl, non-dioxin-like), induced apoptosis in neuronal cell cultures. However, two other reports show apoptosis from the non-dioxin-like PCB 52 (Hwang et al. 2001) and PCB 47, but not the dioxin-like PCB 77 (Howard et al. 2003). Thus, while the acute cell death appears to be due exclusively to *ortho*-substituted PCB congeners, it is likely that dioxin, dioxin-like PCBs, and non-dioxin-like PCBs can induce apoptosis under certain circumstances. The degree to which environmental concentrations of dixoins and/or PCBs act via these pathways is also unclear. Certainly, both dioxins (Hassoun et al. 2000) and PCB mixtures increase oxidative stress (Venkataraman et al. 2006), which is common to both pathways.

ORGANICS THAT ALTER GENE EXPRESSION

Many organic substances alter gene expression, and do so in diverse manners, but there has been relatively little attention to genes that specifically alter neuronal function. Benzene, for example, alters expression of many genes in leukemia cell lines (Sarma et al. 2011). The best-studied pathway is via activation of the aryl hydrocarbon receptor (AhR) by dioxin and dioxin-like compounds including some PCB congeners, some polyaromatic hydrocarbons, and some pesticides. A large number of different genes are either up- or down-regulated by AhR agonists (Mimura and Fujii-Kuriyama 2003; Vezina et al. 2004). But there is an additional complication in that activation of the AhR leads to activation of other pathways that also cause gene induction (Johnson et al. 2004). But gene induction is not limited to agonists of the AhR, as an equally large number of genes are also altered by the non-dioxin-like PCB congener, PCB 153 (Vezina et al. 2004). The altered genes are involved in almost every cell function, and it is difficult to predict how physiological responses will be affected as a consequence of these changes.

The patterns of gene expression are different in different tissues (Maier et al. 2007). Some genes are altered only by single chemicals, but many are altered, even in different directions, by more than one chemical. For example, Lasserre et al. (2009) studied the effects of atrazine and PCBs on protein expression in MCF-7 cells, a human breast cancer cell line. Of 22 differentially expressed proteins, 10 were specific to atrazine, 4 to PCB treatment, and 8 to both atrazine and PCBs. These studies show how complex the interactions may be when there is exposure to mixtures of chemicals.

There is, at present, relatively little information on the role of gene induction as a mechanism of effects of different organics on nervous tissue. Pre- and neonatal exposure to bisphenol A increased mRNA of the dopamine D1 receptor (Suzuki et al. 2003). PCBs alter expression of α-synuclein, synaptophysin, parkin (Malkiewicz et al. 2006), cofilin, and LIM kinase (Tang et al. 2007). Sazonova et al. (2011) exposed rat dams to Aroclor 1254 and investigated brain gene expression in offspring. They reported that 279 genes were altered and that most were due lactational but not gestational exposure. They found only 18 of these genes altered in the same fashion in all brain regions, and suggested that there is a great amount of region specificity. Slotkin and Seidler (2010) investigated gene expression in the PC12 cell line in response to organophostate (chlorpyrifos, diazinon) and organochlorine (dieldrin) pesticides and found that all upregulated genes for neuroactive peptides and their receptors. Chlorpyrifos upregulated genes of cholecystokinin, corticotropin releasing hormone, galanin, neuropeptide Y, neurotensin, preproenkephalin, and tachykinin 1, all peptides very essential for neurodifferentiation. Diazinon and dieldrin had similar actions but were less potent. The authors suggest that these effects on neuropeptides may be important contributors to neurobehavioral actions of a variety of pesticides.

Study of epigenetic changes in neuronal gene expression are recent but important. Wolstenholme et al. (2012) have reported the transgenerational effects of bisphenol A exposure on mRNA in the brain as well as on social behavior in mice. Genes documented as being decreased in the first generation included those for the estrogen receptor α, vasopressin, and oxytocin, and the genes for vasopressin and oxytocin remained decreased up to the F4 generation.

ORGANICS THAT ALTER NEURONAL MIGRATION AND STRUCTURE

As indicated in Chapter 16, ethanol has potent effects on nervous system function, impairing neuronal migration during development. The detailed mechanisms whereby this occurs are uncertain, but it is known that alcohol alters synaptic membrane fluidity (Zerouga and Beaugé 1992) and differentially alters responses to various transmitter substances (Soldo et al. 1994).

Normal dendritic growth of neurons is dependent upon thyroid hormone activity (Smith et al. 2002) and many organic substances interfere with thyroid function (Brucker-Davis 1998; Chapter 9). Kimura-Kuroda et al. (2007) demonstrated that several hydroxylated PCB metabolites inhibit the dendritic development of Purkinje cells and do so by inhibiting thyroid-dependent actions.

Lein et al. (2007) found that perinatal exposure of rat dams to Aroclor 1254 resulted in a pronounced age-related alteration in dendritic length in CA1 hippocampal pyramidal neurons and cerebellar granule cells. At postnatal day (PND), 22 dendritic length was reduced, but at PND 60, it was markedly increased. As reported earlier, Wong et al. (2001) found that PCB 95 increases intracellular calcium by an action at ryanodine receptors. Yang et al. (2009) found that developmental exposure to Aroclor 1254, a commercial PCB mixture, promoted dendritic growth in cerebellar Purkinje and neocortical pyramidal neurons but attenuated or reversed the normal experience-dependent dendritic growth that occurs with training. In a recent study from the same group, Wayman et al. (2012) reported that PCB 95 stimulates dendritic growth in cultured neurons and enhances Ca^{2+} oscillations in somata and dendrites, and that this is mediated via ryanodine receptors. Thus, many physiological functions that are dependent upon intracellular calcium concentration are altered by exposure to at least certain PCB congeners. Wayman et al. (2012) speculate that this action may be related to neurodevelopmental diseases where there are deficits in calcium signaling.

STUDIES OF ORGANICS ON THE CELLULAR BASIS OF LEARNING AND MEMORY

There is a large body of evidence showing that organophosphate pesticides (Prendergast et al. 1998), PCBs (Rice 1999a,b), and many other organic chemicals reduce cognitive function in animals. While the cellular basis of learning and memory is imperfectly understood, an electrophysiological change known as long-term potentiation (LTP) is known to be a central component. LTP is a prolonged increase in the effectiveness of synaptic response at specific synapses following a particular patterned input. The evidence that it is an essential component of learning includes documentation that animals that learn poorly show little or no LTP, and that pharmacological agents that block LTP also block the ability to learn (Lynch 2004).

As for humans (see Chapter 15), there is a strong body of evidence that exposure to PCBs reduces the ability of animals to learn. This has been demonstrated in monkeys (Bowman et al. 1978; Rice 1998) and in rats (Widholm et al. 2001). Interestingly, there appear to be sex differences in that males are more affected on some tasks (Roegge et al. 2000; Widholm et al. 2001), but females on others (Schantz et al. 1995). Eriksson and Fredriksson (1996) investigated learning in mice given a single injection of either PCB 28, PCB 52, PCB 118 (2,3′4,4′5 pentachlorobiphenyl) or PCB 156 (2,3,3′4,4′,5 hexachlorobiphenyl) at 1 day of age. PCBs 118 and 156 are mono-*ortho* PCBs with some dioxin-like activity. They found that PCBs 28 and 52, but not PCBs 118 and 156, caused changes in spontaneous behavior, learning, and memory. These observations indicate that non-dioxin-like PCBs are most important in altering neurobehavior.

This raises the question of whether PCBs alter LTP. Several laboratories have demonstrated that the gestational/lactational exposure to PCBs does reduce LTP (Altmann et al. 1995; Gilbert and Crofton 1999; Hussain et al. 2000). Figure 17.3 (from Hussain et al. 2000) shows LTP measured in the CA1 region of the hippocampus of 30-day rat pups exposed to one di-*ortho* PCB congener, PCB 153, through gestation and lactation. Compared to unexposed

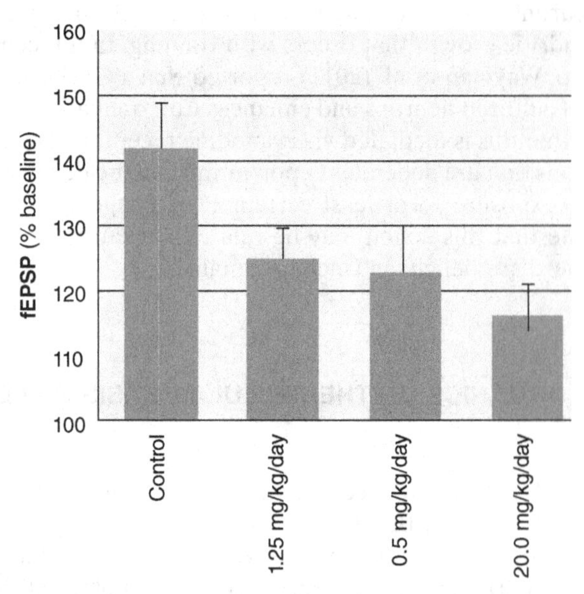

Figure 17.3. Effects of gestational and lactational exposure of rat pups to PCB 153 at various concentrations on the amplitude of the population fast excitatory synaptic potential (fEPSP) in the Schaffer collateral pathway to area CA1 in hippocampal brain slices obtained from the pups at age 30 days. LTP was elicited by application of two 1-minute tetani at 100 Hz. The results show the increase in fEPSP recorded 60 minutes after induction of LTP relative to that before tetanic stimulation (100%). Results were obtained from five animals at each exposure. The result from individual animals was determined as the average obtained from multiple hippocampal slices. The values for 1.25 and 20.0 mg/kg/day are significantly different from the control at the level $p < 0.05$. Reproduced with permission from Hussain et al. 2000.

animals, there is suppression of the monosynaptic response following a tetanic stimulus in hippocampal brain slices, and the LTP suppression is dose dependent.

In vivo exposure of animals is the best model for human exposures, and most investigators have assumed that the suppression of LTP following such studies reflects a development action of the contaminant. However, *in vitro* studies with acute exposure clearly show that there is also a pharmacological action. When hippocampal slices from 30-day rat pups not exposed during development were acutely perfused with a Krebs–Ringer solution containing PCB 153 (approximately 3 nM, prepared from a generator column), there was a similar suppression of LTP after a period of 1 hour (Figure 17.4 [from Hussain et al. 2000]). Later studies found that a coplanar PCB congener, PCB 77, also suppressed LTP in both hippocampal areas CA1 and CA3 when applied acutely to control brain slices (Figure 17.5 [from Ozcan et al. 2004]). However, this suppression occurred at a much higher concentration. These observations suggest that there is both a developmental and pharmacological toxicity of PCBs but do not provide evidence as to the relative importance of either. The observations of Ozcan et al. (2004) indicate that suppression of LTP is not only a result of *ortho*-substituted congeners but that *ortho*-substituted congeners appear to be more potent. Curran et al. (2011) have also presented evidence that both dioxin-like and non-dioxin-like PCBs alter LTP, as well as learning and memory, in a mouse model.

As detailed in Chapter 16, in humans, the reduction in IQ seen with PCBs and other organic chemicals is usually accompanied by changes in behavior.

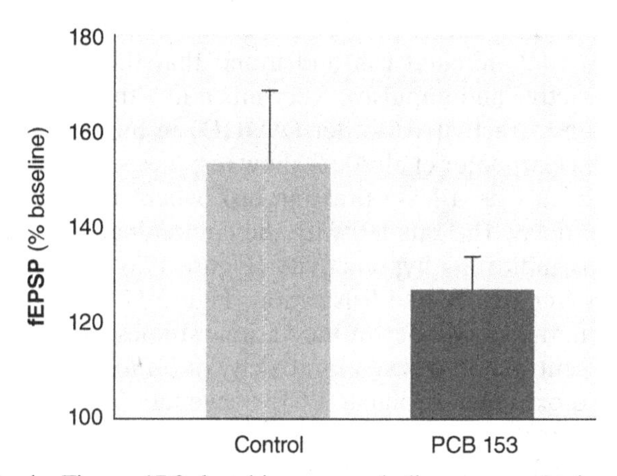

Figure 17.4. As in Figure 17.3, but hippocampal slices were obtained from control animals and the amplitude of the fEPSP monitored 60 minutes after tetanus in control slices ($n = 13$) and slices perfused with a Krebs–Ringer solution equilibrated with PCB 153 via a generator column ($n = 8$), giving a concentration of approximately 3 nM. The reduction in LTP after *in vitro* exposure to PCB 153 is significant at the $p < 0.05$ level. Reproduced with permission from Hussain et al. 2000.

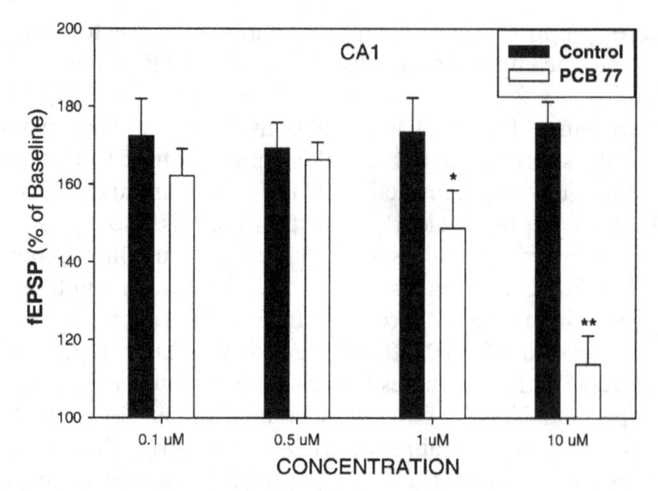

Figure 17.5. As in Figure 17.4 but with perfusion of PCB 77 at various concentrations. PCB 77 solutions were made with DMSO at a concentration of up to 0.05%. DMSO controls for 1 hour did not alter the fEPSP. The results show the dose dependence of reduction of LTP measured 1 hour after the tetanic stimulation. Each bar shows results from a minimum of five different experiments, each the average of measurements from multiple slices obtained from the same animal. * indicates significance at the $p < 0.05$ level; ** indicates significance at the $p < 0.01$ level. Reproduced with permission from Ozcan et al. 2004.

Holene et al. (1998) exposed rats via lactation to PCBs 126 and 153 and 126. Both congeners caused the rats to be hyperactive. Berger et al. (2001) fed young rats PCB-contaminated fish and found that this resulted in the rats becoming hyperactive and impulsive, very much like the syndrome seen with attention deficit/hyperactivity disorder (ADHD) in humans. Figure 17.6 and Figure 17.7 (from Carpenter et al. 2002) show responses of control and fish-fed male and female rats on a lever-pressing task where the reward came only after a 2-minute delay. The rats fed with the contaminated fish pressed more than the controls, indicating hyperactivity (Figure 17.6), and often pressed in bursts, reflecting impatience and frustration (Figure 17.7).

There is a general consistency in the various studies listed earlier showing a much greater central nervous system toxicity of *ortho*-substituted PCB congeners than of dioxin-like, coplanar PCB congeners. This issue was studied directly by Rice (1999a), who drew the conclusion that PCB neurotoxicity was mediated primarily by *ortho*-substituted congeners. There is also considerable evidence that the lower *ortho*-chlorinated congeners have significant toxicity. Lower-chlorinated congeners are more volatile, and therefore exposure can occur by inhalation, but are more rapidly metabolized in the body than are congeners with more chlorines (Matthews and Anderson 1975). However, this indicates that the use of the widely promoted toxic equivalent factors (TEFs),

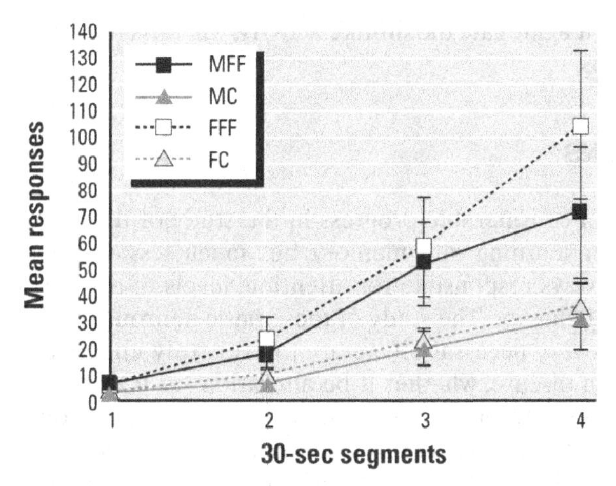

Figure 17.6. Mean responses of male (M) and female (F) rats born to either controls (C) or fish-fed dams (FF) highly contaminated with PCBs during gestation on days 9–19. Offspring were trained and tested at about 60 days of age. The animals earned water by pressing a bar, but the reward was given only after a 2-minute delay. Male and female controls increased the frequency of bar pressing as the 2-minute time was closer, but the offspring of the fish-fed dams pressed at much higher rates, especially in the last minute. These results indicate hyperactivity and impatience. Reproduced with permission from Carpenter et al. 2002.

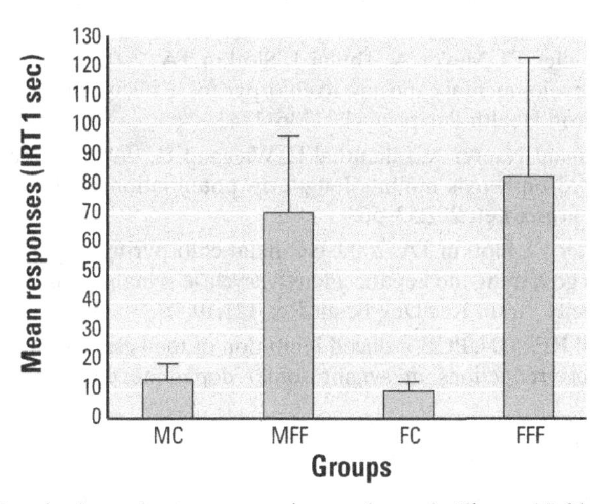

Figure 17.7. Results from the same experiment shown in Figure 17.6 but now plotting the number of burst responses. A burst response is a rapid repetitive pressing of the bar with an inter-response interval of less than 1 second. Both male and female fish-fed animals showed a much greater frequency of burst responses, which is an indication of impulsiveness and frustration. Reproduced with permission from Carpenter et al. 2002.

which represent aggregate dioxin-like activity, will miss most of the neurotoxic effects of PCBs.

CONCLUSIONS

There has been considerable progress in the study of the neuronal pathways responsible for learning and memory but much less understanding of the neuronal pathways responsible for attention, levels of activity, and impulsive and antisocial behavior. The study of these basic neuronal mechanisms is very important and very necessary if we are to ever truly understand the cause of nervous system disease, whether it be alterations in IQ, neurobehavioral diseases such as ADHD, or neurodegenerative diseases. Environmental organic chemicals alter nervous system function at multiple levels, targeting different mechanisms in different neurons. A further complication is that many organic chemicals have actions at multiple sites. However, understanding and preventing the adverse effects of exposure to organic chemicals on the incidence of human neurological diseases require that there be an understanding of the effects of the cellular and molecular levels using the techniques of cellular and animal research. Much more research is necessary to achieve these goals.

REFERENCES

Abdollahi M, Karami-Mohajeri S. 2012. A comprehensive review on experimental and clinical findings in intermediate syndrome caused by organophosphate poisoning. Toxicol Appl Pharmacol 258:309–314.

Aldridge JE, Seidler FJ, Meyer A, Thillai I, Slotkin TA. 2003. Serotonergic systems targeted by developmental exposure to chlorpyrifos: Effects during different critical periods. Environ Health Perspect 111:1736–1743.

Altmann L, Weinand-Haerer A, Lilienthal H, Wiegand H. 1995. Maternal exposure to polychlorinated biphenyls inhibits long-term potentiation in the visual cortex of adult rats. Neurosci Lett 202:53–56.

Auman JT, Seidler FJ, Slotkin TA. 2000. Neonatal chlorpyrifos exposure targets multiple proteins governing the hepatic adenylyl cyclase signaling cascade: Implications for neurotoxicity. Brain Res Dev Brain Res 121:19–27.

Bemis JC, Seegal RF. 2004. PCB-induced inhibition of the vesicular monoamine transporter predicts reductions in synaptosomal dopamine content. Toxicol Sci 80: 288–295.

Berger DF, Lombardo JP, Jeffers PM, Hunt AE, Bush B, Casey A, et al. 2001. Hyperactivity and impulsiveness in rats fed diets supplemented with either Aroclor 1248 or PCB-contaminated St. Lawrence river fish. Behav Brain Res 126:1–11.

Berridge MJ. 1993. Inositol trisphosphate and calcium signaling. Nature 361:315–325.

Birnbaum SG, Yuan PX, Wang M, Vijayraghavan S, Bloom AK, Davis DJ, et al. 2004. Protein kinase C overactivity impairs prefrontal cortical regulation of working memory. Science 306:882–884.

Bowman RE, Heironimus MP, Allen JR. 1978. Correlation of PCB body burden with behavioral toxicology in monkeys. Pharmacol Biochem Behav 9:49–56.

Brooks AI, Chadwick CA, Gelbard HA, Cory-Slechta DA, Federoff HJ. 1999. Paraquat elicited neurobehavorial syndrome caused by dopaminergic neuron loss. Brain Res 823:1–10.

Brucker-Davis F. 1998. Effects of environmental synthetic chemicals on thyroid function. Thyroid 8:827–856.

Cao Z, Shafer TJ, Crofton KM, Gennings C, Murray TF. 2011. Additivity of pyrethroid actions on sodium influx in cerebrocortical neurons in primary culture. Environ Health Perspect 119:1239–1246.

Carpenter DO, Hussain RJ, Berger DF, Lombardo JP, Park HY. 2002. Electrophysiologic and behavioral effects of perinatal and acute exposure of rats to lead and polychlorinated biphenyls. Environ Health Perspect 110(Suppl 3): 377–386.

Casida JE, Quistad GB. 2004. Organophosphate toxicology: Safety aspects of nonacetylcholinesterase secondary targets. Chem Res Toxicol 17:983–998.

Clarkson TW. 1983. Methylmercury toxicity to the mature and developing nervous system: possible mechanisms. In: Biological Aspects of Metals and Metal-Related Diseases (Sarkar B, ed.). New York: Raven Press, pp. 183–197.

Curran CP, Nebert DW, Genter MB, Patel KV, Schaefer TL, Skelton MR, et al. 2011. In utero and lactational exposure to PCBs in mice: Adult offspring show altered learning and memory depending on Cyp1a2 and Ahr genotypes. Environ Health Perspect 119:1286–1293.

DeCaprio AP, Kinney EA, LoPachin RM. 2009. Comparative covalent protein binding of 2,5-hexanedione and 3-acetyl-2,5-hexanedione in the rat. J Toxicol Environ Health A 72:861–869.

Eriksson P, Fredriksson A. 1996. Developmental neurotoxicity of four ortho-substituted polychlorinated biphenyls in the neonatal mouse. Environ Toxicol Pharmacol 1:155–165.

Gant DB, Eldefrawi ME, Eldefrawi AT. 1987. Cyclodiene insecticides inhibit GABAA receptor-regulated chloride transport. Toxicol Appl Pharmacol 88:313–321.

Gilbert ME, Crofton KM. 1999. Developmental exposure to a commercial PCB mixture (Aroclor 1254) produces a persistent impairment in long-term potentiation in the rat dentate gyrus in vivo. Brain Res 850:87–95.

Hassoun EA, Li F, Abushaban A, Stohs SJ. 2000. The relative abilities of TCDD and its congeners to induce oxidative stress in the hepatic and brain tissues of rats after subchronic exposure. Toxicology 145:103–113.

Hille B. 1984. Ionic Channels of Excitable Membranes. Sunderland, MA: Sinauer.

Holene E, Nafstad I, Skaare JU, Sagvolden T. 1998. Behavioural hyperactivity in rats following postnatal exposure to sub-toxic doses of polychlorinated biphenyl congeners 153 and 126. Behav Brain Res 94:213–224.

Hong SJ, Grover CA, Safe SH, Tiffany-Castiglioni E, Frye GD. 1998. Halogenated aromatic hydrocarbons suppress CA1 field excitatory postsynaptic potentials in rat hippocampal slices. Toxicol Appl Pharmacol 148:7–13.

Howard AS, Fitzpatrick R, Pessah I, Kostyniak P, Lein PJ. 2003. Polychlorinated biphenyls induce caspase-dependent cell death in cultured embryonic rat hippocampal but not cortical neurons via activation of the ryanodine receptor. Toxicol Appl Pharmacol 190:72–86.

Hussain RJ, Carpenter DO. 2003. The effects of protein kinase C activity on synaptic transmission in two areas of rat hippocampus. Brain Res 990:28–37.

Hussain RJ, Gyori J, DeCaprio AP, Carpenter DO. 2000. In vivo and in vitro exposure to PCB 153 reduces long-term potentiation. Environ Health Perspect 108:827–831.

Hwang SG, Lee HC, Lee DW, Kim YS, Joo WH, Cho YK, et al. 2001. Induction of apoptic cell death by a p53-independent pathway in neuronal SK-N-MC cells after treatment with 2,2′, 5,5′-tetrachlorobiphenyl. Toxicology 165:179–188.

Inglefield JR, Shafer TJ. 2000. Perturbation by the PCB mixture Aroclor 1254 of GABA(A) receptor-mediated calcium and chloride responses during maturation in vitro of rat neocortical cells. Toxicol Appl Pharmacol 164:184–195.

Inglefield JR, Mundy WR, Meacham CA, Shafer TJ. 2002. Identification of calcium-dependent and -independent signaling pathways involved in polychlorinated biphenyl-induced cyclic AMP-responsive element-binding protein phosphorylation in developing cortical neurons. Neuroscience 115:559–573.

Johnson CD, Balagurunathan Y, Tadesse MG, Falahatpisheh MH, Brun M, Walker MK, et al. 2004. Unraveling gene-gene interactions regulated by ligands of the aryl hydrocarbon receptor. Environ Health Perspect 112:403–412.

Jokanović M, Kosanović M. 2010. Neurotoxic effects in patients poisoned with organophosphorus pesticides. Environ Toxicol Pharmacol 29:195–201.

Kamel F, Hoppin JA. 2004. Association of pesticide exposure with neurologic dysfunction and disease. Environ Health Perspect 112:950–958.

Kanthasamy AG, Kitazawa M, Kanthasamy A, Anantharam V. 2005. Dieldrin-induced neurotoxicity: Relevance to Parkinson's disease pathogenesis. Neurotoxicology 26:701–719.

Katz EJ, Cortes VI, Eldefrawi ME, Eldefrawi AT. 1997. Chlorpyrifos, parathion, and their oxons bind to and desensitize a nicotinic acetylcholine receptor: Relevance to their toxicities. Toxicol Appl Pharmacol 146:227–236.

Keifer MC, Firestone J. 2007. Neurotoxicity of pesticides. J Agromedicine 12:17–25.

Kim KH, Inan SY, Berman RF, Pessah IN. 2009. Excitatory and inhibitory synaptic transmission is differentially influenced by two ortho-substituted polychlorinated biphenyls in the hippocampal slice preparation. Toxicol Appl Pharmacol 237:168–177.

Kimura-Kuroda J, Nagata I, Kuroda Y. 2007. Disrupting effects of hydroxyl-polychlorinated biphenyl (PCB) congeners on neuronal development of cerebellar Purkinje cells: A possible causal factor for developmental brain disorders? Chemosphere 67:S412–S420.

Kitazawa M, Anantharam V, Kanthasamy AG. 2001. Dieldrin-induced oxidative stress and neurochemical changes contribute to apoptopic cell death in dopaminergic cells. Free Radic Biol Med 31:1473–1485.

Kodavanti PR, Ward TR. 1998. Interactive Effects of environmentally relevant polychlorinated biphenyls and dioxins on [3H]phorbol ester binding in rat cerebellar granule cells. Environ Health Perspect 106:479–486.

Kodavanti PR, Shin DS, Tilson HA, Harry GJ. 1993. Comparative effects of two polychlorinated biphenyl congeners on calcium homeostasis in rat cerebellar granule cells. Toxicol Appl Pharmacol 123:97–106.

Kodavanti PR, Derr-Yellin EC, Mundy WR, Shafer TJ, Herr DW, Barone S, Jr., et al. 1998. Repeated exposure of adult rats to Aroclor 1254 causes brain region-specific changes in intracellular Ca^{2+} buffering and protein kinase C activity in the absence of changes in tyrosine hydroxylase. Toxicol Appl Pharmacol 153: 186–198.

Kodavanti PR, Ward TR, Derr-Yellin EC, McKinney JD, Tilson HA. 2003. Increased [3H]phorbol ester binding in rat cerebellar granule cells and inhibition of 45Ca(2+) buffering in rat cerebellum by hydroxylated polychlorinated biphenyls. Neurotoxicology 24:187–198.

Lasserre JP, Fack F, Revets D, Planchon S, Renaut J, Hoffmann L, et al. 2009. Effects of the endocrine disruptors atrazine and PCB 153 on the protein expression of MCF-7 human cells. J Proteome Res 8:5485–5496.

Lein PJ, Yang D, Bachstetter AD, Tilson HA, Harry J, Mervis RF, et al. 2007. Ontogenetic alterations in molecular and structural correlates of dendritic growth after developmental exposure to polychlorinated biphenyls. Environ Health Perspect 115:556–563.

LoPachin RM. 2000. Redefining toxic distal axonopathies. Toxicol Lett 112–113:23–33.

LoPachin RM, DeCaprio AP. 2005. Protein adduct formation as a molecular mechanism in neurotoxicity. Toxicol Sci 86:214–225.

Lynch MA. 2004. Long-term potentiation and memory. Physiol Rev 84:87–136.

Maier MS, Legare ME, Hanneman WH. 2007. The aryl hydrocarbon receptor agonist 3,3′,4,4′,5-pentachlorobiphenyl induces distinct patterns of gene expression between hepatoma and glioma cells: Chromatin remodeling as a mechanism for selective effects. Neurotoxicology 28:594–612.

Malkiewicz K, Mohammed R, Folkesson R, Winblad B, Miroslaw S, Benedikz E, et al. 2006. Polychlorinated biphenyls alter expression of α-synuclein synaptophysin and parkin in the rat brain. Toxicol Lett 161:152–158.

Mariussen E, Fonnum F. 2001. The effect of polychlorinated biphenyls on the high affinity uptake of the neurotransmitters, dopamine, serotonin, glutamate and GABA, into rat brain synaptosomes. Toxicology 159:11–21.

Mariussen E, Andersen JM, Fonnum F. 1999. The effect of polychlorinated biphenyls on the uptake of dopamine and other neurotransmitters into rat brain synaptic vesicles. Toxicol Appl Pharmacol 161:274–282.

Mariussen E, Myhre O, Reistad T, Fonnum F. 2002. The polychlorinated biphenyl mixture Aroclor 1254 induces death of rat cerebellar granule cells: The involvement of the N-methyl-D-asparate receptor and reactive oxygen species. Toxicol Appl Pharmacol 179:137–144.

Matthews HB, Anderson MW. 1975. Effect of chlorination on the distribution and excretion of polychlorinated biphenyls. Drug Metab Dispos 3:371–380.

Mimura J, Fujii-Kuriyama Y. 2003. Functional role of AhR in the expression of toxic effects by TCDD. Biochim Biophys Acta 1619:263–268.

Moriguchi S, Zhao X, Marszalec W, Yeh JZ, Narahashi T. 2007. Effects of ethanol on excitatory and inhibitory synaptic transmission in rat cortical neurons. Alcohol Clin Exp Res 31:89–99.

Narahashi T. 2002. Nerve membrane ion channels as the target site of insecticides. Mini Rev Med Chem 2:419–432.

Olivera BM, Gray WR, Zeikus R, McIntosh JM, Varga J, Rivier J, et al. 1985. Peptide neurotoxins from fish-hunting cone snails. Science 230:1338–1343.

Ozcan M, Yilmaz B, King WM, Carpenter DO. 2004. Hippocampal long-term potentiation (LTP) is reduced by a coplanar PCB congener. Neurotoxicology 25: 981–988.

Pope CN. 1999. Organophosphorus pesticides: Do they all have the same mechanism of toxicity? J Toxicol Environ Health B Crit Rev 2:161–181.

Prendergast MA, Terry AV, Buccafusco JJ. 1998. Effects of chronic, low-level organophosphate exposure on delayed recall, discrimination, and spatial learning in monkeys and rats. Neurotoxicol Teratol 20:115–122.

Priel A, Silberberg SD. 2004. Mechanism of ivermectin facilitation of human P2X4 receptor channels. J Gen Physiol 123:281–293.

Purkerson-Parker S, McDaniel KL, Moser VC. 2001. Dopamine transporter binding in the rat striatum is increased by gestational, perinatal, and adolescent exposure to heptachlor. Toxicol Sci 64:216–223.

Ray DE, Richards PG. 2001. The potential for toxic effects of chronic, low-dose exposure to organophosphates. Toxicol Lett 120:343–351.

Rice DC. 1998. Effects of postnatal exposure of monkeys to a PCB mixture on spatial discrimination reversal and DRL performance. Neurotoxicol Teratol 20: 391–400.

Rice DC. 1999a. Behavioral impairment produced by low-level postnatal PCB exposure in monkeys. Environ Res 80:S113–S121.

Rice DC. 1999b. Effect of exposure to 3,3′,4,4′,5-pentachlorobiphenyl (PCB 126) throughout gestation and lactation on development and spatial delayed alternation performance in rats. Neurotoxicol Teratol 21:59–69.

Roegge CS, Seo BW, Crofton KM, Schantz SL. 2000. Gestational-lactational exposure to Aroclor 1254 impairs radial-arm maze performance in male rats. Toxicol Sci 57:121–130.

Sánchez-Alonso JA, López-Aparicio P, Recio MN, Pérez-Albarsanz MA. 2003. Apoptosis-mediated neurotoxic potential of a planar (PCB 77) and a nonplanar (PCB 153) polychlorinated biphenyl congeners in neuronal cell cultures. Toxicol Lett 144:337–349.

Sanchez-Ramos J, Facca A, Basit A, Song S. 1998. Toxicity of dieldrin for dopaminergic neurons in mesencephalic cultures. Exp Neurol 150:263–271.

Sarma SN, Kim YJ, Ryu JC. 2011. Differential gene expression profiles of human leukemia cell lines exposed to benzene and its metabolites. Environ Toxicol Pharmacol 32:285–295.

Sazonova NA, DasBanerjee T, Middleton FA, Gowtham S, Schuckers S, Faraone SV. 2011. Transcriptome-wide gene expression in a rat model of attention deficit hyperactivity disorder symptoms: Rats developmentally exposed to polychlorinated biphenyls. Am J Med Genet B Neuropsychiatr Genet B156:898–912.

Schantz SL, Moshtaghian J, Ness DK. 1995. Spatial learning deficits in adult rats exposed to ortho-substituted PCB congeners during gestation and lactation. Fundam Appl Toxicol 26:117–126.

Seegal RF, Bush B, Brosch KO. 1991a. Comparison of effects of Aroclors 1016 and 1260 on non-human primate catecholamine function. Toxicology 66:145–163.

Seegal RF, Bush B, Brosch KO. 1991b. Sub-chronic exposure of the adult rat to Aroclor 1254 yields regionally-specific changes in central dopaminergic function. Neurotoxicology 12:55–65.

Seegal RF, Brosch KO, Okoniewski RJ. 1997. Effects of in utero and lactational exposure of the laboratory rat to 2,4,2′,4′- and 3,4,3′,4′-tetrachlorobiphenyl on dopamine function. Toxicol Appl Pharmacol 146:95–103.

Servent D, Fruchart-Gaillard C. 2009. Muscarinic toxins: Tools for the study of the pharmacological and functional properties of muscarinic receptors. J Neurochem 109:1193–1202.

Shafer TJ, Mundy WR, Tilson HA, Kodavanti PR. 1996. Disruption of inositol phosphate accumulation in cerebellar granule cells by polychlorinated biphenyls: A consequence of altered Ca^{2+} homeostasis. Toxicol Appl Pharmacol 141: 448–455.

Slotkin TA, Seidler FJ. 2010. Diverse neurotoxicants converge on gene expression for neuropeptides and their receptors in an in vitro model of neurodifferentiation: Effects of chlorpyrifos, diazinon, dieldrin and divalent nickel in PC12 cells. Brain Res 1353:36–52.

Smith JW, Evans AT, Costall B, Smythe JW. 2002. Thyroid hormones, brain function and cognition: A brief review. Neurosci Biobehav Rev 26:45–60.

Soderlund DM. 2012. Molecular mechanisms of pyrethroid insecticide neurotoxicity: Recent advances. Arch Toxicol 86:165–181.

Soldo BL, Proctor WR, Dunwiddie TV. 1994. Ethanol differentially modulates GABAA receptor-mediated chloride currents in hippocampal, cortical, and septal neurons in rat brain slices. Synapse 18:94–103.

Suzuki T, Mizuo K, Nakazawa H, Funae Y, Fushiki S, Fukushima S, et al. 2003. Prenatal and neonatal exposure to bisphenol-A enhances the central dopamine D1 receptor-mediated action in mice: Enhancement of the methamphetamine-induced abuse state. Neuroscience 117:639–644.

Tan Y, Song R, Lawrence D, Carpenter DO. 2004a. Ortho-substituted but not coplanar PCBs rapidly kill cerebellar granule cells. Toxicol Sci 79:147–156.

Tan Y, Chen C-H, Lawrence D, Carpenter DO. 2004b. Ortho-substituted PCBs kill cells by altering membrane structure. Toxicol Sci 80:54–59.

Tang F, Yan C, Wu S, Li F, Yu Y, Jin X, et al. 2007. Polychlorinated biphenyls disrupt the actin cytoskeleton in hippocampal neurons. Environ Toxicol Pharmacol 23: 140–146.

Vale C, Fonfría E, Bujons J, Messeguer A, Rodríguez-Farré E, Suñol C. 2003. The organochlorine pesticides gamma-hexachlorocyclohexane (lindane), alpha-endosulfan and dieldrin differentially interact with GABA(A) and glycine-gated chloride channels in primary cultures of cerebellar granule cells. Neuroscience 117:397–403.

Venkataraman P, Muthuvel R, Krishnamoorthy G, Arunkumar A, Sridhar M, Srinivasan N, et al. 2006. PCB (Aroclor 1254) enhances oxidative damage in rat brain regions: Protective role of ascorbic acid. Neurotoxicology 28:490–498.

Vezina CM, Walker NJ, Olson JR. 2004. Subchronic exposure to TCDD, PeCDF, PCB126, and PCB153: Effect on hepatic gene expression. Environ Health Perspect 112:1636–1644.

Wayman GA, Bose DD, Yang D, Lesiak A, Bruun D, Impey S, et al. 2012. PCB-95 modulates the calcium-dependent signaling pathway responsible for activity-dependent dendritic growth. Environ Health Perspect 120:1003–1009.

Widholm JJ, Clarkson GB, Strupp BJ, Crofton KM, Seegal RF, Schantz SL. 2001. Spatial reversal learning in Aroclor 1254-exposed rats: Sex-specific deficits in associative ability and inhibitory control. Toxicol Appl Pharmacol 174:188–198.

Wolstenholme JT, Edwards M, Shetty SRJ, Gatewood JD, Taylor JA, Rissman EF, et al. 2012. Gestational exposure to bisphenol A produces transgenerational changes in behaviors and gene expression. Endocrinology 153:3828–3838.

Wong PW, Garcia EF, Pessah IN. 2001. Ortho-substituted PCB95 alters intracellular calcium signaling and causes cellular acidification in PC12 cells by an immunophilin-dependent mechanism. J Neurochem 76:450–463.

Yang D, Kim KH, Phimister A, Bachstetter AD, Ward TR, Stackman RW, et al. 2009. Developmental exposure to polychlorinated biphenyls interferes with experience-dependent dendritic plasticity and ryanodine receptor expression in weanling rats. Environ Health Perspect 117:426–435.

Zerouga M, Beaugé F. 1992. Rat synaptic membrane fluidity parameters after intermittent exposures to ethanol in vivo. Alcohol 9:311–315.

CHAPTER 18

Parkinson's Disease

SAMUEL M. GOLDMAN

ABSTRACT

Background: Parkinson's disease (PD) is a progressive neurodegenerative disorder of aging that affects 1% of the population over age 50. Although it is a systemic disease, the classic motor features of tremor at rest, slowing of movement, muscular rigidity, and impaired balance result primarily from degeneration of dopaminergic neurons in the substantia nigra. Identified genetic causes comprise only a few percent of PD cases, and several lines of research implicate a major etiologic role for environmental factors.

Objectives: This chapter provides a comprehensive review of PD environmental epidemiology and considers the biological plausibility of toxicant associations in the context of underlying pathological mechanisms.

Discussion: Associations between PD and pesticide exposure have been studied for decades. Evidence is strongest for paraquat, rotenone, and organochlorines such as dieldrin and hexachlorohexane (HCH). The solvents trichloroethylene (TCE) and tetrachloroethylene (PERC) have been associated epidemiologically and produce an animal model that closely recapitulates PD pathology. Mechanistic data implicate polychlorinated biphenyls (PCBs), and epidemiological associations in women are fairly consistent, but much more work is needed. Cellular and animal models also support a possible role for metals in PD etiology. Epidemiological evidence is strongest for lead. However, the most etiologically important exposures may occur perinatally or in early childhood and are difficult to measure.

Conclusions: Epidemiological and basic research strongly implicate a major etiologic role for environmental factors in PD, but only a handful of agents have
(*Continued*)

Effects of Persistent and Bioactive Organic Pollutants on Human Health, First Edition.
Edited by David O. Carpenter.

been consistently associated in human populations. Like most chronic diseases, most PD is caused by multiple factors. Future work should investigate interactions between toxicant exposures and genetic variants that affect their metabolism. The effects of combinations of agents at dosages commonly encountered in human populations should be studied in animal models. Future epidemiological studies in humans should focus on large or pooled populations with well-characterized unbiased exposure estimates and should consider specific agents rather than broad compound classes.

INTRODUCTION

Parkinson's disease (PD) is a progressive disabling neurodegenerative disorder affecting up to 1 million Americans, with 50,000–60,000 new cases annually—a figure expected to increase dramatically as the population ages (Parkinson's Disease Foundation 2011). Annual costs in the United States are estimated at $23 billion and a projected $50 billion by 2040 (Huse et al. 2005). Clinical features result primarily from degeneration of pigmented dopaminergic neurons in the substantia nigra pars compacta (Figure 18.1) and include tremor at rest, slowness and paucity of movement (bradykinesia; akinesia), rigidity, and impaired balance. In addition to the classic motor features, there is increasing recognition that PD is a systemic disorder with widespread anatomic involvement and consequent "nonmotor" symptoms. These include hyposmia (loss of sense of smell; associated with pathology in olfactory bulb), cardiac sympathetic denervation (associated with pathology in cardiac autonomic ganglia), constipation (associated with pathology in GI tract autonomic ganglia), depression, and changes in cognitive function (Braak et al. 2003a; Langston 2006). Nonmotor symptoms may occur years or even decades before

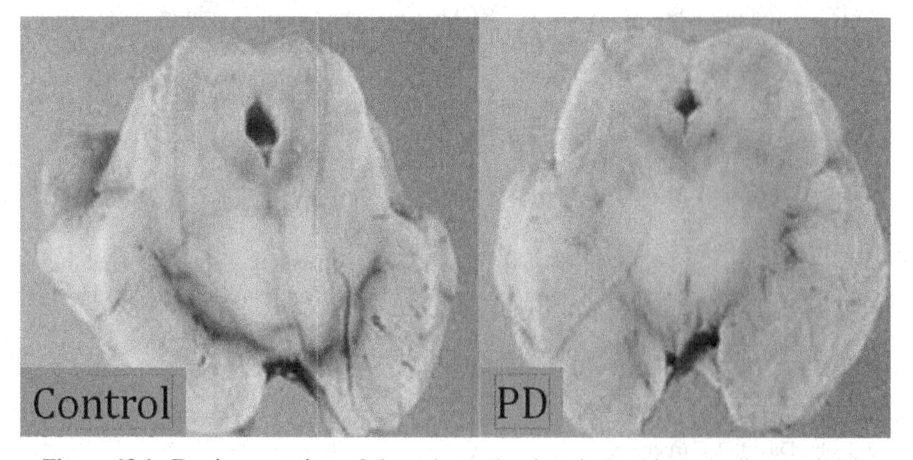

Figure 18.1. Depigmentation of the substantia nigra in Parkinson's disease (PD).

motor features manifest (Ross et al. 2012; Stern et al. 2012), leading some investigators to propose that the causal process may begin in the gut and slowly ascend to the brain via retrograde axonal transport of a toxicant or prion-like process (Braak et al. 2003b; Olanow and McNaught 2011).

There is currently no diagnostic test for PD. Motor signs and symptoms required to meet clinical diagnostic criteria (Gelb et al. 1999) manifest only after 60–80% of striatal dopamine is depleted and neuropathology is highly advanced (Bernheimer et al. 1973). The pathological hallmark of PD is the Lewy body, eosinophilic cytoplasmic inclusions found in dopaminergic neurons of the substantia nigra pars compacta, as well as in other neurons. Lewy bodies, and corresponding Lewy neurites found in axonal processes, consist largely of aggregated alpha-synuclein protein (Spillantini et al. 1998).

The term parkinsonism refers more generally to the motor features seen in PD, and not specifically to the full clinicopathological syndrome of PD. Parkinsonism may result from a variety of causes, including medications (e.g., antipsychotics), vascular insults, acute toxic insults, and other neurodegenerative syndromes. Features of parkinsonism were alluded to as far back as 1000 BC (Gourie-Devi et al. 1991; Manyam 1990), but the complete PD syndrome was not described until the height of the industrial revolution during the early to mid-19th century (Goetz 1986; Parkinson 2002). Competing theories on PD causation date back more than 150 years (Factor et al. 1988; Goetz 1986; Keppel Hesselink 1989) and continue to this day.

THE GENETIC HYPOTHESIS

Polymeropoulos and colleagues reported the first causative genetic mutation in 1997 (Polymeropoulos et al. 1997). Although causal mutations in the gene encoding alpha-synuclein protein (*SNCA*) have subsequently been found to be exceedingly rare, this discovery identified alpha-synuclein protein as playing a major role in typical PD and launched an intensive search for other genetic mutations. To date, at least 15 genes and genetic loci have been associated with PD (Corti et al. 2011), and as statistical power increases, genome-wide association studies (GWASs) continue to identify variants associated with smaller and smaller increases in risk (Nalls et al. 2011). The most common causal genetic variant is *LRRK2* G2019S (Goldwurm et al. 2005). This and several other *LRRK2* mutations are responsible for 1–2% of typical PD, although the frequency is much higher in North African, Ashkenazi Jewish and several other ethnic groups (Correia Guedes et al. 2010).

However, several observations argue against the primacy of genetic determinants of disease. Mutations in single genes explain only a few percent of PD cases (Klein and Ziegler 2011; Lesage and Brice 2009), and the total genetically determined disease liability is estimated at 25% (Do et al. 2011). Even the *LRRK2* G2019S mutation is only 30–40% penetrant (Bonifati 2007), suggesting an important role for environmental factors. In addition, twin

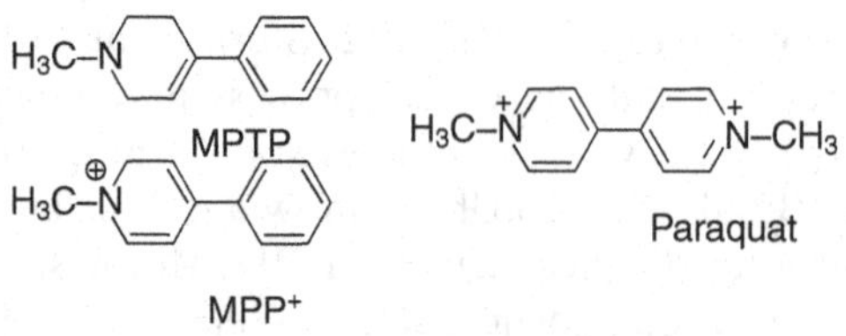

Figure 18.2. Molecular structures of MPTP, MPP^+, and paraquat.

studies have found similar PD concordance in monozygotic and dizygotic twins, further implicating environmental determinants (Tanner et al. 1999; Wirdefeldt et al. 2004).

THE ENVIRONMENTAL HYPOTHESIS

Although there are many case reports of acute parkinsonism resulting from high-dose toxicant exposures, most had atypical clinical and pathological features (Klawans et al. 1982; Pezzoli et al. 1989; Rodier 1955; Tetrud et al. 1994). Langston and colleagues reinvigorated the environmental etiologic hypothesis with their detection of a cluster of intravenous drug users who presented with acute parkinsonism that looked strikingly like typical PD. They identified the causal agent as N-methyl-4-phenyl-1,2,3,6 tetrahydropyridine (MPTP; Figure 18.2), which formed during synthesis of a meperidine analogue (Langston et al. 1983). Like typical PD, these individuals manifested the classic motor signs, responded to standard dopaminergic pharmacological therapies, and had selective loss of nigral dopaminergic neurons. The mechanism of MPTP toxicity has been delineated: MPTP freely crosses the blood–brain barrier and is converted into the proximate toxin, 1-methyl-4-phenylpyridinium (MPP^+), by glial monoamine oxidase B. The presynaptic dopamine transporter (DAT) concentrates MPP^+ in dopaminergic neurons where it inhibits mitochondrial complex I, induces oxidative stress, and causes bioenergetic failure and cell death (Przedborski and Jackson-Lewis 1998; Singer et al. 1988). The MPTP discovery fueled the search for other environmental agents sharing similar molecular structures and toxic mechanisms.

DISEASE MECHANISMS

The rare monogenic forms of PD, in combination with insights from human pathological studies and the MPTP animal model, have helped identify mechanisms likely to play an etiologic role in typical PD. For example, individuals with duplications of *SNCA*, the gene encoding alpha-synuclein protein, have a young-onset, highly virulent form of PD, suggesting that overproduction and/or reduced clearance alpha-synuclein is an important mechanism (Ross et al.

2008). Recapitulating the mechanism of MPTP toxicity, several rare monogenic forms of PD also affect proteins related to mitochondrial function (Martin et al. 2011). Impairment of mitochondrial function, increased oxidative stress, microglial activation and inflammation, and alpha-synuclein protein overproduction and impaired clearance have all been observed in human pathological studies (Henchcliffe and Beal 2008; Schapira 2009). These mechanisms are further implicated in a variety of animal models (Jackson-Lewis et al. 2012). Therefore, environmental toxicants that affect these processes may contribute to disease risk and/or progression.

PD EPIDEMIOLOGY

Challenges

The epidemiological study of PD has many inherent challenges. First, there is no diagnostic test and misdiagnosis is common (Schrag et al. 2002). Second, because PD is primarily a disease of old age, many individuals with early or mild disease die from other causes before motor symptoms manifest, obscuring risk factor associations (Ross et al. 2004, 2006). Third, pathological evidence of a slowly progressive neurodegenerative process suggests that disease-associated environmental exposures may occur years or even decades before diagnosis. Fourth, because PD is relatively rare, prospective study designs are not practical and most studies must rely on a relatively imprecise retrospective recall of exposures. Fifth, considerable variability in clinical course, disease features, and pathology suggests etiologic heterogeneity. Nonetheless, despite these limitations, there has been considerable progress in identifying potential environmental causes of PD.

DESCRIPTIVE EPIDEMIOLOGY: POTENTIAL ETIOLOGIC CLUES

Age and Gender

Age is by far the strongest risk factor for PD, suggesting that the passage of time is necessary for the disease process to unfold. Incidence increases from approximately 2.5 per 100,000 person-years at age 50 to over 100 per 100,000 person-years by age 80 (Van Den Eeden et al. 2003). Incidence in men is 1.5 to 2-fold that of women (Tanner and Goldman 1996). The reason for this disparity is unknown but may reflect hormonal factors or differential exposure to environmental toxicants.

Temporal Trends

If factors related to industrialization cause PD, then incidence should parallel industrial development, and risk should be greatest in persons most directly exposed to those factors. Only a few studies have attempted to assess changes in disease incidence over time. Rajput and colleagues found no change in incidence between 1967 and 1979 using the countywide medical record database in Rochester, Minnesota (Rajput et al. 1984). Rocca et al. (2001) extended the period of observation through 1990 and similarly found no change in incidence. However, secondary causes (e.g., encephalitis, drug induced) included earlier in the study period could have falsely inflated the earlier incidence estimates. Consistent with this premise, a Finnish study found no overall change in incidence between 1971 and 1992, but interestingly saw a 50% increased rate in men and a 30% decreased rate in women (Kuopio et al. 1999). One possible interpretation is that incidence estimates were falsely inflated earlier in the period (for the reasons cited earlier), as reflected in the decreasing rate in women, while in men, there was a true substantial increasing incidence, consistent with an industrial toxicant hypothesis.

Geographic Trends

Although disease clusters have rarely been reported, geographic variation in disease incidence could reflect differential exposure to environmental toxicants. Geographic comparisons are difficult to interpret due to widely varying access to and quality of medical care. Notwithstanding this limitation, several studies suggest that incidence in Africa and (preindustrial) China may be lower than in more industrially developed Western nations (Chen et al. 2001; Okubadejo et al. 2006). Several ecological studies reported increased rates of PD in areas with high concentrations of farming or with the presence of metals or paper-related industries (Barbeau 1986; Rybicki et al. 1993; Willis et al. 2010), and as discussed further, PD may be more common in far northern latitudes.

ANALYTIC EPIDEMIOLOGY: ASSESSING ENVIRONMENTAL RISK FACTORS FOR PD

Lifestyle Factors

Cigarette smoking is consistently associated with a *reduced* risk of PD in dozens of studies around the world. Risk in ever-smokers is approximately half that of never-smokers, with a dose–response relationship (Hernán et al. 2002). Nicotine is neuroprotective in a variety of animal models and may be responsible for the inverse association, but cigarette smoke contains hundreds of additional compounds that may also modify risk. *Caffeine and coffee consumption* are also consistently associated with reduced risk of PD, with magnitude similar to that of smoking, and a dose–response relationship (Hernán et al. 2002). Caffeine may protect dopaminergic neurons through antagonism of the adenosine A2A receptor (Kachroo and Schwarzschild 2012). Other environmental lifestyle factors that may modify PD risk include exercise (decreased risk) (Chen et al. 2005), nonsteroidal anti-inflammatory medications (decreased risk) (Gao et al. 2011; Samii et al. 2009), and dihydropyridine calcium channel blocking drugs (decreased risk) (Marras et al. 2012).

ENVIRONMENTAL TOXICANT ASSOCIATIONS WITH PD

Pesticides

Farming and farming-related occupations and exposures are the most intensively studied environmental risk factors for PD and have been suspected for decades. Farming-related occupations per se have been associated with PD in many, but not all, studies (Tanner and Goldman 1996). However, exposures related to farming may differ dramatically depending on location and time period, the type of farming (field crop, orchard, livestock), and ancillary tasks such as carpentry, painting, and equipment maintenance. In contrast, studies specifically assessing pesticide exposure consistently find an increased risk of PD (Table 18.1). A meta-analysis of 19 studies from North America, Europe, and Asia found a combined odds ratio (OR) of 1.94 (95% confidence interval [CI] = 1.49–2.53) for PD associated with a history of pesticide exposure (Priyadarshi et al. 2000).

Although pesticide use is consistently associated with an increased risk of PD, as summarized in Table 18.1, most studies assessed exposure in a very general way—relying on self-reporting of "ever/never" exposure to "pesticides." In some instances, studies reported functional class (i.e., insecticide, fungicide, herbicide) or application purpose such as eradication of termites, fleas, or weeds. Because most nonprofessionals using pesticides in the home or garden do not know which specific compounds they were exposed to, only rarely have studies assessed exposure to specific compounds or chemical classes (Brown et al. 2006).

TABLE 18.1. Epidemiological Associations of Pesticides and Parkinson's Disease

Study	Location	Design	Cases/ Controls	Exposure	Risk Ratio, 95% CI
Hertzman et al. (1990)	Canada	Case-control	57/122	Pesticide spraying	6.6, $p = 0.031$
				Paraquat	nc, $p = 0.01$
Semchuk et al. (1992, 1993)	Canada	Case-control	130/260	Pesticide application	2.3 (1.3–4.0)
				Herbicide use	3.0 (1.2–7.3)
Hertzman et al. (1994)	Canada	Case-control	127/245	Occupational pesticide exposure	2.0 (1.0–4.1)

(*Continued*)

TABLE 18.1. (*Continued*)

Study	Location	Design	Cases/ Controls	Exposure	Risk Ratio, 95% CI
Seidler et al. (1996)	Germany	Case-control	380/755	Insecticide use, herbicide use	Duration, $p = 0.001$
				Organochlorines	
				Group 1	1.6 (0.4–6.2)
				Group 2	5.8 (1.1–30.4)
Liou et al. (1997)	Taiwan	Case-control	120/240	Exposure to pesticides	2.9 (2.3–3.7)
				Paraquat	3.2 (2.4–4.3)
Petrovitch et al. (2002)	United States	Prospective cohort	116/7,986	Plantation work	1.9 (1.0–3.5)
Baldi et al. (2003)	France	Prospective cohort	24/1,507	Occupational pesticide exposure	5.6 (1.5–22)
Baldereschi et al. (2003)	Italy	Nested case-control	113/4,383	Pesticide-use license	3.7 (1.6–8.6)
Ascherio et al. (2006)	United States	Prospective cohort	413/143,325	Any pesticide exposure	1.8 (1.3–2.5)
Kamel et al. (2007)	United States	Nested case-control	78/55,931	Professional pesticide use	2.3 (1.2–4.5)
				Paraquat	1.8 (1.0–3.4)
				DDT	1.0 (0.6–1.8)
				2.4-D	0.9 (0.5–1.8)
Dick et al. (2007a)	Europe	Case-control	649/1,587	Pesticide use	1.3 (1.02–1.6)
Hancock et al. (2008)	United States	Family study	319/296	Pesticide application	1.6 (1.1–2.3)
				Organochlorines	2.0 (1.1–3.6)
Firestone et al. (2010)	United States	Case-control	404/526	Pesticide worker	1.5 (0.5–4.4)
				DDT	0.8 (0.4–1.6)
				2,4-D	0.8 (0.3–2.0)
Tanner et al. (2009)	United States	Case-control	519/511	Occupational pesticide use	1.9 (1.1–3.2)
				Paraquat	2.8 (0.8–9.7)
				Permethrin	3.1 (0.7–15.8)
				2,4-D	2.6 (1.03–6.5)
Elbaz et al. (2009)	France	Case-control	224/557	Pesticide application	1.8 (1.1–3.1)
				Organochlorines	2.4 (1.2–5.0)
Tanner et al. (2011)	United States	Case-control		Rotenone	2.5 (1.3–4.7)
				Paraquat	2.5 (1.4–4.7)
				Dieldrin	1.6 (0.7–3.3)

CI, confidence interval; nc, not calculable.

Nonetheless, epidemiological studies have convincingly implicated several specific pesticides, and these are discussed further.

Rotenone

Derived from the leaves, roots, and seeds of the cubé plant and other members of the pea family, rotenone-containing plants have been used as pesticides for centuries (Cabras et al. 2002). Most recently, rotenone has been used as an insecticide in agriculture and home gardening, and ubiquitously in household pet products. It has also been used as a piscicide, often in combination with the organochlorine toxaphene, to eradicate invasive fish species in lakes (ATSDR 2010). Rotenone is highly lipophilic and freely enters the brain, where it directly inhibits mitochondrial complex I—the same site as the parkinsonism-causing neurotoxicant MPTP (Greenamyre et al. 1999; Sherer et al. 2007). In animal models, rotenone selectively damages dopaminergic neurons in the substantia nigra and induces a parkinsonian syndrome, which includes bradykinesia, rigidity, and tremor (Cannon et al. 2009). Tanner and colleagues recently reported a significantly increased risk of PD associated with rotenone use in a case-control study (the Farming and Movement Evaluation [FAME] study) nested in an 80,000-person cohort of licensed pesticide applicators and their spouses (OR = 2.5, 95% CI = 1.3–4.7) (Tanner et al. 2011). Although it has a limited environmental half-life and does not bioaccumulate, even relatively brief exposures may cause progressive pathology analogous to that seen in PD (Drolet et al. 2009; Greene et al. 2009; Pan-Montojo and Funk 2010; Pan-Montojo et al. 2010).

Paraquat

The bipyridyl herbicide paraquat has been commercially available since 1962 (Figure 18.1). Used primarily as a defoliant and pre-emergence dessicant, it is one of the most widely used pesticides in the world (Centers for Disease Control and Prevention [CDC] 2006). At least five case-control studies have reported increased PD risk associated with exposure (Hertzman et al. 1990; Kamel et al. 2007; Liou et al. 1997; Tanner et al. 2009, 2011), while two found no association (Firestone et al. 2010; Hertzman et al. 1994). In particular, risk was markedly increased in the FAME study (OR = 2.5, 95% CI = 1.4–4.7) (Tanner et al. 2011), and in a study that used geographic information system (GIS) mapping and pesticide application data to provide unbiased estimates of exposure through well water (Gatto et al. 2009). Providing strong support for a causal relationship with PD, paraquat is a structural analogue of MPTP that generates reactive oxygen species through redox cycling. In animal models, it produces a parkinsonian syndrome that recapitulates many features of PD and causes selective loss of nigrostriatal dopaminergic neurons, mitochondrial impairment, and upregulation and aggregation alpha-synuclein expression (Dinis-Oliveira et al. 2006; Kuter et al. 2010; McCormack et al. 2002). Although paraquat does not bioaccumulate in the food chain, it persists in soil, with reported field half-lives of greater than 1000 days (Staiff et al. 1981; Wauchope et al. 1992).

Organochlorine Pesticides

Organochlorine pesticides include DDT (and its breakdown products, dichlorodiphenyldichloroethylene [DDE] and DDD), aldrin, dieldrin, endrin, endosulfan, hexachlorocyclohexane (HCH) isoforms (e.g., lindane), heptachlor, methoxychlor, chlordane, and others. Although many have been banned since the 1970s, these extremely persistent lipophilic compounds bioaccumulate in the food chain and are still ubiquitous in human tissues (ATSDR 2000). Organochlorines constitute the pesticide class most commonly associated with PD.

Several epidemiological studies found an increased risk of PD associated with exposure to organochlorine pesticides nonspecifically. In a large case-control study in Germany, Seidler et al. (1996) reported an increased risk of PD in subjects ever exposed to an organochlorine pesticide. In a family-based study, Hancock et al. (2008) found a twofold increased risk associated with organochlorine insecticide use (OR = 2.0, 95%CI = 1.1–3.6), but they did not have sufficient power to assess specific agents. Elbaz et al. (2009) reported a significantly increased risk associated with professional use of organochlorine insecticides in a large cohort of French agricultural workers (OR = 2.4, 95%CI = 1.2–5.0). Interestingly, risk was markedly higher in subjects with polymorphisms in the gene encoding p-glycoprotein (P-gp; *ABCB1*) (Dutheil et al. 2010), a transmembrane efflux pump that binds some organochlorines (Sreeramulu et al. 2007).

Dieldrin Dieldrin is a cyclodiene organochlorine insecticide that was commonly used on crops from 1950 to 1970, and to control termites from 1972 to 1987 (ATSDR 2002). Three postmortem studies assessed levels of dieldrin in PD and control brains. Fleming et al. detected dieldrin in 6 of 20 PD brains, but only 1 of 7 Alzheimer's disease (AD) brains and 0 of 14 controls—a highly significant difference (Fleming et al. 1994). Corrigan and colleagues also found significantly higher levels of dieldrin in the caudate nucleus of PD patients compared with normal controls or AD patients (Corrigan et al. 1998). In a subsequent study, these same authors replicated this observation, but importantly, levels of dieldrin were elevated in the substantia nigra but not in the cortex (Corrigan et al. 2000).

Weisskopf et al. assessed PD risk in a 60,000-person prospective Finnish cohort that measured dieldrin levels in blood drawn 20–40 years before disease onset (Weisskopf et al. 2010a). This prospective study design mitigates potential "effect–cause" bias, whereby factors associated with prevalent PD could potentially affect toxicant levels measured after disease onset. Consistent with studies in brain tissue, dieldrin levels in blood were higher in those who subsequently developed PD. Differences were greatest in those over age 66 (OR per interquartile range = 1.7, 95% CI = 1.2–2.4), especially in nonsmokers (OR = 2.6, 95% CI = 1.5–4.4).

Exposure to dieldrin has only been investigated in two epidemiological historical studies. Tanner et al. reported a near-significant association with dieldrin use in male professional pesticide applicators in the FAME study (OR = 1.8, 95% CI = 0.8–4.0) (Tanner et al. 2011), while Kamel et al. found no association (Kamel et al. 2007). However, both studies had limited statistical power.

Animal studies support the human data linking dieldrin and PD. Dieldrin is selectively toxic to dopaminergic neurons in culture (Kitazawa et al. 2001; Sanchez-Ramos et al. 1998) and it depletes brain dopamine in several animal species (Chun et al. 2001; Kanthasamy et al. 2005). Dieldrin also recapitulates many of the etiopathological mechanisms of PD. It increases reactive oxygen species in nigral dopaminergic neurons (Hatcher et al. 2007), activates microglia (Mao and Liu 2008), inhibits mitochondrial oxidative phosphorylation, and reduces ATP levels (Kanthasamy et al. 2005). It also induces fibrillization of alpha-synuclein protein and reduces its degradation (Sun et al. 2005; Uversky et al. 2001b).

Hexachlorohexanes (HCH)

HCHs exist in eight isomeric forms. γ-HCH, also known as lindane, provides most of the insecticidal properties, but an isomeric mixture called "technical-grade" HCH was also sold in the United States (ATSDR 2005).

In the postmortem study described earlier, in addition to dieldrin, Corrigan et al. also found significantly higher levels of γ-HCH (lindane) in the substantia nigra of PD patients. Levels were more than fourfold higher than controls (Corrigan et al. 2000). Richardson et al. measured the serum levels of organochlorine pesticides in subjects with prevalent PD, AD, and controls (Richardson et al. 2009). They detected β-HCH in 76% of PD patients, but only 30% of Alzheimer's and 40% of controls, with ORs of 4.4 and 5.2, respectively. Remarkably, more than 50% of PD patients had levels higher than any subject in the other groups. In a larger follow-up study, subjects with β-HCH levels above the interquartile range were at nearly threefold increased risk of PD (OR = 2.9, 95% = CI 1.8–4.5) (Richardson et al. 2011). However, a large prospective finish study found similar blood levels of β-HCH in cases and controls (Weisskopf et al. 2010a).

Petersen et al. (2008) measured serum concentrations of β-HCH in 79 PD cases and 154 controls residing in the circumpolar Faroe Islands. Residents consume large amounts of sea mammals and whale blubber, which contain high levels of organochlorine compounds (Deutch et al. 2003; Fromberg et al. 1999) (see the section "Polychlorinated Biphenyls (PCBs)"). PD cases had significantly higher levels of β-HCH, especially women (OR = 2.6, 95% CI = 1.03–6.5).

In the single historical epidemiological study that assessed exposure specifically to HCH, lindane was associated with a nonsignificantly increased risk of PD (OR = 1.4, 95% CI = 0.8–2.5).

Animal models provide support for a causal relationship between exposure to HCHs and PD. Lindane significantly decreased brain dopamine in both acute and subchronic dosings in rats (Ortiz Martinez and Martinez-Conde 1995), and markedly increased reactive oxygen species in mouse microglia (Mao and Liu 2008). Lindane also acts synergistically with dieldrin to increase reactive oxygen species, disturb calcium homeostasis, induce mitochondrial dysfunction, and cause death of dopaminergic neurons (Heusinkveld and Westerink 2012; Mao and Liu 2008; Sharma et al. 2010).

DDT and DDE In postmortem studies, Fleming et al. detected DDT and its metabolite DDE in 40 of 41 brains from patients with PD, AD, and controls, but saw no differences between groups (Fleming et al. 1994). Corrigan et al. found higher mean levels of DDE in PD brain than in control brain, but the difference was not significant (Corrigan et al. 1998). A follow-up study found no differences in DDT or DDE in the substantia nigra (Corrigan et al. 2000). Several studies found no difference in serum levels of DDT or DDE in PD cases and controls (Petersen et al. 2008; Richardson et al. 2009; Weisskopf et al. 2010a), although a modestly sized study of Greenland Inuits found significantly higher levels of DDE in PD plasma (42.1 μg/L vs. 15 μg/L, $p = 0.005$) (Koldkjaer et al. 2004).

Historical case-control studies parallel results of tissue studies. Use of DDT was not associated with PD risk in the FAME study of pesticide applicators (OR = 1.1, 95% CI = 0.65–1.7) (Tanner et al. 2011) nor in large population-based case-control studies (Firestone et al. 2010; Kamel et al. 2007).

Mechanistic studies also provide little support for an association of DDT and PD. The sodium channel is the main target of DDT. It also inhibits neuronal ATPases and disrupts calcium homeostasis. However, there is little data regarding effects on dopaminergic toxicity or transmission. Hatcher et al. found that although DDT and its metabolites inhibit the DAT in mouse synaptosomes *in vitro*, there was no evidence of either behavioral or nigral neurochemical changes in mice orally administered DDT or DDE for 30 days (Hatcher et al. 2008).

Other Pesticide Associations

Epidemiological and mechanistic studies provide some support for associations with several other nonorganochlorine pesticides. The dithiocarbamates, alone or in combination with paraquat, have been inconsistently associated with PD risk (Costello et al. 2009; Seidler et al. 1996; Semchuk et al. 1993; Wang et al. 2011), and they produce dopaminergic toxicity in animal models (Thiruchelvam et al. 2000; Zhang et al. 2003). Permethrin is a widely used synthetic pyrethroid insecticide that may inhibit mitochondrial complex I (Gassner et al. 1997). Tanner et al. (2009, 2011) reported a trend toward increased risk associated with permethrin use in two studies, but power was low. 2, 4-Dichlorophenoxyacetic acid (2,4-D) is a component of Agent Orange, an herbicide mixture suggested by the Institute of Medicine to have a possible association with PD (IOM 2008). Exposure was associated with significantly increased risk of PD in a large multicenter case control (OR = 2.59, 95% CI = 1.03, 6.48) (Tanner et al. 2009), but several other studies found no association (Firestone et al. 2010; Kamel et al. 2007; Tanner et al. 2011). In mice, acute exposure to 2,4-D did not produce any observable changes in striatal dopaminergic function (Thiffault et al. 2001).

Gene–Pesticide Interactions

Genetic variability may determine susceptibility to pesticide neurotoxicity. More recent epidemiological studies have begun to consider the joint effects of pesticide exposure and polymorphisms in genes that affect their absorption, distribution, metabolism, and excretion. Metabolic and antioxidant genes that may modify pesticide-associated PD risk include cytochrome P450 2D (*CYP2D6*) (Elbaz et al. 2004), glutathione-S-transferase P1 (*GSTP1*) (Wilk et al. 2006), NADPH dehydrogenase (*NQO1*) (Fong et al. 2007), manganese superoxide dismutase (*SOD2*) (Fong et al. 2007), and paraoxonase 1 (*PON1*) (Manthripragada et al. 2010). Membrane transporters include *MDR1* (multidrug resistance protein) (Zschiedrich et al. 2009) and *ABCB1* (P-gp) (Dutheil et al. 2010).

Pesticide Summary

The bulk of epidemiological, animal, and mechanistic data support a direct causal relationship of pesticides and PD. Few studies have assessed specific pesticides, but existing evidence is strongest for rotenone, paraquat, and the organochlorines dieldrin, γ-HCH (lindane), and β-HCH. Even fewer studies have considered the joint effects of exposure to multiple pesticides or interaction with genetic variants. Future work should investigate PD associations with specific pesticides and gene–pesticide interactions in large, highly exposed cohorts.

SOLVENTS

Although there are anecdotal reports of acute parkinsonism resulting from large solvent exposures (Guggenheim et al. 1971; McCrank and Rabheru 1989; Melamed and Lavy 1977; Pezzoli et al. 1989; Tetrud et al. 1994; Uitti et al. 1994), prior epidemiological studies found inconsistent associations of solvents and PD, with most reporting no association, and others finding increased risk, younger onset, and evidence of nigrostriatal dysfunction (Dick et al. 2007a; Firestone et al. 2010; Hertzman et al. 1994; McDonnell et al. 2003; Pals et al. 2003; Pezzoli et al. 2000; Rango et al. 2006; Seidler et al. 1996; Tanner et al. 2009). However, as with pesticides, few epidemiological studies assessed exposures to specific agents, instead studying solvents as single broad class, often using a self-reported "ever-exposed-to-a-solvent" variable (Dick et al. 2007a; Firestone et al. 2010; Hertzman et al. 1994; McDonnell et al. 2003; Pals et al. 2003; Pezzoli et al. 2000; Rango et al. 2006; Seidler et al. 1996; Tanner et al. 2009).

A major development occurred in 2008 when Gash et al. (2008) reported a cluster of three PD patients with extensive exposure to the solvent trichloroethylene (TCE) in a small manufacturing plant. The patients had worked near each other degreasing metal parts, with dermal and respiratory exposures for more than 25 years. Coworkers were found to perform fine motor tasks significantly slower than controls. These observations are consistent with two prior case reports also linking occupational TCE exposure with PD in four individuals (Guehl et al. 1999; Kochen et al. 2003). Providing further compelling evidence, a recent study in 99 twin pairs discordant for PD found a significant sixfold increased risk in twins occupationally exposed to TCE (OR = 6.1, 95% CI = 1.2–33) (Goldman et al. 2012) and a near-significant association with the structurally similar solvent tetrachloroethylene (PERC) (OR = 10.5, 95% CI = 0.97–113), as well as carbon tetrachloride (CCl_4) (OR = 2.3, 95% CI = 0.9–6.1). Importantly, exposures were assessed in a nonbiased manner by industrial hygienists unaware of subjects' case status. These findings are consistent with prior reports of increased PD risk associated with dry cleaning and textile work (Fall et al. 1999; Schulte et al. 1996; Tüchsen and Jensen 2000).

Supporting the biological plausibility of these associations, TCE recapitulates the key pathological and neurochemical features of PD in a rodent model (Liu et al. 2010). Oral administration to rats for 6 weeks activated microglia, increased markers of oxidative stress, and caused selective dose-dependent loss of dopaminergic neurons in the substantia nigra. In addition, there was marked selective nigral accumulation of alpha-synuclein protein, again paralleling human pathological staging of PD (Braak et al. 2003a). Finally, TCE caused inhibition of mitochondrial complex I, the site of action of the selective parkinsonism-inducing neurotoxin MPTP, and a deficiency seen in typical PD (Di Monte 1991; Sherer et al. 2007).

A proposed proximate toxic species for TCE and PERC is 1-trichloromethyl-1,2,3,4-tetrahydro-β-carboline (TaClo), a potent mitochondrial complex I inhibitor, dopaminergic toxin, and structural analogue of MPTP, which forms in the presence of tryptamine after CYP2E1-mediated oxidation (Figure 18.3) (Bringmann et al. 1995; Heim and Sontag 1997; Riederer et al. 2002). Like

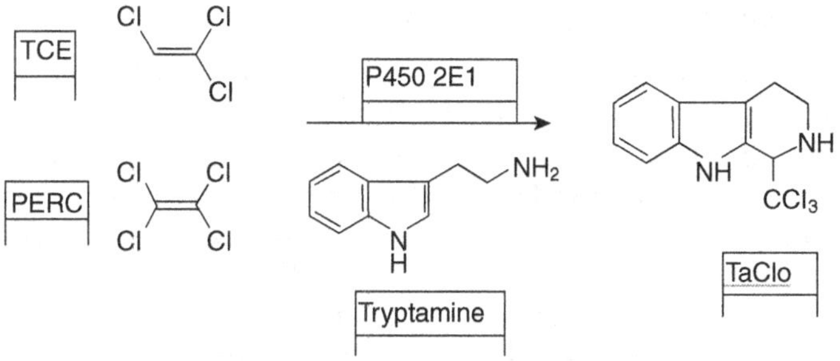

Figure 18.3. TaClo hypothesis of TCE and PERC toxicity.

TCE, both PERC (Lash et al. 2002; Miyazaki and Takano 1983) and CCl_4 (Boer et al. 2009; Manibusan et al. 2007) have been shown to increase markers of oxidative stress, activate microglia, and disrupt mitochondrial function. TCE is used in a wide variety of industrial processes and products. It has been used worldwide since the 1920s as a dry cleaning and degreasing agent and as an additive in common household products, including typewriter correction fluid, adhesives, paints, carpet, and spot removers. Until 1977, it was used as a general anesthetic, skin disinfectant, grain fumigant, and coffee decaffeinating agent (ATSDR 1997b; EPA 2009; IARC 1995; NIOSH 1978; WHO 1997). Today, its most important use is vapor degreasing of metal parts in automotive and metal industries.

TCE is detected in air, soil, food, and human breast milk, and is the most common organic contaminant in groundwater, found in up to 30% of U.S. drinking water supplies (ATSDR 1997b; EPA 2009; Wu and Schaum 2000). Its half-life in fatty tissues is on the order of hours to days (Lash et al. 2000), and although data are lacking, it has low tendency to accumulate in the food chain (Pearson and McConnell 1975). However, it persists for decades in subsurface groundwater (ATSDR 1997b).

PERC was introduced in the 1930s and by the 1960s was the predominant dry cleaning solvent (Andrasik 1990; ATSDR 1997a; EPA 1995). Like TCE, PERC is also widely used for metal cleaning and vapor degreasing. Exposure is primarily through inhalation, and because absorbed PERC is slowly exhaled, workers bring the compound home to their families (Aggazzotti et al. 1994).

Dry cleaning workers chronically exposed to PERC more frequently reported neurological symptoms, had slower reaction times and slower cognitive processing (Ferroni et al. 1992; NIOSH 1976; Seeber 1989), and possibly an increased risk of PD (Fall et al. 1999; Tüchsen and Jensen 2000). Like TCE, PERC has low tendency to accumulate in the food chain, but it may persist in subsurface groundwater for years (ATSDR 1997a; The Interstate Technology and Regulatory Council 2005).

Solvent Summary

The recent cluster of PD in TCE-exposed workers (Gash et al. 2008), the association of occupational TCE and PERC exposure with PD in a population-based twin study (Goldman et al. 2012), and the striking neuropathological similarity of TCE-exposed rodents and human PD (Gash et al. 2008; Guehl et al. 1999; Liu et al. 2010) support a true causal relationship. As evidenced by this work, future epidemiological studies should investigate associations with *specific* solvents and, as for pesticides, should consider the joint effects of solvents and variants in genes involved in their metabolism.

POLYCHLORINATED BIPHENYLS (PCBs)

PCBs were used extensively in industrial applications as lubricants and coolants from 1930 until they were banned in 1977 (ATSDR 2000). Although their stability was a prime advantage for use in industry, this property has led to their persistence in the environment. They cycle between air, water, and soil, and are found worldwide, though global distribution is not uniform. They volatilize in warmer climates, are carried northward by air and sea currents, and condense out and accumulate in cold circumpolar latitudes (Chiu et al. 2004). Because of their lipophilicity, PCBs bioconcentrate in fatty tissues of fish and mammals, especially marine mammals (Carpenter 2006; CDC 2009; Hardell et al. 2010; Welfinger-Smith et al. 2011).

Levels of serum PCBs generally reflect cumulative past exposure, and although PCBs are detectable in virtually the entire population (CDC 2009), levels are especially high in occupationally exposed individuals, such as through the manufacture or industrial use of PCBs (ATSDR 2000), and in populations with large dietary exposures. Inuit people who live at extreme northern latitudes and who consume diets high in fish, marine mammals, meat, and dairy products have consistently been found to have markedly higher levels of PCBs than the general population (ATSDR 2000; Bjerregaard and Hansen 2000; Carpenter et al. 2005; Chiu et al. 2004; Kvalem et al. 2009; Sandau et al. 2000).

PCBs and PD

Paralleling the environmental distribution of PCBs and suggesting a relationship with PD, several studies reported a much higher incidence of PD in peoples who consume large amounts of whale meat and blubber at extreme northern latitudes. In Greenland Inuits, age-adjusted prevalence of PD was 188 per 100,000 persons, and 209 per 100,000 in the Faroe Islands, but only 98 per 100,000 in Denmark and 102 per 100,000 in Norway (Tandberg et al. 1995; Wermuth et al. 1997, 2000, 2002).

Despite these observations, only a handful of epidemiological studies have investigated associations of PCBs and PD. In support of the dietary PCB hypothesis, consumption of whale meat more than twice per month during adulthood was associated with markedly increased risk of PD in a case-control study in the Faroe Islands (OR = 6.5, 95% CI = 3.0–14.1) (Petersen et al. 2008). PD cases had significantly higher levels of PCB 101 in blood, and slightly higher levels of other congeners. However, a study of Greenland Inuits found similar levels of PCBs in PD cases and controls, though levels were high in all subjects (Koldkjaer et al. 2004).

Two postmortem studies compared PCBs in brains from PD patients and controls. Corrigan et al. (2000) found significantly higher levels of congener 153 and a trend toward increased total PCBs in PD caudate nucleus (Corrigan et al. 1998). In a subsequent study, the same authors reported a nonsignificant doubling of total PCBs in the substantia nigra (PD mean 1.4 μg/g lipid, SD 1.4; control mean 0.7, SD 0.9).

Steenland et al. (2006) used National Death Index data to estimate mortality from PD in a cohort of 17,000 workers exposed to PCBs (primarily Arochlor 1254, 1242, and 1016) at plants that produced electrical capacitors from the 1940s to the 1970s. A subset of 400 workers whose blood was tested in the late 1970s had levels 10- to 50-fold higher than community controls. Using air-sample measurements from the 1970s, the investigators constructed a job-exposure matrix (JEM) to estimate cumulative PCB exposures based on plant locations and job tasks. Relative to the U.S. population, workers with high cumulative PCB exposure showed a trend toward increased mortality from PD (standardized mortality ratio [SMR] 2.1, 95% CI = 0.9–4.1). The association was stronger and significant in women (SMR 3.0, 95% CI = 1.1–6.4), possibly due to interaction with estrogen, or higher PCB storage in body fat.

Consistent with the epidemiological evidence of gender-specific PCB toxicity, single-photon emission computed tomography (SPECT) imaging identified an inverse correlation between striatal DAT density and serum PCB levels in female, but not male, capacitor workers. Women with the highest serum PCB levels had a 25% reduction in DAT binding, possibly representing early preclinical disease (Seegal et al. 2010).

Toxicity of PCBs to dopaminergic neurons is well recognized. Recapitulating PD pathology, toxicity primarily affects the substantia nigra, the striatum, and the olfactory tract (Chu et al. 1994, 1998; Seegal et al. 1991, 1994). Specific congeners have been shown to disrupt striatal dopaminergic homeostasis in a variety of cellular, rodent, and primate models—some at levels typically seen in human capacitor workers (Richardson and Miller 2004; Seegal et al. 1991, 1994; Wolff et al. 1982). Dopaminergic toxicity may be greatest for *ortho*-substituted congeners (Mariussen et al. 2001). Most studies observed persistent reductions in levels of dopamine, though some found increased levels. In a recent study, adult rodents exposed to Aroclor 1254 in food for 4 weeks had a significant reduction of dopaminergic neurons in the substantia nigra but a concomitant increase in striatal dopamine that was thought to be a compensatory response. These changes were associated with increases in markers of oxidative stress and lipid peroxidation, possibly due to dysregulation of iron metabolism (Lee et al. 2012). Other proposed mechanisms of toxicity include disruption of dopamine transport due to effects on the DAT or the vesicular monoamine transporter (VMAT) (Caudle et al. 2006; Richardson and Miller 2004), disruption of the blood–brain barrier (Seelbach et al. 2010), and activation of NADPH-oxidase (Myhre et al. 2009).

PCB Summary

Cellular and animal models provide considerable evidence that PCBs are selectively toxic to dopaminergic neurons, supporting the biological plausibility of a causal relationship with PD. Limited, but fairly consistent tissue and historical epidemiological data link PCBs with PD, especially in women. Remarkably, interactions between PCB exposures and variants in genes involved in their metabolism have not been studied. Future research should focus on highly exposed populations, on combined effects with other dopaminergic toxicants and oxidative stressors (Bemis and Seegal 1999), and on gene–PCB interactions. Because the deleterious effects of PCBs may be greater in children, future work should also focus on elucidating relationships between PD and childhood PCB exposures.

METALS

Several lines of evidence implicate metals in PD etiology. Many metals bind neuromelanin, and therefore concentrate specifically in pigmented dopaminergic cells in the substantia nigra (Gerlach et al. 2003; Good et al. 1992; Zecca and Swartz 1993; Zecca et al. 2002). Recapitulating the primary underlying pathology of PD, metals can (1) generate reactive oxygen species that increase cellular stress and damage lipid membranes and DNA (Berg et al. 2004; Stohs and Bagchi 1995), (2) impair mitochondrial function (Huang et al. 2006; Mehta et al. 2006), and (3) enhance aggregation of alpha-synuclein protein (Santner and Uversky 2010; Uversky et al. 2001a). However, as with pesticides and solvents, most epidemiological studies of metals have not assessed specific compounds.

PD may be more common in regions with high concentrations of metallurgic industries (Rybicki et al. 1993; Willis et al. 2010), and several studies of occupational exposure to metals found modestly increased risk. Ever working with metals was nonsignificantly associated with PD (OR = 1.55, 95% CI = 0.78–3.1) in a study that used an industrial hygienist review of manufacturing company records (McDonnell et al. 2003). Pals et al. reported a fourfold increased risk associated with work in "metallurgic activities," but this result may have been confounded by gender differences in case and control groups (Pals et al. 2003). In contrast, occupations that involve working with metals

$$(1)\ Fe^{2+} + H_2O_2 \rightarrow Fe^{3+} + OH\cdot + OH^-$$

$$(2)\ Fe^{3+} + H_2O_2 \rightarrow Fe^{2+} + OOH\cdot + H$$

Figure 18.4. Fenton reaction.

were not associated with PD in large case-control studies that used either self-reported or industrial-hygienist-estimated exposure assessment (Dick et al. 2007b; Feldman et al. 2011; Firestone et al. 2010; Frigerio et al. 2005; Goldman et al. 2005; Kirkey et al. 2001; Tanner et al. 2009).

Iron

Although iron does not bioaccumulate, it can concentrate in soils, water, and air near industrial sources (Polizzi et al. 2007; Smith et al. 1987). Dysregulation of iron metabolism in PD has been recognized for several decades. Numerous postmortem and imaging studies have found increased iron in PD substantia nigra (Dexter et al. 1991; Griffiths and Crossman 1993; Martin 2009; Sofic et al. 1988). Others observed normal levels of total iron, but reduced levels of the iron binding protein ferritin, and concomitant increases in free iron (Dexter et al. 1990; Koziorowski et al. 2007; Wypijewska et al. 2010). Because hydrogen peroxide forms from dopamine metabolism in the substantia nigra (Spina and Cohen 1989), increased free iron can react with hydrogen peroxide through the Fenton reaction to generate highly toxic hydroxyl radicals (Figure 18.4) (Gałazka-Friedman et al. 1996; Halliwell and Gutteridge 1992; Youdim et al. 1993). These reactive oxygen species promote lipid peroxidation and cellular stress specifically in nigral dopaminergic neurons (Dexter et al. 1989a; Wypijewska et al. 2010).

Neonatal mice administered iron at levels equivalent to those found in human infant formula develop progressive midbrain neurodegeneration and enhanced vulnerability to toxic injury (Kaur et al. 2007). However, human epidemiological data on iron exposure are sparse and provide little support for an association with environmental iron exposure. Several studies associated increased dietary iron with PD risk (Johnson et al. 1999; Logroscino et al. 2008; Powers et al. 2003), while others found no association (Logroscino et al. 1998; Miyake et al. 2011). A single study found increased risk associated with occupational exposure to iron in combination with other metals (Gorell et al. 1999), but other studies reported no association (Dick et al. 2007b; Firestone et al. 2010).

Manganese

Manganese is one of the most common earth elements. It exists in elemental form as an oxide or salt, and as organomanganese compounds such as methylmanganese and methylcyclopentadienyl manganese tricarbonyl (MMT) (ATSDR 2008).

Manganese is an essential nutrient and a cofactor in several enzymes, including the mitochondrial antioxidant SOD2 (Robinson 1998). Manganese is used in numerous products and industrial processes. During smelting, it is added to steel to increase hardness, and it is used in some welding rods and other consumables. It is also used in dry-cell batteries, fertilizers, and fungicides (ATSDR 2008). Organic forms are used in fungicides, and in some countries, MMT is used as an antiknock agent in gasoline, replacing lead. Engines that burn fuel with MMT release manganese primarily in inorganic form (Ressler et al. 1999), although occupational exposure to MMT may occur.

Many foods contain manganese, especially whole grains and legumes, and vegetarians may have higher exposures. Some fish have high levels, but there is little magnification in the food chain, and human homeostatic mechanisms minimize the risk of toxicity from food sources. Industrial workers, and persons living in high-density traffic areas where MMT is used may have higher respiratory exposures (ATSDR 2008; Bolté et al. 2004). Direct delivery of inhaled manganese to the brain via olfactory neurons has been shown in rats, although little, if any, reaches the striatum (Dorman et al. 2002). The relevance of this pathway in humans has not been demonstrated.

Manganism Just 20 years after James Parkinson described PD, in 1817, Couper reported a disorder with similar features that resulted from exposure to manganese oxide. Dubbed "manganese crushers" disease, the men developed rapidly progressive leg weakness, salivation, and whispering speech. Many similar reports have been published over the subsequent 170 years in Morocco (Rodier), Egypt (Abdel-Naby and Hassanein 1965), Chile (Cotzias et al. 1968; Schuler et al. 1957), and the United States (Canavan et al. 1934; Casamajor 1913; Cook et al. 1974; Edsall et al. 1919; Greenhouse 1971).

The clinical syndrome now termed *manganism* includes features of parkinsonism, dystonia, and psychiatric disturbances, but differs from PD in several respects. It tends to have an abrupt onset, early gait disturbance, uncommon rest tremor, and responds poorly to dopaminergic therapies effective in PD (Abdel-Naby and Hassanein 1965; Canavan et al. 1934; Casamajor 1913; Cook et al. 1974; Edsall et al. 1919; Fairhall and Neal 1943; Huang et al. 1998; Mena et al. 1967; Rodier 1955; Rosenstock et al. 1971; Schuler et al. 1957; Wang et al. 1989). Pathological studies, and imaging studies in animal models and in smelting factory workers with manganism, identify the primary pathology in the globus pallidus, downstream of the substantia nigra (Guilarte 2010; Perl and Olanow 2007; Shinotoh et al. 1997).

Manganese and PD Several ecological studies have suggested an increased risk or earlier age at PD onset in areas with manganese-related industries (Finkelstein and Jerrett 2007; Rybicki et al. 1993; Willis et al. 2010), although

PD prevalence was not associated with traffic-related pollution in Canada, where MMT was added to gasoline from 1976 until 2004 (Finkelstein and Jerrett 2007).

Epidemiological studies provide little evidence of a relationship between manganese and PD risk. PD patients and controls had similar brain levels of manganese (Dexter et al. 1989b, 1991; Larsen et al. 1981), and most studies found similar serum and urinary levels (Forte et al. 2004; Fukushima et al. 2011; Jiménez-Jiménez et al. 1995). A single study reported that manganese levels were higher in PD cerebrospinal fluid (CSF) (PD 3.3 ng/mL, control 1.9 ng/mL; $p < 0.05$), but analyses were not adjusted for age differences (Hozumi et al. 2011). Most dietary studies of manganese also failed to find an association (Fukushima et al. 2011; Miyake et al. 2011; Powers et al. 2003).

Several studies estimated occupational exposure to manganese in PD cases and controls using unbiased methods. Seidler et al. (1996) found no differences using a JEM in 380 cases and 755 controls. Similarly, industrial hygienist-estimated lifetime exposure was not associated in a study of twins discordant for PD (Goldman et al. 2004) or in a large multicenter study (Dick et al. 2007a). In contrast, Gorell et al. (1997) found an increased risk in subjects exposed to manganese for over 20 years, but this was based on only three subjects.

The relationship between welding-related manganese exposure and PD is highly controversial because of extensive litigation (Kaiser 2003). Racette et al. (2005) reported a high prevalence of PD in welders involved in litigation, but welding was not associated with PD in other studies using mortality, clinic, or industry-based samples, or in studies linking occupational and disease registries (Fored et al. 2006; Fryzek et al. 2005; Marsh and Gula 2006; Park et al. 2006; Stampfer 2009; Tanner et al. 2009).

Despite the lack of epidemiological association, in animal studies, manganese can induce a range of toxic effects potentially related to PD etiology. Manganese is transported into the brain by both facilitated diffusion and active transport. It accumulates in pallidal and striatal mitochondria, and has been shown to disrupt cellular metabolism, activate microglia, and increase inflammatory cytokines (Aschner et al. 2009). *In vitro* and at high concentrations in animal models, it can cause dopaminergic degeneration (Benedetto et al. 2009). However, a recent review concludes that manganese is not likely to produce nigral dopaminergic neuronal degeneration as seen in PD (Guilarte 2010). Instead, its toxicity seems to result largely from the inability to effectively release dopamine from otherwise intact nigrostriatal neurons.

Mercury

Mercury exists in the environment in elemental (metallic), inorganic (mercury salts), and organic forms. Elemental mercury is used in various industrial processes, in thermometers and batteries, and it comprises 50% of dental amalgam used in fillings (ATSDR 1999). Inorganic mercury salts have been used as antiseptics and laxatives, as preservatives, fungicides, and in cosmetics and paints. Organomercury compounds include methyl, dimethyl, ethyl, and phenylmercuric compounds, though methylmercury is the predominant environmental organomercurial. These compounds were commonly used as fungicides for seed grains and in paints until their toxicity was recognized in the 1970s. Mercury has been found in at least half of the Environmental Protection Agency (EPA) National Priorities List (NPL) sites (ATSDR 1999).

Dental amalgams comprise the major source of mercury in most people, although they have not been associated with adverse health effects (Risher and De Rosa 2007). Elemental and inorganic forms constitute the majority of environmental mercury, but microorganisms in soil and water convert inorganic mercury into methylmercury, which accumulates and bioconcentrates in the food chain. Levels are highest in large saltwater fish such as swordfish, and in marine mammals, and therefore in persons who consume large amounts of fish and marine mammals (Muckle et al. 2001).

Metallic mercury enters the body primarily through inhalation. The nervous system is the most sensitive organ to metallic mercury toxicity. Acute and chronic effects include tremor, neuropathy, behavioral, and cognitive changes (Albers et al. 1988; Williamson et al. 1982). Methylmercury enters through gastrointestinal absorption, though other organic forms may have substantial dermal absorption. Much of the human toxicology of methylmercury derives from studies of individuals who consumed highly contaminated fish from Minamata Bay, Japan, from 1930 to 1970 (Japan 2002). Similar to metallic mercury, neurological effects include gait and speech impairment, parasthesias, and behavioral changes. Tremor is also observed but differs from a parkinsonian tremor in character and frequency (Yamanaga 1983). Postmortem studies have found both central and peripheral neuronal degeneration (Eto 1997; Takeuchi 1982).

Epidemiological studies of mercury exposure and PD provide limited support for a possible association, though data are sparse, and the most etiologically important exposures may occur in childhood or *in utero*. In a study of 54 PD and 95 control subjects, the highest tertile of mercury levels in blood was associated with a markedly increased risk of PD (OR = 9.4, 95% CI = 2.5–36), although there was no association with levels in hair (Ngim and Devathasan 1989). In a case-control study in a Faroe Islands population with high consumption of fish and marine mammals, Petersen et al. (2008) found a nonsignificantly increased risk of PD associated with higher blood mercury in women, but a nonsignificantly decreased risk in men.

Two historical case-control studies found modest nonsignificantly increased PD risk associated with occupational exposure to mercury (Ohlson and Hogstedt 1981; Seidler et al. 1996), but others found no association (Gorell et al. 1999; Semchuk et al. 1993). Because of potential exposure to metallic mercury in dental amalgam, several studies assessed PD risk and dental occupations. A large Danish study that linked occupations with neurological diagnoses in more than 5000 dentists and 33,000 dental assistants found no association with

PD (Thygesen et al. 2011). Similarly, PD-related mortality was not elevated in dentists and dental assistants in a U.S. study with more than 33,000 PD deaths (Park et al. 2005).

In contrast to the limited epidemiological data, animal and cellular models lend support for a role of mercury in PD etiology. Mercury has been shown to alter dopaminergic receptors, and mice fed diets supplemented with fish with high levels of methylmercury had decreased striatal dopamine (Bourdineaud et al. 2011; Newland et al. 2008). Mercury has also been shown to disrupt the mitochondrial electron transport chain and to generate reactive oxygen species, to increase release of excitotoxic glutamate, and to promote aggregation of alpha-synuclein protein (Farina et al. 2011; Yamin et al. 2003). In humans, positron emission tomography (PET)-scan-assessed striatal DAT density was inversely related to urine mercury and cumulative occupational exposure to mercury vapor. Therefore, the potential role of mercury in PD etiology deserves more study in well-characterized highly exposed populations.

Lead

Levels of lead in the environment have increased 1000-fold since industrialization. The largest use today is in automobile batteries, solders, and building materials. Because lead does not degrade, the largest exposure risk in the population is due to its prior use in gasoline (until 1995) and paint (until 1978), most of which is now bound to soil (ATSDR 2007). Occupational exposure is common in industries and tasks such as steel smelting, battery manufacturing, and soldering. Exposures occur mainly through ingestion or inhalation.

Lead toxicity has been recognized for centuries and affects virtually all organ systems—especially the developing nervous system. Most toxicity is caused by inorganic forms (salts, oxides) because organic forms, such as in gasoline, quickly oxidize in the environment. Lead is most easily and commonly measured in whole blood, but this reflects a relatively recent exposure. In contrast, lead deposited in bone has a half-life of years (trabecular bone, e.g., patella) to decades (cortical bone, e.g., tibia) and therefore reflects cumulative lifetime exposure. Bone lead is measured using K-shell X-ray fluorescence (KXRF).

Epidemiological associations of PD and lead exposure are inconsistent. Kuhn et al. reported a German case series in which seven of nine postal workers chronically exposed to lead batteries developed PD (Kuhn et al. 1998). A Belgian case-control study found modestly increased risk associated with self-reported exposure (OR = 1.8, 95% CI = 0.8–4.0) (Pals et al. 2003), and hygienist-estimated lead exposure was slightly more common among PD patients in a Michigan health maintenance organization (OR = 1.41, 95% CI = 0.83–2.39)—especially in those exposed for more than 20 years (OR = 2.0, 95% CI = 0.97–4.3) (Gorell et al. 1997). However, lead exposure was not associated in a large study of incident PD that used both self-report and hygienist estimates (Firestone et al. 2010).

Two studies assessed lead levels in bone using KXRF. Coon et al. (2006) combined KXRF and hygienist historical reviews to estimate cumulative lifetime exposures in a population-based sample of 121 PD cases and 414 matched controls. Subjects in the highest quartile of whole body cumulative lead exposure had twice the risk of PD than those in the lowest quartile (OR = 2.3, 95% CI = 1.1–4.5). Weisskopf et al. (2010b) measured bone lead in 330 PD cases and 308 controls. They found no increase in patellar lead but a significantly increased risk in those with the highest quartile of tibial lead (OR = 1.9, 95% CI = 1.0–3.6).

Lead is toxic to dopaminergic neuronal function in animal and cellular models, although it likely does not reduce the number of dopaminergic neurons. It reduces tyrosine hydroxylase activity, alters dopamine release and metabolism, and decreases postsynaptic receptor sensitivity (Kala and Jadhav 1995; Lasley 1992; Pokora et al. 1996; Zuch et al. 1998). Rats fed 250 ppm lead for 3–6 weeks manifested reduced dopaminergic neuronal activity in the substantia nigra (Tavakoli-Nezhad and Pitts 2005). Lead also increases midbrain oxidative stress and lipid peroxidation (Ercal et al. 2001; Sandhir et al. 1994) and strongly enhances fibrillization of alpha-synuclein protein (Yamin et al. 2003).

Other Metals

Other metals have been infrequently investigated. Several studies assessed PD risk associated with copper. *Copper*, like iron, can catalyze a Fenton-like reaction, generating a reactive hydroxyl radical (Lloyd and Phillips 1999). Supporting a relationship with PD, Wilson's disease is a genetic disorder that causes copper accumulation in the basal ganglia, inhibition of mitochondrial complex I, and parkinsonism (Gu et al. 2000; Oder et al. 1991). Hozumi et al. found significantly higher levels of copper in PD CSF than in controls (PD: 18.8 ng/mL, control: 10.2 ng/mL, $p < 0.01$), but analyses were not adjusted for age differences (Hozumi et al. 2011). PD was modestly more common in subjects with hygienist-determined copper exposure (OR = 1.5, 95% CI = 0.9–2.8), especially in those with more than 20 years of exposure (OR = 2.5, 95% = CI 1.1–5.9) (Gorell et al. 1999), but two large case-control studies found no association (Dick et al. 2007a; Firestone et al. 2010), and another study found normal serum levels (Jiménez-Jiménez et al. 1992).

Nickel can also catalyze a Fenton-like reaction (Cavallo et al. 2003; Leonard et al. 2004). Nickel exposure was not associated with PD in a large population-based Swedish twin registry (Feldman et al. 2011).

Zinc levels were higher in PD CSF (Hozumi et al. 2011), and a case-control study found increased risk associated with self-reported exposure, though CIs were very wide (OR = 11.6, 95% CI = 1.5–90.9) (Pals et al. 2003). Conversely, dietary zinc was inversely associated with PD (Miyake et al. 2011); serum levels were normal (Jiménez-Jiménez et al. 1992); and zinc was not associated with PD in case-control studies (Gorell et al. 1997; Seidler et al. 1996).

Metals Summary

Mechanistic data from cellular and animal models support the role of metals in PD etiology. Associations are strongest for mercury and lead, but epidemiological data are inconsistent, perhaps because of very imprecise exposure estimates. Future research should study highly exposed populations. Investigators should consider interactions between metals and variants in genes that code for proteins involved in their transport and binding, and defend against oxidative stress. As demonstrated by studies of early life iron exposure in rodents, perinatal exposures may be particularly important.

PD ASSOCIATIONS WITH OTHER COMPOUNDS

Other environmental pollutants may affect PD mechanistic pathways but have not been studied in humans. For example, *polybrominated diphenyl ethers* (PDBE) bioconcentrates freely enter the brain and alter dopaminergic function in rat synaptosomes (Dreiem et al. 2010). Dioxins have also been shown to affect dopaminergic function (Akahoshi et al. 2009). *Air pollution* increases indices of oxidative stress in the brain (Calderón-Garcidueñas et al. 2008; Campbell et al. 2005), and carbon and metallic *nanoparticulates* may increase aggregation of toxic protein species (Calderón-Garcidueñas et al. 2012; Stern and Johnson 2008).

SUMMARY

PD is caused by multiple factors. Like heart disease or type 2 diabetes, purely genetic or purely environmental causes are rare. However, epidemiological research strongly implicates a major etiologic role for environmental factors in most PD. Pesticides are consistently associated with PD risk, but few studies have assessed specific agents. Mechanistic, animal, and human evidence is strongest for paraquat, rotenone, and organochlorine pesticides such as dieldrin and hexachlorohexane. The solvents TCE and PERC have been associated epidemiologically and produce an animal model that closely recapitulates PD pathology, but results need replication in highly exposed, well-characterized cohorts. Similarly, mechanistic data implicate PCBs, and epidemiological associations are fairly consistent, especially in women, but much more work is needed. Finally, many metals generate free radicals and cause oxidative stress, but, with the exception of lead, exposures are difficult to estimate. In addition, the most etiologically important exposures may occur perinatally or in early childhood.

Exposures to neurotoxicants do not occur in isolation. Future animal models should investigate the effects of combinations of agents at dosages commonly encountered in human populations. Future epidemiological studies in humans should focus on large or pooled populations with well-characterized exposures and should consider interactions with genetic variants that affect metabolism and protect against oxidative stress.

REFERENCES

Abdel-Naby S, Hassanein M. 1965. Neuropsychiatric manifestations of chronic manganese poisoning. J Neurol Neurosurg Psychiatry 28:282–288.

Aggazzotti G, Fantuzzi G, Predieri G, Righi E, Moscardelli S. 1994. Indoor exposure to perchloroethylene (PCE) in individuals living with dry-cleaning workers. Sci Total Environ 156:133–137.

Akahoshi E, Yoshimura S, Uruno S, Ishihara-Sugano M. 2009. Effect of dioxins on regulation of tyrosine hydroxylase gene expression by aryl hydrocarbon receptor: A neurotoxicology study. Environ Health 8:24.

Albers JW, Kallenbach LR, Fine LJ, Langolf GD, Wolfe RA, Donofrio PD, et al. 1988. Neurological abnormalities associated with remote occupational elemental mercury exposure. Ann Neurol 24:651–659.

Andrasik F. 1990. Psychologic and behavioral aspects of chronic headache. Neurol Clin 8:961–976.

Ascherio A, Chen H, Weisskopf MG, O'Reilly E, McCullough ML, Calle EE, et al. 2006. Pesticide exposure and risk for Parkinson's disease. Ann Neurol 60:197–203.

Aschner M, Erikson KM, Herrero Hernández E, Tjalkens R. 2009. Manganese and its role in Parkinson's disease: From transport to neuropathology. Neuromolecular Med 11:252–266.

ATSDR. 1997a. Toxicological Profile for Tetrachloroethylene. Atlanta, GA: P. H. S. U.S. Department of Health and Human Services, Agency for Toxic Substances and Disease Registry.

ATSDR. 1997b. Toxicological Profile for Trichlorethylene. Atlanta, GA: P. H. S. U.S. Department of Health and Human Services, Agency for Toxic Substances and Disease Registry.

ATSDR. 1999. Toxicological Profile for Mercury. Atlanta, GA: P. H. S. U.S. Department of Health and Human Services, Agency for Toxic Substances and Disease Registry.

ATSDR. 2000. Toxicological Profile for Polychlorinated Biphenyls (PCBs). Atlanta, GA: P. H. S. U.S. Department of Health and Human Services, Agency for Toxic Substances and Disease Registry.

ATSDR. 2002. Toxicological Profile for Aldrin/Dieldrin. Atlanta, GA: P. H. S. U.S. Department of Health and Human Services, Agency for Toxic Substances and Disease Registry.

ATSDR. 2005. Toxicological Profile for Alpha-, Beta-, Gamma-, and Delta-hexachlorocyclohexane. Atlanta, GA: P. H. S. U.S. Department of Health and Human Services, Agency for Toxic Substances and Disease Registry.

ATSDR. 2007. Toxicological Profile for Lead. Atlanta, GA: P. H. S. U.S. Department of Health and Human Services, Agency for Toxic Substances and Disease Registry.

ATSDR. 2008. Toxicological Profile for Manganese. Atlanta, GA: P. H. S. U.S. Department of Health and Human Services, Agency for Toxic Substances and Disease Registry.

ATSDR. 2010. Toxicological Profile for Toxaphene. Atlanta, GA: P. H. S. U.S. Department of Health and Human Services, Agency for Toxic Substances and Disease Registry.

Baldereschi M, Di Carlo A, Vanni P, Ghetti A, Carbonin P, Amaducci L, et al. 2003. Lifestyle-related risk factors for Parkinson's disease: A population-based study. Acta Neurol Scand 108: 239–244.

Baldi I, Cantagrel A, Lebailly P, Tison F, Dubroca B, Chrysostome V, et al. 2003. Association between Parkinson's disease and exposure to pesticides in southwestern France. Neuroepidemiology 22:305–310.

Barbeau A. 1986. At the frontiers of the brain—The neurologist and his literature. Union Med Can 115:884–890.

Bemis JC, Seegal RF. 1999. Polychlorinated biphenyls and methylmercury act synergistically to reduce rat brain dopamine content in vitro. Environ Health Perspect 107:879–885.

Benedetto A, Au C, Aschner M. 2009. Manganese-induced dopaminergic neurodegeneration: Insights into mechanisms and genetics shared with Parkinson's disease. Chem Rev 109:4862–4884.

Berg D, Youdim MB, Riederer P. 2004. Redox imbalance. Cell Tissue Res 318:201–213.

Bernheimer H, Birkmayer W, Hornykiewicz O, Jellinger K, Seitelberger F. 1973. Brain dopamine and the syndromes of Parkinson and Huntington. Clinical, morphological and neurochemical correlations. J Neurol Sci 20:415–455.

Bjerregaard P, Hansen JC. 2000. Organochlorines and heavy metals in pregnant women from the Disko Bay area in Greenland. Sci Total Environ 245:195–202.

Boer LA, Panatto JP, Fagundes DA, Bassani C, Jeremias IC, Daufenbach JF, et al. 2009. Inhibition of mitochondrial respiratory chain in the brain of rats after hepatic failure induced by carbon tetrachloride is reversed by antioxidants. Brain Res Bull 80:75–78.

Bolté S, Normandin L, Kennedy G, Zayed J. 2004. Human exposure to respirable manganese in outdoor and indoor air in urban and rural areas. J Toxicol Environ Health A 67:459–467.

Bonifati V. 2007. LRRK2 low-penetrance mutations (Gly2019Ser) and risk alleles (Gly2385Arg)-linking familial and sporadic Parkinson's disease. Neurochem Res 32:1700–1708.

Bourdineaud JP, Fujimura M, Laclau M, Sawada M, Yasutake A. 2011. Deleterious effects in mice of fish-associated methylmercury contained in a diet mimicking the Western populations' average fish consumption. Environ Int 37:303–313.

Braak H, Del Tredici K, Rüb U, de Vos RA, Jansen Steur EN, Braak E. 2003a. Staging of brain pathology related to sporadic Parkinson's disease. Neurobiol Aging 24: 197–211.

Braak H, Rüb U, Gai WP, Del Tredici K. 2003b. Idiopathic Parkinson's disease: Possible routes by which vulnerable neuronal types may be subject to neuroinvasion by an unknown pathogen. J Neural Transm 110:517–536.

Bringmann G, God R, Feineis D, Wesemann W, Riederer P, Rausch WD, et al. 1995. The TaClo concept: 1-trichloromethyl-1,2,3,4-tetrahydro-beta-carboline (TaClo), a new toxin for dopaminergic neurons. J Neural Transm Suppl 46:235–244.

Brown TP, Rumsby PC, Capleton AC, Rushton L, Levy LS. 2006. Pesticides and Parkinson's disease—Is there a link? Environ Health Perspect 114:156–164.

Cabras P, Caboni P, Cabras M, Angioni A, Russo M. 2002. Rotenone residues on olives and in olive oil. J Agric Food Chem 50:2576–2580.

Calderón-Garcidueñas L, Solt AC, Henríquez-Roldán C, Torres-Jardón R, Nuse B, Herritt L, et al. 2008. Long-term air pollution exposure is associated with neuroinflammation, an altered innate immune response, disruption of the blood–brain barrier, ultrafine particulate deposition, and accumulation of amyloid beta-42 and alpha-synuclein in children and young adults. Toxicol Pathol 36:289–310.

Calderón-Garcidueñas L, Kavanaugh M, Block M, D'Angiulli A, Delgado-Chávez R, Torres-Jardón R, et al. 2012. Neuroinflammation, hyperphosphorylated tau, diffuse amyloid plaques, and down-regulation of the cellular prion protein in air pollution exposed children and young adults. J Alzheimers Dis 28:93–107.

Campbell A, Oldham M, Becaria A, Bondy SC, Meacher D, Sioutas C, et al. 2005. Particulate matter in polluted air may increase biomarkers of inflammation in mouse brain. Neurotoxicology 26:133–140.

Canavan MM, Cobb S, Srinker C. 1934. Chronic manganese poisoning. Psychiatry 32:501–512.

Cannon JR, Tapias V, Na HM, Honick AS, Drolet RE, Greenamyre JT. 2009. A highly reproducible rotenone model of Parkinson's disease. Neurobiol Dis 34:279–290.

Carpenter DO. 2006. Polychlorinated biphenyls (PCBs): Routes of exposure and effects on human health. Rev Environ Health 21:1–23.

Carpenter DO, DeCaprio AP, O'Hehir D, Akhtar F, Johnson G, Scrudato RJ, et al. 2005. Polychlorinated biphenyls in serum of the Siberian Yupik people from St. Lawrence Island, Alaska. Int J Circumpolar Health 64:322–335.

Casamajor L. 1913. An unusual form of mineral poisoning affecting the nervous system: Manganese? JAMA 60:646–649.

Caudle WM, Richardson JR, Delea KC, Guillot TS, Wang M, Pennell KD, et al. 2006. Polychlorinated biphenyl-induced reduction of dopamine transporter expression as a precursor to Parkinson's disease-associated dopamine toxicity. Toxicol Sci 92:490–499.

Cavallo D, Ursini CL, Setini A, Chianese C, Piegari P, Perniconi B, et al. 2003. Evaluation of oxidative damage and inhibition of DNA repair in an in vitro study of nickel exposure. Toxicol in Vitro 17:603–607.

CDC. 2006. Facts about Paraquat. Atlanta, GA: Centers for Disease Control and Prevention. Available at http://www.bt.cdc.gov/agent/paraquat/basics/facts.asp [accessed May 14, 2013].

CDC. 2009. Fourth National Report on Human Exposure to Environmental Chemicals. N. C. f. E. H. Centers for Disease Control and Prevention, Division of Laboratory Sciences. Atlanta. Available at http://www.cdc.gov/ExposureReport/index.html [accessed May 14, 2013].

Chen RC, Chang SF, Su CL, Chen TH, Yen MF, Wu HM, et al. 2001. Prevalence, incidence, and mortality of PD: A door-to-door survey in Ilan county, Taiwan. Neurology 57:1679–1686.

Chen H, Zhang SM, Schwarzschild MA, Hernán MA, Ascherio A. 2005. Physical activity and the risk of Parkinson's disease. Neurology 64:664–669.

Chiu A, Beaubier J, Chiu J, Chan L, Gerstenberger S. 2004. Epidemiologic studies of PCB congener profiles in North American fish consuming populations. J Environ Sci Health C Environ Carcinog Ecotoxicol Rev 22:13–36.

Chu I, Villeneuve DC, Yagminas A, LeCavalier P, Poon R, Feeley M, et al. 1994. Subchronic toxicity of 3,′4,4′,5-pentachlorobiphenyl in the rat. I. Clinical, biochemical, hematological, and histopathological changes. Fundam Appl Toxicol 22: 457–468.

Chu I, Poon R, Yagminas A, Lecavalier P, Håkansson H, Valli VE, et al. 1998. Subchronic toxicity of PCB 105 (2,3,3′,4,4′-pentachlorobiphenyl) in rats. J Appl Toxicol 18:285–292.

Chun HS, Gibson GE, DeGiorgio LA, Zhang H, Kidd VJ, Son JH. 2001. Dopaminergic cell death induced by MPP(+), oxidant and specific neurotoxicants shares the common molecular mechanism. J Neurochem 76:1010–1021.

Cook DG, Fahn S, Brait KA. 1974. Chronic manganese intoxication. Arch Neurol 30:59–64.

Coon S, Stark A, Peterson E, Gloi A, Kortsha G, Pounds J, et al. 2006. Whole-body lifetime occupational lead exposure and risk of Parkinson's disease. Environ Health Perspect 114:1872–1876.

Correia Guedes L, Ferreira JJ, Rosa MM, Coelho M, Bonifati V, Sampaio C. 2010. Worldwide frequency of G2019S LRRK2 mutation in Parkinson's disease: A systematic review. Parkinsonism Relat Disord 16:237–242.

Corrigan FM, Murray L, Wyatt CL, Shore RF. 1998. Diorthosubstituted polychlorinated biphenyls in caudate nucleus in Parkinson's disease. Exp Neurol 150:339–342.

Corrigan FM, Wienburg CL, Shore RF, Daniel SE, Mann D. 2000. Organochlorine insecticides in substantia nigra in Parkinson's disease. J Toxicol Environ Health A 59:229–234.

Corti O, Lesage S, Brice A. 2011. What genetics tells us about the causes and mechanisms of Parkinson's disease. Physiol Rev 91:1161–1218.

Costello S, Cockburn M, Bronstein J, Zhang X, Ritz B. 2009. Parkinson's disease and residential exposure to maneb and paraquat from agricultural applications in the central valley of California. Am J Epidemiol 169:919–926.

Cotzias GC, Horiuchi K, Fuenzalida S, Mena I. 1968. Chronic manganese poisoning. Clearance of tissue manganese concentrations with persistence of the neurological picture. Neurology 18:376–382.

Deutch B, Pedersen HS, Jørgensen EC, Hansen JC. 2003. Smoking as a determinant of high organochlorine levels in Greenland. Arch Environ Health 58:30–36.

Dexter DT, Carter CJ, Wells FR, Javoy-Agid F, Agid Y, Lees A, et al. 1989a. Basal lipid peroxidation in substantia nigra is increased in Parkinson's disease. J Neurochem 52:381–389.

Dexter DT, Wells FR, Lees AJ, Agid F, Agid Y, Jenner P, et al. 1989b. Increased nigral iron content and alterations in other metal ions occurring in brain in Parkinson's disease. J Neurochem 52:1830–1836.

Dexter DT, Carayon A, Vidailhet M, Ruberg M, Agid F, Agid Y, et al. 1990. Decreased ferritin levels in brain in Parkinson's disease. J Neurochem 55:16–20.

Dexter DT, Carayon A, Javoy-Agid F, Agid Y, Wells FR, Daniel SE, et al. 1991. Alterations in the levels of iron, ferritin and other trace metals in Parkinson's disease and other neurodegenerative diseases affecting the basal ganglia. Brain 114:1953–1975.

Dick FD, De Palma G, Ahmadi A, Scott NW, Prescott GJ, Bennett J, et al. 2007a. Environmental risk factors for Parkinson's disease and parkinsonism: The Geoparkinson study. Occup Environ Med 64:666–672.

Dick S, Semple S, Dick F, Seaton A. 2007b. Occupational titles as risk factors for Parkinson's disease. Occup Med (Lond) 57:50–56.

Di Monte DA. 1991. Mitochondrial DNA and Parkinson's disease. Neurology 41:38–42.

Dinis-Oliveira RJ, Remião F, Carmo H, Duarte JA, Navarro AS, Bastos ML, et al. 2006. Paraquat exposure as an etiological factor of Parkinson's disease. Neurotoxicology 27:1110–1122.

Do CB, Tung JY, Dorfman E, Kiefer AK, Drabant EM, Francke U, et al. 2011. Web-based genome-wide association study identifies two novel loci and a substantial genetic component for Parkinson's disease. PLoS Genet 7:e1002141.

Dorman DC, Brenneman KA, McElveen AM, Lynch SE, Roberts KC, Wong BA. 2002. Olfactory transport: A direct route of delivery of inhaled manganese phosphate to the rat brain. J Toxicol Environ Health A 65:1493–1511.

Dreiem A, Okoniewski RJ, Brosch KO, Miller VM, Seegal RF. 2010. Polychlorinated biphenyls and polybrominated diphenyl ethers alter striatal dopamine neurochemistry in synaptosomes from developing rats in an additive manner. Toxicol Sci 118:150–159.

Drolet RE, Cannon JR, Montero L, Greenamyre JT. 2009. Chronic rotenone exposure reproduces Parkinson's disease gastrointestinal neuropathology. Neurobiol Dis 36:96–102.

Dutheil F, Beaune P, Tzourio C, Loriot MA, Elbaz A. 2010. Interaction between ABCB1 and professional exposure to organochlorine insecticides in Parkinson's disease. Arch Neurol 67:739–745.

Edsall DL, Wilbur FP, Drinker CK. 1919. The occurrence, course, and prevention of chronic manganese poisoning. J Ind Hyg 1:183–193.

Elbaz A, Levecque C, Clavel J, Vidal JS, Richard F, Amouyel P, et al. 2004. CYP2D6 polymorphism, pesticide exposure, and Parkinson's disease. Ann Neurol 55:430–434.

Elbaz A, Clavel J, Rathouz PJ, Moisan F, Galanaud JP, Delemotte B, et al. 2009. Professional exposure to pesticides and Parkinson's disease. Ann Neurol 66:494–504.

EPA, Office of Compliance. 1995. Profile of the Dry Cleaning Industry. EPA Office of Compliance Sector Notebook Project. Washington, DC, EPA.

EPA. 2009. Iris Toxicological Review of Trichloroethylene. Washington, DC, EPA.

Ercal N, Gurer-Orhan H, Aykin-Burns N. 2001. Toxic metals and oxidative stress part I: Mechanisms involved in metal-induced oxidative damage. Curr Top Med Chem 1:529–539.

Eto K. 1997. Pathology of Minamata disease. Toxicol Pathol 25:614–623.

Factor SA, Sanchez-Ramos J, Weiner WJ. 1988. Trauma as an etiology of Parkinsonism: A historical review of the concept. Mov Disord 3:30–36.

Fairhall LT, Neal PA. 1943. Industrial Manganese Poisoning. Washington, DC: National Institutes of Health.

Fall PA, Fredrikson M, Axelson O, Granérus AK. 1999. Nutritional and occupational factors influencing the risk of Parkinson's disease: A case-control study in southeastern Sweden. Mov Disord 14:28–37.

Farina M, Rocha JB, Aschner M. 2011. Mechanisms of methylmercury-induced neurotoxicity: Evidence from experimental studies. Life Sci 89:555–563.

Feldman AL, Johansson AL, Nise G, Gatz M, Pedersen NL, Wirdefeldt K. 2011. Occupational exposure in parkinsonian disorders: A 43-year prospective cohort study in men. Parkinsonism Relat Disord 17:677–682.

Ferroni C, Selis L, Mutti A, Folli D, Bergamaschi E, Franchini I. 1992. Neurobehavioral and neuroendocrine effects of occupational exposure to perchloroethylene. Neurotoxicology 13:243–247.

Finkelstein MM, Jerrett M. 2007. A study of the relationships between Parkinson's disease and markers of traffic-derived and environmental manganese air pollution in two Canadian cities. Environ Res 104:420–432.

Firestone JA, Lundin JI, Powers KM, Smith-Weller T, Franklin GM, Swanson PD, et al. 2010. Occupational factors and risk of Parkinson's disease: A population-based case-control study. Am J Ind Med 53:217–223.

Fleming L, Mann JB, Bean J, Briggle T, Sanchez-Ramos JR. 1994. Parkinson's disease and brain levels of organochlorine pesticides. Ann Neurol 36:100–103.

Fong CS, Wu RM, Shieh JC, Chao YT, Fu YP, Kuao CL, et al. 2007. Pesticide exposure on southwestern Taiwanese with MnSOD and NQO1 polymorphisms is associated with increased risk of Parkinson's disease. Clin Chim Acta 378:136–141.

Fored CM, Fryzek JP, Brandt L, Nise G, Sjögren B, McLaughlin JK, et al. 2006. Parkinson's disease and other basal ganglia or movement disorders in a large nationwide cohort of Swedish welders. Occup Environ Med 63:135–140.

Forte G, Bocca B, Senofonte O, Petrucci F, Brusa L, Stanzione P, et al. 2004. Trace and major elements in whole blood, serum, cerebrospinal fluid and urine of patients with Parkinson's disease. J Neural Transm 111:1031–1040.

Frigerio R, Elbaz A, Sanft KR, Peterson BJ, Bower JH, Ahlskog JE, et al. 2005. Education and occupations preceding Parkinson's disease: A population-based case-control study. Neurology 65:1575–1583.

Fromberg A, Cleemann M, Carlsen L. 1999. Review on persistent organic pollutants in the environment of Greenland and Faroe Islands. Chemosphere 38:3075–3093.

Fryzek JP, Hansen J, Cohen S, Bonde JP, Llambias MT, Kolstad HA, et al. 2005. A cohort study of Parkinson's disease and other neurodegenerative disorders in Danish welders. J Occup Environ Med 47:466–472.

Fukushima T, Tan X, Luo Y, Kanda H. 2011. Serum vitamins and heavy metals in blood and urine, and the correlations among them in Parkinson's disease patients in China. Neuroepidemiology 36:240–244.

Gałazka-Friedman J, Bauminger ER, Friedman A, Barcikowska M, Hechel D, Nowik I. 1996. Iron in parkinsonian and control substantia nigra—A Mössbauer spectroscopy study. Mov Disord 11:8–16.

Gao X, Chen H, Schwarzschild MA, Ascherio A. 2011. Use of ibuprofen and risk of Parkinson's disease. Neurology 76:863–869.

Gash DM, Rutland K, Hudson NL, Sullivan PG, Bing G, Cass WA, et al. 2008. Trichloroethylene: Parkinsonism and complex 1 mitochondrial neurotoxicity. Ann Neurol 63:184–192.

Gassner B, Wüthrich A, Scholtysik G, Solioz M. 1997. The pyrethroids permethrin and cyhalothrin are potent inhibitors of the mitochondrial complex I. J Pharmacol Exp Ther 281:855–860.

Gatto NM, Cockburn M, Bronstein J, Manthripragada AD, Ritz B. 2009. Well-water consumption and Parkinson's disease in rural California. Environ Health Perspect 117:1912–1918.

Gelb DJ, Oliver E, Gilman S. 1999. Diagnostic criteria for Parkinson's disease. Arch Neurol 56:33–39.

Gerlach M, Double KL, Ben-Shachar D, Zecca L, Youdim MB, Riederer P. 2003. Neuromelanin and its interaction with iron as a potential risk factor for dopaminergic neurodegeneration underlying Parkinson's disease. Neurotox Res 5:35–44.

Goetz CG. 1986. Charcot on Parkinson's disease. Mov Disord 1:27–32.

Goldman SM, Quinlan P, Smith AR, Langston JW, Tanner CM. 2004. Manganese exposure and risk of Parkinson's disease in twins. Mov Disord 19:S162.

Goldman SM, Tanner CM, Olanow CW, Watts RL, Field RD, Langston JW. 2005. Occupation and parkinsonism in three movement disorders clinics. Neurology 65:1430–1435.

Goldman SM, Quinlan PJ, Ross GW, Marras C, Meng C, Bhudhikanok GS, et al. 2012. Solvent exposures and Parkinson's disease risk in twins. Ann Neurol 71:776–784.

Goldwurm S, Di Fonzo A, Simons EJ, Rohé CF, Zini M, Canesi M, et al. 2005. The G6055A (G2019S) mutation in LRRK2 is frequent in both early and late onset Parkinson's disease and originates from a common ancestor. J Med Genet 42:e65.

Good PF, Olanow CW, Perl DP. 1992. Neuromelanin-containing neurons of the substantia nigra accumulate iron and aluminum in Parkinson's disease: A LAMMA study. Brain Res 593:343–346.

Gorell JM, Johnson CC, Rybicki BA, Peterson EL, Kortsha GX, Brown GG, et al. 1997. Occupational exposures to metals as risk factors for Parkinson's disease. Neurology 48:650–658.

Gorell JM, Johnson CC, Rybicki BA, Peterson EL, Kortsha GX, Brown GG, et al. 1999. Occupational exposure to manganese, copper, lead, iron, mercury and zinc and the risk of Parkinson's disease. Neurotoxicology 20:239–247.

Gourie-Devi M, Ramu MG, Venkataram BS. 1991. Treatment of Parkinson's disease in "Ayurveda" (ancient Indian system of medicine): Discussion paper. J R Soc Med 84:491–492.

Greenamyre JT, MacKenzie G, Peng TI, Stephans SE. 1999. Mitochondrial dysfunction in Parkinson's disease. Biochem Soc Symp 66:85–97.

Greene JG, Noorian AR, Srinivasan S. 2009. Delayed gastric emptying and enteric nervous system dysfunction in the rotenone model of Parkinson's disease. Exp Neurol 218:154–161.

Greenhouse AH. 1971. Manganese intoxication in the United States. Trans Am Neurol Assoc 96:248–249.

Griffiths PD, Crossman AR. 1993. Distribution of iron in the basal ganglia and neocortex in postmortem tissue in Parkinson's disease and Alzheimer's disease. Dementia 4:61–65.

Gu M, Cooper JM, Butler P, Walker AP, Mistry PK, Dooley JS, et al. 2000. Oxidative-phosphorylation defects in liver of patients with Wilson's disease. Lancet 356:469–474.

Guehl D, Bezard E, Dovero S, Boraud T, Bioulac B, Gross C. 1999. Trichloroethylene and parkinsonism: A human and experimental observation. Eur J Neurol 6:609–611.

Guggenheim MA, Couch JR, Weinberg W. 1971. Motor dysfunction as a permanent complication of methanol ingestion. Presentation of a case with a beneficial response to levodopa treatment. Arch Neurol 24:550–554.

Guilarte TR. 2010. Manganese and Parkinson's disease: A critical review and new findings. Environ Health Perspect 118:1071–1080.

Halliwell B, Gutteridge JM. 1992. Biologically relevant metal ion-dependent hydroxyl radical generation. An update. FEBS Lett 307:108–112.

Hancock DB, Martin ER, Mayhew GM, Stajich JM, Jewett R, Stacy MA, et al. 2008. Pesticide exposure and risk of Parkinson's disease: A family-based case-control study. BMC Neurol 8:6.

Hardell S, Tilander H, Welfinger-Smith G, Burger J, Carpenter DO. 2010. Levels of polychlorinated biphenyls (PCBs) and three organochlorine pesticides in fish from the Aleutian Islands of Alaska. PLoS ONE 5:e12396.

Hatcher JM, Richardson JR, Guillot TS, McCormack AL, Di Monte DA, Jones DP, et al. 2007. Dieldrin exposure induces oxidative damage in the mouse nigrostriatal dopamine system. Exp Neurol 204:619–630.

Hatcher JM, Delea KC, Richardson JR, Pennell KD, Miller GW. 2008. Disruption of dopamine transport by DDT and its metabolites. Neurotoxicology 29:682–690.

Heim C, Sontag KH. 1997. The halogenated tetrahydro-beta-carboline "TaClo": A progressively-acting neurotoxin. J Neural Transm Suppl 50:107–111.

Henchcliffe C, Beal MF. 2008. Mitochondrial biology and oxidative stress in Parkinson's disease pathogenesis. Nat Clin Pract Neurol 4:600–609.

Hernán MA, Takkouche B, Caamaño-Isorna F, Gestal-Otero JJ. 2002. A meta-analysis of coffee drinking, cigarette smoking, and the risk of Parkinson's disease. Ann Neurol 52:276–284.

Hertzman C, Wiens M, Bowering D, Snow B, Calne D. 1990. Parkinson's disease: A case-control study of occupational and environmental risk factors. Am J Ind Med 17:349–355.

Hertzman C, Wiens M, Snow B, Kelly S, Calne D. 1994. A case-control study of Parkinson's disease in a horticultural region of British Columbia. Mov Disord 9:69–75.

Heusinkveld HJ, Westerink RH. 2012. Organochlorine insecticides lindane and dieldrin and their binary mixture disturb calcium homeostasis in dopaminergic PC12 cells. Environ Sci Technol 46:1842–1848.

Hozumi I, Hasegawa T, Honda A, Ozawa K, Hayashi Y, Hashimoto K, et al. 2011. Patterns of levels of biological metals in CSF differ among neurodegenerative diseases. J Neurol Sci 303:95–99.

Huang CC, Chu NS, Lu CS, Chen RS, Calne DB. 1998. Long-term progression in chronic manganism: Ten years of follow-up. Neurology 50:698–700.

Huang XP, O'Brien PJ, Templeton DM. 2006. Mitochondrial involvement in genetically determined transition metal toxicity I. Iron toxicity. Chem Biol Interact 163:68–76.

Huse DM, Schulman K, Orsini L, Castelli-Haley J, Kennedy S, Lenhart G. 2005. Burden of illness in Parkinson's disease. Mov Disord 20:1449–1454.

IARC. 1995. Dry Cleaning, Some Chlorinated Solvents and Other Industrial Chemicals. Summary of Data Reported and Evaluation. IARC Monographs on the Evaluation of Carcinogenic Risks to Humans. Lyon, France, WHO. 63.

IOM. 2008. Veterans and Agent Orange Update, Institute of Medicine. Available at http://www.iom.edu/Reports/2009/Veterans-and-Agent-Orange-Update-2008.aspx [accessed May 14, 2013].

Jackson-Lewis V, Blesa J, Przedborski S. 2012. Animal models of Parkinson's disease. Parkinsonism Relat Disord 18(Suppl 1):S183–S185.

Japan, M o. t. E. G. o. (2002). Minamata Disease: The History and Measures. Tokyo, Japan: E. H. Department.

Jiménez-Jiménez FJ, Fernández-Calle P, Martínez-Vanaclocha M, Herrero E, Molina JA, Vázquez A, et al. 1992. Serum levels of zinc and copper in patients with Parkinson's disease. J Neurol Sci 112:30–33.

Jiménez-Jiménez FJ, Molina JA, Aguilar MV, Arrieta FJ, Jorge-Santamaría A, Cabrera-Valdivia F, et al. 1995. Serum and urinary manganese levels in patients with Parkinson's disease. Acta Neurol Scand 91:317–320.

Johnson CC, Gorell JM, Rybicki BA, Sanders K, Peterson EL. 1999. Adult nutrient intake as a risk factor for Parkinson's disease. Int J Epidemiol 28:1102–1109.

Kachroo A, Schwarzschild MA. 2012. Adenosine A2A receptor gene disruption protects in an α-synuclein model of Parkinson's disease. Ann Neurol 71:278–282.

Kaiser J. 2003. Manganese: A high-octane dispute. Science 300:926–928.

Kala SV, Jadhav AL. 1995. Low level lead exposure decreases in vivo release of dopamine in the rat nucleus accumbens: A microdialysis study. J Neurochem 65:1631–1635.

Kamel F, Tanner C, Umbach D, Hoppin J, Alavanja M, Blair A, et al. 2007. Pesticide exposure and self-reported Parkinson's disease in the agricultural health study. Am J Epidemiol 165:364–374.

Kanthasamy AG, Kitazawa M, Kanthasamy A, Anantharam V. 2005. Dieldrin-induced neurotoxicity: Relevance to Parkinson's disease pathogenesis. Neurotoxicology 26:701–719.

Kaur D, Peng J, Chinta SJ, Rajagopalan S, Di Monte DA, Cherny RA, et al. 2007. Increased murine neonatal iron intake results in Parkinson-like neurodegeneration with age. Neurobiol Aging 28:907–913.

Keppel Hesselink JM. 1989. Trauma as an etiology of parkinsonism: Opinions in the nineteenth century. Mov Disord 4:283–285.

Kirkey KL, Johnson CC, Rybicki BA, Peterson EL, Kortsha GX, Gorell JM. 2001. Occupational categories at risk for Parkinson's disease. Am J Ind Med 39:564–571.

Kitazawa M, Anantharam V, Kanthasamy AG. 2001. Dieldrin-induced oxidative stress and neurochemical changes contribute to apoptopic cell death in dopaminergic cells. Free Radic Biol Med 31:1473–1485.

Klawans HL, Stein RW, Tanner CM, Goetz CG. 1982. A pure parkinsonian syndrome following acute carbon monoxide intoxication. Arch Neurol 39:302–304.

Klein C, Ziegler A. 2011. From GWAS to clinical utility in Parkinson's disease. Lancet 377:613–614.

Kochen W, Kohlmüller D, De Biasi P, Ramsay R. 2003. The endogeneous formation of highly chlorinated tetrahydro-beta-carbolines as a possible causative mechanism in idiopathic Parkinson's disease. Adv Exp Med Biol 527:253–263.

Koldkjaer OG, Wermuth L, Bjerregaard P. 2004. Parkinson's disease among Inuit in Greenland: Organochlorines as risk factors. Int J Circumpolar Health 63(Suppl 2):366–368.

Koziorowski D, Friedman A, Arosio P, Santambrogio P, Dziewulska D. 2007. ELISA reveals a difference in the structure of substantia nigra ferritin in Parkinson's disease and incidental Lewy body compared to control. Parkinsonism Relat Disord 13: 214–218.

Kuhn W, Winkel R, Woitalla D, Meves S, Przuntek H, Müller T. 1998. High prevalence of parkinsonism after occupational exposure to lead-sulfate batteries. Neurology 50:1885–1886.

Kuopio AM, Marttila RJ, Helenius H, Rinne UK. 1999. Environmental risk factors in Parkinson's disease. Mov Disord 14:928–939.

Kuter K, Nowak P, Gołembiowska K, Ossowska K. 2010. Increased reactive oxygen species production in the brain after repeated low-dose pesticide paraquat exposure in rats. A comparison with peripheral tissues. Neurochem Res 35:1121–1130.

Kvalem HE, Knutsen HK, Thomsen C, Haugen M, Stigum H, Brantsaeter AL, et al. 2009. Role of dietary patterns for dioxin and PCB exposure. Mol Nutr Food Res 53:1438–1451.

Langston JW. 2006. The Parkinson's complex: Parkinsonism is just the tip of the iceberg. Ann Neurol 59:591–596.

Langston JW, Ballard P, Tetrud JW, Irwin I. 1983. Chronic Parkinsonism in humans due to a product of meperidine-analog synthesis. Science 219:979–980.

Larsen NA, Pakkenberg H, Damsgaard E, Heydorn K, Wold S. 1981. Distribution of arsenic, manganese, and selenium in the human brain in chronic renal insufficiency, Parkinson's disease, and amyotrophic lateral sclerosis. J Neurol Sci 51: 437–446.

Lash LH, Fisher JW, Lipscomb JC, Parker JC. 2000. Metabolism of trichloroethylene. Environ Health Perspect 108(Suppl 2):177–200.

Lash LH, Qian W, Putt DA, Hueni SE, Elfarra AA, Sicuri AR, et al. 2002. Renal toxicity of perchloroethylene and S-(1,2,2-trichlorovinyl)glutathione in rats and mice: Sex- and species-dependent differences. Toxicol Appl Pharmacol 179:163–171.

Lasley SM. 1992. Regulation of dopaminergic activity, but not tyrosine hydroxylase, is diminished after chronic inorganic lead exposure. Neurotoxicology 13:625–635.

Lee DW, Notter SA, Thiruchelvam M, Dever DP, Fitzpatrick R, Kostyniak PJ, et al. 2012. Subchronic polychlorinated biphenyl (Aroclor 1254) exposure produces oxidative damage and neuronal death of ventral midbrain dopaminergic systems. Toxicol Sci 125:496–508.

Leonard SS, Harris GK, Shi X. 2004. Metal-induced oxidative stress and signal transduction. Free Radic Biol Med 37:1921–1942.

Lesage S, Brice A. 2009. Parkinson's disease: From monogenic forms to genetic susceptibility factors. Hum Mol Genet 18:R48–R59.

Liou HH, Tsai MC, Chen CJ, Jeng JS, Chang YC, Chen SY, et al. 1997. Environmental risk factors and Parkinson's disease: A case-control study in Taiwan. Neurology 48:1583–1588.

Liu M, Choi DY, Hunter RL, Pandya JD, Cass WA, Sullivan PG, et al. 2010. Trichloroethylene induces dopaminergic neurodegeneration in Fisher 344 rats. J Neurochem 112:773–783.

Lloyd DR, Phillips DH. 1999. Oxidative DNA damage mediated by copper(II), iron(II) and nickel(II) fenton reactions: Evidence for site-specific mechanisms in the formation of double-strand breaks, 8-hydroxydeoxyguanosine and putative intrastrand cross-links. Mutat Res 424:23–36.

Logroscino G, Marder K, Graziano J, Freyer G, Slavkovich V, Lojacono N, et al. 1998. Dietary iron, animal fats, and risk of Parkinson's disease. Mov Disord 13(Suppl 1):13–16.

Logroscino G, Gao X, Chen H, Wing A, Ascherio A. 2008. Dietary iron intake and risk of Parkinson's disease. Am J Epidemiol 168:1381–1388.

Manibusan MK, Odin M, Eastmond DA. 2007. Postulated carbon tetrachloride mode of action: A review. J Environ Sci Health C Environ Carcinog Ecotoxicol Rev 25:185–209.

Manthripragada AD, Costello S, Cockburn MG, Bronstein JM, Ritz B. 2010. Paraoxonase 1, agricultural organophosphate exposure, and Parkinson's disease. Epidemiology 21:87–94.

Manyam BV. 1990. Paralysis agitans and levodopa in "Ayurveda": Ancient Indian medical treatise. Mov Disord 5:47–48.

Mao H, Liu B. 2008. Synergistic microglial reactive oxygen species generation induced by pesticides lindane and dieldrin. Neuroreport 19:1317–1320.

Mariussen E, Andersson PL, Tysklind M, Fonnum F. 2001. Effect of polychlorinated biphenyls on the uptake of dopamine into rat brain synaptic vesicles: A structure-activity study. Toxicol Appl Pharmacol 175:176–183.

Marras C, Gruneir A, Rochon P, Wang X, Anderson G, Brotchie J, et al. 2012. Dihydropyridine calcium channel blockers and the progression of parkinsonism. Ann Neurol 71:362–369.

Marsh GM, Gula MJ. 2006. Employment as a welder and Parkinson's disease among heavy equipment manufacturing workers. J Occup Environ Med 48:1031–1046.

Martin WR. 2009. Quantitative estimation of regional brain iron with magnetic resonance imaging. Parkinsonism Relat Disord 15(Suppl 3):S215–S218.

Martin I, Dawson VL, Dawson TM. 2011. Recent advances in the genetics of Parkinson's disease. Annu Rev Genomics Hum Genet 12:301–325.

McCormack AL, Thiruchelvam M, Manning-Bog AB, Thiffault C, Langston JW, Cory-Slechta DA, et al. 2002. Environmental risk factors and Parkinson's disease: Selective degeneration of nigral dopaminergic neurons caused by the herbicide paraquat. Neurobiol Dis 10:119–127.

McCrank E, Rabheru K. 1989. Four cases of progressive supranuclear palsy in patients exposed to organic solvents. Can J Psychiatry 34:934–936.

McDonnell L, Maginnis C, Lewis S, Pickering N, Antoniak M, Hubbard R, et al. 2003. Occupational exposure to solvents and metals and Parkinson's disease. Neurology 61:716–717.

Mehta R, Templeton DM, O'brien PJ. 2006. Mitochondrial involvement in genetically determined transition metal toxicity II. Copper toxicity. Chem Biol Interact 163: 77–85.

Melamed E, Lavy S. 1977. Parkinsonism associated with chronic inhalation of carbon tetrachloride. Lancet 1:1015.

Mena I, Marin O, Fuenzalida S, Cotzias GC. 1967. Chronic manganese poisoning. Clinical picture and manganese turnover. Neurology 17:128–136.

Miyake Y, Tanaka K, Fukushima W, Sasaki S, Kiyohara C, Tsuboi Y, et al. 2011. Dietary intake of metals and risk of Parkinson's disease: A case-control study in Japan. J Neurol Sci 306:98–102.

Miyazaki Y, Takano T. 1983. Impairment of mitochondrial electron transport by tetrachloroethylene. Toxicol Lett 18:163–166.

Muckle G, Ayotte P, Dewailly E, Jacobson SW, Jacobson JL 2001. Determinants of polychlorinated biphenyls and methylmercury exposure in Inuit women of childbearing age. Environ Health Perspect 109:957–963.

Myhre O, Mariussen E, Reistad T, Voie OA, Aarnes H, Fonnum F. 2009. Effects of polychlorinated biphenyls on the neutrophil NADPH oxidase system. Toxicol Lett 187:144–148.

Nalls MA, Plagnol V, Hernandez DG, Sharma M, Sheerin UM, Saad M, et al. 2011. Imputation of sequence variants for identification of genetic risks for Parkinson's disease: A meta-analysis of genome-wide association studies. Lancet 377:641–649.

Newland MC, Paletz EM, Reed MN. 2008. Methylmercury and nutrition: Adult effects of fetal exposure in experimental models. Neurotoxicology 29:783–801.

Ngim CH, Devathasan G. 1989. Epidemiologic study on the association between body burden mercury level and idiopathic Parkinson's disease. Neuroepidemiology 8: 128–141.

NIOSH, Division of Criteria Documentation and Standards Development. 1976. Criteria for Occupational Exposure to Tetrachloroethylene. NIOSH Criteria for a Recommended Standard. Washington, DC: NIOSH, CDC, HHS.

NIOSH, Division of Criteria Documentation and Standards Development. 1978. Special Occupational Hazard Review of Trichloroethylene. Special Occupational Hazard Review with Control Recommendations. Rockville, MD: NIOSH, CDC, DHHS.

Oder W, Grimm G, Kollegger H, Ferenci P, Schneider B, Deecke L. 1991. Neurological and neuropsychiatric spectrum of Wilson's disease: A prospective study of 45 cases. J Neurol 238:281–287.

Ohlson CG, Hogstedt C. 1981. Parkinson's disease and occupational exposure to organic solvents, agricultural chemicals and mercury—A case-referent study. Scand J Work Environ Health 7:252–256.

Okubadejo NU, Bower JH, Rocca WA, Maraganore DM. 2006. Parkinson's disease in Africa: A systematic review of epidemiologic and genetic studies. Mov Disord 21:2150–2156.

Olanow CW, McNaught K. 2011. Parkinson's disease, proteins, and prions: Milestones. Mov Disord 26:1056–1071.

Ortiz Martinez A, Martinez-Conde E. 1995. The neurotoxic effects of lindane at acute and subchronic dosages. Ecotoxicol Environ Saf 30:101–105.

Pals P, Van Everbroeck B, Grubben B, Viaene MK, Dom R, van der Linden C, et al. 2003. Case-control study of environmental risk factors for Parkinson's disease in Belgium. Eur J Epidemiol 18:1133–1142.

Pan-Montojo FJ, Funk RH. 2010. Oral administration of rotenone using a gavage and image analysis of alpha-synuclein inclusions in the enteric nervous system. J Vis Exp 44. DOI: 10.3791/2123.

Pan-Montojo F, Anichtchik O, Dening Y, Knels L, Pursche S, Jung R, et al. 2010. Progression of Parkinson's disease pathology is reproduced by intragastric administration of rotenone in mice. PLoS ONE 5:e8762.

Park RM, Schulte PA, Bowman JD, Walker JT, Bondy SC, Yost MG, et al. 2005. Potential occupational risks for neurodegenerative diseases. Am J Ind Med 48:63–77.

Park J, Yoo CI, Sim CS, Kim JW, Yi Y, Shin YC, et al. 2006. A retrospective cohort study of Parkinson's disease in Korean shipbuilders. Neurotoxicology 27:445–449.

Parkinson J. 2002. An essay on the shaking palsy. 1817. J Neuropsychiatry Clin Neurosci 14:222–236; discussion 222.

Parkinson's Disease Foundation. 2011. Statistics on Parkinson's Disease. Available at http://www.pdf.org/en/parkinson_statistics [accessed May 14, 2013].

Pearson CR, McConnell G. 1975. Chlorinated C1 and C2 hydrocarbons in the marine environment. Proc R Soc Lond B Biol Sci 189:305–332.

Perl DP, Olanow CW. 2007. The neuropathology of manganese-induced Parkinsonism. J Neuropathol Exp Neurol 66:675–682.

Petersen MS, Halling J, Bech S, Wermuth L, Weihe P, Nielsen F, et al. 2008. Impact of dietary exposure to food contaminants on the risk of Parkinson's disease. Neurotoxicology 29:584–590.

Petrovitch H, Ross GW, Abbott RD, Sanderson WT, Sharp DS, Tanner CM, et al. 2002. Plantation work and risk of Parkinson disease in a population-based longitudinal study. Arch Neurol 59:1787–1792.

Pezzoli G, Barbieri S, Ferrante C, Zecchinelli A, Foà V. 1989. Parkinsonism due to n-hexane exposure. Lancet 2:874.

Pezzoli G, Canesi M, Antonini A, Righini A, Perbellini L, Barichella M, et al. 2000. Hydrocarbon exposure and Parkinson's disease. Neurology 55:667–673.

Pokora MJ, Richfield EK, Cory-Slechta DA. 1996. Preferential vulnerability of nucleus accumbens dopamine binding sites to low-level lead exposure: Time course of effects and interactions with chronic dopamine agonist treatments. J Neurochem 67:1540–1550.

Polizzi S, Ferrara M, Bugiani M, Barbero D, Baccolo T. 2007. Aluminium and iron air pollution near an iron casting and aluminium foundry in Turin district (Italy). J Inorg Biochem 101:1339–1343.

Polymeropoulos MH, Lavedan C, Leroy E, Ide SE, Dehejia A, Dutra A, et al. 1997. Mutation in the alpha-synuclein gene identified in families with Parkinson's disease. Science 276:2045–2047.

Powers KM, Smith-Weller T, Franklin GM, Longstreth WT, Jr., Swanson PD, Checkoway H. 2003. Parkinson's disease risks associated with dietary iron, manganese, and other nutrient intakes. Neurology 60:1761–1766.

Priyadarshi A, Khuder SA, Schaub EA, Shrivastava S. 2000. A meta-analysis of Parkinson's disease and exposure to pesticides. Neurotoxicology 21:435–440.

Przedborski S, Jackson-Lewis V. 1998. Mechanisms of MPTP toxicity. Mov Disord 13(Suppl 1):35–38.

Racette BA, Tabbal SD, Jennings D, Good L, Perlmutter JS, Evanoff B. 2005. Prevalence of parkinsonism and relationship to exposure in a large sample of Alabama welders. Neurology 64:230–235.

Rajput AH, Offord KP, Beard CM, Kurland LT. 1984. Epidemiology of parkinsonism: Incidence, classification, and mortality. Ann Neurol 16:278–282.

Rango M, Canesi M, Ghione I, Farabola M, Righini A, Bresolin N, et al. 2006. Parkinson's disease, chronic hydrocarbon exposure and striatal neuronal damage: A 1-H MRS study. Neurotoxicology 27:164–168.

Ressler T, Wong J, Roos J. 1999. Manganese speciation in exhaust particulates of automobiles using MMT-containing gasoline. J Synchrotron Radiat 6:656–658.

Richardson JR, Miller GW. 2004. Acute exposure to Aroclor 1016 or 1260 differentially affects dopamine transporter and vesicular monoamine transporter 2 levels. Toxicol Lett 148:29–40.

Richardson JR, Shalat SL, Buckley B, Winnik B, O'Suilleabhain P, Diaz-Arrastia R, et al. 2009. Elevated serum pesticide levels and risk of Parkinson's disease. Arch Neurol 66:870–875.

Richardson JR, Roy A, Shalat SL, Buckley B, Winnik B, Gearing M, et al. 2011. β-Hexachlorocyclohexane levels in serum and risk of Parkinson's disease. Neurotoxicology 32:640–645.

Riederer P, Foley P, Bringmann G, Feineis D, Brückner R, Gerlach M. 2002. Biochemical and pharmacological characterization of 1-trichloromethyl-1,2,3,4-tetrahydro-beta-carboline: A biologically relevant neurotoxin? Eur J Pharmacol 442:1–16.

Risher JF, De Rosa CT. 2007. Inorganic: The other mercury. J Environ Health 70: 9–16.

Robinson BH. 1998. The role of manganese superoxide dismutase in health and disease. J Inherit Metab Dis 21:598–603.

Rocca WA, Bower JH, McDonnell SK, Peterson BJ, Maraganore DM. 2001. Time trends in the incidence of parkinsonism in Olmsted County, Minnesota. Neurology 57:462–467.

Rodier J. 1955. Manganese poisoning in Moroccan miners. Br J Ind Med 12:21–35.

Rosenstock HA, Simons DG, Meyer JS. 1971. Chronic manganism. Neurologic and laboratory studies during treatment with levodopa. JAMA 217:1354–1358.

Ross GW, Petrovitch H, Abbott RD, Nelson J, Markesbery W, Davis D, et al. 2004. Parkinsonian signs and substantia nigra neuron density in decendents elders without PD. Ann Neurol 56:532–539.

Ross GW, Abbott RD, Petrovitch H, Tanner CM, Davis DG, Nelson J, et al. 2006. Association of olfactory dysfunction with incidental Lewy bodies. Mov Disord 21:2062–2067.

Ross OA, Braithwaite AT, Skipper LM, Kachergus J, Hulihan MM, Middleton FA, et al. 2008. Genomic investigation of alpha-synuclein multiplication and parkinsonism. Ann Neurol 63:743–750.

Ross GW, Abbott RD, Petrovitch H, Tanner CM, White LR. 2012. Pre-motor features of Parkinson's disease: The Honolulu-Asia Aging Study experience. Parkinsonism Relat Disord 18(Suppl 1):S199–S202.

Rybicki BA, Johnson CC, Uman J, Gorell JM. 1993. Parkinson's disease mortality and the industrial use of heavy metals in Michigan. Mov Disord 8:87–92.

Samii A, Etminan M, Wiens MO, Jafari S. 2009. NSAID use and the risk of Parkinson's disease: Systematic review and meta-analysis of observational studies. Drugs Aging 26:769–779.

Sanchez-Ramos J, Facca A, Basit A, Song S. 1998. Toxicity of dieldrin for dopaminergic neurons in mesencephalic cultures. Exp Neurol 150:263–271.

Sandau CD, Ayotte P, Dewailly E, Duffe J, Norstrom RJ. 2000. Analysis of hydroxylated metabolites of PCBs (OH-PCBs) and other chlorinated phenolic compounds in whole blood from Canadian Inuit. Environ Health Perspect 108:611–616.

Sandhir R, Julka D, Gill KD. 1994. Lipoperoxidative damage on lead exposure in rat brain and its implications on membrane bound enzymes. Pharmacol Toxicol 74:66–71.

Santner A, Uversky VN. 2010. Metalloproteomics and metal toxicology of α-synuclein. Metallomics 2:378–392.

Schapira AH. 2009. Etiology and pathogenesis of Parkinson's disease. Neurol Clin 27:583–603.

Schrag A, Ben-Shlomo Y, Quinn N. 2002. How valid is the clinical diagnosis of Parkinson's disease in the community? J Neurol Neurosurg Psychiatry 73:529–534.

Schuler P, Oyanguren H, Maturana V, Valenzuela A, Cruze E, Plaza V, et al. 1957. Manganese poisoning; environmental and medical study at a Chilean mine. Ind Med Surg 26:167–173.

Schulte PA, Burnett CA, Boeniger MF, Johnson J. 1996. Neurodegenerative diseases: Occupational occurrence and potential risk factors, 1982 through 1991. Am J Public Health 86:1281–1288.

Seeber A. 1989. Neurobehavioral toxicity of long-term exposure to tetrachloroethylene. Neurotoxicol Teratol 11:579–583.

Seegal RF, Bush B, Brosch KO. 1991. Sub-chronic exposure of the adult rat to Aroclor 1254 yields regionally-specific changes in central dopaminergic function. Neurotoxicology 12:55–65.

Seegal RF, Bush B, Brosch KO. 1994. Decreases in dopamine concentrations in adult, non-human primate brain persist following removal from polychlorinated biphenyls. Toxicology 86:71–87.

Seegal RF, Marek KL, Seibyl JP, Jennings DL, Molho ES, Higgins DS, et al. 2010. Occupational exposure to PCBs reduces striatal dopamine transporter densities only in women: A beta-CIT imaging study. Neurobiol Dis 38:219–225.

Seelbach M, Chen L, Powell A, Choi YJ, Zhang B, Hennig B, et al. 2010. Polychlorinated biphenyls disrupt blood–brain barrier integrity and promote brain metastasis formation. Environ Health Perspect 118:479–484.

Seidler A, Hellenbrand W, Robra BP, Vieregge P, Nischan P, Joerg J, et al. 1996. Possible environmental, occupational, and other etiologic factors for Parkinson's disease: A case-control study in Germany. Neurology 46:1275–1284.

Semchuk KM, Love EJ, Lee RG. 1992. Parkinson's disease and exposure to agricultural work and pesticide chemicals. Neurology 42:1328–1335.

Semchuk KM, Love EJ, Lee RG. 1993. Parkinson's disease: A test of the multifactorial etiologic hypothesis. Neurology 43:1173–1180.

Sharma H, Zhang P, Barber DS, Liu B. 2010. Organochlorine pesticides dieldrin and lindane induce cooperative toxicity in dopaminergic neurons: Role of oxidative stress. Neurotoxicology 31:215–222.

Sherer TB, Richardson JR, Testa CM, Seo BB, Panov AV, Yagi T, et al. 2007. Mechanism of toxicity of pesticides acting at complex I: Relevance to environmental etiologies of Parkinson's disease. J Neurochem 100:1469–1479.

Shinotoh H, Snow BJ, Chu NS, Huang CC, Lu CS, Lee C, et al. 1997. Presynaptic and postsynaptic striatal dopaminergic function in patients with manganese intoxication: A positron emission tomography study. Neurology 48:1053–1056.

Singer TP, Ramsay RR, McKeown K, Trevor A, Castagnoli NE, Jr. 1988. Mechanism of the neurotoxicity of 1-methyl-4-phenylpyridinium (MPP+), the toxic bioactivation product of 1-methyl-4-phenyl-1,2,3,6-tetrahydropyridine (MPTP). Toxicology 49:17–23.

Smith GH, Williams FL, Lloyd OL. 1987. Respiratory cancer and air pollution from iron foundries in a Scottish town: An epidemiological and environmental study. Br J Ind Med 44:795–802.

Sofic E, Riederer P, Heinsen H, Beckmann H, Reynolds GP, Hebenstreit G, et al. 1988. Increased iron (III) and total iron content in post mortem substantia nigra of parkinsonian brain. J Neural Transm 74:199–205.

Spillantini MG, Crowther RA, Jakes R, Hasegawa M, Goedert M. 1998. alpha-Synuclein in filamentous inclusions of Lewy bodies from Parkinson's disease and dementia with Lewy bodies. Proc Natl Acad Sci U S A 95:6469–6473.

Spina MB, Cohen G. 1989. Dopamine turnover and glutathione oxidation: Implications for Parkinson's disease. Proc Natl Acad Sci U S A 86:1398–1400.

Sreeramulu K, Liu R, Sharom FJ. 2007. Interaction of insecticides with mammalian P-glycoprotein and their effect on its transport function. Biochim Biophys Acta 1768:1750–1757.

Staiff DC, Butler LC, Davis JE. 1981. A field study of the chemical degradation of paraquat dichloride following simulated spillage on soil. Bull Environ Contam Toxicol 26:16–21.

Stampfer MJ. 2009. Welding occupations and mortality from Parkinson's disease and other neurodegenerative diseases among United States men, 1985–1999. J Occup Environ Hyg 6:267–272.

Steenland K, Hein MJ, Cassinelli RT, 2nd, Prince MM, Nilsen NB, Whelan EA, et al. 2006. Polychlorinated biphenyls and neurodegenerative disease mortality in an occupational cohort. Epidemiology 17:8–13.

Stern ST, Johnson DN. 2008. Role for nanomaterial-autophagy interaction in neurodegenerative disease. Autophagy 4:1097–1100.

Stern MB, Lang A, Poewe W. 2012. Toward a redefinition of Parkinson's disease. Mov Disord 27:54–60.

Stohs SJ, Bagchi D. 1995. Oxidative mechanisms in the toxicity of metal ions. Free Radic Biol Med 18:321–336.

Sun F, Anantharam V, Latchoumycandane C, Kanthasamy A, Kanthasamy AG. 2005. Dieldrin induces ubiquitin-proteasome dysfunction in alpha-synuclein overexpressing dopaminergic neuronal cells and enhances susceptibility to apoptotic cell death. J Pharmacol Exp Ther 315:69–79.

Takeuchi T. 1982. Pathology of Minamata disease. With special reference to its pathogenesis. Acta Pathol Jpn 32(Suppl 1):73–99.

Tandberg E, Larsen JP, Nessler EG, Riise T, Aarli JA. 1995. The epidemiology of Parkinson's disease in the county of Rogaland, Norway. Mov Disord 10:541–549.

Tanner CM, Goldman SM. 1996. Epidemiology of Parkinson's disease. Neurol Clin 14:317–335.

Tanner CM, Ottman R, Goldman SM, Ellenberg J, Chan P, Mayeux R, et al. 1999. Parkinson's disease in twins: An etiologic study. JAMA 281:341–346.

Tanner CM, Ross GW, Jewell SA, Hauser RA, Jankovic J, Factor SA, et al. 2009. Occupation and risk of parkinsonism: A multicenter case-control study. Arch Neurol 66:1106–1113.

Tanner CM, Kamel F, Ross GW, Hoppin JA, Goldman SM, Korell M, et al. 2011. Rotenone, paraquat, and Parkinson's disease. Environ Health Perspect 119: 866–872.

Tavakoli-Nezhad M, Pitts DK. 2005. Postnatal inorganic lead exposure reduces midbrain dopaminergic impulse flow and decreases dopamine D1 receptor sensitivity in nucleus accumbens neurons. J Pharmacol Exp Ther 312:1280–1288.

Tetrud JW, Langston JW, Irwin I, Snow B. 1994. Parkinsonism caused by petroleum waste ingestion. Neurology 44:1051–1054.

The Interstate Technology and Regulatory Council. 2005. Perchlorate: Overview of Issues, Status, and Remedial Options. Washington, DC: Interstate Technology & Regulatory Council.

Thiffault C, Langston WJ, Di Monte DA. 2001. Acute exposure to organochlorine pesticides does not affect striatal dopamine in mice. Neurotox Res 3:537–543.

Thiruchelvam M, Richfield EK, Baggs RB, Tank AW, Cory-Slechta DA. 2000. The nigrostriatal dopaminergic system as a preferential target of repeated exposures to combined paraquat and maneb: Implications for Parkinson's disease. J Neurosci 20:9207–9214.

Thygesen LC, Flachs EM, Hanehøj K, Kjuus H, Juel K. 2011. Hospital admissions for neurological and renal diseases among dentists and dental assistants occupationally exposed to mercury. Occup Environ Med 68:895–901.

Tüchsen F, Jensen AA. 2000. Agricultural work and the risk of Parkinson's disease in Denmark, 1981–1993. Scand J Work Environ Health 26:359–362.

Uitti RJ, Snow BJ, Shinotoh H, Vingerhoets FJ, Hayward M, Hashimoto S, et al. 1994. Parkinsonism induced by solvent abuse. Ann Neurol 35:616–619.

Uversky VN, Li J, Fink AL. 2001a. Metal-triggered structural transformations, aggregation, and fibrillation of human alpha-synuclein. A possible molecular NK between Parkinson's disease and heavy metal exposure. J Biol Chem 276:44284–44296.

Uversky VN, Li J, Fink AL. 2001b. Pesticides directly accelerate the rate of alpha-synuclein fibril formation: A possible factor in Parkinson's disease. FEBS Lett 500:105–108.

Van Den Eeden SK, Tanner CM, Bernstein AL, Fross RD, Leimpeter A, Bloch DA, et al. 2003. Incidence of Parkinson's disease: Variation by age, gender, and race/ethnicity. Am J Epidemiol 157:1015–1022.

Wang JD, Huang CC, Hwang YH, Chiang JR, Lin JM, Chen JS. 1989. Manganese induced parkinsonism: an outbreak due to an unrepaired ventilation control system in a ferromanganese smelter. Br J Ind Med 46:856–859.

Wang A, Costello S, Cockburn M, Zhang X, Bronstein J, Ritz B. 2011. Parkinson's disease risk from ambient exposure to pesticides. Eur J Epidemiol 26:547–555.

Wauchope RD, Buttler TM, Hornsby AG, Augustijn-Beckers PW, Burt JP. 1992. The SCS/ARS/CES pesticide properties database for environmental decision-making. Rev Environ Contam Toxicol 123:1–155.

Weisskopf MG, Knekt P, O'Reilly EJ, Lyytinen J, Reunanen A, Laden F, et al. 2010a. Persistent organochlorine pesticides in serum and risk of Parkinson's disease. Neurology 74:1055–1061.

Weisskopf MG, Weuve J, Nie H, Saint-Hilaire MH, Sudarsky L, Simon DK, et al. 2010b. Association of cumulative lead exposure with Parkinson's disease. Environ Health Perspect 118:1609–1613.

Welfinger-Smith G, Minholz JL, Byrne S, Waghiyi V, Gologergen J, Kava J, et al. 2011. Organochlorine and metal contaminants in traditional foods from St. Lawrence Island, Alaska. J Toxicol Environ Health A 74:1195–1214.

Wermuth L, Joensen P, Bünger N, Jeune B. 1997. High prevalence of Parkinson's disease in the Faroe Islands. Neurology 49:426–432.

Wermuth L, von Weitzel-Mudersbach P, Jeune B. 2000. A two-fold difference in the age-adjusted prevalences of Parkinson's disease between the island of Als and the Faroe Islands. Eur J Neurol 7:655–660.

Wermuth L, Pakkenberg H, Jeune B. 2002. High age-adjusted prevalence of Parkinson's disease among Inuits in Greenland. Neurology 58:1422–1425.

WHO. 1997. Monographs on the Evaluation of Carcinogenic Risks to Humans, World Health Organization, International Agency for Research on Cancer (IARC). 63.

Wilk JB, Tobin JE, Suchowersky O, Shill HA, Klein C, Wooten GF, et al. 2006. Herbicide exposure modifies GSTP1 haplotype association to Parkinson onset age: The GenePD Study. Neurology 67:2206–2210.

Williamson AM, Teo RK, Sanderson J. 1982. Occupational mercury exposure and its consequences for behaviour. Int Arch Occup Environ Health 50:273–286.

Willis AW, Evanoff BA, Lian M, Galarza A, Wegrzyn A, Schootman M, et al. 2010. Metal emissions and urban incident Parkinson's disease: A community health study of Medicare beneficiaries by using geographic information systems. Am J Epidemiol 172:1357–1363.

Wirdefeldt K, Gatz M, Schalling M, Pedersen NL. 2004. No evidence for heritability of Parkinson's disease in Swedish twins. Neurology 63:305–311.

Wolff MS, Fischbein A, Thornton J, Rice C, Lilis R, Selikoff IJ. 1982. Body burden of polychlorinated biphenyls among persons employed in capacitor manufacturing. Int Arch Occup Environ Health 49:199–208.

Wu C, Schaum J. 2000. Exposure assessment of trichloroethylene. Environ Health Perspect 108(Suppl 2):359–363.

Wypijewska A, Galazka-Friedman J, Bauminger ER, Wszolek ZK, Schweitzer KJ, Dickson DW, et al. 2010. Iron and reactive oxygen species activity in parkinsonian substantia nigra. Parkinsonism Relat Disord 16:329–333.

Yamanaga H. 1983. Quantitative analysis of tremor in Minamata disease. Tohoku J Exp Med 141:13–22.

Yamin G, Glaser CB, Uversky VN, Fink AL. 2003. Certain metals trigger fibrillation of methionine-oxidized alpha-synuclein. J Biol Chem 278:27630–27635.

Youdim MB, Ben-Shachar D, Eshel G, Finberg JP, Riederer P. 1993. The neurotoxicity of iron and nitric oxide. Relevance to the etiology of Parkinson's disease. Adv Neurol 60:259–266.

Zecca L, Swartz HM. 1993. Total and paramagnetic metals in human substantia nigra and its neuromelanin. J Neural Transm Park Dis Dement Sect 5:203–213.

Zecca L, Tampellini D, Gatti A, Crippa R, Eisner M, Sulzer D, et al. 2002. The neuromelanin of human substantia nigra and its interaction with metals. J Neural Transm 109:663–672.

Zhang J, Fitsanakis VA, Gu G, Jing D, Ao M, Amarnath V, et al. 2003. Manganese ethylene-bis-dithiocarbamate and selective dopaminergic neurodegeneration in rat: A link through mitochondrial dysfunction. J Neurochem 84:336–346.

Zschiedrich K, König IR, Brüggemann N, Kock N, Kasten M, Leenders KL, et al. 2009. MDR1 variants and risk of Parkinson's disease. Association with pesticide exposure? J Neurol 256:115–120.

Zuch CL, O'Mara DJ, Cory-Slechta DA. 1998. Low-level lead exposure selectively enhances dopamine overflow in nucleus accumbens: An in vivo electrochemistry time course assessment. Toxicol Appl Pharmacol 150:174–185.

CHAPTER 19

Psychiatric Effects of Organic Chemical Exposure

JAMES S. BROWN, JR.

ABSTRACT

Background: The literature describing the psychiatric effects of organic chemicals traces back to the 19th century, and current knowledge in the field reflects clinical experiences on an international scale.

Objectives: Organic chemicals are categorized in this review as persistent or nonpersistent in the environment, and this chapter will review the evidence that both types can cause mental illness.

Results: Persistent organic pollutants (POPs), especially dioxin, polychlorinated biphenyls (PCBs), and chlordane, are known to cause depression and anxiety following chronic exposures. Compared to POPs, nonpersistent organic chemicals cause more severe psychiatric symptoms after both acute and chronic exposures. Psychiatric reactions to the nonpersistent substances, notably organophosphates (OPs), organometals, and solvents, include not only depression and anxiety but also mania, psychosis, and aggression.

Conclusions: Clinicians should remain aware that psychiatric symptoms can arise from toxic chemicals in diverse situations including terrorist attacks with chemical agents, mass chemical disasters in industrial or community settings, individual chemical accidents, and intentional solvent inhalation. Emerging evidence also indicates that prenatal exposure to organic compounds adversely affects neurodevelopment in humans and may be associated with later risk of mental illness.

Effects of Persistent and Bioactive Organic Pollutants on Human Health, First Edition.
Edited by David O. Carpenter.

INTRODUCTION

This chapter focuses on the association of organic pollutants with mental or psychiatric illness. Preceding chapters in this book discuss the relationship between organic pollutant exposure and cognitive and attention problems especially in children. To some extent, the chapters on intelligence and neurodevelopment apply to any discussion of mental illness as clinicians frequently encounter comorbidity of cognitive and attention problems with emotional and thought disorders. Although this chapter discusses how exposures may cause psychiatric symptoms such as depression, anxiety, mania, and psychosis, these conditions rarely occur in isolation from abnormalities of cognition and/or attention.

Rather than symptoms alone, what distinguishes the literature of this chapter from previously discussed neurodevelopmental issues is that the exposures described here usually occur after prenatal/perinatal stages or early childhood. This distinction, however, reflects only the current literature, not a barrier to relating prenatal exposures to mental illness. Three lines of evidence support the possibility that prenatal chemical exposures might increase the risk of mental illness. First, an increasing awareness is apparent in the medical literature that prenatal programming can influence adult-onset diseases (Wintour and Owens 2006). Second, cognitive disability, attention deficit/hyperactivity disorder, and autism, conditions that may be influenced by prenatal exposure, are also associated with schizophrenia, bipolar disorder, and various mood abnormalities (Arnold et al. 2011; King and Lord 2011; Nettelbladt et al. 2009). Third, as discussed further, recent studies have shown that increased risk of mental illness results from prenatal exposure to two organic compounds (methylmercury and tetrachloroethylene [TCE]) and lead of unspecified origin (i.e., inorganic vs. organic).

To describe the current state of knowledge regarding the issues mentioned earlier, the author first divides organic chemical exposures according to two possible times of exposure: childhood/adulthood versus prenatal. The author previously reviewed the known associations of psychiatric conditions in children and adults with organic chemical exposures in 2002 (Brown 2002) and 2007 (Brown 2007). The current review summarizes the previous reviews and updates the information published since 2007. Table 19.1 summarizes the psychiatric effects of organic chemical exposures discussed in this and in the two previous reviews.

This review further categorizes the material to examine the psychiatric issues resulting from persistent organic pollutants (POPs) versus non-POPs. POPs are defined here as those 12 chemicals or groups of chemicals formally defined as POPs plus an additional 12 chemicals listed as potential POPs by the Stockholm Convention in 2008 (Harrad 2010). Chemicals formally defined as POPs are aldrin, chlordane, dichlorodiphenyltrichloroethane (DDT), dieldrin, endrin, heptachlor, hexachlorobenzene, mirex, polychlorinated biphenyls (PCBs), dioxins, dibenzofurans, and toxaphene (Harrad 2010, p. 2).

TABLE 19.1. Psychiatric Symptoms Attributed to Chemical Exposures Occurring in Childhood and Adulthood

	PCB/D	OC	OP	OL	OM	OT	SV
Mood							
Depression	X	X	X	X	X	X	X
Anxiety	X	X	X	X	X	X	X
Labile mood	X	X	X	X		X	X
Irritability	X	X	X	X		X	X
Behavior							
Agitation	X	X	X	X		X	X
Suicidal ideation	X	X	X	X			X
Mania				X		X	X
Aggression				X		X	X
Perceptual							
Hallucinations		X	X	X		X	X
Paranoia			X	X		X	X
Delusions				X		X	X
Cognitive							
Confusion	X	X	X	X		X	X
Memory loss	X	X	X	X		X	X
Other							
Insomnia	X	X	X	X		X	X
Fatigue	X	X	X			X	X
Decreased libido	X	X	X	X		X	X
Loss of appetite		X	X				X
Nightmares		X	X				X
Excessive dreaming			X				
Dissociation			X				

PCB/D, polychlorinated biphenyl/dioxin (includes Agent Orange); OC, organochlorine; OL, organic lead compound; OM, organic mercury compound; OT, organotin; SV, solvent.

PSYCHIATRIC EFFECTS OF EXPOSURES BEGINNING IN CHILDHOOD AND ADULTHOOD

POPs

Summarizing the literature on the psychiatric effects of POPs, the majority of groundbreaking reports were published in the 20th century. The 21st century literature agrees with earlier discoveries, and the majority of mental symptoms from these chemicals have been long known. The well-known psychiatric effects of POPs are divided for discussion according to two subclasses generally defined by the severity of psychiatric reactions associated with them. POPs in the PCB/dioxin-related class typically cause less severe, but still clinically important, psychiatric reactions compared with POPs in the organochlorine/pesticide-related group. These differences are approximations and the categories do not necessarily indicate the psychiatric severity of all possible exposures.

Regarding the first and less psychiatrically severe category, recent analyses of adult cohorts confirm an increased risk of mental symptoms especially anxiety and depression from PCB and dioxin exposures (see Table 19.1). A study of PCBs in adult residents of the Upper Hudson River area of New York reported these findings (Fitzgerald et al. 2008), as did a study of military veterans of Operation Ranch Hand in the Vietnam War who were exposed to the dioxin contaminant of the defoliant Agent Orange (Michalek et al. 2003).

The best-known and most widely cited epidemics of PCB poisonings are the Yusho and Yucheng epidemics in Japan and Taiwan, respectively. Adult survivors of these epidemics, who were children at the time of exposure, have not consistently reported abnormalities of mood, but one study of children in the Yucheng epidemic described them as having more "negative mood" than controls (Guo and Hsu 2001). Not all PCB-related POP exposures, however, are associated with mental symptoms. For example, no psychiatric effects from PCDD/Fs and brominated flame retardants, which are listed under POP categories, have been reported.

In contrast to the PCB/dioxin-related POPs, the organochlorine POPs can cause major psychiatric effects (Table 19.1). One exception is the best-known organochlorine, DDT, which has few psychiatric effects despite its other toxic properties. From a psychiatric standpoint, the cyclodiene organochlorines (aldrin, dieldrin, heptachlor, chlordane, endosulfan) are more neurotoxic than DDT or its derivatives. A well-known case of cyclodiene poisoning with subsequent mental symptoms occurred in Hopewell, Virginia, in the 1970s from large-scale industrial poisoning by Kepone (chlordecone) (Cannon et al. 1978). Kepone or chlordecone is recommended for formal listing as a POP under the Stockholm Convention (Harrad 2010). Kepone-exposed individuals in the Hopewell epidemic developed numerous mood complaints similar to major depression with additional cognitive problems and hallucinations (Brown 2002). Many cyclodiene insecticides were banned from use in the United States, and reports of clinical psychiatric effects of these compounds have become rare since the 1980s. Lindane, a cyclodiene currently used for treating head lice, has not been associated with mental symptoms.

Other Organic Pollutants

Organophosphates (OPs) OPs comprise a wide variety of chemicals ranging in toxicity from chemical weapons known as "nerve agents" or "nerve gases" to the insecticides parathion and malathion. The author previously reviewed the psychiatric importance of terrorist and military exposures to chemical weapons including OPs (Brown 2007). The psychiatric symptoms of nerve agents, especially sarin, soman, and tabun, are listed in Table 19.1 under OPs as a group. The psychiatric effects of OPs, including nerve gases, differ only in the doses required to cause psychiatric reactions, which may vary in individuals based on genetic vulnerabilities (Yamasaki et al. 1997).

OP insecticides have worldwide use, and accidental mass poisonings from agricultural applications still occur (Kavalci et al. 2009). By inhibiting acetylcholinesterase in nerve cells, acute human OP exposures can cause major, life-threatening poisonings followed by delayed physiological and mental changes ranging from mild to severe (Table 19.1). Important clinical findings during the last several years include the observation that OPs cause mental disorder persisting 5–10 years after exposure. This was shown in victims of the nerve gas terrorist attacks in the Tokyo subway system in 1994–1995 (Yanagisawa et al. 2006). In that study, researchers determined that persistent mental symptoms were not fully accounted for by posttraumatic stress disorder (PTSD), a psychiatric condition that could confound studies of the toxic effects of such disasters. Symptoms like depressed mood, apathy, forgetfulness, and lack of concentration were related to cognitive dysfunction from neurological damage, not PTSD from the attack.

Imaging studies of the human brain in acute OP poisonings reveal various abnormalities including abnormal single-photon emission computed tomography (SPECT). Distribution of SPECT lesions in these cases may predict neuropsychological sequelae (Mittal et al. 2011; Ozyurt et al. 2008). These neuropsychological changes can remain physiologically measurable after recovery from the intoxication (Dassanayake et al. 2007). Other radiological brain findings after OP poisoning include abnormal magnetic resonance imaging (MRI) such as those reported in both soldiers exposed to low-dose sarin nerve agent during the Gulf War and civilian victims of the Tokyo subway sarin attack (Chao et al. 2011; Heaton et al. 2007; Yamasue et al. 2007).

The psychiatric effects of chronic, low-dose OP exposure remain controversial, but the evidence for an effect is compelling. Studies of agricultural workers and pesticide applicators continue to find that chronic, low-level OP exposure over a period of years may cause depression, anxiety, and cognitive abnormalities (Kamel et al. 2007; Mackenzie Ross et al. 2010; Rohlman et al. 2007). Additional evidence supporting a mental effect from low-level exposure includes neurobehavioral studies of children working as pesticide applicators or living in agricultural communities. In these children, cognitive dysfunction was significantly related to low-dose, chronic OP exposure (Abdel Rasoul et al. 2008; Lizardi et al. 2008), although not all recent studies replicated these findings in adults (Colosio et al. 2009). Investigations of the risk of suicide following OP exposure are also conflicting. One recent study found twice the relative risk of death from suicide and nonvehicular accidents in pesticide applicators who worked with chlorpyrifos compared to controls (Lee et al. 2007). This was not replicated in a study of suicide risk from OP exposure in adult workers with a known history of cholinesterase inhibition (MacFarlane et al. 2011). From the above-mentioned literature, it is evident that the relationship of low-dose, chronic OP exposure to mood changes and risk of suicide has a long controversial history that requires further study (Bushnell and

Moser 2006). Some portion of the inconsistencies described earlier may result from less than optimal biomarkers of exposure chosen for comparison between exposed and nonexposed groups (Rohlman et al. 2011).

Organometals Certain organic compounds containing a metal have psychiatric importance, the most important being lead and mercury. The psychiatric effects of organic lead compounds have become mostly historical since tetraethyl-leaded gasoline was phased out in the 1970s. However, psychosis from ingestion or inhalation of organic lead in fuel stabilizers and antiknock agents can occur as these compounds are used globally (Wills et al. 2010). Clinicians should remain alert for the psychiatric symptoms of organic lead poisoning shown on Table 19.1, even in the United States, as intentional inhalation of lead-containing compounds by substance abusers can result in severe psychotic, delusional, and paranoid states (Brown 2002). Organic lead poisoning causes psychiatric symptoms similar to acute solvent poisoning (see further), which include mania, psychosis, and aggression in the most severe cases.

The well-known but now mostly historical psychiatric manifestations of mercury poisoning known as erethism and acrodynia resulted from inorganic, rather than organic, mercury poisoning (Brown 2002). Compared to inorganic mercury poisonings that still occur (Chan 2011), organic mercury or methylmercury poisoning presents with fewer and less severe psychiatric symptoms (Table 19.1). However, organic mercury exposure *in utero* causes devastating neurological injury as discussed further.

Beginning in the latter half of the 20th century, sources of organic mercury poisoning ranged from contaminated food to large-scale industrial pollution. Survivors of the best-known epidemic developed a condition called "Minamata disease" caused by organic mercury poisoning from industrial contamination of Minamata Bay in Japan in the 1950s (Brown 2002). Children were particularly affected as organic mercury readily crosses the placenta and enters maternal milk, which permitted extensive cerebral damage in the offspring of pregnant and lactating women who lived near Minamata Bay (Brown 2002). The survivors still exhibit psychiatric and neurological sequelae after both pre- and postnatal exposure (Yorifuji et al. 2011). A recent psychiatric follow-up of Minamata survivors found that moderate and severe prenatal exposure resulted in later mood disturbance in contrast to previous studies that indicated only severe exposures had psychiatric effects (Yorifuji et al. 2011). The chronicity of effects after prenatal exposure suggests that prenatal methylmercury can serve as a model for psychiatric illness resulting from prenatal organic chemical exposure.

Investigations of mental development in populations who consumed methylmercury in seafood were also performed (Davidson et al. 2010; Jedrychowski et al. 2007; Strain et al. 2008). Some found no adverse effect from low-level or "background" methylmercury from seafood regardless of whether the consumption was pre- or postnatal (Cao et al. 2010; Davidson et al. 2011; Marques et al. 2012; Strain et al. 2008) Those instances were mild compared to the moderate and severe examples in Minamata, and the negative findings possibly resulted from the neuroprotective effect of long-chain polyunsaturated fatty acids in seafood (Davidson et al. 2011; Marques et al. 2012). Studies of low-level methylmercury usually assess cognitive rather than emotional outcomes so that the risk of mood disorders following mild organic mercury, especially in seafood, remains unclear. Organic mercury in vaccines was also investigated for a purported relationship to autism in children. Initially, there was concern that ethylmercury from thimerosal-containing vaccines caused autism in children, but well-controlled studies found no association between ethylmercury and autism (Aschner and Ceccatelli 2010; Ng et al. 2007; Price et al. 2010).

Organotin compounds such as trimethyltin do have significant neurotoxic properties (Brown 2002).The few known cases of mental disorder from organotin poisoning showed that it causes irreversible neuronal damage and severe psychiatric illness manifested by depression, mania, psychosis, and physical and sexual aggression (Brown 2002). Table 19.1 lists additional psychiatric symptoms of organotin poisoning, which could be underreported and likely rank in psychiatric severity with organolead and solvent poisoning.

Solvents Solvents, a class of chemicals that extract, dissolve, or suspend insoluble materials, include many chemical subclasses of which the most psychiatrically important are carbon disulfide (CS), halogenated hydrocarbons, aromatic hydrocarbons, and mixtures (e.g., gasoline and jet fuel) (Brown 2002). The *Diagnostic and Statistical Manual of Mental Disorders* (DSM-IV-TR) (American Psychiatric Association 2000) of the American Psychiatric Association identifies more solvent-induced mental disorders, especially those related to inhalant abuse, than any other chemical exposure (Brown 2007). These conditions have a broad range of symptoms including psychosis, anxiety, amnesia, and nonspecified mood problems.

CS poisoning, now a rare occurrence with the implementation of modern industrial safety controls, presents with severe psychosis, mania, and often suicide, a "classic" severe CS poisoning syndrome first recognized in the 1850s in rubber workers who experienced heavy CS contact during the vulcanization process (Brown 2002, 2007). CS is not the only solvent with dangerous properties from a psychiatric perspective. Methyl bromide, a fumigant, also causes severe psychiatric symptoms described in case reports of psychosis, mania, and violence following exposures (Brown 2002).

As a group, solvents can cause most of the psychiatric symptoms on Table 19.1, although the particular solvent and dose determines which symptoms occur. Based on the author's subjective impression, the severity of psychiatric impairment induced by various solvents, depending on dose and individual vulnerability, ranges in severity from highest to lowest in the following order: CS, methyl bromide, trichloroethylene, methyl chloride, mixtures (gasoline, jet fuel), ethylene oxide, toluene, 1,1,1-trichloroethane, xylene, styrene, perchloroethylene, or TCE. This informal list does not cover all possible solvents of psychiatric importance but rather reflects solvents of widespread importance, with the largest amounts of supporting literature, and consistently reported psychiatric effects.

As with OPs, the effect of prolonged contact with low-dose solvents on mental function remains debatable. Since the 1970s, the medical literature has described a condition called "chronic solvent-induced encephalopathy" (CSE) in solvent-exposed workers (van Valen et al. 2009). The condition manifests with multiple vague mood and cognitive and somatic complaints. A recent systematic review of the syndrome concluded that CSE does not typically progress once diagnosis is made (van Valen et al. 2009). Whether the condition increases risk for cognitive decline in older ages remains unclear or, in some cases, conflicting (Dick et al. 2010; Nordling et al. 2007, 2010). Some reports suggest CSE actually results from misdiagnosed sleep apnea rather than solvents (Viaene et al. 2009), although a follow-up study found that solvent exposure increases the risk of sleep apnea, which accompanies the increased risk of neurobehavioral changes from the exposure (Godderis et al. 2011). Changes in mood and/or neurocognitive performance may also occur in children exposed to low levels of solvents in either environmental or occupational circumstances (Perera et al. 2011, 2012; Saddik et al. 2009). However, low-level exposure in both children and adults needs further research to clarify its relationship with mental abnormalities.

Inhalant abusers present a special area of concern regarding the psychiatric effects from intentional solvent inhalation. Inhalant abuse is a prevalent and increasing problem on an international scale notably in young people (Garland et al. 2011; Lopez-Quintero and Neumark 2011; Neumark and Bar-Hamburger 2011). Most abused inhalants are solvents, and many chemicals of everyday life are abused: aerosols, glues, paints, polishes, refrigerants, deodorizers, and others. An extensive body of literature substantiates the psychiatric effects of solvent abuse including mood, anxiety, psychotic, and dementia-like disorders (Brown 2007). Acute exposures to many abused solvents, especially toluene, can cause marked damage in more than one area of the brain (Aydin et al. 2009). Toluene, when intentionally inhaled to induce intoxication, can cause serious neuropsychological and neurological damage impacting both gray and white brain matter (Yucel et al. 2008).

As an introduction to the discussion of the prenatal effects of chemicals in the next section, a new approach in solvent research questions whether adult or prenatal exposure increases the risk for developing specific mental illnesses such as schizophrenia and bipolar disorder. One case report, for instance, attributes the onset of schizophrenia to adult occupational solvent exposure (Stein et al. 2010). Two studies that may have seminal significance in the literature reported an association of prenatal TCE exposure with increased risk of specific mental illnesses. One found a relationship between prenatal TCE and schizophrenia (Perrin et al. 2007), and the other reported an increased risk for bipolar disorder, schizophrenia, and PTSD from prenatal TCE (Aschengrau et al. 2012). The actual causes of schizophrenia and bipolar disorder are unknown, and although these studies offer compelling evidence of a relationship between chemicals and psychiatric illness, whether and how such exposures cause these specific mental disorders will require further investigation.

PSYCHIATRIC ISSUES IN PRENATAL CHEMICAL EXPOSURES

Compared to previous decades, the effect of prenatal organic chemical exposure on child neurodevelopment has received considerably more attention in recent years. Research is moving toward the question of whether prenatal chemical exposure increases risk of mental illnesses including schizophrenia and bipolar disorder. Prenatal lead exposure, for example, increases risk for schizophrenia in the offspring (Guilarte et al. 2011), although whether the source of lead can be both inorganic and organic is not clear. As discussed earlier, there is also an epidemiological overlap between abnormal neurodevelopment and mental illness, so the question arises whether chemically related causes of the former can increase risk of the latter. The author has mentioned earlier the association of prenatal TCE as a possible cause of schizophrenia or bipolar disorder, and that prenatal methylmercury may cause depression in the offspring. Since previous chapters of this book address the relationship of neurodevelopment and organic chemical exposure in detail, the following section, summarized on Table 19.2, discusses the primary issues related to this question.

POPs

A recent review and synthesis of PCB literature indicates that prenatal human PCB exposure reduces childhood cognitive performance (Boucher et al. 2009). Some investigators believe this effect operates through impairment of information processing (Boucher et al. 2010; Davidson et al. 2010); others suggest prenatal PCBs reduce attention (Sagiv et al. 2008, 2010) or affect childhood

TABLE 19.2. Prenatal Chemical Exposures of Psychiatric Importance

	Reduced Cognition/ Attention	Mood Change	Mental Illness
Methylmercury	X	X	
Lead			X
Tetrachloroethylene			X
Organophosphates (general)	X		
Solvents (general)	X		
PCBs	X		
Bisphenol A	X		

cognition through endocrine disruption and altered neurotransmitters (Park et al. 2009, 2010). Similar findings are reported for fetal contact with brominated flame retardants (Roze et al. 2009) but not for certain POPs under consideration such as perfluorooctanoate (PFOA) and perfluorooctanesulfonate (PFOS) (Fei and Olsen 2011).

Background or very low levels of PCB can also cause reduced intelligence in offspring (Gray et al. 2005), although inconsistencies between the findings of studies that examined these effects (Lynch et al. 2012) could be explained by the neuroprotective effect of breastfeeding (Jacobson and Jacobson 2001), other socioenvironmental factors (Forns et al. 2012a), birth order (Tatsuta et al. 2012), confounding by simultaneous prenatal exposure to methylmercury (Grandjean et al. 2012), sexually dimorphic effects of PCBs (Sagiv et al. 2012), and variable effects of PCB congeners on different cognitive domains (Forns et al. 2012b,c). Contrary to what might be expected given the prenatal effects of PCBs on cognition and altered brain function as measured by functional MRI in male adolescents who had high prenatal exposures to mixtures of methylmercury and PCBs (White et al. 2011), there are no definitive reports of a relationship between PCBs and autism, although studies of PCB-induced abnormalities in animals do have similarities to autism in humans (Hertz-Picciotto et al. 2008; Jolous-Jamshidi et al. 2010; Kimura-Kuroda et al. 2007).

Other Organic Pollutants

Recent literature also raises concern regarding the prenatal impacts of other organic chemicals on childhood cognitive and/or attention problems. This concern extends to OPs (Bouchard et al. 2011; Engel et al. 2011; Rauh et al. 2011), solvents, and methylmercury (Boucher et al. 2010), although the association is not always consistent for methylmercury. Prenatal exposure to permethrin, a frequent substitute for OP insecticides, does not diminish neurocognitive development in humans, but piperonyl butoxide, a synergist added to boost the potency of permethrin, does impair neurodevelopment (Horton et al. 2011). Contamination of the fetal environment by solvents, often severe in mothers who abuse solvents, can cause the "fetal solvent syndrome" of which cognitive disorder is just one component of the teratogenic effects of solvents (Bowen 2011; Hannigan and Bowen 2010). Background or low-level prenatal solvents may also influence later cognition (Edwards et al. 2010; Fei and Olsen 2011; Perera et al. 2009), and prenatal polycyclic aromatic hydrocarbon exposure may have adverse cognitive or behavioral effects in children (Perera et al. 2012a–c).

Another group of organic chemicals with evidence of developmental neurotoxicity are plastic-related chemicals with endocrine-disrupting properties such as bisphenol A (BPA) and phthalates. Evidence supports a positive relationship between prenatal phthalates and impaired infant and child neurodevelopment, intelligence, and attention (Cho et al. 2010; Engel et al. 2009, 2010; Kim et al. 2009). Some studies report similar associations with BPA

(Braun et al. 2011), while others do not (Yolton et al. 2011). One study reported an association of prenatal BPA exposure with behavioral problems in children especially emotional reactivity and aggression (Perera et al. 2012). The author described elsewhere the similarities between the anatomical, physiological, and hormonal effects of prenatal BPA in animal studies compared to similar findings in schizophrenia (Brown 2009), supporting the recommendation that research on BPA and other endocrine disruptors warrant monitoring for applications in psychiatric research.

CONCLUSION

The evidence indicates that clinicians should remain aware of the potential role that several organic chemicals, both POPs and non-POPs, can play in causing psychiatric symptoms as these chemicals remain a hazard in the 21st century. While research continues to clarify the psychiatric effects of established neurotoxins, attention has turned to the prenatal effects of pollutants. The evidence brings to light the possibilities that prenatal chemicals might increase risk for severe mental illnesses and that abnormalities of cognition and attention induced by prenatal organic chemical exposures could impair future mental well-being. Dependent on the outcome of ongoing research, future findings in this field could evolve to major public health issues.

As awareness of the psychiatric importance of organic substances continues to increase, new chemicals are constantly invented, and new materials may have neurotoxic properties. Nanoparticle neurotoxicology, which includes organic materials, is an emerging field of research with possible psychiatric relevance. Engineered nanoparticles can cross the blood–brain barrier (Hu and Gao 2010), and chronic exposures could cause neurological health effects (Oszlanczi et al. 2010; Simko and Mattsson 2010). The new organometal fuel additive, methylcyclopentadienyl manganese tricarbonyl (MMT), does not have known clinical importance but contains manganese, a well-known inorganic neurotoxin (Brown 2002) The cutting edge of research, however, lies with the prenatal effects of organic chemicals and how they influence developmental programming and adult psychiatric outcomes. These issues will no doubt cause controversy in the future, but the debate will add to the important body of knowledge concerning the effect of chemical exposures on psychiatric illness.

REFERENCES

Abdel Rasoul GM, Abou Salem ME, Mechael AA, Hendy OM, Rohlman DS, Ismail AA. 2008. Effects of occupational pesticide exposure on children applying pesticides. Neurotoxicology 29:833–838.

American Psychiatric Association. 2000. Diagnostic and Statistical Manual of Mental Disorders, 4th edition, Text Revision. Washington, DC: American Psychiatric Association.

Arnold LE, Demeter C, Mount K, Frazier TW, Youngstrom EA, Fristad M, et al. 2011. Pediatric bipolar spectrum disorder and ADHD: Comparison and comorbidity in the LAMS clinical sample. Bipolar Disord 13:509–521.

Aschengrau A, Weinberg JM, Janulewicz PA, Romano ME, Gallagher LG, Winter MR, et al. 2012. Occurrence of mental illness following prenatal and early childhood exposure to tetrachloroethylene (PCE)-contaminated drinking water: A retrospective cohort study. Environ Health 11:2.

Aschner M, Ceccatelli S. 2010. Are neuropathological conditions relevant to ethylmercury exposure? Neurotox Res 18:59–68.

Aydin K, Kircan S, Sarwar S, Okur O, Balaban E. 2009. Smaller gray matter volumes in frontal and parietal cortices of solvent abusers correlate with cognitive deficits. AJNR Am J Neuroradiol 30:1922–1928.

Bouchard MF, Chevrier J, Harley KG, Kogut K, Vedar M, Calderon N, et al. 2011. Prenatal exposure to organophosphate pesticides and IQ in 7-year-old children. Environ Health Perspect 119:1189–1195.

Boucher O, Muckle G, Bastien CH. 2009. Prenatal exposure to polychlorinated biphenyls: A neuropsychologic analysis. Environ Health Perspect 117:7–16.

Boucher O, Bastien CH, Saint-Amour D, Dewailly E, Ayotte P, Jacobson JL, et al. 2010. Prenatal exposure to methylmercury and PCBs affects distinct stages of information processing: An event-related potential study with Inuit children. Neurotoxicology 31:373–384.

Bowen SE. 2011. Two serious and challenging medical complications associated with volatile substance misuse: Sudden sniffing death and fetal solvent syndrome. Subst Use Misuse 46(Suppl 1):68–72.

Braun JM, Kalkbrenner AE, Calafat AM, Yolton K, Ye X, Dietrich KN, et al. 2011. Impact of early-life bisphenol A exposure on behavior and executive function in children. Pediatrics 128:873–882.

Brown JS, Jr. 2002. Environmental and Chemical Toxins and Psychiatric Illness. Washington, DC: American Psychiatric Publishing, Inc.

Brown JS, Jr. 2007. Psychiatric issues in toxic exposures. Psychiatr Clin North Am 30:837–854.

Brown JS, Jr. 2009. Effects of bisphenol-A and other endocrine disruptors compared with abnormalities of schizophrenia: An endocrine-disruption theory of schizophrenia. Schizophr Bull 35:256–278.

Bushnell PJ, Moser VC. 2006. Behavioral toxicity of cholinesterase inhibitors. In: Toxicology of Organophosphate & Carbamate Compounds (Gupta RC, ed.). Amsterdam: Elsevier, pp. 347–360.

Cannon SB, Veazey JM, Jr., Jackson RS, Burse VW, Hayes C, Straub WE, et al. 1978. Epidemic kepone poisoning in chemical workers. Am J Epidemiol 107: 529–537.

Cao Y, Chen A, Jones RL, Radcliffe J, Caldwell KL, Dietrich KN, et al. 2010. Does background postnatal methyl mercury exposure in toddlers affect cognition and behavior? Neurotoxicology 31:1–9.

Chan TY. 2011. Inorganic mercury poisoning associated with skin-lightening cosmetic products. Clin Toxicol (Phila) 49:886–891.

Chao LL, Abadjian L, Hlavin J, Meyerhoff DJ, Weiner MW. 2011. Effects of low-level sarin and cyclosarin exposure and Gulf War Illness on brain structure and function: A study at 4T. Neurotoxicology 32:814–822.

Cho SC, Bhang SY, Hong YC, Shin MS, Kim BN, Kim JW, et al. 2010. Relationship between environmental phthalate exposure and the intelligence of school-age children. Environ Health Perspect 118:1027–1032.

Colosio C, Tiramani M, Brambilla G, Colombi A, Moretto A. 2009. Neurobehavioural effects of pesticides with special focus on organophosphorus compounds: Which is the real size of the problem? Neurotoxicology 30:1155–1161.

Dassanayake T, Weerasinghe V, Dangahadeniya U, Kularatne K, Dawson A, Karalliedde L, et al. 2007. Cognitive processing of visual stimuli in patients with organophosphate insecticide poisoning. Neurology 68:2027–2030.

Davidson PW, Leste A, Benstrong E, Burns CM, Valentin J, Sloane-Reeves J, et al. 2010. Fish consumption, mercury exposure, and their associations with scholastic achievement in the Seychelles Child Development Study. Neurotoxicology 31:439–447.

Davidson PW, Cory-Slechta DA, Thurston SW, Huang LS, Shamlaye CF, Gunzler D, et al. 2011. Fish consumption and prenatal methylmercury exposure: Cognitive and behavioral outcomes in the main cohort at 17 years from the Seychelles child development study. Neurotoxicology 32:711–717.

Dick FD, Bourne VJ, Semple SE, Fox HC, Miller BG, Deary IJ, et al. 2010. Solvent exposure and cognitive ability at age 67: A follow-up study of the 1947 Scottish Mental Survey. Occup Environ Med 67:401–407.

Edwards SC, Jedrychowski W, Butscher M, Camann D, Kieltyka A, Mroz E, et al. 2010. Prenatal exposure to airborne polycyclic aromatic hydrocarbons and children's intelligence at 5 years of age in a prospective cohort study in Poland. Environ Health Perspect 118:1326–1331.

Engel SM, Zhu C, Berkowitz GS, Calafat AM, Silva MJ, Miodovnik A, et al. 2009. Prenatal phthalate exposure and performance on the Neonatal Behavioral Assessment Scale in a multiethnic birth cohort. Neurotoxicology 30:522–528.

Engel SM, Miodovnik A, Canfield RL, Zhu C, Silva MJ, Calafat AM, et al. 2010. Prenatal phthalate exposure is associated with childhood behavior and executive functioning. Environ Health Perspect 118:565–571.

Engel SM, Wetmur J, Chen J, Zhu C, Barr DB, Canfield RL, et al. 2011. Prenatal exposure to organophosphates, paraoxonase 1, and cognitive development in childhood. Environ Health Perspect 119:1182–1188.

Fei C, Olsen J. 2011. Prenatal exposure to perfluorinated chemicals and behavioral or coordination problems at age 7 years. Environ Health Perspect 119:573–578.

Fitzgerald EF, Belanger EE, Gomez MI, Cayo M, McCaffrey RJ, Seegal RF, et al. 2008. Polychlorinated biphenyl exposure and neuropsychological status among older residents of upper Hudson River communities. Environ Health Perspect 116:209–215.

Forns J, Torrent M, Garcia-Esteban R, Caceres A, Pilar Gomila M, Martinez D, et al. 2012a. Longitudinal association between early life socio-environmental factors and attention function at the age 11 years. Environ Res 117:54–59.

Forns J, Torrent M, Garcia-Esteban R, Grellier J, Gascon M, Julvez J, et al. 2012b. Prenatal exposure to polychlorinated biphenyls and child neuropsychological development in 4-year-olds: An analysis per congener and specific cognitive domain. Sci Total Environ 432C:338–343.

Forns J, Lertxundi N, Aranbarri A, Murcia M, Gascon M, Martinez D, et al. 2012c. Prenatal exposure to organochlorine compounds and neuropsychological development up to two years of life. Environ Int 45:72–77.

Garland EL, Howard MO, Vaughn MG, Perron BE. 2011. Volatile substance misuse in the United States. Subst Use Misuse 46(Suppl 1):8–20.

Godderis L, Dours G, Laire G, Viaene MK. 2011. Sleep apnoeas and neurobehavioral effects in solvent exposed workers. Int J Hyg Environ Health 214:66–70.

Grandjean P, Weihe P, Nielsen F, Heinzow B, Debes F, Budtz-Jorgensen E. 2012. Neurobehavioral deficits at age 7 years associated with prenatal exposure to toxicants from maternal seafood diet. Neurotoxicol Teratol 34:466–472.

Gray KA, Klebanoff MA, Brock JW, Zhou H, Darden R, Needham L, et al. 2005. *In utero* exposure to background levels of polychlorinated biphenyls and cognitive functioning among school-age children. Am J Epidemiol 162:17–26.

Guilarte TR, Opler M, Pletnikov M. 2011. Is lead exposure in early life an environmental risk factor for schizophrenia? Neurobiological connections and testable hypotheses. Neurotoxicology 33:560–574.

Guo YL, Hsu CC. 2001. Yucheng and Yusho: The effects of toxic oil in developing humans in Asia. In: PCBs. Recent Advances in Environmental Toxicology and Health Effects (Robertson LW, Hansen LG, eds.). Lexington, KY: University Press of Kentucky, pp. 137–141.

Hannigan JH, Bowen SE. 2010. Reproductive toxicology and teratology of abused toluene. Syst Biol Reprod Med 56:184–200.

Harrad S. 2010. Beyond the Stockholm Convention: An introduction to current issues and future challenges in POPs research. In: Persistent Organic Pollutants (Harrad S, ed.). Chichester: Wiley, pp. 1–4.

Heaton KJ, Palumbo CL, Proctor SP, Killiany RJ, Yurgelun-Todd DA, White RF. 2007. Quantitative magnetic resonance brain imaging in US army veterans of the 1991 Gulf War potentially exposed to sarin and cyclosarin. Neurotoxicology 28:761–769.

Hertz-Picciotto I, Park HY, Dostal M, Kocan A, Trnovec T, Sram R. 2008. Prenatal exposures to persistent and non-persistent organic compounds and effects on immune system development. Basic Clin Pharmacol Toxicol 102:146–154.

Horton MK, Rundle A, Camann DE, Boyd BD, Rauh VA, Whyatt RM. 2011. Impact of prenatal exposure to piperonyl butoxide and permethrin on 36-month neurodevelopment. Pediatrics 127:e699–e706.

Hu YL, Gao JQ. 2010. Potential neurotoxicity of nanoparticles. Int J Pharm 394: 115–121.

Jacobson JL, Jacobson SW. 2001. Developmental effects of PCBs in the fish eater cohort studies. In: PCBs. Recent Advances in Environmental Toxicology and Health Effects (Robertson LW, Hansen LG, eds.). Lexington, KY: University Press of Kentucky, pp. 127–136.

Jedrychowski W, Perera F, Jankowski J, Rauh V, Flak E, Caldwell KL, et al. 2007. Fish consumption in pregnancy, cord blood mercury level and cognitive and psychomotor development of infants followed over the first three years of life: Krakow epidemiologic study. Environ Int 33:1057–1062.

Jolous-Jamshidi B, Cromwell HC, McFarland AM, Meserve LA. 2010. Perinatal exposure to polychlorinated biphenyls alters social behaviors in rats. Toxicol Lett 199:136–143.

Kamel F, Engel LS, Gladen BC, Hoppin JA, Alavanja MC, Sandler DP. 2007. Neurologic symptoms in licensed pesticide applicators in the Agricultural Health Study. Hum Exp Toxicol 26:243–250.

Kavalci C, Durukan P, Ozer M, Cevik Y, Kavalci G. 2009. Organophosphate poisoning due to a wheat bagel. Intern Med 48:85–88.

Kim BN, Cho SC, Kim Y, Shin MS, Yoo HJ, Kim JW, et al. 2009. Phthalates exposure and attention-deficit/hyperactivity disorder in school-age children. Biol Psychiatry 66:958–963.

Kimura-Kuroda J, Nagata I, Kuroda Y. 2007. Disrupting effects of hydroxy-polychlorinated biphenyl (PCB) congeners on neuronal development of cerebellar Purkinje cells: A possible causal factor for developmental brain disorders? Chemosphere 67:S412–S420.

King BH, Lord C. 2011. Is schizophrenia on the autism spectrum? Brain Res 1380: 34–41.

Lee WJ, Alavanja MC, Hoppin JA, Rusiecki JA, Kamel F, Blair A, et al. 2007. Mortality among pesticide applicators exposed to chlorpyrifos in the Agricultural Health Study. Environ Health Perspect 115:528–534.

Lizardi PS, O'Rourke MK, Morris RJ. 2008. The effects of organophosphate pesticide exposure on Hispanic children's cognitive and behavioral functioning. J Pediatr Psychol 33:91–101.

Lopez-Quintero C, Neumark Y. 2011. The epidemiology of volatile substance misuse among school children in Bogota, Colombia. Subst Use Misuse 46(Suppl 1): 50–56.

Lynch CD, Jackson LW, Kostyniak PJ, McGuinness BM, Louis GM. 2012. The effect of prenatal and postnatal exposure to polychlorinated biphenyls and child neurodevelopment at age twenty four months. Reprod Toxicol 34(3):451–456.

MacFarlane E, Simpson P, Benke G, Sim MR. 2011. Suicide in Australian pesticide-exposed workers. Occup Med (Lond) 61:259–264.

Mackenzie Ross SJ, Brewin CR, Curran HV, Furlong CE, Abraham-Smith KM, Harrison V. 2010. Neuropsychological and psychiatric functioning in sheep farmers exposed to low levels of organophosphate pesticides. Neurotoxicol Teratol 32: 452–459.

Marques RC, Dorea JG, Leao RS, Dos Santos VG, Bueno L, Marques RC, et al. 2012. Role of methylmercury exposure (from fish consumption) on growth and neurodevelopment of children under 5 years of age living in a transitioning (tin-mining) area of the western Amazon, Brazil. Arch Environ Contam Toxicol 62:341–350.

Michalek JE, Barrett DH, Morris RD, Jackson WG, Jr. 2003. Serum dioxin and psychological functioning in U.S. Air Force veterans of the Vietnam War. Mil Med 168: 153–159.

Mittal T, Gupta N, Kohli A, Bhalla A, Singh B, Singh S. 2011. Correlation of defects in regional cerebral blood flow determined by 99mTc SPECT with residual neurocognitive testing abnormalities during and 3 months post exposure in acutely poisoned patients with organophosphates. Clin Toxicol (Phila) 49:464–470.

Nettelbladt P, Goth M, Bogren M, Mattisson C. 2009. Risk of mental disorders in subjects with intellectual disability in the Lundby cohort 1947–97. Nord J Psychiatry 63:316–321.

Neumark Y, Bar-Hamburger R. 2011. Volatile substance misuse among youth in Israel: Results of a national school survey. Subst Use Misuse 46(Suppl 1):21–26.

Ng DK, Chan CH, Soo MT, Lee RS. 2007. Low-level chronic mercury exposure in children and adolescents: Meta-analysis. Pediatr Int 49:80–87.

Nordling NL, Barregard L, Sallsten G, Hagberg S. 2007. Self-reported symptoms and their effects on cognitive functioning in workers with past exposure to solvent-based glues: An 18-year follow-up. Int Arch Occup Environ Health 81:69–79.

Nordling NL, Karlson B, Nise G, Malmberg B, Orbaek P. 2010. Delayed manifestations of CNS effects in formerly exposed printers—A 20-year follow-up. Neurotoxicol Teratol 32:620–626.

Oszlanczi G, Vezer T, Sarkozi L, Horvath E, Konya Z, Papp A. 2010. Functional neurotoxicity of Mn-containing nanoparticles in rats. Ecotoxicol Environ Saf 73: 2004–2009.

Ozyurt G, Kaya FN, Kahveci F, Alper E. 2008. Comparison of SPECT findings and neuropsychological sequelae in carbon monoxide and organophosphate poisoning. Clin Toxicol (Phila) 46:218–221.

Park HY, Park JS, Sovcikova E, Kocan A, Linderholm L, Bergman A, et al. 2009. Exposure to hydroxylated polychlorinated biphenyls (OH-PCBs) in the prenatal period and subsequent neurodevelopment in eastern Slovakia. Environ Health Perspect 117:1600–1606.

Park HY, Hertz-Picciotto I, Sovcikova E, Kocan A, Drobna B, Trnovec T. 2010. Neurodevelopmental toxicity of prenatal polychlorinated biphenyls (PCBs) by chemical structure and activity: A birth cohort study. Environ Health 9:51.

Perera FP, Li Z, Whyatt R, Hoepner L, Wang S, Camann D, et al. 2009. Prenatal airborne polycyclic aromatic hydrocarbon exposure and child IQ at age 5 years. Pediatrics 124:e195–e202.

Perera FP, Wang S, Vishnevetsky J, Zhang B, Cole KJ, Tang D, et al. 2011. Polycyclic aromatic hydrocarbons-aromatic DNA adducts in cord blood and behavior scores in New York City children. Environ Health Perspect 119:1176–1181.

Perera F, Li TY, Lin C, Tang D. 2012a. Effects of prenatal polycyclic aromatic hydrocarbon exposure and environmental tobacco smoke on child IQ in a Chinese cohort. Environ Res 114:40–46.

Perera FP, Tang D, Wang S, Vishnevetsky J, Zhang B, Diaz D, et al. 2012b. Prenatal polycyclic aromatic hydrocarbon (PAH) exposure and child behavior at age 6–7 years. Environ Health Perspect 120:921–926.

Perera F, Vishnevetsky J, Herbstman JB, Calafat AM, Xiong W, Rauh V, et al. 2012c. Prenatal bisphenol A exposure and child behavior in an inner city cohort. Environ Health Perspect 120(8):1190–1194.

Perrin MC, Opler MG, Harlap S, Harkavy-Friedman J, Kleinhaus K, Nahon D, et al. 2007. Tetrachloroethylene exposure and risk of schizophrenia: Offspring of dry cleaners in a population birth cohort, preliminary findings. Schizophr Res 90:251–254.

Price CS, Thompson WW, Goodson B, Weintraub ES, Croen LA, Hinrichsen VL, et al. 2010. Prenatal and infant exposure to thimerosal from vaccines and immunoglobulins and risk of autism. Pediatrics 126:656–664.

Rauh V, Arunajadai S, Horton M, Perera F, Hoepner L, Barr DB, et al. 2011. Seven-year neurodevelopmental scores and prenatal exposure to chlorpyrifos, a common agricultural pesticide. Environ Health Perspect 119:1196–1201.

Rohlman DS, Lasarev M, Anger WK, Scherer J, Stupfel J, McCauley L. 2007. Neurobehavioral performance of adult and adolescent agricultural workers. Neurotoxicology 28:374–380.

Rohlman DS, Anger WK, Lein PJ. 2011. Correlating neurobehavioral performance with biomarkers of organophosphorous pesticide exposure. Neurotoxicology 32: 268–276.

Roze E, Meijer L, Bakker A, Van Braeckel KN, Sauer PJ, Bos AF. 2009. Prenatal exposure to organohalogens, including brominated flame retardants, influences motor, cognitive, and behavioral performance at school age. Environ Health Perspect 117:1953–1958.

Saddik B, Williamson A, Black D, Nuwayhid I. 2009. Neurobehavioral impairment in children occupationally exposed to mixed organic solvents. Neurotoxicology 30: 1166–1171.

Sagiv SK, Nugent JK, Brazelton TB, Choi AL, Tolbert PE, Altshul LM, et al. 2008. Prenatal organochlorine exposure and measures of behavior in infancy using the Neonatal Behavioral Assessment Scale (NBAS). Environ Health Perspect 116: 666–673.

Sagiv SK, Thurston SW, Bellinger DC, Tolbert PE, Altshul LM, Korrick SA. 2010. Prenatal organochlorine exposure and behaviors associated with attention deficit hyperactivity disorder in school-aged children. Am J Epidemiol 171:593–601.

Sagiv SK, Thurston SW, Bellinger DC, Altshul LM, Korrick SA. 2012. Neuropsychological measures of attention and impulse control among 8-year-old children exposed prenatally to organochlorines. Environ Health Perspect 120:904–909.

Simko M, Mattsson MO. 2010. Risks from accidental exposures to engineered nanoparticles and neurological health effects: A critical review. Part Fibre Toxicol 7:42.

Stein Y, Finkelstein Y, Levy-Nativ O, Bonne O, Aschner M, Richter ED. 2010. Exposure and susceptibility: Schizophrenia in a young man following prolonged high exposures to organic solvents. Neurotoxicology 31:603–607.

Strain JJ, Davidson PW, Bonham MP, Duffy EM, Stokes-Riner A, Thurston SW, et al. 2008. Associations of maternal long-chain polyunsaturated fatty acids, methyl mercury, and infant development in the Seychelles Child Development Nutrition Study. Neurotoxicology 29:776–782.

Tatsuta N, Nakai K, Murata K, Suzuki K, Iwai-Shimada M, Yaginuma-Sakurai K, et al. 2012. Prenatal exposures to environmental chemicals and birth order as risk factors for child behavior problems. Environ Res 114:47–52.

van Valen E, Wekking E, van der Laan G, Sprangers M, van Dijk F. 2009. The course of chronic solvent induced encephalopathy: A systematic review. Neurotoxicology 30:1172–1186.

Viaene M, Vermeir G, Godderis L. 2009. Sleep disturbances and occupational exposure to solvents. Sleep Med Rev 13:235–243.

White RF, Palumbo CL, Yurgelun-Todd DA, Heaton KJ, Weihe P, Debes F, et al. 2011. Functional MRI approach to developmental methymercury and polychlorinated biphenyl neurotoxicity. Neurotoxicology 6:975–980.

Wills BK, Christensen J, Mazzoncini J, Miller M. 2010. Severe neurotoxicity following ingestion of tetraethyl lead. J Med Toxicol 6:31–34.

Wintour EM, Owens JA. 2006. Early Life Origins of Health and Disease. New York: Springer Science+Business Media.

Yamasaki Y, Sakamoto K, Watada H, Kajimoto Y, Hori M. 1997. The Arg192 isoform of paraoxonase with low sarin-hydrolyzing activity is dominant in the Japanese. Hum Genet 101:67–68.

Yamasue H, Abe O, Kasai K, Suga M, Iwanami A, Yamada H, et al. 2007. Human brain structural change related to acute single exposure to sarin. Ann Neurol 61:37–46.

Yanagisawa N, Morita H, Nakajima T. 2006. Sarin experiences in Japan: Acute toxicity and long-term effects. J Neurol Sci 249:76–85.

Yolton K, Xu Y, Strauss D, Altaye M, Calafat AM, Khoury J. 2011. Prenatal exposure to bisphenol A and phthalates and infant neurobehavior. Neurotoxicol Teratol 33:558–566.

Yorifuji T, Tsuda T, Inoue S, Takao S, Harada M. 2011. Long-term exposure to methylmercury and psychiatric symptoms in residents of Minamata, Japan. Environ Int 37:907–913.

Yucel M, Takagi M, Walterfang M, Lubman DI. 2008. Toluene misuse and long-term harms: A systematic review of the neuropsychological and neuroimaging literature. Neurosci Biobehav Rev 32:910–926.

CHAPTER 20

Growth and Development

LAWRENCE M. SCHELL, MIA V. GALLO, and KRISTOPHER K. BURNITZ

ABSTRACT

Background: Physical growth in prenatal and postnatal life is an indicator of the health of the individual. The population-level pattern of physical growth reflects the population's health as well. The effects of exposure to organic pollutants can be assessed in terms of the impact on the pattern of growth and development.

Objective: To evaluate the evidence presented in peer-reviewed publications that prenatal or postnatal exposure to organic pollutants has an effect on the pattern of physical growth, and to make recommendations regarding future work.

Discussion: Many studies of newborns, children, and youth demonstrate relationships between organic pollutant exposures and subsequent growth, although the effects range from negative to positive values, and some studies do not find relationships. There is no consistent effect of all organic pollutants. The diversity in results could be attributed to (1) differences in the toxicants, (2) differences in the exposure timing, (3) differences in the biological media used for exposure assessment, (4) the accuracy of exposure measurement, (5) gender differences in susceptibility, and/or (6) differences in susceptibility due to growth stage. In addition, the usual threats to internal validity such as uncontrolled covariates are especially problematic because growth is sensitive to many social and demographic factors.

Conclusion: Too many studies have found relationships between organic pollutants and some aspect of growth, including size at birth or height in postnatal life, to ignore the probability that growth is sensitive to such pollutants. Research is needed that will clarify prenatal and postnatal exposure contributions to physical growth. Additionally, investigations should examine effects on growth velocity as well as size for age.

Effects of Persistent and Bioactive Organic Pollutants on Human Health, First Edition.
Edited by David O. Carpenter.

INTRODUCTION

Since 1980, there have been a number of reviews on child growth and development in relation to several pollutants by this author and others, the most recent of which was published in 2009 (Karmaus 2006; Schell 1991; Schell et al. 2009). This chapter will review the literature and focus on work produced since the most recent review. However, it also provides an analysis of the research on growth and organic toxicants from the perspective of auxology (the science of human growth and development) with the purpose of identifying approaches and specific methods that can be strengthened to produce a clearer picture of toxicant effects on growth and development.

AUXOLOGY

Good physical growth is one of the most common signs of good health among subadults of plants and animals (Schell 1997; Schell et al. 1992; Tanner 1978). In humans, positive growth from one pediatric visit to the next is one of the most certain expectations of parents and clinicians. Thus, unexpected reduced physical growth of an individual child is a cause for some concern and follow-up. In toxicology, impaired growth is a classic sign of toxicity in subadult experimental animals, just as weight loss is among adult animals. Toxicity testing of novel materials in the laboratory is calibrated to levels that are not able to reduce growth because such reductions are an obvious sign of toxicity. Growth is a necessary signal of health but is not sufficient to confirm health. In other words, children growing poorly are signaling a health problem, but children growing well are not necessarily healthy in every respect.

James Tanner, a leader in the study of child growth for decades, defined auxological epidemiology as the use of growth data to discover populations in suboptimal environments. He dated the origin of this application to the social reformers of the mid-19th century (Tanner 1981). The application of growth data to the problem of toxicant exposure is a simple extension of this tradition. The scientific study of human growth has developed over the past nearly 200 years, and subtleties in the study of growth are sometimes not understood by researchers coming to this area of research from other disciplines with different foci and methods (e.g., epidemiology, pediatrics, toxicology, public health). For this reason, when reviewing studies of child growth conducted chiefly by nonauxologists, it is necessary to use this opportunity to blend in knowledge from auxology for maximum benefit.

Therefore, this chapter is a critique of our current knowledge of organic pollutants and child growth rather than a review of each and every study ever done. Those studies done more recently than other published reviews are considered in some detail as are especially influential ones. This critique seeks to identify the weaknesses in the studies conducted to date in order to create a path for more informative work in the future.

The Meaning of Diversity in Results Concerning Growth and Development

The studies to date exhibit great diversity in many important characteristics: the compounds investigated, the methods of exposure assessment, the timing of exposure, and the age ranges of the sample children. All of these factors are capable of creating a lack of agreement in the results.

The conclusion easily reached by reviewing these studies is that we have incomplete knowledge of the influence of many compounds, especially organic pollutants, on the physical growth of children and youth. Further, it is now clear that the investigation of these relationships is extremely complex and requires extraordinary control over extraneous factors, sophisticated assessment of exposure in terms of amount and timing, and realistic basis for assessing nonstandard dose–response relationships.

Another possible conclusion from this diversity is that the investigations are flawed in ways that are largely unrecognizable at present. This leads to the additional conclusion that some investigations have produced spurious results (although it is not clear which ones). While there can be little doubt that investigations have weaknesses, it is also necessary to recognize that the organic compounds studied are considered endocrine disruptive and that myriad effects are possible. A considerable portion of the studies reviewed show evidence of alterations in the action or level of sex steroids and thyroid hormones. Given that growth is driven by these (in addition to others such as insulin-lie growth factor-1 [IGF-1] and growth hormone [GH]), it is reasonable that effects of the organic compounds could differ by sex and by age as the influences of endogenous sex steroids and thyroid hormones vary with stage of growth. In other words, differing effects by sex can be expected, and different effects from the same compound at different stages of growth can be expected also. The kind of precise concordance of results that some expect in studies of human samples may be a more reasonable expectation in laboratory studies where control over extraneous factors is more completely exercised.

Current Weaknesses in the Literature on Growth and Organics

Although the study of growth in relation to environmental factors involves at a minimum an analysis of both growth distance and velocity, few studies have considered the importance of the difference. "Distance" or size may summarize the history of an individual's growth; it does not reflect very well the current status of the individual. Appropriate size for age at one point in time may be the result of past exemplary growth but recent cessation. The cessation would be worrying and it can be missed easily by focusing on size for age. Formerly well-growing children would have to cease growing for several months before crossing a major centile line (e.g., from the 50th to the 25th centile) of a U.S. growth chart (e.g., using the CDC growth charts, an

18-month-old boy at the 97th centile of height would have to stop growing for 6 months before reaching the 50th centile; he would have to stop growing for 16 months before reaching the fifth centile and possibly attracting clinical attention). Growth velocity is a better indicator of current growth status. True velocity requires multiple measures, and this is rare in studies of growth and toxicants and is entirely absent in studies of large, nationally representative samples.

As growth is ecosensitive, the study of growth in relation to organic compounds also requires the measurement of multiple targets and covariates including other nonorganic toxicants. Earlier work tested relationships between growth parameters and one toxicant, and only sometimes more. More recent studies tend to have included information on a greater range of toxicants, though usually all of the same type that can be measured in one laboratory analysis (e.g., dichlorodiphenyldichloroethylene [DDE], polychlorinated biphenyls [PCBs], and hexachlorobenzene [HCB]). Given the findings from studies of other outcomes that some effects are especially pronounced through the range of lower exposures (e.g., of the effect of lead on mental performance in children), the inability to control or to model low exposures of the co-occurring but nontargeted toxicants raises some concern regarding uncontrolled confounding. For example, in studies focusing on organic pollutants and growth, the measurement of lead is rarely included despite the fact that lead has been associated with later age at menarche in youth and slower growth in children (Denham et al. 2005; Selevan et al. 2003; Wu et al. 2003). This often reflects the focus of the investigators, which, in this area of research, tends to be defined by toxicant type (metals vs. organics and within organics by family of compound, etc.). Thus, researchers who focus on air pollution might not consider exposure (dietary) to organics, and those focusing on organics may overlook metals. Clearly, it is impossible to measure the entire anthropogenic environment, but it is not a simple matter to determine the appropriate place to stop when studying prenatal or child growth as both are highly ecosensitive outcomes having many environmental influences.

Another important source of diversity is in the timing of exposure and the accuracy of the exposure measurement. At the very least, chronic and acute exposures can be expected to contribute different effects on growth. Thus, results from studies of Yusho and Yucheng sufferers (described later in this chapter) are usually presented separately from reviews of effects that may follow from chronic exposure. The stage of development of the child is a related factor. Critical period theory, which states that effects of exposure at different stages of development can produce different effects, is now well established through studies of thalidomide, alcohol, mercury, and diethylstilbestrol, to name some prominent examples. Although the criticality of the ovular and embryonic stages of human development is well recognized, heightened sensitivity to environmental factors can be found in the postnatal life as well. Suspected or confirmed critical periods in postnatal life include the first few months after birth when endocrine regulation becomes stabilized, early childhood (adiposity rebound), and adolescence (Cameron and Demerath 2002). Thus, it is prudent when analyzing growth to consider exposures in any of these postnatal stages as sensitive stages though not exactly equivalent in potential to produce effects on growth and development.

Variation in measurement methods and the interpretation of levels contributes to diversity also. The chronicity of PCB and other lipophilic, persistent organocompounds suggests that levels measured at different times may stand for levels at other times. Thus, third trimester measurements may be a fairly accurate estimate of first trimester exposure, and prenatal exposure may stand well for early postnatal exposure. The length of time between serum sampling and point estimate of toxicant level may be a factor. However, long-term longitudinal studies to model changes in levels across developmental stages have not been conducted to date. Research designs are retrospective and exposure assessment is often problematic and involves many assumptions. With chronic exposures, the point at which damage occurred is impossible to determine since exposure is long lasting. Thus, there could be critical periods of sensitivity, but given the chronicity of exposure and of serum levels, these cannot be detected. Frequently, the earliest exposure is interpreted as the causal one, yet this assumption needs to be established. A chronic exposure coupled with a sensitive period in early postnatal life can be missed.

Given the significant and genuine sources of diversity possible in studies of growth and development (exposure timing, stage of development at exposure, stage of development at size measurement, exposure measurement accuracy, and sex), conclusions concerning effects on growth are limited by these factors. In fact, over-generalization of results disregards those limits. A major source of confusion created by results that appear to conflict, or at least are not concordant, is due to investigators' lack of specificity in summarizing the effects they have discovered. Statements such as, "compound X was unrelated to child growth" are vague, too broad, and too often interpreted globally. Such statements of effect or lack of effect, on the growth of children are common in abstracts and in the discussion or concluding sections of research articles. Instead, a statement that is more reflective of the particulars of the study, such as "compound X over the range of A to B has such and such a relationship to a particular dimension of size or growth at the age of Z" recognizes the limitations of the sampling, the range of exposure, and the age range of the subjects. As growth occurs through very recognizable phases (infancy, childhood, puberty, and postpubertal phases) that are directed by different endocrines, usually in combination (IGF, GH, sex steroids, as well as thyroid hormones), generalizing to more than one such stage from observations in any other stage is unwarranted and, to the point, adds to the apparent heterogeneity of results among studies.

These observations from the perspective of auxology are apparent in the review that follows. This review focuses on linear growth (length and height) as other chapters of this volume address sexual maturation, overweight, and obesity. These topics are touched upon only when highly relevant to interpret results on linear growth. Not all types of organic compound exposures are dealt with owing to the lack of information on some despite the level of interest currently, but ones with relatively substantial investigations are included.

INEXACTLY CHARACTERIZED EXPOSURES

Yusho and Yucheng

Studies of growth velocity following acute exposure to organic compounds are very rare and such studies deserve special attention. Exposure to a mix of organic chemicals occurred in two food poisoning incidents that affected large numbers of children. The first occurred in Japan in 1968, affecting some 1600 residents. The ensuing disease was called Yusho in Japanese. A second occurrence in Taiwan in 1979 was strikingly similar in origin to the first and was termed Yucheng (or Yu-Cheng). In both cases, persons consumed rice oil that had been contaminated with a variety of polychlorobiphenyl compounds chiefly from the thermal degradation of PCBs including polychlorinated dibenzofurans (PCDFs), dibenzodioxins, tetrachlorodibenzodioxin, PCBs, and polychlorinated quaterphenyls (Kuratsune 1980; Masuda and Yoshimura 1984). It is not known exactly how much rice oil each person consumed; thus, individuals' exposures are not known precisely. However, the exposure was severe judging by the presence of several symptoms including the dermatological sign of chloracne, and there is no evidence of exposure to anything other than organic compounds. Furthermore, postnatal exposure to children occurred, allowing investigators to estimate the direct effects of the mixture on child growth without prenatal exposure, as well as the effects of prenatal exposure as registered in the children born to mothers who had consumed the rice oil either while pregnant or before becoming pregnant. In both cases of rice oil poisoning, a national registry of victims was established, and these were used in the follow-up studies.

A very common and well-established consequence of prenatal exposure in Yusho and Yucheng is small size at birth (Funatsu et al. 1971; Taki et al. 1969; Yamashita and Hayashi 1985; Yoshimura 1974). Children born to Yucheng mothers exhibited a higher percentage of low birth weights, lower birth length, and retarded growth (Rogan et al. 1988; Yen et al. 1989, 1994), similar to Yusho children (Yamashita and Hayashi 1985). A recent retrospective examination of 190 births to Yusho patients reported maternal levels of non-*ortho*-PCBs significantly correlated with a reduction in birth weight in all Yusho patients of nearly 162 g in males but not in females (Tsukimori et al. 2012).

Japan administers growth examinations of children at regular and frequent intervals through the public health system. An early report compared height and weight increments of 23 boys and 19 girls of school age to 719 healthy controls. Height increments before (1967–1968) and after (1968–1969) the poisoning were compared to increments by control children. Boys exposed through their consumption of rice oil grew in length and weight more slowly than unexposed children; for example, velocity was reduced. The picture was less clear in girls despite generally smaller postpoisoning increments in the exposed girls (Yoshimura 1971).

The routine physical examinations in 1970 and 1971 provided data on growth increments well after the rice oil poisoning (Yoshimura and Ikeda 1978). Sample sizes are small: Yusho boys' age-specific sample sizes ranged from 18 to 45, and girls' from 17 to 43. Boys' increments in height in the years following the poisoning were not significantly different from the increment obtained in the year of the poisoning. Thus, the yearly rates of growth as indicated by the increments in the 2 years following the poisoning were not significantly different from the rate in the year of the poisoning. This indicates that there was no catch-up growth after the poisoning stopped. Similarly in girls, the height increment in the year after the poisoning was not different from the increment in the year of the poisoning. Two years later, the increment was greater, and though the probability level was not less than 0.05, it was below 0.10. Body weight increments of both girls and boys did not differ across the years of study.

A small number of children were available for comparison of increments from before the poisoning (Yoshimura and Ikeda 1978). Among the boys ($n = 9$), the mean height increment z-score before the poisoning was –0.50; it became substantially smaller in the year of the poisoning, reaching –1.76 and then reduced to –0.58 and –0.67 in the following 2 years. While the small number precludes statistical testing, the mean values suggest that height velocity was similar to prepoisoning levels after 2 years but was not appreciably greater, indicating that there was no catch-up growth, only a partial return to prepoisoning rates. In girls ($n = 7$), 2 years after the poisoning, z-scores of height increments became positive; that is, they are far larger than the prepoisoning increments, suggesting that some catch-up growth did occur. In conclusion, Yusho boys' and girls' growth rates slowed significantly in the year of the poisoning and did not recover in the 2 years afterward, although the rates did appear less retarded.

In Taiwan, Yucheng patients numbered 1978 (Yen et al. 1994). Reduced postnatal growth was observed among the Yucheng children exposed *in utero*. Yucheng children born between 6/1978 and 3/1985 ($n = 117$) were examined in 4/1985, at an average age of 32 months, and were compared to matched controls ($n = 108$). Yucheng children were 97% of the heights of controls and 93% of the weights adjusted for sex and age (Rogan et al. 1988). In 2/1991 when Yucheng children were between 6 and 12 years of age, 55 children from the original registered cohort were examined as were matched controls (Guo et al. 1994a,b). Heights of affected children were 3.1 cm less than controls, a statistically significant difference. Weights were not different. However, the difference in height was confined to firstborn children after the poisoning. When the comparison was limited to later-born children, heights of cases and controls did not differ. Exposure probably contributed to the difference in effects as firstborn children had, on average, 1.57 times the level of later-born ones.

In 1997–1999, girls between 13 and 19 years of age who were the offspring of exposed mothers were examined for menstrual characteristics (Yang et al. 2005). This report of a comparison of Yucheng girls ($n = 27$) and controls ($n = 21$) also included a description of the heights for the two groups, though uncorrected for other factors. Heights of Yucheng girls averaged 154.8 cm (±0.9), while the average for controls was 157.0 (=/–1.1) a difference of 2.2 cm, though a nonsignificant difference ($p = 0.13$).

In conclusion, Japanese children exposed in postnatal life to contaminated rice oil experienced reduced growth as indicated by reduced velocity of growth. Over the following 2 years, growth velocity climbed, though not to pre-exposure levels, indicating that catch-up growth did not occur, as this would have involved higher than pre-exposure growth velocities. The studies of Yusho children are among the few of postnatal exposure and true growth velocity, not size for age. Studies of Taiwanese children have focused on prenatal exposure and have found long-lasting sequelae of exposure including reduced height, weight, and other parameters. Growth velocity has not been examined as of this writing. The finding that Taiwanese children who were later in the birth order showed lesser or no effects bears interpretation. The reduced effects may be due to their having received less exposure by virtue of being born after siblings exposed *in utero* probably reducing mothers' toxicant burden. This suggests that a relatively high level of exposure causes the observed postnatal effects. Alternatively, it could be that the effects of poisoning on the mother's ability to nurture *in utero* are repaired by the time later-born children are developing *in utero*.

Love Canal

Love Canal is a residential area near the city of Niagara Falls, New York, where chemical dumping created one of the most significant hazardous waste sites in the United States. Discovered in 1978, it also was one of the earliest discovered, and its investigation taught the public health community much about the proper way to deal with public concerns regarding hazardous waste exposure. One source of chronic uncertainty about the risk was the poor knowledge of the compound mixture comprising the exposure; it has never been completely identified. At the time of the crisis, a study of air quality in nearby homes discovered evidence of benzene, toluene, chloroform, trichloroethylene, tetrachloroethylene, hexane, and xyelenes (NYSDOH 1978). Recently, an analysis of sera from persons who had resided in the Love Canal area ($n = 373$) in 1978–1979 revealed three compounds in most participants: 1,2,4-trichlorobenzene, beta-hexachlorocyclohexane, and 1,2-dichlorobenze. However, the relationship of these 373 participants to the births and children whose growth and size have been studied is not known (Kielb et al. 2010).

Prenatal growth has been examined in several studies using different sampling strategies. The most definitive of these, performed by the New York State Department of Health, found that birth weights were reduced according to some measures of exposure (again, it is not clear who were the most exposed) and that exposure close to the time of chemical dumping at Love Canal showed the largest effects (Vianna and Polan 1984). Effects on length at birth were not described and presumably were not analyzed. A recent analysis found evidence that families living closest to the canal had a higher frequency of low birth weight. Again, length at birth was not considered (Austin et al. 2011).

Postnatal growth at Love Canal was examined, though exposures of individual children could not be determined. Residence and percent of life lived in the Love Canal area were the proxy for chemical exposure. Children's heights were transformed into percentiles of National Center for Health Statistics (NCHS) reference values for sex and age. The average height percentile of children from Love Canal was 50.8 versus 53 for controls, a nonsignificant difference. Children who had been born in Love Canal and lived 75% or more of their lives in the Love Canal area ($n = 170$) averaged 46.6, significantly less than the average height percentile of 53.3 for control area children. Weight percentiles did not differ nor did weight for height.

Despite the lack of information on amounts and types of toxicant burdens, the analyses of growth at Love Canal do show that postnatal growth can be affected. Evidence for diminution of prenatal growth (weight at birth) is stronger as there have been several analyses with similar results, but no analyses to date of length at birth or the relationship of diminished size at birth to later growth.

PCBs and Polybrominated Biphenyls (PBBs)

PCBs are in a class of organohalogens that have been classified as endocrine disruptors, exhibiting hormonal activity *in vivo*, with the potential to interact and disturb normal physiological functions (Hansen 1998; Safe 2004). It has become increasingly evident that different PCB congeners vary in the type and strength of effect they have on outcomes (Stewart et al. 1999; Wolff and Landrigan 2002). Different PCB congeners have been found to be weakly estrogenic, antiestrogenic, and antiandrogenic (Wolff et al. 1997), and can affect thyroid hormone function (Schell et al. 2008). Many studies have not presented findings that are congener specific, although this factor is being given increasing attention (Brouwer et al. 1999; Chen and Hsu 1994; Patandin et al. 1998; Seo et al. 1995; Shain et al. 1991). While PCBs and PBBs were banned by the mid-1970s in the United States, these halogenated and chlorinated organic compounds have a wide range of biochemical properties and are well established in modern society by their use as solvents and heat exchange fluids.

An important determinant of child health and survival is birth weight, and low birth weight has been associated with increased infant mortality and childhood morbidity (McCormick 1985). Studies of populations with PCB burdens not from acute poisoning episodes but from chronic exposures have produced mixed results (Dar et al. 1992; Fein et al. 1984; Grandjean et al. 2003; Rogan et al. 1986, 1987). Higher prenatal PCB exposure has been associated with reduced size at birth from 22 to 500 g (Fein et al. 1984; Hertz-Picciotto et al. 2000; Murphy et al. 2010; Patandin et al. 1998; Rylander et al. 1998; Taylor et al. 1984, 1989).

These associations are often sex specific and/or involve only some of the dimensions included in a given study. For example, Hertz-Picciotto et al. (2000) found PCBs to be inversely related to birth weight and length only among males, not females. This finding is similar to that of Tsukimori et al. (2012), suggesting that males may be more susceptible to growth restriction induced by *in utero* toxicant levels regardless of their mother's exposure level. A large data-driven project examined over 200,000 births among a PCB-exposed group of mothers identified by zip code and detected a decrease of 21 g in birth weight with an increased risk of over 6% for giving birth to a male infant of low birth weight (Baibergenova et al. 2003). However, this study was limited by a group measurement of PCBs rather than an individual measurement.

The earliest study of chronic PCB exposure and human growth came from a cohort of mothers and offspring who consumed fish from Lake Michigan, Michigan (Fein et al. 1984). Neonates with cord serum levels of 5.0 ng/mL or greater had decreased birth weight of approximately 160–190 g and smaller head circumference compared with children with lower levels (Fein et al. 1984), suggesting a dose-dependent form. Also, these effects could be attributable to confounding by contaminants in fish since mothers consumed moderate amounts of Lake Michigan fish for an average of 16 years. Birth weight, but not gestational length, was reduced by approximately 471 g in 55 neonates born to women in the highest versus the lowest tercile of preconception anti-estrogenic PCB concentration, yet no such decrement was observed in relation to estrogenic PCB concentration (Murphy et al. 2010). Similar results were reported from a large study comprising 12 European cohorts that measured PCB 153 (used as a biomarker of PCB exposure) in maternal and cord blood and breast milk samples from nearly 8000 women (Govarts et al. 2011). Independent of gestational age and after controlling for a number of confounders, a 150-g decline in birth weight was observed with a corresponding 1-μg/L increase in cord serum concentration of PCB 153 (Govarts et al. 2011). A study of infants in the Netherlands found a reduction of approximately 165 g and delayed growth for 3 months correlated with higher PCB levels measured in cord plasma and mother's blood (Patandin et al. 1997, 1998). A recent study of Great Lakes fish eaters on the effect of PCBs and DDE on birth weight found a significant decrease (~500 g) in birth weight of infants born to mothers with PCB levels of 25 μg/L or more, irrespective of smoking status (Karmaus and Zhu 2004). In nonsmoking participants, this effect was still present even in women with low levels of PCB exposure (<5 μg/dL). A 153-g difference in birth weight was found in 51 children born to women with high occupational exposure to PCBs, resulting in nearly 8% under 2501 g at birth (Taylor et al. 1984). This effect remained after adjustment for gestational age, and further examination indicated that this reduction in birth weights was primarily the result of the shortening of gestational age in the high-exposure group (Taylor et al. 1984). In contrast, one study evaluated PCBs by maternal recall of fish consumption in the past year reported an increase in birth weight by approximately 4 g if the mother gained less than 34 lb during the term of her pregnancy, yet if the mother's gestational weight gain exceeded 34 lb, a substantially smaller 0.86-g increase was noted (Dar et al. 1992).

In general, these studies indicate that prenatal exposure to PCBs can cause lower birth weight and shorter gestation. It has been well established that PCBs have an efficient transplacental transfer (Covaci et al. 2002). Better evaluation of the magnitude of exposure to organic pollutants is necessary given the potential of differential effects on human outcomes between acute exposures versus long-term background exposure.

In postnatal life, the effect on growth of lactational exposure has been the subject of several investigations. In a North Carolina cohort, breast-fed infants weighed less than bottle-fed ones by 1 year of age (Rogan et al. 1986, 1987). However, breastfeeding itself may have been responsible for the slower growth rate (Rogan et al. 1986, 1987) since bottle-fed infants usually add more weight than breast-fed ones, especially after 6 months of age (Bergmann et al. 2003). Recently, Grandjean et al. (2003) reported that lactational exposure to PCBs and mercury was associated with reductions of 0.59 kg and 1.5 cm in weight and height, respectively, at 18 months of age.

Less research has been conducted into the pattern of development after early childhood of individuals exposed to PCBs either prenatally or as young children. Only two studies could be found that examined the effects of prenatal PCB exposure on postnatal growth, one on both boys and girls (Jacobson et al. 1990), and the other only in girls (Blanck et al. 2002). The former found no association between PCBs and height in 285 4-year-old children born to fish-eating mothers (77% of whom ate at least 11 kg of fish over a 6-year period) while controlling for other environmental exposures (PBBs, DDT, and Pb). Likewise, after estimating PBB and PCB (measured as Aroclor 1254) exposure during pregnancy in 308 women accidentally exposed through ingestion of contaminated food, no relation between these toxicants and their daughters' self-reported height was observed approximately 5–15 years later (Blanck et al. 2002). However, the authors did not control an unmeasured source of confounding: Maternal exposure to the metabolite of DDT, 2,2-bis(*p*-chlorophenyl)-1,1-dichloroethylene (DDE), was often found to be highly correlated with PCBs, established in fish-eating populations such as this, and was found to affect growth in childhood and early adolescence (Gladen et al. 2000; Karmaus et al. 2002; Verhulst et al. 2009).

Similarly, no relationship of prenatal PCB levels to height was found in 176 adult daughters of women enrolled in the Michigan fish eaters cohort measured some 25 years later with a mean age of 62 years (Karmaus et al. 2009), nor in a prospective study of over 340 8- to 10-year-old German children whose heights were measured at three different intervals (Karmaus et al. 2002). Follow-up on 594 children from a large North Carolina cohort found no association between measured transplacental (maternal, cord, and placental blood) or lactational (breast milk) exposure to PCBs and height in either boys or girls, controlling for maturation (Gladen et al. 2000). No relationship between PCB 153 and height at 4 and 7 years of age was found in a smaller study of children born to Swedish fishermen's wives who consumed fatty fish from the Baltic Sea, yet weight was negatively impacted at these same ages among those children with a normal birth weight (Rylander et al. 1998). A large study of 1712 children with measured prenatal toxicant levels (their mother's serum from pregnancy) found no association between PCBs and children's height at 1, 4, and 7 years of age (Ribas-Fito et al. 2006). In short, several significant studies have not found associations between some PCB congeners and some dimension of size at birth.

On the other hand, decreased length was reported from an analysis of postnatal growth of 207 infants exposed *in utero* (cord blood measurements) who had not been breast-fed. Length was decreased from 1 to 3 months of age in association with an increase in four predominantly persistent PCBs (PCBs 118, 138, 153, and 180). Yet, there was a negative effect on growth rates for body length among breast-fed infants between 3 and 7 months of age (Patandin et al. 1998), indicative of a postnatal effect of PCB exposure through breast milk.

While these studies have observed decreased growth in relation to PCB levels, increased height in relation to PCB levels has also been found. In a study of 399 mother/child pairs selected from a nationally representative sample (Child Health and Development Study), conducted in the 1960s when PCB use was ongoing worldwide, increased height was found. In proportion to their maternal PCB levels (Σ of nine PCBs# 105, 110, 118, 137, 138, 153, 170, 180, and 187), 5-year-old girls had greater standing height and sitting height (trunk size) after controlling for other toxicants and other significant covariates (Hertz-Picciotto et al. 2005).

Likewise, a large Flemish study examined background levels of toxicants in relation to body size in nearly 1700 adolescents between 14 and 15 years of age. Serum levels of PCB 118, a highly persistent dioxin-like congener, exhibited a positive association with height in boys, but not in girls, after controlling for maturational stage (Dhooge et al. 2010). A smaller study of 150 African American mother/child pairs tested the relationship of maternal serum levels of PCBs grouped by structure and reported an increased height in girls. In a population with a low prevalence of breastfeeding (16%), girls exhibited increased height at 4, 7, and 17 years in relation to prenatal exposure to only the non-*ortho*-substituted and non-dioxin-like PCBs, by a 2% height increase in 4 and 7-year-olds when doubling maternal serum concentration (Lamb et al. 2006).

A careful analysis of PCB congeners revealed a reduction in prepubescent girls' height, which was observed in relation to eight of the measured mono-*ortho*-substituted congeners and one of the di-*ortho*-congeners (PCB 101), a marker of continuing exposure given its short half-life. This congener was also related to increased height in prepubescent boys, as was two tri-*ortho*-substituted congeners, but only marginally (Lamb et al. 2006).

After a 3-year follow-up restricted to 473 Russian boys of 8 and 9 years of age, the majority of whom were breast-fed (86%), higher peripubertal serum PCBs were associated with a significant reduction in height *z*-scores with increasing age (Burns et al. 2011). Furthermore, annual height velocity decreased by 0.19 cm/year, yet neither the extent nor the duration of peak height velocity could be measured as these boys had not reached full maturation by the age of 12 years (Burns et al. 2011).

The effects of PCBs and/or PBBs on birth length and childhood height vary between studies since the 1990s and often are due to variances in exposure assessment. The discrepancies between these studies on height may center on differences in growth measurements, exposure, and population characteristics. In summary, the data on PCBs reported to date suggest that the greatest risks to development are associated with exposure during the prenatal period. The effects reported on growth postnatally suggest possible impairment during a critical period of neuroendocrine or other central nervous system mechanisms regulating postnatal growth.

Dioxins

The polychlorinated dibenzo-*p*-dioxins (PCDDs) comprise a number of structurally similar compounds. Better known simply as dioxins, they are also often referred to as 2,3,7,8-tetrachlorodibenzo-*p*-dioxin (TCDD), PCDDs, and PCDFs, formed during combustion processes such as waste incineration, forest fires, backyard burning, paper pulp bleaching, and herbicide manufacturing (Agency for Toxic Substances Disease Registry 2004). Dioxins are considered highly toxic chemicals with the potential to cause cancer (Agency for Toxic Substances Disease Registry 2004; Miller 2004; Webster and Commoner 2003), reproductive disorders (Eskenazi and Kimmell 1995; Eskenazi et al. 2000, 2002a,b; Lundqvist et al. 2006; Warner et al. 2004), and developmental alterations (Lundqvist et al. 2006; Patandin et al. 1999; Sauer et al. 1994), to interfere with the immune system (Kaneko et al. 2006; Leijs et al. 2009), and to alter hormonal status (Aoki 2001; Goodman et al. 2010; Lundqvist et al. 2006; Matsuura et al. 2001). Human exposure is mainly through food (primarily meat, fish, shellfish, and dairy products), and because of their chemical stability and their ability to be stored in fat, they are not readily metabolized (Amakura et al. 2003; Svensson et al. 1991; Turrio-Baldassarri et al. 2009).

The influence of transplacental exposure to high levels of dioxins from accidental exposure (Yusho/Yu-Cheng) on prenatal and postnatal growth and development has been described earlier in this chapter. However, reports on the influence of low-level exposure to PCDDs/Fs and TCDDs on birth size and later physical growth have been less frequent and are inconsistent.

In a report of occupationally exposed mothers to wood preservatives containing PCDD/Fs and octachlorodioxins (OCDDs), birth length and weight were reduced significantly (by 150 g and 2 cm, respectively) in comparison to nonexposed newborns (Karmaus and Wolf 1995). Another report measured PCDDs/Fs in breast milk from 167 lactating mothers from two different areas in Finland (Vartiainen et al. 1998). Birth weight was negatively correlated with dioxin toxic equivalents (TEQs) of PCDDs/Fs in all singleton births, but more so in boys; when restricted to only primiparae, this relationship disappeared (Vartiainen et al. 1998). A statistically significant, yet moderate, inverse association between OCDD, but not PCDDs/Fs, and birth weight was found in 207 Japanese breast-fed infants (Tajimi et al. 2005). A more recent project examined the relationship between newborn size at birth and dioxin isomers in 42 breast milk samples as an indicator of maternal exposure (Nishijo et al. 2008). Controlling for confounders including child sex, gestational age, and parity, head circumference at birth, and TCDD, as well as PCDF, there was no correlation with either birth weight or length (Nishijo et al. 2008). As a proxy for direct dioxin measurement, maternal consumption of at least four fatty-fish meals per month from the Baltic Sea was found to have an increased risk (odds ratio [OR] = 1.9) of having a low-birth weight infant, and if restricted to boys, the odds increased to over three times that of women who ate less fish per month (Rylander et al. 1996). A study of 514 Japanese mother/child pairs found a 272.7-g decrease in birth weight with a 10-fold increase in total PCDF levels as measured in maternal sera (Konishi et al. 2009).

Reports on later growth in children and adolescents in relation to dioxins are limited. Findings from a follow-up study of 92 mother/child pairs exposed to dioxins *in utero* concluded that overall height was significantly increased in relation to maternal PCDD/F TEQ levels at age 2 and 5 years after adjustment for thyroid and GHs (Su et al. 2010). Gender differences were noted for this effect in females at age 2, but not at age 5 years. Further examination determined that overall thyroid and GHs were also significantly higher with respect to higher dioxin levels, suggesting a broad effect of *in utero* exposure to PCDD/Fs on growth and development in preschool children (Su et al. 2010). Higher quintiles of PCDD/Fs concentrations were found to have a nonlinear association with Russian boys' height z-scores, but not with height velocity (Burns et al. 2012). A positive correlation between height (adjusted for age), as well as head circumference, and current serum PCDD/F TEQs was noted in 33 14- to 19-year-olds from the Netherlands (Leijs et al. 2008).

In summary, though there is a paucity of data on growth in children exposed pre- or postnatally to PCDD/Fs and TCDD, it is clear that there are some significant relationships with these toxicants when at background levels. There is some evidence of gender-specific effects. Exposure to endocrine-disrupting chemicals during critical windows of child and adolescent development may result in permanent alterations not only in physical growth but also potentially in the hormonal mechanisms of growth.

DDT/DDE

DDE is a known endocrine disruptor that possesses antiandrogenic properties and is a metabolite of DDT. DDT, a pesticide in popular use throughout the world prior to the 1970s, can be found in water and agricultural contexts. Exposure to DDE may impact normal height growth throughout the entirety of childhood.

As levels of toxicants, including DDE, can be measured from different tissues, care must be taken to identify how DDE was assessed. Toxicological studies have utilized maternal serum levels, placental tissue levels, and cord blood levels to analyze the effect DDE has on birth length and growth in infants. A Saudi Arabian study of maternal and placental concentrations of DDE and the impact on infant length found no significant relationship existed between maternal DDE levels and birth length (Al-Saleh et al. 2012). Similarly, two separate studies of maternal serum DDE levels and infant height growth in Mexican infants showed no significant relationship (Cupul-Uicab et al. 2010; Garced et al. 2012). Other recent studies analyzing toxicants in New York infants found no relationships between maternal serum DDE and birth length as well (Wolff et al. 2007). Utilizing maternal serum appears to be unrelated to birth length in older research articles as well, even in populations born during peak DDT use (Farhang et al. 2005; Jusko et al. 2006; Longnecker et al. 2001).

Utilizing other methods of determining DDE level such as the use of cord blood, placental tissue, or lactation has found similar nonsignificant results between DDE and birth length. A Spanish study utilizing cord blood samples found that DDE was not a significant predictor of birth length (Lopez-Espinosa et al. 2011). A study of Flemish individuals found that DDE was unrelated to lower standard deviation scores for length between 1 and 3 years of age. Similar nonsignificant relationships have been found between cord blood and birth length in older studies as well (Sagiv et al. 2007). However, the use of placental tissue found that placental concentrations of DDE were negatively correlated to birth length (Al-Saleh et al. 2012), although this finding is not universal (Rogan et al. 1986). DDE levels determined from breast milk do not appear to affect length growth in breast-fed infants (Pan et al. 2010). Based on these findings, it appears that DDE does not have a strong effect on prenatal or infant linear growth, regardless of the tissue used for analysis.

This result is surprising though, as many studies have found associations between DDE levels and birth weight and infant weight growth (Jusko et al. 2006; Lamb et al. 2006; Mendez et al. 2011). Weight and body mass index (BMI) are easier to measure in newborns and infants and are more regularly available for research purposes. Increased weight and BMI are also associated with higher DDE levels at older ages (Dhooge et al. 2010, 2011; Karmaus et al. 2009). These findings are fairly consistent, though not universal (Cupul-Uicab et al. 2010). This suggests that DDE's impact on weight utilizes separate biological pathways compared to the potential impacts on height.

Relationships between DDE levels and height are more apparent in older populations with several studies detecting significant relationships. A study of transplacental exposure to toxicants found that elevated levels of transplantational DDE were associated with increased height in males only in a population born after peak DDT usage (Gladen et al. 2000). A multicenter U.S. study of children born in the 1950s and 1960s, a period of peak DDT usage, found that extremely high maternal DDE levels were associated with decreased height in females only at ages 1, 4, and 7, with increased reductions in height at subsequent ages (Ribas-Fito et al. 2006). Older studies of German 8-year-old females (Karmaus 2002) and North Carolinian adolescent males (Gladen et al. 2000) found that higher DDE was significantly associated with shorter height. More recently, a study of Russian adolescent males found no reduced height in males with the highest levels of DDE (Burns et al. 2012).

However, some studies have found no associations between DDE and adolescent height. An adolescent male sample drawn from Philadelphia born during the mid-20th century found no relationship between maternal DDE and anthropometric measures, including height (Gladen et al. 2004). A recent study of Flemish 14- and 15-year-olds found that high-serum DDE was negatively related to BMI in both males and females, but no association was found between DDE and increased height (Dhooge et al. 2010). Studies of adult levels of DDE and height have also found no significant relationships (Karmaus et al. 2009). Thus, it appears that while some evidence exists for DDE impacting postnatal height, the relationship between DDE and height remains unclear.

Though DDE does not impact adolescent height universally, many studies have found significant relationships between adolescent maturation and DDE levels. Higher DDE is associated with faster and earlier pubertal maturation in studies of prenatal exposure (Vasiliu et al. 2004) and body burden measures (Dhooge et al. 2010). The observation that adolescent height is not commonly affected by DDE but that sexual maturation is routinely affected is surprising and requires further investigation.

There are two potential explanations for this phenomenon. Height growth is controlled by GH and IGF-1 during childhood and increases further with the release of sex steroids during puberty. As DDE is antiandrogenic, its impact on height at these ages would be minimal during childhood as the sex steroids are at very low levels and do not affect growth much prior to puberty. However, DDE may impact the GH/IGF-1 axis during this period. Studies of the impact of DDE on IGF-1 and GH are severely limited in humans, although a study of Spanish adolescents and young adults found lowered IGF-1 in young males is associated with higher levels of DDE (Zumbado et al. 2010). DDE is associated with lower testosterone levels in adult humans (Blanco-Muñoz et al. 2011; Langer et al. 2012), but relationships among adolescents are nonexistent (Dhooge et al. 2010; Gladen et al. 2000). Further studies of the effects DDE on hormone levels in children and adolescents are necessary in order to better understand these relationships.

In summary, although DDE has known endocrine-disrupting properties, the impact on height appears to be small. Regardless of the age of the participants under investigation or the method of determining the level of DDE, only a few significant relationships have been found with height. Furthermore, it should be noted that even in studies where DDE levels are high, relationships to height are rarely identified (Cupul-Uicab et al. 2010; Dhooge et al. 2010; Jusko et al. 2006). Compared to the other toxicants discussed, DDE's chemical properties do not appear to greatly impact the biological processes responsible for height independent of the effects on maturation.

CONCLUSION

The studies reviewed here demonstrate a range of relationships between one or more features of physical growth and an exposure that was measured in some way. In short, there is considerable variety in both outcomes and exposures, their content, and the rigor of their assessment. Studies that have found alterations of growth suggest that the compound is capable of altering basic pathways in the control of physical growth and development. This in itself can be taken as a sign of some degree of toxicity. The interpretation is more complex when the exposure is related to enhanced growth. While the relationship of reduced growth to health outcomes has been established over many years (e.g., with birth weight), relationships of artificially enhanced growth to health outcomes have not been studied. Thus, the meaning of the relationship of DDE to greater heights is not at all clear, although it is clearly an alteration of the usual growth pattern and may be viewed as a negative effect for that reason alone.

The diversity of outcomes is related to the wide range of growth outcomes that can be studied, the very different stages of growth that exist in the pattern of human growth with different hormonal controls, and unfortunately, to the variation in exposure assessment, and to the wide range of extraneous influences on growth that are difficult to control in studies of free-ranging human populations. Especially difficult is the interpretation of birth weight's role in interpreting effects in postnatal life. Is birth weight in the causal pathway to postnatal effects or not? The answer affects how multivariate models are employed and their ability to register effects on postnatal growth and development.

Future work would make great strides if it dealt with growth as growth velocity rather than size alone. Additionally, the study of subadults who were exposed in postnatal life would clarify the issue of whether only prenatal exposure produces effects on postnatal growth and development. The continued study of prenatal exposures is worthwhile and warranted owing to the sensitivity of prenatal development, but such study cannot clarify whether all postnatal effects are due to prenatal exposure. Finally, greater consideration of the ecosensitivity of growth is required. This should result in better control for other influences on growth such as other toxicants as well as aspects of the social context and character of those studied. In this way, we might be able to delineate and assess the effects of some toxicants on children.

REFERENCES

Agency for Toxic Substances Disease Registry. 2004. ToxFAQs February 1999 What are CDDs? U.S. Department of Health and Human Services.

Al-Saleh I, Al-Doush I, Alsabbaheen A, Mohamed GED, Rabbah A. 2012. Levels of DDT and its metabolites in placenta, maternal and cord blood and their potential influence on neonatal anthropometric measures. Sci Total Environ 416:62–74.

Amakura Y, Tsutsumi T, Sasaki K, Maitani T. 2003. Levels and congener distributions of PCDDs, PCDFs and Co-PCBs in Japanese retail fresh and frozen vegetables. Shokuhin Eiseigaku Zasshi 44:294–302.

Aoki Y. 2001. Polychlorinated biphenyls, polychlorinated dibenzo-p-dioxins, and polychlorinated dibenzofurans as endocrine disrupters—What we have learned from Yusho disease. Environ Res 86:2–11.

Austin AA, Fitzgerald EF, Pantea CI, Gensburg LJ, Kim NK, Stark AD, et al. 2011. Reproductive outcomes among former Love Canal residents, Niagara Falls, New York. Environ Res 111:693–701.

Baibergenova A, Kudyakov R, Zdeb M, Carpenter DO. 2003. Low birth weight and residential proximity to PCB-contaminated waste sites. Environ Health Perspect 111:1352–1357.

Bergmann KE, Bergmann RL, Von Kries R, Bohm O, Richter R, Dudenhausen JW, et al. 2003. Early determinants of childhood overweight and adiposity in a birth cohort study: Role of breast-feeding. Int J Obes Relat Metab Disord 27: 162–172.

Blanck HM, Marcus M, Rubin C, Tolbert PE, Hertzberg VS, Henderson AK, et al. 2002. Growth in girls exposed in utero and postnatally to polybrominated biphenyls and polychlorinated biphenyls. Epidemiology 13:205–210.

Blanco-Muñoz J, Lacasaña M, Aguilar-Garduño C, Rodríguez-Barranco M, Bassol S, Cebrián ME, et al. 2011. Effect of exposure to p,p′-DDE on male hormone profile in Mexican flower growers. Occup Environ Med 69:5–11.

Brouwer A, Longnecker MP, Birnbaum LS, Cogliano J, Kostyniak P, Moore J, et al. 1999. Characterization of potential endocrine-related health effects at low-dose levels of exposure to PCBs. Environ Health Perspect 107(Suppl 4):639–649.

Burns JS, Williams PL, Sergeyev O, Korrick S, Lee MM, Revich B, et al. 2011. Serum dioxins and polychlorinated biphenyls are associated with growth among Russian boys. Pediatrics 127:e59–e68.

Burns JS, Williams PL, Sergeyev O, Korrick SA, Lee MM, Revich B, et al. 2012. Serum concentrations of organochlorine pesticides and growth among Russian boys. Environ Health Perspect 120:303–308.

Cameron N, Demerath EW. 2002. Critical periods in human growth and their relationship to diseases of aging. Yearb Phys Anthropol 45(Suppl 35):159–184.

Chen WJ, Hsu CC. 1994. Effects of prenatal exposure to PCBs on the neurological functioning of children: A neuropsychological and neurophysiological study. Dev Med Child Neurol 36:312–320.

Covaci A, Jorens P, Jacquemyn Y, Schepens P. 2002. Distribution of PCBs and organochlorine pesticides in umbilical cord and maternal serum. Sci Total Environ 298:45–53.

Cupul-Uicab LA, Hernandez-Avila M, Terrazas-Medina EA, Pennell ML, Longnecker MP. 2010. Prenatal exposure to the major DDT metabolite 1,1-dichloro-2,2-bis (p-chlorophenyl)ethylene (DDE) and growth in boys from Mexico. Environ Res 110:595–603.

Dar E, Kanarek MS, Anderson HA, Sonzogni WC. 1992. Fish consumption and reproductive outcomes in Green Bay, Wisconsin. Environ Res 59:189–201.

Denham M, Schell LM, Deane G, Gallo MV, Ravenscroft J, DeCaprio AP, et al. 2005. Relationship of lead, mercury, mirex, dichlorodiphenyldichloroethylene, hexachlorobenzene, and polychlorinated biphenyls to timing of menarche among Akwesasne Mohawk girls. Pediatrics 115:e127–e134.

Dhooge W, Den Hond E, Koppen G, Bruckers L, Nelen V, Van De Mieroop E, et al. 2010. Internal exposure to pollutants and body size in Flemish adolescents and adults: Associations and dose–response relationships. Environ Int 36:330–337.

Dhooge W, Den Hond E, Bruckers L, Schoeters G, Nelen V, van de Mieroop E, et al. 2011. Internal exposure to pollutants and sex hormone levels in Flemish male adolescents in a cross-sectional study: Associations and dose-response relationships. J Expo Sci Environ Epidemiol 21:224–233.

Eskenazi B, Kimmell G. 1995. Workshop on perinatal exposure to dioxin-like compounds. II. Reproductive effects. Environ Health Perspect 103(Suppl 2):143–147.

Eskenazi B, Mocarelli P, Warner M, Samuels S, Vercellini P, Olive D, et al. 2000. Seveso Women's Health Study: A study of the effects of 2,3,7,8-tetrachlorodibenzo-p-dioxin on reproductive health. Chemosphere 40:1247–1253.

Eskenazi B, Mocarelli P, Warner M, Samuels S, Vercellini P, Olive D, et al. 2002a. Serum dioxin concentrations and endometriosis: A cohort study in Seveso, Italy. Environ Health Perspect 110:629–634.

Eskenazi B, Warner M, Mocarelli P, Samuels S, Needham LL, Patterson DG, Jr., et al. 2002b. Serum dioxin concentrations and menstrual cycle characteristics. Am J Epidemiol 156:383–392.

Farhang L, Weintraub JM, Petreas M, Eskenazi B, Bhatia R. 2005. Association of DDT and DDE with birth weight and length of gestation in the Child Health and Development Studies, 1959–1967. Am J Epidemiol 162:717–725.

Fein GG, Jacobson JL, Jacobson SW, Schwartz PM, Dowler JK. 1984. Prenatal exposure to polychlorinated biphenyls: Effects on birth size and gestational age. J Pediatr 105:315–320.

Funatsu I, Yamashita F, Yoshikane T, Funatsu T, Ito Y, Tsugawa S, et al. 1971. A chlorobiphenyl induced fetopathy. Fukuoka Igaku Zasshi 62:139–149.

Garced S, Torres-Sánchez L, Cebrián ME, Claudio L, López-Carrillo L. 2012. Prenatal dichlorodiphenyldichloroethylene (DDE) exposure and child growth during the first year of life. Environ Res 113:58–62.

Gladen BC, Ragan NB, Rogan WJ. 2000. Pubertal growth and development and prenatal and lactational exposure to polychlorinated biphenyls and dichlorodiphenyl dichloroethene. J Pediatr 136:490–496.

Gladen BC, Klebanoff MA, Hediger ML, Katz SH, Barr DB, Davis MD, et al. 2004. Prenatal DDT exposure in relation to anthropometric and pubertal measures in adolescent males. Environ Health Perspect 112:1761–1767.

Goodman JE, Kerper LE, Boyce CP, Prueitt RL, Rhomberg LR. 2010. Weight-of-evidence analysis of human exposures to dioxins and dioxin-like compounds and associations with thyroid hormone levels during early development. Regul Toxicol Pharmacol 58:79–99.

Govarts E, Nieuwenhuijsen M, Schoeters G, Ballester F, Bloemen K, de Boer M, et al. 2011. Birth weight and prenatal exposure to polychlorinated biphenyls (PCBs) and dichlorodiphenyldichloroethylene (DDE): A meta-analysis within 12 European Birth Cohorts. Environ Health Perspect 120:162–170.

Grandjean P, Budtz-Jorgensen E, Steuerwald U, Heinzow B, Needham LL, Jorgensen PJ, et al. 2003. Attenuated growth of breast-fed children exposed to increased concentrations of methylmercury and polychlorinated biphenyls. FASEB J 17:699–701.

Guo YL, Chen YC, Yu ML, Hsu CC. 1994a. Early development of Yu-Cheng children born seven to twelve years after the Taiwan PCB outbreak. Chemosphere 29:2395–2404.

Guo YL, Lin CJ, Yao WJ, Ryan JJ, Hsu CC. 1994b. Musculoskeletal changes in children prenatally exposed to polychlorinated biphenyls and related compounds (Yu-Cheng children). J Toxicol Environ Health 41:83–93.

Hansen LG. 1998. Stepping backward to improve assessment of PCB congener toxicities. Environ Health Perspect 106(Suppl 1):171–189.

Hertz-Picciotto I, Keller J, Willman E, James R, Teplin S, Charles MJ. 2000. Fetal and early childhood growth in relation to prenatal PCB and organochlorine pesticide exposures. Organohalogen Compd 48:163–166.

Hertz-Picciotto I, Charles MJ, James RA, Keller JA, Willman E, Teplin S. 2005. *In utero* polychlorinated biphenyl exposures in relation to fetal and early childhood growth. Epidemiology 16:648–656.

Jacobson JL, Jacobson SW, Humphrey HE. 1990. Effects of exposure to PCBs and related compounds on growth and activity in children. Neurotoxicol Teratol 12: 319–326.

Jusko TA, Koepsell TD, Baker RJ, Greenfield TA, Willman EJ, Charles MJ, et al. 2006. Maternal DDT exposures in relation to fetal and 5-year growth. Epidemiology 17:692–700.

Kaneko H, Matsui E, Shinoda S, Kawamoto N, Nakamura Y, Uehara R, et al. 2006. Effects of dioxins on the quantitative levels of immune components in infants. Toxicol Ind Health 22:131–136.

Karmaus W. 2002. Growth in girls exposed in utero and postnatally to polybrominated biphenyls and polychlorinated biphenyls (letter). Epidemiology 13:604.

Karmaus W. 2006. Commentary: Halogenated organic compounds and child's growth: A growing public health problem. Int J Epidemiol 35:858–861.

Karmaus W, Wolf N. 1995. Reduced birthweight and length in the offspring of females exposed to PCDFs, PCP, and lindane. Environ Health Perspect 103:1120–1125.

Karmaus W, Zhu X. 2004. Maternal concentration of polychlorinated biphenyls and dichlorodiphenyl dichlorethylene and birth weight in Michigan fish eaters: A cohort study. Environ Health 3:1–9.

Karmaus W, Asakevich S, Indurkhya A, Witten J, Kruse H. 2002. Childhood growth and exposure to dichlorodiphenyl dichloroethene and polychlorinated biphenyls. J Pediatr 140:33–39.

Karmaus W, Osuch JR, Eneli I, Mudd LM, Zhang J, Mikucki D, et al. 2009. Maternal levels of dichlorodiphenyl-dichloroethylene (DDE) may increase weight and body mass index in adult female offspring. Occup Environ Med 66:143–149.

Kielb CL, Pantea CI, Gensburg LJ, Jansing RL, Hwang S-A, Stark AD, et al. 2010. Concentrations of selected organochlorines and chlorobenzenes in the serum of former Love Canal residents, Niagara Falls, New York. Environ Res 110: 220–225.

Konishi K, Sasaki S, Kato S, Ban S, Washino N, Kajiwara J, et al. 2009. Prenatal exposure to PCDDs/PCDFs and dioxin-like PCBs in relation to birth weight. Environ Res 109:906–913.

Kuratsune M. 1980. Yusho. In: Halogenated Biphenyls, Terphenyls, Naphthalnenes, Dibenzodioxins and Related Products (Kimbrough RD, ed.). New York: Elsevier/ North-Holland, pp. 287–302.

Lamb MR, Taylor S, Liu X, Wolff MS, Borrell L, Matte TD, et al. 2006. Prenatal exposure to polychlorinated biphenyls and postnatal growth: A structural analysis. Environ Health Perspect 114:779–785.

Langer P, Kocan A, Drobna B, Susienkova K, Radikova Z, Huckova M, et al. 2012. Blood testosterone in middle aged males heavily exposed to endocrine disruptors is decreasing more with HCB and p,p′-DDE related to BMI and lipids, but not with Sigma15PCBs. Endocr Regul 46:51–59.

Leijs MM, Koppe JG, Olie K, van Aalderen WM, Voogt P, Vulsma T, et al. 2008. Delayed initiation of breast development in girls with higher prenatal dioxin exposure; a longitudinal cohort study. Chemosphere 73:999–1004.

Leijs MM, Koppe JG, Olie K, van Aalderen WM, de Voogt P, ten Tusscher GW. 2009. Effects of dioxins, PCBs, and PBDEs on immunology and hematology in adolescents. Environ Sci Technol 43:7946–7951.

Longnecker MP, Klebanoff MA, Zhou H, Brock JW. 2001. Association between maternal serum concentration of the DDT metabolite DDE and preterm and small-for-gestational-age babies at birth. Lancet 358:110–114.

Lopez-Espinosa MJ, Murcia M, Iniguez C, Vizcaino E, Llop S, Vioque J, et al. 2011. Prenatal exposure to organochlorine compounds and birth size. Pediatrics 128: e127–e134.

Lundqvist C, Zuurbier M, Leijs M, Johansson C, Ceccatelli S, Saunders M, et al. 2006. The effects of PCBs and dioxins on child health. Acta Paediatr Suppl 95:55–64.

Masuda Y, Yoshimura H. 1984. Polychlorinated biphenyls and dibenzofurans in patients with yusho and their toxicological significance: A review. Am J Ind Med 5:31–44.

Matsuura N, Uchiyama T, Tada H, Nakamura Y, Kondo N, Morita M, et al. 2001. Effects of dioxins and polychlorinated biphenyls (PCBs) on thyroid function in infants born in Japan—The second report from research on environmental health. Chemosphere 45:1167–1171.

McCormick MC. 1985. The contribution of low birth weight to infant mortality and childhood morbidity. N Engl J Med 312:82–90.

Mendez MA, Garcia-Esteban R, Guxens M, Vrijheid M, Kogevinas M, Goni F, et al. 2011. Prenatal organochlorine compound exposure, rapid weight gain, and overweight in infancy. Environ Health Perspect 119:272–278.

Miller RW. 2004. How environmental hazards in childhood have been discovered: Carcinogens, teratogens, neurotoxicants, and others. Pediatrics 113(4 Suppl):945–951.

Murphy LE, Gollenberg AL, Buck Louis GM, Kostyniak PJ, Sundaram R. 2010. Maternal serum preconception polychlorinated biphenyl concentrations and infant birth weight. Environ Health Perspect 118:297–302.

Nishijo M, Tawara K, Nakagawa H, Honda R, Kido T, Nishijo H, et al. 2008. 2,3,7,8-Tetrachlorodibenzo-p-dioxin in maternal breast milk and newborn head circumference. J Expo Sci Environ Epidemiol 18:246–251.

NYSDOH (New York State Department of Health). 1978, Love Canal: Public health time bomb. A special report to the Governor and Legislature. Albany, NY.

Pan IJ, Daniels JL, Herring AH, Rogan WJ, Siega-Riz AM, Goldman BD, et al. 2010. Lactational exposure to polychlorinated biphenyls, dichlorodiphenyltrichloroethane, and dichlorodiphenyldichloroethylene and infant growth: An analysis of the Pregnancy, Infection, and Nutrition Babies Study. Paediatr Perinat Epidemiol 24:262–271.

Patandin S, Koopman-Esseboom C, Weisglas-Kuperus N, Sauer PJJ. 1997. Birth weight and growth in Dutch newborns exposed to background levels of PCBs and dioxins. Organohalogen Compd 34:447–450.

Patandin S, Koopman-Esseboom C, De Ridder MAJ, Weisglas-Kuperus N, Sauer PJJ. 1998. Effects of environmental exposure to polychlorinated biphenyls and dioxins on birth size and growth in Dutch children. Pediatr Res 44:538–545.

Patandin S, Lanting CI, Mulder PG, Boersma ER, Sauer PJ, Weisglas-Kuperus N. 1999. Effects of environmental exposure to polychlorinated biphenyls and dioxins on cognitive abilities in Dutch children at 42 months of age. J Pediatr 134:33–41.

Ribas-Fito N, Gladen BC, Brock JW, Klebanoff MA, Longnecker MP. 2006. Prenatal exposure to 1,1-dichloro-2,2-bis (p-chlorophenyl)ethylene (p,p′-DDE) in relation to child growth. Int J Epidemiol 35:853–858.

Rogan WJ, Gladen BC, McKinney JD, Carreras N, Hardy P, Thullen J, et al. 1986. Neonatal effects of transplacental exposure to PCBs and DDE. J Pediatr 109:335–341.

Rogan WJ, Gladen BC, McKinney JD, Carreras N, Hardy P, Thullen J, et al. 1987. Polychlorinated biphenyls (PCBs) and dichlorodiphenyl dichloroethene (DDE) in human milk: Effects on growth, morbidity, and duration of lactation. Am J Public Health 77:1294–1297.

Rogan WJ, Gladen BC, Hung K-L, Koong S-L, Shih L-Y, Taylor JS, et al. 1988. Congenital poisoning by polychlorinated biphenyls and their contaminants in Taiwan. Science 241:334–336.

Rylander L, Stromberg U, Hagmar L. 1996. Dietary intake of fish contaminated with persistent organochlorine compounds in relation to low birthweight. Scand J Work Environ Health 22:260–266.

Rylander L, Strömberg U, Dyremark E, Ostman C, Nilsson-Ehle P, Hagmar L. 1998. Polychlorinated biphenyls in blood plasma among Swedish female fish consumers in relation to low birth weight. Am J Epidemiol 147:493–502.

Safe SH. 2004. Endocrine disruptors and human health: Is there a problem. Toxicology 205:3–10.

Sagiv SK, Tolbert PE, Altshul LM, Korrick SA. 2007. Organochlorine exposures during pregnancy and infant size at birth. Epidemiology 18:120–129.

Sauer PJJ, Huisman M, Koopman-Esseboom C, Morse DC, Smits-van Prooije AE, van de Berg KJ, et al. 1994. Effects of polychlorinated biphenyls (PCBs) and dioxins on growth and development. Hum Exp Toxicol 13:900–906.

Schell LM. 1991. Effects of pollutants on human prenatal and postnatal growth: Noise, lead, polychlorinated compounds and toxic wastes. Yearb Phys Anthropol 34(Suppl 13):157–188.

Schell LM. 1997. Using patterns of child growth and development to assess community wide effects of low-level exposure to toxic materials. In: Hazardous Waste: Impact on Human and Ecological Health (Johnson BL, Xintara C, Andrews JS, eds.). Princeton, NJ: Princeton Scientific Publishing Co., pp. 478–483.

Schell LM, Madan M, Davidson GK. 1992. Auxological epidemiology and methods for the study of effects of pollution. Acta Med Auxol (Milano) 24:181–188.

Schell LM, Gallo MV, Denham M, Ravenscroft J, DeCaprio AP, Carpenter DO. 2008. Relationship of thyroid hormone levels to levels of polychlorinated biphenyls, lead, p,p′-DDE, and other toxicants in Akwesasne Mohawk youth. Environ Health Perspect 116:806–813.

Schell LM, Gallo MV, Ravenscroft J. 2009. Environmental influences on human growth and development: Historical review and case study of contemporary influences. Ann Hum Biol 36:459–477.

Selevan SG, Rice DC, Hogan KA, Euling SY, Pfahles-Hutchens A, Bethel J. 2003. Blood lead concentration and delayed puberty in girls. N Engl J Med 348:1527–1536.

Seo BW, Li M-H, Hansen LG, Moore RW, Peterson RE, Schantz SL. 1995. Effects of gestational and lactational exposure to coplanar PCB congeners or TCDD on thyroid hormone concentrations in weanling rats. Toxicol Lett 78:253–262.

Shain W, Bush B, Seegal R. 1991. Neurotoxicity of polychlorinated biphenyls: Structure-activity relationship of individual congeners. Toxicol Appl Pharmacol 111:33–42.

Stewart P, Darvill T, Lonky E, Reihman J, Pagano J, Bush B. 1999. Assessment of prenatal exposure to PCBs from maternal consumption of Great Lakes fish: An analysis of PCB pattern and concentration. Environ Res 80(2 Pt 2):S87–S96.

Su PH, Chen JY, Chen JW, Wang SL. 2010. Growth and thyroid function in children with in utero exposure to dioxin: A 5-year follow-up study. Pediatr Res 67:205–210.

Svensson BG, Nilsson A, Hansson M, Rappe C, Akesson B, Skerfving S. 1991. Exposure to dioxins and dibenzofurans through the consumption of fish. N Engl J Med 324: 8–12.

Tajimi M, Uehara R, Watanabe M, Oki I, Ojima T, Nakamura Y. 2005. Relationship of PCDD/F and Co-PCB concentrations in breast milk with infant birthweights in Tokyo, Japan. Chemosphere 61:383–388.

Taki I, Hisanaga S, Amagase Y. 1969. Report on Yusho (chlorobiphenyls poisoning). Pregnant women and their fetuses. Hukuoka Acta Med 60:471–474.

Tanner JM. 1978. Foetus into Man: Physical Growth from Conception to Maturity. Cambridge, MA: Harvard University Press.

Tanner JM. 1981. A History of the Study of Human Growth. London: Cambridge University Press.

Taylor PR, Lawrence CE, Hwang H-L, Paulson AS. 1984. Polychlorinated biphenyls: Influence on birthweight and gestation. Am J Public Health 74:1153–1154.

Taylor PR, Stelma JM, Lawrence CE. 1989. The relation of polychlorinated biphenyls to birth weight and gestational age in the offspring of occupationally exposed mothers. Am J Epidemiol 129:395–406.

Tsukimori K, Uchi H, Mitoma C, Yasukawa F, Chiba T, Todaka T, et al. 2012. Maternal exposure to high levels of dioxins in relation to birth weight in women affected by Yusho disease. Environ Int 38:79–86.

Turrio-Baldassarri L, Alivernini S, Carasi S, Casella M, Fuselli S, Iacovella N, et al. 2009. PCB, PCDD and PCDF contamination of food of animal origin as the effect of soil pollution and the cause of human exposure in Brescia. Chemosphere 76:278–285.

Vartiainen T, Jaakkola JJK, Saarikoski S, Tuomisto J. 1998. Birth weight and sex of children and the correlation to the body burden of PCDDs/PCDFs and PCBs of the mother. Environ Health Perspect 106:61–66.

Vasiliu O, Muttineni J, Karmaus W. 2004. *In utero* exposure to organochlorines and age at menarche. Hum Reprod 19:1506–1512.

Verhulst SL, Nelen V, Hond ED, Koppen G, Beunckens C, Vael C, et al. 2009. Intrauterine exposure to environmental pollutants and body mass index during the first 3 years of life. Environ Health Perspect 117:122–126.

Vianna NJ, Polan AK. 1984. Incidence of low birth weight among Love Canal residents. Science 226:1217–1219.

Warner M, Samuels S, Mocarelli P, Gerthoux PM, Needham L, Patterson DG, Jr., et al. 2004. Serum dioxin concentrations and age at menarche. Environ Health Perspect 112:1289–1292.

Webster TF, Commoner B. 2003. Overview: The dioxin debate. In: Dioxins and Health (Arnold Schecke, ed.). Hoboken, NJ: John Wiley & Sons, Inc., pp. 1–53.

Wolff MS, Landrigan PJ. 2002. Organochlorine chemicals and children's health. J Pediatr 140:10–13.

Wolff MS, Camann D, Gammon M, Stellman SD. 1997. Proposed PCB congener groupings for epidemiological studies. Environ Health Perspect 105:13–14.

Wolff MS, Anderson HA, Britton JA, Rothman N. 2007. Pharmacokinetic variability and modern epidemiology—The example of dichlorodiphenyltrichloroethane, body mass index, and birth cohort. Cancer Epidemiol Biomarkers Prev 16:1925–1930.

Wu T, Buck GM, Mendola P. 2003. Blood lead levels and sexual maturation in U.S. girls: The third national health and nutrition examination survey, 1988–1994. Environ Health Perspect 111:737–741.

Yamashita F, Hayashi M. 1985. Fetal PCB syndrome: Clinical features, intrauterine growth retardation and possible alteration in calcium metabolism. Environ Health Perspect 59:41–45.

Yang C-Y, Yu M-L, Guo H-R, Lai T-J, Hsu C-C, Lambert G, et al. 2005. The endocrine and reproductive function of the female Yucheng adolescents prenatally exposed to PCBs/PCDFs. Chemosphere 61:355–360.

Yen YY, Lan SJ, Ko YC, Chen CJ. 1989. Follow-up study of reproductive hazards of multiparous women consuming PCBs-contaminated rice oil in Taiwan. Bull Environ Contam Toxicol 43:647–655.

Yen YY, Lan SJ, Yang CY, Wang HH, Chen CN, Hsieh CC. 1994. Follow-up study of intrauterine growth of transplacental Yu-Cheng babies in Taiwan. Bull Environ Contam Toxicol 53:633–641.

Yoshimura T. 1971. A case control study on growth of school children with "Yusho." Fukuoka Igaku Zasshi 62:109–116.

Yoshimura T. 1974. Epidemiological study on Yusho babies born to mothers who had consumed oil contaminated by PCB. Fukuoka Igaku Zasshi 65:74–80.

Yoshimura T, Ikeda M. 1978. Growth of school children with polychlorinated biphenyl poisoning or yusho. Environ Res 17:416–425.

Zumbado M, Luzardo OP, Lara PC, Alvarez-Leon EE, Losada A, Apolinario R, et al. 2010. Insulin-like growth factor-I (IGF-I) serum concentrations in healthy children and adolescents: Relationship to level of contamination by DDT-derivative pesticides. Growth Horm IGF Res 20:63–67.

CHAPTER 21

How Much Human Disease Is Caused by Exposure to Organic Chemicals?

DAVID O. CARPENTER

ABSTRACT

Background: This book has consisted of chapters dealing with different diseases and the role that exposure to organic chemicals plays in risk. The goal of this concluding chapter is to try to put the overall issue into perspective.

Significance: Because in modern society organic chemicals are found in food, water, air, and consumer products, everyone is exposed to one degree or another. Thus, it is critical to understand what are the hazards and then what steps can be taken to reduce exposure and risk.

Discussion: Organic chemicals act in many different ways to influence cellular processes in almost every organ of the body. While the mechanisms of action are incompletely known, some act by disrupting endocrine and neuronal function and many act by gene induction.

Conclusion: Exposure to organic chemicals influences susceptibility to almost every major category of human disease. While it is not possible to answer the question of how much human disease is caused by exposure to organic chemicals, it is clear that the answer is larger than is usually appreciated.

INTRODUCTION

As clearly shown by chapters in this book, organic chemicals have actions on many different organ systems, including the brain, immune system, endocrine systems, heart, bones, and joints. In addition, exposure is associated with increased risk of developing many different diseases, including cancer, heart

Effects of Persistent and Bioactive Organic Pollutants on Human Health, First Edition.
Edited by David O. Carpenter.

disease and hypertension, diabetes, respiratory disease, infections, learning disabilities, neurodegenerative diseases, and even mental illness. The goal of this concluding chapter is to attempt to generalize the role of exposure to organics within the broader context of human disease.

We can conclude that every organ in the human body is affected by exposure to organic chemicals. While there have not been specific chapters on the gastrointestinal system, the liver, or the kidney, many organics are absorbed through the gastrointestinal tract, metabolized by the liver, and either the parent compounds or its metabolites excreted in part through the kidney. Except for diseases that are solely a function of genetics, organic chemicals contribute to some degree to almost all of the diseases that result in human morbidity and mortality. In some cases, these effects are mediated through specific receptor-binding sites, as is the case for some of the endocrine disruptors, while others are a consequence of gene induction and gene–environment interactions. Some exposures early in life result in development of disease much later, while other exposures result in more immediate disease. As concluded by Khoury et al. (2005), "environmental factors play an important role in the etiology of almost all human diseases." This is not to say that the environmental factors are the sole cause of all human disease, but rather that through gene–environment interactions, environmental exposures must be considered in order to understand the development of disease (Hunter 2005).

Ideally, it would be great if we could identify a percentage of human disease that is a consequence of exposure to organic chemicals. Several reports have attempted to determine what portion of the global burden of disease is secondary to environmental factors (Briggs 2003; Prüss-Üstün and Corvalán 2006; Prüss-Ustün and Corvalán 2007; Prüss-Ustün et al. 2008, 2011; Smith et al. 1999). These reports have focused sources of exposure such as sanitation, indoor and outdoor air pollution, and occupational exposures. These analyses are by their nature conservative and for the most part have considered exposures from only some sources. Most of these publications have concluded that environmental factors cause only a relatively small percentage of human diseases. For example, Prüss-Ustün et al. (2011) conclude that 8.3% of deaths and 5.7% of disability-adjusted life years (DALYs) are attributable to environmental exposures. But they also clearly state "Chemicals with known health effects, such as dioxins, cadmium, mercury or chronic exposure to pesticides could not be included in this article due to incomplete data and information." Thus, exposures that come from daily, nonoccupational sources are ignored. This author's conclusion is that these reports grossly underestimate the contribution of environmental exposures, and specifically exposure to organic chemicals, to human disease.

THE IMPORTANCE OF GENE–ENVIRONMENT INTERACTIONS

There are some diseases that are clearly caused by genetics. But most are not due solely to genetic factors. Lichtenstein et al. (2000), by a study of cancers in Scandinavian twins, conclude that "inherited genetic factors make a minor contribution to susceptibility to most types of neoplasms." In a study of environmental contributions to cancer, Boffetta and Nyberg of IARC (2003) consider only exposure to asbestos, outdoor air pollution including residence near major industrial emission sources, environmental tobacco smoke, indoor radon, other sources of indoor air pollution, arsenic in drinking water, chlorination by-products in drinking water, and other drinking water pollutants, and conclude that there are "relatively small relative risks of cancer following exposure to environmental carcinogens." The problem with this conclusion is that it ignores many sources to known and probable human carcinogens, particularly those in food and personal care products. Newby and Howard (2006) commented that anecdotal evidence suggests that malignant disease was rare in preindustrial societies, which is consistent with the conclusion that environmental exposures are important. The report of Boffetta and Nyberg (2003) also ignores early-life exposure to carcinogens, causing cancer later in life. Barton et al. (2005) reviewed animal studies comparing juvenile to adult cancer potencies and found a geometric mean ratio of 11, so early-life exposure is extremely important in understanding the relationship between organic chemical exposure and cancer. Nor does the analysis of Boffetta and Nyberg (2003) consider the impact of inflammation caused by environmental exposures (see Chapter 13) on risk of development of cancer (Aggarwal and Gehlot 2009).

The last several years have seen an enormous growth in our understanding of gene–environment interactions, and these interactions have not been considered in previous attempts to determine the extent of environmentally induced disease. Many chemical contaminants alter the expression of hundreds of different genes, often genes regulating so many different cellular functions that it is not possible to trace the exact pathway leading from exposure to a particular disease. For example, 2,3,7,8-tetrachlorodibenzo-*p*-dioxin (TCDD) alters expression by a factor of greater than 2 of 310 genes in human hepatoma cells (Puga et al. 2000), including genes involved in calcium regulation, kinases, phosphatases, transcription factors, cardiovascular and respiratory function, cell cycle regulation, apoptosis, cell adhesion, cancer, metastasis, protein traffic and membrane integrity, drug metabolism, DNA stability, and others. With such a great variety of cell functions influenced by just one organic chemical, it is not surprising that many different physiological functions are altered by TCDD. Figure 21.1 (from Johnson et al. 2004) shows the enormous complexity of the networks activated by organic chemicals that bind to the aryl hydrocarbon receptor (AhR), like TCDD. When one views the diversity of genes identified as being altered by this one pathway, it is not surprising that many different cellular functions are changed. Other studies have demonstrated that only some of the gene expression effects of TCDD are mediated by the AhR, contrary to general belief (Akintobi et al. 2007), and that the patterns of gene expression are organ specific (Volz et al. 2005). Thus, even the complexity shown in Figure 21.1 may not completely identify the variety of ways in which TCDD and other AhR ligands can alter function.

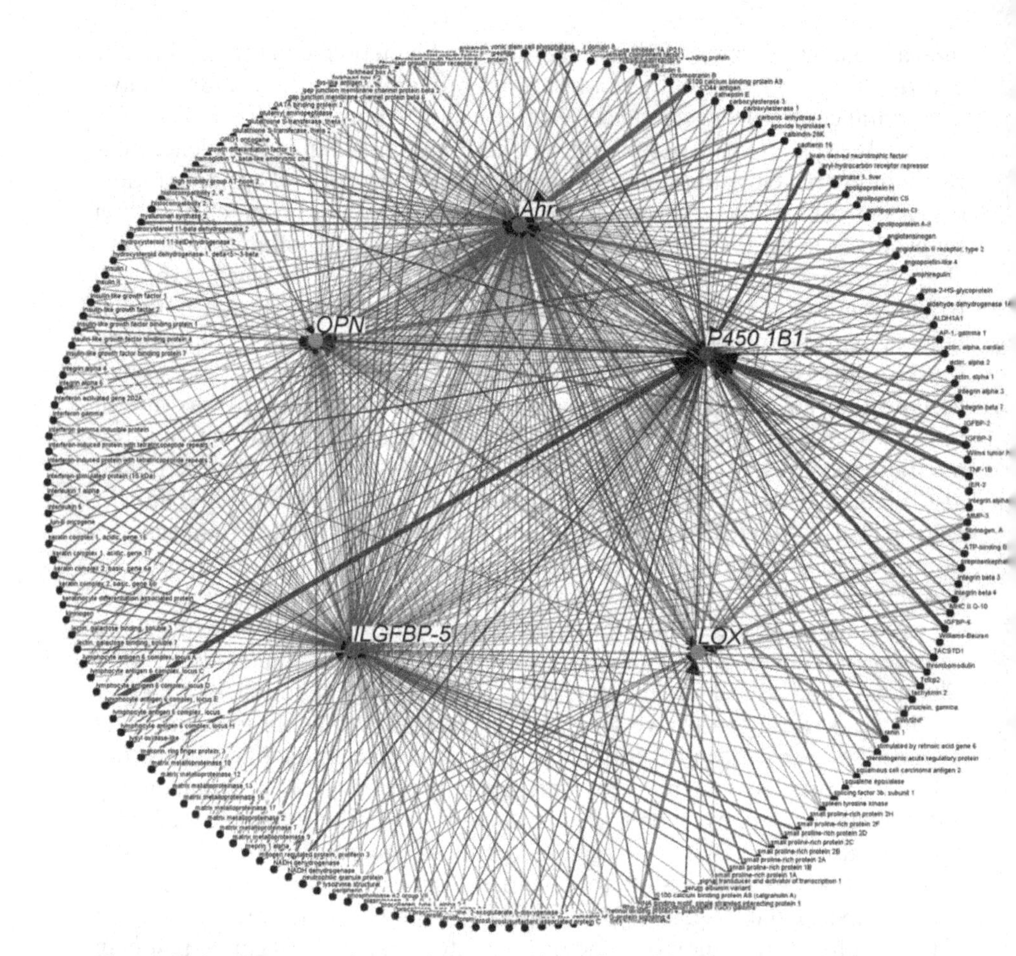

Figure 21.1. Gene–gene interaction networks activated by ligands of the aryl hydrocarbon receptor (AhR). All three gene combinations for each target that met the cutoff of COD > 0.9 and error < 0.5 were individually plotted using a program developed by Breitkreutz et al. (2003). The thickness of the line denotes the selection frequency for individual genes for each target. Figure and legend reproduced from Johnson et al. (2004).

We also know much more about polymorphisms, and how small differences in DNA structure between individuals greatly alter susceptibility to disease secondary to environmental exposures. However, Manolio et al. (2009) lament the fact that current genome-wide studies and the enormous increase in knowledge of polymorphorisms have only explained relatively small increments in risk for a great variety of human disease. This does not mean, however, that the ultimate understanding of gene–environment interactions will not at some time lead to a more complete understanding of what causes disease. For example, it has long been known that individuals differ in their genetic susceptibility to smoking-induced cancer (Spitz et al. 1999), and a number of the specific genotypes that confer susceptibility have been identified. While detailed pathways between exposure to chemicals that are associated with increases in risks of cancer, diabetes, heart disease, or an altered immune system may not be clear, the lack of a detailed mechanistic pathway should not be a barrier to identification of the disease being at least in part "environmental."

Epigenetic effects of organic chemicals add to the complexity as well, and such changes almost certainly contribute to the cause of some diseases that occur in families but are not explained by purely genetic factors. Epigenetic changes occur without alteration of DNA sequence, and occur through a variety of processes including DNA methylation, histone modifications, non-coding RNA, and networks of transcription factors (Bonasio et al. 2010; Kormaz et al. 2011), and these changes can be passed from one generation to the next. There is increasing evidence that a variety of classes of environmental contaminants, including peroxisome proliferators, air pollutants, endocrine-disrupting, and reproductive toxicants, cause epigenetic changes (Baccarelli and Bollati 2009) and that these changes may underlie diseases later in life resulting from early-life exposures (Dolinoy and Jirtle 2008; Dolinoy et al. 2007; Jirtle and Skinner 2007). Thus, epigenetics may be the basis for transgenerational changes in everything from learning and behavior (Weaver et al. 2004) to chronic diseases of older age, like obesity (Ng et al. 2010). It has been proposed that epigenetic changes are the reason for the differences in development of cancers in identical twins (Weidman et al. 2007). Rusiecki et al. (2008) reported that there is global DNA hypomethylation associated with elevated concentrations of persistent organic pollutants (POPs) in Greenlandic Inuits. While the true extent of organic chemical-induced epigenetic changes is still very unclear, there is little question but that this is an important mechanism whereby these contaminants alter risk of disease.

BARRIERS IN QUANTITATING THE BURDEN OF DISEASE FROM ORGANIC CHEMICALS

It is not possible to adequately quantitate how much disease is caused by exposure to organic chemicals for several reasons. First, as indicated earlier, we do not yet understand the magnitude of gene–environment interactions. In addition, exposure assessment is very difficult. We know more about the health consequences of exposure to persistent organics primarily because their persistence allows measurements of concentrations in blood and adipose tissue that reflect exposure over a relatively long period of time, but most organics are not so persistent. However, this does not mean that they do not cause disease. We have ways of monitoring recent exposure of nonpersistent organics by looking at serum levels and metabolites and, in some cases, urinary excretion of metabolites. But such exposure assessment is only a measure of recent exposure, not overall exposure over time.

Duration of exposure is a very important factor. Many organics are so omnipresent in our air, food, and water that we are exposed to low concentrations on a continuous basis. For others, exposures occur less continuously and at varied intensities, making monitoring of long-term exposure very difficult. This complicates exposure assessment greatly, particularly for less persistent organics.

Time between exposure and the development of disease is another very important factor that makes exposure assessment difficult. We know, for example, that the latency between exposure to asbestos and development of lung cancer, asbestosis, and mesothelioma may be as long as 20–30 years (Selikoff and Seidman 1991), and mortality from asbestosis peaks with a latency of 40–45 years (Lehman et al. 2008). McLaughlin et al. (1995) followed U.S. veterans for 26 years in relation to risk of 19 different types of cancer and reported that "the tendency for risks to diminish with time was generally slight and apparent only for certain cancers." Brain cancer risk following exposure to ionizing radiation has been found to exist for a lifetime (Preston et al. 2002). Other cancers, such as leukemia, may develop with a much shorter latency. Little is known about the latency between exposure and elevated risk of many other diseases.

Barker (2004) was the first to report that low birth weight was a major risk factor for development of chronic adult diseases such as cardiovascular disease and diabetes. There is increasing evidence that early-life exposure to environmental contaminants can result in elevated risk of a variety of diseases much later in life (Barouki et al. 2012; Gluckman and Hanson 2004; Grandjean 2008), probably by epigenetic mechanisms (Gluckman et al. 2008). Cupul-Uicab et al. (2012) have shown that *in utero* exposure to tobacco smoke increases the risk of obesity, hypertension, and gestational diabetes in women of reproductive age. Prenatal exposure to a great variety of organic chemicals has been associated with elevated risk of obesity, including polychlorinated biphenyls (PCBs), 2,2-bis(*p*-chlorophenyl)-1,1-dichloroethylene (DDE), DDT (Mendez et al. 2011; Valvi et al. 2012; Verhulst et al. 2009), hexachlorobenzene (Smink et al. 2008), and perfluorooctanoate (Halldorsson et al. 2012). Obesity, in turn, is a major risk factor for many other diseases, including diabetes, cardiovascular disease, osteoarthritis, and others.

Another major difficulty in assessing the degree to which organic chemicals influence the incidence of human disease stems from the fact that we are all exposed to chemical mixtures, usually not individual compounds (Rice et al. 2008; Sarigiannis and Hansen 2012; Sexton and Hattis 2007). Biomonitoring studies demonstrate that most of us are exposed to hundreds of different chemicals. Exposure to two or more chemicals may show additive, less than additive, or even synergistic effects (Carpenter et al. 2002). While most mixtures studied to date have shown either additive or antagonistic effects, some have shown synergism (Evans et al. 2012; Parvez et al. 2009). Humans are exposed to so many different chemicals, often at concentrations too low to demonstrate a biological effect when only one chemical is monitored.

Rajapakse et al. (2002) showed, however, that simultaneous exposure to low concentrations of xenoestrogens, below the concentrations that would have biological effects from single exposures, gave significant actions from the mixture (Rajapakse et al. 2002). But when considering the vast number of chemicals to which we are exposed, the great number of genes induced by many of these chemicals, and the multiple organ systems impacted by many individual chemicals, the challenge of quantitatively assessing the overall impact on disease risk becomes impossible.

A number of human diseases are often considered to be "lifestyle" diseases in that they are reflections of diet, habits such as smoking, drug, or alcohol consumption, and patterns of exercise. Behavioral and lifestyle factors vary greatly with socioeconomic status, which is a major consideration in health status. There is no question that diet, habits, and exercise influence susceptibility to disease. However, these factors have major influences on the degree of exposure to organic chemicals. For example, risk of colorectal cancer is known to be elevated by smoking, diabetes, obesity, and high meat intake (Huxley et al. 2009) and reduced by consumption of fruit and vegetables (van Duijnhoven et al. 2009). But as documented in various chapters in this book, smoking, diabetes, obesity, and high meat consumption all are associated with elevations in intake of POPs, and elevations in intake of fruits and vegetables may be simply a marker of reduced intake of animal fats containing POPs. Thus, diet, like smoking and substance abuse, is a central consideration regarding the intake of organic chemicals and should certainly not be dismissed as unrelated to exposure.

Many organics perturb immune system function (see Chapter 13). Some suppress immune system function, increasing risk of infection and cancer. But when a person gets an infection, it is almost never identified as being, at least in part, a consequence of exposure to an organic chemical. Not only can some organics directly increase the risk of infectious disease by immune suppression but there is also now clear evidence that perinatal exposure to some organic chemicals reduces the ability of the body to develop antibodies to vaccines, and thus greatly increases the risk of development of viral diseases that other members of the population are protected against via immunization. Other chemicals increase risk of asthma or allergies. While we know of many organics that trigger asthma attacks, the possibility that exposure increases the risk of development of the disease is rarely noted. Thus, almost every disease risk is altered by immune system function, and immune system function is in turn altered by exposure to many different organic chemicals. However, it is not possible to determine, for example, what percentage of infectious disease is a consequence of exposure to organics, but certainly some is, and it may be significant.

There is yet another complication in determining what percent of human disease is caused by exposure to organic chemicals. There is developing evidence that the dose–response relationships between exposure and disease is not always linear. It has been a rule of toxicology that more is worse, or "the poison is in the dose." However, a number of investigators have confirmed that this is not always the case (Fagin 2012). In some circumstances, the dose–response curves may be inverted, nonlinear, and there are effects at very low concentrations that are no longer found or are different from those at higher concentrations. Nonlinear dose–response relations appear to be common among endocrine-disruptive chemicals (Calabrese 2008; Lee et al. 2010; Welshons et al. 2003). In at least some cases, low-dose neurotransmitter effects are mediated through different receptors than those activated at higher concentrations (Ayrapetyan and Carpenter 1991). This process of low-dose, nonlinear dose–response curves has been called "hormesis," and while it is incompletely understood, the implications may be very significant. The results of Lee et al. (2010) demonstrating associations between low concentrations of POPs with risks of diabetes that disappear at higher concentrations suggest the possibility that the association has been missed for many years because in previous investigations, the lowest-exposed group already had maximal effect.

Figure 21.2 shows a pie chart indicating the worldwide burden of disease, measured as DALYs (data from the World Health Organization [WHO]). This is helpful in beginning to understand the possible role of exposure to organic chemicals in relation to overall disease. The various chapters in this book have documented the influence of organics on violent and antisocial behavior, respiratory diseases, cardiovascular disease, neuropsychiatric conditions, diabetes mellitus, malignant neoplasms, perinatal conditions, and susceptibility to infections. While no one has specifically implicated exposure to organics in susceptibility to HIV/AIDS, all the evidence that the immune system is "set" by a combination of genetics and exposures suggests that even here, there may be influences of exposure. Organics play a major role in endocrine disorders and skin diseases and may play important roles in other morbidities such as congenital anomalies. Thus, as stated by Khoury et al. (2005) "environmental factors play an important role in the etiology of almost all human diseases." Exposure to organic chemicals is among the most important of all these environmental factors.

It is very difficult to prevent disease if one does not understand what causes disease. The lack of general understanding within the medical community of how important exposure to organic chemicals is to the development of human disease is a major barrier. But it is especially a barrier to the development of strategies for the prevention of the disease in the first place. The best way to prevent most of the diseases discussed earlier is to prevent exposure to the environmental agents that contribute to the cause of the disease, whether that be an organic chemical in food, water, or air or a behavior that results in an increased exposure to organic chemicals, such as smoking cigarettes or drinking excessive amounts of alcohol. It is important that the medical community and the general public have greater understanding of those chemicals in our food supply, drinking water, air, and consumer products that are carcinogenic or act to increase the risk of other diseases, and become informed on how to reduce exposure and thus reduce risk. It is common knowledge that "an ounce

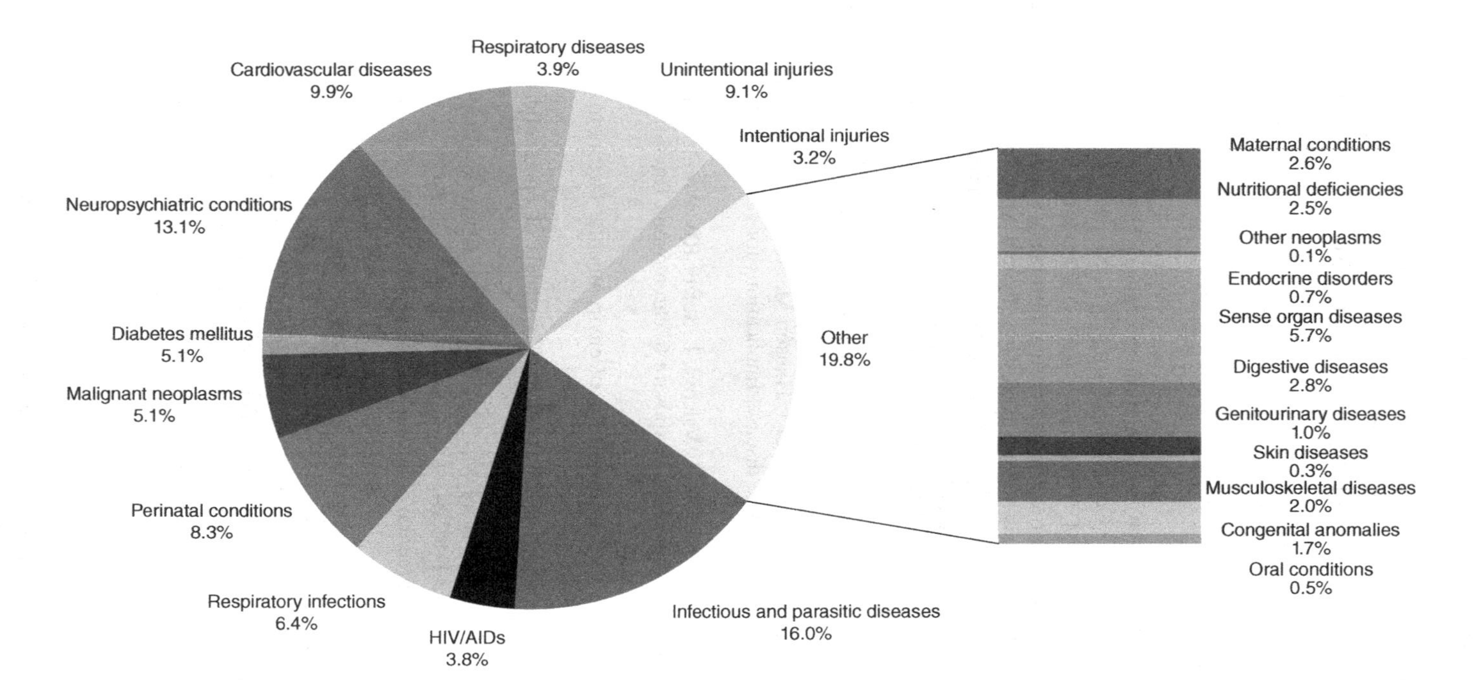

Figure 21.2. Figure distribution of burden by cause (all ages), 2004. Adapted from WHO (2008).

of prevention is worth a pound of cure," but the things that our mothers taught us are often not implemented in practice. Effective action requires an understanding of the magnitude of the problem but also its causes and underlying mechanisms.

REFERENCES

Aggarwal BB, Gehlot P. 2009. Inflammation and cancer: How friendly is the relationship for cancer patients? Curr Opin Pharmacol 9:351–369.

Akintobi AM, Villano CM, White LA. 2007. 2,3,7,8-Tetrachlorodibenzo-p-dioxin (TCDD) exposure of normal human dermal fibroblasts results in AhR-dependent and -independent changes in gene expression. Toxicol Appl Pharmacol 220:9–17.

Ayrapetyan S, Carpenter D. 1991. Very low concentrations of acetylcholine and GABA modulate transmitter responses. Neuroreport 2:563–565.

Baccarelli A, Bollati V. 2009. Epigenetics and environmental chemicals. Curr Opin Pediatr 21:243–251.

Barker DJ. 2004. The developmental origins of adult disease. J Am Coll Nutr 23:588S–595S.

Barouki R, Gluckman PD, Grandjean P, Hanson M, Heindel JJ. 2012. Developmental origins of non-communicable disease: Implications for research and public health. Environ Health 11:42.

Barton HA, Cogliano VJ, Flowers L, Valcovic L, Setzer RW, Woodruff TJ. 2005. Assessing susceptibility from early-life exposure to carcinogens. Environ Health Perspect 113:1125–1133.

Boffetta P, Nyberg F. 2003. Contribution of environmental factors to cancer risk. Br Med Bull 68:71–94.

Bonasio R, Tu S, Reinberg D. 2010. Molecular signals of epigenetic states. Science 330:612–616.

Breitkreutz BJ, Stark C, Tyers M. 2003. Osprey: A network visualization system. Genome Biol 4:R22.1–R22.4.

Briggs D. 2003. Environmental pollution and the global burden of disease. Br Med Bull 68:1–24.

Calabrese EJ. 2008. Hormesis: Why it is important to toxicology and toxicologists. Environ Toxicol Chem 27:1451–1474.

Carpenter DO, Nguyen T, Le L. 2002. Understanding the human health effects of chemical mixtures. Environ Health Perspect 110(Suppl 1):25–42.

Cupul-Uicab LA, Skjaerven R, Haug K, Travlos GS, Wilson RE, Eggesbø M, et al. 2012. Exposure to tobacco smoke in utero and subsequent plasma lipids, ApoB and CRP among adult women in the MoBa cohort. Environ Health Perspect 120: 1532–1537.

Dolinoy DC, Jirtle RL. 2008. Environmental epigenomics in human health and disease. Environ Mol Mutagen 49:4–8.

Dolinoy DC, Das R, Weidman JR, Jirtle RL. 2007. Metastable epialleles, imprinting, and the fetal origins of adult diseases. Pediatr Res 61:30R–37R.

van Duijnhoven FJ, Bueno-De-Mesquita HB, Ferrari P, Jenab M, Boshuizen HC, Ros MM, et al. 2009. Fruit, vegetables, and colorectal cancer risk: The European Prospective Investigation into Cancer and Nutrition. Am J Clin Nutr 89: 1441–1452.

Evans RM, Scholze M, Kortenkamp A. 2012. Additive mixture effects of estrogenic chemicals in human cell-based assays can be influenced by inclusion of chemicals with differing effect profiles. PLoS One 7:e43606. DOI: 10.1371/journal.pone.0043606.

Fagin D. 2012. The learning curve. Nature 490:462–465.

Gluckman PD, Hanson MA. 2004. Living with the past: Evolution, development, and patterns of disease. Science 305:1733–1736.

Gluckman PD, Hanson MA, Cooper C, Thornburg KL. 2008. Effect of in utero and early-life conditions on adult health and disease. N Engl J Med 359:61–73.

Grandjean P. 2008. Late insights into early origins of disease. Basic Clin Pharmacol Toxicol 102:94–99.

Halldorsson TI, Rytter D, Haug LS, Bech BH, Danielsen I, Becher G, et al. 2012. Prenatal exposure to perfluorooctanoate and risk of overweight at 20 years of age: A prospective cohort study. Environ Health Perspect 120:668–673.

Hunter DJ. 2005. Gene-environment interactions in human diseases. Nat Rev Genet 6:287–298.

Huxley RR, Ansary-Moghaddam A, Clifton P, Czernichow S, Parr CL, Woodward M. 2009. The impact of dietary and lifestyle risk factors on risk of colorectal cancer: A quantitative overview of the epidemiological evidence. Int J Cancer 125: 171–180.

Jirtle RL, Skinner MK. 2007. Environmental epigenomics and disease susceptibility. Nat Rev Genet 8:253–262.

Johnson CD, Balagurunathan Y, Tadesse MG, Falahatpisheh MH, Brun M, Walker MK, et al. 2004. Unraveling gene-gene interactions regulated by ligands of the aryl hydrocarbon receptor. Environ Health Perspect 112:403–412.

Khoury MJ, Davis R, Gwinn M, Lindegren ML, Yoon P. 2005. Do we need genomic research for the prevention of common diseases with environmental causes? Am J Epidemiol 161:799–805.

Kormaz A, Manchester LC, Topal T, Ma S, Tan D-X, Reiter RJ. 2011. Epigenetic mechanisms in human physiology and diseases. J Exp Integr Med 1:139–147.

Lee DH, Steffes MW, Sjödin A, Jones RS, Needham LL, Jacobs DR, Jr. 2010. Low dose of some persistent organic pollutants predicts type 2 diabetes: A nested case-control study. Environ Health Perspect 118:1235–1242.

Lehman EJ, Hein MJ, Estill CF. 2008. Proportionate mortality study of the United Association of Journeymen and Apprentices of the Plumbing and Pipe Fitting Industry. Am J Ind Med 51:950–963.

Lichtenstein P, Holm NV, Verkasalo PK, Iliadou A, Kaprio J, Koskenvuo M, et al. 2000. Environmental and heritable factors in the causation of cancer—Analyses of cohorts of twins from Sweden, Denmark, and Finland. N Engl J Med 343:78–85.

Manolio TA, Collins FS, Cox NJ, Goldstein DB, Hindorff LA, Hunter DJ, et al. 2009. Finding the missing heritability of complex diseases. Nature 461:747–753.

McLaughlin JK, Hrubec Z, Blot WJ, Fraumeni JF, Jr. 1995. Smoking and cancer mortality among U.S. veterans: A 26-year follow-up. Int J Cancer 60:190–193.

Mendez MA, Garcia-Esteban R, Guxens M, Vrijheid M, Kogevinas M, Goñi F, et al. 2011. Prenatal organochlorine compound exposure, rapid weight gain, and overweight in infancy. Environ Health Perspect 119:272–278.

Newby JA, Howard CV. 2006. Environmental influences in cancer aetiology. J Nutr Environ Med 15:56–114.

Ng SF, Lin RC, Laybutt DR, Barres R, Owens JA, Morris MJ. 2010. Chronic high-fat diet in fathers programs β-cell dysfunction in female rat offspring. Nature 467: 963–966.

Parvez S, Venkataraman C, Mukherji S. 2009. Nature and prevalence of non-additive toxic effects in industrially relevant mixtures of organic chemicals. Chemosphere 75:1429–1439.

Preston DL, Ron E, Yonehara S, Kobuke T, Fujii H, Kishikawa M, et al. 2002. Tumors of the nervous system and pituitary gland associated with atomic bomb radiation exposure. J Natl Cancer Inst 94:1555–1663.

Prüss-Üstün A, Corvalán C. 2006. Preventing Disease through Healthy Environments. Towards an Estimate of the Environmental Burden of Disease. Geneva: World Health Organization.

Prüss-Ustün A, Corvalán C. 2007. How much disease burden can be prevented by environmental interventions? Epidemiology 18:167–178.

Prüss-Ustün A, Bonjour S, Corvalán C. 2008. The impact of the environment on health by country: A meta-synthesis. Environ Health 7:7.

Prüss-Ustün A, Vickers C, Haefliger P, Bertollini R. 2011. Knowns and unknowns on burden of disease due to chemicals: A systematic review. Environ Health 10:9.

Puga A, Maier A, Medvedovic M. 2000. The transcriptional signature of dioxin in human hepatoma HepG2 cells. Biochem Pharmacol 60:1129–1142.

Rajapakse N, Silva E, Kortenkamp A. 2002. Combining xenoestrogens at levels below individual no-observed-effect concentrations dramatically enhances steroid hormone action. Environ Health Perspect 110:917–921.

Rice G, MacDonell M, Hertzberg RC, Teuschler L, Picel K, Butler J, et al. 2008. An approach for assessing human exposures to chemical mixtures in the environment. Toxicol Appl Pharmacol 233:126–136.

Rusiecki JA, Baccarelli A, Bollati V, Tarantini L, Moore LE, Bonefeld-Jorgensen EC. 2008. Global DNA hypomethylation is associated with high serum-persistent organic pollutants in Greenlandic Inuit. Environ Health Perspect 116:1547–1552.

Sarigiannis DA, Hansen U. 2012. Considering the cumulative risk of mixtures of chemicals—A challenge for policy makers. Environ Health 11(Suppl 1):S18.

Selikoff IJ, Seidman H. 1991. Asbestos-associated deaths among insulation workers in the United States and Canada, 1967–1987. Ann N Y Acad Sci 643:1–14.

Sexton K, Hattis D. 2007. Assessing cumulative health risks from exposure to environmental mixtures—Three fundamental questions. Environ Health Perspect 115: 825–832.

Smink A, Ribas-Fito N, Garcia R, Torrent M, Mendez MA, Grimalt JO, et al. 2008. Exposure to hexachlorobenzene during pregnancy increases the risk of overweight in children aged 6 years. Acta Paediatr 97:1465–1469.

Smith KR, Corvalán CF, Kjellström T. 1999. How much global ill health is attributable to environmental factors? Epidemiology 10:573–584.

Spitz MR, Wei Q, Li G, Wu X. 1999. Genetic susceptibility to tobacco carcinogenesis. Cancer Invest 17:645–659.

Valvi D, Mendez MA, Martinez D, Grimalt JO, Torrent M, Sunyer J, et al. 2012. Prenatal concentrations of polychlorinated biphenyls, DDE, and DDT and overweight in children: A prospective birth cohort study. Environ Health Perspect 120:451–457.

Verhulst SL, Nelen V, Hond ED, Koppen G, Beunckens C, Vael C, et al. 2009. Intrauterine exposure to environmental pollutants and body mass index during the first 3 years of life. Environ Health Perspect 117:122–126.

Volz DC, Bencic DC, Hinton DE, Law JM, Kullman SW. 2005. 2,3,7,8-Tetrachlorodibenzo-p-dioxin (TCDD) induces organ- specific differential gene expression in male Japanese medaka (*Oryzias latipes*). Toxicol Sci 85:572–584.

Weaver IC, Cervoni N, Champagne FA, D'Alessio AC, Sharma S, Seckl JR. 2004. Epigenetic programming by maternal behavior. Nat Neurosci 7:847–854.

Weidman JR, Dolinoy DC, Murphy SK, Jirtle RL. 2007. Cancer susceptibility: Epigenetic manifestation of environmental exposures. Cancer J 13:9–16.

Welshons WV, Thayer KA, Judy BM, Taylor JA, Curran EM, vom Saal FS. 2003. Large effects from small exposures. I. Mechanisms for endocrine-disrupting chemicals with estrogenic activity. Environ Health Perspect 111:994–1006.

WHO. 2008. The global burden of disease, 2004 update. Available at http://www.who.int/healthinfo/global_burden_disease/en/index.html [accessed December 14, 2012].

INDEX

Effects of Persistent and Bioactive Organic Pollutants on Human Health, First Edition.
Edited by David O. Carpenter.

Printed in the United States
By Bookmasters